Progress in Plasmas and Gas Electronics

Volume 1

Progress in Plasmas and Gas Electronics

Volume 1

Edited by

R. Rompe and M. Steenbeck

With 211 Figures and 80 Tables

AKADEMIE-VERLAG · BERLIN

1975

Translated from the original German
„Ergebnisse der Plasmaphysik und der Gaselektronik",
Band 1,

published by Akademie-Verlag, with revisions and additions.

Erschienen im Akademie-Verlag, 108 Berlin, Leipziger Str. 3 – 4

Lizenznummer: 202 · 100/573/74
Gesamtherstellung: VEB Druckerei „Thomas Müntzer", 582 Bad Langensalza
Bestellnummer: 761 745 4 (55 70/1E) · LSV 1195
Printed in GDR
EVP 115,–

Foreword

Plasma physics has assumed considerable proportions in recent times. This applies both to exploratory research and technical application; new phenomena are discovered almost every day in the field of plasma physics and related fields of gas electronics, theoretically interpreted and, in some cases, applied in practice. While in other fields of contemporary physics one occasionally hears that following a phase of rapid development we are now approaching a calmer era, this does not generally hold true for plasma physics. The extraordinary multiplicity of the field — physics, geophysics, astrophysics, chemistry, electronics are all contributing their share — complicates as a matter of course adequate arrangement of the individual results and justifies the need for a complete, systematic representation. In view of the rapid pace of development, representations of this kind must be realizable in a relatively short time to make them useful.

Contemplating this task it seems that it is hardly possible to give a coherent representation of our present knowledge of plasma physics and its technical application in the conventional form of a text-book or handbook.

It is remarkable that attempts to give comprehensive representations of plasma physics, which were still possible some 25 to 30 years ago, have no longer been made in recent years, apart from literature dealing with special fusion problems, which, despite its up-to-date nature, is still rather one-sided.

Plasma physics is the physics of the most frequent aggregate state occurring in nature. The universe is immersed in a sea of matter in plasma state featuring a wide variety of the characteristic parameters. Only some celestial bodies, such as our own planet, where temperatures of several hundred °K are encountered at high densities, have condensed phases — this is but a small fraction. To represent plasma physics only from the point of view of fusion systems would mean to omit broad fields of knowledge. This would also be the disadvantage of trying to elaborate a representation of plasma physics solely on the basis of astronomic phenomena occurring in the cosmos. The representation of plasma physics must be approached differently, by taking into account the latest advances in the field of physics.

The courage to attempt comprehensive representations of plasma physics, which many physicists possessed twenty-five to thirty years ago — we may consider ourselves to be among their number — originated in the situation extant in physics of that time characterized by the permeation of the classical gas discharge physics by advances in the fields of quantum physics of electrons, atoms and ions. Its charm lay in the possibility of fundamental simplification making it possible to comprehend the physics of gas discharge. Essentially, it was a matter of viewing all phenomena in plasma from the point of view of the kinetics of its components, a task which, as a whole, could be fulfilled. The predominant approach was based on the individual particles; the binary impacts primarily determined the interpretation of the phenomenon. Since this method of treatment has been verified in a wide range of circumstances, it could be applied successfully to plasmas of astrophysics, of the ionosphere, of sparks and arc discharges, and of gas discharges at low pressure. It is also suitable for many problems of plasma chemistry, whose importance is increasing nowadays. This approach has resulted in extraordinary methodical progress and clarified the common physical foundation in the great majority of plasmas. Although the work was confined to the classical realm in the field of plasma, this progress has been comparable and took place nearly concurrently with the development in the field of solid state physics, which began after the studies conducted by SOMMERFELD and BLOCH in the field of the one-electron theory of metals, semiconductors and insulators.

If the kinetics of homogeneous isotropic plasmas are approached on the basis of discrete particles, two physically significant problems must necessarily recede into the background: the question of various collective interactions of the plasma components, of typical multi-particle problems and the question of the morphology of plasma, of the outer form assumed by plasma for various reasons. In special cases one speaks of the "problem of support" — a problem of a very general nature with which we are confronted in the cosmos as well as in the laboratory.

But it is precisely these two aspects that give plasma its identity as a special medium, comparable perhaps with liquid being characterized by a free surface and its greater degree of disarrangement than observed in crystalline structures. A great deal has been written about the fundamental significance of these two questions in monographs appearing in the past ten years. But the initiates will not be surprised to hear that the time is not nearly ripe for answering these two questions. Particularly in this area, a great deal of persevering and purposive work must still be done in order to arrive at tangible results.

Multi-particle problems have become part of the group of major problems in the theoretical foundations of modern physics. It will suffice to mention the extraordinarily successful work being done in this connection in the fields of superconductivity, superfluidity, states of excitation of solids (exciton, polaron, plasmon) and of the theory of atomic nuclei.

Our German series entitled Ergebnisse der Plasmaphysik und der Gaselektronik appears as a number of self-contained volumes. Specific areas of study that have, in our opinion, attained the necessary degree of maturity will be summarized by researchers who are themselves actively working in the respective fields. The emphasis will be on topical areas of research and technology that have an important bearing on the development of plasma physics in the broadest sense. We believe that although it is not possible to give a systematic overall survey of the field today, we can at least formulate the major problems in research and application.

We ask all colleagues interested in promoting the development of plasma physics and gas electronics along these lines to cooperate with us, be it as authors – we do not presume to be able to find all those best suited for this purpose – or as readers of the contributions by their constructive criticism.

Robert Rompe
Max Steenbeck

Contents

Spectroscopic Methods of Plasma Diagnostics

by W. Neumann

Chapter 1. Introduction . . . 3
Chapter 2. Method for Determining the Emission Coefficients and Line Profiles . . . 5
Chapter 3. Classification of Spectra . . . 87
Chapter 4. Calculation of the Composition of a Thermal Plasma with a Boltzmann Distribution of the Electronic Levels . . . 95
Chapter 5. Determination of Plasma Parameters via the Intensity of Spectral Lines, emitted by Atoms and Ions in an Optically Thin Layer. Plasma in Thermal Equilibrium . . . 119
Chapter 6. Determination of Temperatures and Particle Densities of Plasmas Displaying Appreciable Absorption . . . 198
Chapter 7. Determination of Plasma Parameters from the Band Spectrum of Diatomic Molecules . . . 237
Chapter 8. Determination of Particle Densities in Plasmas by Means of the Refractive Index. Laser Diagnostics . . . 258
Chapter 9. Determination of Plasma Parameters by Means of Continuous Radiation 289
Chapter 10. Determination of Plasma Parameters from Line Profiles. Fading of Spectral Lines at the Series Limit and the Problem of the Reduction in the Ionization Energy . . . 324
Chapter 11. Validity Limits of the Methods of Thermal Spectroscopy discussed in Chapters 5 to 10. Some Special Methods of Nonthermal Spectroscopy . . . 404
Bibliography . . . 436

Magnetohydrodynamic Waves

(Waves in Ideal Media)

by K. Baumgärtel and G. Wallis

Chapter 1. Introduction . . . 513
Chapter 2. Propagation of Plane Waves in Homogeneous Media . . . 516
Chapter 3. Interaction of MHD-Waves with Surfaces of Discontinuity. Propagation in Inhomogeneous Media . . . 541
Chapter 4. Propagation in Inhomogeneous Media . . . 563
Chapter 5. Appendix: Matrices . . . 582
Bibliography . . . 589

Validity Conditions for Local Thermodynamic Equilibrium

by H. W. Drawin

Chapter 1. Introduction . . . 593
Chapter 2. Partial L.T.E. in Time-Independent and Homogeneous Optically Thin Plasmas . . . 597
Chapter 3. Complete L.T.E. in Time-Independent and Homogeneous Optically Thin Plasmas . . . 606
Chapter 4. Influence of Resonance Absorption on the Establishment of Complete L.T.E. in Homogeneous Stationary Plasmas . . . 608
Chapter 5. Influence of Resonance Absorption on the Establishment of Partial L.T.E. in Homogeneous Stationary Plasmas . . . 611
Chapter 6. Influence of Heavy Particle Collisions on the Equilibrium Population of Highly Excited Levels . . . 614
Chapter 7. Validity of Partial L.T.E. in Quasi-Stationary and Quasi-Homogeneous Plasmas (optically thin and thick) . . . 617
Chapter 8. Relaxation Times for the Establishment of Partial and Complete L.T.E. in Homogeneous Transient Plasmas . . . 631
Chapter 9. Relaxation Lengths for the Establishment of Partial and Complete L.T.E. in Stationary Non-Uniform Plasmas . . . 637
Chapter 10. Perturbation of the Electron Temperature by Externally Applied Electric Fields . . . 646
Chapter 11. Calculation of Optical Reduction Factors Λ_{ji} and Λ_i . . . 648
Chapter 12. Concluding Remarks . . . 657
Bibliography . . . 658

Subject Index . . . 661

W. Neumann

Spectroscopic Methods of Plasma Diagnostics

CHAPTER I

Introduction

The investigation of plasmas (ionized gases) has come during the last two decades into the focus of technological and scientific interest. Of special interest are high- and low-pressure plasmas, produced by steady and pulsed electric discharges, by adiabatic and shock-wave compression, by focusing of strong electromagnetic radiation, by thermal heating, by chemical processes, and by injection of energetic particles into magnetic bottles.

Plasmas find application in lighting, welding, cutting, melting, and heat treatment of materials used at high temperatures. Furthermore, plasmas are encountered in gaselectronic devices, in circuit breakers, in high-temperature chemistry, in devices for conversion of heat into electric energy, such as thermionic converters and MHD-generators, and also in research on controlled thermonuclear reactions in high-temperature plasmas, in which nuclear energy is to be converted into other, technologically useful forms of energy.

For the description and explanation of phenomena occurring in a plasma, it is necessary to determine as many parameters of the plasma as possible with the help of as many as possible mutually independent methods, as accurately as possible. Among the relevant parameters are the material composition of the plasma, the temperature and density of the different plasma components, the magnetic and electric field strengths, and the flow velocities of the individual components of the plasma.

To the numerous methods of plasma diagnostics, which were developed for the purpose of determining the plasma parameters, belong the methods of plasma spectroscopy. Their application has the advantage over other methods in that the plasma remains practically undisturbed.

At the present time the useful range of wavelengths for plasma spectroscopy extends from the infrared (maximum wavelength 0.1 cm) to the vacuum-ultraviolet (wavelengths down to a few Å). Shortwave spectroscopy is of special interest in the study of high-temperature plasmas with temperatures of the order of 10^6 °K and higher.

The spectroscopic methods were primarily developed for the determination of the plasma parameters in thermal plasmas, i.e., plasmas in a local thermodynamic equilibrium. Most methods, however, can also be applied in those cases where the equilibrium is disturbed due to the difference in temperature of the different plasma components and when certain conditions are met: each plasma component must have a Maxwellian velocity distribution and the electron states must be occupied with electrons in accordance with the Boltzmann distribution. In plasmas containing molecules the occupation of the rotational and vibrational terms must also correspond to a Boltzmann distribution.

Several spectroscopic methods are known which are valid even though neither the Maxwell distribution nor the Boltzmann distribution occur. To these methods belong those for the determination of density, whereby the refractive index is measured.

Besides these, there are special methods which are only applicable to low-pressure plasmas with an extremely disturbed Boltzmann distribution and in which the line intensities emitted from relatively low levels are measured.

In the following discussion we shall first deal with the general measuring and computational methods for the determination of line intensities and line profiles (Chapter 2). In Chapter 3 we shall briefly treat the classification of spectra. Chapter 4 deals with the calculation of the plasma composition in a thermal atomic plasma or in a plasma in which chemical reactions take place.

In Chapters 5—10 are discussed in detail those plasma diagnostic methods, which were developed under the assumption of thermal equilibrium. In these methods the intensities in the line spectrum and in the continuous spectrum of emission and absorption spectra are measured, the refractive index is determined, and the line profiles and line shifts are obtained.

In Chapter 11 the limits of application of the above methods are stated and certain specific methods are mentioned in which the plasma parameters in low-pressure plasmas with non-Boltzmann electron distributions can be determined.

CHAPTER 2

Method for Determining the Emission Coefficients and Line Profiles

2.1. Spectroscopic Instrumentation

Spectroscopic devices are used to provide photographic or electric recording of the spectra and to determine the line intensities and line profiles. With the help of prisms, gratings, and interferometers the light can be analyzed. Where a resolution of 0.1 Å is sufficient, prism spectrographs can be used*. When higher resolution is requisite, grating spectrographs or interference devices are necessary which require a preresolution of the rays by means of a monochromator, prism spectrograph, or grating spectrograph. For grating spectrographs the interfering overlapping orders must be eliminated by filters or by a prism. The prism devices are distinguished from the grating spectrograph in that a higher light intensity is obtained in the former. If stray light is unwanted as it may bring serious errors into the experimental results especially when working in the ultraviolet spectral region, then for its reduction those spectral devices must be used which have a high linear dispersion or include two light dispersing systems in series (for example double monochromators). Among the last kind of instruments there are to mention double monochromators enclosing a prism and grating in series, giving by this means a relative high dispersion over a relative broad wavelength region.

Within the frame of this section, only basic types of spectral devices are dealt with. For more details see for instance [2.52]. About special devices used for high-speed spectroscopical work see section 2.5.

2.1.1. Prism Spectrographs

With a prism spectrograph a spectral range of 0.11 to 40 μm can be covered. The individual spectral ranges for different prism materials are listed in Table 2.1.

The optical image is effected by means of glass lenses, in the visual range. In the other spectral ranges lenses made of the above materials can be employed. By combining several of these substances (e.g., quartz and lithium fluoride) colour-correcting lenses can be produced. Mirror lenses are universally suitable for optical images within the total spectral range covered with prism substances.

* Regarding the two main forms of spectral devices – spectrographs and monochromators there is no principal difference in their working principle, therefore we restrict our attention in the following, if not otherwise stated, to spectrographs only.

Table 2.1. Spectral ranges for different prism materials

Material	Chemical formula	Wavelength range
Lithium fluoride	LiF	0.11 – 6 μm
Calcium fluoride	CaF_2	0.12 – 9 μm
Quartz	SiO_2	0.185 – 3.5 μm
Rock salt	NaCl	0.20 – 17 μm
Potassium bromide	KBr	0.21 – 28 μm
Potassium iodide	KJ	0.25 – 31 μm
Flint glass		0.34 – 1.0 μm
Sylvite	KCl	0.38 – 21[1]) μm
Thallium bromoiodide	TlBrJ	0.50 – 40 μm

[1]) Absorption bands at 3.2 and 7.1 μm.

The intensity of the spectrograph depends essentially on the relative aperture $1/o$ of the camera lens (o = focal length:effective diameter). Relative apertures of 1:2 to 1:30 are typical. To photograph low-intensity light waves or short-time processes high-power spectrographs are preferable, provided the generally small linear dispersion and the lower resolution of the high-power instruments can be allowed for.

The linear dispersion $\frac{\Delta\lambda}{\Delta x}$ is usually given in Å/mm and is dependent on the wave mode. The relationship between the wavelength λ and the position on the plane of the photographic plate where this wavelength appears is given by the Hartmann interpolation formula. Excluding the vacuum ultraviolet it has the form ([2.53]* p. 438)

$$\lambda - \lambda_0 = \frac{c}{x - x_0}. \tag{2.1}$$

The constants λ_0, c and x_0 in this case apply only to a small spectral range and can be determined for this range from 3 measured lines with known wavelengths.

For the vacuum-ultraviolet the Hartmann dispersion formula ([2.53], page 438) becomes

$$\lambda - \lambda_0 = \left(\frac{c}{x - x_0}\right)^{0.5}. \tag{2.2}$$

The resolution of the spectrograph is given by

$$\Delta\lambda_{\text{theoret.}} = \frac{1.7\, o\, \lambda \cdot \Delta\lambda}{\sin\beta \cdot \Delta x}, \tag{2.3}$$

where o denotes as explained above the f/number of the camera lens and β the angle of inclination of the photographic plate to the ray direction. To obtain the maximum resolving power it is suitable to choose the width of the entrance slit as follows

$$B = o_{\text{coll.}} \cdot \lambda, \tag{2.4}$$

* For references see page 436.

where o_{coll} is the f/number of the collimator. When the slit width is too small the resolution is worsened due to diffraction at the slit, and too large slit widths cause pure geometrical blurring in the plane of the photographic plate.

The resolution can be written after transforming (2.3) as follows

$$\Delta\lambda_{\text{theoret.}} = \frac{1.7\,\lambda}{b}\,\frac{\Delta\lambda}{\Delta\delta} \tag{2.5}$$

where b denotes the utilizable diameter of the cameralens. This depends on the size of the prism, and $\frac{\Delta\lambda}{\Delta\delta}$ is the angular dispersion, which is connected to the linear dispersion $\frac{\Delta\lambda}{\Delta x}$ by

$$\frac{\Delta\lambda}{\Delta x} = \frac{1}{f}\,\frac{\Delta\lambda}{\Delta\delta}\sin\beta\,. \tag{2.6}$$

If a high linear dispersion is desired for a given prism material, i.e., if $\frac{\Delta\lambda}{\Delta x}$ is to be small, the largest possible focal length of the camera lens is used and several prisms are inserted in series. When high resolution is desired, the largest possible prisms are employed and likewise several prisms are inserted in series. In the last case the angular dispersion is increased as well (i.e., $\frac{\Delta\lambda}{\Delta\delta}$ becomes smaller).

A further improvement of resolving power and dispersion is attained, if the light after passing the prism set is directed back by means of the "autocollimation mirror" and passes the same prism set once again in the opposite direction. This principle is found within autocollimation spectrographs. Before using this type of spectrograph, it has to be taken in mind that as a result of unwanted

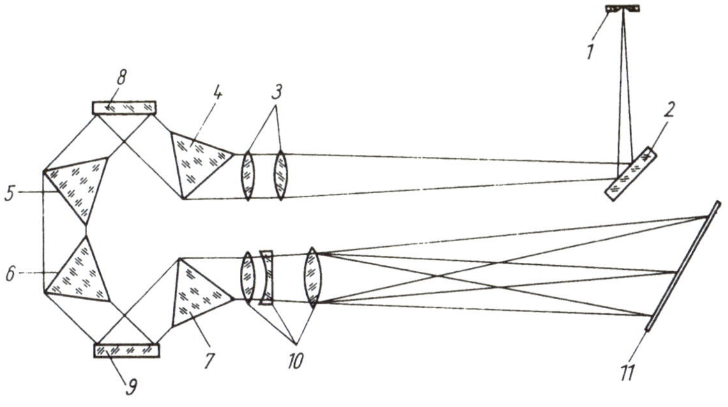

Figure 2.1. Optical diagram of the 4-prism quartz spectrograph Fuess 110 H (R. Fuess, Berlin, prospectus)

1 — slit, *2* — plane mirror, *3* — collimator objective, *4, 5, 6, 7* — prisms, *8, 9* — plane mirror, *10* — camera objective, *11* — plate

reflection by the prism surfaces there will exist a disturbing spectrum superimposed to the main spectrum. Then the former spectrum must be shifted relative to the last by tilting the autocollimation mirror.

An example of a prism spectroscopic device is given in Figure 2.1, which illustrates the optical diagram of a 4-prism quartz spectrograph. Below we list some commercial spectroscopic apparatus.

Table 2.2. Data on spectroscopic apparatus

Spectrograph type Manufacturer	Spectral range [Å]	f_{camera} [cm]	°	λ [Å]	$\frac{\Delta\lambda}{\Delta x}$ [Å/mm]	$\lambda_{theoret.}$ [Å]
3-prism spectrograph	3600	12	2.3	5000	105	0.2
VEB Carl Zeiss Jena	–	27	5.5	5000	47	0.2
with glass prisms	10000	130 (Autom.)	26	5000	4.9	0.1
4-prism quartz spectrograph	2100	95	ca. 16	3000	5	0.04
110 H	–			5000	25	0.3
R. Fuess Berlin	8,000					
Quartz prism spectrograph	2000	50	11	3000	13.5	0.08
Q 24	–			5000	58	0.5
VEB Carl Zeiss Jena	5,800					
Mirror monochromator with	P: 2000	40	8.4			
prisms (P)	–					
and grating (G)	400000					
Type SPM 2	G: 2000					
VEB Carl Zeiss Jena	-10000					
Plane-grating spectrograph	2000	207.5	28.5	3000	7.4	0.1
PGS 2	–			5000	7.4	0.2
VEB Carl Zeiss Jena	28000	207.5	28.5	3000	3.7	0.05
	[1])	[2])		5000	3.7	0.1

[1]) These data correspond to a grating with 650 lines/mm and to the first order of diffraction.
[2]) The data in the two last lines correspond to a ray path where the radiation is twice diffracted by the grating.

2.1.2. Grating Spectrographs

The most widely used spectrographs of this type are plane-grating spectrographs with a reflecting grating, in which the design is such that the greater part of the radiation is reflected into a certain angle (Blaze-gratings). The favoured wavelength within a given spectral order is called the "Blaze-wavelength". Figure 2.2a illustrates the optical diagram of such a spectrograph with an Ebert mount, manufactured by VEB Carl Zeiss Jena; Figure 2.2b shows the external view of the instrument.

Rowland gratings, which do not require additional optical components, since they are shaped like concave mirrors, are also used, and they are of high value for vacuum ultraviolet work. A shortcoming is the astigmatic image obtained when usual grating mounts are employed. As a measure of the light intensity one can again use the relative aperture $1/o$ (effective diameter: focal length of the camera object lens). However, it should be stressed that the light intensity decreases toward higher orders.

As mentioned above, the overlapping of several orders can be prevented by preresolving the radiation by means of filters or monochromators.

The linear dispersion for the case in which the rays are incident normal to the plane of the plate, is given by

$$\frac{d\lambda}{dx} = \frac{d \cdot \cos\beta}{f n} \qquad (2.7)$$

where d is the distance between adjacent grating rules (grating constant), β the angle between the normal to the grating and the ray diffracted by the grating, f the focal length of the camera objective, and n the order of the spectrum.

If the incident radiation is normal to the grating, the condition for obtaining diffraction maxima is given by the inequality

$$\frac{n\lambda}{d} < 1 . \qquad (2.8)$$

Taking $n = 1$ we get for the maximal wavelength the relation

$$\lambda_{max} = d \qquad \text{(for normal incidence)} .$$

For oblique incidence λ_{max} can become twice the above value. Accordingly, a grating with 1000 lines/mm in the most extreme case can be used for a maximum wavelength of 20,000 Å. Radiation of longer wavelength (infrared radiation) can therefore be diffracted only by gratings with larger intergroove spacings.

The theoretical resolving power is given by

$$\frac{\lambda}{\Delta\lambda_{\text{theoret.}}} = n N \qquad (2.9)$$

where N is the total number of illuminated rulings of the grating.

According to equation (2.9) it is possible to increase the resolution for a given grating by using a higher order n, but this procedure is limited, if not in consequence of the decreasing light intensity (which can be countered by using especially adapted Blaze gratings [2.15a]) then in consequence of formula (2.8), giving for the maximal resolving power of a grating the relation

$$\frac{\lambda}{\Delta\lambda_{\text{theoret.}}} = n N \le \frac{dN}{\lambda}$$

where dN is the effective breadth of the grating. With $dN = 10$ cm (medium size grating) and $\lambda = 5000$ Å we get $\left(\frac{\lambda}{\Delta\lambda}\right)_{max} = 200{,}000$, $\Delta\lambda_{min} = 0.025$ Å.

According to these statements, with a medium-size grating spectrograph working in a high enough spectral order it is possible to measure spectral line width down to 0.2 Å with sufficient accuracy [2.32]. To resolve profiles of narrower spectral lines, large spectrographs or interference devices (see section 2.1.3) have to be used.

For short wavelengths (below 1000 Å) only reflection gratings can be used. Since the reflecting metals have low reflecting powers below 1000 Å, glass gratings are mostly employed. For wavelengths below 200 Å (soft X-rays) the radiation must fall upon the grating at grazing incidence. The radiation is then totally reflected, since the index of refraction for glass in this wavelength range is smaller than 1.

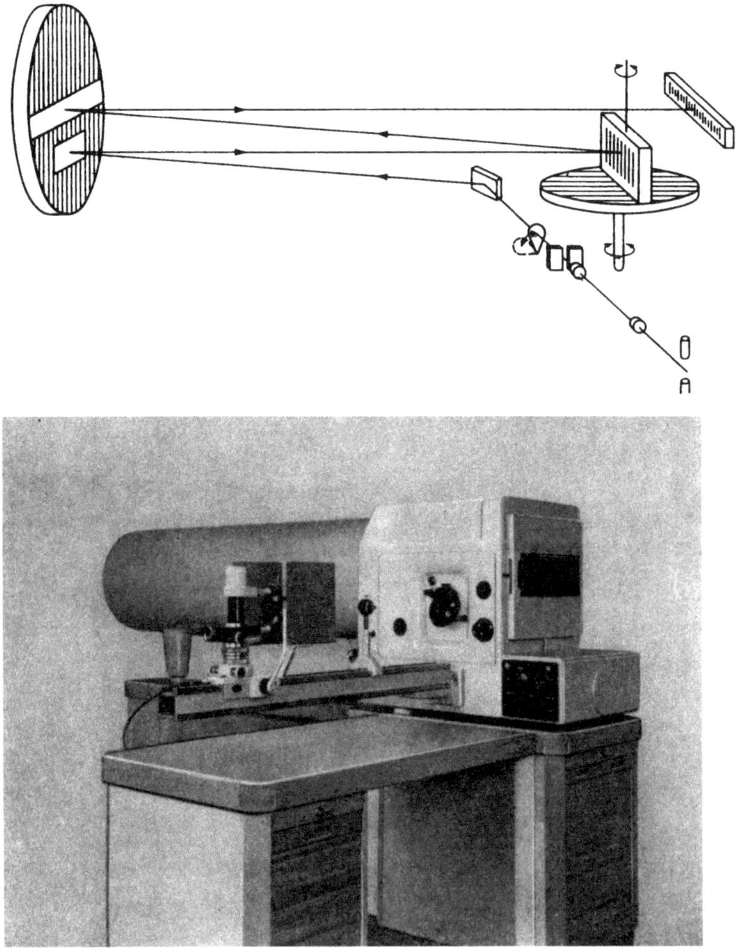

a)

b)

Figure 2.2. Ray path (a) and externed view (b) of the of the 2-m plane-grating spectrograph PGS 2 of VEB Carl Zeiss Jena

B. Edlèn [2.35] and F. Tyrén [2.125c] obtained the following relationship between the angle of incidence α and the shortest wavelength λ_{min}:

α (in °)	0	60	80	85.6	89
λ_{min} (in Å)	320	160	75	53	7

At an angle of incidence of 89° it was possible to observe a spectrum down to 16 Å in the first order [2.39] or down to 10 Å respectively 7 Å when higher spectral orders are used [2.125b], [2.125c].

Figure 2.3 gives a schematic representation of a vacuum spectrograph for the case of grazing incidence.

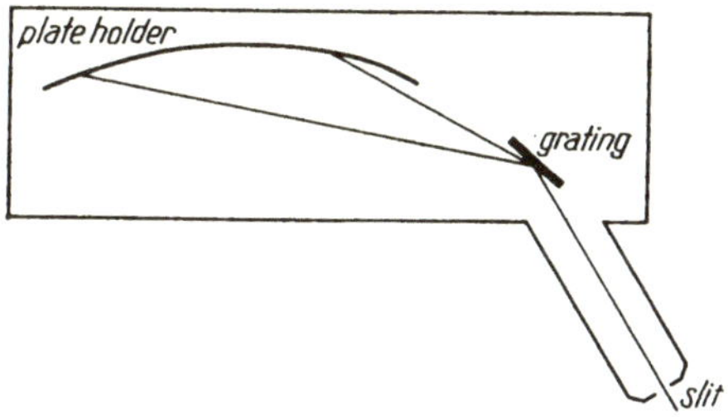

Figure 2.3. Schematic representation of a vacuum spectrograph for grazing incidence (Harrison, et al. [2.52], p. 538)

In order to separate the different orders which normally overlap in such a grating, an auxiliary mirror can be placed in front of the entrance slit, which (depending on the angle of incidence) does not let the short-wave radiation pass through and whereby the limit of total reflection is exploited (see Astoin et al. [2.4]). For more detail concerning vacuum ultraviolet work see [2.35a] and [2.81a].

A special type of diffracting grating useful for the determination of spectrum line profiles is the Echelle-grating which differs from an ordinary diffraction grating in its large intergroove spacing. To get a sufficient number of interfering rays, a high spectral order must be chosen. P. M. Griffin et al. [2.48] used for the determination of the profiles of the spectrum lines C III 4647, 4650 and 4651 Å emitted from a high temperature plasma a grating with 8 rulings per mm and 600 rulings total. The resolving power in the 479th order was 250,000. The Echelle grating was used without pre-dispersal.

2.1.3. Interferometric Devices

These devices are used among other things to resolve lines whose half-width is less than 0.2 Å. The most popular is the Fabry-Perot interferometer also known under the name air plate interferometer. Its working principle will be demonstrated by means of figure 2.4 in which the corresponding ray path is schematically shown. There is an air layer between the two mirror-coated glass plates P_1 and P_2. The reflectivity r of the semitransmissive reflecting films S_1 and S_2 in general lies between 0.85 and 0.90. The number of interfering rays is (see [2.94], p. 132)

$$N_i = \frac{\pi \sqrt{r}}{1 - r}. \tag{2.10}$$

With the usual metal coatings a number $N_i \approx 30$ will be obtained. Higher numbers of interfering rays within variable but limited wavelength ranges are

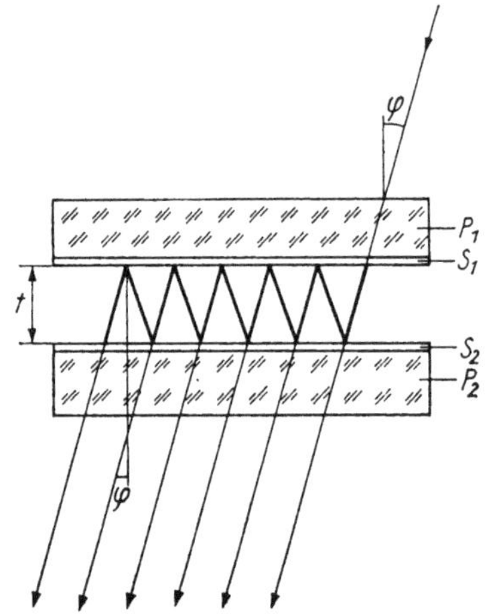

Figure 2.4. Ray paths in a Fabry-Perot interferometer (Nagibina and Prokofiev [2.94], p. 132)

P_1 and P_2 – glass plates with reflecting coatings S_1 and S_2, respectively

obtainable when instead of single metal coatings dielectric multiple layers are used, the absorption losses of which may be as low as 1% and the reflection coefficients as high as 0.97 [2.105a], [2.120a]. In consequence of this, the resulting absorption losses are lower than those obtained with a metal coated interferometer.

To eliminate unwanted reflections, the uncoated surfaces of the interferometer plates are inclined relative to the reflecting surfaces by an angle of about 5′. For further construction details see for instance [2.94] p. 137.

To obtain the dispersion relation, it must be considered that rays with wavelength λ interfere constructively only for those directions for which

$$2\,t \cos \varphi = m\,\lambda\,, \tag{2.11}$$

where m is a whole number. If a convex lens of focal length f is placed behind the interferometer, the rays of equal φ will interfere in the focal plane and form circular fringes. m is the order of the interference fringe. The diameter D and the order m are connected by (see [2.52], p. 559)

$$m = \frac{2t}{\lambda} + \frac{2tD^2}{8\lambda f^2} \tag{2.12}$$

The lowest order is m_0 and corresponds to the fringe with the smallest diameter $D_0\left(m_0 \geqq \frac{2t}{\lambda}\right)$.

The dispersion (change in the fringe diameter caused by change of wavelength) is given by

$$\frac{\mathrm{d}D}{\mathrm{d}\lambda} = \frac{4f^2 \cos\varphi}{D\lambda}. \tag{2.13}$$

When $\cos\varphi \approx 1$, the free wavelength distance between two orders is

$$\Delta\lambda \approx \frac{\lambda^2}{2t}. \tag{2.14}$$

When working with the Fabry-Perot interferometer, a preresolution of the spectrum is necessary with a resolving power better than $\Delta\lambda$. For an interferometer with $t = 1$ cm, the required preresolution is 0.1 Å for $\lambda = 5000$ Å. For isolated spectral lines with a total line width smaller than $\Delta\lambda$, the demands on the preresolution are not so high.

For realizing preresolution, the Fabry-Perot interferometer can be placed before a spectrograph as shown in Figure 2.5, so that parallel ray bundles pass

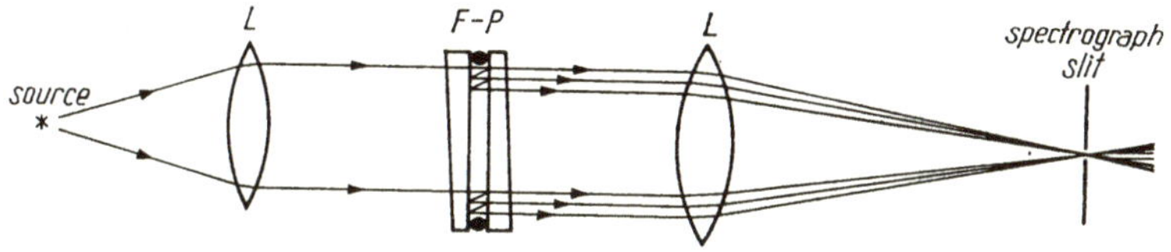

Figure 2.5. Fabry-Perot interferometer arrangement before the spectrograph slit (Harrison et al. [2.52], Figure 20.10)

through the interferometer. Thus, after proper adjustment in the plane of the slit, an interference pattern as shown in Figure 2.6 appears. Bergstedt [2.14] utilized a different arrangement with converging ray paths in order to uniformly illuminate the slit of the spectrograph with radiation from an iron arc. Figure 2.7 shows the resulting interference pattern.

Figure 2.8 illustrates a system in which the interferometer is placed between the collimator lens and the prismatic system inside the prism spectrograph.

The theoretical resolving power of the Fabry-Perot interferometer is according to Meissner [2.26] given by

$$\frac{\lambda}{\Delta\lambda} = \frac{5.96 \cdot t \cdot r^{1/2}}{\lambda(1 - r)}. \tag{2.15}$$

Thereby it is accepted that the reflecting surfaces are so perfectly plane parallel and polished that the distance t between them after adjustment of the interferometer plates is constant within an error limit of

$$\Delta t \leq \frac{\lambda}{2 N_i}. \tag{2.16}$$

Otherwise the resolving power is determined by the surface roughness.

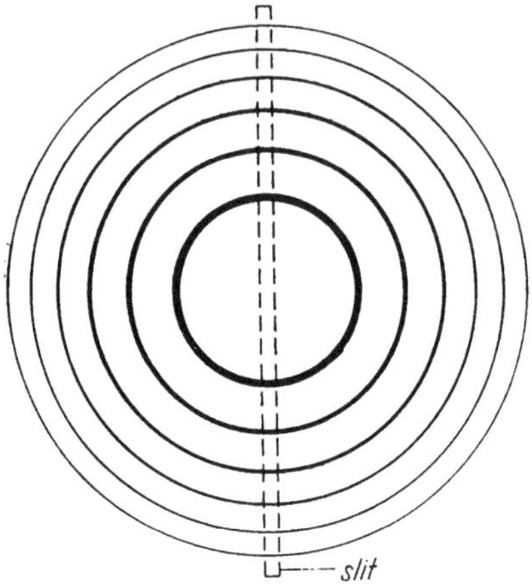

Figure 2.6. Interference pattern on the spectrograph slit for monochromatic incident radiation (Harrison et al. [2.52], Figure 20.11)

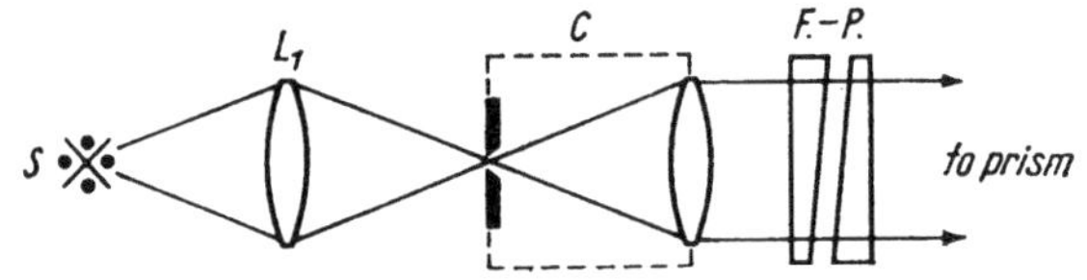

Figure 2.8. Arrangement of Fabry-Perot interferometer inside a spectrograph. (C. Candler [2.26], p. 297)

For illustration, with $t = 1$ cm, $r = 0.90$ and $\lambda = 5000$ Å, a resolving power of $\lambda/\Delta\lambda \approx 10^6$ is obtained which is considerably higher than the resolving power of medium size spectrographs ($\leq 2 \times 10^5$).

The interference fringes have a profile when highly monochromatic light is observed, which, expressed in angular units ([2.26], p. 303 and [2.94], p. 133), is

$$\frac{I}{I_0} = \frac{1}{1 + \frac{4r}{(1-r)^2} \sin^2 \frac{2\pi t \cos\varphi}{\lambda}} \tag{2.17}$$

with a half-width h' and in terms of wavelength units (see [2.14])

$$h' \approx \frac{\lambda^2 (1-r)}{2\pi t \sqrt{r}}. \tag{2.18}$$

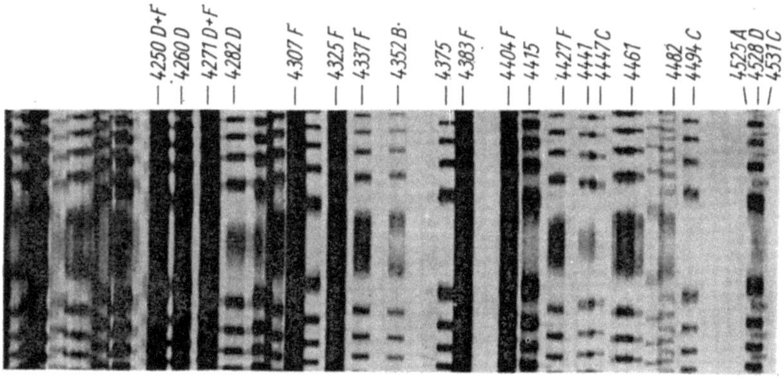

Figure 2.7. Interferometer photograph of an iron spectrum; mirror distance about 6 mm, wavelengths in Å (Bergstedt [2.14])

If the half-width of the line, whose profile is to be measured, is large compared to h', then except for the normal coordinate correction (transformation of the D-scale to the λ-scale and conversion of $I(D)$ into $I(\lambda)$ via the relationship $I(D)\,dD = I(\lambda)\,d\lambda$) no further corrections are necessary. Otherwise, the measured line profiles which are transferred to $\lambda(I)$ must be freed of the influence of the apparatus [2.11], [2.42]. This is further discussed in section 2.9.

From the other types of interferometric devices useful for plasmadiagnostical work there are to mention the wedge interferometer [2.22a] and a special interferometric system worked out by Weizel and Meister [2.131] by means of which spectral line profiles can be determined by measuring the contrast between the interference fringes.

A very high spectral resolution (of the order of 10^{-3} Å) can also be obtained with correlation-spectroscopical methods. For details we refer to the literature [2.3b], [2.27d], [2.86a].

For a detailed treatment of interferometric systems see also Korolev [2.70].

2.1.4. Crystal Spectrometer

In the range of soft X-rays (below 16 Å) the crystal spectrometer with a Beryllium monocrystal is best suited. Bearden et al. [2.13] and Jahoda et al. [2.61] used this device to photograph the spectrum of a Scylla discharge (magnetically compressed high-temperature plasma) in Helium (see Figure 2.9).

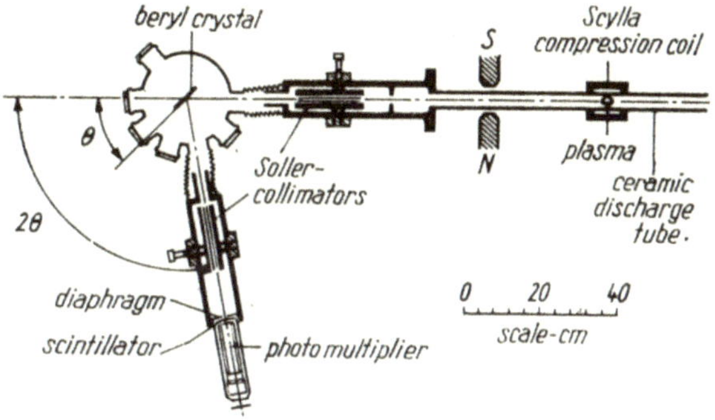

Figure 2.9. Crystal spectrometer for photographing plasma spectra in the range from 6 to 16 Å (Jahoda et al. [2.61])

Before it became customary to employ crystal spectrometers to photograph plasma spectra, various absorbers of different thickness in combination with an appropriate radiation detector were used in order to realize spectral resolution (Jahoda et al. [2.60]).

2.2. Wavelength Calibration of Spectroscopic Instruments

The calibration of wavelengths is effected by employing light sources, which insofar as possible emit atomic spectrum lines of only one element, with the aid of a spectral table or spectral atlas. As light sources spectral lamps, iron arcs, Geissler tubes, and hollow-cathode discharges can be used.

For an initial calibration it is appropriate to photograph a mercury spectrum. It is usually sufficient to subsequently improve the wavelength calibration by means of an iron spectrum.

Gaps in the calibrating curves can be filled by interpolation. For grating spectral devices linear interpolation is sufficient while for prism spectral devices the Hartmann-interpolation formula (2.1) and (2.2) may be applied.

Below we give a list of spectral atlases and spectral tables which are useful for the identification of spectra.

LIST OF SPECTRAL ATLASES

1. J. M. Eder and E. Valenta, "Atlas typischer Spektren", Verlag A. Hölder, Vienna, 1911.
2. J. Bardet, "Atlas de Spectre d'Arc", G. Doin, Paris 1926. Spectral range 2500 to 3500 Å.
3. W. Grotian, "Graphische Darstellung der Spektren von Atomen und Ionen mit 1, 2, und 3 Valenzelektronen", Springer-Verlag, Berlin 1928, Vol. 1 and 2.
4. F. Löwe, "Atlas der letzten Linien der wichtigsten Elemente", Steinkopff-Verlag Dresden 1928; 43 elements, 2200 – 4700 Å.
5. W. J. Crook, Metallurgical spectrum analysis, Stanford University Press, Stanford 1935. Atlas contains iron lines 5671 – 2794 Å.
6. A. Gatterer and J. Junkes, "Atlas der Restlinien von 30 chemischen Elementen", Specola Vaticana 1937; 2200 – 8000 Å.
7. M. Minnaert, G. F. W. Mulders, and J. Houtgast, Photometric Atlas of the Solar Spectrum, Amsterdam 1940.
8. W. R. Brode, Chemical Spectroscopy, Wiley & Sons, New York 1943; 2310 – 5090 Å.
9. A. Gatterer and J. Junkes, "Spektren der seltenen Erden", Specola Vaticana 1945; 2265 – 7600 Å.
10. S. K. Kalinin, A. A. Yanvel', and L. E. Naimark, "Atlas of the Arc and Spark Spectrum of Iron", Metallurgizdat, Moscow 1953.
11. S. K. Kalinin, L. E. Naimark, V. L. Marzuvanov, and K. I. M. Ismagulova, "Atlas of Spectral Lines for a Glass Spectrograph", Gosgeoltechizdat, Moscow 1956.
12. Q 24 – Iron Spectrum of 2227 – 4555 Å. Published by VEB Carl Zeiss Jena, Verlag Gustav Fischer, Jena 1957.
13. A. I. Alekseev, I. G. Grinman, S. K. Kalinin, Yu. A. Kushnikov, and V. L. Marzuvanov, "Atlas of the Hg-Spectrum", Akad. Nauk Kazakhstan SSR, Alma Ata 1959. 226 lines in the range from 2240 – 10140 Å.
14. S. K. Kalinin, A. A. Yanvel', A. I. Alekseeva, V. L. Marzuvanov, and L. E. Naimark, "Atlas of Spectral Lines for Quartz Spectrographs", Gosgeoltechizdat, Moscow 1959.
15. S. K. Kalinin, V. L. Marzuvanov, T. B. Bekbaulova, "Atlas of Spectral Lines in the Range 2095 – 1940 Å", Akad. Nauk Kazakhstan SSR, Alma Ata 1960.
16. R. Intonti and A. Taddeucci, Spettro di arco del ferro righe analitiche di 55 elementi (The Spectrum of the Iron Arc and sensitive Lines of 55 Elements), Milan, Assoc. Ital. Metallurg., Centro Ric. Spettrochim., 1961.
17. H. Jaffe, Atlas of analysis lines, Hilger & Watts Ltd., London 1962.

LIST OF SPECTRAL TABLES

1. H. Kayser, "Handbuch der Spektroskopie", Verlag Hirzel, Leipzig. Vol. 5, 1910: 20,000 lines of 45 elements A—N, including air lines and some band heads; range: 1030—39100 Å. Vol. 6, 1912: 50,000 lines of 41 elements Na—ZR and iron; contains also principal lines, range: 1850—9000 Å.
2. P. Auger et al., "Données numériques spectroscopie", Gauthier-Villars, Paris 1910 to 1936. 150,000 lines of 76 elements and some bands, range: 2000—10,000 Å.
3. F. Exner and E. Haschek, "Die Spektren der Elemente bei normalen Druck", Verlag Deuticke, Leipzig and Vienna 1911. Vol. 2: 60,000 lines of 67 elements and 214 band heads, range: 2200—6800 Å. Vol. 3: 56,000 lines of 78 elements and 107 band heads, range: 2200—6800 Å.
4. F.Paschen and R.Götze,"Seriengrenze der Linienspektren",Springer-Verlag, Berlin 1922.
5. H. Kayser and H. Konen, "Handbuch der Spektroskopie", Verlag Hirzel, Leipzig. Vol. 7, Part 1—3, 1923—1934: 54,000 lines of 49 elements, range: 70—75,000 Å; Vol. 8, 1932: 22,000 lines of 19 elements, Ag—Cu, range: 250—29,300 Å.
6. W. Weizel, "Handbuch der Experimentalphysik", Supplement I, Akad. Verlagsges., Leipzig 1931. Band data for about 150 molecules.
7. Landolt-Börnstein, "Phys.-chem. Tabellen", Supplement I, p. 336; Vol II, p. 529; Vol. III, p. 763, Springer-Verlag, Berlin 1931—1935. 10,000 lines of 81 elements, including some band heads and tables of the most intense and last lines. Range: 291—74,360Å.
8. W. Jevons, Report on Band Spectra of Diatomic Molecules. Cambridge Univ. Press 1932. Band data for 142 molecules.
9. G. Scheibe and C. F. Lindström, "Tabellen der Funken- und Bogenspektren des Eisens etc.", R. Fuess, Berlin-Steglitz 1933. Range 2300—3700 Å.
10. G. R. Harrison, MIT, Wavelength tables with intensities in arc, spark and discharge tube of more than 100,000 spectrum lines. Wiley & Sons, New York 1939. 109,275 lines, some air lines and band heads. Range: 2000—10,000 Å.
11. H. Kayser and R. Ritschl, "Tabelle der Hauptlinien der Linienspektren aller Elemente", Springer-Verlag, Berlin 1939; reprint 1968. 27,000 lines of 88 elements, 33—90,850 Å.
12. C. E. Moore, A multiplet table of astrophysical interest. Pt. I: Table of multiplets; Pt. II: Finding list of all lines in the table of multiplets. Contr. Princeton Univ. Obs. Nr. 20, 1945.
13. C. E. Moore, An ultraviolet multiplet table, Circular 488, sect. 1—5, Nat. Bur. Stand., Washington 1950—1962.
13a. Atlas des longueurs d'onde caractéristiques des bandes d'émission et d'absorption des mólecules diatomiques. Hermann & Cie, Dépositaires, Paris 1952.
14. G. Milazzo, Spectrum tables used for wavelength identification in the Schuman region. Instituto di Sanita, Rome, ca. 1958.
15. J. A. Norris, Wavelength table of rare earth elements and associated elements including zirconium, thorium, hafnium and tellurium. Office of Techn. Serv., Dep. Commerce, Washington 1960.
16. A. N. Zaidel', V. K. Prokofiev, and S. M. Raiskii, "Tables of Spectral Lines", Fizmatgiz, Moscow 1962.
17. J. Kuba, Hlavni čáry emisni spektrálni analysys koincidencemi od 2000—10,000 Å (principal Lines of Emission Spectrum Analysis with Coincidences from 2000—10,000 Å) Prague, Publ. House of the Czechoslovakia Acad. Sci., 1962.
18. R. W. B. Pearse and A. G. Gaydon, The Identification of Molecular Spectra. Chapman & Hall Ltd., London 1963.
19. R. L. Kelly, A Table of Emission Lines in the Vacuum Ultraviolet for all elements (6—2000 Å). UCRL-5612, Univ. Calif., Lawrence Radiat. Lab., 1960.
20. F. R. Kessler and H. Hauser, Quecksilberemissionslinien zur Monochromatoreichung im nahen Ultrarot. Z. angew. Phys. 17 (1964) 280.

2.3. Radiation Detectors

To these belong photographic plates, electrical detectors, and photochemical radiation detectors. The photographic plates have the advantage that a two-dimensional photograph can be obtained (one dimension is the wavelength and the other is either the spatial coordinate or time). A disadvantage is that the conversion of the blackening density to intensity is troublesome and not very accurate.

Photoelectric receivers have the advantage that the reading is proportional to the intensity of the radiation. Furthermore, by using the principle of light chopping in combination with phase selective amplification of the output signal, a high sensitiveness of light detection in the case of stationary light sources is obtainable.

2.3.1. Photographic Plates and Films

We shall give a short survey on the spectral ranges to which the photographic plates apply and refer to the data in the literature.

For the infrared spectral range with wavelengths exceeding 12,000 Å there are no photographic materials. But if we trace out an image on a fluorescent

Table 2.3. Photographic material for diverse wavelength ranges

Type of plate or film	Wavelength range Å [1])	Reference
Eastman-Kodak Company, diverse kinds of spectral plates, altogether	2000 – 12,000	[2.52] p. 164
ORWO spectral plates inclusive infrared plates	2300 – 10,800	prospectus
ORWO H_α-plate and H_α-film, low Schwarzschild effect	2300 – 6,900	prospectus
УФ-ГИКФИ and РЛ-1Л, sensibilized with Natriumsalicylate	150 – 2,200	[2.127]
Airphoto film (USSR, 1400 ГОСТ), sensibilized with Natriumsalicylate	at least 150 – 2,200	[2.106]
Airphoto film "ПАНХРОМ 10", sensibilized with machine oil	150 – 2,100	[2.64]
Agfa and Kodak X-ray films, sensibilized with Na-Salicylate	500 – 2,000	[2.47]
ORWO Schumann plate UV2	40 – 5,000	prospectus
НИКФИ – special film	23 – 113	[2.83]
Kodak S.W.R. film	at least 16 – 50	[2.39]
Ilford Q-Plate	9 – 950	[2.44a]

[1]) Given is the minimum range of sensitivity of the corresponding photographic material.

screen (e.g., with a NaCl-lens), which fluoresces over its entire surface and whose fluorescence intensity decreases with increasing temperature, a negative image of the body is formed on the screen, which can be determined photographically [2.34]. A like effect is obtained with a liquid crystal screen the colour of which changes above a critical temperature [2.64a].

2.3.2. Electrical Radiation Detectors

These will refer to those radiation detectors, which transform the incident radiation in such a way that it is indicated electrically. In table 2.4 we give the spectral ranges, which the various radiation detectors cover.

Nonselective radiation detectors (radiation thermocouples, bolometers) are particularly well suited for the determination of total radiative powers of plasmas [2.58], [2.97], [2.124]. By appropriate construction the frequency response is so fast (up to 20 kHz for bolometers, see [2.14c]) that alternating light technique can be applied which guarants in the stationary case a high sensitivity of detection.

For an absolute calibration of these detectors, electrical light sources with water-cooled walls can be used, whose total radiative power is determined by the difference between the electrical power and the power loss due to the cooling water. According to Neumann [2.97], the tubular XF 3000 Xenon high-pressure lamp is well suited as a light source.

From the group of selective radiation detectors the photomultiplier is the most appropriate for spectroscopic-plasmadiagnostical work because of its high sensitivy and fast time response (generally 0.1 μs [2.125a]). In special arrangements a time resolution of 10^{-9} s is attainable, see [2.46b], and also [2.3a].

The spectral response of the various types of photocathode-window-combinations is classified by the symbols S1···S20 and can be derived from [2.125a]. The upper wavelength limit is 12,000 Å. The lower limit of 1100 Å is given by the cutoff of transparency when using LiF-windows.

With growing limiting wavelength of the photocathode, that part of the dark current rises which originates from the thermoemission. The dark current which is of statistical nature [2.72] can be lowered 1. by reducing the light sensitive area of the photocathode by means of magnetic deflection [2.42a], [2.127b] and 2. by cooling of the photocathode. Some constructional details of photomultiplier coolers are given in [2.114c], [2.119a], [2.119c], [2.127b], [2.131c].

For wavelength below 5000–6000 Å the photocathode can be cooled down to −190 °C (temperature of liquid nitrogen), see [2.114c], [2.127b]. For longer wavelength the reduction in sensitivity of the photocathode due to the reduction of cathode temperature counteracts the reduction of dark current. Thus an optimal signal: noise ratio is obtained for cathode temperatures between −10 and −50 °C, see [2.42a], [2.127b], [2.136a].

For the detection of extremely weak light intensities it is advantageous to use the photomultiplier as quantum counter, see [2.114c].

When working with higher light intensities, care must be taken that the linearity of the output signal is preserved [2.83a].

After extremely intensive illumination of the photocathode (for instance when the photomultiplier is directly illuminated by daylight) reversible changes of sensitivity has been observed, lasting about a few hours (see, for example [2.53a]).

If radiation has to be measured in the vacuum ultraviolet or soft X-ray region, the radiation can be transformed by means of a fluorescent sheet or a scintillator crystal before arriving the photomultiplier giving a constant quantum efficiency over a relative broad spectral region. In the case of a fluorescent sheet of Na-Salicylate, the constancy of quantum efficiency has been confirmed for a wavelength region 800—2500 Å by Johnson, Watanabe and Tousey [2.61c].

For other vacuum UV detector systems see [2.81b], [2.125a] and the citation in table 2.4.

Table 2.4. Electrical radiation detectors

Type of radiation detector	Wavelength range [Å]	Reference
Thermocouple and bolometer, windowless	non selective	[2.89]
Photomultiplier without window	<5— 1400	[2.56], [2.44a], [2.49a]
Photomultiplier with scintillator	at least 1—16	[2.13], [2.59b], [2.81]
Photomultiplier coated with Na-Salicylate	200— 6000	[2.108], [2.114]
Photoelectrical tungsten UV detector	<300— 1200	[2.104]
Counter tube with foil window	10— 120	[2.91b]
I_2 Counter tube wit LiF window	1080— 1317	[2.31]
Ionization chamber	vacuum UV	[2.40]
Thermocouple, bolometer and GOLAY cell with diverse windows, see table 2.1	1100—400000	
Photomultiplier with window	1100— 12000	[2.53], [2.125a],
Selenium photovoltaic cell, VEB Carl Zeiss Jena	2000— 12000	prospectus
Phototube, Preßler, Leipzig	2000— 12000	prospectus
Photoconductive detector, VEB Carl Zeiss Jena	<4000— 35000	prospectus
InAs Photodiode, rise time 1/20 ns	8000— 34000	[2.17c]

In addition, we note that image converters can also be used as electrical radiation detectors. As a rule they are employed in such a way that the spectrum is projected on the image converter cathode and appears on a fluorescent screen. It is profitable that the spectrum in the infrared and ultraviolet spectral range can be directly observed [2.30]. Furthermore, we can increase the light intensity which is important for photographing low-intensity spectra.

2.3.3. Chemical Radiation Detectors

These are particularly suited for the determination of the radiative power in the vacuum ultraviolet spectral range. The radiation enters a reaction chamber containing a still or flowing molecular gas. The quantum yield of the reaction when the energy of the light quanta appreciably exceeds the threshold energy is independent of the energy, i.e., constant. Feates et al. [2.40] used a cell filled with CO_2 to detect the radiation in the range from 1100 to 1400 Å. The radiation entered the cell via an LiF-window. The quantum yield of the reaction $CO_2 - CO + O$, for which values between 1.9 and 0.98 dissociated molecules per light quantum are given in [2.50], [2.62], was taken by Feates and co-workers to be 1.0. The methods for setting up the measurement equipment and measuring the reaction yield will be discussed in detail.

2.4. Standard Sources

By comparing the radiation of the investigated source and that of a standard source absolute values can be obtained for the integral intensities of spectral lines or for the intensities per unit frequency, respectively.

It is a prerequisite for the standard sources that their intensity is constant in time for every wavelength that will be used. Furthermore, the spectral intensity distribution must be known in absolute measure (for pure line emitters it is sufficient to know the integral intensity of the diverse spectrum lines).

Below we shall treat a few types of standard sources.

2.4.1. Black Body Radiators

Black body radiators are usually employed in the form of a cavity radiator. This may consist of an electrically heated resistance body with suitable interior geometry in such a way that the black body emits the radiation according to the Planck radiation law.

Thus, the radiation which is emitted per unit surface area, frequency, and solid angle normal to the radiating surface (hole in cavity) is given by

$$B_\nu = \frac{2h\nu^3}{c^2} \frac{1}{e^{h\nu/kT} - 1} = \frac{1.472 \times 10^{-47} \nu^3}{e^{\frac{4.79 \cdot 10^{-11} \nu}{T}} - 1} \tag{2.19a}$$

(B_ν in erg cm^{-2} sr^{-1}, ν in s^{-1}, T in °K).

If the spectral intensity is taken per unit wavelength instead of unit frequency, we have

$$B_\lambda = \frac{2hc^2}{\lambda^5} \frac{1}{e^{\frac{hc}{\lambda kT}} - 1} = \frac{1.19\ 10^{-5} \lambda^{-5}}{e^{\frac{1.438}{\lambda T}} - 1} \tag{2.19b}$$

(B_λ in erg s^{-1} cm^{-3} sr^{-1}, λ in cm, T in °K).

Among black body radiators are some which work at relatively low temperatures and others for which the working temperature is relatively high (e.g., above 3000 °K). The latter in general have a shorter lifetime. An advantage, however, is that these, especially in the ultraviolet spectral range, emit much more strongly than black bodies, which are operated at relatively low temperatures.

Figure 2.10 shows a black body radiator designed by Lummer and Kurlbaum (see F. Henning [2.55]). It consists of an inner tube A of Marquardt material or magnesia with built-in diaphragms. The diaphragms on the right have apertures through which a thermocouple can be introduced into the chamber. The diaphragms on the left are built in such a way that only the radiation from the middle chamber K is observed. The cavity inside is blackened with iron, chromium, or cobalt oxide. Platinum ribbon is wound around the ceramic body and electrically heated. The ceramic tubes B and C serve as heat insulators. This type of black body radiator is suitable up to a working temperature of 1500 °C.

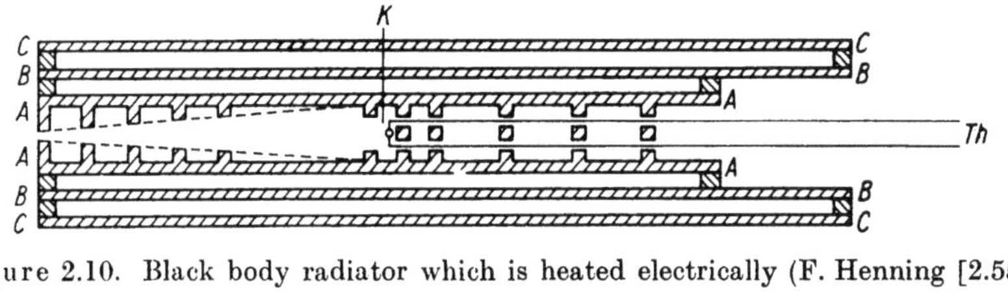

Figure 2.10. Black body radiator which is heated electrically (F. Henning [2.55])
A – tube of Marquardt material, B, C – heat-insulating ceramic tubes, Th – thermocouple, K – emission chamber

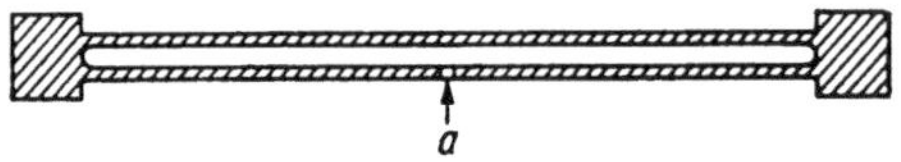

Figure 2.11. Schematic diagram of a black body radiator, consisting of an electrically heated metal tube with a small hole a, through which the radiation can be observed (Harrison et al. [2.52], p. 172)

Figure 2.11 illustrates a simple black body radiator, consisting of an electrically heated metal tube with a hole on one side through which the black body radiation can be observed. Solid metal rods, in which a small hole on the side can be drilled through which the radiation is observed, can also be used as black body radiators, provided the depth of the hole is several times larger than the diameter of the hole (Euler and Ludwig [2.38], p. 140).

Mehltretter [2.88] constructed a tungsten tube furnace, which can be used as a black body radiator up to a temperature of 2900 °K.

Quinn und Barber [2.106b] used a tungsten tube of 2 mm inner diameter mounted within a lamp bulb and obtained a working temperature up to 2700 °C.

Higher temperatures can be obtained with electrical heated carbon tubes. A construction for working temperature up to 3500 °K is given by Dahm [2.29a] who used an inductively heated carbon tube.

Anacker and Mannkopff [2.1a] developed a black body radiator, consisting of an electrically heated carbon tube with a slit on the side. This radiator can work up to the sublimation temperature of carbon (4000 °K). The life-time of the carbon tube which is clamped between water-cooled holders amounts, however, only to a few minutes. The reproducibility of the source at the sublimation temperature is very good.

The anode crater of the carbon arc

A very simple standard source is the positive crater of a homogeneous carbon arc, when the current of the arc is maintained just below the hissing temperature. According to Null and Lozier [2.99] this crater is a black body radiator, i.e., it possesses an absorptivity of about 1. Of course the overlapping CN-emission bands in the range from 3700—3900 Å exert a disturbing effect, which can be suppressed when the arc is burned in CO_2 (Euler [2.37]) or in Ar [2.38a], [2.61a].

The electrodes are placed in such a way that the anode lies horizontal and the cathode either points with the tip upwards or is inclined at about 30° to the normal so that both electrodes form an obtuse angle. Before stable values of the luminous density of the anode crater are obtained after the discharge is triggered the arc must be burned about 2 min.

To increase the stability of the arc, it is advisable to stabilize the discharge current, for example, by inserting screen sections or barretters in the discharge path.

For every choice of the electrode material there are definite differences in the luminous density behavior. Null and Lozier established that when employing graphite anodes the maximum luminous density in the center of the crater occurs immediately before the hissing point (current intensity at which the arc starts to hiss), whereas in the case of carbon anodes the maximum crater luminance occurs at a current intensity, which lies 10% below the hissing point. However, it is difficult to maintain this current intensity with accuracy. Thus, the application of graphite anodes is more suitable.

For anodes the authors recommend the use of 1/4 in. diameter "National Grade SPK", produced by the National Carbon Co. USA: A suitable cathode is the "National Standard Purity AGKS" graphite cathode (1/8 inch diameter). The working current intensity is around 12.3—13.3 A. The authors found that in the range between 3000 and 42,000 Å the radiation follows the Planck's radiation law for a temperature of 3800 °K.

Using the same type of electrodes, Hattenburg [2.53b] found that the radiation intensity is comparable to that of a black body radiator at a temperature of 3792 °K with the exception that at some wavelength bands the emission is somewhat higher. For the true crater temperature a value of 3810 °K was found [2.117 a].

Euler [2.37] recommends the use of the spectroscopic carbon Ringsdorf Spektral RW II (5 mm diameter) for anodes. Any kind of homogeneous carbon can be taken for the cathode, such as for example, thin copperplated carbon (4 mm diameter). Euler obtained the value 3806 °K for the black body temperature. According to his measurements the actual temperature of $T = 3996$ °K is appreciably higher than the above value, and the surface of the anode has a wavelength-dependent absorption coefficient (0.717 for 2500 Å up to 0.80 for 18,000 Å). Anacker and Mannkopff [2.2] obtained nearly the same values for the absorption coefficient of carbon. These results contradict the latest results of Null and Lozier, who obtained an absorption coefficient of almost 1 for the arc with the RW II-carbon (see also [2.17 a])*. In addition, Null and Lozier specified smaller values than those of Euler for the intensities in the spectral range <6000 Å.

In order to clear this discrepancy, Magdeburg and Schley [2.83 b], [2.83 c] performed some measurements with an improved technique. They were able to show that the data of Null and Lozier are more realistic than those of Euler. For the true maximal temperature of the crater of the RW II anode of diameter 6.35 mm they found a value of about 3960 °K at a current of 8.6 A which was about 1.4 A below the hissing current. The authors found that maximal stability of radiation intensity is obtained at a somewhat lower current intensity of 7.3 A, and for this current they have tabulated the datas for the emissivity and for the spectral intensity in the spectral range 0.25—15 μm. The contribution of the continuous radiation from the arc plasma to the total spectral intensity is included in the tabulation. For $\lambda > 0.3\,\mu$m this contribution is $<3\%$, while for $\lambda < 0.25\,\mu$m this contribution is in excess of 20%. Because this contribution is sensitive against changes in geometry, the carbon crater ceases to be a radiation standard for wavelengths $\lambda < 0.25\,\mu$m.

2.4.2. Surface Sources

We shall first consider the tungsten filament lamp which is unsurpassed in the simplicity of its operation. The radiation is emitted by an electrically heated tungsten filament with dimensions between 20×2 and 8×0.5 mm. The intensity is given by the Kirchhoff-formula

$$I_\nu = A_\nu B_\nu \quad \text{or} \quad I_\lambda = A_\lambda B_\lambda \tag{2.20}$$

where B_ν or B_λ, respectively, is obtained from (2.19) and A_ν or A_λ, respectively, is the wavelength-dependent absorptivity of the tungsten filament. The filament is enclosed in a glass bulb, which is either evacuated or filled with an inert

* In [2.17 a] an actual temperature of 3830 °K and a wavelength-independent absorptivity $A_\nu = 0.97$ were assumed for the RW II-crater.

gas. Better designs have a window, through which the radiation passes. The rear side of the bulb must be slanted, so that the radiation reflected from this side will not pass along the principal ray path. Figure 2.12 illustrates such a tungsten-filament lamp.

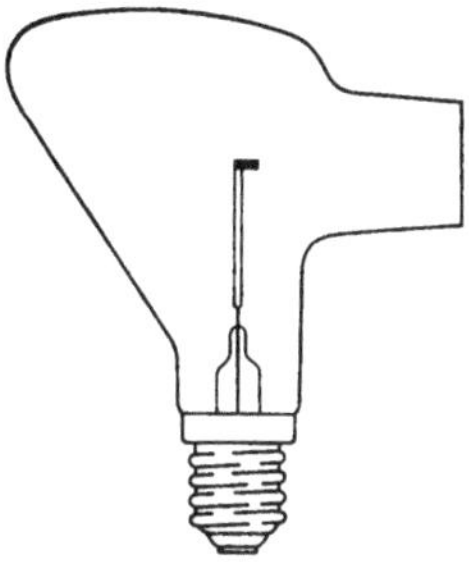

Figure 2.12. Tungsten-filament lamp with fused-on plane window (Willenberg [2.134])

Designs employing glass bulbs are suitable only for wavelengths between 4000 and 20,000 Å. Versions with quartz windows display an absorption beginning at a wavelength below 2500 Å and thus can be still used for wavelengths somewhat below 2000 Å, when allowing for the wavelength-dependent absorption of the wall. Fused quartz transmits in the infrared up to 35,000 Å. Information on the transmissivity of windows can be found, for example, in [2.38] (p. 155 and 156).

The transmissivity of the bulb and the window is always less than 1 because of losses due to reflection. The reflection losses depend on the refractive index and are given by

$$R = (n - 1)^2/(n + 1)^2 . \tag{2.21}$$

For indexes between 1.42 and 1.74 the reflection loss amounts to 3.02 and 7.29%, respectively ([2.38], p. 157), assuming normal incidence of the radiation. The transmission factor for several interfaces (number $= x$), through which the radiation must successively penetrate is

$$D = (1 - R)^x . \tag{2.22}$$

An additional decrease in the transmissivity occurs, because tungsten or tungsten nitride can evaporate on the bulb of the filament lamp and on the exit window. This film can already be present when the filament lamp is supplied by the manufacturer from old stock. It is therefore advisable in each case to check the transmission factors of the bulb and of the window before measurements are started. Filament lamps with a symmetrical bulb or with two adjacent windows (W_2 in Figure 2.13) are irradiated with a second filament (W_1 in Figure 2.13) or an incandescent lamp, and the radiation transmitted by the lamp is measured with a photovoltaic cell-galvanometer system. Then the irradiated filament lamp is removed from the radiation path and the unimpaired radiation incident on the photocell is measured. It is thereby important to be sure that all the light which passes the filament lamp in the middle of the radiation path is detected by the light-sensitive element of the photocell. The light sensitivity of the photovoltaic cell must be equally distributed over the receiving surface. Further, the rays are to pass the filament lamp in the immediate vicinity of the tungsten filament, but no radiation is allowed to fall on the filament itself. From the galvanometer deflections A_1 and A_2 with and without the filament lamp in the radiation path, respectively, the transmission factor of the window and the bulb is given by

$$D = (A_1/A_2)^{1/2} . \tag{2.23}$$

Figure 2.13 gives a measuring system for the determination of the transmissivity of the filament lamp.

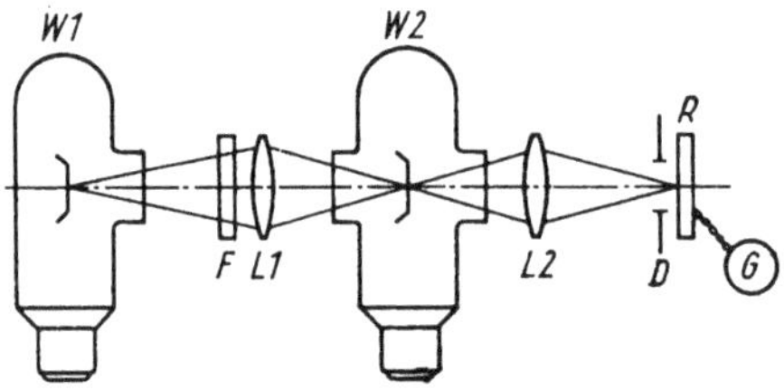

Figure 2.13. Measuring system for determining the transmission factor of the window and glass bulb of a tungsten-filament lamp (Euler and Ludwig [2.38])

W1 – auxiliary lamp, *F* – filter, *L1* and *L2* – lenses, *W2* – lamp to be measured, – *D* – diaphragm, *R* – radiation detector with galvanometer *G*

The transmission factor of the bulb and window should be determined for several wavelengths. For this purpose different interference filters F or other monochromatic filters are inserted in the radiation path.

When working with filament lamps it should be ensured that it is always mounted in the same position when used as a standard source and when determining the temperature of the tungsten-filament. This is especially desirable in the case of gas-filled lamps in which an additional cooling of the tungsten filament results due to convection. For lamps operating on direct current, the filament temperature can depend on the current direction. Thus, the filament lamp must be operated and calibrated under the same polarization conditions. By convention, the positive pole is to be connected to the outer part of the lamp socket [2.59a].

It should be ensured that the direct current intensity is maintained constant. In order to maintain the current intensity manually it is absolutely necessary to use a precision device which gives accurate readings.

To get the spectral energy distribution of the tungsten filament, the black body temperature S_1 is first measured as with a partial radiation pyrometer (wavelength $\lambda = 6500$ Å). The actual temperature T of the tungsten filament (often designated by W in photometric literature) is found via Wien's radiation law

$$B_\nu = \frac{2\,h\,\nu^3}{c^2}\,\mathrm{e}^{-h\nu/kT}, \tag{2.24}$$

Kirchhoff's law (equation 2.20), the transmission factor D_1 of the bulb or window, and the spectral emissivity A_1^* of the tungsten filament, and is given by

$$\frac{1}{T} = \frac{1}{S_1} + \frac{\lambda_1}{c_2}\ln(A_1\,D_1) \tag{2.25}$$

where

$$c_2 = h\,c/k = 1.438 \cdot 10^8\ \text{Å} \cdot {}^\circ\mathrm{K}\,. \tag{2.26}$$

After the temperature T is known, the effective intensity of the tungsten-filament lamp for an arbitrary wavelength is defined by

$$I_\nu = A_\nu\,D_\nu\,B_\nu \tag{2.27}$$

where the spectral emissivity $A_\nu = A_\lambda$ of the tungsten filament depends on the temperature. This has been measured by various investigators ([2.51], [2.57], [2.77], [2.78], [2.128] and [2.129]). We shall use the A_ν-value measured by Hamaker and De Vos in Tables 2.5 and 2.6 ([2.89]), p. 65 and [2.38], p. 151).

A comparison of the values of Tables 2.5 and 2.6 shows that the measured values in some cases deviate by more than 1% from one another. It is assumed that these deviations originate in the different quality of the individual tungsten filaments used in different measurements.

Larrabee [2.77], [2.78] also gives tabulated values for the emissivity of tungsten between 3500 and 8000 Å in the temperature range from 1600 to 2400 °K. Moreover, he found the following analytical relationship between the emissivity A, the temperature T, and the wavelength λ (λ in units of 1000 Å, T in °K):

1. For the range from 3500 to 4500 Å, $T = 1600-2400$ °K:

   $$A = 0.6075 - 0.03000\,\lambda - 0.3265 \cdot 10^{-4}\,T + 0.05900 \cdot 10^{-4}\,\lambda\,T;$$

2. for the range from 4500 to 6800 Å, $T = 1600-2400$ °K:

   $$A = 0.4655 + 0.001558\,\lambda + 0.2675 \cdot 10^{-4}\,T - 0.07305 \cdot 10^{-4}\,\lambda\,T;$$

3. for the range from 6800 to 8000 Å and $T = 1600-2400$ °K:

   $$A = 0.6552 - 0.2633\,\lambda - 0.7333 \cdot 10^{-4}\,T + 0.7417 \cdot 10^{-4}\,\lambda\,T\,.$$

A tungsten surface source with a large reproducibility of the emissivity was constructed by Blau, Martin, and Chaffee [2.16]. It consists of a tungsten monocrystal, whose surface, which emits the radiation we are interested in, is the

* For A_1 initially an approximate value, for example, $A_1(\lambda_1, S)$ is taken.

Table 2.5. Spectral emissivity A_ν of tungsten for different absolute temperatures T [2.89] p. 65 (measurements of Hamaker)

λ in mμ	$T =$ 1000	1200	1400	1600	1800	2000	2200	2400	2600	2800	3000 °K
230	0.414	0.411	0.408	0.406	0.403	0.400	0.398	0.395	0.392	0.390	0.387
240	0.440	0.438	0.435	0.432	0.429	0.427	0.424	0.421	0.418	0.416	0.413
250	0.470	0.467	0.465	0.462	0.459	0.456	0.453	0.450	0.448	0.445	0.442
275	0.494	0.491	0.488	0.485	0.482	0.478	0.475	0.472	0.469	0.466	0.463
300	0.496	0.493	0.491	0.488	0.486	0.483	0.481	0.478	0.475	0.473	0.470
325	0.481	0.479	0.477	0.476	0.474	0.472	0.470	0.469	0.467	0.465	0.463
350	0.472	0.471	0.470	0.469	0.468	0.467	0.466	0.465	0.464	0.463	0.462
375	0.479	0.478	0.477	0.476	0.475	0.473	0.472	0.471	0.470	0.469	0.468
400	0.487	0.484	0.482	0.479	0.476	0.474	0.471	0.468	0.466	0.463	0.460
425	0.484	0.481	0.477	0.473	0.470	0.467	0.463	0.460	0.457	0.453	0.450
450	0.480	0.477	0.473	0.470	0.466	0.463	0.459	0.456	0.452	0.449	0.445
500	0.468	0.465	0.462	0.459	0.456	0.453	0.450	0.447	0.444	0.441	0.438
550	0.460	0.458	0.456	0.453	0.451	0.448	0.446	0.443	0.441	0.439	0.436
600	0.453	0.451	0.449	0.447	0.445	0.443	0.441	0.438	0.436	0.434	0.432
650	0.446	0.444	0.442	0.440	0.438	0.436	0.434	0.432	0.430	0.428	0.426
700	0.442	0.440	0.438	0.436	0.434	0.431	0.429	0.427	0.425	0.423	0.420
750	0.443	0.438	0.434	0.430	0.426	0.422	0.418	0.414	0.410	0.405	0.401
800	0.441	0.434	0.427	0.420	0.412	0.405	0.398	0.391	0.384	0.377	0.370
850	0.436	0.428	0.419	0.410	0.401	0.393	0.384	0.375	0.366	0.358	0.349
900	0.426	0.416	0.407	0.397	0.388	0.378	0.369	0.359	0.350	0.340	0 331
950	0.415	0.406	0.397	0.388	0.379	0.370	0.361	0.352	0.343	0.334	0.325
1000	0.396	0.388	0.380	0.372	0.364	0.355	0.347	0.339	0.331	0.323	0.315
1050	0.375	0.368	0.361	0.354	0.348	0.341	0.334	0.327	0.320	0.313	0.306
1100	–	–	–	0.345	0.334	0.322	0.311	0.299	0.288	0.276	0.265
1200	–	–	–	0.317	0.308	0.298	0.289	0.279	0.270	0.260	0.251
1300	–	–	–	0.295	0.288	0.280	0.273	0.266	0.258	0.251	0.244
1400	–	–	–	0.278	0.273	0.268	0.263	0.258	0.253	0.247	0.242
1500	–	–	–	0.264	0.261	0.258	0.256	0.253	0.250	0.247	0.244
1600	–	–	–	0.252	0.251	0.251	0.250	0.249	0.249	0.248	0.247
1700	–	–	–	0.242	0.243	0.245	0.246	0.247	0.248	0.250	0.251
1800	–	–	–	0.233	0.236	0.240	0.243	0.246	0.249	0.253	0.256
1900	–	–	–	0.225	0.230	0.234	0.239	0.243	0.248	0.252	0.257
2000	–	–	–	0.217	0.223	0.228	0.234	0.239	0.245	0.251	0.256

(110)-plane with the highest atom density and thus at higher working temperatures (until 3000 °K) is not subjected to thermal etching but remains smooth. The heating up to the piece of tungsten is effected by a high-frequency system with a field concentrator. The reproducibility of the emissivity is excellent.

In the study is further described an interesting method for determining the emissivity of tungsten.

Table 2.6. Spectral emissivity A_ν of tungsten for different absolute temperatures T [2.38], p. 151 (measurements of De Vos)

λ in mμ	T = 1600	1800	2000	2200	2400	2600	2800	2900 °K PTB
250	(0.447)	(0.441)	(0.435)	(0.428)	0.423	0.417	0.411	–
300	0.482	0.478	0.474	0.470	0.465	0.460	0.456	0.424
350	0.479	0.476	0.473	0.470	0.467	0.465	0.461	0.446
400	0.481	0.477	0.474	0.471	0.467	0.464	0.461	0.453
450	0.474	0.471	0.467	0.464	0.460	0.457	0.454	0.450
500	0.469	0.465	0.462	0.458	0.455	0.451	0.448	0.443
550	0.464	0.460	0.456	0.453	0.450	0.446	0.443	0.439
600	0.455	0.451	0.448	0.444	0.441	0.437	0.434	0.436
650	0.449	0.446	0.442	0.438	0.434	0.430	0.427	0.430
700	0.444	0.440	0.436	0.432	0.427	0.423	0.419	0.423
750	0.438	0.433	0.428	0.423	0.418	0.414	0.410	0.412
800	0.431	0.425	0.420	0.414	0.409	0.404	0.402	0.400
900	0.413	0.407	0.401	0.395	0.391	0.387	0.383	0.385
1000	0.390	0.386	0.381	0.377	0.373	0.370	0.367	0.369
1100	0.367	0.364	0.361	0.358	0.356	0.354	0.352	0.353
1200	0.334	0.333	0.331	0.330	0.329	0.328	0.328	0.329
1300	(0.322)	(0.322)	0.323	0.323	0.324	0.324	0.325	0.324
1400	(0.300)	0.303	0.305	0.307	0.310	0.311	0.313	0.311
1500	(0.281)	(0.284)	0.288	0.291	0.296	0.299	0.302	0.298
1600	(0.264)	(0.268)	0.273	0.278	0.284	0.288	0.292	0.287
1800	(0.234)	(0.241)	0.247	0.255	0.262	0.268	0.275	0.265
2000	(0.210)	(0.219)	0.227	0.235	0.243	0.251	0.259	0.244
2200	(0.190)	(0.201)	0.210	0.218	0.228	0.236	0.244	0.228
2400	(0.176)	(0.187)	0.196	0.206	0.215	0.224	0.234	0.215
2600	(0.164)	(0.175)	0.185	0.195	0.205	0.214	0.224	0.205

2.4.3. Gaseous Discharges as Standard Sources

Gaseous discharges can be used as standard sources provided the discharge tubes can be reproducibly manufactured and insofar as it is also ensured that the emitted radiation, both in regard to the absolute energy emitted and the intensity as well as with regard to the spectral composition of the radiation, give reproducible values. Before employing this source the intensity must be measured as a function of frequency or wavelength.

We differentiate between two main types of gasdischarge radiation sources: Those which are useful for the quartz-optical region and those which are useful for the vacuum ultraviolet region.

2.4.3.1. Gaseous discharges as standard sources for the quartz-optical region

In 1937, Krefft, Rössler, and Rüttenauer [2.74] proposed the use of a wall-stabilized high-pressure mercury arc of suitable design as a standard source (UV-standard). In [2.109] Rössler gives the irradiation intensity F for 50 lines between 2302 and 17193 Å at a distancce of 1 m from the lamp in units of $\mu W/cm^2$. Furthermore, the irradiation intensity for continuous radiation is given in units of $\mu W/cm^2$ Å. Also indications are given on how to obtain data on the intensity I of the radiation source by converting the above values, and some values for I are given. The conversion factor is a function of the excitation potential of the lines or the continuum. For the continuum we can assume an excitation potential of 8.6. V.

In [2.110] are given values for the irradiation intensity in the infrared spectral range at a distance of 1 m from the lamp for Hg-lines in the range from 1 to 4 μm and for continuous radiation in the range from 1–8.5 μm.

Newer measurements of the spectral intensity distribution of the UV-standard have been published by Zschaeck [2.136c] for the spectral region 0.21–10 μm (continuum) respectively 0.230–2.32 μm (spectrum lines).

According to Wende [2.131a], [2.131b], a cascade arc burning in argon at atmospheric pressure can be used as radiation standard of high reproducibility ($\pm 2\%$) for the spectral region 2500–6750 Å. In contrast to the mercury arc, the continuous spectrum is more pronounced. Constructive details are given, together with data for the continuous spectral intensity for end-on observation. For $\lambda = 2500$ Å, the continuous intensity of the arc under investigation surpasses that of the carbon arc crater by a factor of 270.

For spectral intensity calibrations when high-speed photographs are taken of the spectra of intensely radiating plasmas, pulse discharge which produce a high radiation density, are of particular importance.

Rath, Manthey and Bötticher investigating a short-arc-pulsed discharge in argon with an initial pressure of 60 atm, have published datas for the spectral intensity distribution between 2200 and 4500 Å. The pulse shape was rectangular, the pulse length was 1.9 ms. The discharge emits practically continuous radiation characterized by a distribution temperature of 8200 °K. The intensity is reproducible within 10–15%. Gabriel [2.43] designed a tubular pulse lamp, through which a squarewave current pulse of 1200 A flowed in a period of 15 μs. Over a time interval of 15 μs the intensity is constant. The emission corresponds to that of a black body at $T = 26{,}000$ °K, which is verified by comparison of the radiation with a calibrated tungsten-filament lamp.

Using a capillary discharge tube filled with argon with an initially pressure of 20 atm, Manthey and Bötticher [2.86b] obtained within the wavelength intervall 2000–6000 Å black body radiation characterized by a temperature of $3 \cdot 10^4$ °K, lasting at a constant level for 2 μs.

Bogen et al. [2.17b] discharged a capacitor through a polyethylene channel of 0.5 mm diameter and 8 cm channel length and found for the spectral region from near infrared to near ultraviolet black body radiation with a temperature $3.5 \cdot 10^5$ °K$-4.5 \cdot 10^5$ °K, lasting for 60 ns. The reproducibility of intensity was $\pm 4\%$.

Further details on pulse lamps can be found in the comprehensive paper of Marschak [2.87]. Likewise, certain information on pulsed light sources can be found in section 6.1 in connection with the line-reversal method.

2.4.3.2. Gaseous discharges as standard sources for the vacuum ultraviolet region. Secondary standards

From the stationary types of vacuum ultraviolet light sources, there is at first to mention the well known hydrogen discharge tube (hydrogen lamp) which operates at a gas pressure of a few Torr. The lamp is suitable for use in the spectral range from 1100 to 2500 Å [2.52] p. 191, [2.65], [2.80], [2.127c]. In the range from 1100 to 1800 Å the spectrum is composed of bands, whereas the spectral intensity in the range from 1800 to 2500 Å is uniform.

In general the lamps are used as light sources for absorption analysis. When employed as standard sources the wavelength dependence of the intensity must be measured beforehand in an absolute scale with a radiation detector, which is nonselective in the ultraviolet spectral range. The hydrogen lamps can be employed in closed form with quartz windows or, if the lamp is to be connected to a vacuum spectrometer, as an open system in conjunction with diaphragms and pressure stage sections [2.127c]. In this case the lamp, which is usually water-cooled, operates in flowing hydrogen (Figure 2.14).

Feates et al. [2.40] employed a stationary argon-neon discharge as a light source for the vacuum ultraviolet. The discharge takes place in a capillary of 2 mm inner diameter. The electrode supports and the capillary are water-cooled. The pressure of the lamp filling is about 10 Torr. The lamp can be

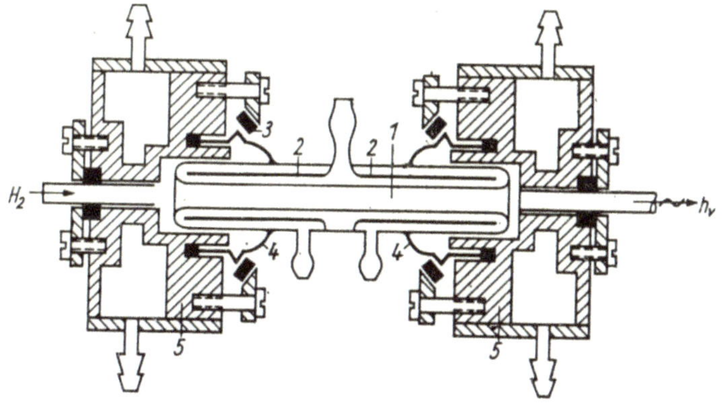

Figure 2.14. Hydrogen discharge tube (Vilesov et al. [2.127c])

1 — discharge capillary, 2 — water supply pipes, 3 — rubber sealing, 4 — quartz collars, 5 — electrodes

operated steadily at an electrical power absorption of up to 6 kW. Supplying the lamp with a LiF-window, a spectral range between 1100 and 1400 Å, for which the lamp emits intense radiation, can be utilized.

Samson and Liebl [2.114], [2.114a] proposed a duoplasmatron-low-pressure discharge in hydrogen or argon as an ultraviolet source for absorption measurements. The relevant wavelength range in the windowless version comprised 500—2000 Å. To determine the spectral energy distribution a multiplier arrangement with a sodium salicylate fluorescent coating can be used, giving at least in the spectral region from 800 to 2000 Å a constant quantum yield.

Tanaka et al. [2.122] produced with condensor discharges at a working pressure of about 800 Torr high-intensity inert gas continua in the short-wave ultraviolet. The connection of the discharge tube to the vacuum spectrograph was effected via a diaphragm which is only opened when the photograph is taken, so that little gas is allowed into the vacuum spectrograph. The utilizable wavelength range lies between 500 and 2000 Å.

An X-ray source for producing spectrum lines in the region from 10 to 120 Å is described in [2.91b].

2.4.3.3. Gaseous discharges as standard sources for the vacuum ultraviolet region. Absolute standards

The predicate "absolute standard" in connection with discharge sources is interpreted in the sense that some characteristic emission datas of the discharge source in the vacuum ultraviolet region will be definitely known in an absolute scale after a few characteristic quantities of the discharge source has been measured by using visible or near-visible spectroscopic technique.

Boldt [2.18] produced black body radiation in the range of the hydrogen line $L_\alpha \lambda = 1216$ Å with a high-current cascade arc. The working pressure was about 1 atm. A cascade arc is shown in Figure 2.15. Via 1, 2, and 3 gas (usually argon) is introduced, while via Z a second gas, for example, hydrogen, is led into the arc chamber. The contraction of the nozzle between the supply channels 2 and 3 ensures that the hydrogen cannot reach the vicinity of the arc cathodes where sharp temperature gradients exist. Thus, when observing through the cathode the self-absorption of the lines of the second gas is eliminated.

The cascade arc can be connected to a vacuum system when a multiple air lock (system of diaphragms with pressure stages) is inserted.

The intensity of the L_α radiation which is optically thick is given by the Planck formula (2.19). The precision with which the emitted radiation in the black body range can be indicated (20%) depends on the precision with which the plasma temperature (for argon as the main gas $T = 11{,}900$ °K) is determined.

An analogous radiation standard in which the spectrum lines N I 1493 respectively 1743 Å give black radiation is described by Streißle [2.121c].

Griffin and McWhirter [2.49] pointed out a further method (branching ratio method) to obtain line radiation of known intensity. In the vacuum ultraviolet they used the fact that optically thin emitted lines which possess a common upper level are related to one another by a constant intensity ratio. If one of the lines lies in the visible range, its intensity can be measured by

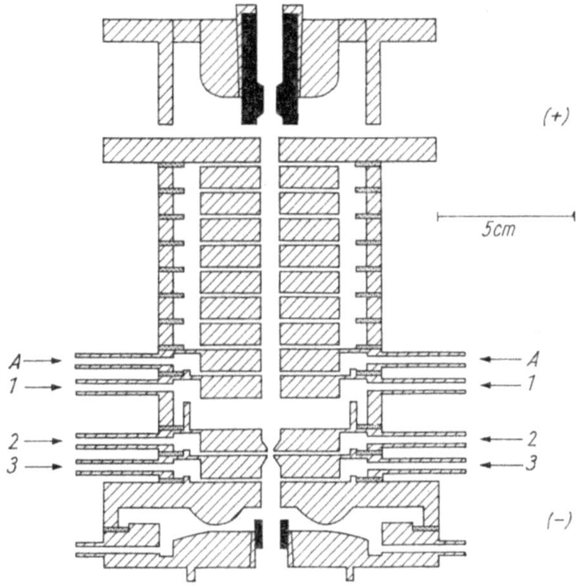

Figure 2.15. Cascade arc for the production of black body radiation in the range of spectral lines (after Boldt [2.18])

1, *2*, and *3* – Gas supply channels, *A* – supply channel of the second gas

conventional means, for example, by employing a tungsten filament lamp. When the intensity ratio is known, the intensity of the second line, which may lie in the vacuum ultraviolet range, is also fixed.

The experiments of Griffin and McWhirter are specifically based on the two hydrogen lines Lyman $\beta, \lambda = 1025$ Å and Balmer $\alpha, \lambda = 6563$ Å, which are emitted by a zeta-discharge. The ratio of the radiated quanta per unit time is $I_{L_\beta} : I_{H_\alpha} = 1.25$.

An analogeous calibration technique is used by van Eck and de Heer [2.34d]. The characteristic line pair is He I 5016 and 537 Å originating from the transitions $3\,^1P \to 2\,^1S$ and $3\,^1P \to 1\,^1S$. The radiation is emitted from helium atoms originating from a 30 keV ion ray which had been neutralized in a target gas (H_2 or Ne) by charge-exchange.

An extension of the branching ratio method to the band spectrum of CO molecules was performed by Aarts, de Heer et al. [2.1], [2.114b].

A similar calibration principle proposed by Conrads [2.27c] is based on the fact that the continuous emission coefficient of a plasma consisting of hydrogen-like ions and electrons is well defined when the plasma parameters are known. The light source under consideration is a pulsed plasma produced by a condensor discharge through a LiH-tube of 20 mm length and 2 mm inner diameter. In the spectral region from 40 to 100 Å, the spectral emission is mainly due to radiative recombination processes occurring by collisions between the hydrogen like Li IV ions and the plasma electrons. While the electron density N_e (maximal value $= 6.10^{18}$ cm^{-3}) and the electron temperature T_e (maximal value 30 000 °K) can be measured by conventional spectroscopic means, the intensity of the continuous radiation is found by calculation.

In [2.136b], the synchrotron radiation is used as a calibrating radiation source for a wavelength region down to a few multiples of 10 Å.

2.5. Special Apparatus or Systems for High-speed Spectroscopy

In this section we do not consider those systems, in which the spectrally resolved radiation is incident upon a radiation detector (e.g., secondary electron multiplier) and is recorded as a function of time, unless several wavelengths are simultaneously recorded.

Next we shall discuss those systems for which there is a wider spectral range in which an instantaneous photograph can be taken. Then spectroscopic apparatus with time resolution will be dealt with, and subsequently, spectroscopic framing cameras. Furthermore, we shall touch upon a few recording techniques, among which those in which the radiation for several wavelengths is simultaneously recorded by an oscillograph. Finally, we shall deal with systems effecting a high-speed recording of spectral line profiles.

2.5.1. Instantaneous Recording of Spectra

By making instantaneous photographs of spectra one can obtain a general view of the spectral intensity distribution, resolved in terms of one spatial coordinate, at a definite instant of time. When we are dealing with the photography of a pulse-triggered plasma, which is reproducible, it becomes possible through repeated recording of instantaneous spectra at different instants of time after the plasma starts to radiate to obtain a general idea of the spectral energy distribution as a function of a spatial coordinate as well as a function of time.

Wurster [2.135] constructed a mechanical high-speed shutter, which blocks the light after a period of time of 30—40 μs upon arrival of an electric signal in the analyzing device. It essentially consists of a slider, which is activated by the shock wave originating from a wire explosion, subsequently causing the closure of a slit.

Camm [2.25] designed a mechanical high-speed shutter, which can be mounted directly on to a spectrograph slit. It consists of a slider with a 20 μm wide slit. The slider blocks the light path in the rest position (see Figure 2.16). If now a current pulse is sent through a hairpin-shaped chrome-nickel wire by a thyratron-triggered capacitor discharge, the thermal expansion causes the wire to push the slider in such a way that the slit in the slider passes across the spectrograph slit (slit width 20 μm) and effectively allows free passage of

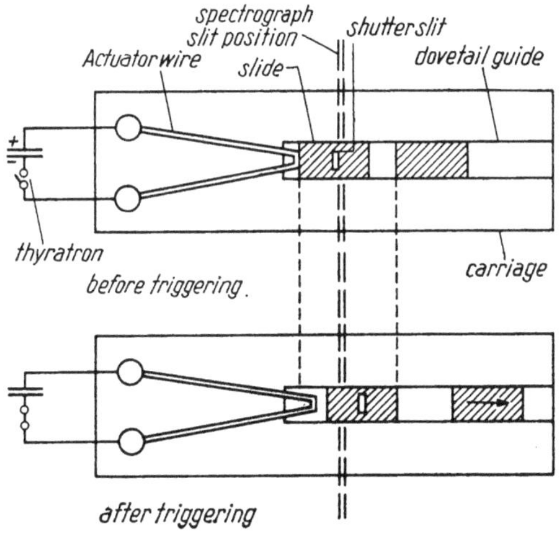

Figure 2.16. High-speed shutter for spectrography (after Camm [2.25]). The actuation results from a thermal expansion of an electrically short-time heated chrome-nickel wire. Top: before exposure; bottom: after exposure

light for a period of time of 5 μs. The slider is not connected to the chrome-nickel wire and must be slid back into the rest position after the process is completed. The time delay between the onset of the capacitor discharge and the shutter opening time is about 20 μs. The electromagnetic driven high speed shutter constructed by Avellone et al. [2.4a] allows a minimum exposure of 5 μs with a delay time of 40 μs.

Clayton [2.27b] describes an analogous system in which the shutter is driven by the action of an electrically exploded metal foil and moves over the spectrographic slit. The delay time is 19 μs, the exposure time 2 μs.

A shutter including an electromagnetic driven metal strip is described in [2.30b], [2.105b]. In [2.30b], the shutter closes the light pass 10 μs after triggering. In [2.105b], the light pass is opened after a delay time of 100 μs for a operating time of 35 μs.

Another type of light-interrupting shutter allowing a greater aperture consists of two glass plates, between which a lattice of lead wire is placed. After the wire is electrically exploded, the light transmission is practically stopped [2.34b].

Weber [2.130] reported on a mechanical shutter for exposure times of $1-20\,\mu s$. It consists of two fixed and one movable prism. These are arranged in such a manner that under normal conditions the light cannot pass through the prisms due to total reflection. However, if the middle prism is brought into contact for a short time with the other two, the light is transmitted through the prism system as if through a plane parallel plate. The short exposure time is achieved in that the movable prism is brought to collide against the fixed prisms and is then deflected after contact.

Wiese [2.132] designed a high-speed spectrograph shutter in which the basic part is a slit S_1 which rotates about a vertical axis. The exposure is realized when the image of S_1, obtained by means of Lens L_2 (Figure 2.17),

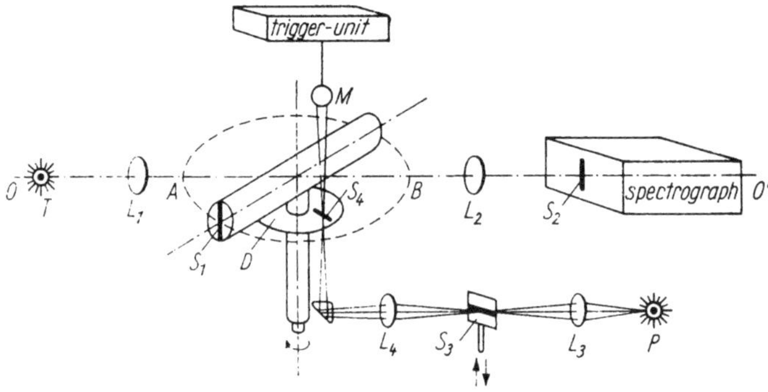

Figure 2.17. Mechanical high-speed spectrograph shutter (after Wiese [2.132])
T – light source, S_1 – rotating slit, S_2 – spectrograph slit, L_1 and L_2 – image-forming lenses. The second ray path associated with the source P serves for synchronization during the photographing of short-time plasmas

moves across the spectrograph slit S_2. To illuminate the slit, an image of the plasma T is formed via the lens L_2 either at point A or point B, past which the slit must move.

The total opening time t of the shutter is given by

$$t = \frac{d_1 + a\,d_2}{2\,\pi\,n\,l}\,, \tag{2.28}$$

where d_1 and d_2 respectively denote the slit width of S_1 and S_2, $1/a$ the magnification factor for the image of S_1 on S_2, n the number of revolutions per sec of the cylinder, on which the slit S is fixed, and l the distance of S_1 from the axis of rotation.

In the limiting case the following values can be chosen: $d_1 = d_2 = 10\,\mu m$, $a = 1$, $n = 330\,s^{-1}$, $l = 15$ cm. This gives a shortest possible exposure time of approximately $6 \cdot 10^{-8}$ s.

In a similar way we can obtain instantaneous photographs of spectra by inserting a rotating disk between the plasma and the entrance slit of the spectrograph [2.7], [2.13b], [2.22b], [2.96], [2.115], [2.116]. This disk has a radial slit whose long sides are (insofar as possible) not precisely parallel but slanted in such a way that they converge to the axis of rotation (constant aperture). Such an arrangement is schematically shown in Figure 2.18 (Schneider [2.115]). The plasma is imaged via L_1 on the rotating slit S_1, which in turn is imaged via L_2 on the spectrograph slit S_2. The ray path for synchronizing the onset of the pulse discharge to be investigated with the angular arrangement of the rotating disk is also shown in the figure. Formula (2.28) holds also for the exposure time in this case, when l is the distance from the middle of the slit to the axis of rotation. Schneider gave the following typical values for the

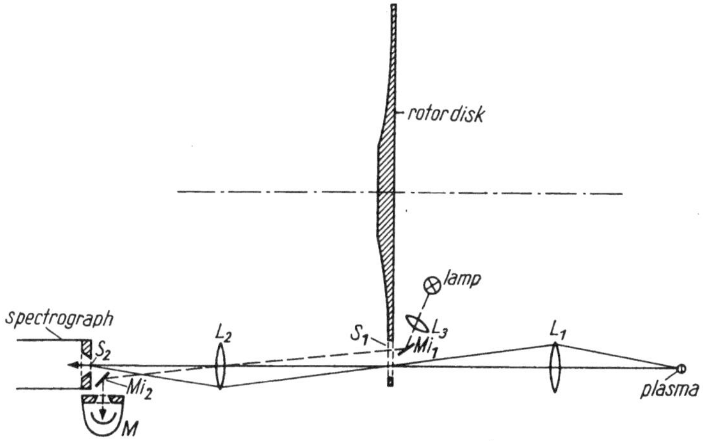

Figure 2.18. Mechanical high-speed spectrograph shutter with rotating slit S_1 (after Schneider [2.115])

Principal ray path: plasma — L_1 — S_1 — L_2 — entrance slit S_2 of the spectrograph. Ray path for synchronization of the onset of the discharge of the pulsed-produced plasma with the phase relationship of the rotating slit: lamp — L_3 — $Mi1$ — L_2 — $Mi2$ — photocell or multiplier M

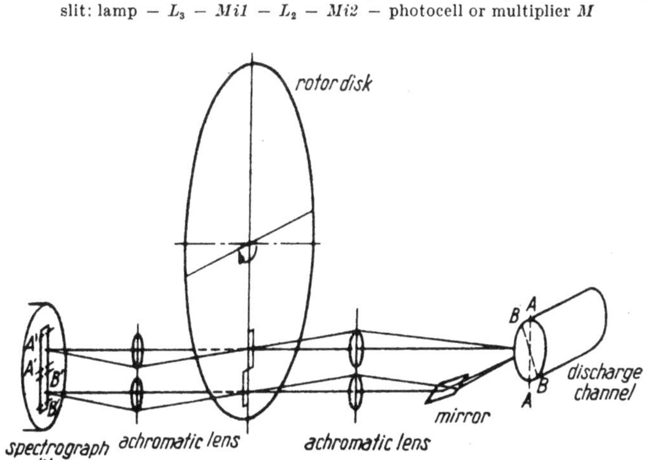

Figure 2.19. Mechanical high-speed shutter for the production of successive instantaneous spectra (after Schneider and Mailänder [2.116])

shutter data: $l = 10$ cm, $n = 333\ s^{-1}$, $d_1 = d_2 = 20\ \mu m$, $a = 1$, and obtained an exposure time of 10^{-7} s. By decreasing the slit widths by a factor of 2 a lowering of the shutter time by the same factor is possible.

Schneider and Mailänder [2.116] constructed a rotating-disk system with 2 slits, which are somewhat displaced with respect to one another in such a way that 1) the rays are transmitted by the second slit somewhat later, and 2) the image of the second slit is formed either above or below that of the first slit on the spectrograph slit. In order that the same portions of the plasma can be photographed, four lenses instead of two are used for forming the optical image (Figure 2.19). With this arrangement a plasma spectrum can be produced at two different instances of time on the same plate.

To prevent double exposures, in systems with rotating slits we usually make use of a second mechanical shutter, which transmits the radiation only during a single rotation of the rotating slit. The second slit can be omitted only if the plasma radiates at the most for the length of time of a full revolution of the slit, or when the plasma has a periodic intensity and oscillates synchronously with the rotating slit.

With central shutters the shortest shutter times that can be achieved are 1/250 s. If shorter switching times are required, one can refer back to an arrangement of Schneider and Mailänder [2.116], which is schematically shown in Figure 2.20. Besides the slit disk there is a second switching disk with

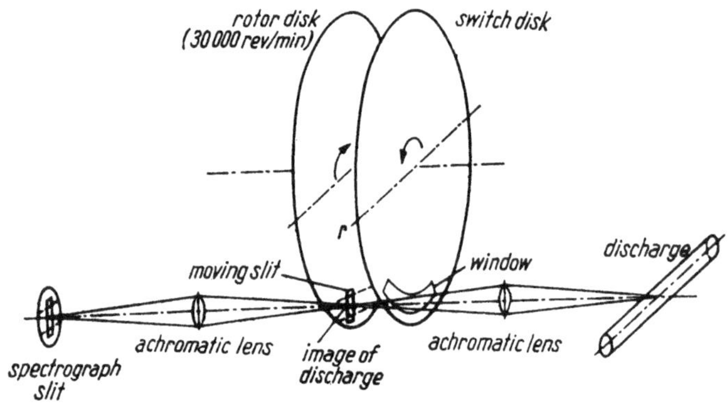

Figure 2.20. High-speed spectrographic shutter with a second mechanical shutter which opens once (after Schneider and Mailänder [2.116])

a window section coaxial to the former. Just before exposure the selector disk is given a sudden push by a strong spring and is subsequently retarded after half a revolution by a corresponding mechanism. The width of the window (10 mm) is chosen in such a way that during the time in which the window lets the radiation through, the image of the rotating slit passes across the spectrograph slit only once. The synchronization circuit can be found in [2.116].

A high-speed shutter with a rotating mirror was designed by Bardócz [2.8]. Here, the light source *Sp* behind the slit S_1 is imaged via a rotating concave mirror (or via a rotating plane mirror-lens combination) on the slit S_2 (entrance slit of the spectrograph) (see Figure 2.21, in which the complementary ray path for synchronization is also shown). Expression (2.28) gives the shutter

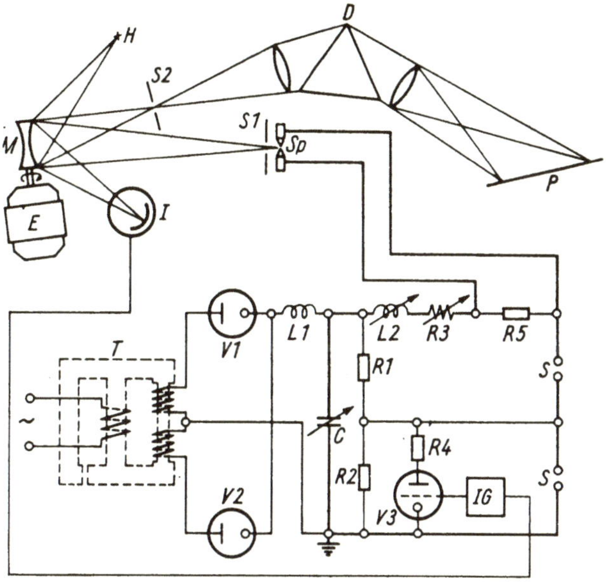

Figure 2.21. High-speed spectrograph shutter with rotating concave mirror (after Bardócz [2.8])

The slit S_1 in front of the spark plasma *Sp* is imaged via the rotating concave mirror *M* on the plane of the entrance slit S_2 of the spectrograph. The synchronization circuit is also shown

time, when l is taken as twice the distance between the rotating mirror M and the spectrograph slit S_2. For suitably large l and n, extremely short exposure times can be achieved.

Van Calker and Tacke [2.24] used a rotating prism as the high-speed shutter for a spectrograph. They did not employ a second slit and the image of the plasma (spark discharge) was directly projected on the spectrograph slit via a lens and the rotating prism (Figure 2.22). With such an arrangement van Calker and Tacke took photographs of spectra of periodically recurring spark plasmas in different phase relationships.

A Kerr-cell shutter was employed by Beams [2.12] as well by Larenz and Dunnington [2.76]. Such systems have the disadvantage that the Kerr fluid absorbs, i.e., the light is never completely excluded from a shutter of this type. Furthermore, when using nitrobenzene ultraviolet rays are no longer transmitted. In addition, the polarizers absorb radiation, especially when foil polarizers are employed. Kerr cells are rarely used in high-speed spectroscopy,

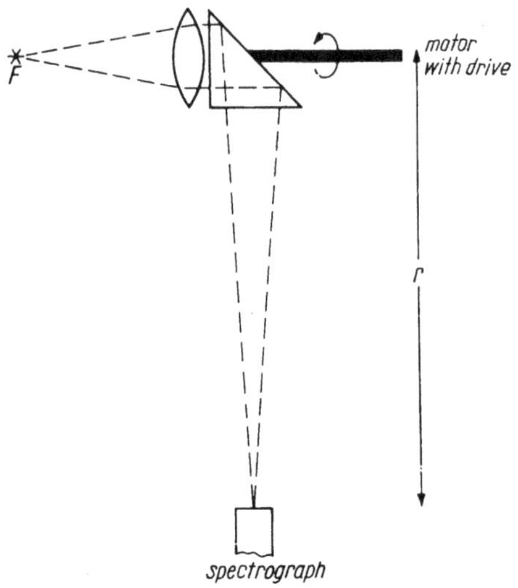

Figure 2.22. High-speed spectrograph shutter with rotating prism (after Van Calker and Tacke [2.24])

The image of the process at F is formed after reflection at the hypotenuse plane of the 90 -prism on a circle of radius r

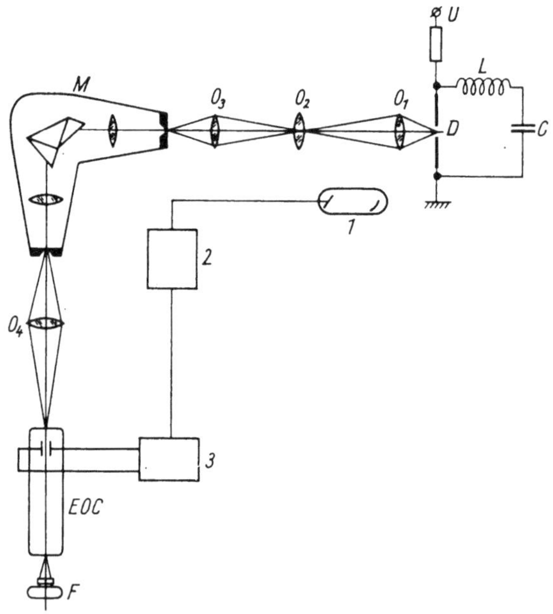

Figure 2.23. Instantaneous photograph of a spectral section by means of an electron-optical image converter (after Balashov et al. [2.5])

D – pulse-produced plasma, M – monochromator with wide exit slit, EOC – electron-optical image converter, F – photographic camera, 1 – multiplier for detection of the synchronizing light signal, 2 – adjustable timedelay circuit, 3 – square-voltage generator since the time resolution of 10^{-8} s can also be achieved by means of a rotating mirror system.

Very short times of exposure can be achieved with the use of an electron-optical image converter (EOC). Figure 2.23 shows such a system used by Balashov et al. [2.5]. A section of the spectrum which appears in the wide exit slit of a monochromator, is imaged upon the photocathode of the EOC (Figure 2.23). In the normal state the electron rays starting from the photocathode are deflected toward the side. An enlarged and a more intense image of the spectrum appears on the screen only when a square-pulse voltage is applied to the deflection capacitor. From the screen a photograph of the spectrum can be taken.

The intensifying action of the image converter tube allows to get exposure times down to 20 ns [2.84a], [2.115b] which can hardly be obtained by conventional methods.

Additionally intensifying the spectrum allows one to get spectra of higher spectral resolution, is possible, when the intensity distribution of the image on the luminescent screen is recorded in the direction of spectral dispersion by means of a photomultiplier-scanning-equipment [2.115b] (see also section 2.5.4). To get more time for registration, the application of an afterglowing luminescent screen is essential.

A spectrum of higher intensity can be obtained also with a system in which an image intensifier is set behind the image converter. Thus an additionally intensifying factor of 10^6 is obtainable [2.81b].

2.5.2. Continuous Time-resolved Spectra

By employing suitable devices, of which we shall enumerate 6 in this section, spectra can be photographed in such a way that blackening density can be read off as a function of wavelength as well as of time. We shall now discuss the various devices of this type.

Type 1. A rotating mirror serves as the time-resolving element (see [2.6], [2.8], [2.9], [2.29], and [2.43]). The ray path in a typical experimental arrangement of this type is shown in Figure 2.24 [2.29]. The image of the plasma F is formed on a horizontal slit, the image of which is formed via the concave mirror M_1, the plane mirror M_2, and the rotating plane mirror RM on a surface tangent to the entrance slit of the spectrograph S_2, so that the image of S_1, which lies diagonally across S_2, passes in front of slit S_2. In this way a time-resolved spectrum such as shown in Figure 2.25 is obtained.

Instead of a rotating plane mirror a rotating concave mirror can also be used [2.6], [2.8], which simultaneously ensures the formation of the optical image.

For objects which emit low-intensity light, the spectrum can be repeatedly taken at the same point, provided the plasma can be generated reproducibly and the synchronization is sufficiently accurate. A time resolution of 0.3 μs for repeated exposures can be directly obtained. For objects emitting intense light a higher time resolution is possible. Gabriel [2.43] obtained a time resolution of 10^{-7} s.

Type 2. Here the time-resolving element consists of a disk with beveled, spiral-like or several concentric slits, which are distributed in stepped fashion around the disk [2.8], [2.84], [2.113]. The disk rotates as a rule directly in front of the spectrograph slit. In this manner the slits pass in front of the

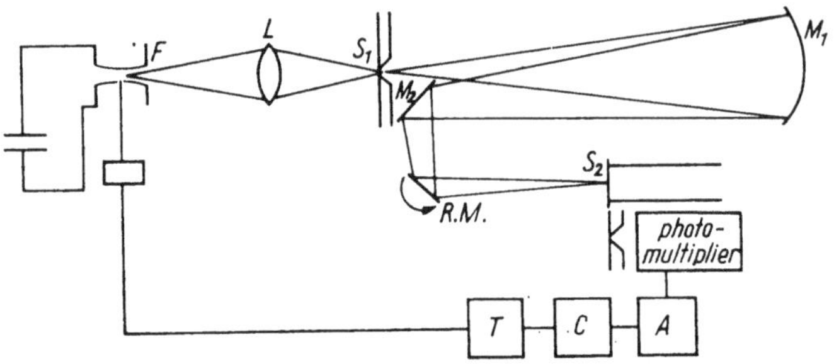

Figure 2.24. Time-resolved spectroscopy by means of a rotating mirror (after Curzon and Greig [2.29])

F – light source, *RM* – rotating mirror, S_2 – entrance slit of the spectrograph, *A* – amplifer, *C* – control mechanism, *T* – trigger system

spectrograph slit, so that different points of the slit are successively exposed at different heights. Similarly as in type 1 the spectrum appears resolved according to time on the spectroscopic plate. When applying stepped slits (Figure 2.26) one obtains time markings on the time-resolved spectrum. With type 2-arrangements a time resolution of up to 2 μs can be attained.

Types 3—5. The time-resolving element is located behind the camera lens of the spectrograph. The entrance slit of the spectrograph is maintained at a small slit height.

In type 3 the film in the photographic plane of the spectrograph, in which the spectrum is formed, moves linearly and at right angles to the dispersion direction (see Figure 2.27a). A corresponding system, which gives a time resolution up to 10^{-6} s, is described in [2.21].

In type 4 (Figure 2.27b) the image of the spectrum is formed on a rotating film inside a film drum [2.21], [2.91], [2.112], [2.123]. A time resolution of 10^{-7} s is guaranteed with this system. A typical film drum speed is 10,000 rev/min. The diameter of the drum is of the order of a few tens of centimeters.

In type 5 (Figure 2.27c) the image of the spectrum is formed via a camera lens and a rotating mirror on a circular film at rest [2.9b], [2.21], [2.44], [2.118]. When employing sufficiently rapid running motors to drive the rotating mirror (of the order of 1000 rev/sec) a time resolution of up to 10^{-8} s can be reached.

Type 6. The image of the spectrum is projected on the photocathode of an mage converter and then photographed from the screen. Inside the image

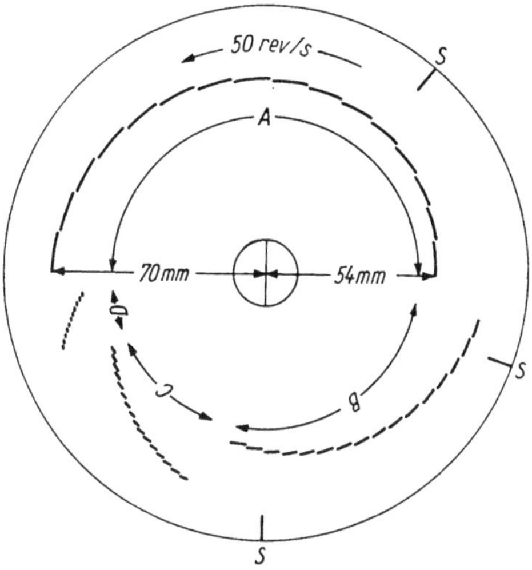

Figure 2.26. Rotating disk with step-like slits for the production of time-resolved spectra with time markings (after McK. Nobbs [2.84])

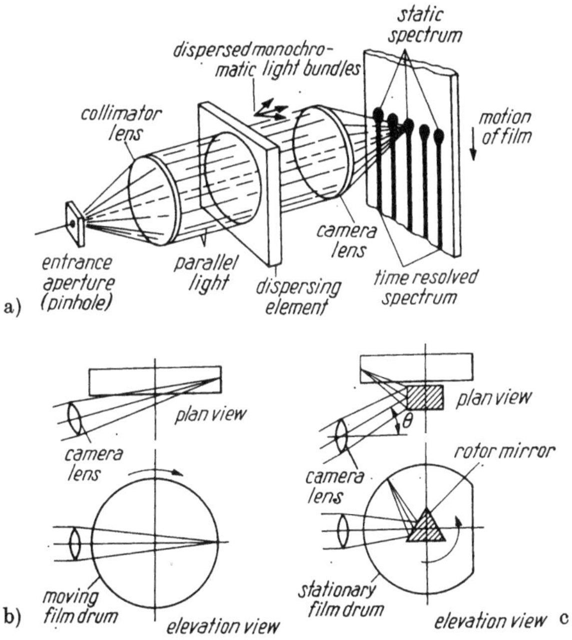

Figure 2.27. Certain types of time-resolving spectrographs. The time-resolving elements are located behind the camera lens (after Bovey and Gabriel [2.21])

a) film moving in a plane; b) rotating film; c) film movement by means of a rotating mirror

converter is located a magnetic quadrupole lens which allows the scale of the image to be changed independently in both directions. For example, in the dispersion direction the spectrum can be enlarged and in the direction perpendicular thereto it can be so strongly contracted that without other manipulation a hairline-type spectrum is formed. The time resolution is effected by deflection of the electron rays inside the image converter, so that the hairline-type spectrum on the fluorescent screen shifts with time in a direction perpendicular to the dispersion direction [2.23].

An improved time-scanning system is described by Liu and Steinberg [2.81 b], in which the time-resolved spectrum appearing on the luminescent screen is intensified by means of a second image converter. An intensifying factor of up to 10^6 can be obtained.

2.5.3. Spectroscopic Framing Devices

These devices are used to photograph in short time sequences spatially resolved spectra with a short exposure time. They consist of a combination of a spectrograph and a high-speed framing camera [2.20], [2.123]. The spectrograph can be placed either in front or behind the high-speed camera. The ray diagram for the former system [2.20] is given in Figure 2.28. The image of

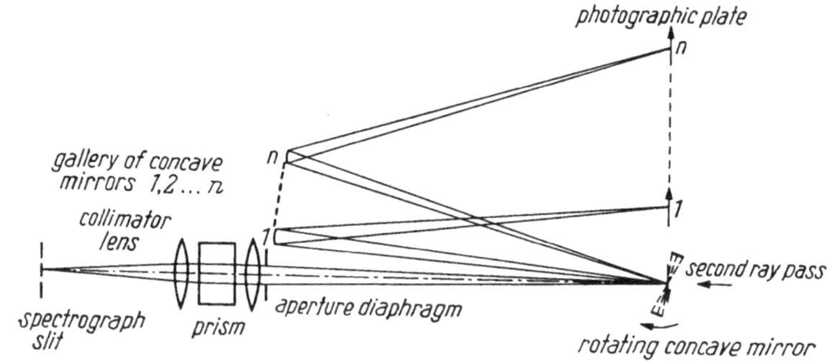

Figure 2.28. Ray diagram in a spectroscopic framing device (after Bortfeldt [2.20])

the spectrum is projected on a rotating concave mirror and then successively imaged via a system of concave mirrors on the photographic plate at n different discrete points.

In another version the radiation is first transmitted through a time-resolving element, and then the spectrum is analyzed (Figure 2.29). The image of the discharge is formed via n objective concave mirrors on n field concave mirrors. By means of a lens the image is formed on the spectrograph slit. Between the lens and the slit diaphragm is located a rotating plane mirror, which lets the field mirror images or the plasma images formed in the field mirrors pass one by one in front of the spectrograph slit. In this manner, several spectra are

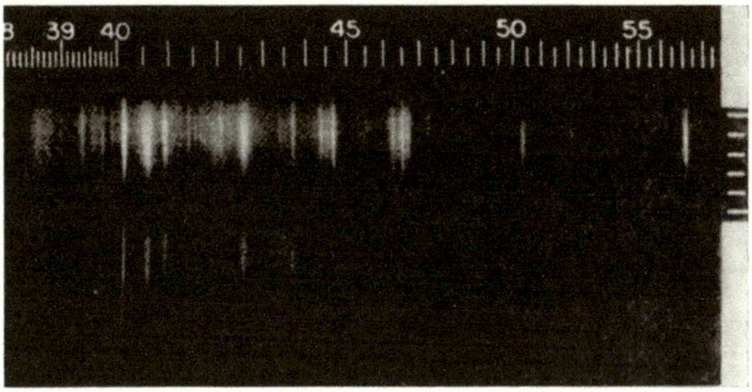

Figure 2.25. Time-resolved spectrum (after Curzon and Greig [2.29]).

Time scale: 1.1 μs per time marking

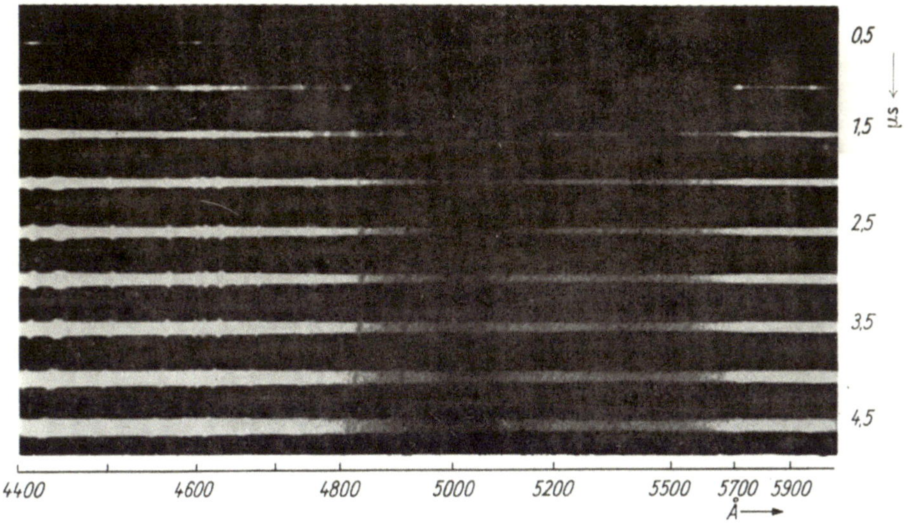

Figure 2.30. Spectroscopic kinematographic photograph of an exploding wire in argon (after Bortfeldt [2.20])

formed one after another at different plate heights. In the extreme case a series of spectra can be obtained with a sequence of 10^{-8} s. Figure 2.30 shows a photograph obtained by the last-mentioned method (exploding wire in argon).

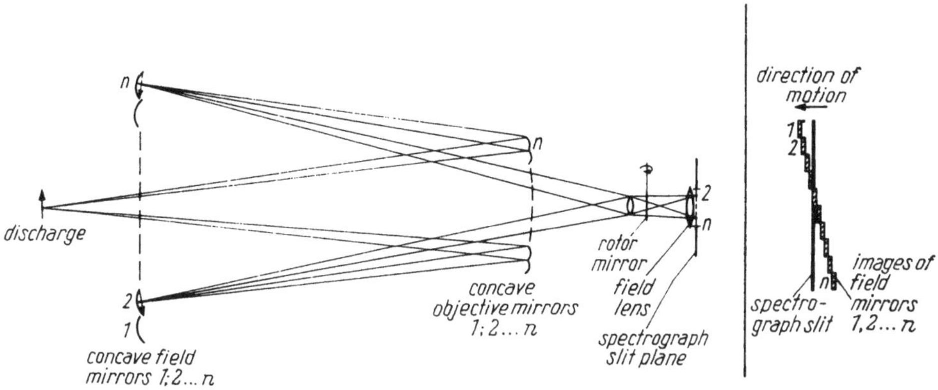

Figure 2.29. Spectroscopic framing device with a spectrograph located behind (after J. Bortfeldt [2.20])

2.5.4. Special Recording Methods for Spectroscopic Studies

2.5.4.1. Production of monochromatic pictures with a spectrograph

In the device described by Schneider [2.115a], the image of the plasma under investigation is horizontally swept over the entrance slit of the spectrograph. Behind the exit slit there is moving a photographic film also in the horizontal direction. By suitable adjustment of the film speed a monochromatic picture of the plasma object on the film is obtained.

2.5.4.2. High speed recording with spatial respectively spectral resolution

In order to determine the radial intensity distribution Schneider [2.115] formed the image of a vertical, axially symmetric discharge via a rotating mirror in such a way that the image passed in front of the entrance slit of the spectrograph in a short time (about 10^{-5} s) in the horizontal direction. Instead of the photographic plate an exit slit with a photomultiplier is used, so that the radial intensity distribution for a spectral wavelength can be recorded on an oscillograph. For horizontal plasmas the image of the plasma must be rotated through 90° with the help of two suitably arranged mirrors.

If one attaches importance to absolute simultaneity when reording the radial intensity distribution we can obtain a spectrum section photographed with a short exposure time (e.g., 10^{-7} s) via a spectrograph-image converter combination (see section 2.5.1), when the image converter is employed as a high-speed electronic shutter. The fluorescent screen displays afterglow, so that a longer time is available, in which the spectrum can be scanned photoelectrically via a rotating mirror and a multiplier (for example, in the manner whereby the radial intensity distribution for a given spectral position (spectral line) appears on the oscillograph screen). Figure 2.31 schematically illustrates such a detection system.

When the axis of the rotating mirror is turned by an angle of 90°, the last arrangement can also be used to rapidly record the spectral energy distribution on an oscillograph, whereby a spatial resolution is foregone (see also [2.115b]).

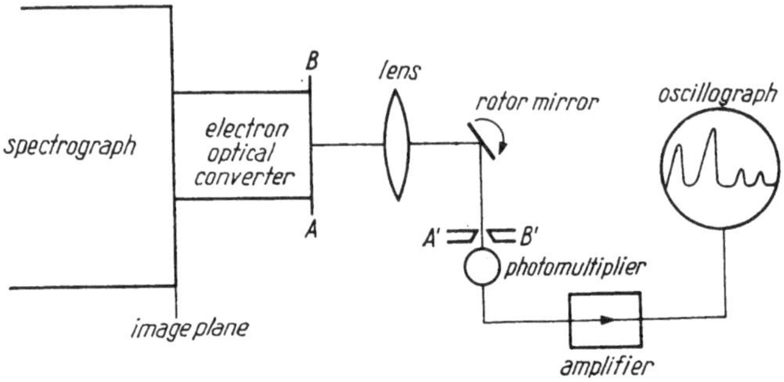

Figure 2.31. Device for recording the radial intensity distribution of a spectrum obtained by an image converter. The image converter is used as a high-speed shutter device. The spectrum which is displayed on the afterglowing screen is scanned photoelectrically (after Schneider et al. [2.117])

An improved, similar spectrum recording system in which the image converter and the photomultiplier are coupled directly forming a single unit in the manner that no luminescent screen is needed, is described in [2.93a].

If somewhat more time is at our disposal, high-speed recording of the spectral energy distribution can be performed with the help of a rotating disk in the position where the spectrum occurs [2.69]. The rotating disk has a spiral-like

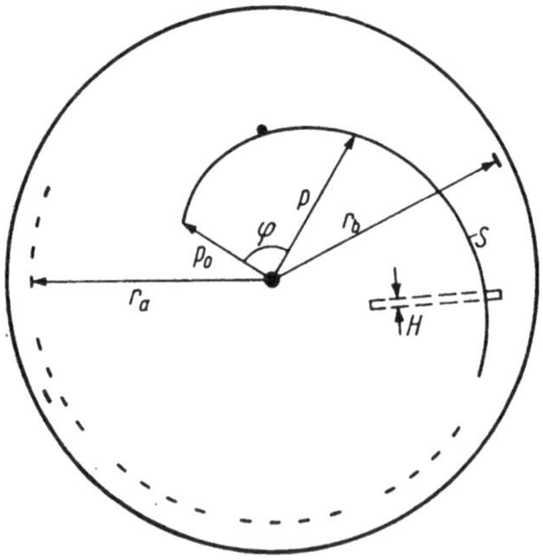

Figure 2.32. Rotating disk in the image plane of a spectrograph with a spiral-like slit for high-speed recording of the spectral energy distribution (after Koltsov [2.69])

S – spiral slit, H – horizontal slit

slit (see Figure 2.32), behind which a linear slit is located whose longitudinal side extends in the direction of the dispersion. Behind the slit there is a photoelectric detector (e.g., a photomultiplier) with a recording arrangement (such as a multiplier-oscillograph system). With a suitable shape of the spiral a recording curve with a linear wavelength scale can be obtained. The total recording time depends among other things on the number of revolutions of the disk. A typical value is $\Delta t = 1/100$ s.

Another high-speed spectral scanning device consists of a monochromator which includes in the interior a rotating mirror as an optical element, so that the spectrum image is moved in dispersion direction across the exit slit. The signal from the photomultiplier, placed behind the exit slit, is displayed by an oscilloscope. This device is especially useful for the recording of spectral line profiles. A typical recording time is 10^{-7} s [2.14b], [2.27a].

A similar improved device which allows to record a spectral line profile repeatedly with a repetition time of about 3 μs is described by Hill [2.56a].

In [2.35c] another system for high-speed recording of spectral line profiles is presented in which the rotating mirror is placed outside the spectrograph. A section of the spectrum which appears at the output plane of the spectrograph is projected on a slit behind which a photomultiplier coupled to an oscilloscope is placed. When the mirror rotates, the spectrum image is swept across the slit with a speed of 0.11 Å/μs, thus a high-speed record of spectral intensity distribution is obtained.

2.5.4.3. Multichannel systems for spectrum registration (Polychromators)

The application of polychromators enables us to simultaneously record the spectral intensity of a plasma as a function of the time for different wavelengths. As a rule 7 to 10 channel arrangements are used, i.e., the spectral intensity for 7 to 10 different wavelengths is simultaneously recorded.

Apart from the cases where it is of interest to simultaneously record the time dependence of the intensity for different spectral lines or different wavelengths of the continuous spectrum, polychromators are mainly used in those cases where line profiles are to be determined [2.2a], [2.14a], [2.21], [2.91a], [2.100], [2.120].

Polychromators are spectrographs or monochromators having a high linear dispersion, and whereby in the image plane various sliding slits are usually arranged which let through only a small range of wavelengths, each of which falls on a multiplier and is recorded as a function of time. For the recording of line profiles one can no longer employ normal slits in conjunction with multipliers, since these require too much space. The problem of incorporating the multiplier is solved by means of light guides. Glass fibres (see Bovey and Gabriel [2.21]) or improvised flat pieces of plexiglas (see Spillman et al. [2.120]) and also thin glassfoils [2.91a] are suitable for this purpose. The light guides conduct the radiation from the image plane of the spectrograph to the multipliers. Due to the flexibility of the light guides, the arrangement of the photomultiplier involves no difficulties.

In Spillman's version of a polychromator with 9 channels the image of the spectral section occurring in the image plane of the monochromator is enlarged via a cylindrical lens (plexiglas cylinder) in the dispersion direction on a plane, which is formed by the forepart of the light guides consisting of 1.6-mm thick plane pieces of plexiglas (Figure 2.33). The magnified spectrum has a linear dispersion of about 0.3 Å/mm, so that the light guides, whose width determines the resolution, can guarantee an absolute resolution of 0.5 Å.

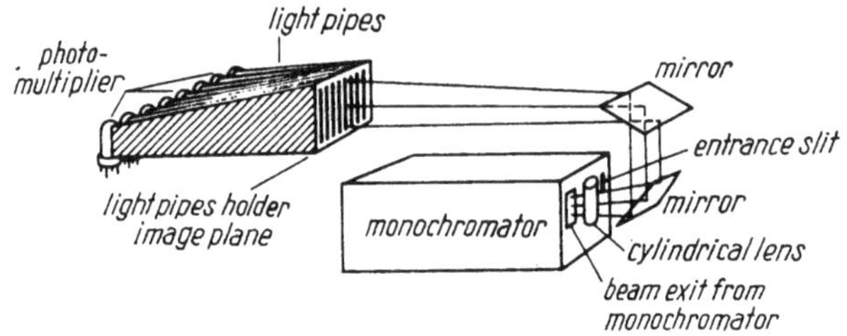

Figure 2.33. Polychromator with plexiglas photoconductors for the determination of time-resolved line profiles (after Spillman et al. [2.120])

Bovey and Gabriel [2.21] obtained a spectral resolution of 0.15 Å upon employing a suitable number of 50-μm glass fibres as light guides.

Using 0.2 mm thick glass foils, Miyamoto et al. [2.91a] obtained a wavelength resolution of about 0.05 Å.

To avoid the restriction to longer wavelengths given by the absorption of ultraviolet radiation in the light guides made from glass or from plexiglass, Bernstein [2.14a] has constructed a polychromator whose image plane is faced by a set of 0.15 mm thick plane mirrors which split the radiation within a given wavelength intervall into 8 separate channels, ending at the cathodes of 8 photomultipliers. Mirror optics are used throughout. A wavelength resolution of 0.12 Å was obtained.

The wavelength resolution of the polychromators is sufficient to determine the profiles of spectral lines emitted in high temperature plasmas because in these plasmas nearly all the spectral lines emitted by elements with an atomic number which is not too high are so strongly broadened due to the Doppler effect that the line width comes into the order of 1 Å. Moreover, the resolution of polychromators is sufficient to record Stark-profiles of hydrogenic Balmer lines for electron densities in excess of 10^{12} cm^{-3} (see [2.61b]).

If the recording of the spectral intensity is to be carried out only for two closely spaced wavelengths, we can insert a prism beam splitter in the image plane of the spectrograph. The prism reflects all the radiation in a certain spectral range for $\lambda > \lambda_0$ in one, for $\lambda < \lambda_0$ in another direction, and in this manner also enables the insertion of the slits and multipliers (see, for example [2.113]).

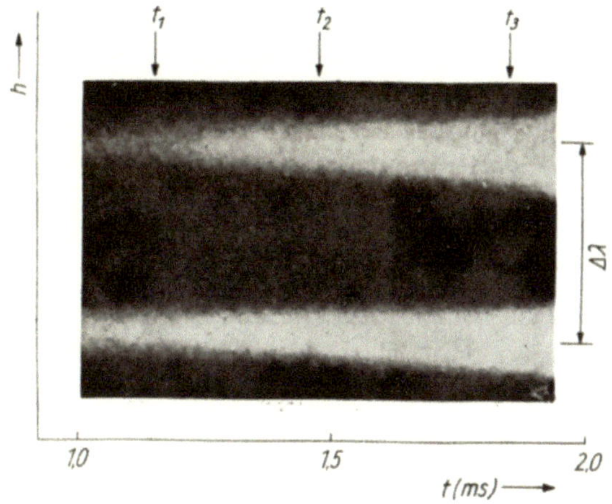

Figure 2.34. Time-resolved Fabry-Perot interference fringes of the Hg I 4047 Å spectral line. Two interference fringes are visible, from whose profile the true line profile can be determined (after Malyshev et al. [2.86])

A light splitting system using as beam splitter a semitransparent mirror, is described in [2.127d].

According to Vanhauvermeiren [2.127a], light splitting into three channels can be achieved by the use of a semitransparent mirror in series with a prismatic light splitter.

The problem of inserting 3 multipliers behind 3 slits located close to one another was solved by Hearn, Jones, and Ramsden [2.54] by letting the light ray through the middle slit fall directly upon a multiplier, while behind the other slits deflecting prisms were placed, which deflected the light coming through these two slits to both sides, where it fell upon two other multipliers. This system was used to record the time variations of the Doppler width of spectral lines in a high-temperature plasma.

2.5.4.4. High-speed recording of line profiles by means of a Fabry-Perot interferometer

The Fabry-Perot interferometer in this case is usually crossed with a spectrograph (see Figure 2.5). In the exit plane of the spectrograph an interference pattern appears, as shown in Figure 2.7. Each spectral line displays a pattern of interference fringes transverse to the dispersion direction of the spectrograph, and photometric recording of this pattern gives the line profile.

In order to achieve high-speed recording of the line profile a spectral line in the image plane of the spectrograph is firstly delimited by means of a vertical slit, and then by means of a horizontal slit a narrow range of the interference pattern, which corresponds to a well-defined narrow wavelength interval, is delimited. Now the light path between the interferometer plates is rapidly changed. This can be accomplished either by attaching one of the interferometer plates to a barium titanate oscillator (Cooper and Greig [2.28], [2.28a], [2.104a]) or by rapidly changing the gas pressure in the space between the plates [2.15], [2.73]. Bradley [2.22] constructed an interferometer, in which the interplate separation is altered by purely mechanical means. This causes a shift in the interference fringe pattern, so that successively radiation of different wavelengths falls upon a photomultiplier and is recorded by an oscillograph. After a few geometrical corrections of the recorded curve the line profile is obtained.

The recording time of such systems is very short. Cooper and Greig give a minimum obtainable recording time of 10^{-7} s for the Barium-titanate oscillator system.

In some other arrangements the light path in the Fabry-Perot interferometer is not altered. The fringe pattern is stationary.

In the device used by Mewe and DeVries [2.88a], the wavelength scanning of the transmitted radiation is obtained by the vertical movement of a horicontal slit placed in front of the entrance slit of the spectrograph upon which the interference rings are projected. In cases in which only the half width of the recorded spectral line is of interest, a recording time as short as $2 \cdot 10^{-8}$ s is obtainable. To record the line profile in all details, a longer recording time is needed.

In the device described in [2.85], [2.86], the spectral line which shows the interference structure in vertical direction, is isolated by means of a perpendicular slit in the image plane of the spectrograph or by means of the exit slit of the monochromator, which is used instead of the spectrograph. Then by projection an image of the line is formed on the photocathode of an image converter. A second image of the line is formed via the deflection system in the image converter on the fluorescent screen and is split in the course of the deflection time. Figure 2.34 shows a corresponding phosphor-screen photograph of the Hg-line $\lambda = 4047$ Å, which was emitted by a pulsed high-pressure Hg-lamp. On the photograph the increase in the width and the total intensity of the line in the relevant time interval are clearly visible. By photometry of the photograph the time dependence of the line width and the total intensity can be determined. The time resolution is essentially limited by the light intensity of the object to be photographed. Malyshev et al. [2.86] obtained a time resolution of 10 μs.

Bradley [2.22] designed a similar apparatus for the high-speed recording of line profiles, in which the image of the interference rings is first formed on the photocathode of an image converter. By means of a system of magnetic coils, through which a periodically varying current flows, the scale of the image of the circular fringe system on the fluorescent screen could be periodically varied. Behind the screen is placed a ring shaped diaphragm for delimiting a narrow wavelength-range in an interference ring. Due to the periodic variation of the magnification, the fringe system shifts in such a way that the line profile is recorded with a photomultiplier placed behind the ring slit and connected to an oscillograph. The recording system employed by Bradley has a time resolution of 0.3 ms.

A Fabry-Perot interferometer coupled with a multichannel photoelectric detection system is described in [2.34c] and [2.40a]. In [2.34c], the inner zone of the interference ring system is projected on a set of concentric, circular mirrors of equal area. The reflection angle of each mirror is tilted against the others in a manner that the radiation selected out by each of the concentric mirrors reaches 8 separate photomultipliers. The spectral resolution of this system is better than 0.2 Å.

In [2.40a], the inner zone of the interference fringe system is projected on a set of ten concentric fibre optical light guides of equal area, ending at 10 separate photomultipliers the photocurrents of which are recorded by oscillographic means.

2.6. Intensity Measurements with Photographic Plates

When photographing spectral sections on spectroscopic plates the next task is to convert the blackening density at given points on the spectroscopic plate to spectral intensities per unit frequency I_ν or to radiant energy per unit frequency and solid angle, which is emitted per cm^2 plasma surface area,

Usually this is done by photographing a standard source with the same optical arrangement as that for the spectroscopic photograph of the plasma to be measured. Furthermore, the blackening marks must be photographed, while between the source, which is to be imaged on the spectroscopic arrangement, and the spectrograph slit filters or similar means of attenuation are inserted which enable a measurable light attenuation. The filters should be, insofar as possible, nonselective.

All photographs should be taken, if possible, at the same exposure time to avoid the Schwarzschild effect. A consequence of this effect is that the blackening density S is no longer determined by the product of the spectral intensity I_ν and the exposure time t.

If spectroscopic photographs are to be taken of very intense luminous plasmas in a very short exposure time, it can occur that for this exposure time the standard source either produces no blackening at all or merely a too weak, no longer reproducible blackening. In such a case we must make use of a secondary standard source. In general this is a pulsed light source, which produces a sufficiently strong spectral intensity in the relevant spectral range. The intensity of this intermediate source must be determined by comparison with the actual standard source. A longer exposure time can be chosen when resorting to this second measure, while the secondary source emits radiation at a constant rate.

2.6.1. Principles of Intensity Determinations from Photographic Blackening

In the following we shall make the general assumption that we require only one standard source by means of which the blackening markings can also be photographed. Then for the determination of the absolute spectral intensities the following measures must be taken:

1. The plasma under investigation is photographed with an exposure time t. The blackening density S on the plate is either a function of ν (frequency) and x (spatial coordinate transverse to the direction of observation) or a function of ν and t, depending on whether an instantaneous photograph or a time-resolved photograph is taken.

   The blackening density S is defined by

$$S = -\lg T\,, \tag{2.29}$$

   where T denotes the transmittancy of the photographic layer (T is the ratio of the light transmittance at the blackened point of the photographic material to the light transmittance at an unblackened point).

2. To produce absolute blackening markings the image of the standard source with spectral intensity $I_{\nu\,0}$ is formed on the spectroscopic device by inserting filters with transmission factors $D_1, \ldots, D_n$ with the same optical arrangement. The spectrally resolved image falls upon the same plate or film as

the image of the plasma under investigation. Blackening densities $S_1, \ldots, S_n$ are obtained, which are functions of the frequency ν.

3. The characteristic film curve $S = f(\log I_\nu)$ is plotted (Figure 2.35) for a given frequency ν for the points $(I_{\nu 1} = D_1 \cdot I_{\nu 0}, S_1), \ldots, (I_{\nu n} = D_n \cdot I_{\nu 0}, S_n)$.
4. From this curve the correlation between S and I_ν is obtained for the investigated plasma.

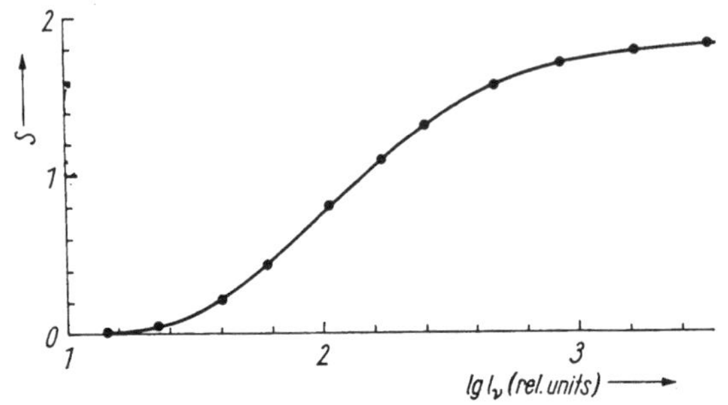

Figure 2.35. Characteristic film curve $S = f(I_\nu)$

2.6.2. Developing of the Photographic Material

When developing the photomaterial special care must be taken that the reagent reacts uniformly at all points of the material. The reagent should be made to flow over the photographic emulsion. This was accomplished by Török [2.125] for films by means of a device whereby the film with the emulsion facing toward the outside is wound on a drum, which is placed in a cylindrical vessel half-filled with the reagent (Figure 2.36). When the vessel rotates about its horizontal axis, the developer streams along the film surface.

During the developing of plates in a basin it is recommended to stroke the surface of the plates with a badger-hair brush. In this manner the reagent on the surface of the plates is permanently removed (see Honerjäger-Sohm and Kaiser [2.59]). Otherwise the Eberhard effect will occur. This happens when the reagent is more quickly consumed at the more strongly blackened points than at those less strongly blackened. If the reagent is not removed fast enough from the more strongly blackened points the blackening in the vicinity of these points will diminish and be rendered irreproducible.

Instead of the badger-hair brush strips of coarse woollen cloth can also be used [2.34a].

In a second method the reagent is driven over the photographic emulsion by swinging the developer bath unperiodically about a horizontal axis which lies perpendicular to the direction of dispersion.

According to the third method (see [2.30c]), at a distance of 0.32 cm above the photographic emulsion there is placed a lattice substituted from vertically

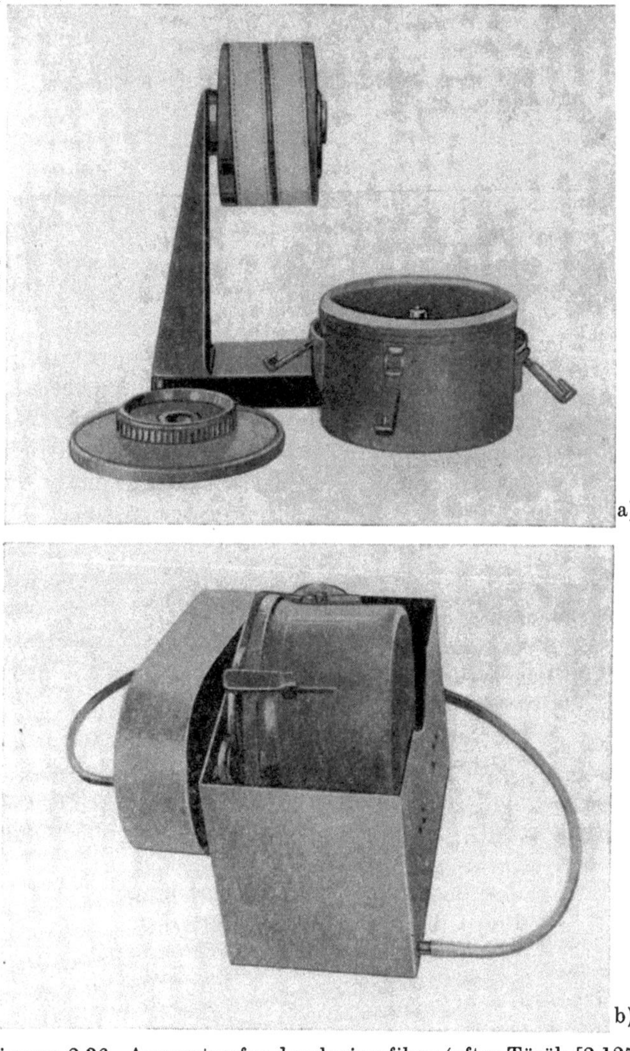

a)

b)

Figure 2.36. Apparatus for developing films (after Török [2.125])

positioned steel blades. When this lattice moves horizontally in a direction perpendicular to the blade surfaces, the photographic emulsion is "hydraulically brushed".

2.6.3. Adjustment of Slit Widths

Usually the blackening distribution on the spectroscopic plate or the film is determined with the help of a microphotometer. (We shall not discuss here the stereoscopic density technique [2.79]). In this method the image of the spectrum is formed in the plane of the photometer slit. If the blackening density of spectral lines is photographed, special care should be taken that the width of the photometer slit is appreciably smaller than the width of the image of the spectral line in the plane of the slit.

The width of the entrance slit of the spectral arrangement is chosen as follows. If the intensity of the continuous radiation is to be determined, then the slit width can be chosen proportionally large. The upper limit of the slit width is determined by the fact that the radiation in the image plane of the spectrograph at each point is not strictly monochromatic, but is composed of a finite large wavelength range $\lambda_1 < \lambda < \lambda_2$ which broadens with increasing slit width. The slit width may be at the utmost so large that the spectral intensities of the source under investigation, and also of the standard source, can be considered as practically constant over this wavelength range. Moreover, the continuum must not be distorted by neighboring lines.

If total intensities of spectral lines are to be determined, then it is recommended to choose the slit width as large as possible so that the image of the spectral line on the photographic plate will be as broad as possible and the density measurement rendered easier. If the "line width due to the apparatus" exceeds the true line width when the slit is wide, then a small range of constant blackening S_{max} (Figure 2.37 b) is obtained for the middle of the line. This blackening density S_{max} depends only on the total intensity of the spectral line and the intensity of the continuous background, and no longer on the profile.

A limit for the slit width is given by the condition that the line studied may not extend to a neighboring line and photographically merge with it.

To determine line profiles from a spectrograph, the width of the entrance slit must be chosen as narrow as possible (Figure 2.37 a), otherwise the line profile will be too strongly distorted, which makes corrections for the apparatus more difficult.

When working in conjunction with an interferometer system the slit width has an upper limit due to the high demands imposed on the preresolution of the spectrum.

2.6.4. Straightening of the Characteristic Film Curve

The characteristic film curve $S = f(\log I_\nu)$ is usually curved (see Figure 2.38).

To straighten the curve, the so-called Seidel or W-transformation can be applied (see [2.59])

$$W = \log\left(\frac{1}{T} - 1\right) \tag{2.30}$$

(W is a single-valued function of the blackening density S in accordance with (2.29) and (2.30)) and W can be expressed as a function of $\log I_\nu$ (Figure 2.38). Thus, in many cases we see that the bottom part of the curve is now less curved toward the other side. The curvature is practically removed when a suitably

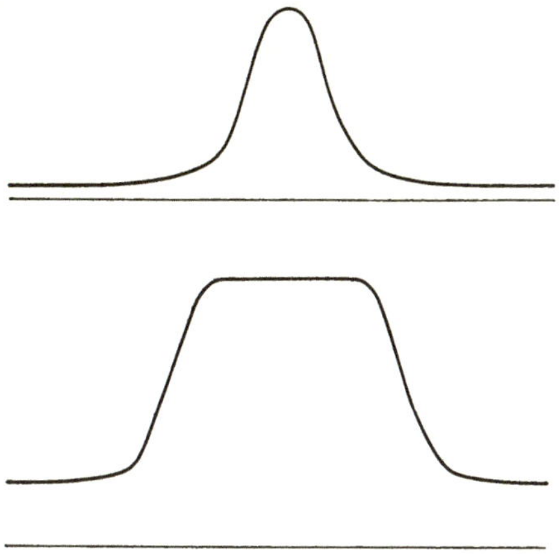

Figure 2.37. Blackening density profile of a spectral line (schematic)
a) with a narrow, b) with a wide entrance slit

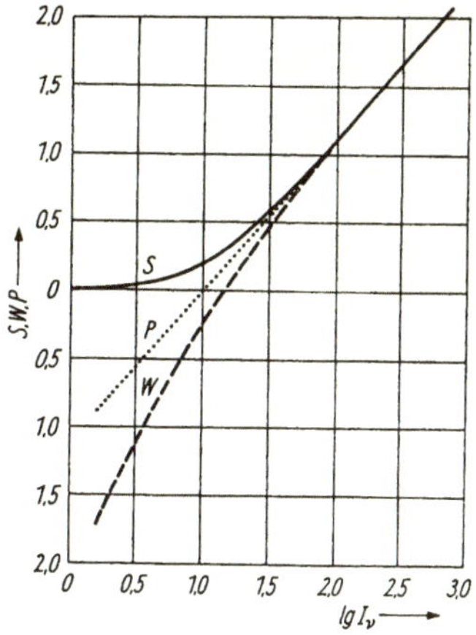

Figure 2.38. Characteristic film curve of an Agfa spectroscopic plate blue-extra-hard at 2800 Å with P- and W-transformation (after Honerjäger-Sohm and Kaiser [2.59])

chosen intermediate value between W and S is selected as a measure for the blackening density, i.e., when a partial Seidel transformation is applied. In the simplest case the value

$$P = \frac{S + W}{2} \tag{2.31}$$

can be chosen, which again is uniquely related to S (P-transformation). Figure 2.38 illustrates the special case for which a straight-line characteristic film curve $P = f(\log I_\nu)$ can be obtained in the range of weak blackening densities.

If this transformation is not satisfactory, then a better fit can be attempted by means of a slightly modified partial Seidel-transformation

$$P_a = \frac{S + a\,W}{1 + a}. \tag{2.32}$$

Then a in general will depend on the wavelength. In plasma spectroscopy only the simple transformations (2.30) and (2.31) enter into consideration, whereas the more complicated transformations (see in particular Török [2.125]) are mainly applicable to quantitative spectral analysis.

2.6.5. The Schwarzschild Effect

As a rule all photographic materials exhibit this effect in the visible spectral range, i.e., for a given frequency ν the blackening S is no longer a function of the product $I_\nu \cdot t$, where I_ν is the spectral radiation intensity per cm² incident upon the photographic emulsion, but depends explicitly on I_ν as well as on t.

Figure 2.39 shows a set of complete characteristic film curves S = const for Ilford HP3-film, which was exposed to radiation of wavelength $\lambda = 5770$ Å. From the figure we can deduce in which way the Schwarzschild effect works. Accordingly, the slightest amount of radiation $I_\nu\, t$ to produce a given blackening is required when the exposure time is about 10^{-2} s (minimum of the curves S = const).

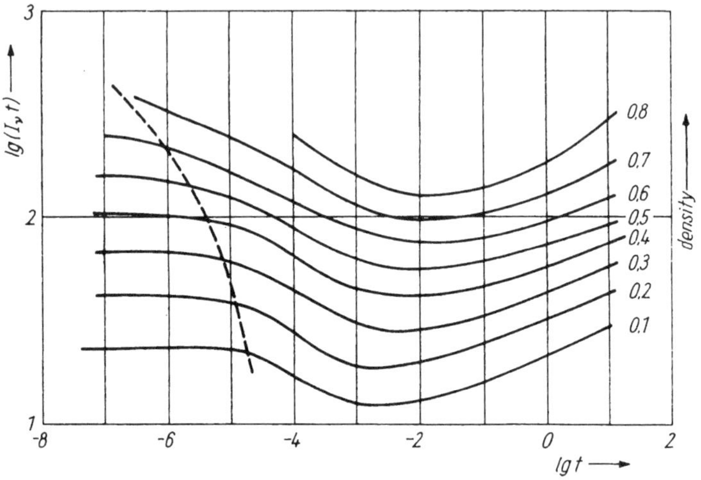

Figure 2.39. A set of complete characteristic film curves obtained with Ilford HP3-film for $\lambda = 5770$ Å (after Gabriel [2.43])

Except at this minimum we can express the correlation between the spectral intensity I_ν, the exposure time t, and the blackening S by

$$S = F(I_\nu\, t^p)\,, \tag{2.33}$$

where p is the Schwarzschild exponent. This is less than 1 for exposure times exceeding 10^{-2} s. Typical values of p for exposure times between 1 and 20 s are 0.8—0.9 (see Jürgens [2.63]).

For exposure times below 10^{-5} s (the exact time limit is denoted on Figure 2.39 by a dotted line) the reciprocity theorem is again valid. This fact is of great significance for high-speed spectroscopy, when extremely short exposure times (e.g., 10^{-7} s) are desired. It is then sufficient to produce blackening marks by using an intermediate standard source (pulse lamp, see Gabriel [2.43]) at exposure times of about 10^{-5} s. This restricts the requirements imposed on the intermediate source.

It is of great interest here to produce photographic materials which exhibit a slight Schwarzschild effect. In Agfa-Wolfen* photographic materials for

* Current trademark ORWO.

Raman-spectroscopy as well as for H_α-films and -plates a slight Schwarzschild effect is displayed.

Castle et al. [2.27] investigated this effect in 6 different types of Kodak film of very high sensitivy, which were developed with normal and very strong reagent. Figure 2.40 gives the curves S = const for the heavily developed films

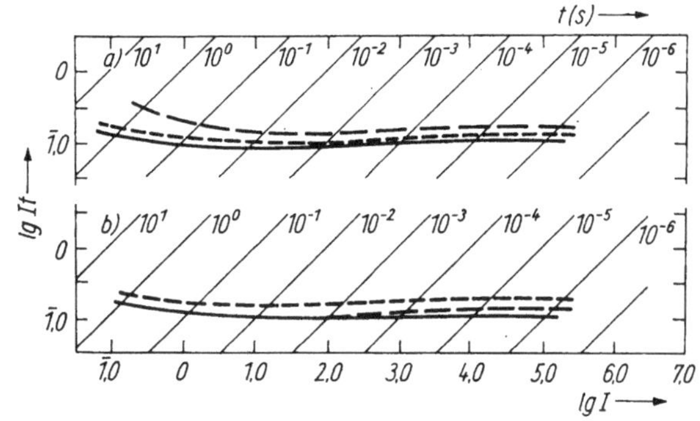

Figure 2.40. Deviations from the reciprocity theorem (after Castle et al. [2.27])

The curves S = const are given for the following types of Kodak film when heavily developed: a) — Kodak Spectroscopic film, Type I–D(2); — — — Kodak Tri-X Aerial Recon Film; - - - - - Kodak Royal Pan Sheet Film. b) — Kodak Royal Ortho Sheet Film; — — — Kodak Tri-X Negative Film; - - - - - Kodak Linagraph Pan Film

in the log t versus log I diagram. The lines t = const are also given. From the diagram we see that the deviation from the reciprocity theorem is at the most 25%. When using standard reagents a larger deviation can be expected. It thus follows that the calibration photographs taken with the standard source for a suitable choice of photographic material can be carried out with exposure times above 10^{-5} s.

It should be still mentioned that when working with X-rays the reciprocity theorem is always valid ([2.3], page 59).

2.6.6. **Slit-width Effect**

Hitherto it was commonly accepted that the photographic density for a given photographic plate, given exposure time and given wavelength band is a function of the irradiation intensity only.

As Funk* has shown, this is not longer true when the photographic density exceeds a value of 2. Under this condition it was found that the shape of the characteristic curves derived from spectral line blackening datas, show a dependence from the spectrograph slit width explicitely. For a given irradiation intensity, the photographic density grows with growing slit width. Saturation

* private communication.

is obtained for slit widths greater than 100 μm. Then the characteristic curve is identical with that obtained when a continuous radiator is used as calibrating light source.

To avoid the slit width effect, the exposition rate must be reduced to an extent that the photographic density is kept below 2.

2.6.7. Measures for Producing a Measurable Drop in the Light Intensity

A measurable drop in the intensity of the radiation, which is incident upon the spectroscope, is necessary for constructing characteristic film curves, and further for the matching of the intensity between the investigated source and a standard source.

The means by which the intensity drop is effected should be non-selective, insofar as possible. This simplifies the investigation. A drop in the intensity can be produced among other things by the following:

1. Rotating sector diaphragms,
2. Suitably placed diaphragms,
3. Wire gauzes,
4. Filters,
5. Reflecting optical surfaces.

We shall briefly discuss these means of attenuating radiation and indicate their range of application.

2.6.7.1. Rotating sector diaphragms

These consist in their simplest form of a rotating disk, in which a sector with aperture angle α is cut out. The rotating sector acts as a nonselective filter with transmission factor

$$D = \alpha^0/360^\circ , \tag{2.34}$$

for the time-integrated radiation, which passes through the aperture of the sector.

Modern versions are constructed in such a manner that the aperture angle α can be changed during operation [2.121]. Moreover, not one but several sectors are usually employed.

The sector diaphragms can also be constructed so that they act as a stepped filter. These are nearly always placed in front of the spectrograph slit in such a way that the axis of rotation is located below the slit. The sector is divided into different radial ranges with different aperture angles $\alpha_1 \cdots \alpha_n$, so that the radiation quantity passing through each of the corresponding sections of the spectrograph slit is proportional to $\alpha_1 \cdots \alpha_n$, assuming an uniform illumination of the region in front of the slit. Figure 2.41 illustrates such a rotating sector with 5 different apertures.

Rotating sectors are thus only applicable for intensity attenuation when the total exposure time of the process to be recorded is so long, that during this time at least 100 openings and interruptions of the light path result. Then the inaccuracy in the amount of light transmitted is less than 1%.

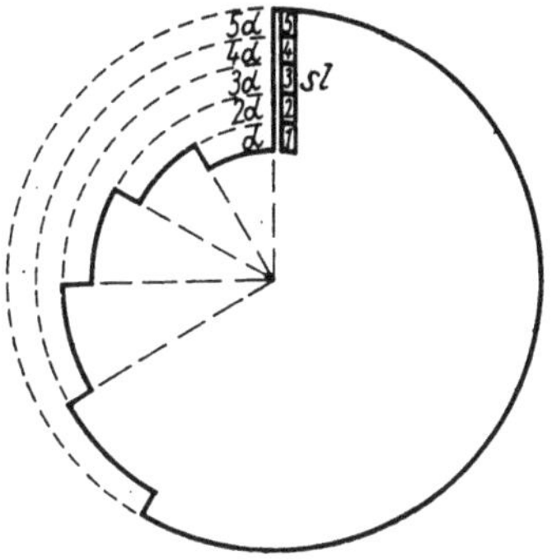

Figure 2.41. Rotating step sector (schematic) with 5 aperture angles increasing according to a given law (after Löwe [2.82], page 115)

Another condition for the application of the rotating sector is that the frequency of the light interruptions should be at least 100 s^{-1}. This is connected with the Schwarzschild effect and the intermittency effect ([2.3], page 175). When using the rotating sector to produce a drop in the intensity the actual exposure time is shortened. If only this effect were active, the photographic densities obtained for the case $p < 1$ would be too large. In contrast hereto the intermittency effect has the result that the photographic effect of the intermittent light is less than that for the continuous light.

Both effects mutually compensate one another at higher frequencies of the light interruptions. Honerjäger-Sohm and Kaiser [2.59] give 90 s^{-1} as a suitable working frequency. Lang [2.75] found that the exact limiting frequency depends on the type of film, which, however, can be used without hesitation above an intermittency frequency of 100 Hz of the rotating sector.

2.6.7.2. Intensity attenuation by diaphragms

These can be used when one is certain that a wavelength-dependent intensity distortion of the spectrum does not result due to diffraction effects.

The most convenient are diaphragms in the principal plane of objectives, which uniformly lower the intensity of the transmitted light. However, as a rule, these cannot be employed since the already small relative aperture of the spectrograph (in most cases of the order of 1:16) is hereby reduced below that which is permissible. The result would be a line broadening on the photographic plate due to diffraction.

The step diaphragm designed by Hansen (Figure 2.42) serves as a steplike light absorber. It must be uniformly illuminated and its image is formed via an objective lens on the spectrograph slit. Further, a cylindrical lens is located between the step diaphragm and the objective. Its axis is directed along that of the slit, which forms the image of the step diaphragm in the

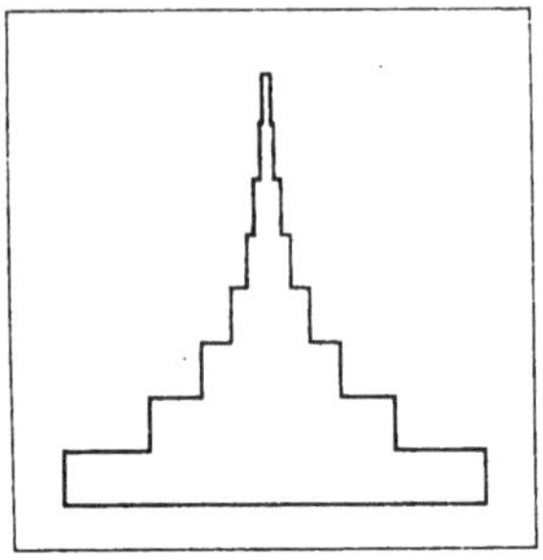

Figure 2.42. Step diaphragm after Hansen (Angerer and Joos [2.3])

objective. Thus, the rectangular apertures of the diaphragm are divided into long streaks, whose brightness in the center is proportional to the width of each aperture.

Light absorption, which is effected in a continuous fashion over the spectrograph slit, can be achieved with the diaphragm system of Hirsch and Schön (Figure 2.43). An uniformly illuminated surface to the right is imaged via

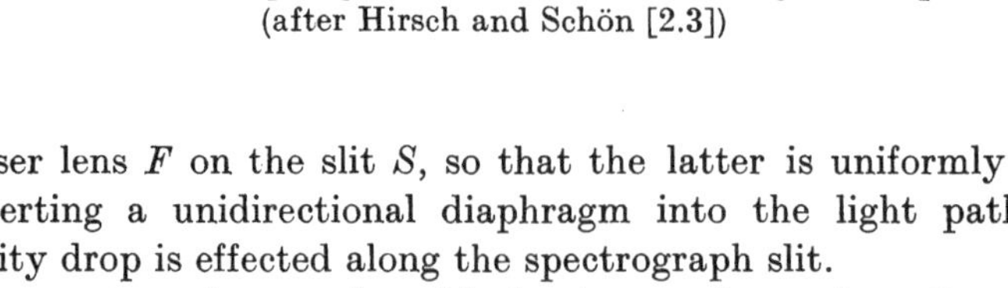

Figure 2.43. Diaphragm system for continuous light absorption (after Hirsch and Schön [2.3])

the condenser lens F on the slit S, so that the latter is uniformly illuminated. By inserting a unidirectional diaphragm into the light path a regular light intensity drop is effected along the spectrograph slit.

An arrangement used to produce blackening marks, such as those produced by the diffraction pattern of a slit on the photographic plate whereby the calculable intensity distribution is the basis for the calibration, was designed by Paetzold [2.103].

2.6.7.3. Light attenuation by means of wire grids

In principle, these can be inserted at an arbitrary point along the ray path. They are suitable for light attenuation only when their transmissivity is experimentally determined. Wire grids are to be used in that wavelength range, where the light absorption is independent of the wavelength.

Newman and Binder [2.98] employed wire grids of stainless steel for light absorption. By combining several grids a transmission range of over 6 orders of magnitude can be attained. Newman and Binder confirmed that the light absorption was independent of the wavelength in the range up to at least 14 μm, by using wire grids with mesh widths of 40 μm.

Using the etching technique, Low [2.81 c] prepared a wire mesh screen to give a transmission coefficient which is a linear function of the coordinate x along the screen.

2.6.7.4. Light absorption by filters

Filters have the advantage that they absorb the incident light uniformly over their entire cross section. However, in most cases the wavelength dependence of the absorption factor must be allowed for. This is, for example, the case for the Schott grey filters (NG-filters).

Filters having a weak wavelength dependence of the absorption factor can be produced by depositing a metallic film, for example, platinum or rhodium, on a quartz plate. Zoned filters as well as homogeneous absorption filters can be made. By combining several metallic films one can obtain a practically nonselective absorption in the spectral range between 3000 Å and 2.3 μm ([2.3], p. 175). For further possibilities of light absorption by means of filters see [2.3].

2.6.7.5. Light attenuation by partially reflecting surfaces

Light attenuation is obtained when the light is reflected by a partially reflecting optical surface. The latter may be either a thin mirror-coated glass-plate or in the simplest case a pure uncoated glass plate [2.50 a]. Care must be taken that unwanted reflection from the back surface is directed outside the main ray path.

2.6.8. **Photographic Line Width**

If the spectroscopic photographs are given ample exposure, it usually occurs that some of the lines appearing on the photographic emulsion are strongly broadened [2.92]. Clearly, such spectral lines cannot ba analyzed in normal fashion, since for this purpose shorter exposure times or less intensely illuminated photographs must be used.

On the other hand these overexposed photographs are of interest for quantitative spectral analysis. The concentration of the element can be determined on the basis of the photographic width of these lines. For more details on this subject see Körber and Röllig [2.68]. For plasmadiagnostic work, the mentioned effect is of little interest.

2.7. Radiation Density Measurements With Photoelectric Detectors

Let us consider the case, in which the image of the source under investigation is formed on the entrance slit of a monochromator. The photoelectric detector can be placed behind the exit slit. Let the photocurrent i correspond to the frequency ν at which the monochromator is set.

To obtain the absolute radiation densities from the photocurrent i, the image of a standard source with spectral intensity $I_{\nu 0}$ is likewise formed on the entrance slit of the monochromator, whereby exactly the same ray path is followed. Let the photocurrent at the photoelectric detector be i_0 for the frequency ν. Then the radiation density of the source under investigation is given by

$$I_\nu = \frac{i}{i_0} \cdot I_{\nu 0} \,. \qquad (2.35)$$

This expression is completely valid when the frequency ν falls in a range, in which the source emits only continuous radiation. If the measurements are extended (mostly in the form of a recording) over a frequency range in which a spectral line occurs, then the latter is recorded in distorted fashion. The distortion has no significance, however, when the integral line intensity is determined.

2.8. Determination of Integral Line Intensities

2.8.1. Integral Intensity of Isolated Spectral Lines

As a rule, we start from a recording curve $I(\nu)$ of the spectral intensity distribution. This curve can be obtained by photographic means (characteristic film curve) (see section 2.6), or by means of photoelectric recording (see section 2.7).

In general the spectral line is overlapped by continuous radiation (see Figure 2.44). Denoting the frequencies, between which the spectral line in the recording curve appears, by ν_1 and ν_2, the integral line intensity I is given by

$$I = \int_{\nu_1}^{\nu_2} I_\nu \, \mathrm{d}\nu - (\nu_2 - \nu_1) \cdot \frac{I_{\nu 1} + I_{\nu 2}}{2} \,. \qquad (2.36)$$

In many applications, the procedure of obtaining the total intensity of a spectral line over the recording curve is too complicated.

In section 2.6.3 we mentioned that when the entrance slit is sufficiently wide the intensity distribution of the radiation in the image plane is smeared out in such a fashion that all the spectral lines have a flat plateau in the middle

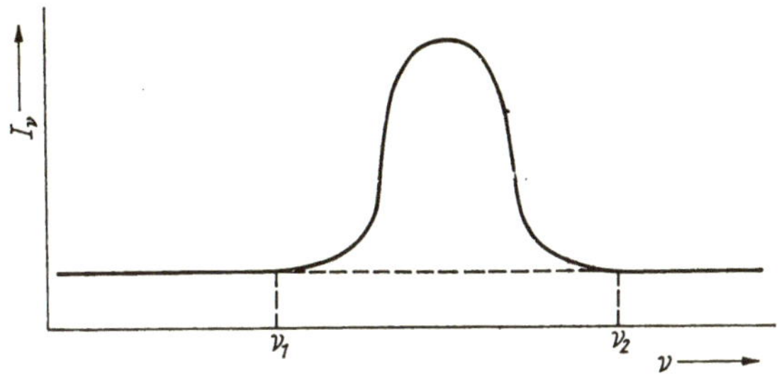

Figure 2.44. Curve to determine the integral line intensity I from the spectroscopically analyzed intensity I_ν (see text)

(see Figure 2.37 b). This plateau appears also in the recording curves when working with monochromators having a wide entrance slit and a narrow exit slit. (It is also possible to work with a monochromator having a wide exit slit. However, then the width of the entrance slit may not exceed a certain limit, which is set in such a way that the image of the apparatively broadened spectral line in the plane of the exit slit is markedly narrower than the exit slit itself.)

We shall denote the spectral intensity corresponding to the flat plateau in the recording curve $I_\nu = f(\nu)$ by $I_{\nu m}$ (see Figure 2.45) and assume that the

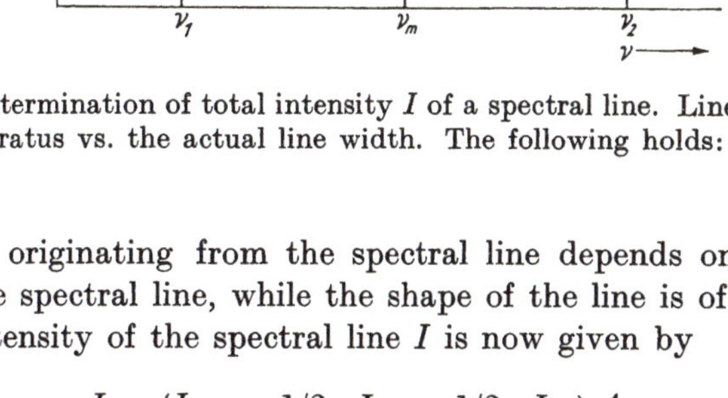

Figure 2.45. Determination of total intensity I of a spectral line. Line width broadened by the apparatus vs. the actual line width. The following holds: $I = I_{\nu m} \cdot \Delta\nu$

portion of $I_{\nu m}$ originating from the spectral line depends only on the total intensity of the spectral line, while the shape of the line is of no importance.

The total intensity of the spectral line I is now given by

$$I = (I_{\nu m} - 1/2 \cdot I_{\nu 1} - 1/2 \cdot I_{\nu 2})\, \Delta\nu\,. \qquad (2.37)$$

ν_1 and ν_2 are again two frequencies, between which the spectral line occurs. $\Delta\nu$, the line width due to the apparatus, is given by

$$\Delta\nu = \frac{\int_{\nu_1}^{\nu_2} I \, d\nu - \frac{1}{2}(\nu_2 - \nu_1)(I_{\nu\,1} + I_{\nu\,2})}{I_{\nu\,m} - \frac{1}{2} I_{\nu\,1} - \frac{1}{2} I_{\nu\,2}} . \tag{2.38}$$

$\Delta\nu$ need only be determined by recording once over a given frequency range in the neighborhood of the spectral line. After this we only require measurement values for 1 to 3 discrete frequencies for the determination of I from (2.37).

1. The measured value $I_{\nu\,m}$ is sufficient when the continuum can be disregarded.
2. If the continuum on both sides of the line is equally strong ($I_{\nu\,1} = I_{\nu\,2}$), then only the measured values $I_{\nu\,m}$ and $I_{\nu\,1}$ are required.
3. When the continuum is not the same on both sides of the line, the 3 measured values $I_{\nu\,m}$, $I_{\nu\,1}$, and $I_{\nu\,2}$ are required,

If the widths of the entrance slit and the exit slit of the monochromator expressed in wavelength units are identical, then the recording curves show no plateau, and the simplified procedure of obtaining integral line intensites can be applied only in those cases when the true line width is a small fraction of the line width caused by apparatus. In these cases, formulas (2.37) and (2.38) hold again when $I_{\nu\,m}$ is interpreted as the maximum value of the recording curve $I_\nu(\nu)$.

2.8.2. Intensity of Overlapping Spectral Lines

We consider the problem of determining integral line intensities of two partially overlapping spectral lines.

In order to keep the overlapping at a low level, narrow slit widths must be used when recording the spectral intensity distribution I_ν. From the recording curve which is shown schematically in Figure 2.46, an attempt should be made

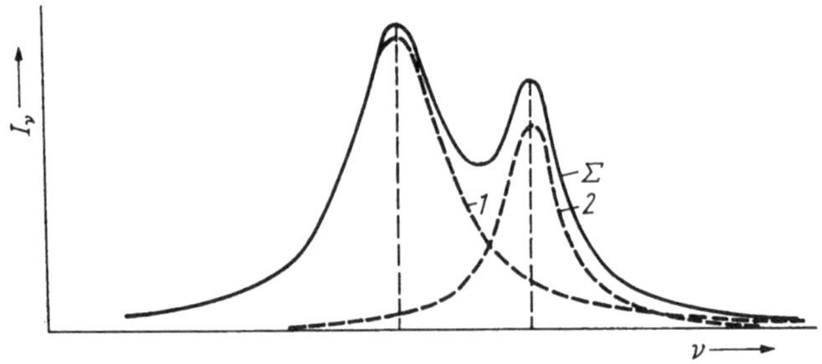

Figure 2.46. Determination of the individual intensities of two overlapping lines by resolving the total profile (Σ) into the two component parts *1* and *2* (schematic)

to resolve the two component parts to which the lines correspond. This is only possible, when at least one of the lines is symmetrically broadened. The separation is easily effected when the lines have an approximate dispersion profile. The total intensities of the two spectral lines I_1 and I_2 are obtained by integrating over the partial profiles.

Wiese and Shumaker [2.133] considered in detail the case, in which an intense spectral line (O I 6157 Å), whose integral intensity was to be determined, is distorted by a neighboring weak line (O I 6106 Å), and that besides the two lines a continuous backgrund is superposed. The lines were emitted from a wall-stabilized high-current arc and were strongly broadened. The recording curve is shown in Figure 2.47.

In carrying out this task it is important to exactly determine the intensity of the continuous backgrund. In a first approximation, the background is arbitrarly drawn as a straight line (dotted) and the line at 1657 Å is graphically freed of the influence of the neighboring line at 6106 Å. Thus, we obtain the dashed profile curve, which together with the dotted line forms a closed curve. This profile has a half-width of $2\,\Delta\lambda^*_{1/2}$ and a base width $2\,\Delta\lambda^*$. The area under this curve is

$$I^* = \int_{-\Delta\lambda^*}^{+\Delta\lambda^*} I^*_\lambda \,\mathrm{d}\Delta\lambda\,. \tag{2.39}$$

The line is then fitted to a dispersion profile

$$I_\lambda = C\,[1 + (\Delta\lambda/\Delta\lambda_{1/2})^2]^{-1} \tag{2.40}$$

in such a manner that the measured half-width and the measured line profile area are taken as the basis of the dispersion profile. For this we must first transform the half half-width $\Delta\lambda^*_{1/2}$ to the somewhat larger half half-width $\Delta\lambda_{1/2}$ of the dispersion profile. The transformation is

$$\frac{\Delta\lambda^*}{\Delta\lambda_{1/2}} = [\Delta\lambda^*/\Delta\lambda^*_{1/2})^2 - 2]^{1/2}\,. \tag{2.41}$$

The second transformation is obtained, due to the fact that the line area I of the dispersion profile is larger than the area I^* of the constructed line profile:

$$\frac{I}{I^*} = \frac{\int_{-\infty}^{+\infty} I_\lambda \,\mathrm{d}\Delta\lambda}{\int_{-\Delta\lambda^*}^{\Delta\lambda^*} I^*_\lambda \,\mathrm{d}\Delta\lambda} = \frac{\pi}{2}\left[\arctan\frac{\Delta\lambda^*}{\Delta\lambda_{1/2}} - \left(\frac{\Delta\lambda_{1/2}}{\Delta\lambda^*} + \frac{\Delta\lambda^*}{\Delta\lambda_{1/2}}\right)^{-1}\right]^{-1} = \gamma\,. \tag{2.42}$$

Both corrective transformations are obtained on the basis of the fact that the background of the dispersion profile lies deeper than the constructed line background.

Thus, the task of determining the line intensity I is fulfilled. The dispersion profile is defined by I and $\Delta\lambda_{1/2}$ (the constant C in (2.40) is $I/(\pi\,\Delta\lambda_{1/2})$ and is designated by a dashed line in Figure 2.47. The dash-dot line indicates the corrected continuous background.

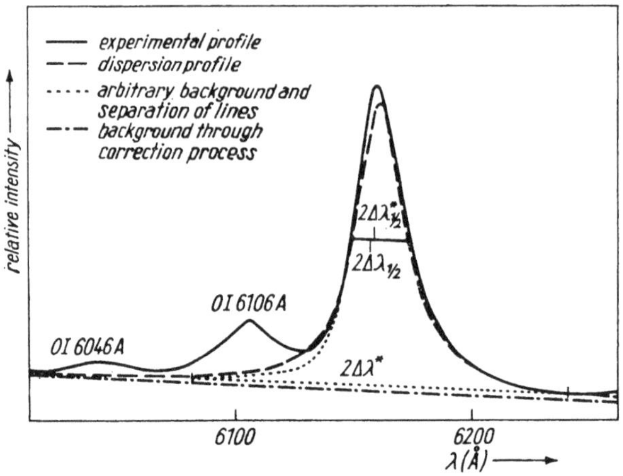

Figure 2.47. Profile determination of the O I 6157 line, emitted by a 100 A-arc and distorted by a neighboring line (after Wiese and Shumaker [2.133])

2.9. Proper Reproduction of Line Profiles distorted by the Apparatus

Spectral lines are usually reproduced in distorted fashion. We shall consider in this section those distortions, which are due to the apparatus. There are further sources of distortion, such as those caused by self-absorption. These are the subject of discussion in sections 2.10.2 and 6.3.2.

2.9.1. General Methods

The distortion due to the apparatus is a consequence of the imperfect resolution of the spectroscopic device. The causes lie in finite slit widths, diffraction at the objective stops, light scattering, and radiation diffusion in photographic plates.

Denoting the true spectral intensity distribution by $w(x')$, and that measured by the spectroscopic device by $s(x)$, we have the following expression

$$s(x) = \int_{-\infty}^{+\infty} a\,(x - x')\, w(x')\, \mathrm{d}x' \tag{2.43}$$

where $a\,(x - x')$ is the apparatus profile. This is normalized as follows

$$\int_{-\infty}^{+\infty} a(x)\, \mathrm{d}x = 1\,.$$

The correction of the measured profile $s(x)$ is effected by solving the integral equation in (2.43), whereby the apparatus profile $a\,(x - x')$ must be known (the determination of the apparatus function will be discussed later).

To solve (2.43) many methods are available.

A Fourier integral method is given in [2.106] and [2.111]. This method can be modified, introducing Fourier series instead of Fourier integrals [2.97 a].

A direct numerical correction method was developed by Bondarev [2.19], who applied step functions for the functions s, a, and w, and reduced the problem of reestablishing of the original line profile to the problem of solving a set of algebraic equations. He limited himself to the case of a symmetric line profile, but the given method can be extended to the case of nonsymmetric profiles [2.97 a].

A comprehensive discussion of a correction method for the case, in which the apparatus function can be represented as a dispersion profile, can be found in [2.45] (Genkin and Rumyantsev).

A time saving iteration method for resolving the profile was developed by Drawin [2.30 a]. An improved version of it [2.97 b] will be demonstrated here.

At first, we replace the function $w(x')$ in (2.43) by a point function

$$w(x')_{\text{approx}} = w(x')\,\delta\,(x' - x_k)\,, \qquad x_k = x_0 + k\,\Delta x\,. \tag{2.44}$$

Regarding all x-dependent quantities only for discrete values x_l, we get instead of (2.43):

$$s(x_l) = \sum_{k=0}^{N} w(x_k)\,A\,(x_l - x_k)\,, \qquad \begin{array}{l} x_l = x_0 + l\,\Delta x\,, \\ l = -\,M \cdots N + M\,. \end{array} \tag{2.45}$$

$2\,M \cdot \Delta x$ is the full width of the instrumental profile function $A\,(x_l - x_k)$. The latter differs from $a\,(x_l - x_k)$ only by a numerical factor of order 1 because the normalizing condition has the modified form

$$\sum_{m=-M}^{+M} A(\Delta x \cdot m) = 1\,. \tag{2.46}$$

When the instrumental function $A\,(x_l - x_k)$ is known, then $w(x_k)$ can be derived from $s(x_l)$ by the following iteration procedure.

For this purpose, equation (2.45) is transformed into

$$w(x_l) = s(x_l) + w(x_l) - \sum_{k=0}^{N} w(x_k)\,A\,(x_l - x_k)\,. \tag{2.47}$$

In the first iteration step, in the right hand side of equation (2.47) w is substituted by s, leading to

$$w^{(1)}(x_l) = 2\,s(x_l) - \sum_{k=0}^{N} s(x_k)\,A\,(x_l - x_k)\,. \tag{2.48}$$

In most cases, the first iteration step is sufficient. If not, the function $w(x_l)$ respectively $w(x_k)$ on the right hand side of (2.47) must be substituted by the first iterated function $w^{(1)}(x_l)$ respectively $w^{(1)}(x_k)$ to get the second iteration step.

Before we treat the Voigt functions for the approximation of the functions s, a, and w, we should say more about the determination of the apparatus function

$a(x - x')$, which hitherto has been assumed as known. To do this three possibilities present themselves. In one of these the apparatus function can be experimentally determined by spectroscopically recording a very narrow line. In this way the recorded line profile $s(x)$ is to whithin a normalization factor equal to the apparatus function $a(x)$. Suitable sources are cooled low-pressure lamps. No resonance lines should be used. Furthermore, spectral lines, emitted by atomic beams perpendicular to the direction of motion, are suitable as well as, insofar as available, laser rays.

Secondly, the apparatus function $a(x)$ can be determined in the case of non negligible test line width on the basis of (2.43) from a record of the apparent line profile $s(x)$ of the test line, if the true profile of this line $w(x)$ is known.

Thirdly, the apparatus function can be theoretically calculated. Moser [2.93] and Mielenz [2.90] did this by calculating the apparatus profile determined by diffraction in a prism spectrograph, whereby the influence of prism absorption is allowed for. For details on the apparatus profile determined by diffraction in spectroscopic devices with diffraction gratings, see Unsöld [2.126, p. 250]. Fricke [2.42], Semenov et al. [2.119], and Bayer-Helms [2.11] studied apparatus profiles of Fabry-Perot interferometers.

To obtain higher accuracy it is always advisable to determine the apparatus function experimentally.

2.9.2. Correction of Line Profiles by Means of Voigt Functions

We shall now consider the two-parameter Voigt-function. This is represented as a folding of a dispersion distribution $f_1(x)$ of half-half-width β_1 and a Doppler distribution $f_2(x)$ of half $1/e$-width β_2:

$$V(x|\beta_1, \beta_2) = \int_{-\infty}^{+\infty} f_1(x - x') f_2(x') \, dx' , \tag{2.49}$$

where

$$f_1(x) = \frac{1}{\pi} \frac{\beta_1}{\beta_1^2 + x^2} , \tag{2.50}$$

$$f_2(x) = \frac{1}{\sqrt{\pi}\,\beta_2} \cdot e^{-(x/\beta_2)^2} . \tag{2.51}$$

All 3 functions are normalized, i.e.,

$$\int_{-\infty}^{+\infty} V(x) \, dx = 1 , \quad \int_{-\infty}^{+\infty} f_1(x) \, dx = 1 , \quad \int_{-\infty}^{+\infty} f_2(x) \, dx = 1 .$$

By means of the Voigt-functions the correction problem (see (2.43))

$$s(x) = \int_{-\infty}^{+\infty} a(x - y) \, w(y) \, dy \tag{2.52}$$

is readily solved, insofar the functions s and a can be represented as or approximated by Voigt-functions. Then w is also a Voigt-function.

Let us write

$$s(x) = V(x|\beta_1(s), \beta_2(s)) , \tag{2.53}$$

$$a(x) = V(x|\beta_1(a), \beta_2(a)) , \tag{2.54}$$

$$w(x) = V(x|\beta_1(w), \beta_2(w)) . \tag{2.55}$$

Then the following relationship holds between the parameters of the Voigt-functions

$$\left. \begin{aligned} \beta_1(s) &= \beta_1(a) + \beta_1(w) , \\ \beta_2^2(s) &= \beta_2^2(a) + \beta_2^2(w) . \end{aligned} \right\} \tag{2.56}$$

If we therefore approximate the apparent, measured line profile $s(x)$ and the apparatus profile $a(x)$ by Voigt-functions with the parameters $\beta_1(s)$, $\beta_1(a)$, $\beta_2(s)$, and $\beta_2(a)$, we obtain from (2.56) the parameters $\beta_1(w)$ and $\beta_2(w)$ for the Voigt-function, which properly reproduces the true line profile in approximate fashion. The main task is in approximating the measured functions $s(x)$ and $a(x)$ by Voigt-functions.

This is made easier by the fact, that the shape of the Voigt-functions depend explicitly only on the ratio β_1/β_2. If β_1 and β_2 are given, the Voigt-function $V(x|\beta_1, \beta_2)$ can be immediately plotted (see Figure 2.48 after Unsöld [2.126], p. 259). From this Voigt-profile we read off a central ordinate c and a full half-width h.

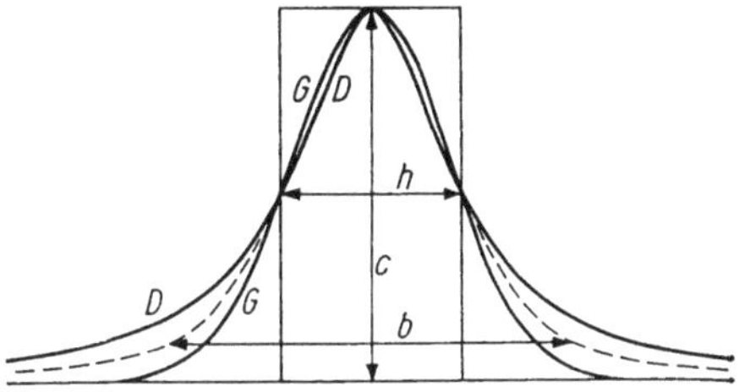

Figure 2.48. Voigt-profile (Unsöld [2.126], p. 259)

G – Gaussian profile, D – dispersion profile, and intermediate profile (dotted line) with $\beta_1/h = 0.25$

If we now take the ordinate as referring to $c = 1$ and the corresponding full widths b to $h = 1$, then we see from Table 2.7 ([2.126], p. 259), that the correlation between the ordinate values (listed for ordinate values from 0.01 to 0.8) and the reduced full widths b/h are determined by the shape factor β_1/β_2 only. On the left-hand side of the table are further listed the quantities β_1/h, β_2/h, β_2^2/h^2, associated with β_1/β_2, and the quantity p, defined by

$$A = p\,h\,c , \tag{2.57}$$

where A is the area under the Voigt-profile.

The correlations between the parameters of the Voigt-profiles can be derived also from Figure 2.49.

To find the best Voigt-profile for a given profile, the following procedure can be followed:

1. Take the central coordinate of the given profile to be c, the full line width at $0.5\,c$ (= half-width h), as well as the full width at $0.1\,c$ ($b_{0.1}$)

2. With the found value of $b_{0.1}/h$ we refer to the graph (Figure 2.49), from which the associated values β_1/h, β_2/h, and p can be taken. Since h and c are known, one can find the sought parameters (β_1 and β_2) of the Voigt-function V from the above quantities. The quantity p, which specifies with (2.57) the area under the Voigt-profile, can be useful when normalizing the Voigt-function

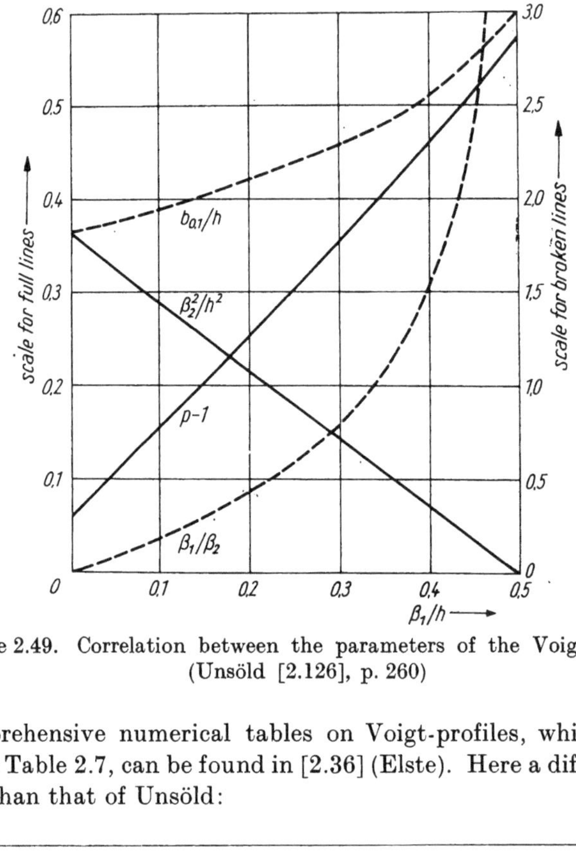

Figure 2.49. Correlation between the parameters of the Voigt-profile (Unsöld [2.126], p. 260)

More comprehensive numerical tables on Voigt-profiles, which in principle correspond to Table 2.7, can be found in [2.36] (Elste). Here a different notation is employed than that of Unsöld:

Unsölds notation [2.126] (used in this text)	β_1/h	β_1/β_2	β_2/h	β_2^2/h^2	p	b	b/h
Elstess notation [236]	b_1	α	b_2	b_2^2	p	$2\,\Delta\lambda$	u

A tabulation of Voigt profiles is also given in [2.11 a].

Not all profiles can be approximated by Voigt-profiles.

Table 2.7. Data on Voigt-profiles [2.126]

β_1/h	β_1/β_2	β_2/h	β_2^2/h^2	p	0.8	0.7	0.6	0.5	0.4	0.3	0.2	0.1	0.05	0.02	0.01
Parameters					Ordinate (referred to $c = 1$) — full line width b/full half-width h										
Gaussian profile															
0.000	0.00	0.60	0.36	1.06	0.57	0.72	0.86	1.00	1.15	1.32	1.52	1.82	2.08	2.38	2.58
0.025	0.04	0.59	0.34	1.08	0.56	0.72	0.86	1.00	1.15	1.33	1.53	1.84	2.12	2.49	2.82
0.050	0.09	0.57	0.32	1.11	0.56	0.71	0.86	1.00	1.15	1.33	1.54	1.87	2.19	2.63	3.13
0.075	0.14	0.55	0.31	1.13	0.56	0.71	0.86	1.00	1.16	1.33	1.56	1.90	2.25	2.79	3.56
0.100	0.19	0.54	0.29	1.16	0.56	0.71	0.86	1.00	1.16	1.34	1.57	1.94	2.34	3.00	4.08
0.125	0.24	0.52	0.27	1.18	0.56	0.71	0.86	1.00	1.17	1.34	1.59	1.98	2.42	3.24	4.58
0.150	0.30	0.50	0.25	1.20	0.55	0.71	0.85	1.00	1.17	1.35	1.60	2.02	2.54	3.52	5.05
0.175	0.36	0.48	0.23	1.23	0.55	0.70	0.85	1.00	1.17	1.36	1.62	2.06	2.64	3.80	5.50
0.200	0.43	0.46	0.21	1.25	0.55	0.70	0.85	1.00	1.18	1.37	1.64	2.10	2.75	4.14	5.96
0.225	0.51	0.44	0.20	1.28	0.54	0.70	0.85	1.00	1.18	1.38	1.66	2.15	2.87	4.44	6.40
0.250	0.59	0.42	0.18	1.30	0.54	0.70	0.84	1.00	1.18	1.39	1.68	2.19	2.98	4.73	6.78
0.275	0.69	0.40	0.16	1.33	0.53	0.69	0.84	1.00	1.19	1.40	1.71	2.24	3.12	5.03	7.15
0.300	0.79	0.38	0.14	1.35	0.53	0.69	0.84	1.00	1.19	1.41	1.74	2.29	3.26	5.32	7.52
0.325	0.92	0.35	0.12	1.38	0.53	0.68	0.84	1.00	1.19	1.42	1.77	2.34	3.39	5.57	7.86
0.350	1.07	0.33	0.11	1.40	0.52	0.68	0.84	1.00	1.20	1.44	1.81	2.40	3.54	5.83	8.21
0.375	1.26	0.30	0.09	1.43	0.52	0.68	0.83	1.00	1.20	1.45	1.85	2.46	3.70	6.07	8.55
0.400	1.50	0.27	0.07	1.45	0.52	0.67	0.83	1.00	1.21	1.47	1.88	2.54	3.85	6.30	8.86
0.425	1.83	0.23	0.05	1.48	0.51	0.67	0.83	1.00	1.21	1.48	1.92	2.64	4.00	6.55	9.18
0.450	2.38	0.19	0.04	1.51	0.51	0.66	0.82	1.00	1.22	1.50	1.96	2.74	4.13	6.76	9.50
0.475	3.54	0.13	0.02	1.54	0.51	0.66	0.82	1.00	1.22	1.52	1.98	2.87	4.25	6.92	9.77
0.500	∞	0.00	0.00	1.57	0.50	0.66	0.82	1.00	1.22	1.53	2.00	3.00	4.36	7.00	9.95
dispersion profile															

The range of application of the Voigt-function according to Elste [2.36] can be nevertheless extended, due to the fact that many line profiles can be represented as the sum or difference of two Voigt-functions. Asymmetric profile can also be approximated in this manner. The corrected profile can again be represented as the sum or difference of two Voigt-functions. For details we refer the reader to the relevant literature.

2.10. Transformation of the Spectral Intensity Distribution $I(x)$ at Right Angles to the Axis to the Radial Distribution of the Emission Coefficient $\varepsilon(r)$ for a Source having Rotational Symmetry (Inversion)

2.10.1. Inversion without Allowance for Plasma Absorption

Disregarding the absorption is permissible in an optically thin plasma, as long as the absorptivity remains below 5%. In applying the below-described method the radiating plasma should therefore be investigated for self-absorption deficiency (see section 2.12).

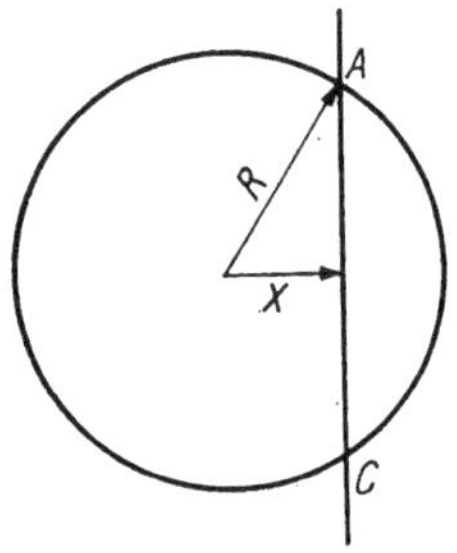

Figure 2.50. Conversion of spectral intensity distribution $I(x)$ to volume intensity distribution $\varepsilon(r)$. The spectral intensity is equal to the volume intensity integrated along the line of sight AC (Nestor and Olsen [2.95])

If self-absorption is neglected, the following relationship exists between the measured radial intensity distribution $I(x)$ at right angles to the axis of rotation and the radial distribution of the emission coefficient $\varepsilon(r)$ (see Figure 2.50)

$$I(x) = 2 \int_x^R \frac{\varepsilon(r)\, r\, \mathrm{d}r}{\sqrt{r^2 - x^2}} . \tag{2.58}$$

Equation (2.58) is an integral equation for $\varepsilon(r)$ and can be transformed after a suitable substitution into an Abel integral equation (see Schmeidler [2.114d]). The solution is

$$\varepsilon(r) = -\frac{1}{\pi} \int_r^R \frac{I'(x)\, \mathrm{d}x}{\sqrt{x^2 - r^2}} . \tag{2.59}$$

able 2.8. Coefficient scheme $\dot{K}(n, m)$ for transformation of intensity distribution $\dot{I}(x)$ at right angles to the axis to the rad
The tabulized quantit

m \ n	0	1	2	3	4	5	6	7	8	9	10	11	12	13	14
0	6366	—													
1	2122	3676	—												
2	1273	1396	2847	—											
3	909	950	1117	2406	—										
4	707	726	791	958	2122	—									
5	579	589	622	692	852	1919	—								
6	490	496	515	553	623	775	1766	—							
7	424	428	440	463	502	571	716	1644	—						
8	374	377	385	400	425	464	530	668	1544	—					
9	335	337	343	353	370	394	433	497	629	1461	—				
10	303	305	309	316	328	345	370	407	470	596	1389	—			
11	277	278	281	287	295	307	325	349	386	446	568	1327	—		
12	255	255	258	262	269	278	290	308	332	368	426	543	1273	—	
13	236	236	238	242	247	254	263	276	293	317	352	408	521	1225	--
14	220	220	222	224	228	234	241	251	263	280	303	338	393	502	1182
15	205	206	207	209	213	217	223	230	240	252	269	292	325	379	485
16	193	193	194	196	199	202	207	213	221	230	243	259	281	314	366
17	182	182	183	185	187	190	194	198	205	212	222	234	250	272	304
18	172	172	173	174	176	179	182	186	191	197	205	214	226	242	264
19	163	163	164	165	167	169	172	175	179	184	190	198	207	219	235
20	155	155	156	157	158	160	162	165	169	173	178	184	192	201	213
21	148	148	149	150	151	152	154	157	160	163	167	172	178	186	195
22	141	142	142	143	144	145	147	149	151	154	158	162	167	173	181
23	135	136	136	137	137	139	140	142	144	147	150	153	158	163	169
24	130	130	130	131	132	133	134	136	137	140	142	145	149	153	158
25	125	125	125	126	126	127	128	130	131	133	136	138	141	145	149
26	120	120	120	121	122	122	123	125	126	128	130	132	135	138	141
27	116	116	116	116	117	118	119	120	121	122	124	126	129	131	134
28	112	112	112	112	113	113	114	115	116	118	119	121	123	126	128
29	108	108	108	108	109	109	110	111	112	113	115	116	118	120	123
30	104	104	105	105	105	107	106	107	108	109	110	112	114	115	117
31	101	101	101	102	102	102	103	104	104	105	107	108	109	111	113
32	98	98	98	98	99	99	99	100	101	102	103	104	105	107	109

…ation of the emission coefficient $\varepsilon(r)$ for a source having rotational symmetry according to the formulas (2.63) and (2.64…
…scales in units of 10^{-4}

…6	17	18	19	20	21	22	23	24	25	26	27	28	29	30	31	32
…3	—															
…5	1077	—														
[illegible]	442	1047	—													
[illegible]	335	430	1019	—												
[illegible]	278	326	419	994	—											
[illegible]	242	271	318	409	971	—										
[illegible]	216	236	265	310	399	949	—									
[illegible]	197	211	230	259	303	391	929	—								
[illegible]	180	192	206	225	253	297	382	909	—							
[illegible]	168	176	187	201	220	247	290	374	891	—						
	157	164	172	183	197	216	242	285	367	874	—					
	147	153	160	169	179	193	211	238	279	360	858	—				
	139	144	150	157	165	176	189	207	233	274	354	843	—			
	132	136	141	147	154	162	172	186	204	229	269	347	829	—		
	126	129	133	138	144	151	159	169	182	200	225	264	342	815	—	
	120	123	127	131	136	141	148	156	166	179	186	221	260	336	802	—
	115	118	121	124	128	133	139	145	153	163	176	193	217	256	331	790

Here $I'(x)$ is the derivative of the measured spectral intensity distribution at right angles to the axis at x. R is the radius, for which the intensity is practically zero.

Numerous methods exist for carrying out the calculations in accordance with (2.59). We have chosen one of these, which is described by Nestor and Olsen [2.95] and stems from Gooderum and Wood [2.46].

We divide the integration range from $x = 0$ into N intervals of length a (see Figure 2.51). The mth interval lies in the range $x_m \leq x \leq x_{m+1}$ where $x_m = m \cdot a$.

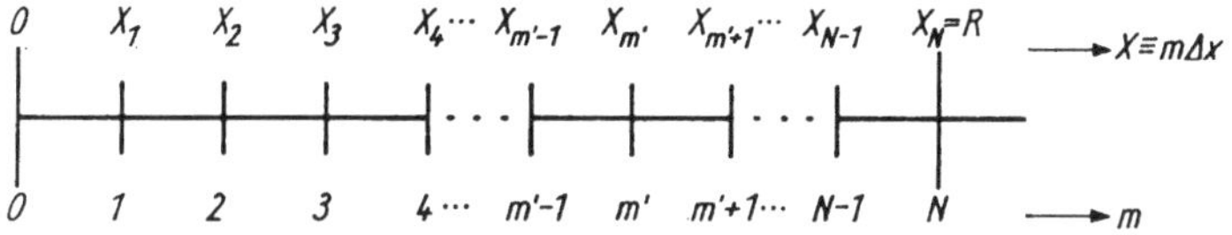

Figure 2.51. Subdivision of the range of integration for x into N intervals of width a

After the transformation $r^2 = v$ and $x^2 = u$, equation (2.59) becomes

$$\varepsilon[r(v)] = -\frac{1}{\pi} \int_v^{R^2} \frac{I'(u)\,\mathrm{d}u}{(u - v)^{1/2}}\,. \tag{2.60}$$

The integral is now separated into subintegrals in accordance with the subdivision of the interval. For every interval $I(u)$ is taken to be a linear function of u. We have for the discrete values $r_n = n\,a$

$$\varepsilon_n = f(a\,n) = -\frac{1}{\pi} \sum_{m=n}^{N-1} I'_m(u) \int_{(a\,m)^2}^{[a(m+1)]^2} \frac{\mathrm{d}u}{[u - (a\,n)^2]^{1/2}} \tag{2.61}$$

where

$$I'_m(u) = \frac{I_{m+1}(u) - I_m(u)}{a^2\,[(m + 1)^2 - m^2]}\,. \tag{2.62}$$

Now we shall carry out the integration and transform back into the original coordinates. The result is

$$\varepsilon_n = \frac{1}{a} \sum_{m=n}^{N-1} K(n, m) \cdot \Delta I(m) \tag{2.63}$$

where

$$\Delta I(m) = I(m) - I\,m(+ 1) \tag{2.64}$$

and

$$K(n, m) = \frac{2}{\pi}\,\frac{[(m + 1)^2 - n^2]^{1/2} - [m^2 - n^2]^{1/2}}{2\,m + 1}\,. \tag{2.65}$$

The corresponding values ε_n can be quite easily calculated from (2.63) even with a mechanical calculator. Table 2.8 gives the coefficient scheme $K(n, m)$ which is used for this purpose. In general, one arrives at a solution for a subdivision of the variation range into 20 to 30 intervals.

Minimum calculation time is needed when for the calculation of (2.63) an analogue computer is used [2.13a].

Before calculation, the ΔI_m-values have to be smoothed. A separate smoothing procedure is omitted when the values ΔI_m are derived from a graphical smoothed $I(x)$ plot.

A numerical procedure which includes smoothing manipulation by itself, is given by Barr [2.9a]. Instead of [2.59], $\varepsilon(r)$ is obtained using the relations

$$\varepsilon(r) = -\frac{1}{2\pi r}\frac{\mathrm{d}F}{\mathrm{d}r}, \tag{2.66}$$

$$F(r) = 2\int_r^R \frac{I(x)\,x\,\mathrm{d}x}{(x^2 - r^2)^{1/2}}. \tag{2.67}$$

The function $F(r)$ is naturally smoothed due to integration principle of its construction, while $\varepsilon(r)$ is smoothed by graphical differentiation of $F(r)$.

A refined strictly numerical procedure which gives smoothed values for $\varepsilon(r)$ from measured values $I(x)$ immediately, is given by Cremers and Birkebak [2.28b]. This procedure suitable for automatic machine calculation, includes the division of the x-axis into 5 intervals and the approximation of the measured function $I(x)$ in each intervall by a polynomal of degree 4 with respect to x respectively x^2.

For a special discussion about optimal smoothening of Abel-inverted intensities see Gorenflo [2.46b].

In the last period there has been constructed automatically working photoelectric spectral devices in which the procedures of wavelength-scanning, recording of the intensity cross-distribution and Abel-inversion (by means of analogous computing circuits) are combined [2.103a], [2.119b], [2.136]. With these devices, recording curves of emission coefficients for some given values of the radial coordinate r as function of wavelength can be obtained.

When the plasma is viewed side-on through a cylindrical glass tube, the observed intensity cross profile is distorted. For correcting procedures in this case see [2.22c], [2.66a], [2.121b].

2.10.2. Inversion with Allowance for Plasma Absorption, Determination of Plasma Absorption Coefficients

The following arguments now apply only to the radiation in a narrow wavelength range, in which the radiation intensity hardly varies, i.e., we are concerned with continuous radiation or with a sufficiently well-resolved spectral line. The methods mentioned enable radial distributions of emission coefficients per unit frequency to be determined.

To obtain radial distributions of emission coefficients integrated over the spectral line, it is in general necessary first to determine radial distributions of emission coefficients per unit frequency for different frequencies along the spectral line and subsequently to integrate over the frequency.

For weak to moderate self-absorption, inversion formulas can also be given which allow integral line intensities to be transformed immediately into integral emission coefficients.

The inversion methods are of further significance for the proper reproduction of spectral line profiles $\varepsilon_\nu = f(\nu)$ for different values of r, which are calculated from the profiles $I_\nu = f'(\nu)$ recorded for different x, which are distorted due to self-absorption.

For weak to moderate self-absorption the inversion formulas can also indicate the intensities integrated over the range of a spectral line.

In general, for the inversion procedure, besides data for the intensity cross-distribution $I(x)$, also data for the radial distribution of the absorption coefficient $\varkappa(r)$ are required. The latter can be obtained from measurements of the absorptivity-cross-distribution $A(x)$ as will be shown in section 2.10.2.1.

Section 2.10.2.2. deals with an one-step inversion procedure proposed by Olsen, useful for all degrees of self absorption, while in section 2.10.2.3. are dealt with iterative inversion formulas derived by Elder et al. The number of iteration steps depends from the degree of self absorption.

2.10.2.1. Determination of local absorption coefficients $\varkappa(r)$ from side-on measurements of the absorptivity

To obtain $\varkappa(r)$ in the case of axialsymmetric plasma sources, we follow a procedure proposed by Freeman and Katz [2.41].

The plasma and the measuring system are shown schematically in Figure 2.52. First the intensity distribution $I(x)$ is measured without using a mirror (Figure 2.52) and then the intensity distribution $K(x)$ is measured for the case when a mirror having a reflectivity f is placed behind the plasma*. Since the source for given x radiates equally in the forward direction (direction of observation) and in the backward direction (direction of the mirror), the following expression holds

$$K(x) = I(x) + f \cdot I(x) \cdot \exp\left\{-\int\limits_{-(R^2-x^2)^{1/2}}^{(R^2-x^2)^{1/2}} \varkappa(r)\, \mathrm{d}y\right\}, \qquad r^2 = x^2 + y^2 . \quad (2.68)$$

From this we obtain

$$\ln \frac{1}{1 - A(x)} = \int\limits_{-(R^2-x^2)^{1/2}}^{(R^2-x^2)^{1/2}} \varkappa(r)\, \mathrm{d}y , \qquad r^2 = x^2 + y^2 . \quad (2.69)$$

* Optically satisfactory windows may be located between the source and the mirror.

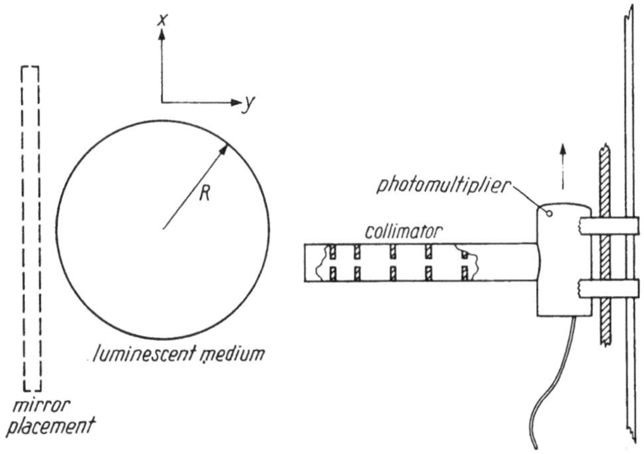

Figure 2.52. Experimental layout for the determination of the radial absorption coefficient $\varkappa(r)$ and the emission coefficient $\varepsilon(r)$ from the measured lateral distribution $I(x)$ and $K(x)$ with and without reflection, respectively. See text for explanation (after Freeman and Katz [2.41])

$A(x)$ is the absorptivity of the plasma, and is given by

$$A(x) = 1 - \frac{1}{f}\left(\frac{K(x)}{I(x)} - 1\right). \tag{2.70}$$

Introducing the expression

$$F(x) = \ln \frac{1}{1 - A(x)} \tag{2.71}$$

we obtain after a small conversion for $\varkappa(r)$ the Abel integral equation

$$F(x) = 2 \int_x^R \frac{\varkappa(r)\,\mathrm{d}r}{\sqrt{r^2 - x^2}}. \tag{2.72}$$

The solution is

$$\varkappa(r) = -\frac{1}{\pi} \int_x^R \frac{F'(x)\,\mathrm{d}x}{\sqrt{x^2 - r^2}}. \tag{2.73}$$

Formula (2.73) corresponds to formula (2.59) and can be solved by the same numerical procedure.

Finally, it should be noted the quantity $A(x)$ can also be determined without the use of mirrors when the plasma is irradiated by means of a second light source (see section 6.2.1.), whose light is suitably modulated (chopped-light method, Bauer [2.10]).

2.10.2.2. Inversion of measured intensity datas into local emission coefficients for absorbing plasmas. The Olsen inversion [2.10], [2.102]

Let a source having rotational symmetry extend from $r = 0$ to $r = R$. We divide the variation range for r into N intervals of width r_0 (see Figure 2.53). Let the nth interval lie in the range $r_n \leq r \leq r_{n+1}$, where $r_n = n\, r_0$.

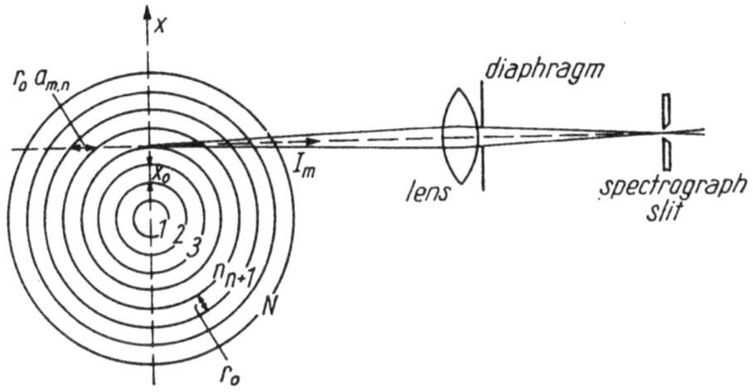

Figure 2.53. System for the inversion of intensities in the case of an absorbing source (after Olsen [2.101])

The geometrical path length, which the radiation covers in each interval n, varies with the position of the ray $x_m = m\, r_0$ (along the line of sight), and is designated by $r_0\, a_{m\,n}$. The associated optical depth is

$$\tau_{m\,n} = \frac{\varkappa_n + \varkappa_{n+1}}{2}\, r_0\, a_{m\,n}\,, \tag{2.74}$$

where $\varkappa_n$ and $\varkappa_{n+1}$ denote the absorption coefficients at the points r_n and r_{n+1}, respectively, and $a_{m\,n}$ the normalized geometrical path length:

$$a_{m\,n} = [(n+1)^2 - m^2]^{1/2} - [n^2 - m^2]^{1/2}\,. \tag{2.75}$$

By using expression (2.74) for the optical depth in each interval, we can obtain for the lateral distribution of the spectral intensity $I(x)$, taken for discrete values $x = x_m$ for a source, whose radial distribution of emission coefficient is $\varepsilon(r)$ and whose radial distribution of the absorption coefficient is $\varkappa(r)$, the expression

$$I_m = \sum_{n=m}^{N} \frac{\varepsilon_n + \varepsilon_{n+1}}{2} \cdot A_{m,\,n} \tag{2.76}$$

where

$$A_{m,\,n} = \frac{2\,(1 - \mathrm{e}^{-\tau_{m\,n}})}{\varkappa_n + \varkappa_{n+1}} \left[1 + \mathrm{e}^{-\tau_{m\,n}} \prod_{l=m}^{n-1} \mathrm{e}^{-2\,\tau_{m\,l}}\right] \prod_{l=n+1}^{N} \mathrm{e}^{-\tau_{m\,l}}\,. \tag{2.77}$$

$A_{m,\,n}$ is a triangular matrix, which can be analytically transformed into the inverse matrix $B_{n\,m}$

$$B_{n\,m} = A_{m\,n}^{-1}\,. \tag{2.78}$$

The inversion equation for the determination of the volume intensity distribution $\varepsilon(r)$ from the spectral intensity distribution $I(x)$ now becomes

$$\frac{\varepsilon_n + \varepsilon_{n+1}}{2} = \sum_{n=m}^{N} I_m B_{n,m} . \tag{2.79}$$

The prerequisite for applying this equation is that the radial distribution of the absorption coefficient $\varkappa(r)$ is known. In some cases, a direct determination of $\varkappa(r)$ is possible (see the foregoing section), while in others it is not (for example, when the plasma is enclosed in a cylindrical quartz tube with nonideally transmission properties of the envelope leading to a disturbance of the absorption measurements). Then the inversion problem may be solved by an iterative procedure, beginning with an estimated distribution for $\varkappa(r)$ (we may also choose $\varkappa(r) = 0$ initially).

Thus, we obtain for $\varepsilon(r)$ a zeroth approximation. Further measures are aimed at determining whether we are dealing with the radiation of a continuum or with that in the range of a spectral line, and whether the temperature distribution is already known or should be first determined from the volume intensity to be calculated.

We shall consider the case, in which the temperature distribution is to be determined from the volume intensity ε_ν of the continuum. Let the relationship between ε_ν and T be known.

We can now carry out the inversion of $I(x)$ to $\varepsilon_\nu^{(0)}(r)$ by assuming that the absorption coefficient $\varkappa = 0$. The corresponding temperature distribution will be designated by $T^{(0)}(r)$. Furthermore, we can obtain with the help of Kirchhoff's law, which is valid for local thermal equilibrium, a zeroth approximation $\varkappa^{(0)}$ for the absorptoin coefficient $\varkappa$ according to the equation

$$\varkappa^{(0)} = \frac{\varepsilon_\nu^{(0)}}{B_\nu(T^{(0)})} \tag{2.80}$$

(B_ν is the Planck function (see (2.19))).

Applying the values $\varkappa^{(0)}(r)$ a second inversion can be performed, whereby $\varepsilon_\nu(r)$, $T(r)$, and $\varkappa(r)$ are obtained as first approximations. Should the first approximation not suffice, a second approximation can always be made.

Let us consider the case where the temperature distribution is to be determined from the integral emission coefficient of a spectral line with known transition probability $A_{n \leftarrow m}$.

A complication arises from the wavelength dependence of the absorption coefficient across the spectral line profile.

If the degree of self absorption is not too great, then a mean absorption coefficient can be introduced. Assuming for the spectral line a dispersion profile with a full half-width $2\,\beta$ (in frequency units) we obtain for the absorption coefficient taken with respect to the radiation of the line integrated over frequency

$$\varkappa(T) = \frac{g_m A_{mn} c^2 (1 - e^{-h\nu/kT})}{16 \pi \nu^2 \beta} \frac{n(T)}{Q(T)} e^{-U_n/kT} . \tag{2.81}$$

U_n is the excitation energy of the lower level, n the number of particles per cm^3, which emit the relevant spectral line, $Q(T)$ the partition function, and 2β the full half-width of the spectral line, which also depends on temperature. In the first approximation the full half-width can be considered as constant and can be found from the spatially integrated spectral line profile. The inversion follows in the same manner as for continuous radiation: first of all we determine $\varepsilon^{(0)}(r)$ from $I(x)$ without allowing for absorption. We then calculate $T^{(0)}(r)$, and subsequently from (2.81) $\varkappa^{(0)}(r)$. With the values $I(x)$ and $\varkappa^{(0)}(r)$ we carry out another inversion and obtain the better values $\varepsilon^{(1)}(r)$, $T^{(1)}(r)$, and $\varkappa^{(1)}(r)$.

See Olsen [2.102] for a more precise determination of $\varepsilon(r)$ and $\varkappa(r)$ for the case when the radial temperature distribution $T(r)$ is known and the transition probability is the wanted quantity.

2.10.2.3. Inversion of measured intensity data into local emission coefficients for absorbing plasmas. Iterative inversion after Elder et al.

In this section we follow an iterative inversion procedure proposed by Elder et al. [2.35b] to derive the radial distribution of local emission coefficient $\varepsilon(r)$ from the measured cross distribution of intensity $I(x)$ and absorptivity $A(x)$. The convergence of this procedure has been proofed by Birkeland and Oss [2.15b].

At first, we write down the formula for the intensity $I(x)$ at the cross coordinate x

$$I(x) = \int\limits_{-\sqrt{R^2-x^2}}^{\sqrt{R^2-x^2}} \mathrm{d}y\, \varepsilon(r) \cdot \mathrm{e}^{-\left(\int\limits_{y}^{\sqrt{r^2-x^2}} \varkappa(\varrho)\,\mathrm{d}\eta\right)}, \qquad \begin{aligned} \varrho^2 &= x^2 + \eta^2 \\ r^2 &= yx^2 + y^2 \end{aligned} \tag{2.82}$$

respectively

$$I(x) = \mathrm{e}^{-\int\limits_0^{\sqrt{R^2-x^3}} \varkappa\,\mathrm{d}y} \cdot \int\limits_0^{\sqrt{R^2-x^2}} \varepsilon(r) \left(\mathrm{e}^{\int\limits_0^{y} \varkappa(\varrho)\,\mathrm{d}\eta} + \mathrm{e}^{-\int\limits_0^{y} \varkappa(\varrho)\,\mathrm{d}\eta} \right) \mathrm{d}y\,. \tag{2.83}$$

From this we obtain

$$\frac{I(x)}{T^{1/2}(x)} = 2 \int\limits_x^R \frac{\varepsilon(r) \cosh G(x, \mathrm{r}) \cdot r\,\mathrm{d}r}{(r^2 - x^2)^{1/2}}, \tag{2.84}$$

in which

$$T(x) = \mathrm{e}^{-2\int\limits_x^R \frac{\varkappa(r)\, r\,\mathrm{d}r}{(r^2-x^2)^{1/2}}} \tag{2.85}$$

denotes the transmittancy of the plasma, which is connected with the absorptivity by the relation

$$T(x) = 1 - A(x)\,,$$

while the quantity G is given by

$$G(x, r) = \int\limits_x^r \frac{\varkappa(\varrho)\,\varrho\,\mathrm{d}\varrho}{(\varrho^2 - x^2)^{1/2}}\,. \tag{2.86}$$

If the transmittancy is greater than 80%, then the term cosh $G(x, r)$ in (2.84) can be substituted by 1, and we have the usual Abel integral equation which can be solved as shown in section 2.10.1. For this region of transmittancy, Freeman and Katz [2.41] obtained the same result.

For transmittancies below 80%, a solution of (2.84) can be obtained in form of a series

$$\varepsilon(r) = \varepsilon^{(0)}(r) + \varepsilon^{(1)}(r) + \varepsilon^{(2)}(r) + \cdots \tag{2.87}$$

where the terms are given by

$$2\int_x^R \frac{\varepsilon^{(0)}(r)\, r\, \mathrm{d}r}{(r^2 - x^2)^{1/2}} = \frac{I(x)}{T^{1/2}(x)} \tag{2.88}$$

and by the recursive formula

$$\int_x^R \frac{\varepsilon^{(n)}(r)\, r\, \mathrm{d}r}{(r^2 - x^2)^{1/2}} = -\int_x^R \varepsilon^{(n-1)}(r)\, \bigl(1 - \cosh G(x, r)\bigr) \frac{r\, \mathrm{d}r}{(r^2 - x^2)^{1/2}} \tag{2.89}$$

which is like (2.88) and Abel integral equation and can be solved according to the procedure given in section 2.10.1 after the integration of the expression in the right hand side of equation (2.89) had been carried out.

According to [2.35 b], the term cosh $G(x, r)$ can be substituted by $1 + \frac{1}{2}\, G^2(x, r)$, if the transmittancy of the plasma is at least 10%.

A refined numerical procedure for evaluating $\varepsilon(r)$ which is based upon a polynomal approximation of the measured quantities, is given in [2.15 b].

2.11. Direct Determination of Emission Coefficients with Optical Probes

According to Stojanoff [2.121 a], local emission coefficients in inhomogeneous plasmas of not too small size can be determined directly by means of an optical probe which is quickly moved through the plasma. The principle of construction can be seen from the upper part of Figure 2.54. Main part of the probe is a stainless steel tube with 1.5 mm inner diameter. The inner surface is blackened. One side of the tube is coupled to a flexible light guide which is inserted a short distance into the probe tube. Close in front of the open end of the tube is placed a blackened obstacle (probe cap). Probe tube and probe cap as a whole are moved quickly in the direction C through the plasma (see Figure 2.54). Simultaneously the signal of the plasma radiation from volume D after passing the light guide and an interference filter with low bandwidth is picked up by a photomultiplier and recorded by an oscillograph. The signal is proportional to the local emission coefficient.

By repetitive interchanging the blackened probe cap with a reflecting one, absorption coefficients can be determined (compare section 2.10.2.1.).

With the double probe system shown in Figure 2.54 below, a higher signal to noise ratio is obtainable.

Using an improved optical probe system, Stojanoff [2.121 a] obtained automatic records of the spatial temperature distribution in a plasma. The temperature was derived from the intensity ratio of two spectral lines with different excitation energy (two-line method, see section 5.2.2.).

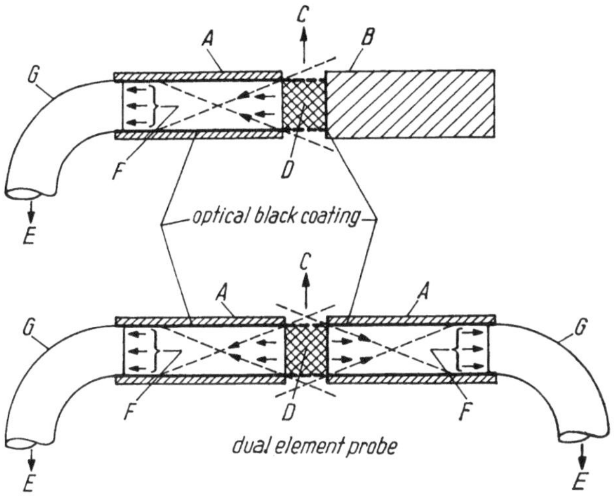

Figure 2.54. Optical probe as means for the direct determination of local emission and absorption coefficients. After Stojanoff [2.121 a])

2.12. Analysis of the Emitting Plasma for Self-absorption Deficiency

In many cases for the evaluation of spectroscopic measurements methods are used which are based on an optically thin medium. Before applying such methods it is necessary to check whether the plasma has a low absorptivity A ($\leq 5\%$).

In this connection, methods, in which the plasma absorption is directly measured, are of significance. These methods were already discussed in section 2.10.2.1. The absorptivity can also be calculated from approximation formulas (see section 2.12.1). Finally, we shall discuss the experimental check for low absorptivity (section 2.12.2).

2.12.1 Approximation Formulas for the Absorptivity A_ν

Since for a low absorptivity the approximation

$$A_\nu = \varkappa_\nu l \tag{2.90}$$

is valid, it follows from (2.80) that the absorptivity is given by

$$A_\nu = \frac{\varepsilon_\nu l}{B_\nu(T)} . \tag{2.91}$$

Here ε_ν denotes the emission coefficient and B_ν the Planck function (see (2.19)). When applying (2.91) it is assumed that ε_ν and T are at least approximately known. See Chapter 9 for the calculation of ε_ν in the case of a continuous spectrum.

In the case of spectral lines, the spectral emission coefficient depends from two quantities: from the integral line emission coefficient and from the line profile.

The integral line emission coefficient is according to chapter 5, given by

$$\varepsilon = \int_{\text{(line)}} \varepsilon_\nu \, \mathrm{d}\nu = \frac{A_{n\,m}\, g_m\, n(T)\, h\,\nu\, \mathrm{e}^{-U_m/kT}}{4\,\pi\, Q(T)},$$

where

$$U_m - U_n = h\,\nu\,.$$

Here $A_{n\,m}$ denotes the transition probability, m the higher level, n the lower level, g_m the statistical weight of the higher level, $Q(T)$ the partition function, and $n(T)$ the number density of ions or atoms which emit the relevant spectral line.

For the case of the dispersion distribution (2.50) with β_1 = half the half-width in frequency units we have for the absorptivity at the line center

$$A_{\nu\,\max} = \varkappa_{\nu\,\max}\, l = \frac{A_{n\,m}\, g_m\, n(T)\, \mathrm{e}^{-U_n/kT}\,(1 - \mathrm{e}^{-h\nu/kT})\, l}{Q(T)\, 8\,\pi^2\,\nu^2\,\beta_1} \cdot c^2 \tag{2.92}$$

and in the case of a Doppler distribution with β_2 = half 1/e-width (see equation (2.51))

$$A_{\nu\,\max} = \varkappa_{\nu\,\max}\, l = \frac{A_{n\,m}\, g_m\, n(T)\, \mathrm{e}^{-U_n/kT}\,(1 - \mathrm{e}^{-h\nu/kT})\, l}{Q(T)\, 8\,\pi^{3/2}\,\nu^2\,\beta_2} \cdot c^2\,. \tag{2.93}$$

When applying equations (2.92) and (2.93) it is assumed that the width of the relevant spectral line (β_1 or β_2) is known and that there exists information on the transition probability. For estimates these quantities need only be approximately known.

If a line width due to smallness can not be determined with a spectroscopic device, it is still always possible to establish a lower limit (usually the Doppler width). The half 1/e-width in frequency units is

$$\beta_2 = \frac{\nu}{c}\,\xi_0$$

with

$$\xi_0^2 = \frac{2\,kT}{M},$$

where M is the weight of the ion or atom, which emits the relevant spectral line. Then for an estimate of $A_{\nu\,\max}$ we simply have the following inequality

$$A_{\nu\,\max} \leq \frac{A_{n\,m}\, g_m\, n(T)\, \mathrm{e}^{-U_n/kT}\,(1 - \mathrm{e}^{-h\nu/kT})\, l}{Q(\mathrm{T})\, 8\,\pi^{3/2}\,\beta_2} \cdot c^2\,. \tag{2.94}$$

When a continuous background of appreciable intensity is superimposed on the spectral line, the portion of the continuous absorptivity is added to the absorptivity of the spectral line.

2.12.2. Experimental Check for Low Self-absorption

The basis of our considerations lies in the expression for the emission from a homogeneous layer of length l

$$I_\nu = (1 - e^{-\varkappa_\nu l}) B_\nu , \tag{2.95}$$

which for an optically thin layer ($\varkappa_\nu l \ll 1$) can be replaced by an expression proportional to l

$$I_\nu \approx \varkappa_\nu l B_\nu . \tag{2.96}$$

Deviations between (2.95) and (2.96) of 5% or more begin to occur when $\varkappa_{\nu l} > 0.1$.

Equations (2.95) and (2.96) are also valid as an approximation for inhomogeneous plasmas, when T is understood as the temperature in the center of the plasma and for l a suitable value is taken (effective length of plasma). (These interpretations are not allowed in the case of resonance lines).

These equations are based on the assumption that the radiation can be spectrally resolved. This is not always the case, such as for example, when the radiation of a narrow spectral line is examined. Now, instead of (2.95) and (2.96), we have

$$I = B_\nu \int_{\text{(line)}} (1 - e^{-\varkappa_\nu l}) \, d\nu \tag{2.97}$$

as well as

$$I \approx \varkappa l B_\nu \tag{2.98}$$

for small optical layer thicknesses $\varkappa_{\nu \max} l \ll 1$. I here denotes the integral line intensity and $\varkappa$ the mean absorption coefficient.

Based upon equations (2.95)—(2.98), we shall now examine some criteria for low self-absorption and in connection herewith some practicle examples.

Criterion 1. The radiation is optically thin when it can be verified that the spectral intensity I_ν or the total intensity I of a spectral line is proportional to the plasma length l or increases linearly with l.

This criterion was used by Knopp et al. [2.66] to verify that the line A I 4158.6 Å from an argon plasma jet is laterally emitted from an optically thin layer.

For this purpose the line intensity was first measured in the usual manner, after which the line intensity was measured when the length of the source l was doubled. The doubling of l was effected by placing a mirror behind the plasma ray. The intensity with the mirror was 95% higher than that without. Assuming a loss due to reflection $(1 - f)$ of a few percent, it follows from (2.70) that the absorptivity of the plasma becomes vanishingly small.

When employing wall-stabilized arcs and observing along the axis it is possible to measurably alter the plasma length l. In this case it can be very readily confirmed whether I_ν or I varies linearly with l.

Criterion 2. We consider two sources with source lengths l_1 and l_2, respectively, having similar temperature distributions. Let us put $l_2 > l_1$.

The exact values of l_1 and l_2 are not required. Let us examine the spectrally resolved radiation $I_{\nu 1}$ and $I_{\nu 2}$ for the two frequencies ν_1 and ν_2, respectively, whose separation lies within the range where the Planck function B_ν can be taken to be constant. Say $I_{\nu 1} < I_{\nu 2}$. Then as a condition for an optically thin plasma for both frequencies and source lengths we have

$$\varkappa_\nu l \ll 1 \quad \text{when} \quad \frac{I\,\nu_1(l_1)}{I\,\nu_2(l_1)} = \frac{I\,\nu_1(l_2)}{I\,\nu_2(l_2)} \,. \tag{2.99}$$

Criterion 3. Again we consider two sources with lengths l_1 and l_2, where $l_1 < l_2$, and a sufficiently narrow spectral range in which B_ν can be taken as constant. Denoting $I_{\nu\,\max}(l_2)$ as the maximum spectral intensity in the relevant spectral range for a source length l_2, the following inequalities hold

$$\varkappa_\nu l_{1,2} < 0.1 \quad \text{when} \quad I_\nu(l_{1,2}) < 0.1 \cdot I_{\nu\,\max}(l_2) \,. \tag{2.100}$$

Let us consider an example.

To examine the optical thickness of the hydrogen line H_α in a water stabilized electric arc, Jürgens [2.63] recorded the profile of this line once while observing along the axial direction (end-on, $l = l_2$) and once in a direction at right angles

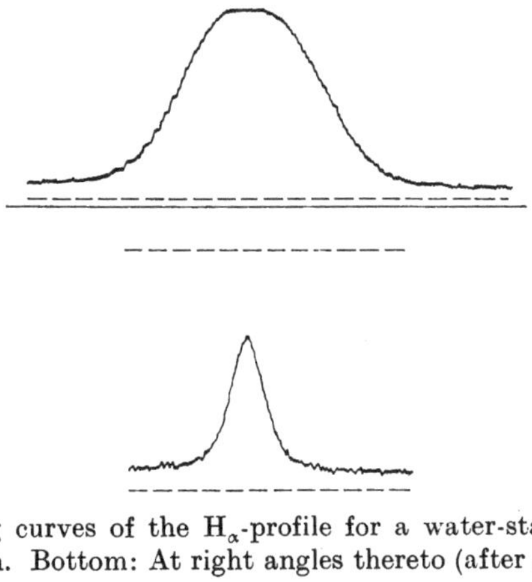

Figure 2.55. Recording curves of the H_α-profile for a water-stabilized arc. Top: taken in axial direction. Bottom: At right angles thereto (after Jürgens [2.63])

thereto (side-on, $l = l_1$) (see Figure 2.55). If the lines were optically thin for both plasma lengths l, then according to (2.99) the two line profiles would have to be proportional to one another. However, the recording shows that the profile taken along the axial direction is flattened in the center.

If we plot $I_\nu(l_1)$ against $I_\nu(l_2)$ for different wavelengths within the range of the H_α-line (Figure 2.56), we see that for small intensities (wings) linearity exists, while in the region of larger intensities saturation occurs for $I_\nu(l_2)$ $\left(I_\nu(l_2) = B_\nu\right.$ at the center of the line$\left.\right)$. This means that for the wings the plasma is optically thin for both directions of observation.

Further, we can state, owing to the validity of expression (2.100), that the side-on observed spectral line as a whole is optically thin in this direction.

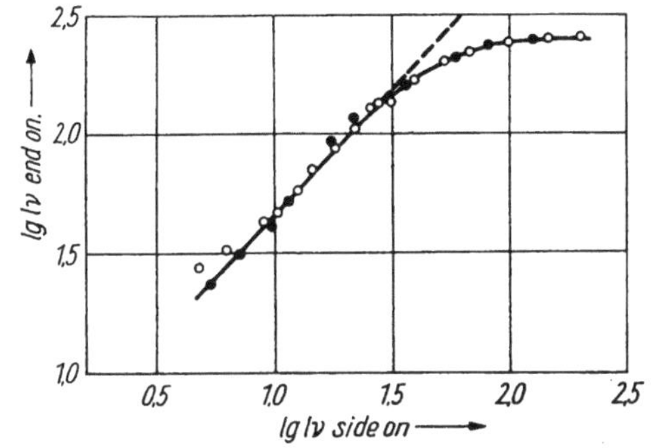

Figure 2.56. The spectral intensity I_ν, taken along the axial direction (end-on), as a function of the intensity I_ν, taken side-on, for frequencies within the range of the H_α-line (Jürgens [2.63])

○ – red wing, ● – blue wing

Criterion 4. Two unresolved spectral lines with intensities I_1 and I_2 and different absorption coefficients are radiated practically free of self-absorption when the ratio of their intensities is independent of the length l of the emitting plasma in the direction of observation.

This criterion can in general be applied to two lines with equal upper levels (see Pearce [2.105]) when they have different intensities (different A_{nm}-values). Since the upper level is decisive for the line width, the latter is practically the same for both lines (an exception are those lines in which the lower level n is a resonance level). Thus it follows from (2.92) and (2.93), that both lines posses different absorption coefficients $\varkappa_{\nu\,\max}$, which we had assumed to begin with.

Criterion 5. An unresolved spectral line of intensity I_1 optically thin when the ratio of I_1 and the spectral intensity I_ν of the continuum in the neighborhood of the line is independent of the length l of the source.

On the basis of this criterion Neumann [2.97] verified that the spectral lines of Xe I, which were emitted by an XF 3000 xenon high-pressure lamp at right angles to the arc axis, originated from an optically thin layer for wavelengths below 8000 Å, while the lines close to the infrared spectral range displayed considerable self-absorption. To prove this the intensities for two different observation directions were measured: once along the normal to the arc axis, and

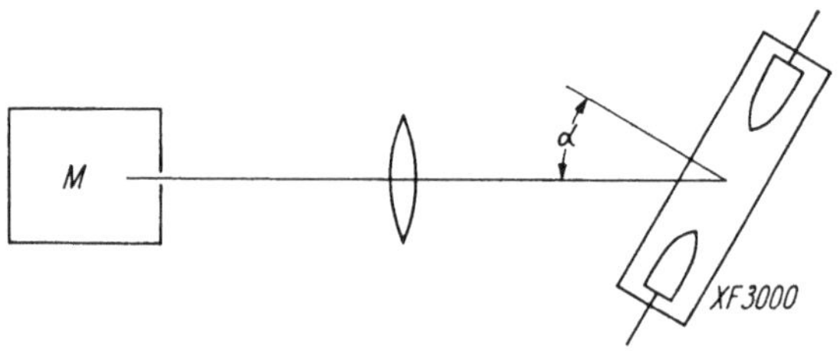

Figure 2.57. Experimental layout for checking the self-absorption of the radiation emitted from a wall-stabilized xenon high-pressure arc

M – monochromator with photocell and galvanometer

once in a direction making an angle α with the normal to the arc axis (see Figure 2.57). The lengths l, from which the radiation reached the measuring device, are thus in the ratio $\cos\alpha : 1$.

CHAPTER 3

Classification of Spectra

3.1. Atomic Spectra

We shall give a brief treatment of the nomenclature of electron states of atoms and atomic ions and discuss the classification of the transitions between these states, insofar as this has bearing on plasma diagnostic methods.

In the framework of our considerations we shall consider the spectra of atomic ions in the same sense as atomic spectra, since the former are essentially subject to the same laws as neutral atoms.

For a given atom of given charge number we need only consider the electrons occupying the outer shell (designated by the principal quantum number n) for the classification of the electron states. The following principal quantum numbers are associated with the following shells:

$n =$	1	2	3	4	5	6	7
shell:	K	L	M	N	O	P	Q

The state of each individual electron is first of all designated by two quantum numbers: the principal quantum number n and the orbital quantum number l. The latter gives the orbital angular momentum of the electron. The following scheme gives the value of the orbital quantum number for a specific electron designation:

$l =$	0	1	2	3	4	5	6	7	8	9	10	11	12	13	14	15
electron designation	s	p	d	f	g	h	i	k	l	m	n	o	q	r	t	u

Further, each electron has a spin quantum number $\pm 1/2$ which is associated with the spin angular momentum.

In order to describe the state of the electron cloud as a whole we must have precise knowledge on the quantum numbers for each individual emission electron. Fortunately, there exist certain couplings between these quantum numbers, so that a state of the electron cloud can be described by fewer parameters.

Most frequently LS (or Russell-Saunders) coupling occurs. In this case all the individual orbital quantum numbers l add up vectorially to form the total orbital angular momentum L, which describes the whole electron system, and the resultant of the spin quantum numbers of the individual electrons ($\pm 1/2$) gives the spin quantum number S for the whole system. The resultant of L and S gives the total angular momentum quantum number J. If a magnetic field is applied the projection of J on this field is designated as the magnetic quantum number M.

The values of L associated with the individual total electron states are as follows:

$L =$	0	1	2	3	4	5	6	7	8	9	10	11	12	13	14	15
designation of the total electron state	S	P	D	F	G	H	I	K	L	M	N	O	Q	R	T	U

The following table gives the atomic states respectively group of states and their characterization by quantum numbers (see Allen [3.1]):

concept	characterized by	statistical weight	occurrences upon transition
state	n, L, S, J, M	1	component of a spectral line
level = group of states	n, L, S, J	$2J + 1$	spectral line
term = group of levels	n, L, S	$(2S + 1)(2L + 1)$	multiplet
polyad = group of terms with common parent term having multiplicity S			super multiplet
configuration	n and l stated for all electrons		transition scheme

The above nomenclature is based upon LS coupling. This will be adhered to even when LS coupling is not strictly observed.

Here we should point out that frequently JJ coupling also occurs. In this case the resultants of the quantum numbers l and s of the individual electrons give the quantum numbers j, the resultant of which gives the total angular momentum quantum number J of the electron system.

We still have to introduce some further definitions: a *term series* denotes a series of terms with given quantum numbers L and S, whereby the principal quantum number n of each successive term increases by 1. A *parent term* is the ground state of the next highest ionization stage of the ion created by the removal of an electron to which a certain term series converges. The inert gases, for example, have several parent terms.

To illustrate the symbolism we shall take a typical designation for an atomic level, whereby for the sake of completeness we shall include the electron configuration.

$$\text{symbol: } 2\,p^3\,{}^4S^0_{1^1/_2}$$

this is interpreted as:

- 2 principal quantum number of the outer electrons = 2, L-shell,
- p^3 there are 3 outer electrons in the states $l = 1$,
- ^{4} multiplicity of the level $= 4 = 2S + 1$ or $S = 1^1/_2$,
- S total orbital quantum number = 0 (not to be confused with the spin quantum number S),

$1^1/_2$ $J = 1^1/_2$. The statistical weight is $(2J + 1) = 4$,

^{0} the parity of the level is odd. For even levels the designation ^{0} is left out. The designation even or odd is based on the sum of the orbital quantum numbers of the electron configuration.

In a magnetic field the levels split into $(2J + 1)$ states, which are normally not indicated by any symbols.

For many of the complicated atomic systems, such as, for example, the inert gases neon, argon, etc., *LS* coupling no longer occurs. In Ne, Ar, etc., the levels are paired and are described by a somewhat different symbolism. Let us consider the symbol for a certain level pair of argon

$$3\,p^5\,(^2P^0_{1^1/_2})\,4\,d[1/2]^0\,.$$

This reads as follows:

$3\,p^5\,4\,d$ the electron configuration of the unfilled shells,

$(^2P^0_{1^1/_2})$ the parent term of the next highest ionization stage of the ion, to which the corresponding term series converges,

1/2 (in square brackets) the mean value of the two J-values ($J = 0$ and $J = 1$) of the level pair,

^{0} (on the upper right of the square bracket) the symbol for an odd level (see above).

In the general case the energy of a level depends on all of the quantum numbers of the emission electrons explicitly (in the case of *LS* coupling the energy depends on n, L, S, and J). In hydrogen and hydrogen like ions, a degeneracy appears. Here, without allowing for the fine structure, the energy depends only on the principal quantum number n. By applying strong electric and magnetic fields the degeneracy disappears.

Selection rules: Any transitions between two levels must take place in accordance with the following selection rules

1. $\Delta L = \pm 1\,,$ (3.1)
2. $\Delta J = 0, \pm 1$ (except $J = 0 \rightarrow J = 0$) . (3.2)

The selection rule for the magnetic quantum number M is

3. $\Delta M = 0, \pm 1\,,$ (3.3)
4. even parity goes over into odd parity and vice versa.

3.2. Classification of Molecular Spectra

Molecular spectra are more complicated than atomic spectra, because the energy of the excited levels is dependent upon both the rotational and vibrational energy of the molecule.

This is evident when expressing the energy U of the level in the following way

$$U = U_e + U_v + U_r \tag{3.4}$$

where U_e is the electron component, U_v the vibrational component, and U_r the rotational component of the total energy U of the level.

We shall separately consider the energies and the quantum states associated with these components. For our purposes we shall limit the discussion to diatomic molecules.

3.2.1. Electron Quantum States

An electron state is determined by the electrons in the outer shells of those atoms which make up the molecule. In particular, the term energy depends on the quantum numbers n, l, and λ, where λ is the projection of the angular momentum quantum number l of each electron along the internuclear axis. The sum of all λ-values is the so-called axial quantum number Λ, which is the parameter primarily determining the state. The states associated with the axial quantum numbers Λ are designated by Σ, Π, Δ, . . . We have the following scheme

axial quantum number Λ	0	1	2	3	4
designation of molecular state	Σ	Π	Δ	Φ	Γ

The axial quantum number Λ can assume the integral values 0, 1, 2, . . . , L, when L is the resultant angular momentum of all the emission electrons.

The spin quantum number Σ is obtained by projecting all spin quantum numbers s of the individual electrons on the internuclear axis.* Here, in contrast with atoms, Σ can also assume negative values, i.e., it can have the values S, $S - 1, \ldots, 1 - S, -S$, when S designates the resultant spin quantum number of the outer electrons. In accordance with the $2S + 1$ different values for Σ the quantity $2S + 1$ is called the multiplicity.

Adding the quantum numbers Λ and Σ we get the quantum number of the total angular momentum:

$$\Omega = |\Lambda + \Sigma| .$$

Now we are in a position to characterize a term. On the upper left appears the multiplicity $(2S + 1)$, the upper-case Greek letter denotes the axial quantum number Λ; for example,

$$^2\Pi$$

signifies a term with multiplicity $2\,(S = 1/2)$ and axial quantum number $\Lambda = 1$.

* In a number of cases which will be not considered here, the resultant spin vector S is not coupled with the internuclear axis; then the symbols Σ and Λ losse their meanings (see Herzberg [3.3] p. 221).

Each term splits into $(2S + 1)$ levels, whose energies lie close to one another. The levels are characterized firstly by the multiplicity (upper left), secondly by the axial quantum number, which again is designated by an upper-case Greek letter, and thirdly by the quantum number Ω of the projection of the total angular momentum along the molecular axis. This quantum number is indicated by a lower index, next to the symbol for the axial quantum number.

As an example let us consider the splitting of the $^2\Pi$ term into the two levels

$$^2\Pi_{1/2} \quad \text{und} \quad ^2\Pi_{3/2} \, .$$

Here

2 is the multiplicity ,

Π states that $\Lambda = 1$.

The values 1/2 and 3/2 are the values for $\Omega = |\Lambda + \Sigma|$ ($\Lambda = 1$, $\Sigma = +1/2$ and $-1/2$).

The designation of the terms is made more complete by introducing several additional letters and symbols.

1. A letter in front of the term symbol signifies the term:
   X = ground term.
   A, B, C terms have the same multiplicity as the ground term; a, b, c, . . . are terms having different multiplicity. The order A, B, C, . . . and a, b, c, . . . corresponds to the order of the principal quantum numbers $n = 1, 2, 3$, from lower to high term energies.
2. The symbol $+$ or $-$ on the upper right next to the term symbol signifies whether the eigenfunction preserves or changes sign upon reflection in the plane through the internuclear axis.
3. The symbol g (gerade = even) or u (ungerade = odd) on the lower right next to the term symbol refers to the symmetry property of the eigenfunction with respect to reflection upon the center of symmetry for molecules made up of the same nuclei; g denotes no change of sign, while u indicates a sign change.

Let us specify the designation of the term

$$b^1\,\Sigma_g^+ \, .$$

The letter b indicates that the multiplicity of the term is different from that of the ground term (lower-case letter). 1 denotes the multiplicity. Σ specifies that the axial quantum number is 0, $+$ and g point to the fact that the eigenfunctions preserve sign upon reflection in a plane through the internuclear axis and upon reflection with respect to the center of symmetry.

The statistical weight of a term is in the case of Σ-terms ($\Lambda = 0$) equal to its multiplicity $2S + 1$, for other terms ($\Lambda \neq 0$) it is equal to $2(S + 1)$ (Λ-type doubling).

The selection rules for transitions between electron terms and between electron levels are

$$\Delta\Lambda = 0, \pm 1, \Delta\Omega = 0, \pm 1 \, , \quad \text{and} \quad \Delta\Sigma = 0 \tag{3.5}$$

For the Σ-terms the rule takes on the form

$$\Delta S = 0 \,. \tag{3.6}$$

The following selection rules hold for the symmetry properties

$$+ \leftrightarrow - \quad \text{and} \quad g \leftrightarrow u \,. \tag{3.7}$$

Details on molecular terms in diatomic molecules can be found in [3.3] (Herzberg).

3.2.2. Vibrational States

The vibrational energy U_v corresponds to that of an anharmonic oscillator.

$$U_v = h\,c\,\omega_e\,(v + 1/2) - h\,c\,\omega_e\,x_e\,(v + 1/2)^2 + h\,c\,\omega_e\,y_e\,(v + 1/2)^3 + \cdots \,. \tag{3.8}$$

where $c\,\omega_e$ is the fundamental frequency of the oscillator, v the oscillation quantum number (0, 1, 2, . . .), x_e and y_e are measures of the anharmonicity. The quantities ω_e, x_e, and y_e are, as the symbol e signifies, dependent on the electron term. The relevant tabulated values can be found in [3.3] (Herzberg).

The statistical weight of all vibrational states is

$$g_v = 1 \,. \tag{3.9}$$

There are no selection rules for vibrational transitions. All vibrational transitions are allowable. However, the line intensities decrease with increasing vibrational quantum number v' and in general also with increasing difference $|v' - v''|$ between the vibrational quantum numbers of the upper and lower vibrational states which are denoted by v' and v''.

3.2.3. Rotational States

The rotational energy U_r is given by the expression

$$U_r = h\,c\,B_v\,J\,(J + 1) - D_v\,J^2\,(J + 1)^2 + \cdots \tag{3.10}$$

where

$$B_v = B_e - \alpha_e\,(v + 1/2) + \cdots \tag{3.11}$$

(see Herzberg [3.3], p. 179). J is the rotational quantum number. The quantities B_e, α_e, and D_v are listed, among other places, in [3.3].

The statistical weight of the rotational states is normally defined by

$$g_r = 2\,J + 1 \,. \tag{3.12}$$

However, if the molecules are composed of the same nuclei, every second rotational state is unoccupied altogether or it is less heavily occupied. In this case the statistical weight g_r must be multiplied by the factor Φ_p accounting for nuclear spin degeneracy, the value of which alternates with growing rotational quantum number between

$$\Phi_{p,\,\mathrm{min}} = \frac{I}{2\,I+1}\,, \qquad \Phi_{p,\,\mathrm{max}} = \frac{I+1}{2\,I+1} \tag{3.13}$$

when I denotes the nuclear spin quantum number.

If (in the case of heteronuclear molecules) the two nuclei have nuclear spins I_1 and I_2, respectively, the statistical weight is increased by a factor of $(2\,I_1 + 1) \times \times (2\,I_2 + 1)$. When computing the material composition the nuclear spin need not be allowed for, since it cancels out in the corresponding equations.

In the case of homonuclear molecules, the statistical weight is increased by the factor $\frac{1}{2}(2\,I + 1)^2$, where the term 1/2 allows for the nuclear spin degeneracy. When computing the composition, the term 1/2 does not cancel out but must be considered, by multiplying the corresponding partition function calculated without allowing for nuclear spin degeneracy by a factor of 1/2 (see section 4.2.1.).

The selection rule for rotational transitions is given by

$$\Delta J = 0, \pm 1\,. \tag{3.14}$$

This does not hold for pure rotational spectra, but only for band spectra.

The line series, for which $\Delta J = 0$, is called the Q branch, the series with $\Delta J = +1$ is the positive or R branch, and the series with $\Delta J = -1$ is the negative or P branch. ΔJ is defined as $\Delta J = J' - J''$ (′ signifies the upper, ″ the lower level).

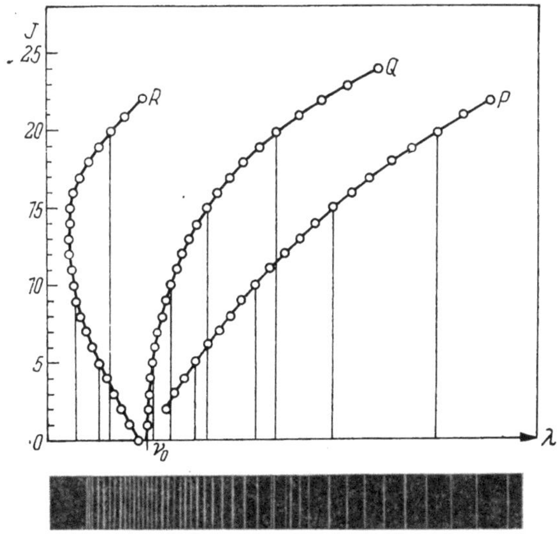

Figure 3.1. Division of spectral lines in a molecular band into the P, Q, and R branch (Finkelnburg [3.2], p. 404)

A band is the collection of all lines (lines of the Q, R, and P branch) which belong to one electron transition and one specific vibrational transition.

Figure 3.1 serves as an example for the division of lines within a band. Below a picture of a band is given, upon which the lines of the P, Q, and R branches are graphically represented as a function of the lower rotational quantum number J'' (designated by J in the diagram). This is a so-called Fortrat diagram.

For a specified wavelength the curve of the R branch displays first a negative trend and then passes over into a positive trend. Here the lines are crowded together and a sharp boundary of the band occurs, which is called the band head. The positions of the most important band heads are listed in tables (see section 2.2).

CHAPTER 4

Calculation of the Composition of a Thermal Plasma with a Boltzmann Distribution of the Electronic Levels

The evaluation of spectroscopic measurements in many cases requires that the component composition of the plasma is known as a function of temperature and pressure. This composition can be calculated under the assumption of thermal equilibrium. No additional assumption is needed if the plasma is substituted from only one element. Otherwise, additional information is needed about the elemental composition of the plasma.

The calculation is relative simple under the assumption that the relative concentrations c_n of the plasma constituents (for definition see footnote on page 107) are constant. This assumption holds for example in a homogeneous plasma or in a plasma with heavy intermixing. In quiet or laminar streaming plasmas, diffusive demixing processes bring complication into the calculation of plasma composition. For details see for example Frie [4.12b] who calculated the composition of a SF_6-Plasma with allowance for reactive demixing processes and Hermel [4.16a] who investigated a H_2-He plasma and found demixing of the plasma constituents due to thermodiffusion. For this process, a time constant of 20–40 µs was found. About demixing correlated with current transport see for example Morgenroth [4.25a].

In this chapter we shall treat corresponding methods for the calculation of the particle number densities without considering demixing processes: a) for the case in which there are no molecules in the plasma (atomic plasma), and b) for the case where the plasma contains molecules and molecular ions (chemically reacting plasma).

4.1. Calculation of the Component Composition of an Atomic Plasma

The plasma will consist of

1. electrons with density n_e,
2. atoms and ions of the element A ($A = 1, \ldots, s$) having charge $z\,e$ ($z = 0, \ldots, t$) and density n_z^A.

Our task is to determine a total of $(t + 1)\,s + 1$ unknowns (n_e and n_z^A) for a given temperature T, pressure p, and material composition of the plasma. We must therefore set up an equal number of equations. Let us start with a system of t Saha equations

$$\frac{n_{z+1}^A n_e}{n_z^A} = 2\left(\frac{2\pi m kT}{h^2}\right)^{3/2} \frac{Q_{z+1}^A}{Q_z^A} e^{-\frac{U_{z,\,\mathrm{eff}}^A}{kT}} = S_z^A(T)\,, \quad z = 0, \ldots, t-1\,, \tag{4.1}$$

where Q_z^A denotes the contribution of the internal degree of freedom of the ion respectively atom, designated by A and z, to the partition function, and $U_{z,\,\mathrm{eff}}^A$ the energy required to ionize the above ion (in what follows we shall consider the atoms as ions). Both quantities depend to a certain degree on temperature and the charge carrier concentration as we shall see later in section 4.1.1.

Tables of calculated values for the Saha function $S_z^A(T)$ for 38 elements and maximal 5 ionization steps can be found in the book of Drawin and Felenbok [4.10b]. Parameter in these tables is the lowering of ionization energy, the precise value of which can be found by iteration.

The next equation at our disposal is the quasi neutrality definition

$$n_e = \sum_{A=1}^{s} \sum_{z=1}^{t} z\, n_z^A\,. \tag{4.2}$$

When we treat the plasma as an ideal gas, which is allowable when the average distance between the plasma components is large compared to the mean diameter of the latter, we obtain as a further equation the pressure balance*

$$p = kT\left(n_e + \sum_{A=1}^{s} \sum_{z=0}^{t} n_z^A\right). \tag{4.3}$$

Finally, we have a system of $s - 1$ equations, which describe the material composition of the plasma:

$$\sum_{z=0}^{t} n_z^1 : \sum_{z=0}^{t} n_z^2 : \cdots : \sum_{z=0}^{t} n_z^s = c_1 : c_2 \cdots : c_s\,. \tag{4.4}$$

c_A denotes the given relative contribution of the element A ($A = 1, \ldots, s$) to the total number of heavy particles.

As can be seen, equations (4.1)—(4.4) form a nonlinear system of $(t + 1)\,s + 1$ equations with the same number of unknowns. The solution is usually obtained by iteration, starting from a zeroth approximation for n_e.

4.1.1. Calculation of the Effective Ionization Energy $U_{z,\,\mathrm{eff}}^A$ and the Contribution Q_z^A of the Internal Degree of Freedom to the Partition Function

To solve equations (4.1)—(4.4) more information on the quantities $U_{z,\,\mathrm{eff}}^A$ and Q_z^A is required. The effective ionization energy $U_{z,\,\mathrm{eff}}^A$ is that energy which is required to remove an electron from an ion (with charge number z and of the

* It must be mentioned that in plasmas embedded in a strong magnetic field the total pressure p varies with the spatial coordinate (see, for example, [4.10a], [4.19f]).

species A) in the ground state and to bring this electron into such a state that it is no longer bound to the rest ion (designated by $z + 1$ and A), but can freely move around in the plasma.

In rarefied plasmas ($n_e < 10^{15}\,\text{cm}^{-3}$) the effective ionization energy is practically the same as that given in tables for isolated ions in the ground state. The spectroscopically determined ionization energy U_z^A is listed in Table 4.1 (see section 4.1.2) for the common elements.

In high-pressure plasmas ($n_e > 10^{15}\,\text{cm}^{-3}$) the removal of an electron from the rest ion requires less energy than A_z^A due to the presence of electrostatic microfields in the plasma:

$$U_{z,\,\text{eff}}^A = U_z^A - \Delta U_z^A\,. \tag{4.5}$$

In accordance with calculations carried out by Ecker and Kröll [4.12] with the help of statistical methods the reduction ΔU_z^A in the ionization energy can be described by the expressions

$$\Delta U_z^A = \frac{e^2\,\varkappa}{\varepsilon}\,(z+1)\,, \qquad n \leq n_{cr}\,, \tag{4.6}$$

$$\Delta U_z^A = \frac{C\,e^2}{\varepsilon\,r_0}\,(z+1)\,, \qquad n > n_{cr} \tag{4.7}$$

($\varepsilon = 1$ in the electrostatic cgs-system). Accordlingly, reduction of the ionization energy for charge carrier densities n below the critical density n_{cr} is described by a polarization expression (4.6) and for higher carrier densities by an expression (4.7), which is analogous to the expression for the electrostatic lattice energy.

The critical carrier density n_{cr} is defined as that density below (above) which the mutual electrostatic potential energy of the plasma particles is smaller (larger) than their mean thermal energy. n_{cr} is given by the expression

$$n_{cr} = \left(\frac{3}{4\,\pi}\right)\left[\frac{kT}{(e\,t)^2}\right]^3, \tag{4.8}$$

where t is the maximum number of electrons separated from an atom in the plasma.

In equation (4.6) $\varkappa$ denotes the reciprocal Debye length, which is given by

$$\varkappa = \left(\frac{4\,\pi}{\varepsilon} \sum_{\substack{A=1,\,\ldots,\,s\\ z=1,\,\ldots,\,t}} \frac{n_z^A\,e^2\,z^2}{kT}\right)^{1/2}. \tag{4.9}$$

The r_0 in (4.7) signifies the mean distance between two charge carriers, and is written as

$$r_0 = \left(\frac{3}{4\,\pi\,n}\right)^{1/3}, \tag{4.10}$$

where n is the total number of carrier per unit volume:

$$n = n_e + \sum_{\substack{A=1,\,\ldots,\,s\\ z=1,\,\ldots,\,t}} n_z^A\,. \tag{4.11}$$

The constant C in (4.7) is given by

$$C = 2.2\left[\sum_{(A,\,z)} n_z^A\, e^2 z^2/(\varepsilon kT)\right]^{1/2} n_{cr}^{-1/3}\,. \qquad (4.12)$$

For a plasma, consisting of similar atoms, electrons, and simply charged ions, the constant becomes $C = 1.22$. The expression for the reduction in the ionization energy then simplifies to

$$\left.\begin{aligned} \Delta U &= 2.09 \cdot 10^{-8}\,(n_e/T)^{1/2}\,, && n_e \leq 2.55 \cdot 10^7\, T^3\,,\\ \Delta U &= 3.58 \cdot 10^{-7}\,(n_e)^{1/3}\,, && n_e \leq 2.55 \cdot 10^7\, T^3\\ &(\Delta U \text{ in eV})\,. \end{aligned}\right\} \qquad (4.13)$$

For this case the reduction in the ionization energy is illustrated by the diagram in Figure 4.1, which was taken from a work of Ecker and Kröll [4.12]. Note: $n = 2\, n_e$.

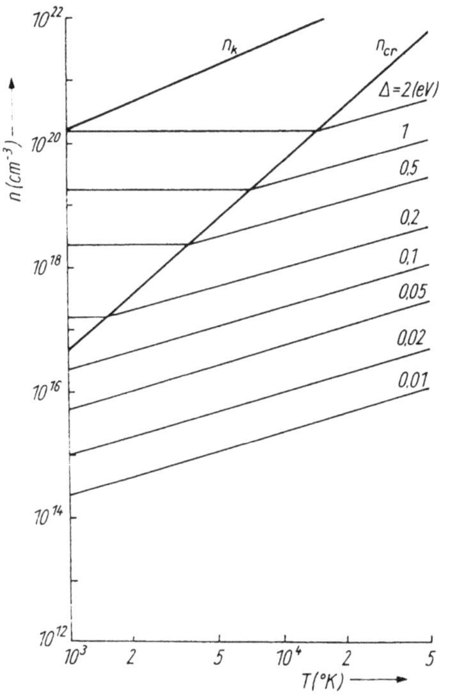

Figure 4.1. Reduction in the ionization energy ΔU as a function of n ($n = 2\, n_e$) and T for a plasma containing electrons, atoms, and simply charged ions (after Ecker and Kröll [4.12]). n_{cr} is the critical charge carrier density, n_k gives the limit up to which the electrons can be described classically

From this figure we can deduce, among other things, that the critical density for $T > 10{,}000$ °K lies above 10^{19} cm^{-3}. This means that for most laboratory plasmas with pressures below 100 atm we can employ equation (4.13) for the ionization energy.

Next, we indicate the contribution Q_z^A of the internal degree of freedom of the plasma particle type, characterized by A and z, to the partition function:

$$Q_z^A = \sum_{U_{z,r}^A = 0}^{U_{z,Q}^A} g_{z,r}^A \, e^{-\frac{U_{z,r}^A}{kT}} . \tag{4.14}$$

Here, $g_{z,r}^A$ denotes the statistical weight of the electron term with excitation energy $U_{z,r}^A$. (In this case, as in what follows, we shall designate all discrete energy states, insofar as this does not lead to any confusion, by terms.) The sum must be taken over all terms with energy values between 0 and $U_{z,Q}^A$. For a rough estimate it is allowed to substitute for the quantity $U_{z,Q}^A$ the quantity $U_z^A - 1.2\,(n_e/10^{18}\ \mathrm{cm}^{-3})^{1/3}$ which is of the order of the effective ionization energy (see section 10.8.2).

Numerical data on the term energies U and statistical weights g ($g = 2\,J + 1$) can be found in tables compiled by Moore [4.25]. However, Moore does not give any data on terms of elements with atomic numbers 58–71, 85, 87, 90ff. Some details on these terms can be found in [4.21] (Landolt-Börnstein), with the exception of the elements with atomic numbers 61, 85, 87, 93ff.

Because the tables mentioned above include not all existing levels, the missing levels must be found by extrapolation (see, for example [4.23d]).

4.1.2. **Practical Examples, Simplifications**

As an example for finding the solution when calculating the plasma composition of a complex plasma we give Figure 4.2, which depicts the different particle number densities of a completely dissociated water-vapor plasma for $p = 1$ atm and $T = 5{,}000-50{,}000$ °K.

In the case of a complex plasma we must solve the system of equations (4.1)–(4.4), which is not possible in closed form. Olsen [4.30], however, found a closed form for the calculation of the particle concentration for a simple atomic plasma for the case of simple to complex ionization. At the same time the ideal gas law $p = \sum n\,kT$ is extended by the Debye-Hückel pressure corrections, which allow for the cohesive pressure of the plasma. Corrections for volume (in the sense of the van der Waals equation of state) were not allowed for.

The system of equations linking the total pressure p, the electron pressure p_e, and the temperature T is given by

$$p = p_e \left[\frac{1 + \sum_{z=0}^{t-1} (z+2)\, C_z}{\sum_{z=0}^{t-1} (z+1)\, C_z} \right] - \frac{e^3 \sqrt{\pi}}{3\,(kT)^2} \left[p_e \frac{\sum_{z=0}^{t-1} (z+2)(z+1)\, C_z}{\sum_{z=0}^{t-1} (z+1)\, C_z} \right]^{3/2} \tag{4.15}$$

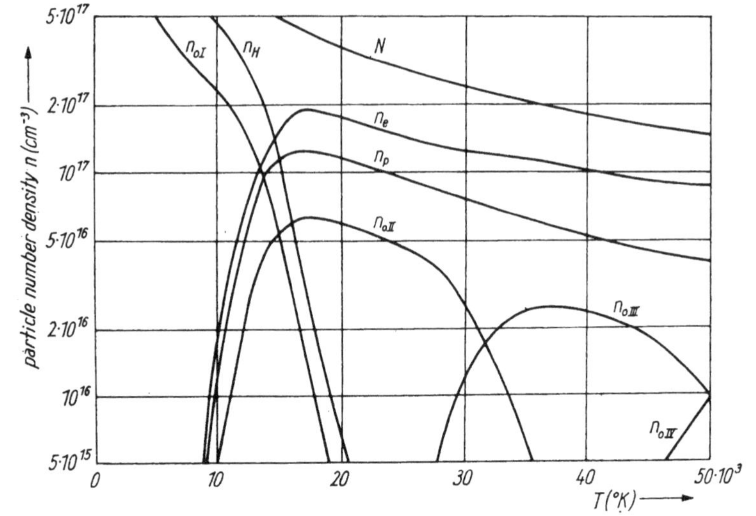

Figure 4.2. Particle densities per cm^{-3} as a function of temperature for water vapor as the starting substance. $p = 1$ atm. N = total number of particles per cm^3 (after Burhorn, Maecker, and Peters [4.5])

with

$$C_z = \prod_{i=0}^{z} \frac{F_i}{p_e}, \tag{4.16}$$

$$F_i = 2 \left(\frac{2\pi m_e}{h^2}\right)^{3/2} k^{5/2} \frac{Q_{i+1}}{Q_i} T^{5/2} \mathrm{e}^{-(U_i - \Delta U_i)/kT} \tag{4.17}$$

and (for the case $n < n_{cr}$)

$$\Delta U_i = (i+1)\, 2 \frac{\sqrt{\pi}}{kT} e^3 \left[\frac{p_e \sum_{z=0}^{t-1} (z+2)(z+1) C_z}{\sum_{z=0}^{t-1} (z+1) C_z}\right]^{1/2}. \tag{4.18}$$

The partition function

$$Q_i = \sum_r g_{z,r}\, \mathrm{e}^{-\frac{U_{i,r}}{kT}} \tag{4.19}$$

is taken over all terms with excitation energies up to the effective ionization energy $U_i - \Delta U_i$.

The solution of (4.15)—(4.19) is effected iteratively, whereby first estimated values $\Delta U_i^{(0)}$ are taken for the reduction in the ionization energy U_i, and a zero approximation $p_e^{(0)}$ for the partial electron pressure p_e is obtained for given values of p and T. On the basis of $p_e^{(0)}$ and $\Delta U_i^{(0)}$ a better value $\Delta U_i^{(1)}$ for the reduction in the ionization energy can be calculated from equation (4.18). With this a better p_e-value, $p_e^{(1)}$, can be determined from the above system of equations (4.15)—(4.17) and (4.19). Further iteration is usually no longer required.

Accordingly, the particle concentrations are obtained from the expressions

$$n_e = p_e/kT\,, \qquad n_0 = n_e\Big/\left(\sum_{z=0}^{t-1} (z+1)\, C_z\right) \quad \text{and} \quad n_{z+1} = n_0\, C_z\,. \qquad (4.20)$$

When estimates are made it is often sufficient to replace the partition function by the statistical weight $g^A_{z,0}$ of the deepest lying multipet. The error thus involved is usually small for a small degree of ionization.

For this purpose we list in Table 4.1 the statistical weights $g^A_{z,0}$ of the deepest multiplet for different elements along with their respective ionization energies U^A_z. The tabulation procedure corresponds to that of Unsöld [4.41]. The symbol ** next to the statistical weight signifies that the term immediately following the ground term lies within a distance of 1 eV (corresponding to 8068 cm^{-1}). The symbol * means that the term immediately following the ground term lies within a distance of 2 eV (wave number 16,136 cm^{-1}). The data in the table were taken from [4.2a], [4.11], [4.13c], [4.19], [4.21], [4.23b], [4.25], [4.39], and [4.41].

Numerical values of the complete partition function of 39 elements and their ions, given as function of temperature T and lowering of ionization energy ΔU_i, can be found in [4.10b].

Up to this point in our discussion we have not considered the case in which negative ions occur in the plasma. This happens when the plasma contains electronegative gases, such as hydrogen, oxygen, and chlorine. No essential changes have to be introduced into equations (4.1)—(4.4) for the calculation of the ionization equilibrium. Only additional terms with $z = -1$ appear and the quasineutrality condition assumes a slightly different form and in the case of a simple plasma reduces to $n_+ = n_e + n_-$.

For details on the calculation of the reduction in the ionization energy in a plasma containing negative ions, see Ecker and Kröll [4.12].

Saha functions $S(T)$ for a hydrogen plasma containing negative ions are given by Drawin and Felenbok [4.10b] p. 227.

Boldt [4.3] studied the case where negative ions are formed by attachment of electrons to excited atoms.

4.2. Calculation of the Component Composition of a Chemically reacting Plasma as a Function of Temperature and Pressure

When calculating the equilibrium concentrations we must consider the chemical reactions taking place between the neutral plasma components. Also, we need only limit ourselves to the case of a simply ionized thermal plasma, since in more highly ionized plasma there are practically no molecules present.

Table 4.1. Ionization energy U (in eV) and statistical weight g of basic atomic and ionic terms, $Z = 1 - 60$

Z	Element	g_0	U_0	g_1	U_1	g_2	U_2	g_3	U_3	g_4	U_4	g_5	U_5	g_6	U_6	g_7	U_7	g_8	U_8
1	H	2	13,598																
2	He	1	24.587	2	54.416														
3	Li	2*	5.392	1	75.638	2	122.451												
4	Be	1	9.322	2	18.211	1	153.888	2	217.713										
5	B	6	8.298	1	25.155	2	37.930	1	259.363	2	340.217								
6	C	9*	11.260	6	24.382	1	47.883	2	64.492	1	392.084	2	489.980						
7	N	4	14.532	9*	29.600	6	47.438	1	77.471	2	97.888	1	552.063	2	667.029				
8	O	9*	13.618	4	35.117	9	54.900	6	77.413	1	113.902	2	138.115	1	739.327	2	871.39		
9	F	6	17.422	9	34.970	4	62.661	9	87.10	6	114.242	1	157.156	2	185.185	1	953.90	2	1103.09
10	Ne	1	21.564	6	41.08	9	63.45	4	97.10	9	126.19	6	157.93	1	207.26	2	239.10	1	1195.82
11	Na	2	5.139	1	47.302	6	71.74	9	98.905	4	138.401	9	172.138	6	208.496	1	264.20	2	299.85
12	Mg	1	7.646	2	15.035	1	80.137	6	109.32	9	141.267	4	186.539	9	224.955	6	265.93	1	327.98
13	Al	6	5.986	1	18.828	2	28.447	1	119.987	6	153.81	9	190.46	4	241.439	9	284.60	6	330.24
14	Si	9**	8.151	6	16.345	1	33.492	2	45.141	1	166.77	6	205.16	9	246.47	4	303.15	9	351.07
15	P	4	10.486	9*	19.725	6	30.163	1	51.48	2	65.023	1	220.47	6	263.37	9	309.33	4	371.69
16	S	9*	10.360	4*	23.411	9*	35.0	6	47.304	1	72.68	2	88.049	1	281.05	6	328.43	9	379.05
17	Cl	7	12.967	9*	23.80	4	39.914	9*	53.5	6	67.65	1	97.03	2	114.193	1	348.48	6	400.3
18	Ar	1	15.759	6	27.629	9*	40.91	4	59.69	9	75.2	6	91.17	1	124.5	2	143.44	1	422.8
19	K	2*	4.341	1	31.71	6	45.81	9	61.11	4	82.66	9	100.0	6	117.82	1	155.1	2	175.79
20	Ca	1*	6.113	2*	11.871	1	51.18	6	67.27	9	84.50	4	108.78	9	128.0	6	147.6	1	188.8
21	Sc	10*	6.540	15**	12.800	10	24.758	1	73.93	6	91.9	9	111.0	4	138.0	9	159.0	6	180.5
22	Ti	21**	6.820	28**	13.577	21*	27.47	10	43.247	1	99.87	6	119.7	9	140.7	4	170.4	9	193.2
23	V	28**	6.740	25**	14.65	28*	29.32	21*	46.6	10	65.23	1	128.95	6	150.6	9	173.4	4	205.8
24	Cr	7**	6.765	6*	16.497	25*	30.96	28*	49.1	21*	69.3	10	90.60	1	161.14	6	184.7	9	209.3
25	Mn	6	7.434	7*	15.640	6	33.70	25	51.2	28	72.4	21*	95	10	119.27	1	196.46	6	221.8
26	Fe	25**	7.870	30**	16.182	25	30.651	6	54.8	25	75.0	28	99	21	125	10	151.18	1	234.87
27	Co	28**	7.864	21**	17.056	28*	33.50	25	51.3	6	79.5	25	102	28	129	21	157	10	186.3
28	Ni	21**	7.635	10*	18.152	21	35.17		54.9	25	75.5	6	108	25	133	28	162	21	193
29	Cu	2*	7.726	1	20.291	10	36.83		57.1		79.9		103	6	139		166		199
30	Zn	1	9.394	2	17.964	1	39.71	10	59.4		82.6		108		134		174		203

Table 4.1. (Continuation)

Z	Element	g_0	U_0	g_1	U_1	g_2	U_2	g_3	U_3	g_4	U_4	g_5	U_5	g_6	U_6	g_7	U_7	g_8	U_8
31	Ga	6	6.00	1	20.51	2	30.70	1	64.2		(90)		(118)		(149)		(183)		(231)
32	Ge	9**	7.88	6	15.93	1	34.21	2	45.7	1	93.4		(123)		(155)		(189)		(226)
33	As	4*	9.81	9*	18.63	6	28.34	1	50.1	2	62.6	1	127.5		(160)		(196)		(234)
34	Se	9*	9.75	4*	21.5	9*	32	6	43	1	68	2	82	1	155		(202)		(241)
35	Br	6	11.84	9*	21.6	4*	35.9	9	47.3	6	59.7	1	88.6	2	103	1	193	10	(248)
36	Kr	1	13.996	6	24.56	9*	36.9	4	(68)		(66)		(80)		(110)		(127)	1	(256.6)
37	Rb	2*	4.176	1	27.5	6	40		(80)		(71)		(86)		(102)		(134)	2	(153)
38	Sr	1*	5.692	2*	11.027	1	(43)	6	57		(72)		(93)		(109)		(126)		(161)
39	Y	10*	6.38	1**	12.23	10**	20.5		(62)	6	77		94		(117)		(135)		(152)
40	Zr	21**	6.84	28**	13.13	21**	22.98	10	34.33		(83)	6	99		(118)		(143)		(163)
41	Nb	30**	6.88	25**	14.32	28	25.04	21**	38.3	10	50	1	103	6	125		(145)		(172)
42	Mo	7*	7.10	6*	16.15	25	27.13	28	46.4	21*	61.2	10	68	1	126	6	153		(174)
43	Tc	6**	7.28	7**	15.26		30.5		37.7		60.0		(72)		(94)		(162)		(184)
44	Ru	35**	7.364	28*	16.76	25	28.46		(47)		(63)		(81)		(100)		(119)		(194)
45	Rh	28**	7.46	21*	18.07	28*	31.05		(46)		(67)		(85)		(105)		(126)		(174)
46	Pd	1**	8.33	10	19.42	21*	32.92		(49)		(66)		(90)		(111)		(132)		(155)
47	Ag	2	7.574	1	21.48	10	34.82		(52)		(70)		(89)		(116)		(139)		(162)
48	Cd	1	8.991	2	16.904	1	37.47	10	(55)		(73)		(94)		(115)		(146)		(170)
49	In	6	5.785	1	18.86	2	28.03	1	54.4	10	(77)		(98)		(121)		(144)		(178)
50	Sn	9*	7.342	6	14.628	1	30.49	2	40.72	1	72.3	10	(103)		(126)		(151)		(176)
51	Sb	4*	8.639	9*	16.5	6	25.3	1	44.1	2	56	1	108		(132)		(157)		(184)
52	Te	9*	9.01	4*	18.6	9	31	6	38	1	60	2	72	1	137		(164)		(192)
53	I	6	10.454	9*	19.09	4	33	9	(42)		(52)	1	(77)	2	(90)	1	170		(200)
54	Xe	1	12.127	6	21.2	9	32.1	4*	(46)		(76)		(68)		(96)		(110)		(204.7)
55	Cs	2*	3.893	1	25.1	6	(34.6)		(51)		(58)		(74)		(86)		(117)		(132)
56	Ba	1*	5.210	2**	10.001		(37)	6	(49)		(62)		(80)		(93)		(106)		(140)
57	La	10**	5.61	21**	11.43	10	19.17		(52)		(66)		(80)		(100)		(114)		(128)
58	Ce	33	5.60	44	14.8	(9)	19.5	14	36.715		(76)		(85)		(100)		(122)		(137)
59	Pr		5.48				23.2						(89)		(106)		(122)		(146)
60	Nd	65	5.53												(111)		(129)		(147)

Table 4.1. (Continuation)

Z	Element	g_9	U_9	g_{10}	U_{10}	g_{11}	U_{11}	g_{12}	U_{12}	g_{13}	U_{13}	g_{14}	U_{14}	g_{15}	U_{15}	g_{16}	u_{16}	g_{17}	u_{17}	g_{18}	u_{18}
10	Ne	2	1362.16																		
11	Na	1	1465.11	2	1648.66																
12	Mg	2	367.46	1	1761.82	2	1962.6														
13	Al	1	398.62	2	441.94	1	2086.0	2	2304.1												
14	Si	6	401.42	1	476.13	2	523.3	1	2437.7	2	2673.1										
15	P	9	424.39	6	479.50	1	560.5	2	611.6	1	2816.9	2	3069.8								
16	S	4	447.08	9	504.58	6	564.5	1	651.9	2	706.8	1	3223.8		3494						
17	Cl	9	455.61	4	529.34	9	591.7	6	656.4	1	750.1	2	809.0		3658		3946				
18	A	6	479.1	9	539.04	4	618.5	9	685.7	6	755.3	1	855.3		918		4121		4426		
19	K	1	503.9	6	564.7	9	629.4	4	714.6	9	786.6	6	861.1		968		1034		4611		4934
20	Ca	2	211.26	1	591.9	6	657.2	9	726.6	4	817.6	9	894.5		974		1087		1157		5129
21	Sc	1	225.6	2	249.85	1	686.8	6	756.7	9	830.8	4	927.5		1009		1094		1213		1288
22	Ti	6	216.5	1	265.5	2	291.6	1	788.6	6	863.1	9	941.9		1044		1131		1221		1346
23	V	9	230.5	6	255.7	1	308.6	2	336.4	1	897.3	6	976		1060		1168		1260		1355
24	Cr	4	244.4		270.8		298.0	1	354.8	2	384.4	1	1013		1097		1185		1299		1396
25	Mn	9	248.3		286.0		314.4		343.6	1	404.1	2	435.6	1	1136		1224		1317		1437
26	Fe	6	262.0	9	290.3		330.8	9	361.0	6	392.2	1	457	2	490	1	1265		1358		1456
27	Co	1	276.4	6	305		336		379		411		444	1	512	2	547	1	1402		1500
28	Ni	10	224.6		321.0	6	352		384		430	9	464		499		571	2	608		1546
29	Cu		232		266.1		368.8		401		435		484		520		557		633	2	672
30	Zn		238		274		310.8		419.7		454		490		542		579		619		698
31	Ga		(271)																		
32	Ge		(280)																		
33	As		(274)																		
34	Se		(282)																		
35	Br		(291)																		

Table 4.1. (Continuation)

Z	Element	g_9	U_9	g_{10}	U_{10}	g_{11}	U_{11}	g_{12}	U_{12}	g_{13}	U_{13}	g_{14}	U_{14}	g_{15}	U_{15}	g_{16}	U_{16}	g_{17}	U_{17}	g_{18}	U_{18}
36	Kr		(300)																		
37	Rb	1	277																		
38	Sr		(182)	1	324																
39	Y		(191)			1	374														
40	Zr		(181)																		
41	Nb		(193)																		
42	Mo		(204)																		
43	Tc		(206)																		
44	Ru		(217)																		
45	Rh		(228)																		
46	Pd		(178)																		
47	Ag		(187)																		
48	Cd		(195)																		
49	In		(204)																		
50	Sn		(213)																		
51	Sb		(211)																		
52	Te		(220)																		
53	I		(229)																		
54	Xe		(238)																		
55	Cs		(246)																		
56	Ba		(156)																		
57	La		(165)																		
58	Ce		(152)																		
59	Pr		(162)																		
60	Nd		(171)																		

Table 4.1. Ionization energy U (in eV) and statistical weight g of basic atomic and ionic terms. $Z = 61-92$

Z	Element	g_0	U_0	g_1	U_1	g_2	U_2	g_3	U_3	g_4	U_4	g_5	U_5	U_6	U_7	U_8	U_9
61	Pm																
62	Sm	49	6.6		11.4												
63	Eu	8	5.64	9**	11.4												
64	Gd	45	6.7	50**	(12)												
65	Tb	(66)	5.98														
66	Dy		6.8														
67	Ho																
68	Er		6.08														
69	Tm		5.81														
70	Yb	1	7.1	2	12.11												
71	Lu	10	6.1	1*	14.7	2	(19)										
72	Hf	21	7	10**	14.9		23.3		33.1								
73	Ta	28	7.88	35**	16.2		(22)		(33)		(45)						
74	W	25	7.98	30**	17.7		(24)		(35)		(48)		(61)				
75	Re	6	7.87	7*	16.6		(26)		(38)		(51)		(65)	(79)			
76	Os	25	8.7	30**	17		(25)		(40)		(54)		(68)	(83)	(99)		
77	Ir	28	9		(17)		(27)		(39)		(57)		(72)	(88)	(104)	(121)	
78	Pt	15	9.0	10**	18.56		(29)		(41)		(55)		(75)	(92)	(109)	(127)	(146)
79	Au	2	9.22	1*	20.5		(30)		(44)		(58)		(73)	(96)	(114)	(133)	(153)
80	Hg	1	10.43	2	18.751	1	34.2	28**	72		(82)			(94)	(120)	(139)	(159)
81	Tl	6	6.106	1	20.42	2	29.8	1	50.7		(64)		(81)	(98)	(117)	(145)	(166)
82	Pb	9	7.415	6	15.028	1	31.93	2	42.31	1	68.8		(84)	(103)	(122)	(142)	(173)
83	Bi	4	7.287	9	16.68	6	25.56	6	45.3	2	56.0	1	88.3	(107)	(127)	(148)	(169)
84	Po	9	8.43		(19)		(28)		(38)		(61)		(73)	(112)	(132)	(154)	(176)
85	At		9.2		(18)		(30)		(41)		(51)		(78)	(91)	(138)	(160)	(183)
86	Rn	1	10.746		(20)		(30)		(44)		(55)		(67)	(97)	(111)	(166)	(190)
87	Fr				(22)		(32)		(43)		(59)		(71)	(84)	(117)	(133)	(197)
88	Ra	1*	5.277	2*	10.144		(34)		(46)		(59)		(76)	(89)	(103)	(140)	(156)
89	Ac	10*	6.9	1**	12.1	2**	(20)		(49)		(62)		(76)	(95)	(109)	(123)	(164)
90	Th	21	6.95		(11.5)	33**	29.4				(65)		(80)	(94)	(115)	(130)	(145)
91	Pa												(84)	(100)	(115)	(138)	(154)
92	U	85	4											(104)	(121)	(137)	(162)

Thus, we must consider the following plasma components the number of which is $(2\,r + 1)$:

1. r neutral plasma components K_i, $i = 1, \ldots, r$,
2. r simply ionized plasma components K_i^+, $i = 1, \ldots, r$,
3. electrons e.

We shall consider s dissociation reactions of the type

$$K_l + K_m \leftarrow K_n \,. \tag{4.21}$$

between the neutral plasma components. The energies of dissociation associated with these reactions can be denoted by $D_{l+m\leftarrow n}$.

For given temperature, pressure, and material composition of the plasma the task now consists of determining the $2\,r + 1$ partial pressures p_i, p_i^+, and p_e.

First of all, we have s reaction equations at our disposal:

$$\frac{p_l\,p_m}{p_n} = \frac{Q_l\,Q_m}{Q_n}\,\frac{\left(2\,\pi\,\dfrac{m_l\,m_m}{m_n}\right)^{3/2}}{h^3}\,(kT)^{5/2}\,\mathrm{e}^{-\frac{D_{l+m\leftarrow n}}{kT}}\,. \tag{4.22}$$

The right-hand side is also designated by K_p in the literature and is practically a function of temperature only. We shall treat the calculation of K_p from thermodynamic functions in section 4.2.2. Q_l, Q_m, and Q_n are the partition functions of the respective plasma components.

Further we have r Saha equations for the determination of the ionization equilibrium:

$$\frac{p_i^+\,p_e}{p_i} = \frac{2\,Q_i^+}{Q_i}\cdot\frac{(2\,\pi\,m_e)^{3/2}}{h^3}\,(kT)^{5/2}\,\mathrm{e}^{-\frac{U_{\mathrm{i,\,eff}}}{kT}}\,. \tag{4.23}$$

From quasineutrality considerations we also have

$$p_e = \sum_{i=1}^{r} p_i^+\,, \tag{4.24}$$

and the expressions for the total pressure

$$p = p_e + \sum_{i=1}^{r} (p_i + p_i^+)\,. \tag{4.25}$$

Finally, we still have to write the expression for the material composition of the plasma.

Let c_{ij} be a number denoting how often the plasma constituent* L_j ($j = 1, \ldots, r - s$) occurs in the plasma component K_i or K_i^+ ($i = 1, \ldots, r$). Then the condition for a constant material composition becomes

$$\sum_{i=1}^{r} (p_i + p_i^+)\,c_{i1} : \sum_{i=1}^{r} (p_i + p_i^+)\,c_{i2} : \cdots : \sum_{i=1}^{r} (p_i + p_i^+)\,c_{i,r-s} = c_1 : c_2 \cdots : c_{r-s}\,, \tag{4.26}$$

* By plasma constituents are meant those heavy particles which are not further dissociated, independent of whether they are electrically neutral or charged. See [4.13].

where the relative numbers of plasma constituents $c_1, \ldots, c_{r-s}$ are known beforehand. Expression (4.26) contains $r - s - 1$ mutually independent equations.

With equations (4.22)–(4.26) we have $2r + 1$ equations at our disposal, with which we can determine the $2r + 1$ unknown partial pressures. Once these are known we can readily obtain the particle densities n_i etc., with the help of the expression $p_i = n_i k T$ for an ideal gas.

We need not go into deviations from the ideal gas law. These only begin to play a role at pressure exceeding 100 atm.

Let us briefly outline the procedure for setting up the system of equations for the component composition of a reacting plasma, such as, for example, argon-free air, consisting of 21% oxygen and 79% nitrogen.

The plasma is composed of

1. 5 neutral plasma components N, O, N_2, NO, O_2,
2. 5 simply ionized plasma components N^+, O^+, N_2^+, NO^+, O_2^+,
3. electrons.

The plasma constituents are N and O.

We can specify 3 dissociative reactions:

$$\mathrm{N} + \mathrm{N} \leftarrow N_2\,, \qquad \mathrm{O} + \mathrm{O} \leftarrow \mathrm{O}_2\,, \qquad \mathrm{N} + \mathrm{O} \leftarrow \mathrm{NO}\,,$$

for which we must calculate the three expressions

$$\frac{p_{\mathrm{N}}^2}{p_{\mathrm{N}_2}} = K_{p\,1}(T)\,, \qquad \frac{p_{\mathrm{O}}^2}{p_{\mathrm{O}_2}} = K_{p\,2}(T)\,, \qquad \frac{p_{\mathrm{N}}\, p_{\mathrm{O}}}{p_{\mathrm{NO}}} = K_{p\,3}(T) \tag{4.27}$$

in accordance with (4.22). In addition, we have 5 Saha equations (equation (4.23))

$$\left.\begin{aligned} &\frac{p_{\mathrm{N}^+}\cdot p_e}{p_{\mathrm{N}}} = f_1(T, n_e)\,, \quad \frac{p_{\mathrm{O}^+}\cdot p_e}{p_{\mathrm{O}}} = f_2(T, n_e)\,, \quad \frac{p_{\mathrm{N}_2^+}\cdot p_e}{p_{\mathrm{N}_2}} = f_3(T, n_e)\,,\\ &\frac{p_{\mathrm{NO}^+}\cdot p_e}{p_{\mathrm{NO}}} = f_4(T, n_e)\,, \quad \frac{p_{\mathrm{O}_2^+}\cdot p_e}{p_{\mathrm{O}_2}} = f_5(T, n_e)\,. \end{aligned}\right\} \tag{4.28}$$

The quasineutrality condition is

$$p_e = p_{\mathrm{N}^+} + p_{\mathrm{O}^+} + p_{\mathrm{N}_2^+} + p_{\mathrm{NO}^+} + p_{\mathrm{O}_2^+}\,. \tag{4.29}$$

For the total pressure we have

$$p = 2\,p_e + p_{\mathrm{N}} + p_{\mathrm{O}} + p_{\mathrm{N}_2} + p_{\mathrm{NO}} + p_{\mathrm{O}_2}\,. \tag{4.30}$$

Finally, the expression for the material composition of the plasma is given by

$$\frac{(p_{\mathrm{N}} + p_{\mathrm{N}^+}) + 2\,(p_{\mathrm{N}_2} + p_{\mathrm{N}_2^+}) + (p_{\mathrm{NO}} + p_{\mathrm{NO}^+})}{(p_{\mathrm{O}} + p_{\mathrm{O}^+}) + (p_{\mathrm{NO}} + p_{\mathrm{NO}^+}) + 2\,(p_{\mathrm{O}_2} + p_{\mathrm{O}_2^+})} = \frac{79}{21}\,. \tag{4.31}$$

Now we need only indicate the principle involved in solving the system 4.27)–(4.31), thus saving ourselves the trouble in solving the system (4.22) to 4.26), which is analogous to the principle outlined below.

Instead of p and T we shall assume that p_e and T are known. Then we have with (4.27), (4.29), and (4.31) five algebraic equations in terms of the partial pressures of the five neutral plasma components. Now we can find the partial pressures. Finally, p is given by (4.30).

Carrying through the calculations for different p_e-values when T is constant, we obtain

$$p = f(T, p_e)\,,$$

from which the expression

$$p_e = f(T, p) \tag{4.32}$$

results by graphical means or numerical inversion. Via this equation and for known values of T and p we are at last in a position to determine all the desired partial pressures (and particle densities with the help of the expression $n_x = p_x/kT$).

At this stage we still have to consider the calculation of the internal partition functions of molecules.

4.2.1. Calculation of the Internal Partition Functions of Molecules

Essentially, we shall follow the procedure due to Burhorn and Wienecke [4.6].*
The internal partition function is given by the expression

$$Q_i = \sum_e \sum_r \sum_v g_e\, e^{-U_e/kT}\, g_r\, e^{-U_r/kT}\, g_v\, e^{-U_v/kT}\,, \tag{4.33}$$

where U_e, U_r, and U_v respectively denote the electron, rotational, and vibrational contribution to the internal energy, and g_e, g_r, and g_v are the respectively statistic weigths of the terms.

Carrying out the summation over the electron states explicitly, we obtain

$$\begin{aligned} Q_i &= g_{e\,0}\, Q_{rv}^{(0)} + g_{e\,1}\, e^{-U_1/kT}\, Q_{rv}^{(1)} + g_{e\,z}\, e^{-U_2/kT}\, Q_{rv}^{(2)} + \cdots \\ &= Q_0 + Q_1 + Q_2 + \cdots \end{aligned} \tag{4.34}$$

where

$$Q_{rv}^{(i)} = \sum_r \sum_v g_r\, e^{-U_r/kT}\, g_v\, e^{-U_v/kT}\,. \tag{4.35}$$

The decisive contribution to the total partition function is in nearly all cases made by the ground state of the electron, while the higher electron terms usually (exception: O_2) become excited at a higher temperature and even then only yield corrective contributions to the partition function. For this reason we are justified in employing the model of the rigid rotator and harmonic oscillator for the higher electron terms.

The partial partition function $Q_{rv}^{(0)}$ associated with the ground state of the electron must be determined explicitly.

* For alternative procedures, see [4.31 a].

To begin with, we have

$$\left.\begin{aligned} U_r &= h\,c\,\{B_v\,J\,(J+1) - D_v\,J^2\,(J+1)^2\}\,,\\ B_v &= B_e - \alpha_e\,(v+1/2)\,,\\ g_r &= 2\,J+1\,,\\ U_v &= h\,c\,\{\omega_e\,v - x_e\,\omega_e\,(v^2+v)\}\,, \qquad g_v = 1\,. \end{aligned}\right\} \tag{4.36}$$

In the above expressions J denotes the rotational quantum number, v the vibrational quantum number, and α_e, B_e, ω_e, x_e and D_v are constants depending of the electron state (D_v is dependent also on v).

If, while keeping v constant, the summation over the rotational states in (4.35) is replaced by integration, we finnaly obtain the component of the partition function corresponding to the ground state of the electron:

$$Q_0 = g_{e\,0} \sum_v \mathrm{e}^{-U_v/kT} \left[\frac{kT}{h\,c\,D_v} + \frac{2\,D_v}{B_v}\left(\frac{kT}{h\,c\,B_v}\right)^2 + 12\left(\frac{B_v}{B_v}\right)^2\left(\frac{kT}{h\,c\,B_v}\right)^3\right]. \tag{4.37}$$

The contributions of the higher electron states to the partition function can be written in simplified form:

$$Q_1 + Q_2 + \cdots = g_{e\,1}\,\mathrm{e}^{-U_{e\,1}/kT}\,Q_r^{(1)}\,Q_v^{(1)} + g_{e\,2}\,\mathrm{e}^{-U_{e\,2}/kT}\,Q_r^{(2)}\,Q_v^{(2)} + \cdots \tag{4.38}$$

where

$$Q_r = \frac{kT}{h\,c\,B_v}\,, \qquad Q_v = \frac{1}{1-\mathrm{e}^{-h\,c\,\omega_e/kT}} \tag{4.39}$$

see Herzberg [4.17], p. 125 and 468). The total partition function is obtained by addition of (4.37) and (4.38).

When determining the partition functions the following factors should be allowed for (see section 3.2):

The statistical weight g_e of the electron terms is in the case of Σ-terms equal to the multiplicity of the term. This is indicated by an index on the upper left of the term symbol. Example: $g_e(^1\Sigma) = 1$.

If the molecules consist of two atoms of the same species (e.g., N_2), then from symmetry considerations either the even or the odd rotational states will be depleted, or an additional alternating factor will have to be introduced into their statistical weights. A consequence hereof is that the total partition function must be divided by the symmetry number 2.

When we are not dealing with the Σ-electron states, the corresponding contribution of the electron to the partition function must be multiplied by a factor of 2, due to the Λ-doubling of the rotational terms.

4.2.2. Determination of Chemical Equilibria from Tabulated Thermodynamic Functions

In as far as there are suitable tabulated values, the chemical equilibrium can be determined from the thermodynamic data. These can be found experimentally or can be calculated from data of molecular spectroscopy.

Let us consider the reaction (Godnev [4.15])

$$a\,A + b\,B \longleftrightarrow l\,L + m\,M\,. \tag{4.40}$$

Here, A, B, L, and M denote the atoms or molecules in the plasma. a, b, l, and m are the respective molecule numbers. The law of mass action gives

$$\frac{p_L^l\, p_M^m}{p_A^a\, p_B^b} = K_p \tag{4.41}$$

where p_L, p_N, p_A, and p_B are the respective partial pressures.

The equilibrium constant K_p is obtained from the expression

$$\ln K_p = -\frac{\Delta G^0}{RT}, \tag{4.42}$$

where ΔG^0 is the change in the free enthalpy of the reaction under normal conditions (for each reactant a partial pressure of 1 atm is assumed, which is indicated by the upper index ^{0}). R is the gas constant ($R = 1.9860$ cal/°K mole).

The tables in most cases contain data on the reduced free enthalpy of the reactants

$$\Pi_T^0 = \frac{G_T^0 - E_0^0}{T} = \frac{G_T^0 - H_0^0}{T} = \frac{H_T^0 - H_0^0}{T} - S_T^0 \tag{4.43}$$

as well as the energy of reaction at absolute zero

$$Q_0 = -\,\Delta H_0^0 \tag{4.44}$$

or the energy of reaction at room temperature

$$Q_{298} = -\,\Delta H_{298}^0\,. \tag{4.45}$$

This is connected to the energy of reaction at absolute zero via the expression

$$Q_{298} = Q_0 - \Delta\,(H_{298}^0 - H_0^0)\,, \tag{4.46}$$

where $H_{298}^0 - H_0^0 = h_{298}^0$ is generally tabulated. For K_p we finally have

$$\ln K_p = -\frac{1}{R}\left[\Delta\Pi_T^0 - \frac{Q_0}{T}\right]. \tag{4.47}$$

The heat of reaction at absolute zero is positive when for a low temperature the reactants on the righ hand side in (4.40) dominate. Otherwise it is negative. $\Delta\Pi_T^0$ is given by

$$\Delta\Pi_T^0 = l\,\Pi_T^0(L) + m\,\Pi_T^0(M) - a\,\Pi_T^0(A) - b\,\Pi_T^0(B)\,. \tag{4.48}$$

We shall illustrate the calculation of the chemical equilibrium by means of an example. Let us consider the dissociation

$$2\,H \leftrightarrow H_2 \tag{4.49}$$

at a temperature $T = 3000$ °K, i.e., we are interested in finding the equilibrium constant

$$K_p = p_{H_2} : (p_H)^2\,. \tag{4.50}$$

Table 4.2. List of diatomic molecules and molecular ions for which spectroscopic data exist together with their respective energies of dissociation D_0^0 and ionization U_i. Only those molecules are listed for which the value of D_0^0 exceeds about 3.5 eV

Molecul	D_0^0 [eV]	U_i [eV]	Reference[1])	Molecul	D_0^0 [eV]	U_i [eV]	Reference
AlBr	4.55		41a	CeO	7.7		17
AlCl	5.05		41a	CrCl	3.75		41a
AlF	6.78	9.0	41a	CrO	4.4−5.3		41a
AlI	3.77		41a	CsBr	4.1−4.3		41a
AlO	5.0−5.9		41a	CsCl	4.4−4.6		14, 17, 41a
AlS	3.4−3.5		41a				
As_2	3.94	11.0	41a	CsF	5.0−5.3		41a
As_2^+	2.4		17	CsI	3.4−3.6		41a
AsN	6.5		17	CuBr	3.4		41a
AsO	4.9		41a	CuCl	3.7		41a
AsO^+	7.6		41a	CuO	4.9		14, 17, 41a
AuCl	3.5		17	FeCl	3.6		41a
B_2	3.6		17	FeO	4.2−4.8		17, 41a
BBr	4.47	9.25	41a	GaBr	4.3−4.4		41a
BCl	5.1	10.4	41a	GaCl	5.0		41a
BCl^+	3.03		41a	GaI	3.5−3.9		41a
BF	8.02		41a	GaO	2.5−3.6		41a
BH	3.0−3.4	10.1	41a	GdO	5.9−6.4		17, 41a.
BD	<3.55		17	GeC	4.7		41a
BH^+	1.3		17, 41a.	GeCl	3.5−4.0		17, 41a.
BN	4.0		41a	GeF	4.9	7.3	17, 41a.
BO	7.4−8.0		41a	GeO	6.9		17
BS	6.0		41a	GeS	5.6		17
BaF	3.8		17	GeSe	5.0		41a
BaO	5.95		41a	GeTe	4.1		41a
BeCl	4.76		41a	H_2	4.476	15.42	17, 19.
BeF	6.95		41a	HD	4.511		17
BeO	4.6−4.8		41a	D_2	4.553		17
BiO	3.7		41a	HT	4.524		17
C_2	6.2		41a	T_2	4.588		17
C_2^+	5.5		41a	DT	4.570		41a
CCl	3.3−3.5		41a	H_2^+	2.648		17
CF	4.69	8.9−13.8	41a	H_2^-	0.15		41a
CH	3.47	11.1	11	HBr	3.75	13.2	17, 19.
CD	3.52		17	HBr^+	4.0		41a
CH^+	3.6		17	HCl	4.43	13.8	17, 19.
CN	7.5−8.4	14.5−15.1	41a	DCl	4.481		17
CN^+	4.55		41a	HCl^+	4.71		41a
CO	11.0	14.0	41a	DCl^+	4.53		17
CO^+	8.36		41a	HF	4.55	15.8	41a
CP	6.0		41a	HS	<3.8		17
CS	7.6−7.8	10.7	41a	InBr	4.0		41a
CSe	6.8		17	InCl	4.4−4.5		41a
CaO	5.0		41a	InF	5.5		41a
CaS	5.2		41a	InI	3.4		41a
CdO	<3.8		41a	KBr	3.96		17
CdS	<3.9		17	KCl	4.42		17

Table 4.2. (Continuation)

Molecul	D_0^0 [eV]	U_i [eV]	Reference	Molecul	D_0^0 [eV]	U_i [eV]	Reference
KF	5.0		14, 17.	PrO	7.6		17, 41a.
KI	3.3–3.5		41a	RbBr	4.0		14, 17.
LaO	8.0–8.2		41a	RbCl	4.5–4.7		14, 17, 41a
LiBr	4.3		41a	RbF	5.0–5.2		41a
LiCl	4.8–4.9		41a	RbI	3.4–3.5		41a
LiF	5.95		14, 17.	S_2	4.3	8.3	41a
LiI	3.58	8.55	17, 41a.	SH	3.5		41a
LiK	5.895		17	SO	5.36		41a
LiO	3.4		41a	Sb_2	3.7	8.4–9.5	17, 41a.
LiRb	6.421		17	SbCl	3.7		41a
LuO	5.3		17	SbF	4.2		17
MgCl	3.9		41a	SbN	4.8		17
MgF	4.2–5.2		17, 41a.	SbO	4.4		41a
MgO	4.3		41a	ScF	5.0		41a
MnCl	3.3–3.7		41a	ScO	7		17
MnF	3.9		17	SeO	3.5		14, 17.
MnO	4.2–4.4		17, 41a.	SiBr	3.7		17
MoO	5.0		41a	SiC	4.5		41a
N_2	9.756	15.58–15.60	17, 41a.	SiCl	4.0		17
N_2^+	8.7		41a	SiF	5.4	7.3	41a
NH	3.6	13.10	41a	SiF^+	6.3		41a
ND	3.9		41a	SiN	4.5		17
NO	6.49	9.25–9.26	17, 41a	SiO	8.3	10.8	41a
NO^+	10.6–10.9		17, 40, 41a.	SiO^+	5.7		41a
NS	5.0		41a	SiS	6.4		41a
NaBr	3.85		17	SiSe	5.8		17
NaCl	4.25		41a	SiTe	5.3		41a
NaF	4.9–5.2		41a	SnCl	3.6		17
NbO	7.8		41a	SnF	3.9		17
NiBr	3.7		41a	SnO	5.8		41a
NiCl	3.3–3.8		41a	SnS	4.8		41a
NiO	4.2		41a	SnSe	4.6		17
O_2	5.080	12.1–12.2	17, 41a.	SnTe	4.1		41a
O_2^+	6.65		41a	SrF	3.5		17
OH	4.35	13.18	17, 41a.	SrO	4.8		41a
OD	4.39		17	TaO	8.4–9.3		41a
OT	7.9		41a	ThO	8.5		41a
OH^+	4.97		41a	TiO	6.9		17
OH^-	4.67		41a	TlBr	3.4		41a
P_2	5.031	11.8	17, 41a.	TlCl	3.75		17
PF	4.6		41a	TlF	4.75		14, 17.
PN	7.1		41a	UO	7.8		41a
PO	6.2		17	VO	6.4		14, 17.
PbF	3.47		17	WO	6.7–7.2		41a
PbO	4.1–4.3		17, 41a.	YO	7.2		41a
PbS	4.7		17	ZnO	<4.0		41a
PbSe	4.7		17	ZnS	4.2		41a
PbTe	3.5	8.2	17, 41a.	ZrO	7.8		17

[1]) References given in short symbols, for example "17" stands for [4.17].

The heat of reaction is given by [4.24]

$$Q_0 = 103{,}240 \text{ cal/mole} . \quad (4.51)$$

The respective reduced free enthalpies are

$$\Pi^0_{3000}(\text{H}) = -33.89 \text{ cal/°K} \cdot \text{mole} , \quad (4.52)$$

$$\Pi^0_{3000}(\text{H}_2) = -40.72 \text{ cal/°K} \cdot \text{mole} , \quad (4.53)$$

we thus obtain

$$\Delta\Pi^0_{3000} = (-40.72 + 67.78)\text{cal/°K} \cdot \text{mole} = 27.06 \text{ cal/°K} \cdot \text{mole} . \quad (4.54)$$

From (4.47) we finally get

$$\ln K_p = 3.70; \qquad K_p = 40.3 . \quad (4.55)$$

Table 4.3. Bibliographic references on partition functions[1])

a) partition functions of atomic systems

Atomic system	T-range [°K]	Range of pressure and density, respectively	Reference
Ar I – Ar III	5000 – 19,500	1 atm	Popenoe, Shumaker [4.33a]
Ar I – Ar IV	5000 – 20,000	1 atm	Goldman [4.15a]
Ar I – Ar IV	5000 – 30,000	1 atm	Olsen [4.28]
Ar I – Ar V	4000 – 50,000	1 atm	Bögershausen, Krempl [4.2b]
Ar I – Ar VI	4000 – 50,000	1 atm	Olsen [4.30a]
Ar I – Ar IV	5000 – 35,000	1.1 atm	Olsen [4.29]
Ar I – Ar VI	4000 – 34,500	1.1 atm	Olsen [4.30b]
Ar I – Ar V	5000 – 35,000	0.1 – 5 atm	Drellishak et al. [4.10d]
Cs I	2400 – 7,900	$n_{max} = 7 - 24$[2])	Samuilov, Olevinski [4.39a]
Fe I – Fe III	0 – 15,000		Czernikhovski [4.9d]
H I	3500 – 5,800	$N_e = (3 - 3000)\, 10^{15}$	Eberhagen, Lünow [4.10e]
H I	$10^4 - 10^7$	$\varrho/\varrho_0 = 10^{-5} - 10$	Oppenheim, Haffmann [4.30c]
Hg I – Hg III	1000 – 15,000	10 Torr – 50 atm	Irmer, Worm [4.19b]
N I, N II	500 – 15,000	$\varrho/\varrho_0 = 10^{-5} - 10$	Peng, Pindroh [4.31]
N I, N II	5000 – 30,000	0.2 – 100 atm	Martinek [4.23f]
N I – N III	1000 – 12,000	1 – 50 atm	Irmer, Worm [4.19a]
N I – N IV	2000 – 35,000	0.1 – 5 atm	Dresllishak [4.10c]
O I, O II	500 – 15,000	$\varrho/\varrho_0 = 10^{-5} - 10$	Peng, Pindroh [4.31]
Ti I – Ti VIII	0 – 15,000		Czernichovski [4.9c]
Xe I	10,000, 20,000		McChesney, Jones [4.23d]

[1]) A tabulation of atomic partition functions $Q(T)$ of 38 elements in different ionization stages, in which the parameter is not the pressure p but the lowering of ionization energy ΔU_i, is given by Drawin and Felenbok [4,10b]. Boumans ([4.3b] p. 94) has tabulated partition functions for 57 elements inclusive single charged ions for two temperatures (5000 and 6000 °K). The summation was broken up by 3.3 eV and 3.9 eV, respectively (for Cs I by 3.7 eV). de Galan et al. [4.13b] give polynomal approximations of partition functions for 73 elements inclusive single charged positive ions for the temperature range 1500 – 7000 °K.

[2]) n_{max} = limiting quantum number up to which the partitition function is taken.

4.3. Important Bibliographic References required for the Determination of Plasma Compositions

We shall consider

1. a list of those diatomic molecules and molecular ions together with their respective energies of dissociation D_0^0, for which spectroscopic data exist and for which the value of D_0^0 exceeds about 3.5 eV (Table 4.2). Molecules with lower energies of dissociation are not treated here, since they are usually completely dissociated in the plasma.

In conjunction with this we introduce

2. a list of bibliographic references on partition functions (Table 4.3).

3. Table 4.4 contains bibliographic references for the equilibrium constants K_p of some chemical reactions as well as data on the energy of reaction at absolute zero Q_0.

Table 4.5 lists bibliographic references for tabulated plasma compositions. The results of different authors deviate from one author to another; for example, this occurs in the case of an air plasma. We shall return to this in the footnotes of Table 4.5.

Table 4.3. Bibliographic references on partition functions

b) partition functions of diatomic molecules

Molecule	T-range [°K]	Reference
C_2	2000– 6,000	Clementi [4.9b]
C_2	1000–10,000	Burhorn, Wienecke [4.6]
CN	1000–10,000	Burhorn, Wienecke [4.6]
CO	1000–10,000	Burhorn, Wienecke [4.6]
CO^+	1000– 8,000	Kloss [4.19e]
H_2	1000–10,000	Burhorn, Wienecke [4.6]
N_2	1000–10,000	Burhorn, Wienecke [4.6]
N_2	1000–12,000	Irmer, Worm [4.19a]
N_2	500–15,000	Peng, Pindroh [4.31]
N_2	2000–35,000	Drellishak et al. [4.10c]
N_2^+	1000– 8,000	Kloss [4.19e]
N_2^+	1000–12,000	Irmer, Worm [4.19a]
N_2^+	2000–35,000	Drellishak et al. [4.10c]
NO	1000–10,000	Burhorn, Wienecke [4.6]
NO	500–15,000	Peng, Pindroh [4.31]
NO^+	1000– 8,000	Kloss [4.19e]
NO^+	500–15,000	Peng, Pindroh [4.31]
OH	1000–10,000	Baumann [4.2]
O_2	500–15,000	Peng, Pindroh [4.31]
O_2	1000–35,000	Drellishak et al. [4.10c]
O_2^+	1000–35,000	Drellishak et al. [4.10c]

Table 4.4. Bibliographic references on the equilibrium constants of some chemical reactions insofar respective data exist for $T > 3000$ °K and insofar these reactions have not reached the end stage at 3000 °K. Meanings of the symbols: Q_0 — energy of reaction at absolute zero, C_{gr} — carbon in form of graphite, C, C_2, C_3 — carbon atoms and molecules, respectively, in the gas phase

Reaction	Q_0 [eV]	T-range [10^3 °K]	Reference[1])	Remarks
$C_{gr} \to C$	7.4	0.3– 6	23, 27	
$C_2 \to 2\,C$	6.2	1 –12	10, 19j	Q_0 after 41a
$C_3 \to 3\,C$	13.9	1 –12	19j, 24	Q_0 after 41a
$CH \to C + H$	3.47	1 –12	19j, 24	Q_0 after 41a
$CH_2 \to C + H_2$	4.5	1 –12	19j	Q_0 after 41a
$CH_2 \to C + 2\,H$		2 – 6	24	
$CH_3 \to C + 1.5\,H_2$	5.88	1 –12	19j	Q_0 after 41a
$CH_3 \to C + 3\,H$		2 – 6	24	
$CH_4 \to C + 2\,H_2$	6.7	1 –12	19j	Q_0 after 41a
$CH_4 \to C + 4\,H$		2 – 6	24	
$C_2H \to 2\,C + 0.5\,H_2$	9.85	1 –12	19j	Q_0 after 41a
$C_2H_2 \to 2\,C + H_2$	12.4	1 –12	19j	Q_0 after 41a
$C_2H_2 \to 2\,C + 2\,H$		2 – 6	24	
$C_2H_4 \to 2\,C + 2\,H_2$	14.0	1 – 6	19j	Q_0 after 41a
$C_2H_4 \to 2\,C + 4\,H$		2 – 6	24	
$CN \to C + 0.5\,N_2$	3.3	1 –12	19j	
$CN \to C + N$	8.4	1 –12	19j, 24	Q_0 after 41a
$C_2N_2 \to 2\,C + N_2$	11.6	1 – 6	19j	
$C_2N_2 \to 2\,C + 2\,N$	21.3	1 –12	19j, 24, 38	Q_0 after 41a
$C_2N_2 \to 2\,C_{gr} + N_2$	1.4	3 – 4	22	
$CO \to C_{gr} + 1/2\,O_2$	1.2	0.3– 5	23	
$CO_2 \to CO + 1/2\,O_2$	2.9	0.3– 5	23	
$CO_2 \to C_{gr} + O_2$	4.1	0.3–3.5	22	
$F_2 \to 2\,F$	4.12	0 – 5	18	
$HCN \to 0.5\,H_2 + 0.5\,N_2 + C$	6.0	1 – 6	19j	
$HCN \to 0.5\,H_2 + 0.5\,N_2 + C_{gr}$	1.4	0.3– 4	22	
$HCN \to C + H + N$	13.2	6 –12	19j	
$H_2 \to 2\,H$	4.476	1 –12	19j, 191, 22, 23, 27	
$H_2O \to H_2 + 0.5\,O_2$	2.48	0.3– 5	23	
$H_2O \to OH + 0.5\,H_2$	2.90	0.3– 5	23	
$HF \to 0.5\,H_2 + 0.5\,F_2$	4.55	0.3– 5	23	Q_0 after 41a
$NH \to 0.5\,H_2 + 0.5\,N_2$	2.60	1 – 6	19j	
$NH \to H + N$	3.6	2 –12	19j, 24	Q_0 after 41a
$NH_2 \to H_2 + 0.5\,N_2$	2.0	1 – 6	19j	
$NH_2 \to 2\,H + N$	7.3	2 –12	19j, 24	Q_0 after 41a
$NH_3 \to 1.5\,H_2 + 0.5\,N_2$	−0.1	1 – 6	19j	
$NH_3 \to 3\,H + N$	12.0	2 –12	19j, 24	Q_0 after 41a
$N_2 \to 2\,N$	9.756	0.3–15	19j, 22, 23, 27, 31, 37	
$NO \to 0.5\,N_2 + 0.5\,O_2$	0.93	0.3– 6	23	
$NO \to N + O$	6.50	0.3–15	10, 31, 42	
$O_2 \to 2\,O$	5.10	0.3–15	22, 23, 31, 37, 42	
$P_2 \to 2\,P$	5.03	0.3–3.273	10	
$S_2 \to 2\,S$	4.3	0.3– 5	23	Q_0 after 41a

[1]) References are given in short form, for example "19j" instaed of [4.19j].

Table 4.5. Bibliographic references on plasma composition

Medium	Range of pressure or density	Range of T [10^3 °K], Reference
Air	1 atm	2–9 [4.37][1])
Air	1–30 atm	1–30 [4.8][2])
Air	$10^{-2}-10^3$ atm	20–100 [4.27]; 20–150 [4.26][3])
Air	$\varrho/\varrho_0 = 10^{-6}-10^{-2}$	0.09–15 [4.22a]
Air	$\varrho/\varrho_0 = 10^{-5}-10^0$	0.5–15 [4.31]
Air	$\varrho/\varrho_0 = 10^{-6}-10^2$	0.6–3 [4.18a][4])
Air inclusive	$10^{-3}-10^3$ atm	6–12 [4.34][5]), 12–20 [4.35]
Ar		
Air + (0–75%) C_2	0.001–100 Torr	6–100 [4.20a]
Air + 50% C	1 atm	6–100 [4.20b]
Air + 75% C	$10^{-3}-10^2$ atm	20 [4.20b]
Air + (0–50% (mass) CH_2)	1 atm	3–75 [4.23e]
Ar	1 atm	5–19 .5 [4.33a; 5–30 [4.28]; 4–50 [4.30a]
Ar	0.01–1 atm	9–72 [4.1a]
Ar	1.1 atm	4–34.5 [4.30b]; 5–35 [4.29], [4.30]
Ar	0.1–5 atm	5–35 [4.10d]
Ar	$n_e = 10^{17}-10^{19}$ cm^{-3}	5–45 [4.9a]
Ar	$\varrho/\varrho_0 = 10^{-5}-10$	0.5–15 [4.31]
Ar + 5% H_2	1 atm	7–25 [4.12a]
x Ar + y Zn, $y/x = 10^{-9} - \infty$	1 atm	4–50 [4.2c]
C + H_2	1 atm	1–6 [4.19h], [4.19k], [4.191]
CH_2	0.01–1 atm	7–15 [4.27]
C + 2 H_2	1 atm	1–6 [4.191]; 5–50 [4.19k]
C + H_2 + N_2	10^{-4}–100 atm	1–12 [4.19j]
C + H_2 + NH_3	10^{-4}–100 atm	1–12 [4.19j]
CH_4 + 2 O_2 + K_2CO_3 3 (Vol.-% K)	1 atm	1–4 [4.19c], [4.19d]
CO_2	$10^{-4}-10^2$ atm	1–24 [4.36]
CO_2	$10^{-2}-10^3$ atm	1–12 [4.33]
Cl_2	715 Torr	6.5–8.5 [4.25b]
H_2	10^{-2}–10 atm	2–35 [4.1a]
H_2	1–10^4 Torr	10–50 [4.20]
H_2	1–30 atm	1–30 [4.9]
H_2	$10^{-3}-10^6$ atm	1–10 [4.4]
H_2	$10^{-6}-10^6$ atm	1–1000 [4.16]
H_2	$n_{total} = 0.3-300$ 10^{16} cm^{-3}	3.5–350 [4.10e][6])
H_2O	1–30 atm	1–30 [4.9]
He	1 atm	13–19 [4.3a], 1–35 [4.23c] 8–68 [4.2b]
He	0.01–10 atm	12–75 [4.1a]
He	$10^{-4}-10^3$ atm	8–50 [4.23a]
Hg	10^{-4}–1 atm	1–15 [4.39b]
Hg	10 Torr–50 atm	1–15 [4.19b]
Hg + In I + I_2	21.5 atm	1–7 [4.13a]; 3–32 [4.36a]

Table 4.5. (Continuation)

Medium	Range of pressure or density	Range of T [10^3 °K], Reference
N_2	1 atm	4–50 [4.2b]
N_2	0.01–5 atm	2–35 [4.10c]
N_2	1–30 atm	1–30 [4.8]
N_2	1–50 atm	1–12 [4.19a]
N_2	0.2–100 atm	5–30 [4.23f]
N_2	10^{-4}–10^3 atm	1–12 [4.32]
NO	1–30 atm	1–30 [4.8][7])
O_2	0.01–5.0 atm	1–35 [4.10c]
O_2	1–30 atm	1–30 [4.7]
SF_6	1 atm	1–10 [4.36b][8])

[1]) With and without allowance for demixing processes.

[2]) The neglection of NO-ionization causes errors in the calculated values for the electron density in the region of low degrees of ionization. In the contrary, neglection of ionization of O_2 and N_2 seems to be justified.

[3]) The influence of radiation density is allowed for in the calculations of plasma composition.

[4]) O^- and O_2^- are allowed for as plasma components.

[5]) The calculated concentration of NO is lower by a factor 2 than that given by Burhorn and Wienecke [4.8].

[6]) n_{total} = number density of H-nuclei in [cm^{-3}].

[7]) NO-ionization is not allowed for.

[8]) Demixing is not allowed for.

CHAPTER 5

Determination of Plasma Parameters via the Intensity of Spectral Lines, emitted by Atoms and Ions in an Optically Thin Layer. Plasma in Thermal Equilibrium

5.1. Determination of Plasma Parameters from the Absolute Emission Coefficient of Spectral Lines

Starting from the absolute intensity, measured as a function of the space coordinates of a spectral line, emitted by a spatially symmetric plasma, the emission coefficient of the line can be readily found.

The emission coefficient ε_{nm} is expressed by the well-known formula

$$\varepsilon_{nm} = \frac{1}{4\pi} n_m A_{nm} h \nu_{nm} , \tag{5.1}$$

(ε_{nm} in erg · cm^{-3} s^{-1} sr^{-1}, n_m in cm^{-3}, A_{nm} in s^{-1}, ν_{nm} in s^{-1})

where n_m denotes the number of excited particles per cm^3 in the (upper) level m and A_{nm} the probability of a particle making a transition from an upper level m to a lower level n upon the emission of radiation of frequency ν_{nm}. When the emission coefficient ε_{nm} is measured and the transition probability A_{nm} is known the population of the term n_m can be determined. Since we assume a Boltzmann distribution, we obtain with

$$n_m = \frac{g_m n_z^A}{Q_z^A} e^{-U_m/kT} \tag{5.2}$$

the expression

$$\varepsilon_{nm} = \frac{1}{4\pi} \frac{g_m}{Q_z^A} n_z^A A_{nm} h \nu_{nm} e^{-U_m/kT} . \tag{5.3}$$

for the emission coefficient.

Here g_m is the statistical weight of the upper level*, n_z^A the number of particles of type A with charge number z, which emit the spectral line under investigation, Q_z^A the corresponding partition function (for computation see section 4.1.1), and U_m the excitation energy of the upper level. In the case that the gas temperature is not equal to the electron temperature, T must be replaced by T_e in equations (5.2) and (5.3). Below we shall assume a unique plasma temperature.

We shall now formulate problems of plasmaparameter determination which can be solved by the measurement of absolute emission coefficients of spectral lines.

* In the following we assume that g_m is constant. In reality, this is not the case when the excitation energh U_m of level m lies in the neighbourhood of the ionization limit U_i. About the consequences thereof for plasmadiagnostics see section 10.8.2.

5.1.1. Determination of the Plasma Temperature when Pressure and Material Composition are known

Let the pressure p and the material composition of the plasma be known. The temperature T is determined from a measurement of ε_{nm}.

Firstly, the particle concentration n_z^A of the plasma particles, emitting the spectral line ν_{nm}, must be calculated as a function of temperature with the help of the formulas given in section 4.1. Then the emission coefficient of the spectral line is given as a function of temperature by substituting the particle density into (5.3). The plasma temperature T is then found from a plot of $\varepsilon_{nm} = f(T)$ by inserting the measured value for ε_{nm}.

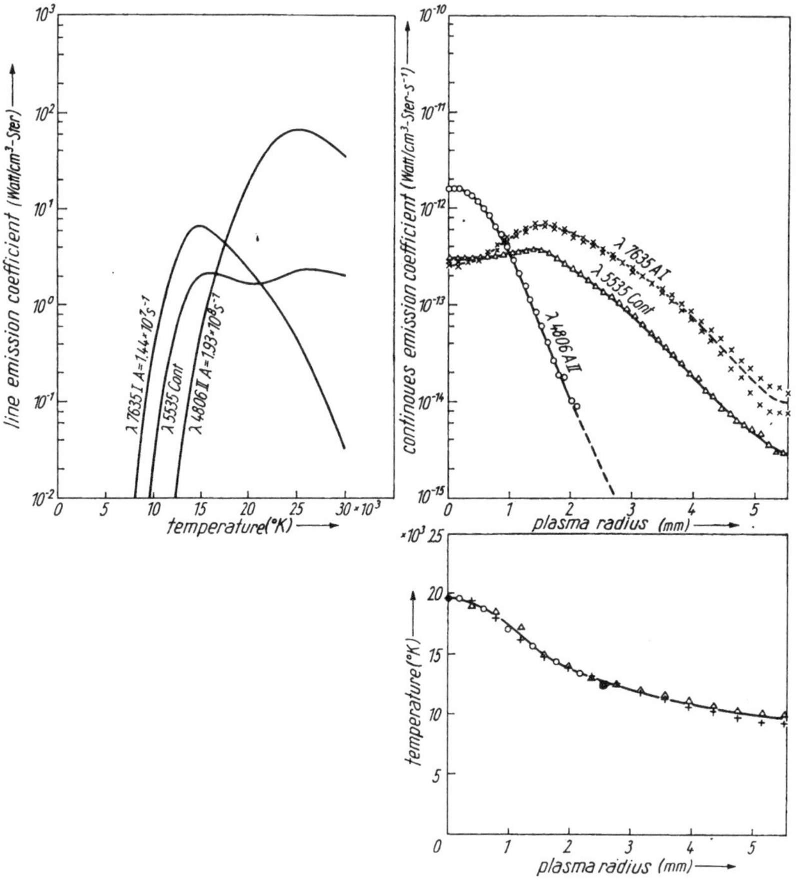

Figure 5.1. Determination of the radial temperature distribution (lower figure) in a plasma of a 1.1 atm, 400-A high-current argon arc from the measured absolute radial intensity distribution (upper right) and the calculated temperature dependence of the absolute intensity (upper left) for 2 spectral lines Ar I 7635 Å and Ar II 4806 Å, and for continuous radiation $\lambda = 5535$ Å (see section 9.5.1) (after Olsen [5.303])

How the radial temperature distribution $T(r)$ is determined from the calculated curves $\varepsilon_{nm} = f(T)$ and the experimentally determined curves $\varepsilon_{nm} = f(r)$ is demonstrated in Figure 5.1 for the case of the radial temperature distribution in a high-current argon arc (current intensity $I = 400$ A, arc length $= 5$ mm) at a distance of 2 mm in front of the cathode. The spectral lines A I 7635 and A II 4806 Å were used.

The figure stems from a work of Olsen [5.303]. It follows from the figure on the upper left that the emission coefficient of each spectral line is maximum for a specific temperature (norm temperature). In the neighborhood of this temperature the temperature cannot be precisely determined with the spectral line in question. However, if we measure the emission coefficient of several spectral lines with different norm temperatures (spectral lines of the same element but different ionization stages, or spectral lines, which are emitted by different elements), the temperature can be measured in a continuous fashion via the norm temperatures.

As a further example we shall discuss the temperature measurement after Jürgens [5.187] in a water-stabilized arc (under conditions of atmospheric pressure and a current intensity of 50 A) by means of the absolute emission

Table 5.1. Temperature determination from absolute intensities of spectral lines when pressure and material composition are known

Plasma object	p [atm]	Emitter	T-range [10^3 °K]	Reference[1])
arc, plasma jet, Ar	1	Ar I	8 – 20	3d, 131a, 131c, 144e, 144f, 184, 254, 254a, 289b, 303, 399, 415b, 455d, 470a.
		Ar II	9 – 20	1b, 122, 184a, 254, 254a, 303, 455d, 455e, 473.
arc, Ar	1.1	Ar I, Ar II	10 – 20	303
arc, Cl	1	Cl I	7 – 8	290c
arc, SF_6	1	F I 6240 Å	9 – 14	290b[2])
arc, Ar + 5% H_2	1	H_α, H_β	12 – 14	183
magnetically confined arc, H_2	0.01 – 0.03	H_β, H_γ	8 – 12	80a
waterstabilized arc	1	$H_\alpha - H_\gamma$	12.6	187
arc, Kr + H_2	0.36 – 1	H_β	10.8 – 11.4	111a
arc, Hg	0.13 – 1	Hg I 5790	5 – 6.3	26b
		He I		84b
arc, plasma jet, N_2	1	N I	9 – 19	58, 184, 399, 400, 415b, 470a.
arc, N_2	1	N II	10 – 13	58
arc, Ne	1	Ne II	21 – 23	111a
arc, O_2	1	O I	8 – 13	84b, 296a, 400.

[1]) References are given in short form, for example, 3d stands for [5.3d].

[2]) Demixing processes are allowed for.

coefficients of the lines H_α, H_β, and H_γ. The measurement gave an axial temperature of $T = 12{,}615 \pm 280$ °K. The smallness of the optical layer is checked by comparing the line profiles for end-on and side-on observations (see section 2.12.2).

A survey of temperature measurements by means of measured absolute emission coefficients of spectral lines is given in Table 5.1.

Day and Griem [5.72a] considered the case of temperature determination in a plasma from absolute line emission coefficients when instead of the pressure p the electron density n_e is known. The object of investigation was a plasma formed in a T-shaped shock tube containing the elements N and He with known concentration ratio. The electron temperature was derived from the sum of the absolute emission coefficients of the spectral lines N II 5676/79/86 Å.

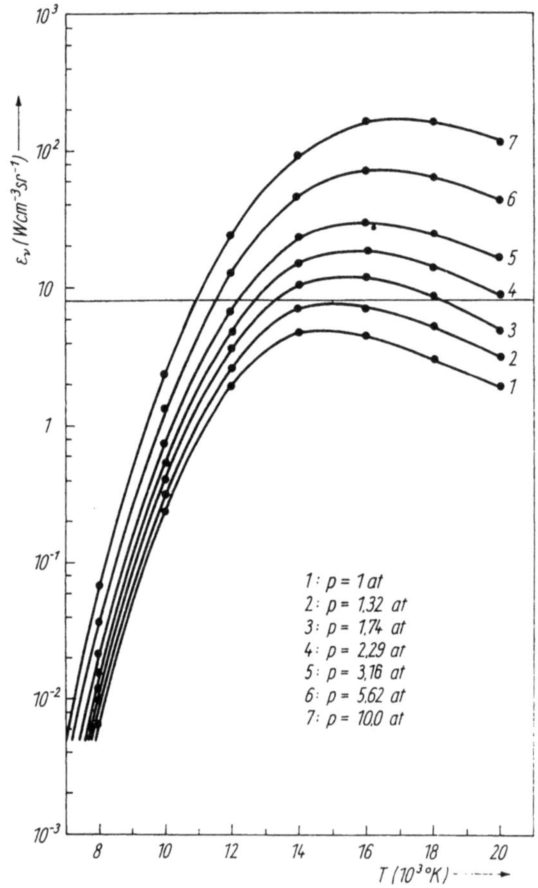

Figure 5.2a. Emission coefficient ε of the spectral line Ar I 7635 Å (a) and Ar II 4806 Å (b) in an argon plasma as a function of temperature and pressure

5.1.2. Determination of the Plasma Temperature and the Plasma Pressure from the Absolute Intensity of Two Spectral Lines when the Material Composition is known

The emission coefficients of the two spectral lines are denoted by ε_1 and ε_2, respectively. Both are functions of temperature and pressure and can be determined according to well-known methods:

$$\varepsilon_1 = f_1(T, p), \qquad \varepsilon_2 = f_2(T, p). \tag{5.4}$$

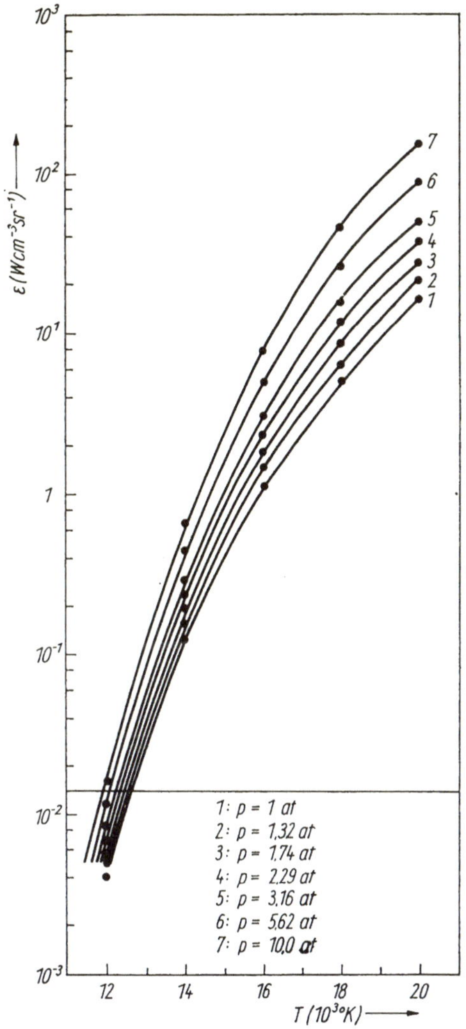

Figure 5.2b

Let ε_1 and ε_2 assume the values a and b in a certain volume element of the plasma. Then, for the determination of T and p we obtain the two equations

$$f_1(T, p) = a\,, \qquad f_2(T, p) = b\,, \tag{5.5}$$

from which the temperature as well as the pressure can be found. Here it must be assumed, however, that the temperature dependence of the intensity is different for both lines.

Let us consider an example.

Say the emission coefficient of the A I-line 7635 Å in a volume element of an argon plasma is found to be 8.0 W cm^{-3} sr^{-1} and that the emission coefficient of the A II line 4807 Å in the same volume element is $1.4 \cdot 10^{-2}$ W cm^{-3} sr^{-1}. In Figure 5.2a and b the emission coefficients of these lines are plotted as a function of p and T. In Figure 5.2a a straight line is drawn at $\varepsilon = 8.0$ W cm^{-3} sr^{-1}, and in Figure 5.2b at $\varepsilon = 1.4 \cdot 10^{-2}$ W cm^{-3} sr^{-1}. The points at which these lines intersect the p = const curves are given below:

A I line

p [atm]	T [°K]
1.74	13230
2.29	12660
3.16	12180
5.62	11480
10.0	10910

A II line

p [atm]	T [°K]
1.0	12610
1.74	12530
2.29	12470
3.16	12350
5.62	12100
10.00	11930

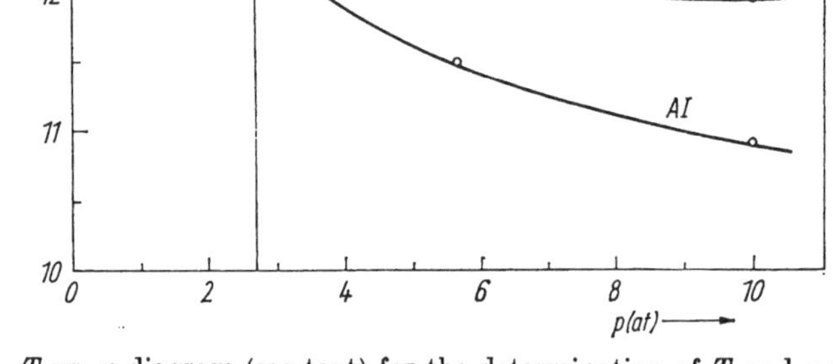

Figure 5.3. T vs. p diagram (see text) for the determination of T and p from the measured line intensities $\varepsilon = 8.0$ W cm^{-3} sr^{-1} (Ar I line 7635 Å) and $\varepsilon = 1.4 \cdot 10^{-2}$ W cm^{-3} sr^{-1} (Ar II line 4806 Å) for an argon plasma. The curves intersect at the point $T = 12{,}400$ °K and $p = 2.7$ atm

From the pairs of values we obtain two $p(T)$-curves, which are shown in Figure 5.3. The point, at which these two curves intersect, gives the sought values

$$T = 12{,}400\ °\mathrm{K}\,, \qquad p = 2.7\ \mathrm{atm}\,.$$

In an analogous manner, Thornton and Cambel [5.421 a] determined the electron temperature and particle number densities in an argon plasma generated in a shock tube via the emission coefficients of the spectral lines A I 4158 and 4259 Å, and A II 4379 and 4579 Å. They obtained typical values of $T_e = (13 - 17) \cdot 10^3\ °\mathrm{K}$, $n_e = (8 - 9) \cdot 10^{16}\ \mathrm{cm}^{-3}$.

See also Lincke and Griem [5.251 d] who determined the electron temperature in a He—H-plasma with known concentration ratio of the substituted elements from the intensity ratio I(He I 5876 Å) : $I(\mathrm{H}_\alpha)$.

5.1.3. Determination of the Material Composition of a Complex Plasma from Absolute Emission Coefficients of Suitable Selected Spectral Lines with known Transition Probabilities

The material composition of a complex plasma can be determined by measuring the absolute intensities of spectral lines, emitted by the different Plasma components.

According to equation (5.3) the emission coefficient of a spectral line, emitted by particles of element A and charge z and when disregarding the pressure dependence of the partition function Q_z^A, is only a function of the temperature T and the concentration n_z^A of the particle characterized by A and z. n_z^A in turn (see section 4.1) is a function of the temperature T, the pressure p, and the material composition of the plasma, which, for a s-component plasma, is defined to within a constant factor by the unknown parameters $c_1, \ldots, c_s$ (element concentrations). The concentrations are normalized as follows:

$$c_1 + c_2 + \cdots + c_s = 1\,. \tag{5.6}$$

Thus, we obtain the following expressions

$$\left.\begin{aligned} \varepsilon_1 &= f_1(p, T, c_1, \ldots, c_s)\,, \\ \varepsilon_2 &= f_2(p, T, c_1, \ldots, c_s)\,, \\ &\vdots \\ \varepsilon_s &= f_s(p, T, c_1, \ldots, c_s)\,, \\ \sum_{n=1}^{s} c_n &= 1\,. \end{aligned}\right\} \tag{5.7}$$

Here, $\varepsilon_1, \ldots, \varepsilon_s$ denote the emission coefficients of the spectral lines, which are respectively emitted by the elements $1, \ldots, s$. If these are known from measurement, then (5.7) represents a system of equations with which, for known pressure p, the temprature T as well as the concentrations $c_1, \ldots, c_s$ can be determined.

Foster [5.107] applied this method to a plasma at 1 atm pressure, which was created in a vortex-fluid stabilized arc chamber, when a swirling water benzene mixture was injected into the chamber. The chemical composition of the medium consists of $C_6H_6 + x\ H_2O$ with unknown mixture ratio x, which was to be determined along with the temperature. To do this the intensities of the lines H_β and O I, $\lambda = 3947$ Å, were calculated as a function of temperature and the mixture ratio x. The isophots were plotted in an x — T diagram (Figure 5.4). Introducing the two measured emission coefficients into the diagram, we obtain the sought quantities x and T from the points where the two isophots intersect.

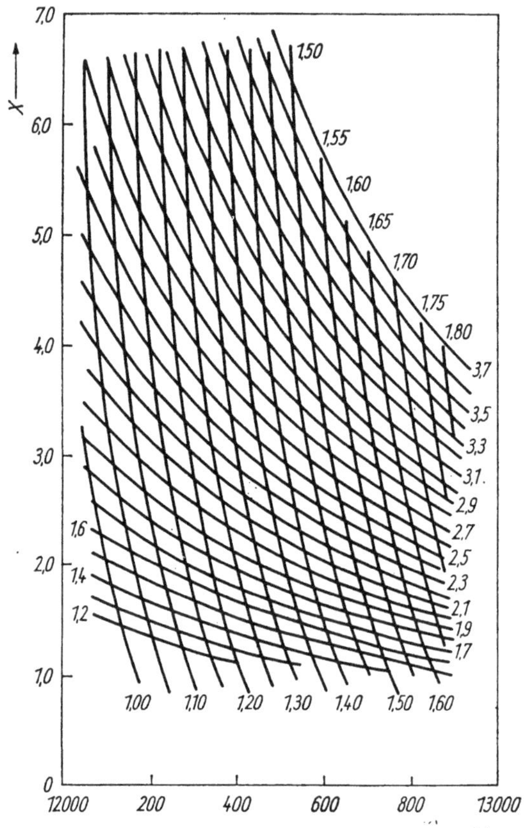

Figure 5.4. Isophot diagram in the x — T coordinate system for the determination of the chemical composition and the temperature of a $C_6H_6 + x\ H_2O$ plasma by means of the absolute intensities of the lines H_β 4860 Å and O I 3947 Å (after Foster [5.107])

The steeper isophots correspond to the H_β line, the flatter ones to the O I 3947 Å line. The numbers beside the isophots indicate the absolute intensity in the following units: [10^8 erg s^{-1} cm^{-2} sr^{-1}] for H_β, [10^5 erg s^{-1} cm^{-2} sr^{-1}] for O I 3947 Å

However, if the pressure is not known, the emission coefficient of an additional spectral line, which is emitted by one of the s elements, must be determined. Among these $s + 1$ spectral lines there have to be at least 2 lines, whose emission coefficients have a quite different dependence on the temperature. It is favorable when the $s + 1$ spectral lines are not all emitted by particles in the same ionization stage.

At this point it should be mentioned that Melchenko [5.267] in an analogous manner determined the temperature T, the effective degree of ionization $\bar{x} = \sum_{i=0}^{t} x_i n_i \Big/ \sum_{i=0}^{t} n_i$, and the particle densities of the components of a plasma at 1 atm, consisting of air, Mg, and Zn, whereby the intensities of the Mg I, Mg II, and Zn I lines had to be measured. The plasma constituents were at the most singly ionized.

Richter [5.367a] has determined the temperature and plasma composition in a cascade arc burning in the binary gas mixtures He—N, H—A, N—A, H—N from absolute emission coefficients of the spectral lines A I 4044 Å, H I 4861 Å (H_β), He I 5876 Å, N I 4935 Å and N II 4613 Å. A temperature in the range of 11,200—13,700 °K was found. Neumann and Serick [5.296b] determined the temperature and the concentration of Cu and Na in a rotating argon arc containing the elements Cu and Na as additives from absolute intensities of the spectral lines (λ in Å) Cu I 5105, 5218, 5220 and Na I 4983. The temperature range was 4000—5600 °K. The pressure was in every case 1 atm.

5.2. Determination of the Plasma Parameters via Relative Intensities of Spectral Lines

5.2.1. Determination of the Particle Density in a Complex Plasma with known Values for the Pressure and Electron Density from Relative Line Intensities (Boldt [5.45])

We shall limit ourselves to the case where the chemical reactions between the plasma constituents can be neglected. Let the plasma be made up of X different constituents K and of Y different components L. (For example, in an $A—CO_2$ plasma, which is completely dissociated, we have the constituents A, C, and O; the components, for temperatures not appreciably exceeding 16,000 °K, are: A I, A II, C I, C II, O I, O II, and e). For the calculation of the individual component densities n_{L_i} $(i = 1, \ldots, Y)$ and the temperature T we have at our disposal

$Y - 1 - X$	Saha equations	
1	quasineutrality condition	(5.8)
1	Dalton's law $p = \sum_{i=1}^{Y} p_i$	

giving a total of $Y + 1 - X$ equations.

The number of unknowns is $Y + 1$, so that we still require $R = X$ remaining equations to determine the densities n_{L_i} and the temperature T.

Assuming that an independent determination of the electron density, for example, via the line broadening, is possible, the number of unknowns reduces by 1 to Y, so that now only $R = X - 1$ equations are missing.

Further assuming that the relative concentrations of the plasma constituents are constant and can be found from the composition of the gas, which is injected into the arc, the $X - 1$ missing equations are given by the system

$$n_{K_1}:n_{K_2}: \ldots :n_{K_X} = c_1:c_2 \ldots :c_X \tag{5.9}$$

where the values n_{K_i}, $i = 1, \ldots, X$ give the concentrations of the constituents.

When diffusion processes lead to demixing (see [5.111]), equation (5.9) no longer holds. Then for the determination of the wanted quantities one may start from relative intensity measurements of properly selected spectral lines and may proceed as follows:

Next we can add to the $Y + 1 - X$ equations N further equations, which connect the relative emission coefficients ε_j of the spectral lines j ($j = 1, \ldots, N$), emitted by N different plasma components, with the particle densities $n_1, \ldots, n_N$ of the N different plasma components. The N equations are

$$\varepsilon_j = K_j n_j f_j(T) , \qquad j = 1, \ldots, N . \tag{5.10}$$

K_j is a constant which depends only on the spectral line, and is essentially the relative transition probability. $f_j(T)$ in essence contains the Boltzmann factor as well as the partition function.

If we know the measured values for the emission coefficient ε_j, we have with (5.10) N further equations at our disposal. However, these equations contain N new unknowns K_j.

To find a solution the measurement of the emission coefficients is not carried out on one plasma but on M different plasmas, which are distinguished by different constituent compositions, whereby always the same constituents are used, but with varying concentrations. If the same optical arrangement is used for the determination of the emission coefficients of the different plasmas, then the K_j-values are equal for all plasmas, while the values of n_{L_i} and T are different for all plasmas.

We obtain a system of equations with

$$M\,(Y - X + 1 + N) \text{ equations in}$$

$$M(Y) + N \qquad \text{unknowns.}$$

The above system can be solved when the number of equations is larger or equal to the number of unknowns, i.e., when

$$M\,(N + 1 - X) \geq N .$$

This can always be achieved with a suitable choice of M and N, for example, by setting $M = N = X =$ number of constituents. At this stage we shall not go into any more detail on this subject and refer the reader to section 5.3.2 (b) and [5.45].

5.2.2. Temperature Determination from the Ratio of the Emission Coefficients of Two Spectral Lines, Emitted by Particles of the Same Element A and Same Charge z (Two-line Method)

According to (5.3) the ratio of the emission coefficients of lines 1 and 2 is given by

$$\frac{\varepsilon_1}{\varepsilon_2} = \frac{g_1}{g_2} \cdot \frac{A_1}{A_2} \cdot \frac{\nu_1}{\nu_2} \cdot \mathrm{e}^{-(U_1 - U_2)/(kT)} . \tag{5.11}$$

It can be seen that this ratio no longer depends on the particle density n_z^A but only on temperature. Determination of the temperature from (5.11) requires that the excitation energies of the lines, their statistical weights, and relative transition probabilities be known. Further the lines must be selected in such a way that the excitation energies of both lines are quite distinct. It is also desirable that the two spectral lines are not so far displaced from one another, then we can reckon under certain conditions with a constant spectral sensitivity of the radiation detector used during measurement. In the latter case it is not necessary to use a standard source, whose relative radiation density usually has to be known with sufficient accuracy for both lines to be measured.

If a plasma does not emit suitable spectral lines which can be used to find the temperature by the above method, this can be effected by admixture of a suitable substance, so that the desired spectral lines are now emitted. The added element is then called a "thermometric element".

Hill [5.171] used this method to determine the temperature in an argon plasma jet from the intensity ratio of the copper lines $\lambda = 5153$ and 5700 Å, shown in Figure 5.5 as a function of temperature. The intensities were measured from the side. Temperatures between 5000 and 6000 °K have been obtained.

Hill did not analyze the radial distribution of the radiation. He did not measure the ratio of the emission coefficients of these lines but rather that of the intensities, i.e., of the emission coefficients integrated along the line of sight. The largest contribution to the intensity is made by the region having the highest emission coefficient. This lies either in the center of the plasma or outside the center when the norm temperature is exceeded. If the plasma temperature remains below the norm temperature (for the copper lines this lies between 8400 and 9700 °K), then the temperature determined from the intensity ratio is characteristic for the region in the center of the plasma. If the norm temperature is reached or exceeded then formally a value is given for the temperature in the line of sight which hardly differs from the norm temperature. Thus, temperature determination by this method no longer has any sense. Therefore, either the radiation must be spatially resolved or suitable spectral lines (e.g., ionic lines) must be used to determine the temperature.

We shall further discuss temperature measurements of an air plasma jet with the help of the lines Cu I 5106 and 5153 Å. (Grechikhin and Shimanovich [5.139]). The measured temperatures (between 6800 and 8300 °K) lay below the norm temperature.

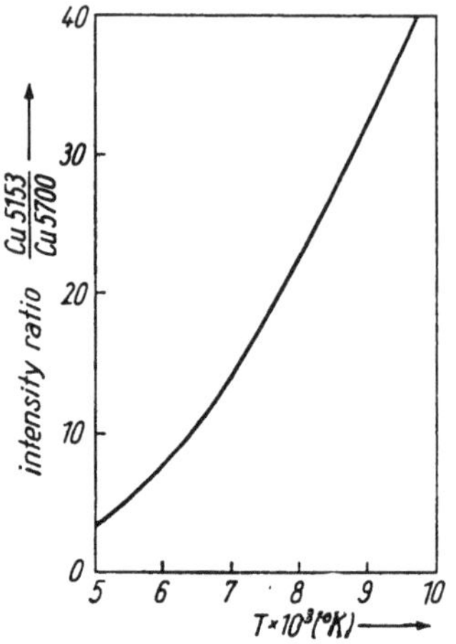

Figure 5.5. Intensity ratio of the copper lines 5153 and 5700 Å as a function of temperature (after Hill [5.171])

When the partial pressure of the thermometer element becomes too high, some lines of this element will be affected by self absorption. Then these lines are not suitable for the purpose of temperature determination.

This case is partially realized in an iron arc. As in all types of free burning metal arcs with eroding electrodes, the partial pressure of the metal and consequently the degree of self absorption grows with growing current intensity, so that especially for high current intensities the thermometer lines must be carefully selected.

Hirschfeld and Weinschenk [5.173] determined the temperature of an iron arc in the current range $I = 30 - 250$ A on the basis of the intensity ratios of 12 Fe I lines in the spectral range from 2577 to 2648 Å with term series $a^5\,F - x^5\,G$ and $a^5\,D - y^3\,D$ and of 16 Fe II lines in the spectral range from 2563 to 2631 Å with term series $a^4\,D - z^4\,P$ and $a^6\,D - z^6\,D$. From these lines different line pairs were taken from which the temperature was found (6000 °K at 150 A) in accordance with equation (5.11). The relative transition probabilities were determined by the above authors from intensity measurements at a 30 A iron arc the temperature of which has been measured by Krysmanski [5.227]. The relative transition probabilities determined by Sobolev [5.404] for the Fe I and Fe II lines in the UV-range could not be used, since these led to discrepancies in the calculations. Accurate data on the degree of self-absorption of the lines employed are unfortunately not given in the work.

5.2.3. Temperature Determination from the Relative Emission Coefficients of Several Spectral Lines of the Same Element and Same Charge z

With this method a more accurate temperature determination is possible than by the method described above in section 5.2.2.

From equation (5.11) we obtain after taking logarithms

$$\lg (\varepsilon\, g^{-1} A^{-1} \nu^{-1}) = \text{const} - \frac{5040 \cdot U_m}{T} \quad (U_m \text{ in eV}, \ T \text{ in °K})\,. \qquad (5.12)$$

If data exist on the statistical weights g, transition probabilities A (these only have to be known in relative measure), and term values U_m, we can calculate the temperature from the measured relative values for the emission coefficients ε. For this purpose $\lg (\varepsilon\, g^{-1} A^{-1} \nu^{-1})$ is plotted as a function of the energy U of the upper level. When self-absorption is negligible the measured values lie along a straight line with slope 5040/T, from which the desired T-value can be obtained.

Morozova and Startsev [5.288] employed this method in determining the temperature from the relative intensities of Fe I lines in an arc burning between two electrodes in air for current intensities between 1 and 6 A. The 16 spectral lines in the range from 3175 to 3355 Å and the 17 lines in the range from 4120 to 4268 Å, which were used, are listed in Table 5.2. The temperature (about 4500 °K) was taken from the graphs of $\lg \frac{I(\lambda/\lambda_0)^3}{gf} = -\frac{5040}{T} U_m + \text{const} = f(U_m)$, which are plotted in Figure 5.6 (see equation (5.44) for the connection between f and A). Note, that I is the spatially unresolved integral line intensity. Figure 5.7 shows that in the center of the arc the given temperatures are too low by 500 °K when the radiation is not spatially resolved. Toward the arc edges these differences become smaller. (According to Krysmanski [5.227] the maximum temperature for a 100-A iron arc is only 200 °K higher than that formally determined from the relative line intensities without spatial resolution.)

As mentioned above, in this method it should be ensured that the selected lines are not affected by self-absorption. This was the case for the lines employed by Morozova and Startsev. On the other hand, however, part of the lines used by Hefferlin [5.155] for finding the temperature of an iron arc of the same type, displayed self-absorption. Hefferlin showed that in this case it is also possible to determine the temperature by applying the theory of self-absorption of spectral lines [5.155], [5.157].

We shall not discuss this any further, since the method is too formal and because in most cases it is possible to select spectral lines free of self-absorption or by slight addition of another element (e.g., copper in the case of an iron arc) other suitable lines can be obtained which do not exhibit self-absorption.

Accordingly, Krysmanski [5.227] found that even for current intensities above 100 A, where most of the Fe lines display self-absorption, there are still a few iron lines with known gf-values of low self-absorption (among which the lines

Table 5.2. Data of Morozova and Startsev [5.288] for temperature determination by means of the relative line intensities of Fe I-lines

Lines in ultraviolet

No.	λ [Å]	Transition	Number of multipet	I	$3 \lg \frac{\lambda}{\lambda_0}$	$\lg(gf)$	U_m [eV]
1	3175.45	$z^7D^0-e^7F$	155	123	0.00	3.63	6.28
2	3178.01	$z^7D^0-f^7D$	156	94	0.00	3.52	6.27
3	3191.66	$a^5D-z^3D^0$	8	117	0.01	1.06	3.87
4	3202.56	$a^1g-w^1F^0$	547	21	0.01	3.43	6.89
5	3205.40	$z^7D^0-e^7F$	155	208	0.01	3.94	6.32
6	3215.94	$z^7D^0-f^7D$	156	227	0.01	3.95	6.30
7	3233.05	$b^3H-x^3I^0$	620	119	0.02	4.47	7.04
8	3234.61	$a^5D-z^3D^0$	8	65	0.02	0.80	3.87
9	3236.22	$a^5D-z^3F^0$	7	113	0.02	1.04	3.86
10	3245.98	$a^5D-z^3D^0$	8	60	0.03	0.83	3.91
11	3257.59	$a^5P-V^5F^0$	90	55	0.03	3.00	5.96
12	3268.23	$a^5P-x^3P^0$	95	29	0.04	2.69	5.99
13	3280.26	$b^3H-x^3I^0$	620	81	0.04	4.35	7.05
14	3323.73	$b^3P-V^3P^0$	379	57	0.06	3.62	6.53
15	3328.87	$b^3H-u^3H^0$	617	50	0.06	4.03	6.96
16	3355.23	$b^3H-u^3H^0$	617	44	0.07	3.99	6.97

Lines in visible spectrum

No.	λ [Å]	Transition	Number of multipet	I	$3 \lg \frac{\lambda}{\lambda_0}$	$\lg(gf)$	U_m [eV]
1	4120.21	$b^3G-z^1H^0$	423	27	0.00	2.97	5.97
2	4121.81	$b^3P-x^3F^0$	356	29	0.00	2.83	5.81
3	4137.00	$a^1P-y^1D^0$	726	50	0.00	3.70	6.38
4	4147.67	$a^3F-z^3G^0$	42	71	0.01	1.80	4.45
5	4153.91	$z^5F^0-f^5F$	695	87	0.01	3.89	6.35
6	4174.92	$a^5F-z^3D^0$	19	31	0.02	0.77	3.87
7	4177.60	$a^5F-z^3F^0$	18	28	0.02	0.71	3.86
8	4184.89	$b^3P-y^3P^0$	355	83	0.02	3.27	5.77
9	4210.35	$z^7D^0-e^7D$	152	136	0.03	3.10	5.40
10	4216.19	$a^5D-z^7P^0$	3	89	0.03	0.27	2.93
11	4217.55	$z^5F^0-e^5G$	693	52	0.03	3.70	6.34
12	4220.35	$c^3P-x^3P^0$	482	20	0.03	2.86	5.98
13	4222.22	$z^7D^0-e^7D$	152	128	0.03	3.05	5.36
14	4225.46	$z^5F^0-e^5G$	693	60	0.03	3.75	6.32
15	4245.26	$b^3P-z^3S^0$	352	40	0.04	2.95	5.75
16	4247,43	$z^5F^0-e^5G$	693	104	0.04	3.92	6.26
17	4267.83	$c^3P-x^3P^0$	482	29	0.05	3.02	5.99

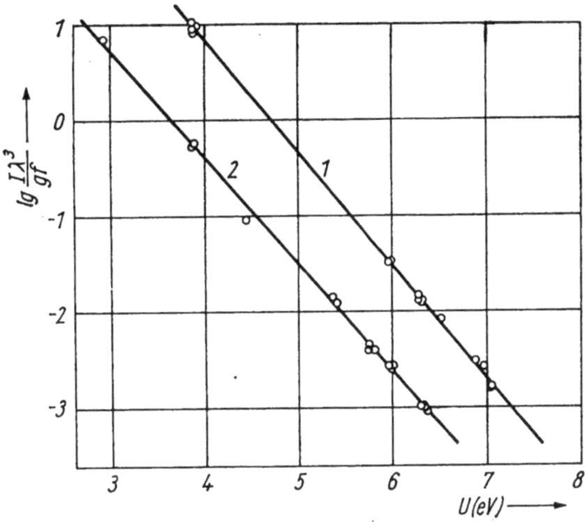

Figure 5.6. Temperature determination from the relative intensities I of 16 FeI lines in the range from 3175 to 3355 Å (curve *1*) and 17 FeI lines in the range from 4120 to 4267 Å (curve *2*) (after Morozova and Startsev [5.288])

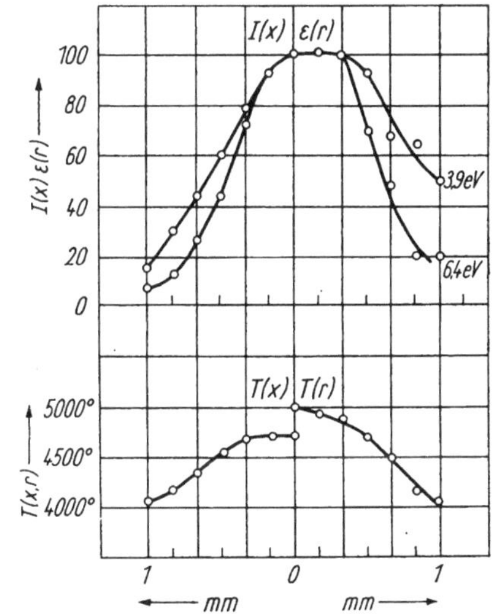

Figure 5.7. Radial temperature distribution of an arc between iron electrodes, determined from relative Fe I line intensities (Morozova and Startsev [5.288])

$T(r)$ — precise temperature, $T(x)$ — formally obtained temperature without spatial resolution of the radiation. Above the relative intensities $I(x)$ and $\varepsilon(r)$ of the lines used for the temperature determination are given for excitation energies of 3.9 and 6.4 eV, respectively

4227, 4247, 4375, and 4427 Å), which are suitable for temperature determination by the above described method.

Table 5.3 lists a number of references in which the temperature determination from relative intensities is described.

Table 5.3. Determination of temperature from relative intensities of spectral lines emitted from atoms or ions of the same ionization stage when relative transition probabilities are known. Abbreviations in column 4: HF. = high frequency discharge, H.P. = high pressure plasma, M.Arc = magnetically confined arc, MD. = medium pressure discharge, Pl. = plasma jet, Pu. = pulse discharge, Sh. = shock wave

Thermometer line λ in [Å]		Excitation energy [eV]	T_e-range [10^3 °K]	Plasma object	Reference[1])
Ag I		5.3–6.7	4.6	Arc, Cu-Air, 1 atm	67a
Al III	4480–5722	17.7–23.5	17 –39	Pu., metal electrodes, p > 1 atm	74a, 74b.
Al VII	240–356	35 –59	200	vacuum spark	246d
Ar I	4198–5606	14.4–15.3	4.1–5.0	afterglowing Pl., Ar + He, 20–140 Torr	302c
Ar II	4000–5010	19 –23	11.6–13	Pl., Ar, p ≤ 1 atm	225b, 255f.
Ar II	3545–5009	19 –23	20 –84	M. Arc, Ar	421b
As I		6.3– 8.4	6.5– 9.5	Arc, Ar + metal	225a
Au I	4792–6278	4.6– 7.7	5.0– 8.4	Arc, C + Ar; N_2 1 atm	40f
Ba I	4726–6110	3.1– 4.2	3	Ar + BaCl	17a
Ba I		1.6– 4.5	4.3– 5.0	Arc, Cu + Air, 1 atm	67a
Ba II	4900–4934	2.5– 5.2	7	HF., Air	214d
C III	4647, 5696		20 –40	Pu., He + Air, p < 1 atm	18f
Ca I		1.9– 6.1	4.2– 4.9	Arc, Cu + Air, 1 atm	67a
Cd I		6.3– 8.0	4.8– 5.1	Arc, Cu + Air, 1 atm	67a
Co I		2.9– 5.7	5.4– 5.5	Arc, Cu + Air, 1 atm	67a
Cr I		3.0– 6.2	4.3	Arc, Cu + Air, 1 atm	67a
Cs I	3611–6723	1.8– 3.8	1.0– 2.5	MD., including Cs	59a, 70c, 139b, 287b.
Cu I	5105–5782	3.8– 6.6	3.5– 8.3	high pressure plasma including Cu	18a, 23, 84a, 86b, 139, 168b, 171, 214d, 277a, 262a, 348a, 349g, 454.
Cu I	2303–3279	3.8– 7.0	6.6–12.6	Pl., Ar	418a
Cu II	2276–2473	8.2–13.6	6.6–12.6	Pl., Ar	418a
Ga I		3.1– 4.3	5.8	Arc, Cu + Air, 1 atm	67a
Fe I	2576–5455	2.4– 7.0	2.4–11.1	HP.	67a, 76, 88a, 112b, 115c, 131d, 155, 157, 173, 214d, 227, 227a, 288, 295a, 336.
Fe II	2562–2631	4.8– 5.8	5 – 8	Arc	173, 436a.
H I	4101–6562	12.1–13.2	6.0–13.3	Arc, H_2O	84a[2]), 187.
He I	3188–3889	23.0–23.7	20	Sh., He	114b
Hg I		7.7– 8.5	6.2	Arc, Cu + Air, 1 atm	67a
Hg I	4339–5769	8.8– 9.5	5.8	MD.	295b
In I		3.0– 4.1	4.5	Arc, Cu – Air, 1 atm	67a

Table 5.3. (Continuation)

Thermometerline λ in [Å]		Excitation energy [eV]	T_e-range [10^3 °K]	Plasma object	Reference[1])
K I		1.6 – 3.4	5.4	Arc, Cu – Air, 1 atm	67a
Mg I	3829 – 5183	5.1 – 7.2	3 – 5.8	HP.	67a, 418d
Mg II	2791 – 2803	4.4 – 8.9	10.0 – 13.4	Arc, Air, 1 atm	421
Mn I		2.2 – 4.5	3.9	Arc, Cu – Air, 1 atm	67a
N II	3995 – 5005	21.1 – 26.2	19 – 50	HP., $p \leq 1$ atm	26a, 311a
N II		21 – 30	20 – 40	MD., He + Air	18f
Ni I		3.5 – 4.6	4.7	Arc, Cu – Air, 1 atm	67a
O I	7774 – 8446	10.7 – 14.1	19 – 34	Pu., Air, 1 atm	311a
O II	4590 – 9676	25.6 – 28.4	15 – 21.7	HP., O + He + H, 1 atm	84b
O II	3739 – 4661	25.6 – 29.6	30 – 44	Air, $p < 1$ atm	26a, 298a
O II	4254 + 4642	25.6 – 34.2	30	He + CO_2, $n_e \approx 10^{17}$ cm^{-3}	369a
O IV	554 – 790	15 – 22	100 – 300	Stellarator, H, $n_e = 10^{13}$ cm^{-3}	172b
Sc I		2.0 – 4.1	5.1	Arc, Cu – Air, 1 atm	67a
Sn I	2850 – 3264	2.5 – 5.4	4 – 9.4	Arc	4b, 131d
Sr I		1.9 – 5.0	4.4	Arc, Cu + Air, 1 atm	67a
Ti I	4301 – 5192	2.4 – 5.0	2.6 – 5.3	HP., $p \geq 1$ atm	67a, 121m, 336
Tl I		3.2 – 5.3	6.4	Arc, Cu + Air, 1 atm	67a
V I		2.2 – 5.2	5.3 – 6.1	Arc, Cu + Air, 1 atm	67a
Zn I	2771 – 3346	3.8 – 8.6	4.8 – 6.6	HP., 1 atm	49a S. 108[3]), 67a, 80b, 131b, 133b, 349g

[1]) References are given in short symbols. For example, 67a stands for [5.67a].

[2]) Instead of the integral line intensity, the center intensity of the spectral line is used as indicator for the temperature.

[3]) Given are calculated diagrams $I_1/I_2 = f(T)$.

5.2.4. Temperature Determination from Relative Intensities of Spectral Lines by the Comparison Method

To find the temperature by this method the relative intensities of at least two spectral lines, emitted by regions with two different temperatures, must be known as well as data on the excitation energies of these lines. On the other hand, however, we do not have to know the transition probabilities of the spectral lines.

Let us consider the ratio $\frac{\varepsilon_m^0}{\varepsilon_m^1}$ of the emission coefficients ε_m^0 and ε_m^1 of a spectral line with excitation energy U_m for the temperatures T_0 and T_1, respectively. This is given in accordance with equation (5.3), assuming weak ionization $\left(Q_z^A \approx \text{const},\ n_z^A \sim \frac{1}{T}\right)$, by

$$\frac{\varepsilon_m^0}{\varepsilon_m^1} = \frac{T_1}{T_0} \, e^{-\frac{U_m}{k}\left(\frac{1}{T_0} - \frac{1}{T_1}\right)}$$

or, taking logarithms

$$\lg \frac{\varepsilon_m^0}{\varepsilon_m^1} = -\lg \left(\frac{T_0}{T_1}\right) + 5040 \cdot U_m \left(\frac{1}{T_1} - \frac{1}{T_0}\right). \tag{5.13}$$

To find T equation (5.13) can be solved either numerically or graphically. This is done by logarithmically plotting the $\varepsilon_m^0/\varepsilon_m^1$-values, obtained from measurements on different spectral lines, as a function of the excitation energy U_m. The plotted points must lie along a straight line, whose slope is given by $5040\left(\frac{1}{T_1} - \frac{1}{T_0}\right)$. The value of T_0/T_1 is obtained from the intercept of the ordinate axis ($U_m = 0$). Now T_0 and T_1 can be found separately (see Figure 5.8).

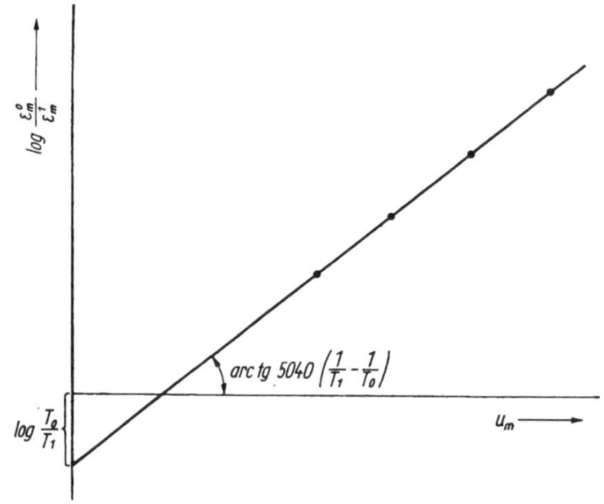

Figure 5.8. Determination of two temperatures T_0 and T_1 from relative emission coefficients ε_m^0 and ε_m^1 for several spectral lines with distinct excitation energy U_m (schematic)

Using the above method Elenbaas [5.90] determined the radial temperature distribution in a wall-stabilized high-pressure mercury arc, where $p = 0.88$ atm, power $= 35$ W/cm, $d_i = 4.1$ cm, mercury filling 12 mg/cm, 0.1% Cd admixture. In the measurement the lines Hg I 4916 and 5770/91 Å as well as the lines Cd I 6438, 5086, 4800, and 4678 Å were used. The line $\lambda = 5461$ Å could not be employed, since it exhibited self-absorption. Temperatures between 5900 °K ($r = 0$) and 4600 °K ($r = 1$ cm) were obtained.

5.2.5. Determination of Temperature and Electron Density from the Relative Intensities of Spectral Lines, emitted by Differently Charged Ions of the Same Element

5.2.5.1. Determination of temperature

The application of this method presupposes that the relative transition probabilities or oscillator strengths of the relevant spectral lines are known. Further, we must know the plasma composition (in the case of complex plasmas) and the pressure or electron density.

Then, from the expression

$$\varepsilon_{n\,m}^{(z)} = \frac{n_z\, g_n^{(z)}\, f_{m\,n}^{(z)}}{Q_z\, \lambda^3}\, \mathrm{e}^{-U_m^{(z)}/(kT)}\, G$$

(the index z refers to the charge of the ion; G is a factor which comprises physical constants), or taking logarithms

$$\lg \frac{\varepsilon\, Q\, \lambda^3}{g\, f\, n_z} = \text{const} - \frac{5040 \cdot U_m}{T} \qquad (U_m \text{ in eV},\ T \text{ in °K}) \qquad (5.14)$$

we obtain a linear relationship between $\lg \frac{\varepsilon\, Q\, \lambda^3}{g\, f\, n_z}$ and U_m (in (5.14) some of the indices have been left out for the sake of simplicity). In this scheme the experimentally determined points must lie along a straight line, from whose slope the temperature can be found.

If the particle number densities n_z entering into formula (5.14) are not known beforehand, then, knowing the the electron density n_e, we can use the Saha equation in the form (4.1) ($n_z = f(n_e, T)$) to calculate the particle densities n_z. Since n_z depends on T the temperature determination can be effected iteratively. First an approximate value for the temperature $T^{(0)}$ is put into the Saha equation. With the particle densities $n_z^{(0)}$ thus found the temperature can be found from equation (5.14) by means of the graphical method. This gives a better temperature approximation $T^{(1)}$, which in turn is substituted into the Saha equation, giving better values for the particle densities $n_z^{(1)}$, and finally the correct temperature is found with (5.14) by the graphical method.

McNally [5.259] employed this method to determine the temperature in a carbon plasma. The plasma was produced by a 1.50-m long high-current carbon arc in a DCX-system, which was magnetically contained along the axis of the discharge vessel. The current intensity was about 250 A. The electron density to be substituted into the Saha equation was found to be 10^{14} cm^{-3} from results of different types of measurements (such as breaking-off the hydrogen lines at the series limit, radiation balance, scattering of corpuscular radiation, microwave transmission measurements, H_δ-line widths). A initial temperature of 23,200 °K (in the first approximation) was substituted into the Saha equation. The intensities of 12 lines were measured, the data of which are given in Table 5.4. The *fg*-values were taken from a work of Allen [5.5].

The excitation temperature was found to be 24,000 °K from the floating point representation (Figure 5.9). Since this temperature was almost in complete agreement with that initially assumed for the calculation of the ion concentrations by means of the Saha equation, it was not necessary to obtain a better temperature value by iteration.

The results indicate that notwithstanding the low electron density of 10^{14} cm^{-3} the excited levels are described by a Boltzmann distribution. Direct verification of the validity of the Saha equation for the case of the C IV—C V ionization equilibrium was provided by the mass-spectroscopically proven fact

Table 5.4. Carbon lines employed by McNally [5.259] for the determination of the excitation temperature

Spectral line, λ [Å]	Excitation energy [eV]	fg
C II 3921	19.49	0.56
C II 4267	20.95	5.0
C II 5145	23.11	0.07
C II 6578	16.33	1.1
C III 3610	42.82	1.8
C III 4070	42.95	10.6
C III 4647	32.19	1.15
C III 5696	34.26	1.2
C IV 3936	58.35	2.1
C IV 4442	58.43	1.8
C IV 4789	58.35	0.4
C IV 5802	39.67	0.57

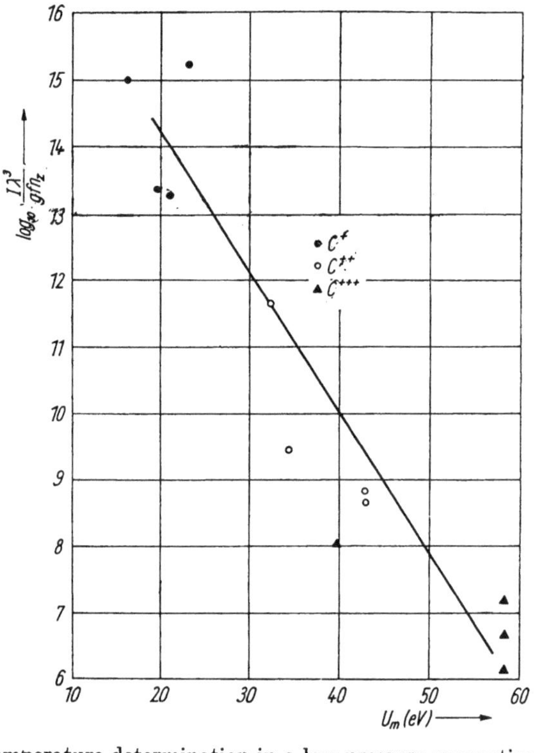

Figure 5.9. Temperature determination in a low-pressure magnetically contained carbon arc for a current intensity of 250 A from the line intensities I of differently charged ions (after McNally [5.259])

The relative particle densities n_z were calculated from the Saha equation for $n_e = 10^{14}$ cm^{-3} and an approximate value $T_e = 23{,}000$ °K. The temperature $T_e = 24{,}000$ °K is obtained from the slope of the straight line

that for $T = 40{,}000$ °K both ion types are represented in nearly equal strength, as expected.

To determine the temperature from the ratio of the emission coefficients ε of 2 spectral lines emitted by the same element in 2 consecutive ionization stage (z and $z + 1$) under the assumption that the electron density is known we start from the formula

$$\frac{\varepsilon^{(z+1)}}{\varepsilon^{(z)}} = \frac{n^{(z+1)}}{n^{(z)}} \frac{g_n^{(z+1)}}{g_n^{(z)}} \frac{f^{(z+1)}}{f^{(z)}} \frac{Q^{(z)}}{Q^{(z+1)}} \frac{\lambda_z^3}{\lambda_{z+1}^3} \cdot \exp\left(\frac{-U_m^{(z+1)} + U_m^{(z)}}{kT}\right),$$

g_n is the statistical weight of the lower level n, Q the partition function.

When the ratio of number densities $n^{(z+1)}/n^{(z)}$ is expressed in terms of the Saha function (4.1), we obtain the resulting equation

$$\frac{\varepsilon^{(z+1)}}{\varepsilon^{(z)}} = \frac{2}{n_e}\left(\frac{2\pi m kT}{h^2}\right)^{3/2} \frac{g_n^{(z+1)}}{g_n^{(z)}} \frac{f^{(z+1)}}{f^{(z)}} \frac{\lambda_z^3}{\lambda_{z+1}^3} \exp\left(\frac{-U_i^{(z)} - U_m^{(z+1)} + U_m^{(z)}}{kT}\right), \quad (5.15)$$

from which T can be obtained.

This method has been used for the determination of T by several authors as can be seen from the following table.

Spectral line used, λ in [Å]	Type of plasma	n_e [10^{16} cm^{-3}]	T [10^3 °K]	Reference[1])
C I 2583 II 2836/37	CH_2, 1 atm		10.7	295c
C I 2583 II 2836/37	C + x H, $x = 1-10$, $p = 0.01-1$ Bar		6–20	225d[2])
C II 4267 III 2296	ring discharge, H_2	1.0–1.4	20–27	90a
He I 3889 II 4686	reflect. shockwave, He–O_2-mixture	13	43	26f
He I 4713 II 4685	pulse discharge, He	2–11	30–40	86, 400b
He I 4471 + 5015 II 4686	pulse disch., He	7–12	32–44	232c
He I 5876 II 3203	shock wave, He	10–25	46–52	114a
O V 2781 VI 3811	pulse disch., H + 0.7% CO_2	3000	160	401b
Ti I, II, various combinations	shock tube, high pressure plasma, CO_2	10–200	8.7–14.5	18g

[1] Short symbols are used, for example 295c stands for [5.295c].

[2] The publication contains diagrams and tabulations $T = f(\mathrm{I(CII)/I(C\,I)})$ with p as parameter.

5.2.5.2. Determination of electron density when the electron temperature is known

As can be seen from formula (5.15), from the ratio of emission coefficients of two spectral lines of the same element in 2 consecutive ionization stages it is possible to derive the electron density n_e if the plasma temperature T is known. The latter quantity can be determined for example by means of the two-line method described in section 5.2.2. Such measurements were performed by Corliss [5.67 b] and also by Goldfarb and Ilina [5.131 b]. In [5.67 b], a copper arc containing diverse metal additives was investigated. The electron density n_e ($1.9 \cdot 10^{13} - 1.6 \cdot 10^{15}$ cm^{-3}) was derived from the intensity ratio of several line pairs emitted by atoms and single charged ions of the elements Ba, Be, Ca, Cd, Mg, Pb, Sc, Sr, Ti, V and Zn, assuming for T a value of 5100 °K. In [5.131 b], a carbon arc in air was investigated. The arc plasma includes metal vapour of diverse elements. From the intensity ratio of the spectral lines Ca I 4226/Ca II 3968 Å an electron density of $n_e = (0.5-4.8) \cdot 10^{15}$ cm^{-3} was found. The temperature was in the range $(5.3-6.6) \cdot 10^3$ °K.

5.2.5.3. Determination of temperature and electron density from intensity ratios of spectral lines emitted by a set of elements in consecutive ionization stages

We rewrite formula (5.15) in the form

$$\frac{\varepsilon^{z+1}}{\varepsilon^{(z)}} = R(z) = \frac{T^{3/2}}{n_e} B(z) \cdot \exp\left(\frac{-U_i^{(z)} - \Delta U_m^{(z)}}{kT}\right)$$

respectively

$$\lg \frac{R(z)}{B(z)} = \lg \left(\frac{T^{3/2}}{n_e}\right) - \frac{5040\,(U_i^{(z)} + \Delta U_m^{(z)})}{T}$$

$$(U_i^{(z)} \text{ and } \Delta U_m^{(z)} \text{ scale in eV, } T \text{ in °K}) .$$

If the quantity $\lg \frac{R(z)}{B(z)}$ is plotted against $5040\,(U_i^{(z)} + \Delta U_m^{(z)})$, then from the slope of the straight line drawn through the coordinate points the temperature T can be derived, while the quantity $\lg (T^{3/2}/n_e)$ respectively n_e is found from the intercept of the ordinate axis.

Berthelot [5.27 d] used the line pairs (λ in Å) (Mg II 2795.53/Mg I 2852.13) and (Ca II 3968.47/Ca I 4226.73 Å) for the determination of the temperature in a carbon arc containing several metal salt additives and obtained $T = (5.3-9.4) \cdot 10^3$ °K. de Galan and Winefordner [5.115 c] determined the temperature $(2.4-3.2) \cdot 10^3$ °K and the electron density ($10^{10}-3.10^{11}$ cm^{-3}) from the intensity ratio of the line pairs (Ca II 3933/Ca I 4227), (Sr II 4077/Sr I 4607) and (Ba II 4554/Ba I 5535). See, for further applications, [5.27 e]. Instead of emission measurements, likewise absorption measurements can be made to obtain the wanted quantities [5.27 e].

5.2.6. **The Norm Temperature Method**

This method is used to find the temperatures of plasmas, whose material composition and pressure (or electron density according to the original work of Fowler and Milne [5.107 e], must be at least approximately known, and was first applied by Larenz ([5.245], [5.246]) to laboratory plasmas. This method has the advantage that the statistical weights and transition probabilities of the spectral lines used do not have to be known. Only the excitation energies of the spectral lines must be given.

The method is based on the following: At constant pressure the emission coefficient ε_{nm} of a spectral line, emitted by particles of the type (z, A), first increases with the temperature, reaches a maximum, and then falls off owing to depletion of particles of type (z, A) in the plasma due to ionization. The temperature, for which the emission coefficient of the corresponding line is maximum, is designated as the norm temperature T^*. This depends on the particular line, pressure, and plasma composition.

In order to employ this method the norm temperature in the plasma must be exceeded.

Let us demonstrate the application of this method for the determination of the radial temperature distribution in a water-stabilized high-power arc (Burhorn et al. [5.59]). First the relative emission coefficient $\varepsilon_{nm}(T)$ of line H_β $\lambda = 4861$ Å, O II $\lambda = 4651$ Å, and O III $\lambda = 3447$ Å is calculated for a stoichiometric water-vapor plasma at $p = 1$ atm as a function of temperature, then normalized by division by $\varepsilon_{nm}(T^*)$, and subsequently plotted on a graph (see Figure 5.10). In the case of the H_β line the solid curve corresponds to the

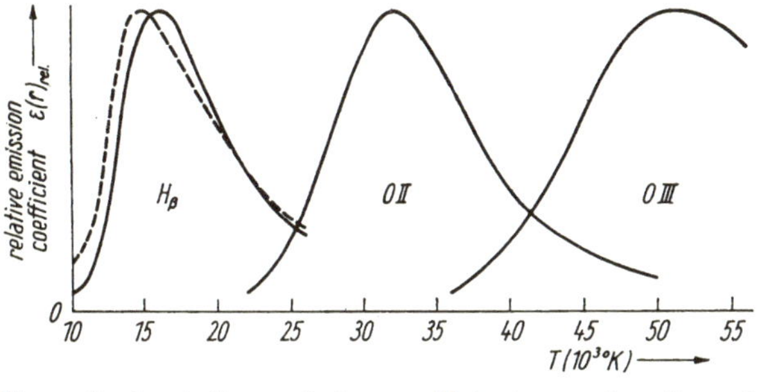

Figure 5.10. Normalized relative emission coefficient as a function of temperature in a water-vapor plasma at $p = 1$ atm for the lines H_β 4860 Å, O II 4651 Å, and O III 3447 Å. In the case of the H_β line the solid curve corresponds to the integral intensity, the dashed curve to the line intensity per unit frequency at the line maximum (after Burhorn, Maecker, and Peters [5.59])

total line intensity, the dashed curve to the line intensity per unit frequency at the maximum. The two curves are distinguished by the fact that the H_β line is broadened in proportion to $n_e^{2/3}$ by the Stark effect and that the ridge height reaches a maximum more quickly (i.e., at a lower temperature) than the total intensity.

Figure 5.11 gives the normalized radial intensity distribution $\varepsilon(r)/\varepsilon(T^*)$ for the above-named 3 lines for the water-stabilized high-power arc, where $I = 1450$A and the channel diameter is 2.3 mm. The maxima of the $\varepsilon(r)$-curves give the norm temperatures, which can be clearly seen in Figure 5.10. The respective intensities and temperatures corresponding to a specific value of the radius are obtained by comparing the ordinates in Figures 5.10 and 5.11. Figure 5.12 gives the resulting radial temperature distribution. The solid curves do not coincide owing to the fact that light dispersion strongly disturbs the measurement when the object measured is small, causing a temperature trend which

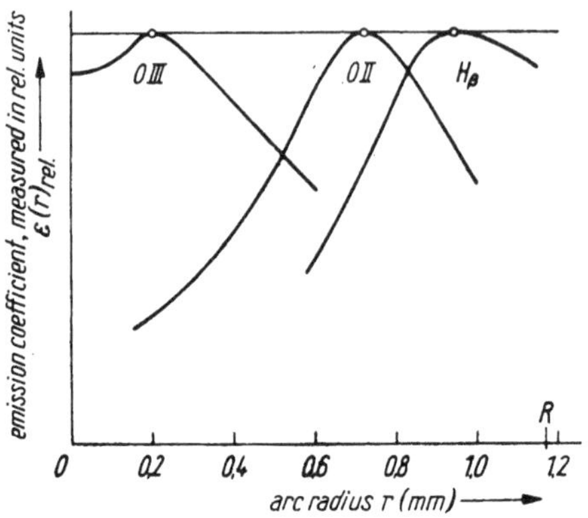

Figure 5.11. Radial intensity distribution $\varepsilon(r)_{\rm rel}$ of the spectral lines H_β, O II 4651 Å, and O III 3447 Å emitted by a water-stabilized high-power arc ($I = 1450$ A, channel diameter $= 2.3$ mm) (after Burhorn, Maecker, and Peters [5.59])

is too smooth. However, the dashed line, which was plotted only on the basis of the norm temperatures and their associated radii (indicated by small circles), clearly represents the true temperature behavior.

This method can also be applied to continuous radiation, if the temperature dependence of the intensity is known (see Larenz [5.246]) and provided that this dependence assumes a maximum at a specific temperature. As an example Figure 5.13 gives the relative spectral intensity of the continuum $\varepsilon_\nu(T)/\varepsilon_\nu(T^*)$ for the wavelength $\lambda = 5600$ Å as a function of temperature for a water-vapor plasma at atmospheric pressure. The intensity maximum lies at $T^* = 18{,}000$ °K.

In Table 5.5 we list norm temperatures for different spectral lines and for different plasmas.

A tabulation of norm temperatures for plasmas of composition $M + x$ Ar, $x = 0 - 10^6$, $p = 1$ atm, where the symbol M stands for Na, Al, Fe, and Zn, can be found in [5.76a]. The spectral lines quoted are diverse lines of Na I, Al I, Al II, Al III, Fe I, Fe II, Zn I, Zn II, C I and C II. The authors tabulated also normtemperatures for plasmas of composition $M + x$ Ar, $x = 0 - 9$, for a total pressure of $p = 1$ atm. The corresponding spectral lines are those of Na I, Fe I and Zn I. While in the first tabulation both the 1. and 2. normtemperature is given, the second tabulation encloses only the 2. normtemperature.

In many cases it is not required to deduce the norm temperatures of spectral lines from intensity-temperature diagrams or to directly calculate the exact

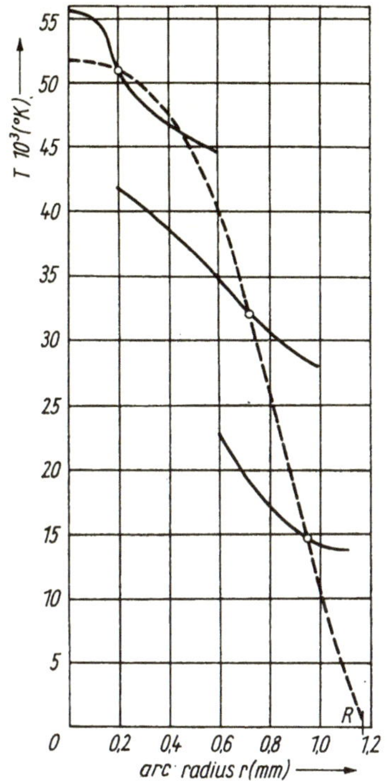

Figure 5.12. Radial temperature distribution in a water-stabilized 1450 A arc (after Burhorn, Maecker, and Peters [5.59])

Dashed curve: temperature dependence extrapolated from the standard temperature (small circles); solid curve: temperature dependences calculated from the relative intensity of 3 spectral lines

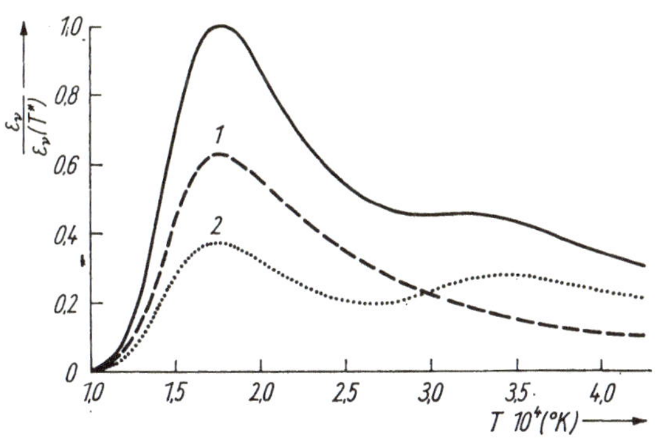

Figure 5.13. Relative spectral intensity $\varepsilon_\nu/\varepsilon_\nu(T^*)$ as a function of temperature for the continuum emitted by a water-vapor plasma, where $p = 1$ atm and $\lambda = 5600$ Å (after Larenz [5.246])

Curves *1* and *2*, respectively, give the contribution of hydrogen and oxygen to the continuum

Table 5.5. Norm temperatures

Line or Continuum λ [A]	Plasma	Norm-temperature [10^3 °K]	Reference[1])
Al I 3961	Al 1 atm	8.1	224
Al I, $U_m = 3.1$ eV	Ar, Air, He + Al, 1 atm	7.9	88b
Al II 2816	Al, 1 atm	18.4	224
Al II, $U_m = 11.8$ eV	Ar, Air, He + Al, 1 atm	17.7	88b
Al III 3601	Al, 1 atm	29.4	224
Al III, $U_m = 17.8$ eV	Ar, Air, He + Al	25	88b
Ar I, div. lines	Ar, 1 atm	15–16	40d, e, 199c, 254a, 256
Ar I 7635, 6965	Ar, 1.1 atm	15	303, 305
Ar, Cont. 5535	Ar, 1.1 atm	16+28	303, 305
Ar II, div. lines	Ar, 1 atm	25–28	40d, e
Ar II 4806	Ar, 1.1 atm	25	303, 305
Ba I 5778	Air + Ba, 1 atm	6.4	421
Ba II 4934	Air + Ba, 1 atm	10.0	421
C I 2478	C, 1 atm	12.8	224
C I 2478	0.99 atm He + 0.01 atm C	10[2])	395
C I, $U_m = 7.7$ eV	Ar, Air, He + C, 1 atm	12.8	88b
C I	H + trace C, $n_e = 10^{16}$ cm^{-3}	10	90a
C II 2512	C, 1 atm	28.5	225
C II 2512	N + trace C, 1 atm	29	225
C II, $U_m = 18.7$ eV	Ar, Air, He + trace C, 1 atm	27	88b
C II, $U_m = 20$ eV	CF_2, 1 atm	22	212d
C II	H + trace C, $n_e = 10^{16}$ cm^{-3}	23	90a
C II	H + trace C, $n_e = 10^{17}$ cm^{-3}	27	90a
C III 2297	C, 1 atm	33.5	224
C III 4651	C, 1 atm	41.8	225
C III 4651	N + trace C, 1 atm	40	225
C III, $U_m = 5.4$ eV	Ar, Air, He + trace C, 1 atm	32	88b
C III, $U_m = 20$ eV	CF_2, 1 atm	33	212d
C III	H + trace C, $n_e = 10^{16}$ cm^{-3}	32	90a
Cu I 5218	Cu, 1 atm	10	395
Cu I 5218	0.99 atm Ar + 0.001 atm Cu	13	395
Cu I 5218	0.99 atm C + 0.01 atm Cu	10	395
Cu I 5218	0.99 atm He + 0.001 atm Cu	8[2])	395
Cu I 5218	0.99 atm Na + 0.01 atm Cu	10	395
Cu I 5218	0.99 atm Ne + 0.01 atm Cu	8[2])	395
Cu I 5105	0.8 atm (Ar + x Cu), $x = 5\cdot 10^{-4}$	5.7[2])	395a
Cu I 5218	see line before	6.0[2])	395a
F II, $U_m = 33-35$ eV	CF_2, 1 atm	35	212d
F III, $U_m = 45-50$ eV	CF_2, 1 atm	55	212d
F IV, $U_m = 56$ eV	CF_2, 1 atm	70	212d
F IV, $U_m = 70$ eV	CF_2, 1 atm	75	212d
F V, $U_m = 70$ eV	CF_2, 1 atm	95	212d
H_β, 4861 A	H_2O, 1 atm	14.6[3])	59
H_β, 4861	H, $n_e = 10^{16}$ cm^{-3}	13.3	90a
H_α, 6563	H, 1 atm	16.5	395
H_α, 6563	0.99 atm C + 0.01 atm H_2	16.5	395

Table 5.5. (Continuation)

Line or Continuum λ [Å]	Plasma	Norm-temperatur [10^3 °K]	Reference
H_α, 6563	0.99 atm He + 0.01 atm H_2	19	395
H Cont., 3500, 4020	H_2O, 1 atm	17	246
H Cont., 5600	H_2O, 1 atm	18	246
He I 3187, 3888	He, 1 atm	24.3	40e
Mg I 4703	Mg, 1 atm	9.2	421
Mg I 5184	Mg, 1 atm	8.5	421
Mg II 2791	Mg, 1 atm	15.9	421
Mg II 2803	Mg, 1 atm	14.4	421
N I 4935	N, 1 atm	15	415b
N II 4613	Air, 1 atm	50	212c
N III 4097	N, 1 atm	48	224
O I 6046	H_2O, 1 atm	16	245
O II 4651	H_2O, 1 atm	32	59
O II 4649	Air, 1 atm	32	212c
O II 3390	Plexiglas	32	48b
O III 3447	H_2O, 1 atm	51	59
O III 3047	Plexiglas, 1 atm	50	48b
O IV 3063	Plexiglas, 1 atm	70	48b
O V 3144	Plexiglas, 1 atm	88	48b
O VI 3811	Plexiglas, 1 atm	109	48b
Pb I 2170; 2393	Pb, 1 atm	9.5; 9.4	224
Pb I 2663; 4168	Pb, 1 atm	8.8; 9.0	224; 427b
Pb I, $U_m = 5.7$ eV	Ar, Air, He, 1 atm + trace Pb	8.5	88b
Pb II 2203; 3786	Pb, 1 atm	17; 18	224; 427b
Pb II, $U_m = 7.2$ eV	Ar, Air, He, 1 atm + trace Pb	15.8	88b
Pb III 2562; 3043	Pb, 1 atm	29; 30	224; 427b
Pb III, $U_m = 23.6$ eV	Ar, Air, He, 1 atm + trace Pb	28.5	88b
Pb IV 2864; 3052	Pb, 1 atm	39.7; 40	224, 427b
Si IV 4089+4116	Si-containing Plasma, $p_e = 1.25 \cdot 10^{-4}$ atm	26.7	224
S I 4695	SF_6, 1 atm	11	290b)[4]
S II 5454	SF_6, 1 atm	13.5	290b[4])

[1]) References are given in short symbols; for example, 212d stands for [5.212d].

[2]) The data correspond to the first of the two maxima.

[3]) The norm temperature is related to the intensity per frequency unit for a frequency corresponding to the line maximum.

[4]) Reactive gas separation processes are allowed for.

norm temperature. If an error of 3% in the norm temperature is permissible, it is sufficient to calculate this with the help of the approximation formula (see Krempl [5.224]).

$$T_z^* = \frac{0.95 \cdot U_z\,[10^3\,°\mathrm{K}]}{1 - 0.33\,\dfrac{U_m}{U_z} + 0.37 \cdot \lg\left(\dfrac{U_z}{10}\right) - 0.14 \cdot \lg\left(\dfrac{g_z}{g_{z+1}}\,p_e^*\right)} \tag{5.16}$$

$$(U_z \text{ in eV},\ p_e^* \text{ in atm})\,.$$

U_z denotes the ionization potential of the ion of charge z, which emits the spectral line under investigation, U_m the excitation energy of the ion, g_z and $g_z + 1$, respectively, the statistical weight of the ground state of the ion with charge z and $z + 1$ ($z = 0$ in the case of atomic lines), and p_e^* the electron pressure at the norm temperature T^*.

We now need an additional relationship connecting p_e^* and p. For a simple plasma we can substitute

$$p_e^* = \frac{2z+1}{2z+3} \cdot p\,, \tag{5.17}$$

into equation (5.16). For a complex plasma with at the most singly ionized particles we can assume

$$p_e^* = \frac{1}{3} \frac{4c}{(1+c)^2} \cdot p\,, \tag{5.18}$$

for the case that the spectral line studied is emitted by an element which yields the major portion of the ions in the plasma. In (5.18) c denotes the concentration of this element based on the total concentration of the heavy particles. In this case we obtain the result that the standard temperature is lower than in the case where (at the same total pressure) the emitting element is the only element occurring in the plasma.

The accuracy of the approximate formulas becomes higher the higher the plasma temperature. This lies herein that for $T > 15{,}000$ °K the ratio p_e/p (except for He), weakly depends of the type of element [5.224]. Therefore, no great erros are involved when the standard temperature for a complex plasma is calculated from (5.16) and (5.17) in this temperature range.

In the case of a gas plasma composed from an element A to which a small amount of an element B with low ionization energy (with respect to the ionization energy of element A) is added, the emission coefficient of an atomic spectral lines emitted by the element B has two maxima, that means, there exist two norm temperatures T_1^* and T_2^*. While the first norm temperature T_1^* can be found from formulas (5.16) and (5.18), the second norm temperature T_2^* can be derived from an implicit formula given by Semenova and Levchenko [5.395a]:

$$1.03 \cdot 10^{-4}\, K\, e^{\xi}\, \xi^{5/2} = \frac{1}{\left(\sqrt{2\,\dfrac{7/2 + \alpha\,\xi}{5/2 + \xi}} - 1\right)^2} - 1$$

with

$$K = \left(\frac{Q_0^A}{Q_1^A}\right) \frac{p}{U_{i(A)}^{5/2}}\,; \qquad \alpha = \frac{U_{i(B)} - U_{m(B)}}{U_{i(A)}}\,; \qquad \xi = \frac{U_{i(A)}}{kT_2^*}$$

$$\left(U_i \text{ (ionization energy) in [eV]},\ p \text{ in [atm]}\right).$$

In the case of a complex plasma the application of the norm temperature method for the determination of temperature becomes difficult when allowing for reactive demixing processes. This case has been considered by Motschmann

[5.290b] for a plasma substituted from SF_6 molecules. For this special case, Motschman found a lowering of the norm temperature for the spectral lines S I 4695 and S II 5454 Å assuming the values $T^* \approx 11{,}000\ °K$ respectively $\approx 13{,}500\ °K$.

A lowering of the norm temperature is also found in those cases where the upper level of the spectral line under consideration has a position close to the ionization limit. For a detailed discussion see section 10.8.2.

The norm temperature method for the determination of the temperature in a rapidly changing plasma can also be applied when a spatial resolution is not possible, for example, in the case of pulsed discharges. If the intensities of spectral lines of different ionization stages are plotted as a function of time, maxima are obtained which designate the norm temperatures. These temperatures are somewhat characteristic of the temperatures in the plasma center.

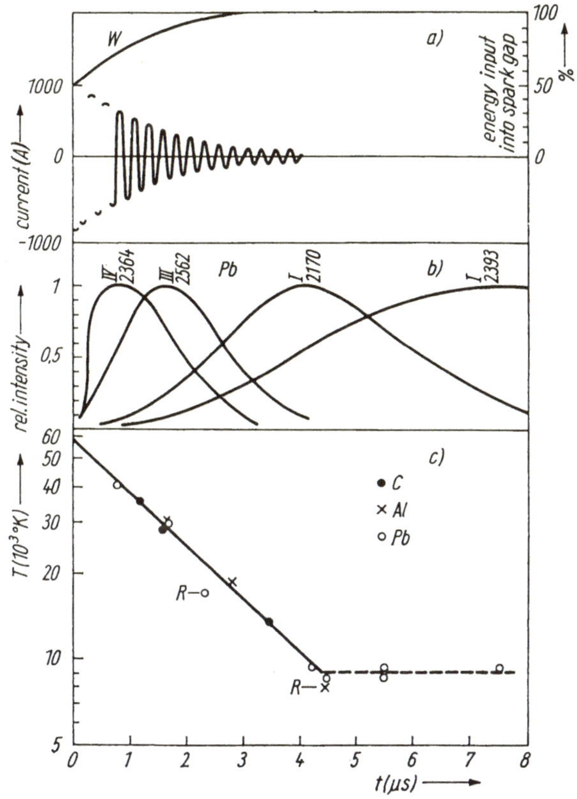

Figure 5.14. Cooling off of a plasma produced by a condenser discharge in air (Krempl [5.224])

a) current dependence and energy release W;
b) rise and falling off of line intensities of the electrode vapor (in this case lead);
c) cooling curve of the plasma.
Abscissa: time in µs with $t = 0$ at spark break-down. R — resonance lines

Krempl [5.224] determined the temperature change with time of a decaying condenser discharge in a plasma, which contained air as the main constituent, by relating the times at which the intensity maxima of different C, Al, and Pb lines occurred to the norm temperatures (see Figure 5.14). The elements C, Al and Pb are supplied from the electrodes.

In table 5.6 we list several references, in which the plasma temperature is determined by the norm temperature method.

As Burhorn [5.58] has shown, the norm temperature method can also be used for temperature determination in plasmas which emit molecular spectra. For the N_2-plasma at 1 atm, the author has calculated the temperature de-

Table 5.6. Temperature determination via the norm temperature method

Plasma	Spectral line or Continuum λ [Å]	T-range in [10^3 °K]	Reference[1])
Ar, 1 atm	Ar I 4158–9657	8– 20	94a, 199c, 212g, 254, 254a, 256 415b.
	Cont. 4810, 4884	10– 14	296
	Ar II 4806, 4880	16– 24	296
	Ar I, II	5– 30	40e, 55a.
Ar 1.1 atm	Ar I 6965, 7635		
	Cont. 5353	10– 28	303, 305.
	Ar II 4806		
Ar + trace	diverse lines		
(C, Al, Zn)	Al I–IV, C II,	7– 27	76b
1 atm	Zn I, Zn II		
CF_2, 1 atm	F II–V, C II, III	80– 80	212d
$C_5H_8O_2$ =	O II 3390, O III 3047,		
(Plexiglass)	O IV 3063, O V 3144;	32–109	48a, 48b.
1 atm	2781, OV VI 3811		
H + additive			
C, O, $n_e = 10^{16}$ cm^{-3}	C II	20– 27	90a
H_2O, 1 atm	H_β, O II, III,	7– 51	59
	Cont. 3500, 4020, 5060;		
	O II 4650	12– 35	246
He, 1 atm	He I 3888	17– 28	40e
Air, 1 atm	N II 4613	22– 35	212c
	O II 4649	24– 35	212c
Air, 1 atm	C I, II, III		
+ traces	Al I, II, III	9– 40	224
C, Al, Pb	Pb I, II, III, IV		
N_2 1 atm	N I 4935	11– 19	415b
Ne 1 atm	Ne II 3334		
	Ne III 2590, Ne IV 2357,	18–100	441a
	V 2265		
SF_6 1 atm	S I 4695		
	S II 5454	11– 14	290b

[1]) References are given in short symbols; for example, 94a stands for [5.94a].

pendence of emission coefficient at the band heads N_2 3371 and N_2^+ 3914 Å. He found a norm temperature of 8000 respectively 8650 °K. See [5.223d], [5.376b], [5.367d] for further applications.

5.3. Determination of Absolute Transition Probabilities and Oscillator Strengths

5.3.1. Determination of Transition Probabilities from the Emission of Spectral Lines in an Optically Thin Layer, using Radiation Sources, for which the Plasma Pressure and the Plasma Composition are known

For this case the emission coefficient of a spectral line as a function of temperature is given by expression (5.3)

$$\varepsilon_{n\,m} = \frac{1}{4\,\pi}\,\frac{g_m}{Q_z^A}\,n_z^A\,A_{n\,m}\,h\,\nu_{n\,m}\,\mathrm{e}^{-U_m/kT}\,,$$

where n_z^A is the number of particles of element A with ionic charge z ($z = 0$ refers to neutral atoms).

Writing (see Olsen [5.304])

$$\varepsilon^*_{n\,m} = \frac{1}{4\,\pi}\,\frac{g_m}{Q_z^A}\,n_z^A\,h\,\nu_{n\,m}\,\mathrm{e}^{-U_m/kT} = f(T) \tag{5.19}$$

we get

$$A_{n\,m} = \frac{\varepsilon_{n\,m}}{\varepsilon^*_{n\,m}}\,. \tag{5.20}$$

If the emission coefficient $\varepsilon_{n\,m}$ of the spectral line is directly measured, according to (5.20) the transition probability appears as the quotient of a measured value and a value determined by the temperature.

In plasmas with rotational symmetry the transition probability can be very accurately found when the norm temperature T^* of the spectral line along the plasma axis is exceeded (Olsen [5.304]). The point where $T = T^*$ can be well localized. The normalized emission coefficient $\varepsilon_{n\,m}(T^*)$ is then fixed with greater accuracy, and $A_{n\,m}$ can be determined with greater precision, as well:

$$A_{n\,m} = \frac{\varepsilon_{n\,m}(T^*)}{\varepsilon^*_{n\,m}(T^*)}\,. \tag{5.21}$$

We shall now consider the case where the standard temperature of the spectral line is not attained. The transition probability $A_{n\,m}$ can again be found from equation (5.3) respectively (5.20), when the temperature T at the point, for which the emission coefficient of the spectral line studied will be determined, is known. The temperature can, for example, be determined by means of the absolute intensity of a second spectral line (see section 5.1.1), when the absolute transition probability of the latter is known.

If this second spectral line is emitted by particles of the same element and at the same ionization stage, the accuracy can be raised even more by employing a graphical method suitably to be applied on axialsymmetric plasmas (Olsen [5.304]). Let us write equation (5.3) in the simplified form

$$\varepsilon_{1,\,2} = K_{1,\,2}\cdot\frac{n}{Q}\,A_{1,\,2}\cdot e^{-U_{1,\,2}/kT}\,. \tag{5.22}$$

The indices 1 and 2 refer to the spectral lines with known transition probability A_1 and unknown transition probability A_2, respectively. n is the number density of particles which emit the spectral lines 1 and 2, Q is the partition function, U_1 and U_2 are the respective excitation energies of the spectral lines. The constants K_1 and K_2 are given by

$$K_{1,2} = \frac{h\,c\,g_{1,2}}{4\,\pi\,\lambda_{1,2}}. \tag{5.23}$$

From (5.3) follows the expression

$$\lg \varepsilon_2 = \frac{U_2}{U_1} \lg \varepsilon_1 + \lg \frac{K_2\,A_2\,n/Q}{(K_1\,A_1\,n/Q)^{U_2/U_1}}. \tag{5.24}$$

If one plots $\lg \varepsilon_2$ as a function of $\lg \varepsilon_1$ after the radial emission coefficients $\varepsilon_1(r)$ and $\varepsilon_2(r)$ are determined, we can draw through the measured points a slightly bent curve, having a slope

$$\frac{\mathrm{d}\lg \varepsilon_2}{\mathrm{d}\lg \varepsilon_1} = \frac{kT^2\,\dfrac{\mathrm{d}n}{\mathrm{d}T} + n\,U_2}{kT^2\,\dfrac{\mathrm{d}n}{\mathrm{d}T} + n\,U_1}. \tag{5.25}$$

When the temperatures belonging to the measured values lie below the norm temperature, the curvature is very low (it vanishes completely when $U_2 = U_1$), and a straight line with slope U_2/U_1 can be drawn through the points. Figure 5.15 gives an example of a $\lg \varepsilon_2$, vs $\lg \varepsilon_1$ plot for A II lines.

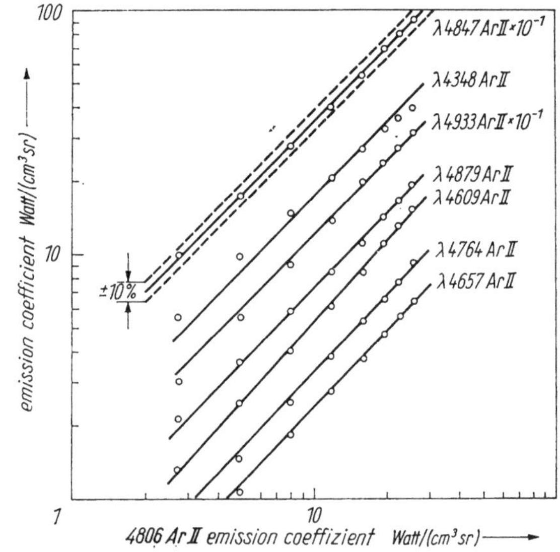

Figure 5.15. Emission coefficient of different Ar II lines plotted logarithmically as a function of the emission coefficient of the Ar II line 4806 Å, also represented on a logarithmic scale. The slopes of the straight lines correspond to the ratios of the excitation energies (after Olsen [5.304])

A_2 can now be found from the intersection of the measured curve with the ordinate. For this we have the expression

$$\varepsilon_1 = 1 \text{ W cm}^{-3}\text{ sr}^{-1}, \qquad \varepsilon_2 = \frac{K_2 A_2 n/Q}{(K_1 A_1 n/Q)^{U_2/U_1}}. \tag{5.26}$$

n/Q depends on T and is known, since T is given by the absolute emission coefficient of line 1, whose transition probability is known. Since n/Q is only slightly dependent on temperature for $T < T^*$, the accuracy with which A_2 is determined from (5.26) is nearly as high as that with which A_1 is known, neglecting errors inherent in the intensity measurements. The accuracy becomes higher the smaller the difference between the excitation energies U_2 and U_1. In the limiting case $U_2 = U_1$, the evaluation of A_2 no longer depends on the temperature.

Suitable plasma sources for the determination of transition probabilities are arc systems, in which the working substance is injected in the form of a gas or liquid (see Figures 5.16—5.19), and shock tubes. With these systems we can determine the transition probabilities of spectral lines of those elements, which at room temperature or somewhat higher temperature exist in the form

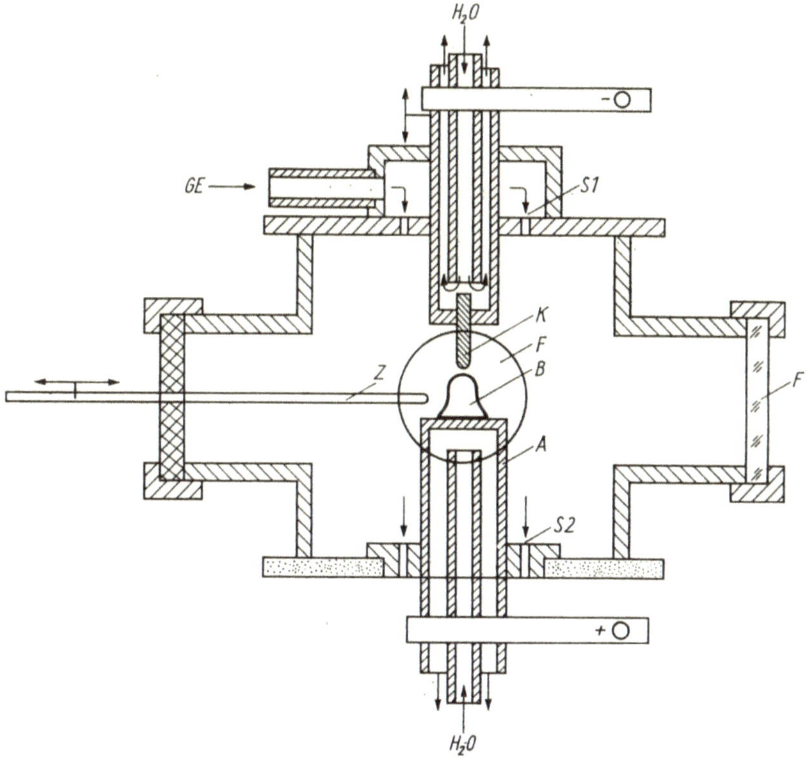

Figure 5.16. Gas-flow stabilized arc (after Neumann [5.296])

K – thoriated tungsten cathode, *A* – copper disc anode, *Z* – ignition rod for *HF*-ignition, *F* – quartz window, *B* – arc. Gas inlet via supply pipes *GE* and screen holes *S1*; gas outlet via screen holes *S2*

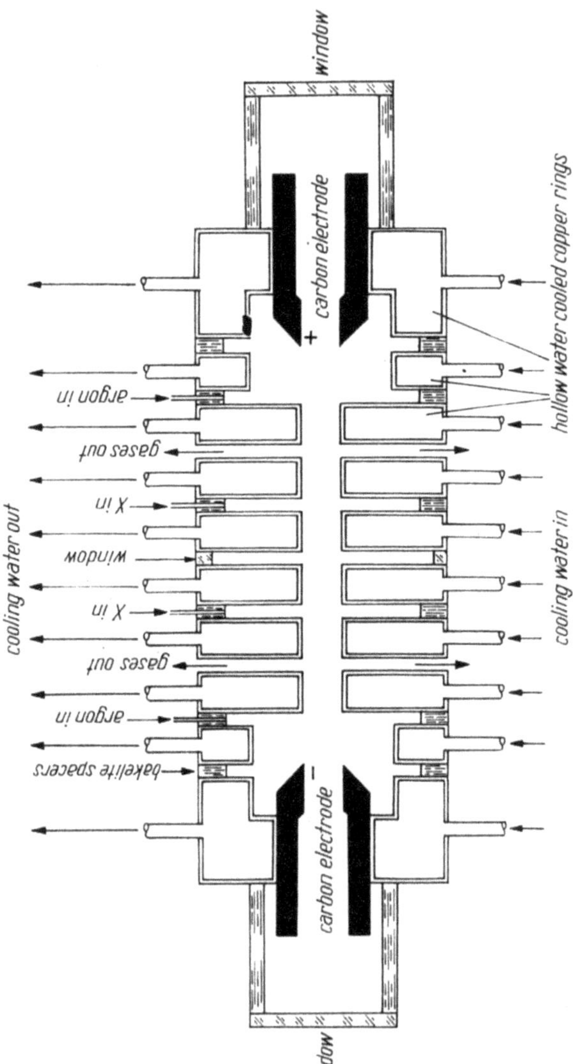

Figure 5.17. Wall-stabilized plate cascade arc. The gas X to be investigated is in the center, the electrodes come into contact only with the buffer gas argon (after Shumaker [5.399])

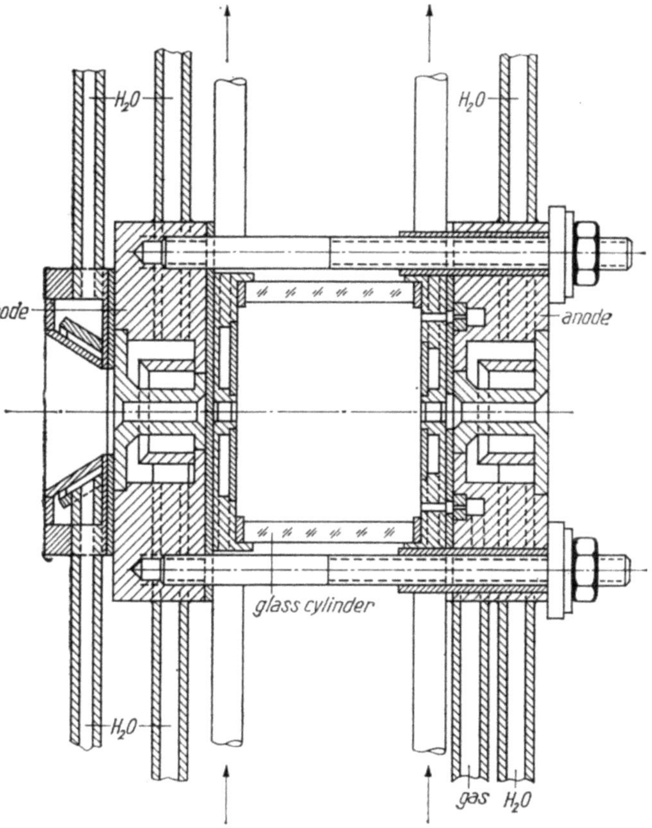

Figure 5.18. Vortex stabilized arc with tangential gas supply. Water-cooled guide plates mounted in front of the electrodes serve to center the arc (after Pfender and Bez [5.349])

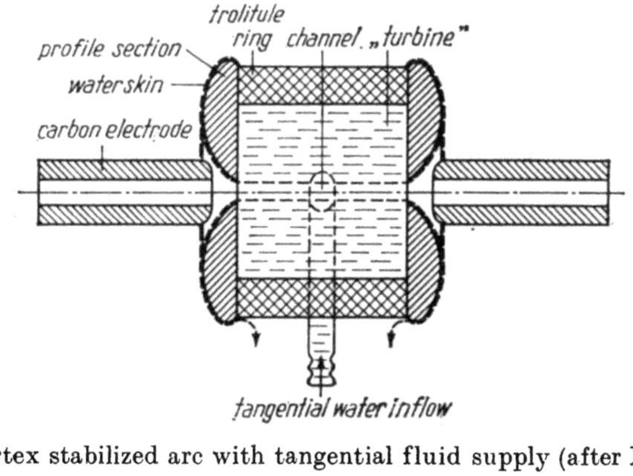

Figure 5.19. Vortex stabilized arc with tangential fluid supply (after Maecker [5.260])

of a gas or liquid, or can be converted into substances of this type by chemical treatment.

To find the absolute transition probabilities in complex plasmas, it is recommended, when possible, to employ vortex stabilized arcs as light sources [5.106]. Here the plasma is rendered more turbulent than in wall-stabilized arcs, and gas separation effects, which in relatively calm plasmas alter the stoichiometric composition of the plasma in a way which is difficult to control (see Frie [5.110]), are here kept to a minimum.

In this connection it may be mentioned that even in pulsed arc plasmas gas separation effects due to ordinary diffusion and thermo diffusion can occur, as has been shown by Hermel [5.164b] for a He-H-plasma, who found for these processes a relaxation time of the order 20 μs.

In shock tubes used as radiation sources, diffusion processes are not so important than in arc sources, because the shock wave plasma has over some distance a high degree of homogeneity. Shock tubes are therefore used in a growing number of cases as radiation source for the determination of transition probabilities of spectral lines emitted by metal atoms which are usually introduced into the plasma device in form of a volatile (for example metalorganic) compound, together with a buffer gas, in most cases Argon. See, for example, [5.54e], [5.302d].

Due to the high degree of plasma homogeneity in the region behind the shock front, at least with respect to a direction perpendicular to the direction of front propagation, the evaluation of transition probability datas from shock tube measurements is relative simple. It requires side-on measurements of spectral line intensities and the knowledge of the optical path length within the radiating volume, the temperature and the number density of radiating atoms. While the latter two quantities can be derived from the initial experimental parameters and the measured shock front velocity by means of the continuity conditions at the shock front (see, for example, Oertel [5.302d], p. 239—242), the temperature can also be found on a pure experimental basis, for example via the line reversal method (see section 6.1).

5.3.2. Determination of Transition Probabilities from the Emission of Spectral Lines in an Optically Thin Layer when the Concentration of the Emitting Element is not known beforehand

The transition probability of the spectral line to be studied is in principle found, as above, by means of equations (5.19) and (5.20). However, the initially unknown quantities T (temperature) and n_z (concentration of particles, emitting the relevant spectral line) must be determined beforehand by other means.

We shall consider four typical cases, for which the determination of the transition probability is effected in this fashion.

Case a. For a mixture of two substances, T and n_z can be found from the absolute intensity of two spectral lines when the pressure is known. The details of this method were already given in section 5.1.3. In this connection we mention a work of Foster [5.107], who employed a turbulent-fluid stabilized arc with water and benzene as working substances to obtain data on the temperature, plasma composition, and the not yet known C I transition probabilities, from the line intensities. The mixture ratio x of the plasma with composition $C_6H_6 + x\, H_2O$ and the plasma temperature were determined from the absolute intensity of the spectral lines H_β and OI 3947 Å.

Case b. The method of Boldt [5.45] for the determination of transition probabilities in complex plasmas with unknown mixture ratio is based in principle on the determination of the plasma composition and temperature from measurements of the relative intensity of spectral lines (see section 5.2.1). After this the absolute transition probabilities are found with the help of (5.19) and (5.20). With this method Boldt determined the transition probabilities of 24 lines and line groups of the plasma component CI in an A-C-O plasma in the spectral range from 1156 to 1765 Å.

The plasma is produced by a cascade arc located between bored carbon electrodes (Figure 5.20). The arc burns essentially in argon, which is supplied via inlets 1—5. Through the gas inlet Z a second gas (CO_2 or O_2) can be injected. This gas only fills the space between the nozzle (where argon is injected via inlet 2) and the anode. In this region 3 different plasmas are produced: 1. a pure argon plasma; 2. an argon plasma with an admixture of O_2, having plasma constituents A and O; 3. an argon plasma with an admixture of CO_2. The plasma constituents in this case are A, O and C. Furthermore, a slight admixture of hydrogen is used, which, however, is of no significance for the composition of the plasma proper. For each of the three plasmas the following was determined by side-on observation via window J: 1. the relative emission coefficient $\varepsilon_{\mathrm{A\,I}}$ of an A I line; 2. the relative emission coefficient $\varepsilon_{\mathrm{O\,I}}$ of an OI line; 3. the relative emission coefficient $\varepsilon_{\mathrm{C\,I}}$ of a C I line. Further, for each of the three plasmas the line width $\Delta\lambda$ of the H_β line is determined by end-on observation through a hole in the anode. Accordingly, we can find the electron density from the expression (see Griem, Kolb, and Shen [5.140])

$$n_e = \beta(\Delta\lambda)^{3/2}\,, \qquad \beta = 3.40 \cdot 10^{14}\ \mathrm{cm}^{-3}\ \text{Å}^{-3/2}\,. \tag{5.27}$$

All 4 quantities are measured with respect to the plasma axis ($r = 0$). We can now calculate the plasma composition for each of the three plasma from the obtained data. For this we have at our disposal the following equations, when we include the electron density n_e in the system of equations:

1. For the Ar-plasma:

$$\left.\begin{aligned} &\frac{n_{\mathrm{A\,II}}^{(1)}}{n_{\mathrm{A\,I}}^{(1)} + n_{\mathrm{A\,II}}^{(1)}} = f_1(n_e^{(1)}, T^{(1)}) \qquad \text{(Saha equation)}\,,\\ &n_{\mathrm{A\,I}}^{(1)} + n_{\mathrm{A\,II}}^{(2)} + n_e^{(1)} = \frac{p}{kT^{(1)}}\,, \qquad p = 1\ \mathrm{atm}\,,\\ &n_e^{(1)} = n_{\mathrm{A\,II}}^{(1)}\,, \qquad \Delta\lambda^{(1)} = f_2(n_e^{(1)})\,, \qquad \varepsilon_{\mathrm{A\,I}}^{(1)} = K_1\, n_{\mathrm{A\,I}}^{(1)} \cdot f_3(T)\,. \end{aligned}\right\} \tag{5.28}$$

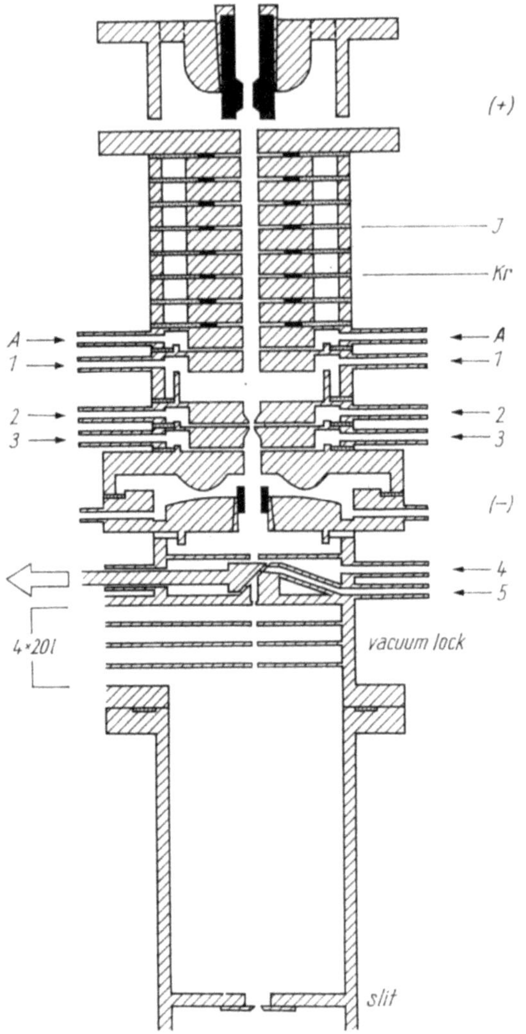

Figure 5.20. Cascade arc for the production of a plasma from composite gases and a vacuum lock for connecting the arc chamber to a vacuum spectrograph (after Boldt [5.45])

1–5 – argon in, *Z* – inlets for second gas, *Kr* – krypton in, *J* – for side-on observation of the arc

We thus have 5 equations in the 5 unknowns $n^{(1)}_{\mathrm{A\,II}}$, $n^{(1)}_{\mathrm{A\,I}}$, $n^{(1)}_e$, $T^{(1)}$, and K_1, of which the first 4 unknowns refer to the A-plasma only, while K_1 (relative transition probability of the A I line) depends only on the measuring arrangement and has identical significance for all 3 plasmas. The system can be solved for all five unknowns.

2. Equations for finding the composition of the A-O_2-plasma:

$$\left.\begin{aligned}
&\frac{n^{(2)}_{\mathrm{A\,II}}}{n^{(2)}_{\mathrm{A\,I}} + n^{(2)}_{\mathrm{A\,II}}} = f_1(n^{(2)}_e, T^{(2)})\,,\\
&\frac{n^{(2)}_{\mathrm{O\,II}}}{n^{(2)}_{\mathrm{O\,I}} + n^{(2)}_{\mathrm{O\,II}}} = f_4(n^{(2)}_e, T^{(2)})\,,\\
&n^{(2)}_{\mathrm{A\,I}} + n^{(2)}_{\mathrm{A\,II}} + n^{(2)}_{\mathrm{O\,I}} + n^{(2)}_{\mathrm{O\,II}} + n^{(2)}_e = \frac{p}{kT^{(2)}}\,, \qquad p = 1\ \mathrm{atm}\,,\\
&n^{(2)}_e = n^{(2)}_{\mathrm{A\,II}} + n^{(2)}_{\mathrm{O\,II}}\,,\\
&\Delta\lambda^{(2)} = f_2(n^{(2)}_e)\,,\\
&\varepsilon^{(2)}_{\mathrm{A\,I}} = K_1\, n^{(2)}_{\mathrm{A\,I}} \cdot f_3(T^{(2)})\,,\\
&\varepsilon^{(2)}_{\mathrm{O\,I}} = K_2\, n^{(2)}_{\mathrm{O\,I}} \cdot f_5(T^{(2)})\,.
\end{aligned}\right\} \quad (5.29)$$

Since K_1 is known, there are 7 equations in the 7 unknowns $n^{(2)}_{\mathrm{A\,II}}$, $n^{(2)}_{\mathrm{A\,I}}$, $n^{(2)}_{\mathrm{O\,II}}$, $n^{(2)}_{\mathrm{O\,I}}$, $n^{(2)}_e$, $T^{(2)}$, and K_2. K_2 again has identical significance for all three plasmas.

3. The system of equations for the A-CO_2-plasma (it is assumed that the mixture ratio between C and O is not constant):

$$\left.\begin{aligned}
&\frac{n^{(3)}_{\mathrm{A\,II}}}{n^{(3)}_{\mathrm{A\,I}} + n^{(3)}_{\mathrm{A\,II}}} = f_1(n^{(3)}_e, T^{(3)})\,,\\
&\frac{n^{(3)}_{\mathrm{O\,II}}}{n^{(3)}_{\mathrm{O\,I}} + n^{(3)}_{\mathrm{O\,II}}} = f_4(n^{(3)}_e, T^{(3)})\,,\\
&\frac{n^{(3)}_{\mathrm{C\,I}}}{n^{(3)}_{\mathrm{C\,I}} + n^{(3)}_{\mathrm{C\,II}}} = f_6(n^{(3)}_e, T^{(3)})\,,\\
&n^{(3)}_{\mathrm{A\,I}} + n^{(3)}_{\mathrm{A\,II}} + n^{(3)}_{\mathrm{O\,I}} + n^{(3)}_{\mathrm{O\,II}} + n^{(3)}_{\mathrm{C\,I}} + n^{(3)}_{\mathrm{C\,II}} + n^{(3)}_e = \frac{p}{kT^{(3)}}\,, \quad p = 1\ \mathrm{atm}\,,\\
&n^{(3)}_e = n^{(3)}_{\mathrm{A\,II}} + n^{(3)}_{\mathrm{O\,II}} + n^{(3)}_{\mathrm{C\,II}}\,,\\
&\Delta\lambda^{(3)} = f_2(n^{(3)}_e)\,,\\
&\varepsilon^{(3)}_{\mathrm{A\,I}} = K_1\, n^{(3)}_{\mathrm{A\,I}} \cdot f_3(T^{(3)})\,,\\
&\varepsilon^{(3)}_{\mathrm{O\,I}} = K_2\, n^{(3)}_{\mathrm{O\,I}} \cdot f_5(T^{(3)})\,,\\
&\varepsilon^{(3)}_{\mathrm{C\,I}} = K_3\, n^{(3)}_{\mathrm{C\,I}} \cdot f_7(T^{(3)})\,.
\end{aligned}\right\} \quad (5.30)$$

Since K_1 and K_2 are known and since measured values exist for $\Delta\lambda^{(3)}$, $\varepsilon^{(3)}_{\mathrm{A\,I}}$, $\varepsilon^{(3)}_{\mathrm{O\,I}}$, and $\varepsilon^{(3)}_{\mathrm{C\,I}}$, the above 9 equations in the 9 unknowns $n^{(3)}_{\mathrm{A\,I}}$, $n^{(3)}_{\mathrm{A\,II}}$, $n^{(3)}_{\mathrm{O\,I}}$, $n^{(3)}_{\mathrm{O\,II}}$, $n^{(3)}_{\mathrm{C\,I}}$, $n^{(3)}_{\mathrm{C\,II}}$, $n^{(3)}_e$, $T^{(3)}$, and K_3, can be solved by simple means.

To determine the oscillator strengths of the CI lines in the vacuum ultraviolet range the spectrum of the A-CO_2-plasma (plasma 3) is photographed end-on, whereby the observation is made through a hole in the cathode. Since the carbon atoms are restricted to the region between the nozzle and the anode and not near the cathode, where the temperature drops to small values, self-absorption of the CI lines could be avoided.

The absolute transition probability can now, in principle, be determined from the absolute line intensities $I_{\mathrm{C\,I}} = l \cdot \varepsilon_{\mathrm{C\,I}}$ (l = length of the A-CO_2-plasma) and the known values for $n_{\mathrm{C\,I}}$ and T (equations (5.19) and (5.20)), when the measuring device is calibrated by means of a standard source in such a way that the measured relative $I_{\mathrm{C\,I}}$-values can be converted to absolute values for $I_{\mathrm{C\,I}}$.

The peculiarity in Boldt's method lies herein that the calibration radiation is produced by the cascade arc itself. For this purpose krypton gas is injected via a gas inlet in such slight amounts, that on the one hand the plasma composition remains practically unaltered, while on the other the radiation of the Kr I line $\lambda = 1236$ Å becomes optically black in the center of the line when viewing end-on through the cathode. Then, in accordance with equation (2.19a), the line intensity per unit frequency is given by the Planck formula

$$I_{\nu\,\mathrm{Abs}} = B_\nu(T)\,. \tag{5.31}$$

With the measured relative values $I_{\mathrm{C\,I}}$ and I_ν we now obtain the absolute value $I_{\mathrm{CI\,Abs}}$ from the expression

$$I_{\mathrm{C\,I\,Abs}} = I_{\mathrm{C\,I\,Rel}} \frac{B_\nu(T)}{I_{\nu\,\mathrm{Rel}}}\,. \tag{5.32}$$

Dividing (5.32) by l gives $\varepsilon_{\mathrm{C\,I\,Abs}}$, and with (5.20) we finally obtain the sought absolute value for $A_{\mathrm{C\,I}}$. Boldt estimated a possible measurement error of 25% in the transition probability.

Case c. Here we shall describe a method for finding the absolute transition probabilities, which can be applied to readily ionizable elements occurring in the plasma, when besides this element only those elements are present, which contribute insignificantly to the ionization. These other elements must either possess a high ionization energy or occur in the plasma only as traces.

The peculiarity of this method is that the number density of emitting particles is derived from the current intensity I and the electric field strength E.

To find the absolute transition probability of a spectral line by this method the absolute emission coefficient of the line must be measured as a function of the radial coordinate. Further, the current intensity, the electric field strength, and the effective cross section for collisions between electrons and atoms must be known. Also, data on the radial temperature distribution $T(r)$ are required.

This method was employed by Eberhagen [5.87] and Eicke [5.89] for the determination of transition probability datas for Sr and Ba lines.

We start with equation (5.3) for the emission coefficient a spectral line, where we shall consider primarily the emission of an atomic line:

$$\varepsilon_{n\,m}(r) = \frac{1}{4\,\pi}\,\frac{g_m}{Q_0^A}\,n_0^A\,A_{n\,m}\,h\,\nu_{n\,m}\,\mathrm{e}^{-U_m/kT}\,. \tag{5.33}$$

Assuming that only the element A is ionized ($n_e = n_1^A$), the Saha equation becomes (we shall now leave out the symbol A for the emitting element)

$$\frac{n_e^2}{n_0} = \frac{2\,Q_1}{Q_0}\,\frac{(2\,\pi\,m)^{3/2}}{h^3}\,[k\,T(r)]^{3/2}\,\mathrm{e}^{-U_i/kT}\,. \tag{5.34}$$

Combining (5.33) and (5.34), we can eliminate n_0 and obtain an expression, linking the electron density, temperature, and intensity of the spectral line in question:

$$n_e^2 = \frac{4\,\pi\,\varepsilon(r)\,Q_1\,F(T)}{g_m\,A_{n\,m}\,h\,\nu_{n\,m}}\,\mathrm{e}^{U_m/kT}\,, \qquad F(T) = \frac{2\,(2\,\pi\,m\,kT)^{3/2}}{h^3}\,\mathrm{e}^{-U_i/kT}\,. \tag{5.35}$$

A further connection between T and n_e is provided by the current transport equation, which is given by Eicke in the form (j = current density)

$$j = \frac{\sqrt{\pi}\,n_e\,e^2\,\lambda_e\,E\,\alpha}{2\sqrt{2\,m\,kT}} \tag{5.36}$$

where $\alpha = 0.75$. Together with the expression for the total current

$$I = 2\,\pi \int_0^R j\,r\,\mathrm{d}r \tag{5.37}$$

we finally have

$$\sqrt{A_{n\,m}} = \frac{\sqrt{8\,\pi}\,C\,\lambda_e\,E}{I\sqrt{g_m\,h\,\nu_{n\,m}}} \int_0^R T(r)^{-1/2}\,\sqrt{Q_1\,F(T)\,\varepsilon_{n\,m}(r)}\,\exp\left(\frac{U_i}{2\,kT}\right) r\,\mathrm{d}r \tag{5.38}$$

where

$$C = \frac{2\,\pi^2\,e^2\,\alpha}{\sqrt{2\,m_e}\,h} = 7.56\cdot 10^{-9}\,[\mathrm{Amp.\,(Volt)^{-1}\cdot cm\cdot (^\circ K)^{1/2}}]\,. \tag{5.39}$$

If we have no further information, we assume

$$\lambda_e = \left(\frac{4\sqrt{2}}{273\,^\circ\mathrm{K}}\right) T\,\lambda_{0\,\mathrm{Gas}}\,[\mathrm{cm}] \tag{5.40}$$

with $\lambda_{0\,\mathrm{Gas}}$ = gas kinetic mean free path for $p = 1$ atm and $T = 273$ °K.

When the temperature is determined, for example, by the two-line method (see section 5.2.2), the absolute transition probability can now be calculated from (5.38).

To obtain reliable values for the transition probability, the mean free path of the electrons λ_e must be known with adequate precision.

In the cases investigated by Eberhagen and Eicke the arc burned essentially in free air, with a slight admixture of the readily ionizable element to be studied, so that only collisions between electrons and air molecules were decisive. Eberhagen gave for this case a value of $\lambda_{0\,\mathrm{Gas}} = 8.0 \cdot 10^{-6}$ cm, independent of the element added.

When the transition probabilities of ionic lines of the element studied are desired (this only applies to singly charged ions), we start with the equation for the emission coefficient $\varepsilon_{n\,m}(r)$ of the ionic line:

$$\varepsilon_{n\,m}(r) = \frac{1}{4\,\pi}\,\frac{g_m}{Q_i}\,n_i\,A_{n\,m}\,h\,\nu_{n\,m}\,\mathrm{e}^{-U_m/kT} \tag{5.41}$$

and the current transport equations (5.36) and (5.37). Combining, we get

$$A_{n\,m} = \frac{2\,\sqrt{\pi}\,C\,E}{I\,g_m\,h\,\nu_{n\,m}} \int\limits_0^R T^{-1/2}\,Q_1(T)\,\lambda_e\,\varepsilon_{n\,m}(r)\,\mathrm{e}^{U_m/kT}\,r\,\mathrm{d}r\,. \tag{5.42}$$

The integral is made up of temperature dependent functions. To solve this integral the temperature must again be known as a function of the radial coordinate r. (See section 5.2.2 for the T-determination by the two-line method.) Eicke [5.89], who determined the transition probabilities for Ba II lines, evaluated the temperature by means of the relative intensities of the Cu lines $\lambda = 5106$, 5218, and 5153 Å. Copper and barium were packed in the form of $CuSO_4$ and $BaCO_3$ together with powdered carbon in bored carbon electrodes, between which the arc burned, and thus succeeded in entering into the arc. Stabilization was effected by a rotating arc system (with the arc enclosed in a rotating glass cylinder, see Schnautz [5.382]) with excellent results.

Case d. The following method given by Berg [5.26e] allows to determine transition probabilities of spectral lines emitted by higher ionized atomic systems.

In order to determine absolute transition probabilities of O III lines, Berg used as radiation source a highly ionized plasma containing the elements He and O, and measured the emission coefficients of four suitably selected spectral lines denoted with εO III), ε(O II), ε(He II), ε(He I). The transition probabilities respectively oscillator strengths of the three last mentioned lines must be known. Then equation (5.15) is applied to the emission coefficients of the line pairs {(O III), (O II)} respectively {(He II), (He I)}. By proper choice of these line pairs the expressions $[U_i^{(z)} + U_{i}^{(z+1)} - U_m^{(z)})]$ in the arguments of the exponential functions are approximately identical (deviations of 1% are allowed). Then the quotient of equations (5.15) applied to the two line pairs is independent from temperature and is given by the expression

$$\frac{\varepsilon(\mathrm{O\,III})}{\varepsilon(\mathrm{O\,II})}\,\frac{\varepsilon(\mathrm{He\,I})}{\varepsilon(\mathrm{He\,II})} = \frac{g_n(\mathrm{O\,III})}{g_n(\mathrm{O\,II})}\cdot\frac{g_n(\mathrm{He\,I})}{g_n(\mathrm{He\,II})}\cdot\frac{f(\mathrm{O\,III})}{f(\mathrm{O\,II})}\cdot\frac{f(\mathrm{He\,I})}{f(\mathrm{He\,II})}\cdot\frac{\lambda^3(\mathrm{O\,II})}{\lambda^3(\mathrm{O\,III})}\cdot\frac{\lambda^3(\mathrm{He\,II})}{\lambda^3(\mathrm{He\,I})}$$

from which the quantity f(O III) can be obtained. Knowing the oscillator strength f, the corresponding transition probability can be derived from formula (5.44) given below.

Knowing the transition probability of one line of the system O III, the transition probabilites of the other lines of the same system can be found using the method described in section 5.3.4. When using this method, the plasma temperature must be known, which can be found for example via the method given in section 5.2.5.3.

5.3.3. Determination of Absolute Transition Probabilities from Relative Ones with the Help of the *f*-Sum Rule

In some cases it is possible to obtain absolute transition probabilities from relative ones by calculating the conversion factors of the Kuhn-Thomas-Reiche *f*-sum rule (see Wobig [5.466] and Tatum [5.419]).

The determination of the relative transition probabilities can be effected, for example, by means of the expression obtained from (5.11)

$$\frac{A_1}{A_2} = \frac{\varepsilon_1 \lambda_1 g_2}{\varepsilon_2 \lambda_2 g_1} \cdot e^{-(U_2 - U_1)/kT} \tag{5.43}$$

which is applied to 1 pair of spectral lines, whose emission coefficients ε_1 and ε_2 have been determined. The statistical weights g_1 and g_2 of the upper level of the lines are assumed to be known. The temperature is also assumed as known or must be measured in some way.

The conversion of the relative transition probabilities A_{rel} to absolute A_{abs} is effected by means of the *f*-sum rule. To apply this rule one must first convert the measured relative transition probabilities into relative absorption oscillator strengths by means of the expression

$$f_{mn} = \frac{g_m}{g_n} A_{nm} \frac{m c^3}{8 \pi^2 e^2 \nu^2} = \frac{g_m}{g_n} \lambda^2 A_{nm} \cdot 1.50 \,. \tag{5.44}$$

The *f*-sum rule for transitions from one term t to all possible higher terms (including the continuum) or lower terms is given by

$$\sum_i g_t f_{ti} = g_t z \tag{5.45}$$

where z is the number of participating emission electrons of the electron system. Usually it is sufficient to consider only one emission electron, so that $z = 1$.

Dividing the transitions into transitions to lower (l) and to higher (h) levels, respectively, we have

$$\sum_l g_t f_{tl} + \sum_h g_t f_{th} = g_t z \,. \tag{5.46}$$

For Ti II, as in the case of elements of the iron group, the most intense lines are emitted due to transitions between the lowest terms of the corresponding series. According to Allen [5.7] their *f*-value is 0.7 that of all transitions to higher terms, including the continuum. Then (5.46) simplifies to

$$\sum t_t = 0.7\, z - \sum_l f_{tl} \,, \tag{5.47}$$

where Σf_t refers to the most energetic transitions. In many cases no transitions to lower levels can be expected, so that $\sum_l f_{tl}$ vanishes.

To obtain the individual f-values from the thus obtained sums, we introduce the relative multiplet intensities (Allen [5.7]):

$$g_t f_t = \frac{1}{\lambda} S(M) . \tag{5.48}$$

For $s - p$ transitions, we have

$$\Sigma g_t f_t = 0{,}233 \cdot g_{ut} . \tag{5.49}$$

The sum must be taken over all lines of a multiplet with the same upper term ($u\ t$). g_{ut} is the statistical weight of the upper term.

For $p - d$ transitions the following holds

$$g_t f_t = 0{,}085 \cdot S(M) , \tag{5.50}$$

and for $d^3 - d^2 p$ transitions

$$g_t f_t = 0{,}026\, S(M) . \tag{5.51}$$

The numerical values for the multiplet strengths are listed by Allen [5.8].

The problem now reduces to finding a factor F. by which the measured relative values $g_t f_t$ are multiplied in such a way that the values calculated by means of the sum formulas $\sum_{\text{cal}} g_t f_t$ and $\underset{\text{cal}}{g_t f_t}$ on the average deviate as little as possible from the values $F \cdot \sum_{\text{meas}} g_t f_t$ and $F \underset{\text{meas}}{g_t f_t}$, respectively. While Tatum [5.419] carried out the approximation only for multiplets with $s - p$ transitions when determining the absolute oscillator strengths of Ti II lines, Wobig [5.466] also included $p - d$ as well as $d^3 - d^2 p$ transitions for the calculation of the conversion factor. The errors, which are involved in finding these $g f$-values, are denoted by a factor 2.

5.3.4. Determination of Absolute Transition Probabilities from Relative Transition Probabilities when the Absolute Transition Probability of at least one Spectral Line of the Element studied is known

We can find relative transition probabilities by applying equation (5.43) to the relevant spectral lines of an element A in an ionization stage z. For this purpose the relative emission coefficients of the spectral lines investigated must be measured, and their statistical weights and excitation energies must be known. Further the temperature must be known in some way or else determined. The thus obtained relative transition probabilities A_{rel} are converted to absolute value A_{abs} by multiplication by a constant factor F. The factor F can be determined, if among the spectral lines, whose relative transition probability was measured, there is at least one for which A_{abs} is known. We shall designate such a line as a calibration line (C) and thus obtain for the conversion factor

$$F = \frac{A_{\text{abs}}(\text{C})}{A_{\text{rel}}(\text{C})} . \tag{5.52}$$

Roder [5.370] employed the above method to find the absolute transition probabilities of Fe I lines. As the radiation source he used a wall-stabilized arc, which essentially burned in argon, which was mixed with 0.2% vaporized iron chloride $FeCl_3$ before injection into the arc. Since gas separation occurred in the arc, the iron concentration in the arc could no longer be calculated, so that a direct method for determining the absolute transition probabilities of iron could not be applied.

He calculated the temperature from the absolute intensities of the lines A I 4158 Å and A II 4348 Å.

After Roder had determined the absolute transition probabilities of Fe I lines by the abovementioned method, he found the absolute transition probabilities of the Fe II lines as well by applying the equation

$$\frac{A_{nm}(\mathrm{Fe}^+)}{A_{n'm'}(\mathrm{Fe})} = \frac{\varepsilon}{\varepsilon'}\,\frac{\lambda}{\lambda'}\,\frac{Q}{Q'}\,\frac{g'_m}{g_m}\,\frac{n_{\mathrm{Fe}}}{n_{\mathrm{Fe}^+}}\,\mathrm{e}^{\frac{U_m - U_{m'}}{kT}} \tag{5.53}$$

to one Fe I line and one Fe II line at a time. Q denotes the partition function. The accented quantities refer to neutral iron. When the temperature is known, the ratio $Q \cdot n_{\mathrm{Fe}} : Q' \cdot n_{\mathrm{Fe}^+}$ is given by the Saha equation (4.1) for iron; the electron density is equal to that of the pure argon plasma, since the iron concentration is so small:

$$\frac{Q\, n_{\mathrm{Fe}}}{Q'\, n_{\mathrm{Fe}^+}} = \frac{1}{2\left(\frac{2\pi m kT}{h^2}\right)^{3/2} \mathrm{e}^{-U_{i,\,\mathrm{eff}}(\mathrm{Fe})/kT}\, n_e} . \tag{5.54}$$

Roder gives an accuracy of 20% for this method, provided there are no errors in the absolute transition probability of the calibration lines. Otherwise the accuracy is reduced.

A difficulty arises when at a given concentration of the element under consideration the calibration line and the line under investigation have quite different intensities resulting in self absorption for the stronger line while the weaker line just overcomes the noise level. Ovechkin et al. [5.330a] avoided this difficulty by supplying the element under consideration to the spectroscopic light source in two different concentrations of known concentration ratio. This ratio is properly adjusted in that way that the stronger line is optically thin for the lower concentration while the weaker line is well observed for the higher concentration. For details see the original literature.

We should still mention that in determining the transition probabilities by means of emitted lines a Boltzmann distribution of the electrons was tacitly assumed. This is usually considered valid when the electron density exceeds a value between 10^{14} and 10^{16} cm^{-3} (see sections 11.1.3 and 11.1.9). Here, it is assumed that an atom, excited in a specific level, experiences, on the average, several inelastic collisions before it emits the excitation energy.

In iron and some other elements there are excited electrons in the inner electron shells. These have an extremely small effective cross section with regard to electron collisions, since these electrons are screened by the outer electrons.

In the case of these electrons we can only expect a Boltzmann distribution when the radiation density is equal to the cavity radiation density. This usually is only the case for a plasma produced in an oven.

Levels, which are hardly influenced by electron collisions, are characterized in that the spectral lines, which correspond to transitions between such levels, display a minimum Stark broadening, while the Doppler broadening predominates (Burhorn [5.57] and Bergstedt [5.27]). As an example of such lines we name the lines Fe I (λ in Å) 4302, 4338, 4346, 4733, 4765, 4832, 4844, 5307, 4352, and 4859 Å. A determination of the transition probability of these lines is as a rule only possible for plasmas produced in ovens. However, the information gained in finding the transition probabilities in this way is of little use.

5.3.5. Determination of Oscillator Strengths by Absorption Measurements on Atoms in an Oven

In ovens steady temperatures of up to about 3000 °K can be produced for a long time. At this temperature only low levels, if at all, of metal atoms are excited, yielding few emission lines (see Carter [5.61]). On the other hand, however, numerous absorption lines appear, whose lower levels lie deep.

To find the oscillator strengths f_{mn} of these lines the gas inside a King furnace (see Figure 5.21) or in a heated absorption tube of length l is irradiated with

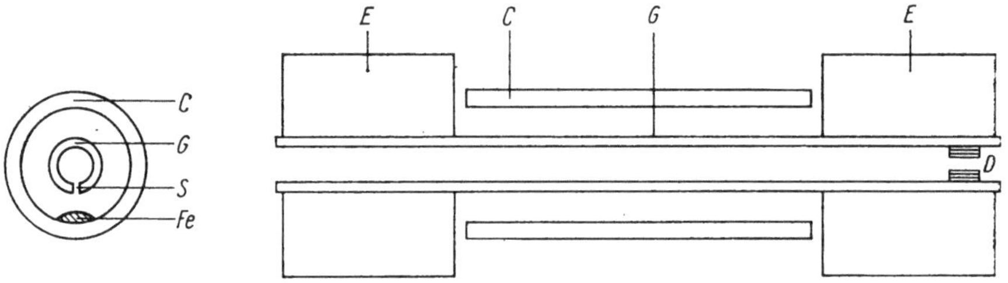

Figure 5.21. King furnace (after Aarts, Harting, and Bakker [5.1a])

E – electrodes, G – electrically heated graphite tube, C – carbon tube, S – slit, D – diaphragm, Fe – Iron-carbon mixture

continuous radiation, such as for example, the radiation from a tungsten filament lamp, and the absorption coefficient k_ν is determined with the help of a high-resolution spectroscopic device as a function of frequency in the range of the spectral line in accordance with the expression

$$I_\nu/I_{\nu_0} = e^{-k_\nu l}. \qquad (5.55)$$

I_{ν_0} is the incident radiation intensity, I_ν the transmitted intensity.

The total absorption coefficient $\int\limits_{(\text{line})} k_\nu \, \mathrm{d}\nu$, which is obtained by integration over the frequency range occupied by the line, is connected with the oscillator strength $f_{m\,n}$ via the expression*

$$\int k_\nu \, \mathrm{d}\nu = \frac{\pi\, e^2}{m\, c}\, n_n f_{m\,n} \tag{5.56}$$

where n_n is the number of absorbing atoms in the quantum state n and can be calculated from T and n_0 (total number of atoms of the element investigated) with the help of the Boltzmann formula (see also (5.2))

$$n_n = \frac{n_0}{Q_0}\, \mathrm{e}^{-U_n/kT}\, g_n\,. \tag{5.57}$$

If n_0 is not accurately known, only relative f-values can be calculated.

To obtain definite values for n_0 and l, King and Stockbarger ([5.208] and [5.210]) used an absorption cell of quartz or pyrex, which is evacuated after filling with the metal (Cd, Cu, or Fe) to be studied and is placed in the center of the oven. n_0 follows from the vapor pressure curve. Since a temperature of 1600 °K could not be surpassed, only absolute f-numbers of resonance lines could be determined, which terminate at the ground level. The absolute f-values are uncertain by a factor of 2 because of the uncertainty in the vapor pressure.

If the frequency dependent absorption power in the range of the spectral line is small compared to 1, we have the expression

$$\int k_\nu \, \mathrm{d}\nu = \frac{1}{l}\, A_\nu\,. \tag{5.58}$$

A_ν is the total absorption of the gas with layer thickness l, expressed in frequency units s^{-1}:

$$A_\nu = \int \frac{I_{\nu_0} - I_\nu}{I_{\nu_0}}\, \mathrm{d}\nu\,. \tag{5.59}$$

Since, in determining A_ν we are only required to measure the intensities integrated over a finite spectral range, the resolution of the spectroscopic device employed does not have to be high.

If the absorptivity of the absorption interval in itself is too large, this can be reduced by introducing a foreign gas into the absorption volume, thus causing a pressure broadening of the spectral line to be studied (see Füchtbauer and Gössler [5.113]). To ensure that the f-numbers are not altered due to the pressure broadening, it is recommended to measure these at several pressures of the foreign gas and, if necessary, to extrapolate them to zero foreign-gas pressure.

When the absorptivity in the center of the line is no longer small compared to 1 (in any case it should not exceed 90%), expression (5.58) no longer holds. If

* Exactly, the right hand side of formula (5.56) has to be multiplied with the factor $(1 - \mathrm{e}^{-h\nu/kT})$ accounting for induced emission. In most cases, this factor is practically 1.

the absorption line displays a Doppler profile, f_{mn} can be determined without the help of high-resolution spectrographs as follows: From the total absorption A_ν and the Doppler width

$$\Delta\nu_D = \frac{\nu}{c}\sqrt{\frac{2\,kT}{M}} \tag{5.60}$$

we calculate the constant C from the relation

$$\frac{A_\nu}{2\,\Delta\nu_D} = \frac{\sqrt{\pi}}{2}\,C\left[1 - \frac{C}{2!\sqrt{2}} + \frac{C^2}{3!\sqrt{3}} - \frac{C^3}{4!\sqrt{4}} + \cdots\right], \quad C < 3 \tag{5.61}$$

(see Unsöld [5.438], p. 290).

We finally obtain

$$f_{mn} = \frac{C\,m\,c\,\Delta\nu_D}{\sqrt{\pi}\,e^2\,n_n\,l}\,. \tag{5.62}$$

5.3.6. Determination of Oscillator Strengths of Resonance Lines by Absorption Measurements on Atomic Beams

An atomic beam, emitted by an atomic-beam oven, serves as the absorber. The principle of the absorption measurement is here the same as that discussed in the previous paragraph.

To determine the particle density n_0 in the beam, the beam is allowed to strike a collector plate and to condensate. If $G(t)$ is the weight increase and $\dot{G}$ its time derivative, we have for the product $n_0\,l$ of the absorbing layer (Kopfermann and Wessel [5.222])

$$n_0\,l = \frac{\dot{G}\,p}{\bar{v}_z\,\mathrm{d}M}\,. \tag{5.63}$$

Here l is the absorption path length across the beam, p a corrective factor, $\bar{v}_z$ the mean beam velocity in the beam direction, d the width of the atomic beam, and M the mass of the individual atoms. $\bar{v}_z$ is given by

$$\bar{v}_z = q\sqrt{\frac{8\,kT}{\pi\,M}} = q\,\bar{v}\,, \tag{5.64}$$

where q depends on the aperture ratio of the atomic beam (1:2.7) and allows for the fact that only the z-components appear in equation (5.64).

The optical layout of the experimental arrangement is given in Figure 5.22. This system has the peculiar feature that a source emitting an iron line (hollow iron cathode) is also used as a radiation source for the study of a beam of iron atoms. The transmitted radiation falls upon a monochromator whose exit slit is so wide, that the entire spectral line is let through by the slit, behind which a photo cell is placed. The total absorption is given by

$$A = \frac{I_0 - I}{I_0}\,, \tag{5.65}$$

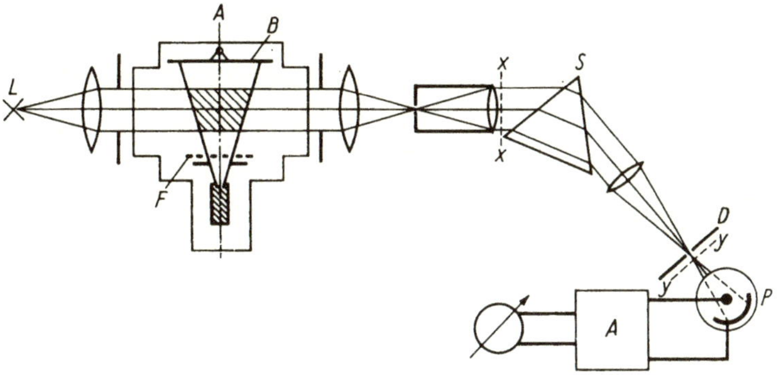

Figure 5.22. Device for determining the absorption oscillator strengths of resonance lines by using an atomic beam (after Kopfermann and Wessel [5.222])

L – light source, *A* – atomic beam apparatus, *B* – scale, *K* – cover, *S* – spectroscopic device, *D* – diaphragm, *P* – photocell, *A* – amplifier

where I_0 and I are the photocurrents for the stopped and transmitted atomic beam, respectively. Assuming a Doppler profile for the emitted as well as for the absorbed line, the constant C is given by (see [5.222])

$$A = C\,r\left[\frac{1}{\sqrt{1+r^2}} - \frac{C}{2!\sqrt{2+r^2}} + \frac{C^2}{3!\sqrt{3+r^2}} - \cdots\right] \tag{5.66}$$

with

$$r = \frac{\Delta\nu_{\mathrm{D}}\,(\text{source})}{\Delta\nu_{\mathrm{D}}\,(\text{absorber})}, \tag{5.67}$$

and f_{mn} is determined from C via (5.62). The Doppler width of the emission line is determined by means of a Fabry-Perot interferometer, while that of the absorption line is calculated from the data on atomic beams. Due to the relatively low temperatures in an atomic beam the levels are practically not thermally excited, and the f-determination is restricted to the resonance lines respectively lines terminating at the ground level.

The accuracy, with which absolute f-values can be obtained by this method, is estimated as 20%.

5.3.7 Oscillator Strengths from the Dispersion in the Vicinity of a Spectral Line

The methods stem from Roshdestvenskii [5.371].

According to the dispersion theory the following equations hold for the refractive index n and for the absorption coefficient k in the vicinity of a spectral line having a dispersion profile (see Unsöld [5.438], p. 272)

$$n - 1 = -\frac{\pi e^2 f_{mn} n_n}{2\,m\,\omega_0}\cdot\frac{2\,\Delta\omega}{(\Delta\omega)^2 + (\gamma/2)^2}, \qquad k = \frac{\pi e^2 f_{mn} n_n}{m\,c}\cdot\frac{\gamma}{(\Delta\omega)^2 + (\gamma/2)^2}. \tag{5.68}$$

The behavior of $n-1$ and k as a function of the circular frequency deviation $\Delta\omega$ from the line center is given in Figure 5.23. For large distances from the line center ($|\Delta\omega| \gg \gamma$) the relative trend of the $(n-1)$-curve is independent

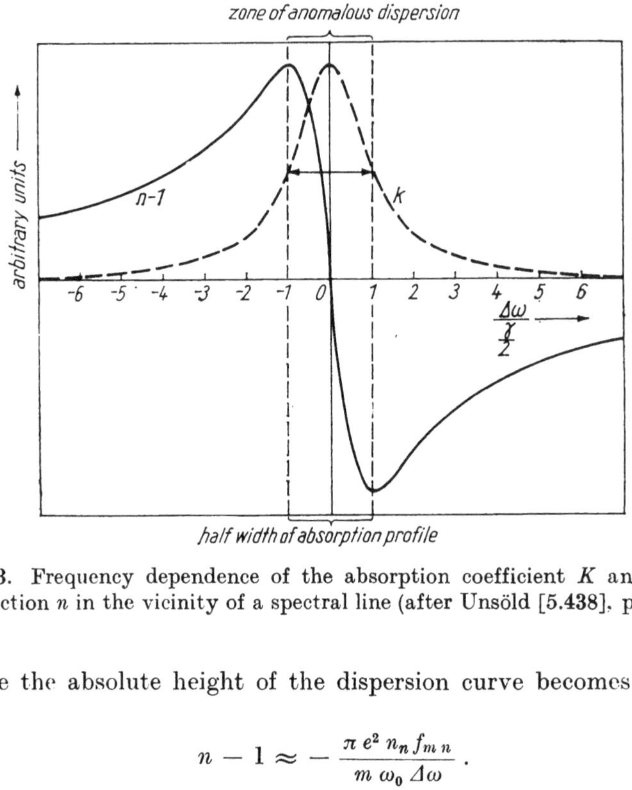

Figure 5.23. Frequency dependence of the absorption coefficient K and index of refraction n in the vicinity of a spectral line (after Unsöld [5.438], p. 273)

of γ, while the absolute height of the dispersion curve becomes proportional to $n_n f_{mn}$:

$$n - 1 \approx -\frac{\pi e^2 n_n f_{mn}}{m \omega_0 \Delta\omega}. \tag{5.69}$$

To determine $f_{mn} n_n$ an interferometric system can be employed in the manner shown in Figure 5.24 (see Ostrovskii and Penkin [5.325]). The light from the continuous radiation source S passes through the interferometer and the entrance slit of a high-resolution spectrograph. In one of the interferometer branches is placed the heated absorption tube A, which contains metal vapor. In the other branch there is an evacuated tube of the same length as well as a compensating plate for producing the "hooks" on the interferogram. Figure 5.25 (see [5.321]) illustrates an example of such an interferogram, on which the "hooks" can be clearly seen. (Hooks denote the maximum and minimum of the interference fringes to left and right, respectively, of the absorption line.) If Δ_{mn} denotes the distance between the hooks, which lie symmetrically with respect to the absorption line (Ostrovskii et all. [5.328]), we have

$$n_n f_{mn} = \frac{\pi m c^2 K_{mn}}{e^2 l \lambda_{mn}^3} \cdot \Delta_{mn}^2, \tag{5.70}$$

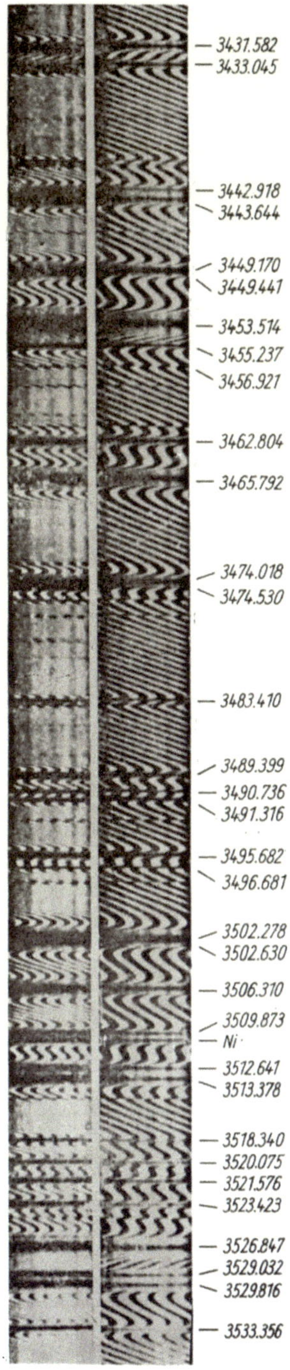

Figure 5.25. Spectro-interferograms with hooks in the neighborhood of Co I lines. Both photographs are distinguished by different oven temperatures and different thicknesses of the compensating plate (after Ostrovskii and Penkin [5.321])

where K_{mn} is a constant of the interferometer depending on the thickness of the compensating plate. l is the length of the absorption path. Knowing the number n_n of absorbing atoms per cm³ (n_n can be calculated from the oven temperature and the vapor pressure curve), the absolute oscillator strengths can be found with low precision from (5.70).

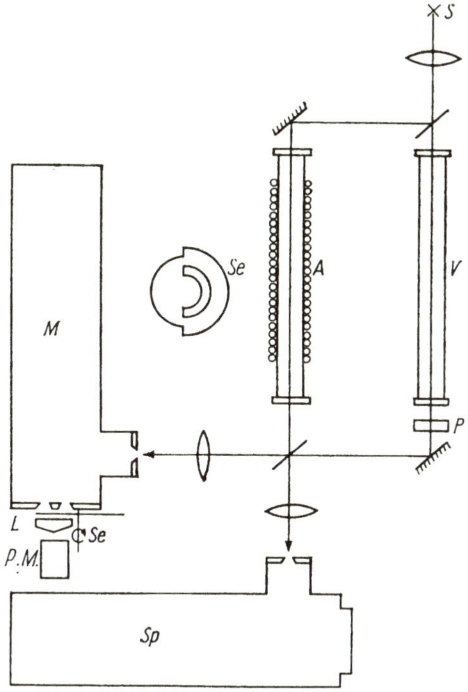

Figure 5.24. Optical layout for determining oscillator strengths (after Ostrovskii and Penkin [5.325])

S — source of continuous radiation, A — heated absorption tube, V — evacuated tube, P — plate for producing the "hooks", Sp — high-resolution spectrograph, M — monochromator, L — beam splitter, Se — rotating sector, $P.M.$ — secondary electron multiplier

According to another version of this method, developed by Ostrovskii, Penkin, and Shabanova [5.328] (see also [5.325]), the oscillator strengths can be directly determined without knowing the exact vapor pressure when, in addition, the total absorption A_λ is measured, which is defined by

$$A_\lambda = \int \frac{I_{\lambda_0} - I_\lambda}{I_{\lambda_0}} \, d\lambda \,, \tag{5.71}$$

where we must integrate over the spectral line. I_{λ_0} is the unabsorbed continuous radiation, I_λ is the radiation attenuated by absorption.

The spectral lines have a profile, which corresponds to a superposition of a Doppler distribution

$$\frac{1}{\sqrt{\pi}\,\Delta\omega_{\mathrm{D}}}\,\mathrm{e}^{-\left(\frac{\Delta\omega}{\Delta\omega_{\mathrm{D}}}\right)^2} \tag{5.72}$$

and a dispersion distribution

$$\frac{\gamma}{2\pi}\,\frac{1}{(\omega-\omega_0)^2+\left(\frac{\gamma}{2}\right)^2} \tag{5.73}$$

where the Doppler width $\Delta\omega_D$ is about 1000 times as large as the damping constant γ. If we observe the spectral line in an absorption spectrum, it then follows that for a small optical depth C in the line center the Doppler profile predominates. The total absorption is proportional to $n_n f_{m\,n}\, l$. For large values of C the total absorption at first increases with growing C only very slowly, until the absorption in the line wings, which depends on the dispersion profile, starts to become appreciable. For $A_\lambda/\Delta\lambda_{\mathrm{D}} \geq 10$ the absorption profile is practically only determined by the dispersion profile [5.321]. Then we have for the damping constant γ

$$n_n f_{m\,n}\, l\,\gamma = \frac{m\,c^3}{e^2\,\lambda^4}\,A_\lambda^2\,. \tag{5.74}$$

From (5.70) and (5.74) we finally obtain

$$\gamma = \frac{c}{\pi\,K_{m\,n}\,\lambda}\left(\frac{A_\lambda}{\Delta_{m\,n}}\right)^2. \tag{5.75}$$

We thus see that the damping constant now only involves the total absorption A_λ and the inter-hook distance $\Delta_{m\,n}$. Both quantities can be measured one after the other or simultaneously. In the latter case γ can be measured with an accuracy of 3% [5.325]. To determine $\Delta_{m\,n}$ and A_λ simultaneously, the interferometric arrangement (Figure 5.24) is supplemented by a monochromator with a double exit slit. One slit is centered on the line, the other on the undisturbed continuum in the vicinity of the line. The radiation passing through one slit is alternately screened by means of a rotating sector and is recorded via a photo-electric radiation detector. We have

$$A_\lambda = \frac{I_0 - I}{I_0}\,\Delta\lambda\,, \tag{5.76}$$

where I_0 and I respectively denote the photocurrents when the undisturbed continuous radiation and that attenuated by absorption fall upon the radiation detector. The wavelength interval $\Delta\lambda$ limited by the slit must be so large that it contains the full absorption line.

The oscillator strengths can be determined as follows from the damping constant γ.

The damping constant is found in accordance with

$$\gamma = \frac{1}{\tau_m} + \frac{1}{\tau_n} + \gamma_{\mathrm{st}} \tag{5.77}$$

from the natural lifetimes of the upper (m) and lower (n) levels as well as from the collision damping constant γ_{st}.

If the transition is studied in the ground state or in a metastable state, we have in the case of no collisions

$$\gamma = \frac{1}{\tau_m} = \sum_{n'} A_{n'm} \tag{5.78}$$

since $1/\tau_n = 0$ and $\gamma_{st} = 0$. Here, the summation is over all n'-values, for which there exist transitions connected with the radiation from the m-state.

If there exists only one transition from level m to level $n' = n$, then we directly have $\gamma = A_{nm}$, whence f_{mn} from equation (5.44). When more transitions exist, then in addition to γ the relative transition probabilities of the lines studied must be known or measured. This can be done, for example, by means of the simple hook method with the help of equations (5.70) and (5.44).

When applying the above method the following conditions must be fulfilled:

1. the quantity $n_n f_{mn} l$ must be sufficiently large, so that the condition $A_\lambda/\Delta\lambda_D \geq 10$ is satisfied.

2. the total concentration n_0 of the metal atoms and the pressure of the residual gas must be sufficiently small, so that the term γ_{st} in (5.77) can be neglected.

In [5.325] it was shown that for an absorption tube length from $l = 50$ cm to 60 cm conditions 1 and 2 can be satisfied for the case where resonance lines of alkaline earth and alkali atoms are studied. This has been verified by determinations of the f-numbers of the resonance lines of Mg I, Ca I, Sr I, Ba I, and Na I by the above described method.

5.3.8. Oscillator Strengths from the Broadening of the Resonance Level

For a sufficiently high atom density the width of the resonance level is appreciable and can exceed the natural linewidth as well as the Doppler width. The resonance broadening depends in accordance with

$$\Delta\omega_{Res} = \frac{6 f e^2 n_0}{m \omega_0} \tag{5.79}$$

(see Fursov, Oganov, and Striganov [5.115]) on the oscillator strength f of the resonance line and the concentration n_0 of the atoms in the ground state of that atomic species, which emits the corresponding resonance line, i.e., foreign atoms are not counted.

The lines corresponding to transitions to a resonance level, in general have the same width (when expressed in frequency units) as the corresponding resonance line, when the pressure is not too low and the degree of ionization is not to high, since the higher level width of the former lines can be neglected compared too the width of the resonance level.

Fursov, Oganov, and Striganov [5.115] applied this method for the determination of the f-number of the argon resonance line $({}^1S_0 - {}^1P_1)$ $\lambda = 1049$ Å.

First the widths, among others, of the lines 4510.7 and 4333.5 Å with transitions $1\,^1P_1 - 3\,^3P_0$ and $1\,^1P_1 - 3\,^3P_2$, which have the resonance level $1\,^1P_1$ as the lower level, were determined for an induced weak-current discharge in argon at pressures between 0.1 and 186 Torr. The widths are of the order of 0.1 Å. The widths of the lines are measured by means of a Fabry-Perot interferometer. n_0 is calculated in accordance with the fixed pressure and an estimated temperature of 1000 °K.

Since there is still a term independent of n_0, which contributes to the line width but is not contained in equation (5.79), equation (5.79) is differentiated with respect to n_0 and used for the f-determination. In this manner a value of $f = 0.5$ was found.

Possible errors besides the measurement inaccuracy originate in the uncertainty in the numerical factor of equation (5.79) as well as in an inaccurate knowledge of the gas temperature.

5.3.9. Transition Probabilities from the Lifetime of Excited Levels

In the method developed by Demtröder [5.74] (see also [5.54]) an atomic beam is produced in an oven. This is irradiated with the high frequency modulated light, emitted by a hollow cathode of the same element. The atomic beam absorbs part of the radiation, and, amongst others, a level m is excited. The fluorescent light, emitted by the excited atoms, is again modulated by the same frequency, but with a phase shift φ. If the time dependence of the primary radiation is given by

$$I_p = C_1\,(1 + a \sin \omega\, t) \tag{5.80}$$

with the modulation factor $a \leq 1$ and modulation frequency $\nu = \frac{\omega}{2\,\pi}$, the secondary radiation is described by

$$I_s = C_2\left(1 + \frac{a}{\sqrt{1 + \omega^2\,\tau_m^2}}\right) \sin\,(\omega\, t - \varphi)\,. \tag{5.81}$$

The phase shift φ between both intensities can be measured with high precision by suitable electronic devices, and is connected with the mean lifetime of the level m via the expression

$$\tan \varphi = \omega\, \tau_m \tag{5.82}$$

from which τ_m can be determined.

We assume that collision damping of the spectral lines in question can be neglected. When several transitions are possible from level m to a lower level n', the transition probabilities $A_{n'm}$, where $n' < m$, are related to the lifetime τ_m in accordance with expression (5.78). If only one transition from m is possible — the one to n — we have

$$\frac{1}{\tau_m} = A_{nm}\,. \tag{5.83}$$

Demtröder modulated the primary beam by means of a standing ultra shock wave. The fluorescence is measured normal to the incident radiation. The transition probabilities and oscillator strengths of transitions to the ground level were determined via the lifetime of the first excited level of neutral Ga, Al, Mg, Tl, and Na. The measurement accuracy was stated as 1—3%. Table 5.7 gives a list of the measurement results.

Table 5.7. Measurement results of Demtröder [5.74] for the oscillator strengths of resonance lines of certain elements

Element	τ_m [10^{-8} s]	Error [%]	Transition	λ [Å] [10^7 s^{-1}]	A_{nm}	f-value	Total f-value
Ga	0.988	1.5	$4\ P_{1/2} - 5\ S_{1/2}$	4032.98	3.60	0.0878	
			$4\ P_{3/2} - 5\ S_{1/2}$	4172.06	6.54	0.0854	0.086
Al	0.643	1.8	$3\ P_{1/2} - 4\ S_{1/2}$	3944.03	5.24	0.122	
			$3\ P_{3/2} - 4\ S_{1/2}$	3961.54	10.30	0.121	0.122
Tl	0.866	3	$6\ P_{1/2} - 7\ S_{1/2}$	3775.72	6.04	0.128	
			$6\ P_{3/2} - 7\ S_{1/2}$	5350.46	5.52	0.118	0.121
Mg	0.329	5	$3\ S_0 \quad -3\ P_1$	2852.13	30.4	1.11	1.11
Na	1.59	1	$3\ S_{1/2} - 3\ P_{1/2}$	5895.93	6.29	0.328	
			$3\ S_{1/2} - 3\ P_{3/2}$	5889.96	6.29	0.654	0.982

The high measurement accuracy lies herein that the phase shift of the high frequency signal is not measured directly, but only after conversion to lower frequency signals by means of a mixer.*

Ziock [5.474], who effected the light modulation via a Kerr cell and directly measured the phaseshift of the high frequency signals, indicated a far greater error (30%) in the transition probability of the iron resonance line 3720 Å, which he measured: $A = 1.8 \cdot 10^7\ \mathrm{s}^{-1} \pm 30\%$.

Heron et al. ([5.165], [5.166]) produced excitation in the $3\ ^1P$, $4^3\ S$, $3^3\ P$, $4^3\ P$, and $3^1\ D$ levels in helium by irradiation with high-speed pulsed 30 eV electron beams (pulse duration of the triangular pulse $= 4 \cdot 10^{-8}$ s, pulse frequency = 10 kHz). The decay of these levels, which follows the exponential law

$$n_m = n_{m\,0}\ \mathrm{e}^{-t/\tau_m} \tag{5.84}$$

is recorded via multiplier-intensity measurements of the lines emitted by the level in question. The precision is increased by using a coincidence circuit, which enables the number of coincidences Z between the irradiating electron pulses and the multiplier pulses to be determined as a function of a variable time lag τ_v of the electron pulse. The lifetime τ_m of the level m can be obtained from the $Z = f(\tau_v)$ curve. The transition probabilities of the lines corresponding

* According to Karstensen [5.189], the high accuracy given by Demtröder is only fictitious, because life time measurements are affected by self absorption or radiation trapping (see Holstein [5.173b], [5.173c]), so that the measured values for the lifetime τ are in excess of the true values. With respect to the resonance levels of Na I, Karstensen was able to show that the τ-values given by Demtröder are at least 11% in excess of the true values.

to transitions from the m-level can now be determined from the lifetime in the manner shown in the previous sections.

The application of this method presupposes a low gas density, so that 1. a small optical depth of the emitted lines is ensured and 2. the collision broadening is small compared to the natural line broadening.

The following table lists the lifetimes of helium levels measured by Heron et al, together with calculated life time datas: (see also Nicholls and Stewart [5.297]). From the data a minimum error of 1.5% for the measured lifetime can be deduced.

Table 5.8. Lifetime of helium levels (τ in 10^{-8} s)

He-level	τ (measured)	τ, calculated [5.21]
3^1 P	7.4 $\pm$ 0.1	7.4
4^3 S	6.75 $\pm$ 0.10	6.4
3^3 P	11.5 $\pm$ 0.5	9.7
4^3 P	15.3 $\pm$ 0.2	13.8
3^3 D	1.0 $\pm$ 0.5	1.39

Brannen et al. ([5.52], [5.53]) determined the lifetime of a level p by measuring the coincidences Z between photons, emitted by the cascade transitions $n \rightarrow p$ and $p \rightarrow m$, when the multiplier signal from the higher transition $n \rightarrow p$ is measurably delayed (time lag $= \tau_v$). The lifetime τ_p of the mean p-level is derived from the slope of the curve $\ln Z = f(\tau_v)$, which is $- 1/\tau_p$. Brannen et al. in this way obtained the value $(1.12 + 0.02) \cdot 10^{-8}$ s for the lifetime of the $7\,^3S_1$ level of mercury. In these experiments, the upper transition corresponds to photorecombination, while the lower transition ($7\,^3S_1 - 6\,^3P_1$) corresponds to the emission of the spectral line 4358.34 Å. A disadvantage is that the radiation from the upper and lower transition, respectively, can fall into two different spectral ranges.

The individual transition probabilities of the lines emitted from the $7\,^3S_1$-level can be calculated in accordance with the methods discussed in section 5.3.7. (That is, the individual transition probabilities result from the lifetime and from relative transition probabilities obtained in a different fashion).

An indirect method for the determination of the life time of excited levels is given by Hanle (see [5.145c], [5.145d], [5.280] chapter V and [5.470g]). He found that the plane of polarization of resonance radiation reemitted by the atoms in a resonance lamp rotates when the lamp is placed in a magnetic field, and that the angle of rotation depends from the magnetic field strength H, from the sought life time τ and from a factor depending from the magnetic splitting constant, but not from the geometry of the radiating volume. Thus, from a measurement of rotation angle the life time of the level under investigation can be derived. See, for details, the literature cited above.

With the beam foil method, life times of spectral levels of atoms as well as of single or multiply charged ions can be determined [5.23b].

5.3.10. Further Methods for the Experimental Determination of Transition Probabilities

It is possible to find the transition probability on the basis of the natural linewidth $\Delta\nu_n$. The latter is defined by the lifetime of the upper and lower level in accordance with the expression $\Delta\nu_n = \gamma/2\pi = \frac{1}{2\pi}\left(\frac{1}{\tau_n} + \frac{1}{\tau_m}\right)$. Under favorable conditions the transition probabilities can be determined on the basis of this expression and the statements made in section 5.3.7.

It should be further mentioned, that the rotation of the plane of polarization of light for a frequency in the vicinity of an absorption line can be used to determine the oscillator strength of this line. Stephenson [5.412] employed this method, which was first developed by Weingeroff [5.453] (see also [5.280] p. 133), to find the oscillator strengths of the resonance lines of Na, K, Li, Rb, and Tl. In this method it is not required that the vapor pressure of the metal be known.

Geiger [5.121j] has determined oscillator strengths of lines emitted from Ne, Ar, and Kr from the differential cross sections corresponding to the excitation of optical transitions for the case when an electron ray of 25 keV particle energy is transmitted through the scattering volume.

5.3.11. Behavior of Oscillator Strengths at High Gas Pressures

Earlier measurements of the total absorption coefficient of the line Hg 2537 Å in the case when the absorption was effected by saturated mercury vapor which together with the admixture was under a high pressure (up to 1300 atm), led to the conclusion that oscillator strengths can be influenced by high pressures. However, Michels et al. [5.271] were able to show that much of the influence of pressure on oscillator strength, as determined from the absorption measurements (see section 5.3.5.) is only fictitious. The reason is that previously it was not payd attention to the fact that the saturation pressure of mercury which is at low pressure a function of temperature only depends in the region of high pressure explicitely on the foreign gas pressure.

Strijland and Nanassy [5.416] by using improved experimental methods, found no indication of a pressure influence on the oscillator strength at all.

Unlike neutral particles, charged particles when appearing in higher concentration can have a pronounced effect on the oscillator strength respectively transition probability. The existence of this effect is indicated by the appearcence of forbidden spectral lines due to the Stark effect caused by the charged plasma particles (see [5.133d], [5.139a] and section 10.7).

In the case of allowed lines, the influence of the microfield on the transition probability [5.133d] is hardly detectable because this effect on line intensity is pronounced only for spectral lines with excitation energy near the ionization limit, and in this case this effect is in general surpassed by the effect of line merging (overlapping of Stark-effect broadened spectral lines with high series members; see [5.181] and section 10.8.1.) in which case intensity measurements are impossible, and by the effect of statistical microfield ionization of high spectral levels, resulting in a pronounced decrease of line intensity. See, for these effects, sections 10.8.2. and 10.8.3.

5.3.12. Calculation of Transition Probabilities

The transition probability A_{nm} and the oscillator strength f_{mn} are respectively connected with the line strength S_{nm} as follows (see [5.438], page 336):

$$A_{nm} = \frac{64\,\pi^2\,\nu^3}{3\,h\,c^3\,g_m}\,S_{nm} = \frac{2.026 \cdot 10^{-6}}{\lambda^3\,g_m}\,\frac{S_{nm}}{e^2\,a_0^2}, \tag{5.85}$$

$$f_{mn} = \frac{8\,\pi^2\,m\,\nu}{3\,h\,e^2\,g_n}\,S_{nm} = \frac{3.037 \cdot 10^{-6}}{\lambda\,g_n}\,\frac{S_{nm}}{e^2\,a_0^2}. \tag{5.86}$$

On the right-hand side of both equations the line strength is expressed in atomic units $e^2\,a_0^2 = 6.457 \cdot 10^{-36}$ electrostatic cgs-units.

The line strength S_{nm} is symmetric with respect to an interchange of the levels m and n.

If the levels m and n split into several individual states, we have

$$S_{nm} = \sum_{i,j} S_{ji}\,. \tag{5.87}$$

Here S_{ji} is the line strength for a component line of the transition $n - m$, which is given by the matrix element [5.297]

$$S_{ji} = |(i\,|P^e|\,j)|^2 \tag{5.88}$$

where

$$(i\,|P^e|\,j) = \int \psi_i^*\,P^e\,\psi_j\,\mathrm{d}\tau\,. \tag{5.89}$$

The dipole moment P^e is defined by

$$P^e = -\sum_s e\,\boldsymbol{r}_s\,, \tag{5.90}$$

where $\boldsymbol{r}_s$ denotes the coordinate of the electron s.

To calculate the transition probabilities we must first solve the Schrödinger equation (see [5.425])

$$H\,\psi = \left(-\frac{1}{2}\sum_{i=1}^{N} \nabla_i^2 + V(r_i)\right)\psi = E\,\psi \tag{5.91}$$

where

$$V(r_i) = -\frac{e^2\,Z}{r_i} + \sum_{\substack{k=1\\k\neq i}}^{N} \frac{e^2}{r_{ik}}\,. \tag{5.92}$$

The potential is made up of the potential of the nucleus and the potential due to the rest electrons at the point where the electron under investigation is located.

With the eigenfunctions ψ_i, ψ_j we can now compute the dipole length matrix

$$M^l_{ij} = \int \psi_i^* \sum_s \boldsymbol{r}'_s \psi_j \, \mathrm{d}\tau' \tag{5.93}$$

which can be replaced by the dipole velocity matrix

$$M^v_{ij} = \frac{\hbar^2}{m\, E_{n\,m}} \int \psi_i^* \sum_s \nabla'_s \psi_j \, \mathrm{d}\tau' \tag{5.94}$$

or by the dipole acceleration matrix

$$M^a_{ij} = \frac{\hbar^2}{m\, E^2_{n\,m}} \int \psi_i^* \left(\sum_s \nabla'_s V \right) \psi_j \, \mathrm{d}\tau' \,. \tag{5.95}$$

In most cases one must be satisfied with approximate wave functions ψ_i and ψ_j. As a measure for the goodness of the approximation the energy of the level is usually taken. For the true eigenfunction this energy assumes an extremum. Of course, the energy in the minimum range is rather insensitive to errors in the wave function.

As a further control may serve the more or less good agreement between expressions (5.93) to (5.95). Which of these expressions yields the best value for the oscillator strength depends on the range of the radial coordinate which gives the best approximation of the computed wave function to the true wave function. For wave functions, which are calculated in accordance with the variational principle

$$\int \psi^* H \psi \, \mathrm{d}\tau = \text{extremum} \,, \qquad \int \psi^* \psi \, \mathrm{d}\tau = 1 \tag{5.96}$$

it is most suitable to carry out the computation via the dipole velocity [5.425].

Accurate eigenfunctions can be obtained for hydrogen and comparable ions as well as for helium (see Hylleraas [5.179]) and helium-like ions, including the negative ion H^- (see Chandrasekhar [5.63]).

For systems with more than two electrons one-electron approximations are used to compute the wave functions.

One of the most important methods for calculating the eigenfunctions is the self-consistent field method (see Hartree [5.149]). The method lies herein, that each electron (i) is considered independently in the field of the nucleus and the remaining electrons, whereby the latter field is averaged over the spherical shells in such a way that it becomes spherically symmetric.

In order to compute the wave function of the ith electron, approximate functions are used for the other electrons. Carrying out the computation for all electrons, we obtain first approximations for all wavefunctions. These yield a first approximation for the self-consistent field, with which the wave functions can be calculated to a second approximation.

The method has in the meantime been improved. Fock [5.104] took into account the exchange effects between electrons having the same spin. Trefftz, Schlüter, Dettmar, and Jörgens [5.427] introduced further correlations between the electrons into the calculation. Biermann and Trefftz [5.38] allowed for the polarization of the other electrons due to the excited electron. In [5.422] the configuration interaction is included.

The error in the transition probabilities for lighter atoms calculated by the self-consistent field method seems, as comparison with experiment has shown, to be about +20%.

As an example we give in Table 5.9 a comparison between measured and computed transition probabilities of Ar II lines (see Table 5 of Olsen [5.28]).

The line intensity can be calculated in a relatively simple way when the field acting on the emission electron is approximated as a Coulomb field (Bates and Damgaard [5.21]). Reliable results on line strengths can be obtained by this method only for transitions within the same multiplet system. Further the upper and lower level must possess the same parent configuration.

Assuming Russel-Saunders coupling, the line strength S can be split into three factors

$$S = S(M) \cdot S(L) \cdot \sigma^2 . \tag{5.97}$$

The first factor $S(M)$ gives the relative strength of the multiplet, the second factor $S(L)$ the relative strength of the line within the multiplet.

$S(M)$ depends on 1. the number k of equivalent eelctrons, 2. the multiplicity, 3. the angular momentum quantum numbers of the individual electrons, and 4. the orbital angular momentum quantum numbers L of the initial and final level. For systems of up to 5 equivalent electrons we list $S(M)$ in the following Table 5.10 (after Allen [5.8]). Further data can be found in [5.127] and [5.128] (Goldberg).

Table 5.9. Transition probabilities of some A II lines. $A_{nm\,\mathrm{exp.}}$: experimentally obtained averaged transition probability (Olsen [5.304]; $A_{nm\,\mathrm{theoret.}}$: transition probability computed by Garstang [5.118])

λ [Å]	A_{nm} exp. [$10^7\,\mathrm{s}^{-1}$]	A_{nm} theoret. [$10^7\,\mathrm{s}^{-1}$]
4348.11	11.5	12.8
4379.74	11.6	11.2
4579.39	7.44	0.10*
4589.93	6.40	10.0
4609.60	9.06	10.7
4637.25	0.64	0.81
4657.94	6.95	0.11*
4764.89	5.40	7.09
4806.90	7.86	7.88
4847.90	0.95	0.81
4879.90	6.59	7.71
4933.24	1.60	2.41
5009.35	1.70	1.39
5062.07	2.20	2.81

Note. The values designated by * were indicated by Garstang as especially sensitive to the choice of the parameter.

Table 5.10. Relative multiplet strengths for systems with 1 to 5 equivalent electrons (after Allen [5.8])

1 electron

s / p	^2P
^2S	6

p / d	^2D
^2P	60

d / f	^2F
^2D	210

f / g	^2G
^2F	504

g / h	^2H
^2G	990

2 electrons

s·s / sp	^3P	^1P
^3S	9	
^1S		3

s^2 / sp	^1P
^1S	6

sp / p·p	^3D	^3P	^3S	^1D	^1P	^1S
^3P	15	9	3			
^1P				5	3	1

sp / p^2	^3P	^1D	^1S
^3P	18		
^1P		10	2

sp / sd	^3D	^1D
^3P	90	
^1P		30

sd / sf	^3F	^1F
^3D	315	
^1D		105

sd / pd	^3F	^3D	^3P	^1F	^1D	^1P
^3D	21	15	9			
^1D				7	5	3

sf / sg	^3G	^1G
^3F	756	
^1F		252

p·p / pd	^3F	^3D	^3P	^1F	^1D	^1P
^3D	126	22·5	1·5			
^3P		67·5	22·5			
^3S			30			
^1D				42	7·5	0·5
^1P					22·5	7·5
^1S						10

p^2 / pd	^3D	^3P	^1F	^1D	^1P
^3P	135	45			
^1D			84	15	1
^1S					20

pd / pf	^3G	^3F	^3D	^1G	^1F	^1D
^3F	405	35	1			
^3D		280	35			
^3P			189			
^1F				135	11·7	0·3
^1D					93·3	11·7
^1P						63

pd / d·d	^3G	^3F	^3D	^3P	^3S	^1G	^1F	^1D	^1P	^1S
^3F	162	42	6							
^3D		84	52·5	13·5						
^3P			31·5	40·5	18					
^1F						54	14	2		
^1D							28	17·5	4·5	
^1P								10·5	13·5	6

pd / d^2	^3F	^3P	^1G	^1D	^1S
^3F	84				
^3D	168	27			
^3P		81			
^1F			108	4	
^1D				35	
^1P				21	12

3 electrons

s·sp / sp·p	^4D	^4P	^4S	^2D	^2P	^2S	^2D	^2P	^2S
^4P	20	12	4						
^2P				10	6	2			
^2P							10	6	2

sp·p / spd	^4F	^4D	^4P	^2F	^2D	^2P	^2F	^2D	^2P
^4D	168	30	2						
^4P		90	30						
^4S			40						
^2D				84	15	1			
^2P					45	15			
^2S						20			
^2D							84	15	1
^2P								45	15
^2S									20

s^2d / spd	^2F	^2D	^2P
^2D	28	20	12

s·sd / spd	^4F	^4D	^4P	^2F	^2D	^2P	^2F	^2D	^2P
^4D	28	20	12						
^2D				14	10	6			
^2D							14	10	6

s^2p / sp^2	^2D	^2P	^2S
^2P	10	18	2

Table 5.10 (cont'd). Relative multiplet strengths

3 electrons

sp^2 \ $p.p^2$	^4D	^4P	^4S	^2F	^2D	^2P	^2D	^2P	^2S	^2P
^4P	20	12	4							
^2D				14	10	6				
^2P							10	6	2	
^2S										6

sp^2 \ p^3	^4S	^2D	^2P
^4P	12		
^2D		15	5
^2P		15	9
^2S			4

spd \ $sd.d$	^4G	^4F	^4D	^4P	^4S	^2G	^2F	^2D	^2P	^2S	^2G	^2F	^2D	^2P	^2S
^4F	216	56	8												
^4D		112	70	18											
^4P			42	54	24										
^2F						108	28	4							
^2D							56	35	9						
^2P								21	27	12					
^2F											108	28	4		
^2D												56	35	9	
^2P													21	27	12

spd \ sd^2	^4F	^4P	^2G	^2F	^2D	^2P	^2S
^4F	112						
^4D	224	36					
^4P		108					
$2\,^2F$			216	56	8		
$2\,^2D$				112	70	18	
$2\,^2P$					42	54	24

sd^2 \ pd^2	^4G	^4F	^4D	^2G	^2F	^2D	^4D	^4P	^4S	^2D	^2P	^2S	^2H	^2G	^2F	^2F	^2D	^2P	^2P
^4F	36	28	20																
^2F				18	14	10													
^4P							20	12	4										
^2P										10	6	2							
^2G													22	18	14				
^2D																14	10	6	
^2S																			6

spd \ p^2d	^2G	^2F	^2D	^2P	^2S	^4F	^4D	^4P	^2F	^2D	^2P	^2F	^2D	^2P
^2F	36	9·3	1·3											
^2D		18·7	11·7	3										
^2P			7	9	4									
^4F						37·3	18·7							
^4D						18·7	3·3	18						
^4P							18	6						
^2F									18·7	9·3				
^2D									9·3	1·7	9			
^2P										9	3			
^2D												9·3	6·7	4

$p^2.p$ \ p^2d	^2G	^2F	^2D	^2P	^2S	^4F	^4D	^4P	^2F	^2D	^2P	^2D
^2F	108	28	4									
^2D		56	35	9								
^2P			21	27	12							
^4D						168	30	2				
^4P							90	30				
^4S								40				
^2D									84	15	1	
^2P										45	15	
^2S											20	
^2P												60

p^2d \ p^2f	^4G	^4F	^4D	^2G	^2F	^2D	^2H	^2G	^2F	^2D	^2P	^2F
^4F	540	47	1·3									
^4D		373	47									
^4P			252									
^2F				270	23	0·7						
^2D					187	23						
^2P						126						
^2G							330	45	3			
^2F								225	63	6		
^2D									144	60	6	
^2P										84	42	
^2S											42	
^2D												210

p^2d \ pd^2	^4G	^4F	$2\,^4D$	^4P	^4S	^2H	$2\,^2G$	$3\,^2F$	$2\,^2D$	$3\,^2P$	$2\,^2S$
^4F	108	233	51								
^4D		75	133	72							
^4P			216	0	72						
^2G						132	108	12			
$2\,^2F$							432	112	58		
$3\,^2D$								338	200	82	
$2\,^2P$									102	108	18
^2S										38	0

p^3 \ p^2d	^4P	$2\,^2F$	$3\,^2D$	$2\,^2P$	^2S
^4S	120				
^2D		210	75	15	
^2P			125	45	10

Table 5.10 (cont'd). Relative multiplet strengths

3 electrons

$d.d^2$ – pd^2

pd^2	$d.d^2$ upper terms	Strengths
^2H	^2I ^2H ^2G	156 53 11
^2G	^2H ^2G ^2F	79 69 32
^2F	^2G ^2F ^2D	28 52 60
^4G	^4H ^4G ^4F	264 81 15
^4F	^4G ^4F ^4D	135 105 40
^4D	^4F ^4D ^4P	48 80 72
^2G	^2H ^2G ^2F	132 40 8
^2F	^2G ^2F ^2D	68 52 20
^2D	^2F ^2D ^2P	24 40 36
^4D	^4F ^4D ^4P	168 30 2
^4P	^4D ^4P	90 30
^4S	^4P	40
^2F	^2G ^2F ^2D	108 28 4
^2D	^2F ^2D ^2P	56 35 9
^2P	^2D ^2P ^2S	21 27 72
^2D	^2F ^2D ^2P	84 15 1
^2P	^2D ^2P	45 15
^2S	^2P	20
^2P	^2D	60

d^3 – pd^2

pd^2	d^3 upper terms	Strengths
^4G	^4F	36
^4F	^4F	252
2^4D	^4F ^4P	216 105
^4P	^4P	48
^4S	^4P	64
^2H	^2H ^2G	79 9
2^2G	^2H ^2G ^2F	317 115 36
3^2F	^2G ^2F 2^2D	200 84 94
3^2D	^2F 2^2D ^2P	132 125 43
3^2P	2^2D ^2P	141 51
^2S	^2P	14

4 electrons

sp^3 – s^2p^2

s^2p^2	sp^3 upper terms	Strengths
^3P	^3D ^3P ^3S	15 9 12
^1D	^1D ^1P	15 5
^1S	^1P	4

spd^2 – s^2d^2

s^2d^2	spd^2 upper terms	Strengths
^3F	^3G ^3F ^3D	54 42 30
^3P	^3D ^3P ^3S	30 18 6
^1G	^1H ^1G ^1F	22 18 14
^1D	^1F ^1D ^1P	14 10 6
^1S	^1P	6

$p \cdot p^3$ – sp^3

sp^3	$p \cdot p^3$ upper terms	Strengths
^3D	^3F ^3D ^3P	21 15 9
^1D	^1F ^1D ^1P	7 5 3
^3P	^3D ^3P ^3S	15 9 3
^1P	^1D ^1P ^1S	5 3 1
^5S	^5P	15
^3S	^3P	9

p^4 – sp^3

sp^3	p^4 upper terms	Strengths
^3D	^3P	15
^3P	^3P	9
^3S	^3P	12
^1D	^1D ^1S	15
^1P	^1D ^1S	5 4

pd^3 – sd^3

sd^3	pd^3 upper terms	Strengths
^5F	^5G ^5F ^5D	45 35 25
^3F	^3G ^3F ^3D	27 21 15
^5P	^5D ^5P ^5S	25 15 5
^3P	^3D ^3P ^3S	15 9 3
^3H	^3I ^3H ^3G	39 33 27
^1H	^1I ^1H ^1G	13 11 9
^3G	^3H ^3G ^3F	33 27 21
^1G	^1H ^1G ^1F	11 9 7

pd^3 – sd^3

sd^3	pd^3 upper terms	Strengths
^3F	^3G ^3F ^3D	27 21 15
^1F	^1G ^1F ^1D	9 7 5
2^3D	2^3F 2^3D 2^3P	42 30 18
2^1D	2^1F 2^1D 2^1P	14 10 6
^3P	^3D ^3P ^3S	15 9 3
^1P	^1D ^1P ^1S	5 3 1

5 electrons

spd^3 – s^2d^3

s^2d^3	spd^3 upper terms	Strengths
^4F	^4G ^4F ^4D	72 56 40
^4P	^4D ^4P ^4S	40 24 8
^2H	^2I ^2H ^2G	52 44 36
^2G	^2H ^2G ^2F	44 36 28
^2F	^2G ^2F ^2D	36 28 20
2^2D	2^2F 2^2D 2^2P	56 40 24
^2P	^2D ^2P ^2S	20 12 4

p^5 – sp^4

sp^4	p^5 upper term	Strengths
^2D	^2P	10
^2P	^2P	18
^2S	^2P	2

p^4d – p^5

p^5	p^4d upper terms	Strengths
^2P	3^2D 2^2P ^2S	190 90 20

The relative line strength $S(L)$ within a multiplet, which is tabulated among others, also by Allen [5.8], depends on the orbital angular momentum quantum numbers L of the initial and final state, on the multiplicity, and on the quantum numbers J for the total angular momentum of the initial and final state. We list $S(L)$ in Table 5.11, where we have restricted ourselves to a formula representation (see Sobelman [5.403 a], p. 372).

Table 5.11. Relative line strength $S(L)$ within a multiplet. The formulas are still valid when the direction of the arrows for the transitions is reversed

Transition	Relative line strength
$L \to L \quad J \to J$	$\frac{\{L(L+1)+J(J+1)-S(S+1)\}^2(2S+1)}{4L(L+1)(2L+1)J(J+1)(2S+1)}$
$L \to L-1 \quad J \to J$	$\frac{(S+L+J+1)(L+J-S)(S+L-J)(S+J-L+1)(2J+1}{4J(J+1)(2L-1)(2L+1)L(2S+1)}$
$L \to L \quad J \to J-1$	$\frac{(S+L+J+1)(L+J-S)(S+J-L)(L+S+1-J)}{4L(L+1)(2L+1)J(2S+1)}$
$L \to L-1 \quad J \to J-1$	$\frac{(L+S+J)(L+S+J+1)(L+J-S-1)(L+J-S)}{4(2L-1)L(2L+1)J(2S+1)}$
$L \to L+1 \quad J \to J-1$	$\frac{(S+J-L-1)(S+J-L)(S+L-J+1)(S+L-J+2)}{4(2L+1)(L+1)(2L+3)J(2S+1)}$

The data on $S(L)$ enable us to calculate the intensity ratios of lines inside a multiplet and in certain cases to compare these with measured intensity ratios. This is, among others, of significance in connection with checking a plasma for low self-absorption (see section 2.12.2).

The last factor in (5.97) is defined by

$$\sigma^2 = \frac{1}{(4l^2-1)}\left(\int_0^\infty R\,R'\,r\,\mathrm{d}r\right)^2, \tag{5.98}$$

where R and R' are the radial components of the wave function of the initial and final state, respectively.

σ can be represented in the form

$$\sigma = \frac{1}{Z}\,\mathcal{F}(n_l^*, l)\,\zeta(n_{l-1}^*, n_l^*, l)\,. \tag{5.99}$$

Z is the charge of the rest ion (for atomic spectra $Z = 1$, for spectra of simply charged ions $Z = 2$, etc.). l is the greater of the two orbital quantum numbers, which differ by 1, and n^* is the effective principal quantum number, defined by

$$n^* = \frac{Z}{(U-U')^{1/2}}, \tag{5.100}$$

where U and U' are the ionization and excitation energy, respectively, in Rydberg units ($R_{\mathrm{H}} = 109{,}678.76\ \mathrm{cm}^{-1}$).

$\mathcal{F}(n_l^*, l)$ and $\zeta(n_{l-1}^*, n_l^*, l)$ are tabulated by Bates and Damgaard [5.21] for l-values of 1, 2, and 3. n_l^* takes on values from 2 to 7, n_{l-1}^* from -4 to $+2.0$ (for $l = 2$ and 3) and from -4 to $+3.9$ (for $l = 1$).

The calculation of line strengths on the basis of the above table involves no difficulties. The accuracy with which the line strengths can be determined by this method practically does not differ from that with which the line strengths are found by the self-consistent field method. The table below, taken from a work of Mastrup and Wiese [5.266], serves as an example of the good agreement between measured oscillator strengths and those calculated after the method of Bates and Damgaard.

Table 5.12. Comparison of measured [5.266] and calculated [5.21] (Bates and Damgaard) oscillator strengths of N II lines

Multiplet	Line Å	f_{meas}	f_{cal}
$3\,s\,^3P_0 - 3\,p\,^3D$	5679.7	0.43	0.36
	5666.6	0.35	0.32
	5676.0	0.45	0.42
	5710.8	0.059	0.063
	5686.2	0.11	0.103
$3\,s\,^3P_0 - 3\,p\,^3P$	4630.5	0.29	0.24
	4613.9	0.067	0.080
	4643.1	0.10	0.079
	4621.4	0.10	0.106
	4601.5	0.15	0.133
	4607.2	0.35	0.32

When applying the Coulomb approximation for the calculation of transition probabilities in the case of more than one equivalent electron, then in the general case the configuration interaction has to be taken into account (Norman [5.302 a]), in order to avoid distinct deviations between calculated and measured transition probability data.

5.3.13. Reference Data on Transition Probabilities

We shall now briefly outline some of the relevant works. Biermann [5.35] compiled important transition probabilities up to 1949; Allen [5.8], as well, tabulated transition probabilities important for astrophysics.

Garstang [5.119] compiled methods and results of transition probability calculations and listed reference data up to 1953 on published data of calculated transition probablities.

Unsöld tabulated transition probabilities and oscillator strengths of hydrogen ([5.438], p. 346ff) and compiled reference data on transition probabilities up to 1954 ([5.438], p. 366ff).

Kolesnikov and Leskov [5.218] listed references for transition probabilities up to 1957. van Regemorter [5.363] drew up a complete bibliography of all oscillator strengths measured up to the end of 1957. Goldberg et al. [5.131] gave a critical survey of the data on transition probabilities, which were known up to 1960.

Levinson and Nikitin [5.251] treat theoretical calculations of line intensities. References on transition probabilities of forbidden and allowed discrete transitions in atoms are contained in the works of Wiese et al. [5.124], [5.267d].

Nicholls and Stewart [5.297] as well as Wiese [5.459] and Foster [5.107a] give a survey on methods of determining transition probabilities.

Corliss and Bozman [5.69] published f-values for 25,000 lines of 70 elements. The light source was a copper arc, whereby the respective element was added in the ratio 1:1000. The temperature of the arc (5100 °K) was assumed to be the same for all admixtures. A radial resolution of the radiation, by means of which the f-values were to be calculated, was not carried out. Kessler and Dobeneck [5.198] pointed out that the listed f-values can involve errors due to simplifications.

After a private communication of Margoshes, quoted in the book of Boumans ([49a], p. 28), the gf-respectively gA-values listed in [5.69] are incorrect for wavelengths below 2500 Å by a factor of about 10.

Griem [5.139e], [5.139f] presents calculated data of oscillator strengths for several light elements, and for cesium.

Wiese et al. [5.461a] have tabulated the most reliable data of transition probabilities given in literature, for the elements H-Ne, and Na-Ca, respectively.

Below we give a survey of measured and calculated transition probabilities (Table 5.13). The table contains the chemical symbol, the chemical element, the atomic number, the ionization state, and references on transition probabilities (and oscillator strengths). Before the reference number (for the sake of simplicity we shall not write the chapter number, i.e., instead of, for example, [5.108] we only write 108, omitting brackets as well) we designate the method used to determine the transition probability by an abbreviation. The methods are indicated as follows:

- A absorption measurements,
- At atomic beam absorption measurements,
- AD method of anomalous dispersion (hook method and similar methods),
- E emission measurements (flame, arc, plasma jet, pulse discharge),
- EK emission measurements with a King furnace,
- ES emission measurements on shock wave tubes,
- Σf transition probability determined via the f-sum rule,
- L lifetime measurements (if the symbol L^* is used, life time data are presented only),
- Li transition probability determined from the line width,
- Q quantum mechanically calculated transition probability,

Table 5.13 Reference data on transition probabilities

Symbol	Element	Z	Ionization State, Method, Reference	
Ag	Silver	47	I	*A*: 29g; *AD*: 182, 345c; *E*: 3, 3a 9, 69, 420; *L**: 55g, 70b; *Q*, Rel: 29b.
			II	*E*: 69.
Al	Aluminium	13	I	*AD*: 299, 315, 332, 333, 341, 344, 345a; *E*: 3, 9, 69, 232a, 441; *E*, Rel: 75a; *L*: 54, 74; *L**: 70a; *Li*: 121i; *Q*: 21, 32, 34, 35, 36, 88, 103a, 139e, 144d, 418c, 438i, 449; *Su*: 145a.
			II	*Q*: 139e, 144d, 163a, 470i; *Su*: 8.
			III	*Q*: 12, 74b, 122b, 139e, 144d; *Su*: 8.
			IV	*Q*: 144d, 191a; *Q*, *Rel*: 292c.
			V	*Q*: 144d, 449.
			VI	*Q*: 144d.
			VII	*Q*: 144d, 246d.
			VIII	*Q*: 144d, 294a.
			IX	*Q*: 144d, 294a, 449.
			X	*Q*: 121h, 144d, 294a.
			XI	*Q*: 294a.
			XII	*Q*: 144d.
Ar	Argon	18	I	*A*: 63f, 362; *AD*: 167, 168; *AD* + $\sum f$: 346a; *AD*, *Rel*: 381; *E*: 2, 48d, 74c, 75, 82, 94a, 111c, 122, 172a, 183, 262e 304, 305, 349c, 367a, 399a, 414a, 455d, 455f; *ES*: 65a; *E*, *Rel*: 1d; $\sum f$: 362; *L*: 91a, 246b, 438r; *L**: 212f, 287a, 312e, 465b; *Li*: 115, 251b, 405e; *Q*: 121d, 144b, 144d, 213, 292b; *Q*, *Rel*: 292c; *So*: 81, 121j; *Su*: 35, 459a.
			II	*E*: 2, 26g, 27f, 225b, 304, 305, 349c, 418b; *E* + *L*: 380e; *E*, *Rel* + *L*: 55j; $\sum f$: 362; *L*: 380a; *L**: 18c, 18e; 59e, *Q*: 12, 118, 139e, 144d, 295, 376a, 400, 405g, 449; *Rel*: 262f.
			III	*E*: 27f; *Q*: 139e, 144d; *Q* + *Su*: 70d; *Su*: 8.
			IV	*Q*: 139e, 144d; *Q* + *Su*: 70d; *Su*: 8.
			V	*Q*: 139e, 144d; *Su*: 8.
			VI	*Q*: 103a, 144d, 449.
			VII + VIII	*Q*: 144d.
			IX	*Q*: 144d, 191a; *Q*, *Rel*: 292c.
			X	*Q*: 144d, 449; *Su*: 8.
			XIII	*Q*: 294a.
			XIV	*Q*: 294a, 449; *Su*: 8.
			XV	*Q*: 121h, 294a.
			XVI	*Q*: 294a.
As	Arsenic	33	I	*E*: 69; *Q*: 121a, 246a.
			II	*Q*: 144c.
			III	*Q*: 121a, 449.
			IV	*Q*: 435a.
			XVII	*Q*: 449.

Table 5.13. Reference data on transition probabilities (Continuation)

Symbol	Element	Z	Ionization State, Method, Reference	
Au	Gold	79	I	*AD*: 345c; *E*: 3a, 3, 40f, 69, 379; *Li*: 55f; *Q*, *Rel*: 29b.
			III	*Q*: 267b.
B	Boron	5	I	*E*: 69; *L*: 27a; *L**: 168a, 246c; *Q*: 46, 47, 123, 139e, 144d, 435b, 449; *Su*: 459a.
			II	*L*: 27a, 246c; *Q*: 47, 121h, 139e, 144d, 438i, 438m, 438n; *Su*: 459a.
			III	*Q*: 139e, 144d, 294a, 438m, 455; *Su*: 461a.
			IV	*Q*: 66a, 144d, 438m, 438n, 455; *Su*: 461a.
Ba	Barium	56	I	*At*: 456; *AD*: 249, 299, 300, 315, 322, 323, 326, 328, 329, 341, 343, 343a, 351, 354, 373; *E*: 3, 69, 89, 132, 219, 226, 247; *E*, *Rel*: 62, 219; *L*: 178a; *Q*: 65, 163a, 438f; *Q*, *Rel*: 211; *Su*: 8, 35.
			II	*AD*: 339, 341; *E*: 3, 69, 89, 226, 364, 405a; *E*, *Rel*: 219, 264; *L**: 55c, 115d, 115e; *Q*: 369b; *Q*, *Rel*: 29c; *Su*: 8.
			IV	*Q*: 449.
Be	Beryllium	4	I	*E*: 3, 69; *L*: 27a; *Q*: 21, 35, 47, 121h, 139e, 144a 144d, 150, 151, 349a, 438i, 438m, 438n, 470d; *Su*: 35, 461a.
			II	*E*: 3, 69; *L*: 27a; *Q*: 8, 103b, 122f, 139e, 144d, 174, 369b, 438i, 438m, 455; *Su*: 461a.
			III	*Q*: 66a, 144d, 438m, 438n, 455c, 470c, 470d.
Bi	Bismuth	83	I	*A*: 254h; *E*: 9, 64, 404; *L**: 70b; *Q*: 121a.
			II	*Q*: 121a, 144c.
			III	*Q*: 121a, 449.
Br	Bromine	35	I	*Q*: 121a, 246a, 449.
			II	*Q*: 121a.
			V	*Q*: 449.
			VI	*Q*: 435a.
C	Carbon	6	I	*E*: 44, 45, 69, 106, 107, 261, 262, 365, 366; *L**: 246c; *Q*: 8, 46, 48, 119e, 139e, 144d, 302, 413, 438i, 455a, 465d, 470, 470d; *Su*: 461a.
			II	*E*: 86a, 261, 262; *E*, *Rel*: *S*: 369a; *L*: 27a; *L**: 246c; *Q*: 8, 12, 35, 36, 46, 47, 139e, 144d, 164a, 174, 438i, 455a, 465d, 449; *Su*: 35, 461a.
			III	*E*: 86a; *L*: 27a; *Q*: 8, 11, 47, 121h, 139e, 144d, 259a, 294a, 349a, 405h, 438m, 438n, 470d; *Su*: 461a.
			IV	*L*: 27b; *Q*: 8, 11, 139e, 144d, 166a, 174, 294a, 438m, 455; *Su*: 461a.

Table 5.13. Reference data on transition probabilities (Continuation)

Symbol	Element	Z	Ionization, State Method, Reference	
C	Carbon	6	V	*Q*: 66a, 144d, 438m, 438n, 455c, 468b; *Su*: 461a.
Ca	Calcium	20	I	*A*: 77a, 306, 307; *At*: 302b; *AD*: 101, 249, 315, 323, 324, 326, 328, 329, 341, 343, 351, 354, 397a; *E*: 3, 6, 62, 69, 192, 223b, 389, 390; *E*, *Rel*: 63b; *L*: 178a, 190b, 254f, 406, 407, 438j, 470f; *L**: 64a; *Li*: 358; *Q*: 16a, 21, 22, 35, 64b, 65, 144a, 152, 163a, 296c, 369b, 416a, 424, 438a, 438d, 438f, 438i; *Rel*: 211; *So*: 329a; *Su*: 5, 8, 200.
			II	*AD*: 299, 341; *E*: 3, 69, 223b; *L**: 115d, 115e, 401c, 401d; *Q*: 8, 21, 22, 35, 81c, 134, 136, 137, 138, 152, 174, 272, 273, 297b, 298, 369b, 393, 426; *Q*, *Rel*: 29c; *So*: 329a; *Su*: 5, 8, 200.
			III	*Q*: 144d; *Q*, *Rel*: 292c.
			IV	*Q*: 449.
			V	*Q*: 144d; *Su*: 8.
			VII	*Q*: 144d.
			VIII	*Q*: 103a, 144d, 174, 449.
			IX, X	*Q*: 144d, 470i.
			XI	*Q*: 144d, 191a; *Q*, *Rel*: 292c.
			XII	*Q*: 144d, 449; *Su*: 8.
			XIII	*Su*: 8.
			XV	*Q*: 173f, 294a; *Su*: 8.
			XVI	*Q*: 294a, 449.
			XVII	*Q*: 121h, 294a.
			XVIII	*Q*: 294a.
Cd	Cadmium	48	I	*A*: 29g, 208, 209, 254h; *AD*: 97, 239, 340, 341, 342, 343; *E*: 3, 69, 164, 247c, 309, 390; *L*: 91, 214, 254e, 405c, 451; *L**: 312c, 438l; *Q*: 21, 121, 121c, 144a, 163a, 211, 438a, 438d, 438f, 438g, 438h; *So*: 230, 231; *Su*: 8, 35, 336.
			II	*E*: 69; *Q*, *Rel*: 29a; *So*: 121k, 121l.
Ce	Cerium	58	II	*E*: 69.
Cl	Chlorine	17	I	*E*: 106, 169, 170, 173a, 288; *Q*: 144d, 246a, 449.
			II	*E*: 106, 170; *Q*: 12, 70d, 144d.
			III, IV	*Q*: 70d, 144d; *Su*: 8.
			V	*Q*: 103a, 144d, 449.
			VI	*Q*: 92.
			VIII	*Q*: 144d, 191a; *Q*, *Rel*: 292c.
			IX	*Q*: 144d, 449.
			X	*Q*: 144d.
			XII	*Q*: 294a.
			XIII	*Q*: 294a, 449.
			XIV	*Q*: 121h, 294a.
			XV	*Q*: 294a.

Table 5.13. Reference data on transition probabilities (Continuation)

Symbol	Element	Z	Ionization State, Method, Reference	
Co	Cobalt	27	I	*A*, *Rel*: 207; *AD*: 315, 321, 341; *E*: 3, 9, 49, 69, 288a, 308; *Q*: 438i; *Su*: 9a.
			II	*E*: 69; *Q*, *Rel*: 143.
			III	*Q*: 448.
			XI + XV	*Q*: 449.
Cr	Chromium	24	I	*A*: 93, 94, 172; *At*: 72; *AD*: 299, 300, 315, 318, 337, 338, 341, 372, 373; *E*: 3, 3a, 9, 10, 63c, 69, 108, 109, 177, 178, 283, 288a; *E*, *Rel*: 178b; *ES*: 59c, 63c, 463, 464, 468a; *L**: 55d; *Q*: 418c; *O*: 63d; *Q*: 418c; *Su*: 7, 8, 9a.
			II	*E*: 3, 10, 69, 378; *ES*: 59c, 62c, 397c, 463, 464, 468a; *Q*: 121a.
			III	*Q*: 448.
			IV	*Q*: 121a, 418c.
			VII	*Q*, *Rel*: 292c.
			VIII	*Q*: 449.
			IX	*Q*: 144d.
			XII	*Q*: 103a, 449.
			XV	*Q*: 191a.
			XVI	*Q*: 449.
			XIX, XX	*Q*: 294a.
Cs	Cesium	55	I	*A*: 50, 51, 114, 387, 442; *AD*: 238, 341, 353, 359; *E*: 1, 3c, 29, 69, 139c, 216, 310, 311; *E*, *Rel*: 111b, 159a, 166d; *L**: 13b, 253a, 253b; *Q*: 13d, 15, 21, 45b, 145e, 293, 369b, 415, 415a, 438a, 468b; *So*: 55h, 274, 286; *Su*: 35.
			II	*AD*: 341.
			III	*Q*: 121a, 449.
Cu	Copper	29	I	*A*: 208, 209, 438q; *A*, *Rel*: 312g; *At*: 24, 72; *AD*: 315, 318, 332; *E*: 3, 3a, 4a, 41, 68, 69, 176, 213b, 252, 253, 367c, 367d, 367f, 389, 390; *E*, *Rel*: 49a (p. 117), 75a; *L**: 55g, 70b; *Li*: 223c; *Q*: 21, 418a, 438i; *Q*, *Rel*: 29b; *So*: 26c, 217, 389; *Su*: 8, 35.
			II	*E*: 69, 213b; *Q*: 121a, 418a.
			XIII, XVII	*Q*: 449.
Dy	Dysprosium	66	II	*E*: 69.
Er	Erbium	68	II	*E*: 69.
Eu	Europium	63	I	*AD*: 214a; *E*: 69; *L*: 243a.
			II	*E*: 69; *Z*: 5.
F	Fluorine	9	I	*Q*: 144d, 290a, 449; *Su*: 461a.
			II	*E*: 212d; *Su*: 461a.
			III	*E*: 212d; *Q*: 46, 144d, 212d, 250; *Su*: 461a.
			IV	*E*: 212d; *Q*: 46, 144d, 212d, 294a, 470; *Su*: 8, 461a.
			V	*E*: 212d; *Q*: 46, 47, 144d, 212d, 294a, 449; *Su*: 461a.

Table 5.13. Reference data on transition probabilities (Continuation)

Symbol	Element	*Z*	Ionization State, Method, Reference	
F	Fluorine	9	VI	*Q*: 121h, 144d, 294a.
			VII	*L*: 27b; *Q*: 144d, 294a, 455; *Su*: 461a.
			VIII	*Q*: 144d, 455c; *Su*: 461a.
Fe	Iron	26	I	*A*: 201, 204, 205, 206, 210, 269, 348, 444; *At*: 24, 222; *AD*: 331, 341, 373, 445; *E*: 1, 3, 9, 69, 69a, 70, 156, 248, 262g, 281, 288a, 289, 334, 348, 370, 404, 443; *EK*: 1a, 61; *E*, *Rel*: 288; *ES*: 59b; *L*: 312, 330, 474; *Q*: 133; *So*: 248a, 248b; *Su*: 7, 8, 9a, 35, 69b, 69c, 316.
			II	*E*: 181, 69, 282, 370, 404; *E*, *Rel*: 288b; *Q*: 119c, 119d, 133c, 142, 174; *Su*, 8, 447.
			III	*Q*: 119a, 446, 448; *Su*: 8.
			IV	*Su*: 8.
			V	*Q*: 119a; *Su*: 8.
			VI	*Q*: 121a; *Su*: 8.
			VII	*Q*: 103b, 121a; *Su*: 8.
			VIII	*Q*: 174.
			IX	*Q*: 95a, 292c.
			X	*Q*: 95a, 120, 449.
			XI	*Q*: 95a, 144d, 225e; *Su*: 8.
			XII	*Q*: 95a.
			XIII	*Q*: 95a, 173f, 225e; *Su*: 8.
			XIV	*Q*: 59, 95a, 103a, 120, 125, 449; *Su*: 8.
			XV	*Q*: 59; *Su*: 8.
			XVI	*Q*: 405h.
			XVII	*Q*: 121b, 191a; *Q*, *Rel*: 292c.
			XVIII	*Q*: 449.
Ga	Gallium	31	I	*A*: 223a, 405f; *AD*: 315, 320, 329, 341, 344, 345a, 345b; *E*: 3a 9, 69; *L*: 54, 74. 330; *L**: 70b; *Q*: 121a, 449; *Rel*: 254b; *Su*: 145a.
			II	*Q*: 163a, 435a.
			XV	*Q*: 449.
Gd	Gadolinium	64	I, II	*E*: 69.
Ge	Germanium	32	I	*AD*: 401; *E*: 3a, 69; *Q*: 121a, 144c, 246a.
			II	*Q*: 121a, 449.
			III	*Q*: 435a.
			XVI	*Q*: 449.
H	Hydrogen	1	I	*A*: 18h, 289a, 402, 403; *AD*: 60; *L*: 17; *L**: 16b, 29f, 64d, 175a; *Q*: 8, 28, 35, 135, 266b, 270, 377, 385, 405d, 418, 438 (p. 345), 438b, 438d, 438m; *Su*: 461a.
He	Helium	2	I	*A*: 54d, 63f, 251a; *AD*: 167, 168; *AD*, *Rel*: 249; *E*: 48e, 232b; *E*, *Rel*: 438p; *ES*: 257; *L*: 166, 199d, 465c; *L**: 18b, 18c, 54c, 107c, 107d, 173d, 173e, 297a, 312b, 312e, 312f; *Li*: 112a, 214c, 251d;

Table 5.13. Reference data on transition probabilities (Continuation)

Symbol	Element	Z	Ionization State, Method, Reference	
He	Helium	2	I	*Q*: 13a (auto-ionizing level), 21, 66a, 86, 115a, 121c, 129, 134a, 144d, 163a, 175, 180, 197, 215, 223e, 255e, 266a, 369, 380, 380c, 427, 438a, 438b, 438d, 438i, 438m, 438n, 440, 440a, 455c, 457, 470b, 470c, 470d; *O*: 40a, 40b, 40c, 56; *Su*: 6, 8, 35, 71, 200, 271a, 272, 336, 435c.
			II	*E*: 86; *L*: 166c; *L**: 186a, 312e, 312f.
Hf	Hafnium	72	I, II	*E*: 69.
Hg	Mercury	80	I	*A*: 38a, 67, 221, 471; *A*, *Rel*: 55b; *A* + *L*: 185a; *AD*: 239, 243, 341, 468; *E*: 3, 3a, 69, 247c, 383, 384, 389, 390; *E*, *Rel*: 9; *L*: 52, 116, 254d, 278, 279, 334a; *L**: 3b, 18j, 18k, 145c, 185b, 247b, 334b, 438l, 470g; *Q*: 21, 121, 121c, 144a, 163a, 255d, 275, 434, 438d, 438f, 438g, 438h; *Rel*: 186, 211; *O*: 18i, 336a; *Su*: 8, 35.
			II	*Q*: 121a.
Ho	Holmium	67	I	*E*: 69.
I	Iodine	53	I	*Q*: 121a, 246a, 449.
In	Indium	49	I	*A*: 64, 223a, 263a, 405f (auto-ionizing-level); *AD*: 315, 320, 329, 341, 344, 345b; *E*: 3a, 69, 232a; *L*: 178a; *L**: 70b; *Q*: 121a, 438c, 438d, 449; *Su*: 145a.
			II	*E*: 69; *Q*: 163a.
			XIV, XVIII	*Q*: 449.
Ir	Iridium	77	I	*E*: 69.
			VI	*Q*: 449.
K	Potassium	19	I	*A*: 77, 80, 154; *AD*: 99, 100, 327, 341, 350, 353, 355, 373; *AD*, *Rel*: 357; *E*: 69, 139d, 159a, 161, 162, 310, 311, 392; *E*, *Rel*: 166d; *L*: 368a, 400c; *L**: 253a; *Q*: 13d, 14, 19, 21, 27c, 30, 31, 32, 33, 35, 36, 66, 139e, 145b, 185, 275, 277, 438a, 438b, 438d, 439, 455b; *O*: 411, 452; *Su*: 8.
			II	*AD*: 300; *Q*: 144d; *Q*, *Rel*: 292c.
			III	*Q*: 139e, 449.
			IV	*Q*: 144d; *Su*: 8.
			VI	*Q*: 144d.
			VII	*Q*: 103a, 144d, 174, 449.
			VIII, IX	*Q*: 144d.
			X	*Q*: 144d, 191a; *Q*, *Rel*: 292c.
			XI	*Q*: 144d, 449.
			XIV	*Q*: 294a.
			XV	*Q*: 294a, 449.

Table 5.13. Reference data on transition probabilities (Continuation)

Symbol	Element	Z	Ionization State, Method, Reference	
K	Potassium	19	XVI	*Q*: 121h, 294a.
			XVII	*Q*: 294a.
Kr	Krypton	36	I	*A*: 63e, 63f, 262b, 346b, 465; *A* + *E*: 262c; *AD* + $\sum f$: 346a; *E*: 111a; *L*: 436b; *L**: 312e, 465b; *Li*: 251b, 438k; *Q*: 81d, 144b, 292b; *O*: 121j, 262c.
			II	*Q*: 121a, 449.
			III	*Q*: 121a.
La	Lanthanum	57	I, II	*E*: 69.
Li	Lithium	3	I	*A*: 410, 428; *AD*: 96, 98, 431, 432, 433; *E*: 69, 84, 199; *L*: 438j; *L**: 190a; *Q*: 15, 21, 103b, 105, 139e, 144d, 145b, 146, 174, 414, 430, 431, 433, 438g, 438h, 438i, 438m, 438o, 455; *O*: 436; *Su*: 8, 35, 98a, 145a, 461a.
			II	*Q*: 66a, 139e, 144d, 163a, 197, 438m, 438n, 455c, 456a, 470c, 470d.
Lu	Lutetium	71	I, II	*E*: 69.
Mg	Magnesium	12	I	*A*: 42, 43; *AD*: 315, 328, 329, 341, 343; *E*: 1, 3, 3a, 4, 6, 69, 195, 196, 330a; *L*: 54, 74, 190b; *L**: 401d; *Li*: 358; *Q*: 16, 21, 35, 38, 121, 121c, 139e, 144a, 144d, 163a, 280, 422, 423, 424, 438a, 438b, 438d, 438f, 438i, 470i; *Rel*: 374; *O*: 79, 254c, 351, 354.
			II	*E*: 3, 69; *L*: 401c; *Q*: 12, 21, 30, 31, 32, 33, 34, 35, 36, 37, 139e, 144d, 174, 369b, 438i.
			III	*Q*: 144d, 191a.
			IV	*Q*: 144d, 191a, 449.
			V, VI	*Q*: 144d.
			VII	*Q*: 144d, 294a.
			VIII	*Q*: 144d, 174, 294a, 449.
			IX	*Q*: 121h, 144d, 294a; *Su*: 8.
			X	*Q*: 144d, 166a, 174, 294a; *Su*: 8.
			XI	*Q*: 144d.
Mn	Manganese	25	I	*At*: 25; *AD*: 299, 300, 315, 316, 317, 318, 341; *E*: 3, 3a, 9, 69, 109, 159, 177, 178, 396, 450; *L**: 254g; *Su*: 9a.
			II	*E*: 69, 396.
			III	*Q*: 448.
			IV	*Q*: 418c.
			V, VI	*Q*: 121a.
			VIII	*Q*, *Rel*: 292c.
			IX	*Q*: 449.
			X	*Q*: 144d.
			XIII	*Q*: 103a, 449.
			XVI	*Q*: 191a.
			XVIII	*Q*: 449.
			XX	*Q*: 294a.

Table 5.13. Reference data on transition probabilities (Continuation)

Symbol	Element	Z	Ionization State, Method, Reference	
Mo	Molyb-denum	42	I	*AD*: 299, 300; *E*: 3a, 69; *E*, *Rel*: 75a.
			II	*E*: 69.
			VIII	*Q*: 449.
N	Nitrogen	7	I	*A*: 289a; *At*: 94b, 95, 349f; *E*: 238a, 290, 367, 400a; *L**: 246c; *Q*: 8, 17b, 17d, 46, 139e, 144d, 193, 193b, 194, 250, 302, 400, 438i, 465d; *Su*: 336, 461a.
			II	*E*: 238a, 265, 266, 401a; *L**: 166b, 168d, 246c, 251c; *Q*: 8, 12, 46, 48, 139e, 144d, 193, 193a, 193b, 255, 413, 438g, 438i, 465d, 470; *Su*: 336, 461a.
			III	*E*, *Rel*: 184b; *L**: 166b, 251c; *Q*: 8, 11, 46, 47, 139e, 174, 193, 193a, 294a, 297c, 438g, 449, 455c; *Su*: 461a.
			IV	*L**: 166b, 251c; *Q*: 8, 11, 47, 121h, 139e, 144d, 193, 193a, 294a, 349a, 405h; *Su*: 461a.
			V	*L*: 27b; *L**: 166b, 251c; *Q*: 8, 139e, 166a, 174, 193, 193a, 294a; *Su*: 461a.
			VI	*Q*: 144d, 455c; *Su*: 461a.
Na	Sodium	11	I	*A*: 242, 273, 386, 428, 429, 430, 435; *AD*: 102, 105a, 235, 236, 237, 266c, 301, 341, 371, 373; *E*: 3, 69, 84, 160, 163, 310, 311; *E*, *Rel*: 49; *L*: 54, 74, 178a, 189, 190, 438j; *L**: 3b, 18m, 70b, 253a; *Li*: 1c; *Q*: 13f, 14, 21, 30, 31, 32, 33, 36, 73, 105, 122a, 139e, 144d, 145b, 174, 185, 347, 352, 368, 375, 417, 433, 438a; 438b, 438g, 438h, 438i, 468b; *O*: 78, 85, 335, 411, 470e; *Su*: 5, 8, 35, 145a.
			II	*Q*: 144b, 144d, 191a; *Q*, *Rel*: 292c.
			III	*Q*: 144d, 191a, 449.
			IV, V	*Q*: 144d, 191a.
			VI	*Q*: 144d, 191a, 294a.
			VII	*Q*: 144d, 174, 191a, 294a, 449.
			VIII	*Q*: 121h, 144d, 191a, 294a.
			IX	*Q*: 144d, 174, 191a, 294a.
			XI	*Q*: 191a.
Nb	Niobium	41	I, II	*E*: 69.
			VII	*Q*: 449.
Nd	Neodymium	60	I, II	*E*: 69.
Ne	Neon	10	I	*A*: 63f, 112, 301b, 388; *A*, *Rel*: 111d; *A*, *E*, *Rel*: 181a; *AD*: 220, 240, 241; $AD + \sum f$: 346a; *E*: 111a; *E*, *Rel*: 81a; *ES*: 214b, 258; *L*: 26d, 141, 301a, 312d; *L**: 18c, 18d, 168c, 212e, 226a, 465b; *Li*: 223, 231a, 251b, 380d; *Q*: 126, 144b, 144d, 191, 225c, 292b, 377a, 394, 398, 437, 438e, 438i; *Q*, *Rel*: 292c;

Table 5.13. Reference data on transition probabilities (Continuation)

Symbol	Element	Z	Ionization, State	Methode, Reference
Ne	Neon	10	I	*O*: 59d, 81b, 121j; *O, Rel*: 394a; *Su*: 35, 461a.
			II	*ES*: 214b; *L*: 380b; *L**: 168c; *Q*: 111a, 117, 118, 139e, 144d, 449; *Su*: 461a.
			III	*L**: 168c; *Q*: 139e, 144d; *Su*: 8, 461a.
			IV	*L**: 29d; *Q*: 139e, 144d; *Su*: 8, 461a.
			V	*L**: 29d; *Q*: 48, 144d, 294a; *Su*: 461a.
			VI	*Q*: 144d, 294a, 449; *Su*: 8, 461a.
			VII	*Q*: 121h, 166a, 294a, 438m, 438n.
			VIII	*L*: 27b; *Q*: 294a, 438m, 455; *Su*: 8, 461a.
			IX	*Q*: 438m, 438n.
			X	*Q*: 438m.
Ni	Nickel	28	I	*A*: 94, 203; *AD*: 331, 341; *E*: 3, 3a, 9, 33, 49a, 68a, 69, 83, 285, 288a, 308; *Q*: 121a, 438i; *Su*: 7, 8, 9a.
			II	*E*: 69, 308; *Q*: 119b, 174; *Q, Rel*: 144; *Su*: 8.
			III	*Su*: 8; *Q*: 119b.
			XI	*Q*: 95a.
			XII	*Q*: 95a, 449; *Su*: 8.
			XIII	*Q*: 95a; *Z*: 8.
			XIV	*Q*: 95a.
			XV	*Su*: 8.
			XVI	*Q*: 449; *Su*: 8.
O	Oxygen	8	I	*A*: 334d; *A* + *E*: 255c; *At*: 289a, 349e, 349f; *E*: 45a, 107, 188, 247a, 404a, 404b, 460, 461; *E, Rel*: 255a; *L*: 115b, 380b; *Q*: 8, 17b, 17d, 65, 115b, 139e, 144d, 193, 193b, 212, 302, 334c, 400, 438a, 438d, 438i, 465d; *Su*: 336, 461a.
			II	*E*: 266, 458; *E, Rel*: 26f; *L**: 29d; *Q*: 8, 12, 13, 17d, 20, 21, 35, 46, 118, 139e, 144d, 153, 193, 193b, 250, 295, 438i, 465d; *Su*: 35, 461a.
			III	*E*: 26e, 26f; *L**: 29d; *Q*: 11, 46, 48, 139e, 144d, 193, 268, 294a, 413, 470; *Su*: 461a.
			IV	*L**: 29d; *Q*: 8, 11, 17d, 46, 47, 139e, 144d, 174, 193, 294a, 449; *Su*: 461a.
			V	*L**: 29e; *Q*: 8, 11, 47, 121h, 139e, 144d, 193, 294a, 349a, 405h; *Su*: 461a.
			VI	*L**: 27b; *Q*: 8, 11, 47, 139e, 144d, 166a, 174, 193, 294a, 455; *Su*: 461a.
			VII	*Q*: 121c, 144d, 455c; *Su*: 461a.
Os	Osmium	76	I	*E*: 69; *Q*: 349d.
			II	*E*: 69.
P	Phosphorus	15	I	*E*: 69; *L**: 380b; *Q*: 70d, 144d; *Su*: 8.
			II	*L**: 380b; *Q*: 70d, 144d.
			III	*Q*: 12, 103a, 144d, 449.
			IV, V	*Q*: 144d, 470i.

Table 5.13. Reference data on transition probabilities (Continuation)

Symbol	Element	Z	Ionization State, Method, Reference	
P	Phosphorus	76	VI	Q: 144d, 191a; Q, Rel: 292c.
			VII	Q: 144d, 449.
			VIII, IX	Q: 144d.
			X	Q: 144d, 294a.
			XI	Q: 294a, 449.
			XII	Q: 121h, 144d, 294a.
			XIII	Q: 294a.
Pb	Lead	82	I	A: 199a, 254h; At: 26, 92; AD: 346, 397; E: 3a, 9, 69; ES: 54e; L: 470h; L*: 70b,378a; Q: 121a,144c,246a; Su: 8.
			II	E: 69, 427b; Q: 121a, 449.
			III, IV	Q: 427b.
Pd	Palladium	46	I	E: 69; L*: 55i.
			II	E: 69.
Pr	Praseodymium	59	II	E: 69.
Po	Polonium	84	I	Q: 121a.
Pt	Platinum	78	I, II	E: 69.
Rb	Rubidium	37	I	A: 311; AD: 130, 341, 353, 359; E: 69, 216, 310, 311, 432; E, Rel: 159a, 166d; L: 55e; L*: 95b, 253a, 361a; Q: 13d, 15, 145b, 185.
			II	Q: 144b.
			III	Q: 121a, 449.
Re	Rhenium	75	I, II	E: 69.
Rh	Rhodium	45	I	E: 3a, 69.
			II	E: 69.
Rn	Radon	86	II	Q: 121a.
Ru	Ruthenium	44	I	E: 3a, 69.
			II	E: 69; Q: 449.
S	Sulphur	16	I	E: 54a, 54b, 107b, 292a; E, Rel: 225a, 255b; L*: 380b; Q: 70d, 144d, 246a, 290a; Su: 8.
			II	E: 54a, 54b, 292a; L*: 380b; Q: 12, 70d, 118, 144d, 290a; Su: 8.
			III	Q: 12, 70d, 144d; Su: 8.
			IV	Q: 103a, 144d, 449; Su: 8.
			V, VI	Q: 144d; Su: 8.
			VII	Q: 144d, 191a; Q, Rel: 292c.
			VIII, IX, X	Q: 144d.
			XI	Q: 294a.
			XII	Q: 294a, 449.
			XIII	Q: 121h, 294a, 449.
			XIV	Q: 294a.
Sb	Antimony	51	I	A: 254h; E: 3a, 69; Q: 121j.
			II	Q: 144c.
			III	Q: 144c, 449.
Sc	Scandium	21	I	AD: 314, 315, 319, 341; E: 69; Q: 8; Su: 7, 9a.
			II	E: 69; Q: 5.
			III	Q: 448, 455b.

Table 5.13. Reference data on transition probabilities (Continuation)

Symbol	Element	Z	Ionization State, Method, Reference	
Sc	Scandium	21	IV	*Q*: 144d; *Q*, *Rel*: 292c.
			V	*Q*: 449.
			VI	*Q*: 144d.
			VIII	*Q*: 144d, 449.
			IX	*Q*: 103a, 144d, 103a, 449.
			X, XI	*Q*: 144d.
			XII	*Q*: 144d, 191a; *Q*, *Rel*: 292c.
			XVI	*Q*: 294a.
			XVII	*Q*: 294a, 449.
			XVIII, XIX	*Q*: 294a.
Se	Selenium	34	I	*E*: 69; *E*, *Rel*: 225a; *Q*: 121a, 246a.
			IV	*Q*: 449.
			V	*Q*: 435a.
			XVIII	*Q*: 449.
Si	Silicium	14	I	*AD*: 401; *E*: 3, 3a, 9, 69, 169, 170, 385a; *E*, *Rel*: 225a; *L**: 380b; *Q*: 8, 17c, 65, 121e, 139e, 144d, 246a, 438i.
			II	*E*: 170, 385a; *L**: 380b; *Q*: 20, 21, 32, 33, 34, 35, 36, 37, 103a, 121g, 139e, 144d, 174, 438i, 449; *O*: 437a.
			III	*Q*: 8, 12, 139e, 144d, 427a, 470i.
			IV	*Q*: 35, 81c, 139e, 144d, 174, 376.
			V	*Q*: 144d, 191a; *Q*, *Rel*: 292c.
			VI, VII, VIII	*Q*: 144d.
			IX	*Q*: 144d, 294a.
			X	*Q*: 120, 144d, 174, 294a, 449.
			XI	*Q*: 121c, 121h, 144d, 294a; *Su*: 8.
			XII	*Q*: 166a, 294a.
Sm	Samarium	62	I, II	*E*: 69.
Sn	Tin	50	I	*AD*: 341, 346; *E*: 3a, 9, 69, 244, 330a; *L*: 470h; *Li*: 294, 358; *Q*: 121a, 144c, 246a.
			II	*Q*: 121a, 449.
Sr	Strontium	38	I	*AD*: 249, 315, 323, 326, 328, 329, 341, 343, 351, 354; *E*: 3, 69, 191, 284, 389, 390; *E*, *Rel*: 62; *L*: 178a, 254f, 470f; *L**: 254i; *Q*: 21, 144a, 163a, 438f; *Q*, *Rel*: 211; *Su*: 8, 35, 336.
			II	*AD*: 299, 300, 341; *E*: 3, 69, 87, 284, 405b; *L**: 115d, 115e; *Q*: 65, 369b; *Q*, *Rel*: 29c; *Su*: 8.
Ta	Tantalum	73	I, II	*E*: 69.
Tb	Terbium	65	I	*E*: 69.
Te	Tellurium	52	I	*A*: 254h; *E*: 69; *Q*: 121a.
			IV	*Q*: 449.
Th	Thorium	90	I	*E*: 69.
			II	*E*: 69; *Q*: 64c.
Ti	Titanium	22	I	*A*: 205, 348, 364a; *AD*: 315, 316, 317, 341; *E*: 69, 109, 147, 147a, 148, 158, 283, 288a, 348, 408, 409, 419; *ES*: 48c; *O*: 397b; *Su*: 7, 336.

Table 5.13. Reference data on transition probabilities (Continuation)

Symbol	Element	Z	Ionization State, Method, Reference	
Ti	Titanium	22	II	*A*: 200, 348; *E*: 3, 69, 147, 283, 348, 408, 409, 419, 466; *ES*: 48c; *Q*: 174, 438i.
			III	*Q*: 121a, 448.
			IV	*Q*: 174, 448.
			V	*Q*: 144d; *Q*, *Rel*: 292c.
			VI	*Q*: 449.
			VII	*Q*: 144d.
			X	*Q*: 103a, 449.
			XI, XII	*Q*: 144d.
			XIII	*Q*: 191a, 449.
			XIV	*Q*: 449.
			XVII, XVIII, XIX, XX	*Q*: 294a.
Tl	Thallium	81	I	*A*: 263, 291, 292; *AD*: 103, 145, 233, 234, 235, 236, 237, 299, 300, 341, 345, 356, 360, 361; *E*: 69; *L*: 74, 115f; *L**: 70b, 133a, 349b, 419a; *Q*: 13e, 21, 121a, 438c, 449; *O*: 229, 230, 411; *Su*: 35, 145a.
			II	*E*: 213a; *Q*: 163a.
Tm	Thulium	69	I, II	*E*: 69.
U	Uranium	92	I, II	*E*: 69.
V	Vanadium	23	I	*A*: 202; *AD*: 315, 321, 341; *E*: 3, 3a, 69, 109, 288a; *Su*: 7, 8, 9a, 336.
			II	*E*: 69; Σf: 8.
			III	*Q*: 448; *Q*, *Rel*: 267a.
			VI	*Q*, *Rel*: 292c.
			VII	*Q*: 449.
			VIII	*Q*: 144d.
			IX	*Q*: 449.
			XIV	*Q*: 103a, 191a.
			XV	*Q*: 449.
			XVIII, XIX, XX	*Q*: 294a.
W	Wolfram	74	I	*E*, *Rel*: 212a.
Xe	Xenon	54	I	*A*: 63a, 63f, 465a; *L*: 13c; *L**: 465b; *Q*: 81d, 144b; *O*: 262d.
			II	*Q*: 121a, 449.
			III	*Q*: 121a.
Y	Yttrium	39	I, II	*E*: 69.
			V	*Q*: 449.
Yb	Ytterbium	70	I	*E*: 69; *L**: 23a.
			II	*E*: 69.
Zn	Zinc	30	I	*A*: 18, 39, 40, 254h; *AD*: 97, 341, 342; *E*: 3a, 69, 247c, 345, 389, 391; *E*, *Rel*: 49a (p. 20), 264; *L*: 254f, 470f; *Q*: 21, 121, 121c, 144a, 163a, 435a, 438f; *Q*, *Rel*: 211; *O*: 55, 232, 405; *Su*: 8, 35.
			II	*E*: 69; *Q*, *Rel*: 29a; *O*: 121k, 121l.
Zr	Zirconium	40	I	*E*: 69.
			II	*E*: 69; *Q*: 174.
			IV	*Q*: 174.

Rel relative transition probability,
O other methods, inclusive comments,
Su summarizing tables.

The transition probability of hydrogenic ions with atomic number Z, such as He II, Li III, Be IV, B V, etc., is not listed, since it can be calculated from the analogous transition probability of the hydrogen atom as follows (Allen [5.8])

$$A_z = Z^4 A_{\mathrm{H}} \quad \text{and} \quad f_z = f_{\mathrm{H}} \,. \tag{5.101}$$

CHAPTER 6

Determination of Temperatures and Particle Densities of Plasmas Displaying Appreciable Absorption

In this chapter we shall treat methods by which the plasma temperature is determined from the emissive and absorptive power in accordance with Kirchhoff's law, provided the latter holds. According to the investigations of Mandel'shtam et al. [6.41] as well as of Bekefi et al. [6.13] this is the case when 1. the electrons possess a Maxwell velocity distribution, 2. the level occupation corresponds to a Boltzmann distribution, and 3. the ionization can be described by the Saha equation. These conditions are satisfied in a plasma in thermal equilibrium.

If the measurements are performed for a small wavelength interval in which the emission or the absorption is effected practically by a single spectrum line only, then for the validity of Kirchhoff's law instead of the strict conditions cited above it is necessary only that the upper (m) and the lower (n) electronic level of the spectrum line are populated according to a relative Boltzmann distribution: $N_m/N_n = (g_m/g_n) \cdot \exp\left((U_n - U_m)/(kT)\right)$ (see section 11.1.3.1).

In order that this condition is satisfied, either the level population must be collision dominated (that means, the excitation rate for transitions $n \to m$ due to particle collisions must exceed the decay rate of level m by radiation) or the plasma must be optically dense in the range of the spectrum line under investigation (see [6.15c], [6.21i–l], [6.46e]).

In order to have a level population determined by electron collisions, the electron density must assume values of at least $10^{12}-10^{14}$ cm^{-3}, depending from the particular line used.

For a low degree of ionization, not too low pressure and not too broad energy gap between levels n and m, a level population satisfying Boltzmann's law can be effected by heavy particle collisions. This has been shown to be the case for Na and Cr resonance lines if the pressure is at least 1 atm for an atomic gas respectively 0.1 atm for a molecular gas [6.28e], [6.31b], [6.47j]. In this case, the measured temperature corresponds to the gas temperature* (in relaxing molecular gases, the measured temperature corresponds to the vibrational temperature [6.32t], due to the strong influence of molecular vibration on level population). For more details concerning the reliability of reversal temperature measurements, see, for example, [6.31b], [6.47j].

* For the pressure range indicated and for the case of stationary plasmas, there will be only an insignificant difference between gas and electron temperature. In this case the measured temperature can be also interpreted as electron temperature.

The relevant measuring methods are discussed in sections 6.1—6.3. In section 6.1 we shall treat homogeneous plasmas, and in section 6.2 inhomogeneous plasmas having rotational symmetry. Bartel's method for determining the temperature in inhomogeneous plasmas having rotational symmetry from the peak intensities of spectral lines displaying self-reversal is discussed in section 6.3. The peculiarity of this method lies herein that direct absorption measurements can be forgone. The method also serves as the basis for the correction of line profiles affected by self-absorption.

In section (6.4) we shall treat the determination of particle densities from the absorption coefficients of resonance lines and from the absorption coefficients for continuous absorption from the ground state.

In the last section (6.5) there will be given a few comments concerning the determination of particle number densities from the intensity of fluorescence radiation.

6.1. Temperature Determination by Means of Kirchhoff's Law for Homogeneous Plasmas (Line-reversal Method)

6.1.1. Fundamentals of the Method

When the plasma is homogeneous (T = const), the temperature can be found from the emissive and absorptive power in accordance with Kirchhoff's law. Let us consider this in the form

$$I_\nu = A_\nu B_\nu \tag{6.1}$$

where

$$B_\nu(T) = \frac{2 h \nu^3}{c^2} \frac{1}{e^{h\nu/kT} - 1} \tag{6.2}$$

is Planck's function (see also equation (2.19)). I_ν is the emissive power or intensity per unit frequency and A_ν is the absorptive power of the plasma at the frequency ν.

The emissive power I_ν can be measured by means of a radiation detector, which was calibrated with a standard. The absorptive power is obtained, for example, from the intensity drop of the radiation passing through the plasma (if D_ν is the spectral transmittance of the plasma, we have $A_\nu = 1 - D_\nu$, since the reflectivity of the plasma can usually be neglected). From I_ν and A_ν we get B_ν via (6.1) and hence the plasma temperature T via (6.2).

This method can be used for the determination of plasma temperature ranging up to some 10,000 °K or more.

In the high temperature region, as reference light source for effecting absorption measurements, a conventional pulse discharge lamp, or, alternatively, a laser light source can be used [6.16a].

The absorptive power can be determined alternatively according to the method given in section 2.10.2.1., in which the radiation from the plasma is measured for two different optical path lengths. The longer optical path length is effected by placing a mirror behind the plasma.

Examples for the application of Kirchhoff's law are given in the publications [6.12a], [6.46d], in which the temperature determination in high pressure discharge lamps has been described.

In low temperature plasma work, it is usual to employ measuring systems of the type shown in principle in Figure 6.1. A source S of continuous radiation with black body temperature T_s is imaged in the Plasma P. This in turn is

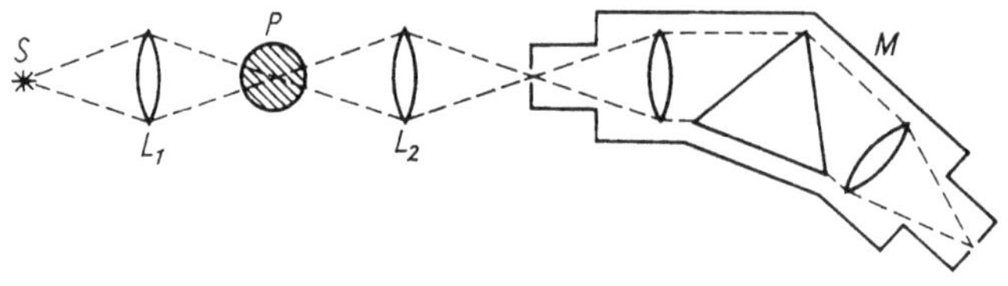

Figure 6.1. Optical layout for determining the plasma temperature by Kirchhoff's law (after Tsuchiya [6.64])

S — continuous radiation source; P — plasma; M — monochromator with measuring system

imaged on the entrance slit of a monochromator M. The radiation to be measured is indicated via the photocurrent i. The following quantities are measured:

- i_1 photocurrent, when only the radiation from S reaches the monochromator,
- i_2 photocurrent, when the plasma radiation as well as that from S weakened by the plasma reaches the measuring device,
- i_3 photocurrent, when only the plasma radiation reaches the monochromator.

If the radiation emitted by S is modulated, it is possible to measure directly the quantity $i_2 - i_3$ (corresponding to the radiation from S, weakened by the plasma) by placing a frequency-selective radiation measuring device behind the monochromator.

Denoting the black body temperature of the continuous source in the frequency range in question by T_s and the plasma temperature by T, we have

$$\left.\begin{aligned} i_1 &= \text{const} \cdot B_\nu(T_s) , \\ i_2 &= \text{const} \cdot B_\nu(T_s)\,(1 - A_\nu) + \text{const} \cdot A_\nu\, B_\nu(T) , \\ i_3 &= \text{const} \cdot A_\nu\, B_\nu(T) . \end{aligned}\right\} \tag{6.3}$$

It thus follows

$$B_\nu(T) = B_\nu(T_s) \cdot \frac{i_3}{i_1 + i_3 - i_2} . \tag{6.4}$$

The plasma temperature is found from $B_\nu(T)$ via (6.2).

Let us now consider the case where the radiation integrated over a finite frequency interval $\Delta\nu$ can be measured. The transmission curve of the monochromator or monochromatic filter is designated by $D(\nu)$. The transmission range should be so small, that in this range Planck's function is practically constant. Then, instead of (6.3) we have for the photocurrents which can be measured

$$\left.\begin{aligned} i_1 &= \text{const} \cdot B_\nu(T_s) \int D(\nu)\,d\nu\,, \\ i_2 &= \text{const} \cdot B_\nu(T_s) \int (1 - A_\nu)\,D(\nu)\,d\nu + \text{const} \cdot B_\nu(T) \int A_\nu\,D(\nu)\,d\nu\,, \\ i_3 &= \text{const} \cdot B_\nu(T) \int A_\nu\,D(\nu)\,d\nu\,. \end{aligned}\right\} \quad (6.5)$$

Rearranging will give us equation (6.4) again. This means that it is also possible to determine the temperature from Kirchhoff's law in a homogeneous plasma, when the measurement is carried out in the range of a spectral line and when the spectroscopic device cannot resolve the line.

In practice it should be ensured that the lens L_1 possesses an aperture which is so large that the delimiting of the light bundles is exclusively effected by the lens L_2. Further, the Light-intensity drop due to lens L_1 and that caused by windows in the ray path (for plasmas in closed chambers) must be allowed for in the system of equations (6.3) and (6.5).

Accurate measurement values can be only obtained when the absorptive power of the plasma is appreciable. By this we understand an absorption of at least 1%. Further the black body temperature T_s of the standard must be of the same order of magnitude as the plasma temperature T. This last condition usually restricts the measurement method to temperatures of about 10,000 °K, which the plasma may not surpass, since for higher temperature there are practically no suitable pulse lamps with reproducible values.

Among suitable light sources are the tungsten filament lamp, the carbon arc crater, high-pressure xenon lamps, as well as the already discussed pulse lamps for the determination of high temperatures (see section 2.4).

Let us now consider the case in which T_s is chosen (e.g., by changing the temperature of the continuous source or by inserting a filter between S and P) in such a way that $i_1 = i_2$. This means that the photocurrent i does not change for a working continuous source when the plasma is placed in the radiation path or removed from it. In other words, the plasma absorbs as much of the continuous radiation as it emits itself. Then, from equation (6.4) we get

$$B_\nu(T) = B_\nu(T_s)\,.$$

The corresponding temperature T is also called the reversal temperature.

The determination of the temperature via the reversal temperature is very practical in those cases where there is sufficient time to perform the measurement. The procedure consists of altering the black body temperature of the continuous source manually or by some suitable control system until $i_1 = i_2$. Many measuring systems are set up in a way that T_s can be directly read off.

In [6.16c], T is derived from the intersectional point of the two curves $i_2 = f(i_1)$ and $i_2 = i_1$, obtained when i_1 and i_2 is measured for various temperatures T_s of the background source.

6.1.2. Increasing the Absorptive Power of a Plasma

If the absorptive power of a plasma is not appreciable, this can be increased, for example, by the addition of sodium. Gaydon and Hurle [6.28e], [6.30] introduced sodium into a shock tube by first leading the gas filling through a heated tube filled with NaCl or NaI, or by leading the gas filling over an electrically heated platinum coil supporting a bead of molten salt.

Other alkali metals and alkaline-earth metals, which exhibit strong line absorption, can also be used for this purpose. The lines in question are Li 6707, K 7664, 7698, Ca 4226, Ba 5535, and Sr 4607 Å. However, none of these elements has advantages over sodium.

Among the volatile compounds lead tetraethyl was once considered as suitable [6.30]. Here a resonance line appears at 2833 Å and a nonresonance line at 4057 Å, where the latter becomes appreciably absorptive only at higher temperatures. The volatile compound $Cr(CO)_6$ is very suitable for increasing the absorptive power of plasmas [6.11], [6.30]. An admixture of 0.05% gives an adequate amount of elementary chromium, so that the absorptive power of the plasma in the range of the lines 4289.7 Å and 5206 Å becomes appreciable. Further, the absorption is high in the range of the CrO band at 6051 Å. For higher temperatures additional lines can be used, for example, Tl 5350 Å, In 4511 Å, Ba II 4554 Å [6.30].

An increase in the absorption occurs also when solid dust particles are present in the gas (for example, in a sooting flame) [6.47]. These assume a somewhat lower temperature than that of the surrounding gas, whose temperature they indicate.

To determine the temperature in argon-helium-nitrogen plasma jets, Tourin [6.62] injected into the working medium admixtures of CO and CO_2, which exhibit strong absorption in the infrared range around 4 μm.

6.1.3. Line-reversal Method

We shall consider the experimental layout shown in Figure 6.1. Let the plasma consist of a flame with a sodium admixture. Depending on the black body temperature at which continuous source S is set, the spectral intensity distribution of the resulting radiation shows the following features: 1. The Na D-line appears as an absorption line, 2. no longer stands out against the continuous background or 3. is observed as an emission line (Figure 6.2). The temperature T_s of the continuous source, for which the line disappears, is called the reversal temperature. This can be visually determined by varying the black body temperature of the radiation source (in the case of a tungsten filament lamp by varying the current, or else by means of light attenuation with filters) until the line in the continuum disappears and $T = T_s$ can be read off (T_s, in the case of the ribbon lamp, is fixed by the lamp current; in the case of a variation in T_s by filtering, it is fixed by the filter setting).

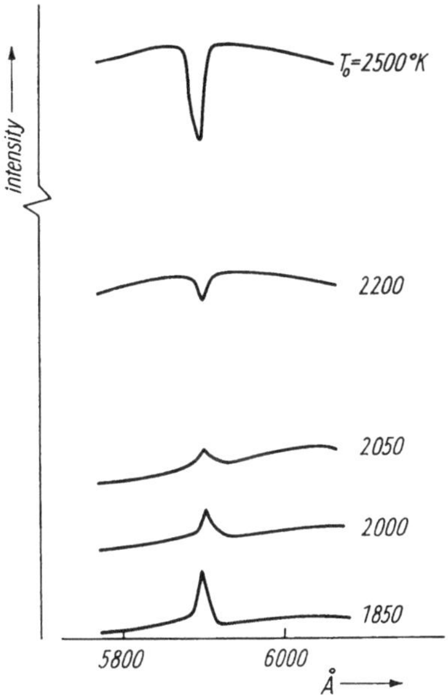

Figure 6.2. Spectroscopic recording of a flame with sodium admixture upon irradiation with continuous light. Depending on the black body temperature at which the continuous source is adjusted an emission line or absorption line appears (after Tsuchiya [6.64])

6.1.4. Plasmas Suitable for Temperature Determination in Accordance with Kirchhoff's Law in the Form (6.1)

Equation (6.1) can be applied, provided the absorbing zones of the plasma which are irradiated, have a uniform temperature.

This is assumed, for example, when measuring the temperature of a shock wave in a shock tube. The observation direction is at right angles to the axis of the tube (see, e.g., [6.27], [6.53]). The condition of homogeneity over the tube cross section is fulfilled when the observations are limited to a region close behind the shock wave front, where the boundary layer along the wall is still thin. The thickness of the boundary layer increases with increasing distance from the shock wave front (see [6.30], p. 70). For a reflected shock wave the boundary layer is sufficiently thin only in the vicinity of the reflecting wall, so that the temperature determination can be carried out only in this region.

Flames in general are inhomogeneous. In this case a formal application of the line reversal method gives incorrect values for the flame center temperature, as shown for example in [6.33a] and [6.47g].

To measure flame temperatures with more accuracy, absorbing admixtures can be added to the flame in such a manner that the absorbing zone has an only small extension in the direction of observation. Figure 6.3 shows a system used by Buchele [6.16]. The Na-compound, which is carried along with the gas, is fed via an injector into the gas upstream from the measuring point. Inside the flame a light track results through which the optical path passes at right angles. The measured reversal temperature characterizes the temperature of

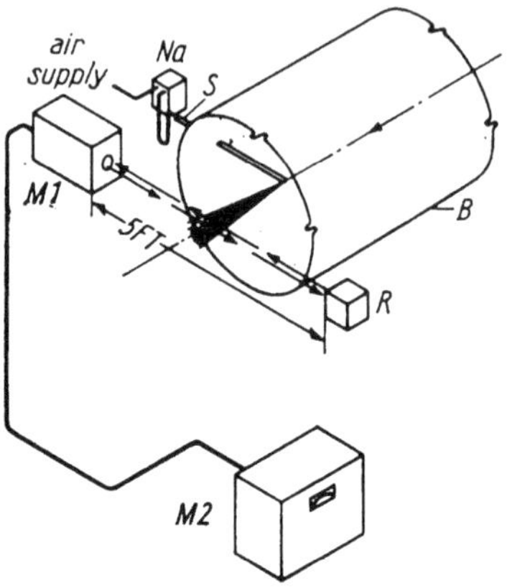

Figure 6.3. System for determining the radial temperature distribution in a flame (after Buchele [6.16])

Na — sodium admixture, *S* — movable injector probe, — *B* — burner, *R* — reflector, M_1 — measuring system with lamp, spectroscope, and photodetector, M_2 — measuring system with current supply, controlling device, and lamp current indicator

the luminous region of the flame. By shifting the injector along the flame cross section the temperature distribution over the cross section of the flame can be measured.

We can also apply the line-reversal method to the middle portion of wall-stabilized cascade arcs, such as used for spectroscopic purposes (see Boldt [6.14] and Figure 2.15, section 2.4), when observations are carried out along the direction of the axis and when the gas near the electrode differs from that in the center of the arc. It should be ensured, however, that the cooler gas layers in front of the electrodes do not produce uncontrollable absorption.

6.1.5. Bichromatic Methods

The bichromatic measuring methods serve as a means of obtaining a rapid temperature determination of plasmas, which exhibit appreciable absorption in the range of a spectral line. The principle of these methods is illustrated in Figure 6.4 (Bauserman [6.12]).

The radiation from a continuous source S with adjustable black body temperature T_s is attenuated due to absorption in the plasma P and reaches the secondary electron multipliers P.M. 1 and P.M. 2 via a beam splitter B and the monochromatic filters F_1 and F_2. The filters are chosen so that F_1 transmits radiation in the range of the spectral line emitted by P, while F_2 only transmits the radiation in the neighborhood of this line. This radiation is equal to the radiation from the continuous source, whereby we assume that the plasma does not display any appreciable absorption in the neighborhood of the line.

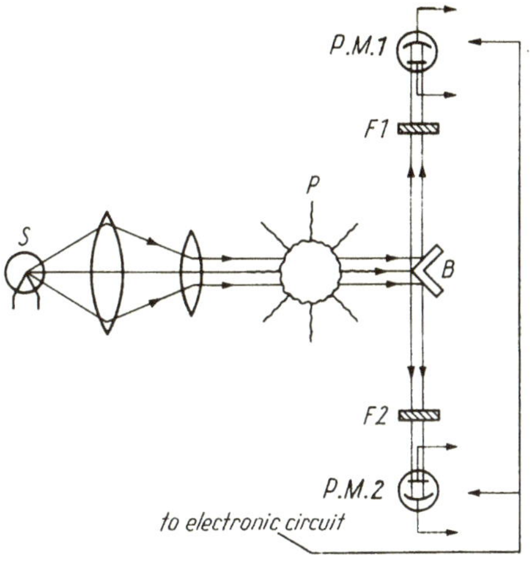

Figure 6.4. Schematic diagram of the system for determining the line-reversal temperature by the bichromatic method (after Bauserman [6.12])

S – continuous radiation source, P – plasma, B – beam splitter, F_1 and F_2 – monochromatic filters, P.M.1 and P.M.2 – secondary electron multipliers

The multiplier sensitivities are adjusted in such a way that they both have the same anode current when the plasma is removed from the radiation path.

If the black body temperature T_s of the continuous source deviates from the flame temperature, the anode currents of the two multipliers will differ during the measuring process. It is possible to introduce a controlling system, utilizing this current difference, which varies the black body temperature of the continuous source until equalization is achieved. The corresponding temperature T_s (reversal temperature) can be read off directly or via a recording arrangement.

To apply the bichromatic method to rapidly varying plasmas, a control for T_s during the measuring process is no longer used and T_s is set at a fixed value [6.12]. While carrying out the measurements the current difference is recorded on an oscillograph as a function of time. If the difference current passes through zero at a certain time t, then at this time the plasma temperature corresponds to $T = T_s$. Given the possibility to reproduce the varying plasma several times, the measurements can be repeated for different settings of T_s and the plasma temperature $T(t)$ can be determined as a function of time. Bauserman [6.12] in this way determined the time dependence of the temperature in burning powders.

Another version of this method in principle consists of inserting a stop in the radiation path between S and P. This enables the measurement of the radiation emitted by the plasma alone. It is thus possible without controlling T_s to determine the plasma temperature T with the help of the equation (6.4), when i_1, i_2, and i_3 have the following meaning:

i_1 = current from P.M. 2, when total radiation is transmitted,
i_2 = current from P.M. 1, when total radiation is transmitted,
i_3 = current from P.M. 1, when radiation from S is blocked.

The time resolution in the best conditions corresponds to the time difference between two light interruptions. A time resolution of 60 kHz can be achieved by using a rotating light chopper [6.43h]. Kerr cells (see [6.78]) enable a time resolution of 10^{-8} s.

When measuring in the infrared range, the measuring speed depends on the inertia of the radiation detector when using bolometers or thermoelements. the minimum time resolution should not be lower than 1/100 s.

Instead of two monochromatic filters a bichromatic spectroscopic device can also be used. For more details on this see Tsuchiya [6.64] and Maleev and Yermakov [6.40].

The roundabout way via the equation (6.4) can be eliminated when T_s is modulated by inserting a light absorber varying steadily with time (polarization foils, of which one rotates [6.47], or a steady modulated Kerr cell) between S and P. The current from P.M. 1 is recorded at those instances of time for which the currents from both multipliers are equal (reversal condition). The temperature T_s associated with this current is then equal to the plasma temperature.

Rössler [6.47] used this principle of modulating T_s for measuring the temperature of rocket flames with a sodium admixture. He imaged a section of the spectrum containing the Na D-line on a drum camera and effected a time resolution of the spectrum. The temperature can be found from the phase angles of the modulated light at the instance of time in which the Na D-line does not stand out against the continuous background.

We should still mention one more method for checking line reversal, whereby the resulting radiation (radiation of the plasma and background) is incident upon the entrance slit of a monochromator, whose exit slit shifts in oscillating fashion and scans the spectral line. The line-reversal is indicated by a photocurrent which is constant in time. When deviations from the line-reversal occur, the radiation received by the multipliers is modulated with time. This modulation can be utilized to adjust T_s until line-reversal occurs.

An interesting biochromatic measuring method, in which no background source is required for the temperature determination, was developed by Bauer [6.10] and applied to the plasma inside a high-pressure xenon lamp. This method is based on observing the spectral intensity I_ν in the center and at the edge of a spectral line, displaying appreciable absorption, for 2 different geometrical layer lengths (in the special case one observes once along the direction of the axis and once at right angles thereto).

Let us write:

l_1 = larger length, l_2 = smaller plasma length,

ω' = frequency at line edge, ω'' = frequence at the line center,

τ = optical depth $= \varkappa\, l$.

The condition

$$\tau_1' < 0.02 < \tau_1'' \tag{6.6}$$

must be fulfilled, i.e., for the line edge the plasma must display a weak absorption in the direction of the larger plasma length, i.e., I_1' must be given with better accuracy than 1% by the approximate expression $\tau_1'\, B_\nu'$.

If we further assume that in the second observation direction the optical depths τ_2' and τ_2'' are both smaller than 0.02, we have

a) for observation in direction 2

$$I_2' = \tau_2'\, B_\nu' \,, \qquad I_2'' = \tau_2''\, B_\nu'' \qquad (B_\nu' \approx B_\nu'' = B_\nu) \,, \tag{6.7}$$

b) in direction 1

$$I_1' = \tau_1'\, B_\nu \,, \qquad I_1'' = (1 - e^{-\tau_1''})\, B_\nu \,. \tag{6.8}$$

Since $\tau_1' = \varkappa'\, l_1$, $\tau_2' = \varkappa'\, l_2$, $\tau_1'' = \varkappa''\, l_1$, and $\tau_2'' = \varkappa''\, l_2$, we get

$$\frac{\tau_1'}{\tau_1''} = \frac{\tau_2'}{\tau_2''} \tag{6.9}$$

and it follows from (6.7) and (6.8) that τ_1'' is given by

$$\frac{1 - e^{-\tau_1''}}{\tau''} = \frac{I_1''\, I_2'}{I_1'\, I_2''} \to \tau_1'' \,. \tag{6.10}$$

If a measurement of the absolute intensity I_1'', in the line center in the direction 1 is made, then $B_\nu(T)$ is given in accordance with (6.8) by

$$B_\nu(T) = \frac{I_1''}{1 - e^{-\tau_1''}} \,. \tag{6.11}$$

with $B_\nu(T)$ we now also know the plasma temperature T.

Bauer considered in more detail the error in determining T in high-pressure xenon lamps due to the temperature drop at the arc edge. This error lies within the permissible limits when the optical depth in the axial direction is smaller than 5. The most accurate measurements are possible for $3 < \tau_1'' < 5$.

6.1.6. The Two-path Method

This method is quite suitable for high-speed recording of plasma temperatures, especially in the case of shock waves (see [6.17], [6.29]), [6.30]). The principle of the method can be explained on the basis of Figure 6.5.

The radiation from the continuous source reaches P.M. 1 via the path L_1, M_1, P, A, L_2, and the interference filter IF. The radiation, which reaches the multiplier P.M. 2 along the second light path via the plasma, must pass in addition through a neutral filter NF. As a consequence the black body temper-

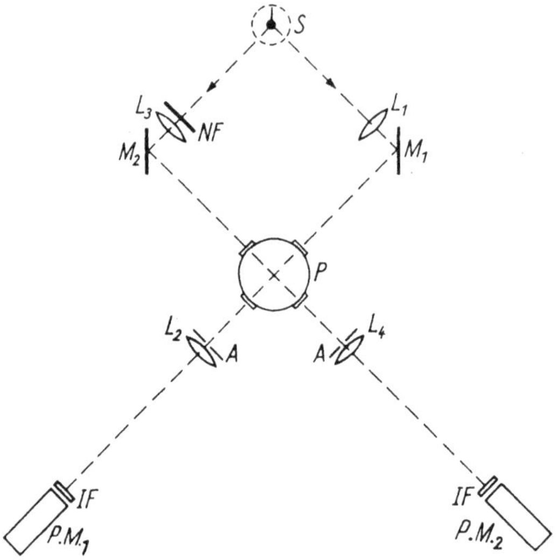

Figure 6.5. Optical layout for high-speed determination of the line-reversal temperature by the two-path method (after Gaydon and Hurle [6.29])

S – source of continuous radiation; NF – neutral filter; P – plasma; A – aperture stop; IF – interference filter; $P.M.$ – secondary electron multiplier

ature T_1 of the source is reduced to a lower temperature T_2. The sensitivity of the multipliers is adjusted in such a manner that the anode currents become equal when the neutral filter is removed.

If the plasma temperature lies between T_1 and T_2, the anode current of P.M. 1 is reduced by a value a compared to the current when the plasma is not present, while the anode current of P.M. 2 increases by a value e. When the difference between T_1 and T_2 is not too large, the plasma temperature is obtained by interpolation:

$$T = T_2 + \frac{e\,(T_1 - T_2)}{a + e}\,. \tag{6.12}$$

Gaydon and Hurle [6.30] indicated an accuracy of ± 30 °K for the case when $T_1 = 2500$ °K and $T_2 = 2350$ °K. By recording the photocurrents e and a on a cathode-ray oscilloscope a time resolution of the order of 1 μs can be achieved without any difficulty.

When T_1 and T_2 differ strongly, or when the plasma temperature T is larger than the black body temperature T_1 of the unattenuated continuous radiation, we must refer back to the system of equations (6.3) and (6.5) to determine T more accurately.

Let us consider the latter case where the plasma temperature T is larger than the black body temperature T_1 of the continuous source. The experimental layout (Figure 6.5) is altered in such a way that instead of the neutral filter a nontransparent disk is employed. The sensitivities of the two multipliers are again adjusted so that both multipliers have the same anode current when the opaque disk is removed.

During measurement only radiation from the plasma reaches P.M. 2, while multiplier P.M. 1 receives radiation from the plasma as well as the continuous background radiation attenuated by the plasma.

Denoting the increase in the anode current of P.M. 1 by i_1, which results when the plasma moves through the measuring range or when in the measuring range the plasma is produced at a point where it did not exist beforehand, and denoting the resulting increase in the anode current of P.M. 2 due to the same circumstances by i_2, we have the following expression for these two quantities (see also equation (6.5))

$$\left.\begin{aligned} i_1 &= K\,B_\nu(T) \int A_\nu\,D(\nu)\,d\nu + K\,B_\nu(T_1) \int A_\nu\,D(\nu)\,d\nu\,, \\ i_2 &= K\,B_\nu(T) \int A_\nu\,D(\nu)\,d\nu\,. \end{aligned}\right\} \tag{6.13}$$

Hence it follows

$$B_\nu(T) = B_\nu(T_1) \cdot \frac{i_2}{i_2 - i_1}\,. \tag{6.14}$$

For the case $e^{-h\nu/kT} \ll 1$ we may use Wien's law instead of the Planck radiation law:

$$B_\nu'(T) = \frac{2\,h\,\nu^3}{c^2} \cdot e^{-h\nu/kT} \qquad [\text{erg cm}^{-2}\,\text{sr}^{-1}]\,. \tag{6.15}$$

The deviation between the Planck radiation law and Wien's Law is less than 1% when the inequality

$$\lambda\,T < 3 \cdot 10^7\ \text{Å} \cdot {}^\circ\text{K} \tag{6.16}$$

holds. (For $T = 3000\,^\circ$K we get, for example, $\lambda < 10{,}000$ Å.)

Then instead of (6.14), we obtain for the temperature

$$e^{-h\nu/kT} = e^{-h\nu/kT_1} \cdot \frac{i_2}{i_2 - i_1}\,. \tag{6.17}$$

If T is greater than T by more than 300 °K, the measurement accuracy of this method begins to drop (see [6.30], p. 168).

6.1.7. Temperature Determination from Black Body Radiation Intensity

In a number of cases the absorptivity of the plasma will be so great that blackbody radiation law can be used to determine the plasma temperature [6.6a], [6.9a], [6.13b], [6.33b].

In the method given by Kandyba [6.33b], the profile of an optical dense line is scanned by means of a Fabry-Perot-interferometer, and from the height of the flat intensity profile in the line center the plasma temperature (black body temperature) is derived.

Basov et al. [6.6a] have determined the radiation temperature-time-dependence for an exploding Cu wire plasma from the spectral intensity at the 3 wavelengths 2400, 3340 and 5000 Å. From the fact that the three $T(x)$ curves lie close together, they concluded that the curves reflect the true temperature-time-profile in the temperature range from 20 to $50 \cdot 10^3$ °K.

Bauder and Plantikow [6.9a] have determined the temperature in a high pressure argon arc at 10 atm from the maxima of the spectral intensity distribution in the region of the red argon lines observed in axial direction. For the temperature they found values ranging from 13 to 18.10^3 °K.

In [6.13b], the temperature in a z-pinch ranging up to 80,000 °K was derived from the spectral intensity in the center of the optical thick He I line at wavelength 5876 Å.

6.1.8. Practical Examples of the Line-reversal Method for Temperature Determination

To illustrate some of the possible applications of the line-reversal method, we have compiled the most important experimental data of some experiments in tabulated form (see Table 6.1). Insofar as known, the measurement accuracies are stated.

The measurement accuracy is on the one hand limited due to experimental and physical causes by the scattering of the physical quantities to be measured, and on the other by the inaccuracy in determining the black body temperature of the continuous source. By no means less significant is the measurement accuracy determined by the statistical fluctuations in the current produced by the photons. For more details see [6.30], p. 168.

6.2. Temperature Determination in Inhomogeneous Plasmas by Kirchhoff's Law

We shall limit ourselves to plasmas having rotational symmetry and treat that case first, in which we can work with light whose spectrum is sufficiently well-resolved, so that within the transmission range of the measuring system the absorption coefficient $\varkappa_\nu$ of the plasma can be considered constant. Then we shall discuss the case in which an entire spectral line falls within the transmission range of the apparatus.

6.2.1. Temperature Determination by Means of Monochromatically Resolved Radiation

The determination of the temperature with monochromatically resolved radiation in inhomogeneous plasmas is based on Kirchhoff's law in the form

$$\frac{\varepsilon_\nu(T)}{\varkappa_\nu(T)} = B_\nu(T) \tag{6.18}$$

Table 6.1. Application of the line reversal method for temperature determination. The symbols in column 3 have the following meanings: R: the temperature is determined while background radiation is adjusted to the reversal point. $E + A$: Background radiation is not adjusted; the temperature is determined from emissivity and absorptivity.
The meanings of the symbols used in column 4 are: P – projection lamp or Pointolite lamp, W – tungsten ribbon lamp, Xe – high pressure Xenon lamp, $Xe(Pu)$ – pulsed discharge in Xenon, C – crater of the carbon arc, Cap – discharge through a textolite capillary. On the right hand side from the short symbol is given the maximum value for the operation temperature T_s (in [10^3 °K]).
"Or" in column 7 means "Order of magnitude".
The references are given as short symbols (for example, 30a instead of [6.30a])

Plasma	Wavelength	Method	Continuous radiator	Measured temperature [10^3 °K]	Precision ± [°K]	Time resolution	Reference
Hefner candle	CO_2 4.25 μm N_2O 4.45 μm CO 4.65 μm	$E + A$	P	1.81 – 2.00	20 – 90	0.1 s	40
Flame	Na D	R	W 2.8, C 2.9	1.6 – 2.5	10 – 25	Or. 0.1 s	16, 25a 62a
Flat oxygen and air flame	4.4 – 6.0 μm	$E + A$		1.5 – 2.6			62a
Pulverized powder-combustion	Na D 5890/96 Å	R	P 3.1	1.54 – 2.4	25	3 ms	12
Solid powder combustion	Na D	R	P 2.8	2.35 – 2.67	30	0.1 s	64
Ar–K channel flow	K I 7664/98 Å	R	W	1.7 – 3.0	80		15c, 47l
Expansion wave, N_2	Na D; Cr I 4254/89 Å	$E + A$	P 3.0	1.4 – 2.8	30		32t, 35c
Expansion wave, Ar	Na D	$E + A$		1.4 – 2.6			32s
Shock wave, fuel–O_2-mixture	Cr I 4254 – 4289 Å CrO 5794	$E + A$	P 2.6	2.2 – 3.3		Or. 1 μs	30a
Shock wave, CO, H_2 + Ar, Ar, 2 CO + 3 O_2	Na D, Cr I 5204/08 Å	$E + A$	P 2.8, Xe	2.15 – 2.94		Or. 1 μs	29

Table 6.1. (Continuation)

Plasma	Wavelength	Method	Continuous radiator	Measured temperature [10^3 °K]	Precision ± [°K]	Time resolution	Reference
Shock wave, N_2	In 4511 Å	$E + A$	K 3.6	2.6 – 3.7	100	Or. 1 μs	17
Shock wave, Air, N_2, Ar	Na D, Ba II 4554 Å	$E + A$	Xe 4.8	2.0 – 4.5	70	Or. 1 μs	26a, 28
Shock wave, Ar	Ca I 4425[1]), 6362 Å	$E + A$	Xe (Pu)	5.8 – 6.8			28d
Reflected shock wave, CO_2, O_2, N_2, Air	Ba II 4554 Å	$E + A$	Xe(Pu) 11	2.5 – 7.0	200–300	Or. 1 μs	35, 52
Shock wave, Ne	Cr I 5208–5328 Å	$E + A$	Xe(Pu)	4.5 – 7.0	<150	Or. 1 μs	16e
Shock wave, Air	N_2^+ 4238 Å	$E + A$	Cap. 30	10 – 13	1000	Or. 1 μs	33h
Shock wave, Ar, Xe, Air, Ar + Air	continuum 0.33–0.56 μm	[2])	without	10 – 90	500–1000	0.1–0.2 μs	78a
Plasma jet, N_2	CO 4.4–4.6 μm	$E + A$	Globar	1.2 – 2.0			61b, 62
Plasma jet, He	continuum 4.05 μm	$E + A$	Globar	4 – 11			62
HF-plasma jet, Air	Mn	R	Xe	3			61
High pressure discharge, Xenon	Xe 8231–9140 Å	[2])	without	8.6 – 8.8			10

[1]) Autoionizing line.
[2]) Temperature is derived from line emission for 2 different sheath thickness.

where

ε_ν = emission coefficient per unit frequency (dimensions: erg $cm^{-3} sr^{-1}$
$\varkappa_\nu$ = absorption coefficient (dimension: $[cm^{-1}]$),
$B_\nu(T)$ = intensity per unit frequency of the black body at temperature T (dimensions: [erg cm^{-2} sr^{-1}], see equation (6.2)).

Equation (6.18) applies to the radiation in a narrow frequency range, in which the variation of the emission and absorption coefficient is insignificant.

The temperature is obtained in accordance with equation (6.18) from the measured emission coefficient ε_ν and absorption coefficient $\varkappa_\nu$.

Since we have already discussed the determination of ε_ν and $\varkappa_\nu$ and their general significance in section 2.10.2, we need not consider this aspect in detail at this stage.

First of all, $\varkappa_\nu(r)$ can be calculated from the absorptive power $A_\nu(x)$ of the plasma by means of (2.71) and (2.73). $A_\nu(x)$ can be found by first measuring the radiation intensity $I_\nu(x)$ directly and then indirectly when a mirror with reflectivity f is placed behind the plasma (measured value $K_\nu(x)$). Then, with the help of (2.70) $A_\nu(x)$ can be found from $I_\nu(x)$, $K_\nu(x)$, and f.

It should be noted that $A_\nu(x)$ can be also determined with high accuracy without using the mirror, when the plasma is irradiated by a background source S of intensity I_{ν_0} (see Figure 6.6). Upon passing through the plasma the

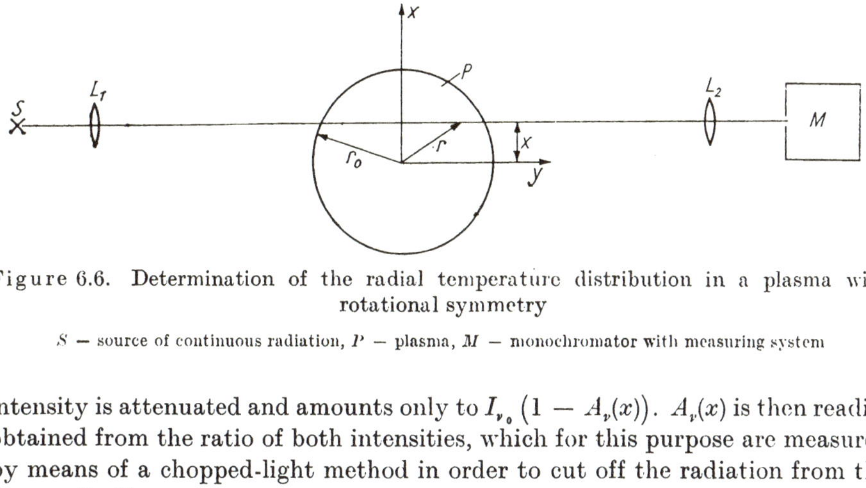

Figure 6.6. Determination of the radial temperature distribution in a plasma with rotational symmetry

S — source of continuous radiation, P — plasma, M — monochromator with measuring system

intensity is attenuated and amounts only to $I_{\nu_0}\left(1 - A_\nu(x)\right)$. $A_\nu(x)$ is then readily obtained from the ratio of both intensities, which for this purpose are measured by means of a chopped-light method in order to cut off the radiation from the plasma itself.

Braun and coworkers [6.13f], (6.15a], [6.15b] used as light source for absorption measurement the central portion of the arc under investigation itself. The lateral scanning for obtaining the radial absorptivity profile was effected by the synchroneously motion of two mirrors. One of them, a concave mirror, reflects the radiation from the arc center back into the arc, where it crosses the arc at the variable axis distance x before it arrives at the detection system.

After $\varkappa_\nu(r)$ is known, $\varepsilon_\nu(r)$ can be found for discrete r-values from the measured intensity distribution $I_\nu(x)$, following the procedures given in sections 2.10.2.2. and 2.10.2.3.

Finally, the radial temperature distribution follows from $\varepsilon_\nu(r)$ and $\varkappa_\nu(r)$ in accordance with (6.18).

Strong and Bundy [6.57] used this method (with a slightly different mathematical formulation) to determine the temperature in the middle of a propane flame. They found a temperature of 2430 °K. The frequency was set in the range of the Na D-lines.

Examples for the determination of the full radial temperature profile in plasmas of rotational symmetry are given in table 6.2.

6.2.2. Temperature Determination by Integrating over a Spectral Line

When the spectral line falls in the transmission range $\Delta\nu$ of the spectroscopic device, Kirchhoff's law takes on the form

$$\frac{\int\limits_{(\nu)} \varepsilon_\nu \, d\nu}{\int\limits_{(\nu)} \varkappa_\nu \, d\nu} = B_\nu(T) . \tag{6.19}$$

The range of integration is identical to the transmission range of the apparatus.

The problem consists of determining the quantities $\int\limits_{(\nu)} \varepsilon_\nu \, d\nu$ and $\int\limits_{(\nu)} \varkappa_\nu \, d\nu$.

To solve this problem we refer to the experimental layout shown schematically in Figure 6.6. The radiation from a continuous source S with intensity $B_\nu(T_s)$ passes through the plasma at a distance x from the axis and reaches the monochromator-measuring system, where the photocurrent i is measured, which is proportional to the intensity I.

We assume that the following intensities are measured:

1. the intensity I_1 from the comparison source S alone:

$$I_1 = B_\nu(T_s) \, \Delta\nu , \tag{6.20}$$

2. the combined intensity from the comparison source and the plasma:

$$I_2 = B_\nu(T_s) \int\limits_{(\nu)} e^{-\int\limits_{-y(x)}^{+y(x)} \varkappa_\nu(r) \, dy} \, d\nu + I_3 \quad \text{where} \quad y(x) = \sqrt{r_0^2 - x^2} , \tag{6.21}$$

and 3. the intensity from the plasma alone:

$$I_3 = \int\limits_{(\nu)} \int\limits_{-y(x)}^{+y(x)} \varkappa_\nu(r) \, B_\nu\big(T(r)\big) \cdot e^{-\int\limits_{y}^{y(x)} \varkappa_\nu(\xi) \, d\xi} \, dy \, d\nu . \tag{6.22}$$

The absorptive power of the plasma (ξ = auxiliary variable in the observation direction) is defined by

$$A(x) = \frac{I_1 - I_2 + I_3}{I_1} \quad \text{or} \quad A(x) = \frac{i_1 - i_2 + i_3}{i_1} .$$

Table 6.2. Application of Kirchhoff's law for the determination of the radial temperature profile

Plasma	Absorber, wavelength	T_{max} [°K]	Background radiator	Wavelength resolution [Å]	Reference
Diffusive propane-air flame, 1 atm	1. carbon dust particles, diam. 0.23 – 1.0 μm 2. Na D, 5890 Å	1900	tungsten ribbon lamp	0.10	[6.35a]
HF discharge, Ar, 1 atm	Ar I 7515, 7635 Å	9500	flash light source, 29,000 °K	0.09	[6.32x]
Xe-compact arc, 2.2 kW, $p > 1$ atm	Xe, 0.3 and 2.7 μm	10,800	carbon crater		[6.33f]
Ar-arc, 3 atm	Ar I 6967 Å	16,000	the arc under investigation itself	0.13	[6.15b]
Hydrogen arc, 1 atm	H_α	17,000	the arc under investigation itself	0.5	[6.13f]

With exception of the case in which the absorption coefficient $\varkappa_\nu$ can be regarded as following a rectangular curve, the determination of $T(r)$ from the measured values $I_1(x)$, $I_2(x)$, and $I_3(x)$ is complicated. We shall here limit ourselves to the cases of weak and strong self-absorption, whereby in the latter case it is assumed that the spectral line has a dispersion profile.

6.2.2.1. Weak self-absorption

We start with equation (6.22) and the plasma absorption $A(x)$, defined by

$$A(x) = \frac{i_1 - i_2 + i_3}{i_1} \tag{6.23}$$

and obtain the expression

$$A(x) = \frac{\int\limits_{(\nu)} \left(1 - e^{-\int_{-y(x)}^{+y(x)} \varkappa_\nu(r)\,dy}\right) d\nu}{\Delta\nu} . \tag{6.24}$$

We can expand the exponential functions in (6.22) and (6.24) in power series and retain only those terms in the integral which contain $\varkappa$ to the first power. We then have

$$I_3 = \int\limits_{(\nu)} \int\limits_{-y(x)}^{+y(x)} \varkappa_\nu(r)\, B_\nu(T(r))\, dy\, d\nu \tag{6.25}$$

and

$$A(x) = \frac{\int\limits_{(\nu)} \int\limits_{-y(x)}^{+y(x)} \varkappa_\nu(r)\, dy\, d\nu}{\Delta\nu} .$$

We introduce the expression

$$\int\limits_{(\nu)} \varkappa_\nu(r)\, d\nu = a(r) \tag{6.26}$$

(integral absorption) and obtain

$$A(x) = \frac{1}{\Delta\nu} \int\limits_{-y(x)}^{+y(x)} a(r)\, dy . \tag{6.27}$$

The solution of this integral equation is of the Abel type (see section 2.10.2) and is given by

$$\frac{a(r)}{\Delta\nu} = -\frac{1}{\pi} \int\limits_r^{r_0} \frac{A'(x)\, dx}{\sqrt{x^2 - r^2}} . \tag{6.28}$$

We now divide (6.25) by (6.20), allowing for (6.26), and obtain

$$I_{31}(x) \equiv \frac{I_3(x)}{I_1(x)} = \frac{\int\limits_{-y(x)}^{+y(x)} \frac{a(r)}{\Delta\nu} B(T(r))\, dy}{B_\nu(T_s)} \tag{6.29}$$

whose solution is

$$B\big(T_\nu(r)\big) = -\frac{B_\nu(T_s)\,\Delta\nu}{\pi\,a(r)} \int\limits_r^{r_0} \frac{I'_{31}(x)\,\mathrm{d}x}{\sqrt{x^2 - r^2}}\,, \tag{6.30}$$

from which $T(r)$ can be determined.

6.2.2.2. Strong self-absorption

Simon [6.51] calculated the line-reversal temperatures of strongly absorbing plasmas having rotational symmetry in the following way.

Starting with equations (6.20)—(6.22) for the measured intensities $I_1(x)$, $I_2(x)$, and $I_3(x)$, consider the case in which for each x the black body temperature T_s of the comparison source is controlled in such a way that line-reversal occurs. The condition for this is $I_1 = I_2$. From this follows the expression

$$B_\nu(T_s) \int\limits_{(\nu)} \left[1 - \mathrm{e}^{-\int_{-y(x)}^{+y(x)} \varkappa_\nu(r)\,\mathrm{d}y}\right] \mathrm{d}\nu = \int\limits_{(\nu)} \int\limits_{-y(x)}^{y(\nu)} \varkappa_\nu(r)\, B_\nu\big(T(r)\big)\, \mathrm{e}^{-\int_{-y}^{y(x)} \varkappa_\nu(r)\,\mathrm{d}\xi}\, \mathrm{d}y\,\mathrm{d}\nu\,. \tag{6.31}$$

For spectral lines with a dispersion profile the following relationship holds approximately for the absorption coefficients (see also equation (5.68))

$$\varkappa_\nu(r) = \frac{e^2\, n_n(r)\, f_{m\,n}\, \gamma(r)}{4\,\pi\, m\, c} \cdot \frac{1}{(\nu - \nu_0)^2}\,. \tag{6.32}$$

Here $\gamma(r)$ is the damping constant or the full half-width of the dispersion curve (in units of circular frequency). n_n is the number of occupied levels in the state n.

Thus, with (6.31) we obtain

$$B_\nu(T_s) = B_\nu\big(T(r = x)\big) \int\limits_{-y(x)}^{+y(x)} \frac{b'(y)}{b'(x)}\,\mathrm{d}y \cdot \frac{\sqrt{2}}{4 \cdot y(x)} \cdot \frac{n_n(x)}{n_n(\bar{\xi})}\, \frac{\gamma(x)}{\gamma(\bar{\xi})}\,. \tag{6.33}$$

For $0 < \bar{\xi} < y(x)$, we have the equation of definition

$$\int\limits_{-y(x)}^{+y(x)} \varkappa_\nu(\xi)\,\mathrm{d}\xi = 2\, y(x)\, \varkappa_\nu(\bar{\xi})\,. \tag{6.34}$$

The quantities $b'(y)$ are the local intensities with self-absorption

$$b'(y) = \int\limits_{(\nu)} \varkappa_\nu(y)\, B_\nu\big(T(r)\big)\, \mathrm{e}^{-\int_y^{y(x)} \varkappa_\nu(\xi)\,\mathrm{d}\xi}\, \mathrm{d}\nu\,. \tag{6.35}$$

Simon further simplified the procedure for calculating $T(r)$. Expression (6.33) can be written in the form

$$B_\nu(T_s) = B_\nu\left(T\,(r = x)\right) \cdot \frac{\beta(x)}{\alpha(x)} \tag{6.36}$$

where $\beta(x)$ and $\alpha(x)$ are corrective factors, which are approximately given by

$$\alpha(x) = \int_{-y(x)}^{+y(x)} \frac{a(r)\, dy}{a\,(r = x)}\,, \tag{6.37}$$

$$\beta(x) = \int_{-y(x)}^{+y(x)} \frac{a(r)\, B_\nu(r)}{a\,(x = r)\, B_\nu\,(x = r)}\, dy\,. \tag{6.38}$$

For the approximate determination of $a(r)$ and $B_\nu(r)$ in (6.37) and (6.38), respectively, we can apply expressions (6.28) and (6.30) with allowance for (6.23), which are derived for the case of weak self-absorption.

6.3. Temperature Determination from the Peak Intensities of Self-reversed Lines and Profile Correction of Self-reversed Lines

6.3.1. Temperature Determination

Quite often one observes self-reversal in spectral lines. This occurs when the absorption coefficient $\varkappa_\nu$ at the line center becomes so large that the optical depth $\tau_\nu = \int_{(y)} \varkappa_\nu\, dy$ of the plasma for this wavelength is large compared to 1.

Self-reversal manifests itself predominantly in strong spectral lines, which in general have a large absorption coefficient. Since the absorption coefficient and the line width are inversely proportional to one another, it follows that for two lines of equal intensity but different width the line with the smaller width will have the greater absorptive power (see (2.92) and (2.93)). The line width, with exception to the lines corresponding to transitions to a resonance level, is given by the width of the upper level. In the case of Stark broadening the line becomes wider as the energy of the upper level approaches the ionization energy. From this it follows that those lines whose upper level lies deep tend toward self-reversal.

To clarify the occurrence of self-reversal in the case of side-on observation of strongly absorbing inhomogeneous plasmas we mention that the main portion of the line radiation reaches the measuring apparatus from ranges with smaller optical depth τ and that the ranges with greater optical depth (e.g., $\tau > 4$) do not contribute to the observed radiation.

In the line center, where the absorption coefficient is large, an appreciable optical depth is already reached for a small geometric path length. The radiation then comes from colder edge layers and has a low intensity in accordance with the Planck radiation law. Approaching the edge from the center of the spectral line, the increasingly deeper lying hotter layers of the plasma contribute to emission, and the observed intensity of the spectral line increases. In the wings the absorption coefficient becomes so small that the total optical depth of the plasma from which radiation reaches the measuring device, is smaller than 1.

In this spectral range the observed intensity begins to drop increasingly as one moves away from the line center. As a consequence two intensity maxima — the so-called reversal peaks — appear on the right and the left between the line center and both wings.

To illustrate the line shapes of self-reversed spectral lines we show the spectral intensity distribution of the radiation of a maximum-pressure mercury lamp of the type OSRAM HBO 200 (Figure 6.7, after Bartels [6.5]). In the figure several self-reversed lines can be recognized.

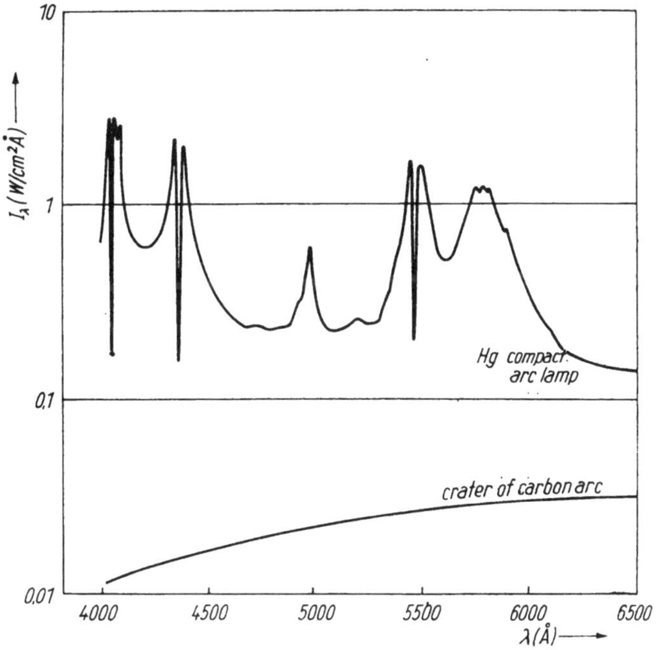

Figure 6.7. Section of the spectrum of a peak-load maximum-pressure mercury arc (after Bartels [6.5])

Bartels showed that it is possible to determine the plasma temperature from the intensity of the reversal peaks. For this the following conditions must be fulfilled:

1. The plasma has rotational symmetry.
2. The ratio of the emission and absorption coefficient (for thermal equilibrium this is the Planck function, otherwise this is the source function) depends only on the axial distance r.
3. There is a Boltzmann equilibrium between the excited spectral levels characterized by the temperature T.
4. In the middle of the line of sight ($y = 0$) T decreases with the square of the coordinate y.
5. The line under investigation is not a resonance line.

In order to derive his method of temperature determination he had to make use of complicated mathematical tools, which, in the final analysis, however, yielded very simple expressions from which the temperature could be readily obtained [6.3], [6.4], [6.6].

We start with the expression for the spectral intensity I_ν when observing at right angles to the axis of a rotationally symmetric plasma along a line of sight, whose distance from the axis is x (we refer to Figure 6.6 without the background source). We have (compare also equation (2.82))

$$I_\nu = \int_{-y(x)}^{+y(x)} B_\nu(r)\, \varkappa_\nu(r)\, \mathrm{e}^{-\int_y^{y(x)} \varkappa_\nu(r)\, \mathrm{d}y'}\, \mathrm{d}y \tag{6.39}$$

$$\text{where} \quad y(x) = \sqrt{r_0^2 - x^2}\,. \tag{6.40}$$

The integral in (6.39) can be reduced to dimensionless form, when the right hand side is expanded by $B_\nu\,(r = x)$ (= maximum value of the Planck function in the center of the line of sight $y = 0$) and the optical depth $\tau_\nu(x)$ of the total irradiated layer:

$$I_\nu = B_\nu\,(r = x)\, \tau_\nu(x) \int_{-y(x)}^{+y(x)} \frac{B_\nu(r)}{B_\nu\,(r = x)} \cdot \frac{\varkappa_\nu(r)}{\tau_\nu(x)} \cdot \mathrm{e}^{-\tau_\nu(x) \int_y^{y(x)} \frac{\varkappa_\nu(r)}{\tau_\nu(x)}\, \mathrm{d}y'}\, \mathrm{d}y\,. \tag{6.41}$$

Next, the dimensionless quantity Φ is introduced, whose significance can be found from the equation of definition

$$I_\nu(x) = \Phi \int_{-y(x)}^{+y(x)} B_\nu(r)\, \varkappa_\nu(r)\, \mathrm{d}y \tag{6.42}$$

in which $\int_{-y(x)}^{+y(x)} B_\nu(r)\, \varkappa_\nu(r)\, \mathrm{d}y$ denotes the total spectral intensity emitted by the plasma along the line of sight $x = \text{const}$. Φ gives the fraction of this intensity which reaches the observer. Φ depends on the distribution $\varkappa_\nu(r)$ and $B_\nu(r)$. For an optically thin plasma $\Phi = 1$.

We thus have

$$I_\nu(x) = B_\nu\,(r = x)\, \tau_\nu(x)\, \Phi \int_{-y(x)}^{+y(x)} \frac{B_\nu(r)}{B_\nu\,(r = x)} \cdot \frac{\varkappa_\nu(r)}{\tau_\nu(x)}\, \mathrm{d}y\,. \tag{6.43}$$

Combining $\tau_\nu(x)\, \Phi$ to a single function $Y(\tau_\nu(x))$ and setting

$$\int_{-y(x)}^{+y(x)} \frac{B_\nu(r)}{B_\nu\,(r = x)} \cdot \frac{\varkappa_\nu(r)}{\tau_\nu(x)}\, \mathrm{d}y = M \tag{6.44}$$

we obtain

$$I_\nu(x) = B_\nu\,(r = x)\, M\, Y(\tau_\nu(x))\,. \tag{6.45}$$

It follows from (6.44) that M corresponds to the fraction by which the irradiated power drops due to the decrease in the temperature with the radius. M depends on the structure of the layer and in the case of a homogeneous layer equals 1.

The quantity $Y(\tau_\nu(x))$ — we shall now designate it by $Y(\tau_0)$ for the sake of simplicity — can be represented by the one-parameter family of curves

$$\overline{Y}(\tau_0, p) = e^{-\tau_0/2} \left\{ \left[\frac{\tau_0}{2} (1 - p) + p \operatorname{sh} \frac{\tau_0}{2} \right] + \frac{1}{\sqrt{p}} \operatorname{sh} \frac{\tau_0 \sqrt{p}}{2} \right\} \tag{6.46}$$

with the parameter

$$p = 12 \frac{\displaystyle\int_{-y(x)}^{+y(x)} \left[\int_0^y \frac{\varkappa(y')}{\tau_0} \, dy' \right]^2 \cdot \frac{B_\nu(r)}{B_\nu(r = x)} \frac{\varkappa_\nu(r)}{\tau_0} \, dy}{\displaystyle\int_{-y(x)}^{+y(x)} \frac{B_\nu(r)}{B_\nu(r = x)} \cdot \frac{\varkappa_\nu(r)}{\tau_0} \, dy} . \tag{6.47}$$

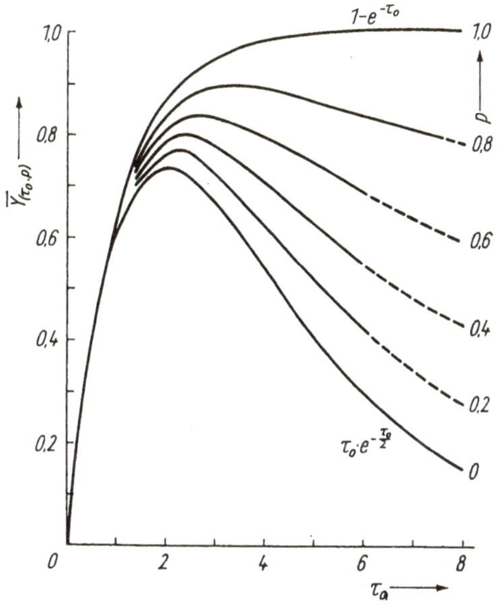

Figure 6.8. The one-parameter family of curves $\overline{Y}(\tau^0, p)$ (after Bartels and Beuchelt [6.6])

The function $\overline{Y}(\tau_0, p)$ is illustrated in Figure 6.8. $p = 1$ (homogeneous layer) and $p = 0$ (luminous filaments in an absorbing sheath) correspond to the two limiting cases:

$$p = 1: \quad \overline{Y} = 1 - e^{-\tau_0} , \tag{6.48}$$

$$p = 0: \quad \overline{Y} = \tau_0 \, e^{-\tau_0/2} . \tag{6.49}$$

Bartels [6.6] showed that the parameter p could be approximately described by the expression

$$p = \frac{6}{\pi} \tan^{-1} \frac{M^2}{\sqrt{1 + 2 M^2}} . \tag{6.50}$$

M can be approximated in 3 ways, according to the different assumptions made.

1. For a spectral line in the atomic spectrum it follows that in the spectral range of the entire spectral line

$$M = \sqrt{\frac{U_n}{U_n + h\nu}} . \tag{6.51}$$

This is based on the assumption, that there is no continuous background. U_n is the excitation energy of the lower level.

2. For all spectral lines of the singly ionized ion spectrum we have

$$M = \sqrt{\frac{\frac{1}{2} U_i + U_n}{\frac{1}{2} U_i + U_n + h\nu}} . \tag{6.52}$$

Here, again, it is assumed that the spectral line is not superposed by a continuum. U_n is the excitation energy of the lower level. U_i is the ionization energy of the neutral atom.

3. When continuous emission (electron-ion continuum) occurs, the excitation energy U_g of the level, which forms the lower boundary of the blurred zone in the level diagram, is introduced. Formally, it plays the same role as the excitation energy of the upper level in atomic spectrum lines. Accordingly it follows that

$$M = \sqrt{\frac{U_g - h\nu}{U_g}} . \tag{6.53}$$

All 3 approximate expressions are valid for the case $kT \ll U_n$ or $kT \ll U_g - h\nu$.

When the line is superimposed on a continuum, intermediate values result in accordance with the portion which the line and the continuum each contribute to absorption.

As an illustration M and p are given in Figure 6.9 as a function of wavelength for the mercury spectrum. From the figure it follows that M and p exhibit only slight variations.

After these preparatory remarks we now return to the temperature determination.

Basing our argument on the following equation for the line intensity

$$I_\nu(x) = B_\nu (r = x) \cdot M \cdot \overline{Y}(\tau_0, p) , \tag{6.54}$$

it follows that the frequency dependence of the intensity is included in the function $\bar{Y}(\tau_0, p)$. As regards the frequency dependence, $\bar{Y}$ as a function of τ_0 is considered an image of the function $I(\nu)$. The maximum of $I(\nu)$ (the peak

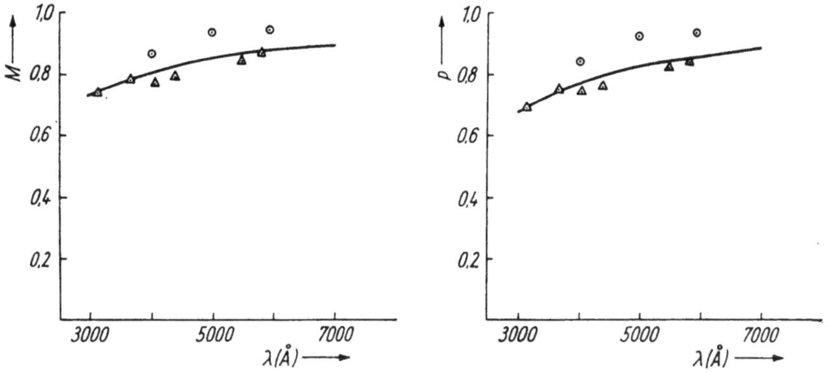

Figure 6.9. M and p for different lines and for the continuum of a mercury arc (after Bartels and Beuchelt [6.6])

The solid curves refer to the continuum, the points designated by △ to lines of the neutral atom, and the points designated by ⊙ to lines of a singly ionized ion

intensity $I_{\nu P}$) is associated with the maximum of the curve $\bar{Y}(\tau_0, p)$. Then from (6.54) it follows

$$I_{\nu P}(x) = B_\nu (r = x) \cdot M \cdot Y_{\max} , \tag{6.55}$$

where M and p are given by (6.50)—(6.53) and $Y_{\max}(p)$ as a function of p can be taken from Figure 6.10. $Y_{\max}$ can also be described by the relationship

$$Y_{\max} = 0.736 + 0.264 \cdot p^2 . \tag{6.56}$$

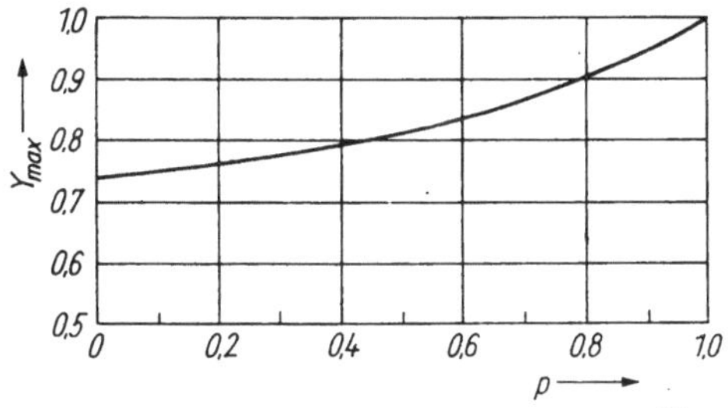

Figure 6.10. The maximum values $Y_{\max}(p)$ of the curves $\bar{Y}(\tau_0, p)$ as a function of p (after Bartels [6.3])

Since the peak intensity $I_{\nu P}(x)$ can be measured, we can also find B_ν $(r = x)$ via (6.55) and thus the temperature T $(r = x)$. Applying equation (6.55) at different distances x of the line of sight from the symmetry axis, we get the radial temperature distribution $T(r)$, though not entirely up to the plasma edge where the optical depth is no longer sufficient to produce self-reversal.

The theory of self-reversal for the determination of the temperature of a discharge cylinder is applied in a somewhat different form to exploding wires (Bartels [6.5]). When the wire is sufficiently thin a cylindrically symmetric plasma is obtained. Since the spectrum is essentially continuous and the optical depth τ practically does not vary with wave length, it is now possible to reproduce the function $Y(\tau)$ in the sense that the radiation is not only observed at right angles to the cylinder axis but also at different angles φ to this normal direction. Due to these rotations the initially unknown optical depth τ_0 is increased by a factor $1/\cos\varphi$. Under these conditions p and M, which initially are also unknown, remain unchanged. Then the relative changes in the intensity corresponds exclusively to the change in the function $Y(\tau_0)$ due to a change in τ_0. Now the measured values should be fit into the $\overline{Y}(\tau_0, p)$ — plot (see Figure 6.11)

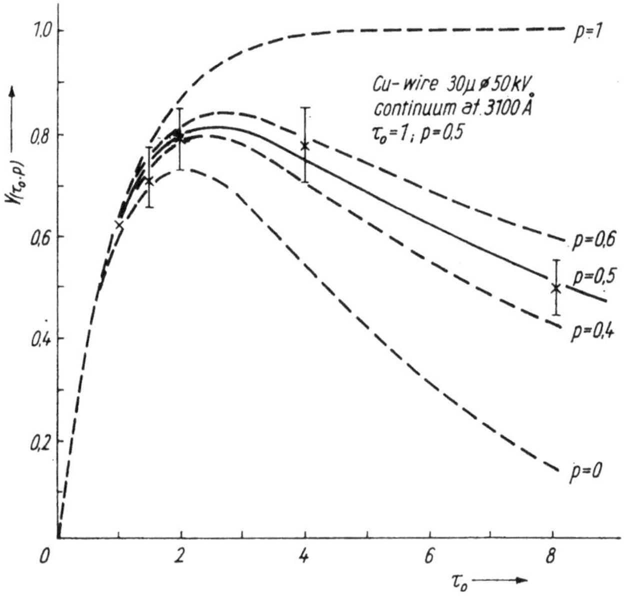

Figure 6.11. Determination of plasma temperature in exploded wires (after Bartels [6.5])

in such a way that a curve $p = \text{const}$ is described with the closest possible approximation. In the example shown $p = 0.5$; in addition, $\tau_0\,(\varphi = 0)$ was found to be 0.9. With p we also know M from (6.50), and T can be determined in the usual fashion (from equation (6.55), when instead of the peak intensity $I_\nu\, P(x)$ the maximum $I_\nu(\varphi)$ of the measured curve is taken).

To demonstrate the possibilities of applying the peak-intensity method in temperature determinations in rotationally symmetric plasmas, we compiled some relevant experimental data in Table 6.3.

6.3.2. Profile Correction of Self-reversed Spectral Lines

We can treat the profile correction of self-reversed spectral lines in connection with the previous formulations. Profile correction in the sense meant here involves the elimination of the influence of self-absorption on the shape of spectral lines.

The profile correction can be carried out with high accuracy. First, the radial temperature distribution $T(r)$ is found by the Bartels method (previous section), then $\varepsilon_\nu(r)$ is determined from $I_\nu(x)$ and $T(r)$ in accordance with the method proposed by Olsen (see section 2.10.2.2) for several frequencies ν in the range of the relevant spectral line. The problem is thereby solved. In many cases it is sufficient to carry out the profile correction in a somewhat more simplified and time-saving fashion (Pingel and Siekmeier [6.46]).

Let us consider the apparent line form $I_\nu(x)$, which results when the plasma is observed side-on at a distance x of the line of sight from the plasma axis. The function $I_\nu/I_{\nu\,\max} = f(\Delta\nu)$ ($\Delta\nu$ = frequency interval from the line center) (see Figure 6.12a) is reproduced by the function $Y(\tau_0)/Y_{\max}$ (Figure 6.12b).

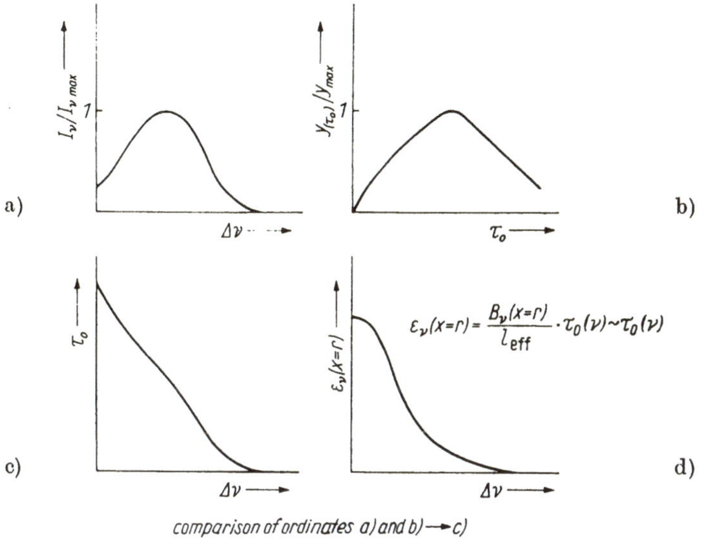

Figure 6.12. Profile correction of self-reversed spectral lines (see text)

Comparing ordinates, we obtain the function $\tau_0 = f(\Delta\nu)$ (Figure 6.12c). Finally, in the approximation of the homogeneous layer

$$\varepsilon_\nu\,(r = x) = \frac{B_\nu(T(r))}{l_{\text{eff}}} \cdot \tau_0(\nu) \sim \tau_0(\nu)\,, \tag{6.57}$$

where $T(r)$ is assumed as known. $\varepsilon_\nu\,(r = x) = f(\nu)$ is the line profile, corrected for self-absorption (see Figure 6.12d) and associated with the emitting volume element at the point $r = x$.

Table 6.3. Application of the Bartels method for temperature determination

Plasma source	Wavelength [Å]	T [°K]	$\pm \Delta T$ [°K]	Reference
arc, K, p = 240 Torr	K I 11771	2650 – 2900		Schmidt [6.47h]
arc, Fe	Fe I 3749, 3815, 3820, 4046	4070 – 5160	200	Startsev and Frish [6.55]
arc, Fe	Fe I 3763, 3767, 3827, 3834	5380	270	Volk-Levanovich [6.66c]
compact arc lamp, Hg	Hg I 3650, 4047, 4358, 5460	7780 – 7900		Bartels [6.5]
compact arc lamp, Hg	Hg I 4047, 4358, 5460	6320 – 7550	30	Volk-Levanovich [6.66b]
compact arc lamp, Hg, pulsed discharge	Hg I 4046, 4358, 5460 Hg II 3983	9000 – 25000		Bartels [6.5]
Xe compact arc lamp, cathodic plasma ball	Xe I 8819	9200 – 9900		Kopec [6.34]
pulsed Xe arc, plasma column	Xe I 8819, 8231	9400 – 10700		Kopec [6.34]
pulsed Xe arc, plasma ball	Xe I 8819, 8231	10200 – 11700		Kopec [6.34]
exploding Cu wire	continuum 3100	63000		Bartels [6.5]
dynamic Z-pinch in He and H	continuum 2000 – 6000	30000 – 120000	10000	Zwicker and Schumacher [6.78c]

From the corrected line profile one finds by integration over frequency the integral line emission coefficient ε.

Knowing the half width of the corrected line profile and the integral emission coefficient, it is possible to determine particle number densities in the plasma, as shown by Grechikhin [6.31a] who has determined the concentration of Li atoms and of electrons in a plasma containing the element Li. In order to eliminate self absorption, he used a numerical procedure somewhat different to that presented here, based on the theory of line profiles in inhomogeneous plasmas in the version of Cowan and Dieke [6.17h] which we will not discuss here. About the differences of this theory to the theory of Bartels see [6.46b], [6.46c].

6.4. Determination of Particle Densities from the Total Absorption Coefficient of Resonance Lines and from the Absorption Coefficient of Continuous Absorption from the Ground State

Neglecting stimulated emission, the total absorption coefficient of a spectral line is given by the expression (compare equation (5.56))

$$\int\limits_{(\nu)} \varkappa_\nu \, d\nu = \frac{\pi e^2}{m c} n_n f_{m n} , \tag{6.58}$$

where n_n is the occupation number of the lower level.

We shall consider the total absorption for resonance lines only and limit ourselves to the case in which essentially only the ground level is occupied, then the occupation number of the latter is practically equal to the number n_0 of those atoms associated with the named spectral line. (That is, if we consider the absorption of the Na-line in a gas containing Na, then n_0 is equal to the number of Na-atoms per cm^3). If we allow the ground term with statistical weight g_0 to be split into several levels with statistical weight $g_{n,0}$, whereby the energies differ only very slightly from one another, we obtain from (6.58)

$$\int\limits_{(\nu)} \varkappa_\nu \, d\nu = \frac{\pi e^2}{m c} \cdot n_0 \cdot \frac{g_{n,0}}{g_0} \cdot f_{m n} . \tag{6.59}$$

The particle density n_0 can now be found from the measured value for $\int\limits_{(\nu)} \varkappa_\nu \, d\nu$ when the oscillator strength $f_{m n}$ is known.

The determination of $\int\limits_{(\nu)} \varkappa_\nu \, d\nu$ was already discussed before. We therefore need only go back to the relevant sections.

We distinguish between the following cases:

1. homogeneous plasma, spectrally resolvable line,
2. homogeneous plasma, spectrally irresolvable line, weak absorption,
3. homogeneous plasma, spectrally irresolvable line, strong absorption,
4. inhomogeneous rotationally symmetric plasma, spectrally resolvable line,
5. inhomogeneous, rotationally symmetric plasma, spectrally irresolvable line, weak absorption.

The last section (6.4.6.) is concerned with possible applications of the absorption method, with inclusion of continuous absorption from the ground level, for the determination of particle number densities.

6.4.1. Determination of the Total Absorption Coefficient $\int_{(\nu)} \varkappa_\nu \, d\nu$ in the Case of a Homogeneous Plasma and Spectrally Resolvable Line

Radiation from an auxiliary source is transmitted through the plasma of length 1. From the unattenuated intensity I_{ν_0} and the attenuated intensity I_ν, the total absorption coefficient

$$\int_{(\nu)} \varkappa_\nu \, d\nu = \frac{1}{l} \int_{(\nu)} \ln \frac{I_{\nu_0}}{I_\nu} \, d\nu \tag{6.60}$$

is derived in accordance with the statements made in section 5.3.5.

6.4.2. Determination of the Total Absorption Coefficient in a Homogeneous Plasma for a Spectrally Irresolvable, Weakly Absorbing Line

We refer back to the statements made in section 5.3.5. From (5.58) and (5.59) it follows that

$$\int_{(\nu)} \varkappa_\nu \, d\nu = \frac{1}{l} \int_{(\nu)} \frac{I_{\nu_0} - I_\nu}{I_{\nu_0}} \, d\nu \,. \tag{6.61}$$

If the intensities are measured photoelectrically and if $\Delta\nu$ is the transmission range of the spectrographic device, within which the spectral line falls, we have instead of (6.61) the following expression

$$\int_{(\nu)} \varkappa_\nu \, d\nu = \frac{\Delta\nu}{l} \, \frac{\int_{(\nu)} I_{\nu_0} \, d\nu - \int_{(\nu)} I_\nu \, d\nu}{\int_{(\nu)} I_{\nu_0} \, d\nu} \,. \tag{6.62}$$

The integrals $\int_{(\nu)} I_{\nu_0} \, d\nu$ and $\int_{(\nu)} I_\nu \, d\nu$ are respectively proportional to the photocurrent when the unattenuated radiation and that attenuated due to plasma absorption fall upon the measuring system. To find the spectral transmission range see section (2.8).

6.4.3. Determination of the Total Absorption Coefficient in a Homogeneous Plasma for a Spectrally Irresolvable, Strongly Absorbing Line

The evaluation of $\int_{(\nu)} \varkappa_\nu \, d\nu$ from $A_\nu = \int_{(\nu)} \frac{I_{\nu_0} - I_\nu}{I_{\nu_0}} \, d\nu$ is possible when the spectral line profile is known, by using for example the curve of growth technique ([6.65a] p. 288). Based on this method, Hiramoto et al. [6.32h] have determined the concentration of potassium in an argon-potassium channel flow used in MHD experiments.

If the spectrum line has a Doppler profile, reference can be made to section 5.3.5. At first, from A_ν and the Doppler width $\Delta\nu_D$, the optical density in the line center C is calculated by means of formula (5.61). Then one obtains

$$\int_{(\nu)} \varkappa_\nu \, \mathrm{d}\nu = \frac{C\sqrt{\pi}}{l} \Delta\nu_D \qquad (6.63)$$

where l is the absorbing length.

Some authors have performed absorption measurements using as back radiation source instead of the usual continuous radiator a line radiator emitting the same line as that under investigation. For the special problems related to this measurement technique see, for example, Torvén [6.61a] and Mitchel and Zemansky [6.42f].

6.4.4. $\int \varkappa_\nu(r) \, \mathrm{d}\nu$ in a Rotationally Symmetric Plasma for a Spectrally Resolvable Line

The absorption coefficient $\varkappa_\nu(r)$ is a function of the radial coordinate and must be separately determined for each frequency within the range of the spectral line.

Next the absorptive power $A_\nu(x)$ of the plasma is obtained as a function of the transverse coordinate x in accordance with the considerations in sections 2.10.2.1 and 6.2. From (2.71) and (2.73) follows $\varkappa_\nu(r)$, and after integrating over ν we finally obtain the sought quantity $\int \varkappa_\nu(r) \, \mathrm{d}\nu$.

6.4.5. $\int \varkappa_\nu(r) \, \mathrm{d}\nu$ in an Inhomogeneous, Rotationally Symmetric Plasma for a Spectrally Irresolvable, Weakly Absorbing Line

Here we refer to the statements in section 6.2.2.1, and in particular to the formulas (6.23), (6.26), and (6.28), from which $\int_{(\nu)} \varkappa_\nu(r) \, \mathrm{d}r$ can be determined.

6.4.6. Possible Applications

The concentrations of atoms (or ions) possessing the lowest first excitation potentials, such as the alkali and alkaline-earth metals, can be determined from the total absorption coefficient of the spectral lines in the visible range of the spectrum. The temperature, however, must be so low that most of the atoms (or ions) are in the ground state.

In the event that the absorptions can be observed in the ultraviolet spectral range, the method can be extended to practically all atomic species.

When the first excited levels lie relatively close to the ground level (as in the case of Ba II where the corresponding excitation energies have the values 0.62 and 0.72 eV, respectively [6.43c]), then the concentration of the atoms or ions in these levels must also be determined, either by absorption technique, or by calculation, to get the total number density of the corresponding atoms or ions. In the latter case, the number densities n_1, n_2 of the excited levels can be derived in a good approximation from the Boltzmann formula $n_{1,2}/n_0 = (g_{1,2}/g_0) \cdot e^{-U_{1,2}/(kT_e)}$ and the measured ground level concentration n_0.

The absorption method can also be applied to those wavelengths, at which only continuous absorption occurs. This corresponds to transitions originating from the ground state of the atom in a continuous state. Then, instead of (6.59), we must allow for the relationship

$$\varkappa_\nu = n_0 \sigma_\nu \tag{6.64}$$

where σ_ν is the effective cross section of an atom for absorbing radiation. The effective cross section is obtained from absorption measurements of a gas with known particle density (see, e.g., Ehler and Weissler [6.25]).

In many works, instead of σ_ν there is given the quantity α, which is identical to the absorption coefficient $\varkappa_\nu$ for normal pressure and temperature (NTP). Avogadro's number is the conversion factor between σ_ν and α:

$$\alpha = 2.687 \cdot 10^{19}\ \mathrm{cm}^{-3} \cdot \sigma\,. \tag{6.65}$$

Some authors present data for the photoabsorption cross section σ_ν in terms of the "oscillator strength per unit frequency" $\mathrm{d}f/\mathrm{d}\nu$ ([6.42b] p. 33). Both quantities are related one to another by the formula

$$\sigma_\nu = \frac{\pi e^2}{m c^2} \frac{\mathrm{d}f}{\mathrm{d}\nu} = 8.82 \cdot 10^{-13} \frac{\mathrm{d}f}{\mathrm{d}\nu}$$

$$(\sigma_\nu \text{ in cm}^2,\ \nu \text{ in s}^{-1})\,.$$

In a gas mixture consisting of N components, the resultant absorption coefficient $\varkappa_\nu$ is given by

$$\varkappa_\nu = \sum_{n=1}^{N} n_n \sigma_\nu^{(n)}\,. \tag{6.66}$$

When the frequency dependencies of the quantities $\sigma_\nu^{(n)}$ differ one from each other, then it is possible from a measurement of $\varkappa_\nu$ at n different frequencies ν_n, $n \geq N$, to derive the individual number densities $n_1 \ldots n_N$, of the gaseous mixture.

The methods described above can also be used to determine the density of molecular gases. In this case, equation (6.59) then refers to the absorption of a whole molecular band, whose lower electron level is a ground level with statistical weight $g_{n,0}$. Accordingly, g_0 denotes the statistical weight of the ground electron term, to which the ground level belongs.

Equation (6.59) therefore only holds when the temperatures are so low (not higher than a few 1000 °C) that essentially only the ground vibrational state is occupied. For details on the decrease in the absorptive power at higher temperatures see Losev and Generalov [6.38].

For elevated temperatures, the number density of molecules in excited vibrational levels gives a substantial contribution to the full number density of molecules. This contribution can be found from integral absorption measurements extended over a number of molecular bands whose lower vibrational levels contribute essentially to the full number density of molecules.

Molecular band absorption technique plays a useful role for the determination of the speed of chemical reaction processes. In [6.1 a], [6.47 i], the OH concentration in diverse flame zones respectively in diverse positions within the shock wave front was monitored by OH band absorption. The relation between $\int\limits_{(\text{band})} \varkappa_\nu \, d\nu$ and n_{OH} was found by calibration. In [6.47 i], the absorption measurements are made at wavelength 3064 Å.

In connection with Cl_2 dissociation studies in shock waves, Jacobs and Giedt [6.32 v] have determined the concentration of Cl_2 from the absorption coefficient at 3640 Å. The temperature dependence of the absorption coefficient was found by calibration.

In [6.33 e], dissociative reaction rates in O_2 and NO, respectively, have been derived from the variation of the corresponding molecule concentrations which was determined from the band absorption at the wavelengths 1470 and 1270 Å, respectively.

For the continuous absorption coefficient of molecular gases due to photodissociation and photoionization from the ground term, the same relation (6.64) holds as that for the absorption coefficient of atomic gases due to photoionization. σ_ν for this purpose is empirically determined from absorption measurements on gases of known composition, temperature, and pressure (see, e.g., Aboud et al. [6.1]).

On the basis of the possible excitations, in particular at higher temperatures, a temperature dependence of the absorption coefficient should not be disregarded. However, the temperature dependence is a function of wave length, this dependence will become less stronger as the wave length decreases, as shown by Soshnikov [6.54] in the case of NO_2-absorption. To eliminate any temperature dependence, quantitative absorption measurements should be carried out, insofar as possible in the ultraviolet spectral range.

In the far vacuum ultraviolet region ($\lambda < 1000$ Å), the absorption is due to photoionization of inner shell electrons and therefore is independent of chemical binding. It depends only from the concentration of atomic nucleus of the element under consideration [6.38 d] and is given by formula (6.66).

As background radiator for absorption measurements in this spectral region, synchrotron radiation sources can be used [6.38 e].

The absorption method is readily employed in density determinations in gas dynamical systems, especially for low gas densities, when other methods (absorption of corpuscular radiation, X-ray absorption, Schlieren and interferometric methods) becomes too insensitive. If the working gas is too weakly absorbing, the absorption can be enhanced by an admixture of a strongly absorbing substance. Suitable admixtures are O_2 (absorption maximum at 1450 Å

Table 6.4. Determination of particle number densities from absorption measurements. In column 4 ("method") is listed the number of the section in which the respective method is described

Plasma	Absorber, wavelength of absorption [Å]	Number density of absorbing element [cm^{-3}]	Method	Reference
acetylene-air-flame	Cu I 2225 – 3274	$2.1 \cdot 10^{10} - 2.5 \cdot 10^{14}$		[6.43g]
H_2O_2-flame + trace Na	Na I 4047, 4358, 5461	$(0.5 - 3.5) \cdot 10^{11}$	6.4.5.	[6.51]
hollow cathode, Au + Ar	Au I 2428, 2676	$(0.6 - 1.1) \cdot 10^{11}$	6.4.3.	[6.34a]
alcohol flame + trace Na	Na I 5895	$1.7 \cdot 10^{12}$	6.4.1.	[6.44]
low pressure discharge, Ar + Cs	Cs I 4555	$(0.15 - 1.3) \cdot 10^{13}$		[6.13i]
low pressure disch., Ar + Cs	Cs I 4555, 8521	$(0.15 - 1.5) \cdot 10^{13}$	absorption at line center and calibration	[6.13h]
d.c. arc containing Na	Na I 5895	$2.1 \cdot 10^{13}$	6.4.1.	[6.44]
low press. disch. Hg	Hg I 2537	$(2 - 8) \cdot 10^{13}$	see [6.61a] and begin of section 6.4	[6.61a]
decaying H-plasma ($7500\,°K < T_e < 20\,000\,°K$)	H 890.8	$(1 - 5) \cdot 10^{15}$	continuous absorption	[6.28b]
Ar-Cs-channel flow	Cs 4555	$10^{14} - 10^{17}$	calibration	[6.66a]
shock heated air or oxygen in wind tunnel	O_2 1400 – 1500	$p < 1$ Torr; qualitative investigation of the flow field		[6.26]

Table 6.5. Bibliographic references on atomic photoionization cross sections. Meanings of the symbols under "Reference": *C* — calculated data) *E* — experimental data, *Su* — summarizing of data. The reference is given in short form, for example 42 e stands for [6.42e]

Atomic species	Reference
Al	*C*: 33g, 38d. *Su*: 42b.
Ar	*C*: 18, 28a, 33g, 38d. *E*: 17a, 28c, 32p, 32r, 37, 38e, 42e, 47b, 47c, 68. *Su*: 19a, 42b.
As	*C*: 38d.
B	*C*: 38d. *Su*: 42b.
Ba	*Su*: 42b.
Be	*C*: 1b, 38d. *Su*: 42b.
Br	*C*: 38d.
Br^-	*C*: 46g.
C	*C*: 38d, 46a. *Su*: 42b.
C^-	*C*: 46g. *E*: 32b. *Su*: 42b.
C^+	*Su*: 42b.
C^{++}, C^{+++}	*C*: 32e.
Ca	*C*: 8, 33g, 38d, 43a. *E*: 33. *Su*: 19a, 42b.
Ca^+	*C*: 16b. *Su*: 42b.
Cd	*E*: 46h. *Su*: 42b.
Cl	*C*: 38d.
Cl^-	*C*: 43b, 46g.
Co, Cr	*C*: 38d.
Cs	*C*: 33g, 38c, 43f, 49, 50c. *E*: 15, 23, 33g, 43. *Su*: 19a, 42b.
F	*C*: 38d. *Su*: 42b.
F^-	*C*: 43b, 46g.
F^+	*Su*: 42b.
Fe	*C*: 38d.
Fe^{13+}	*C*: 74a.
Ca	*C*: 33g, 38d. *Su*: 42b.
Ge	*C*: 38d.
H	*E*: 13c, 13d, 13e. *Su*: 19a, 42b.
H^-	*C*: 38b. *Su*: 42b.
He	*C*: 16b, 28a, 32y, 38c, 38d, 44a, 56a, 66, 75. *E*: 28c, 35b, 37, 38a, 47e. *Su*: 17a, 19a, 42b.
I	*C*: 38c.
I^-	*C*: 46g.
In	*C*: 38c. *E*: 32i. *Su*: 42b.
K	*C*: 16b, 19b, 38c, 38d, 45a, 49, 50c. *E*: 23, 32l. *Su*: 19a, 42b.
K^-	*C*: 43b.
Kr	*C*: 38c, 38d. *E*: 32r, 42e, 45. *Su*: 42b.
Li	*C*: 16b, 30c, 32, 38c, 38d, 44a, 50b, 50c, 56. *E*: 32m, 63, 65. *Su*: 19a, 42b.
Li^-	*C*: 43b.
Li^+	*C*: 13a, 30b, 44a, 56a.
Mg	*C*: 7, 16b, 33g, 38d. *E*: 21. *Su*: 19a, 42b.
Mg^+	*C*: 16b.
Mn, Mo	*C*: 38d.
N	*C*: 9, 28a, 38d. *E*: 17b, 25, 32b. *Su*: 42b.
N^+	*C*: 32y. *Su*: 42b.

Table 6.5. (Continuation)

Atomic species	Reference
N^{++}, N^{+++}	*C*: 32e.
N^{4+}	*C*: 32e, 32u.
Na	*C*: 16b, 33g, 38c, 38d, 44a, 49, 50c. *E*: 32j, 32m. *Su*: 19a, 42b.
Na^-	*C*: 43b.
Na^+	*Su*: 42b.
Nb	*C*: 38d.
Ne	*C*: 24a, 38d, 48, 50, 50a. *E*: 17a, 35b, 36, 47f. *Su*: 19a, 42b.
Ne^+	*C*: 32e. *Su*: 42b.
Ne^{++}, Ne^{3+}, Ne^{4+}	*C*: 32e.
Ni	*C*: 38d.
O	*C*: 9, 16b, 38d. *E*: 16b, 17c. *Su*: 42b.
O^-	*C*: 28c, 46g. *Su*: 42b.
O^+	*C*: 16b. *E*: 32b. *Su*: 42b.
O^{++}, O^{3+}	*C*: 16b, 32e.
O^{4+}	*C*: 32e.
O^{5+}	*C*: 32u.
P	*C*: 38d.
Rb	*C*: 38c, 38d, 49, 50c. *E*: 43. *Su*: 19a, 42b.
Rh, S	*C*: 38d.
S^-	*C*: 46g.
Sb	*C*: 38c.
Sc, Se, Si	*C*: 38d.
Si^-	*C*: 46g.
Si^+	*C*: 16b.
Sn	*C*: 38c.
Sr	*C*: 33g, 38d. *Su*: 42b.
Tc	*C*: 38d.
Te	*C*: 38c.
Ti	*C*: 38d.
Tl	*C*: 33g. *E*: 42. *Su*: 42b.
V	*C*: 38d.
Xe	*C*: 38c. *E*: 13g, 32p, 32r, 42c, 42e, 47d, 72a. *Su*: 42b.
Xe^+	*E*: 13g.
Y	*C*: 38d.
Zn	*C*: 38d. *Su*: 42b.
Zr	*C*: 38d.

with an absorption coefficient $\alpha = 490\ \text{cm}^{-1}$) or mercury, when absorption at a wavelength of 2537 Å is considered (see Winkler [6.77]).

The possibilities of this method are evident in Table 6.4, in which some experimental data are listed. Table 6.5 and 6.6 contain bibliographic references on absorption coefficients of atoms and molecules in the ultraviolet spectral range.

A less recent comprehensive work on UV-absorption by atmospheric gases stems from Ditchburn [6.19]. More recent surveys are given by Wilesov [6.76], Ditchburn and Öpik [6.22], as well as Marr [6.42b], to which we refer the reader for further details.

Table 6.6. Bibliographic references on continuous molecular absorption coefficients. *C*: calculated data; *E*: experimental data; *Su*: summarizing of data. The reference is given in short form (for example, 42d is used instead of [6.42d]

Molecular species	Reference
Anthracene	*E*: 32d.
CF_4	*E*: 17d, 32e.
CH_4	*C*: 18, 38d. *E*: 42d, 68. *Su*: 42b.
C_2H_2	*E*: 42d, 70. *Su*: 42b.
C_2H_4	*E*: 42d, 70. *Su*: 42b.
C_2H_6	*E*: 42d. *Su*: 42b.
CO	*C*: 38d. *E*: 32o, 59, 60. *Su*: 42b.
D_2O	*E*: 2.
H_2	*C*: 33c, 33d. *E*: 13d, 17e, 35d, 68. *Su*: 42b.
H_2O	*C*: 38d (2 H + O). *E*: 2, 17e, 42d, 68. *Su*: 42b.
K_2	*E*: 19b, 23 32l. *Su*: 42b.
N_2	*E*: 17e, 17f, 17g, 32g, 67, 68, 74. *Su*: 42b.
Na_2	*E*: 32j, 32k. *Su*: 42b.
NH_3	*C*: 38d. *E*: 42d, 69, 72. *Su*: 42b.
NO	*C*: 38d (N + O). *E*: 2, 58, 71, 72. *Su*: 42b.
NO_2	*E*: 39.
N_2O	*E*: 2, 71. *Su*: 42b.
O_2	*C*: 32w, 42g. *E*: 1, 17f, 24, 32n, 67, 68, 72a, 73, 77. *Su*: 42b, 77.
O_2^-	*Su*: 42b.
O_3	*Su*: 42b.

6.5. Fluorescence Methods for the Determination of Particle Number Densities

6.5.1. Determination of Particle Number Densities from Resonance Line Scattering

In low pressure plasmas, the effect of resonance line scattering can be used for the determination of particle number densities.

The phenomenon of resonance line scattering can be understood as a two step process. In the first step, the resonance levels are excited by absorption of radiation from an auxiliary radiation source, and in the second step the excited levels are depopulated by radiative deexcitation, that means, by emission of one or more spectrum lines.

From the ratio of the scattered line intensity to the exciting spectral irradiation intensity, the number density of light scattering particles in the electronic ground level can be obtained, when the following two conditions are satisfied:

1. The deexcitation rate due to inelastic collisions must be at least one order of magnitude lower than the reciprocal natural life time of the excited resonance levels.
2. The scattered radiation must arrive the radiation detector without being self absorbed by the plasma.

According to Hofman [6.32i], these conditions are valid in the case of Ba^+-resonance line scattering in a low temperature fully ionized Ba plasma produced by a Q-device, if the ion number density does not exceed a value of 10^{11} cm^{-3}.

For details concerning the absolute scale factor which is influenced by anisotropy effects in magnetized plasmas, see the original reference [6.32i]. For experimental details see also [6.32a], [6.32f], [6.32g].

Because metastable $5D$ levels with low excitation energies exist between the ground level $6S$ and the resonance levels $6P$, the metastable ions contribute merkedly to the total ion number density when the temperature is in excess of 2000 °K. There are two possibilities to find the number density of metastables $n(5\,D)$:

1. The quantity $n(5\,D)$ can be found from a second scattering experiment including the transition $5\,D-6\,P$.
2. If the plasma temperature is known, the number density $n(5\,D)$ can be found in a good approximation via the Boltzmann formula (see also section 6.4.5.).

6.5.2. Determination of Particle Number Densities from Electron Beam Fluorescence Experiments

When an electron beam interacts with a plasma, the plasma particles become excited and emit fluorescence radiation, from the intensity of which the number densities of the diverse kinds of heavy plasma particles (with exception of protons) together with other plasma parameters (for example the plasma temperature, see section 7.3) can be obtained.

This method has been applied in [6.16f], [6.43d], [6.43e], [6.46f], [6.47a], [6.47b] to determine diverse parameters of supersonic wind tunnel flow and to solve problems such as the front structure of shock waves originating from obstacles introduced into the flow. The kinetic energy of the beam electrons are usually in the range of some ten keV.

In Nitrogen, most of the fluorescence radiation is concentrated in the 3914 Å (0—0) band of the 1. negative N_2^+ band system [6.46f]. In Argon, the main radiation is concentrated in lines of Ar I and Ar II, most of them fall in the spectral range from 3000 to 4000 Å. In addition, a series of intense Ar I lines are observed in the spectral region from 7200 to 8500 Å [6.47a]. In Helium, fluorescence is observed for the lines He I 5016 and 4471 Å [6.46f], [6.47a].

Particle number densities in the range from 10^{14} to 10^{16} cm^{-3} are well detectable.

In the process of electron beam interaction with heavy plasma particles, besides of optical line radiation, the emission of X radiation due to bremsstrahlung is observed which also can be used for the determination of plasma density, as shown in [6.78b] for the case of diluted Air-, N_2-, Ne- and Ar plasma objects, excited by 2—90 keV electrons.

CHAPTER 7

Determination of Plasma Parameters from the Band Spectrum of Diatomic Molecules

In this chapter we shall mainly consider the following 3 cases: 1. the determination of the temperature and particle densities from the absolute intensity of individual bands, 2. the determination of the rotational temperature from the relative intensities of the rotational lines within a band, and 3. the determination of the vibrational temperature via the relative intensities of several bands belonging to the same electron transition.

7.1. General Formulation of the Intensities in Spectra of Diatomic Molecules

We are interested in the emission coefficient ε of a line in the band spectrum of a diatomic molecule. Let us consider the transition

$$m\, v'\, J'\, \Lambda' \rightarrow n\, v''\, J''\, \Lambda'' \tag{7.1}$$

where m and n are respectively the upper and lower electron level, v' and v'' the upper and lower vibrational quantum number, J' and J'' the upper and lower rotational quantum number, and, finally, Λ' and Λ'' are the respective term symbols associated with the axial quantum number of the upper and lower electron level.

The emission coefficient (radiative power per unit volume and solid angle) of the above transition is given by the expression

$$\varepsilon_{n\, v''\, J''\, \Lambda'',\, m\, v'\, J'\, \Lambda'} = \frac{1}{4\pi}\, n_{m\, v'\, J'\, \Lambda'}\, h\, \nu\, A_{n\, v''\, J''\, \Lambda'',\, m\, v'\, J'\, \Lambda'} \tag{7.2}$$

or written in short

$$\varepsilon_{v''\, J'',\, v'\, J'} = \frac{1}{4\pi}\, n_{v'\, J'}\, h\, \nu\, A_{v''\, J'',\, v'\, J'} \tag{7.3}$$

where $n_{v'\, J'}$ is the number of molecules per cm³ in the quantum state $m\, v'\, J'$.

The transition probability A can be replaced by the oscillator strength f or the line strength S:

$$A_{v''\, J'',\, v'\, J'} = \frac{8\, \pi^2\, e^2\, \nu^2\, g_n\, (2\, J'' + 1)}{m\, c^3\, g_m\, (2\, J' + 1)} \cdot f_{v'\, J'\, v''\, J''}\, , \tag{7.4}$$

$$A_{v''\, J'',\, v'\, J'} = \frac{64\, \pi^4\, \nu^3}{3\, h\, c^3\, g_m\, (2\, J' + 1)} \cdot S^{m\, v'\, J'\, \Lambda'}_{n\, v''\, J''\, \Lambda''}\, . \tag{7.5}$$

where g_m and g_n are respectively the statistical weight of the upper and lower electronic level. Expressing the emission coefficient in terms of S, we obtain

$$\varepsilon_{v'' J'', v' J'} = \frac{16 \pi^3 \nu^4}{3 c^3 g_m (2 J' + 1)} n_{v' J'} S^{m v' J' \Lambda'}_{n v'' J'' \Lambda''} \tag{7.6}$$

(compare Nicholls and Stewart [7.24]).

The line strength S can be split into two factors:

$$S^{m v' J' \Lambda'}_{n v'' J'' \Lambda''} = S^{J' \Lambda'}_{J'' \Lambda''} p_{v'' v'} \,. \tag{7.7}$$

The first of these, also called the Hönl-London factor, can be represented as follows (see [7.11], p. 208):

For transitions with $\Delta\Lambda = 0$

$$\left.\begin{aligned} S^{J' \Lambda'}_{J'-1, \Lambda'} &= \frac{(J' + \Lambda')(J' - \Lambda')}{J'}, \quad (R\text{-branch}) \qquad S^{J' \Lambda'}_{J' \Lambda'} = \frac{(2 J' + 1) \Lambda'^2}{J'(J' + 1)}, \quad (Q\text{-branch}) \\ S^{J' \Lambda'}_{J'+1, \Lambda'} &= \frac{(J' + 1 + \Lambda')(J' + 1 - \Lambda')}{(J' + 1)} \,. \quad (P\text{-branch}) \end{aligned}\right\} \tag{7.8}$$

For transitions with $\Delta\Lambda = +1$ $(\Lambda' - \Lambda'' = 1)$:

$$\left.\begin{aligned} S^{J' \Lambda'}_{J'-1, \Lambda'-1} &= \frac{(J' + \Lambda')(J' - 1 + \Lambda')}{4 J'}, \quad (R\text{-branch}) \\ S^{J' \Lambda'}_{J', \Lambda'-1} &= \frac{(J' + \Lambda')(J' + 1 - \Lambda')(2 J' + 1)}{4 J'(J' + 1)}, \quad (Q\text{-branch}) \\ S^{J' \Lambda'}_{J'+1, \Lambda'-1} &= \frac{(J' + 1 - \Lambda')(J' + 2 - \Lambda')}{4 (J' + 1)} \,. \quad (P\text{-branch}) \end{aligned}\right\} \tag{7.9}$$

For transitions with $\Delta\Lambda = -1$ $(\Lambda' = \Lambda'' - 1)$:

$$\left.\begin{aligned} S^{J' \Lambda'}_{J'-1, \Lambda'+1} &= \frac{(J' - \Lambda')(J' - 1 - \Lambda')}{4 J'}, \quad (R\text{-branch}) \\ S^{J' \Lambda'}_{J' \Lambda'+1} &= \frac{(J' - \Lambda')(J' + 1 + \Lambda')(2 J' + 1)}{4 J'(J' + 1)}, \quad (Q\text{-branch}) \\ S^{J' \Lambda'}_{J'+1, \Lambda'+1} &= \frac{(J' + 1 + \Lambda')(J' + 2 + \Lambda')}{4 (J' + 1)} \,. \quad (P\text{-branch}) \end{aligned}\right\} \tag{7.10}$$

The factors have a definite meaning in the determination of the rotational temperature, which we shall return to later on.

The following sum rule holds for the S-factors:

$$\sum_{J''} S_{J''\,\Lambda''}^{J'\,\Lambda'} = (2\,J' + 1)\,. \tag{7.11}$$

Thus, the sum intensity of all rotational lines belonging to a band $v'' \leftarrow v'$ is given by

$$\varepsilon_{v''\,v'} = \sum_{J'',\,J'} \varepsilon_{v''\,J''\,v'\,J'} = \frac{16\,\pi^3\,\nu^4}{3\,c^3\,g_m}\,p_{v''\,v'}\,n_{v'}\,, \tag{7.12}$$

where $n_{v'} = \sum_{J'} n_{v'\,J'}$ is the occupation number of the upper vibrational level v'. $p_{v''\,v'}$ is designated as the band strength.

Since (7.12) involves the quantity ν^4, this equation is only valid when the band is so narrow that within the band we can reckon with a constant value of ν which we can take to be $\nu = \frac{1}{h}\,(U_{v'\,0} - U_{v''\,0})$ (U is the level energy).

The sum rule for the quantities $p_{v''\,v'}$ is given by

$$\sum_{v''} p_{v''\,v'} = p_{v'\,v'} \qquad \text{independent of } v'\,, \tag{7.13}$$

which is of significance for determining the vibrational temperature. $p_{v'\,v'}$ depends only on the particular electron transition (compare [7.11], p. 203 and [7.24]).

We should still like to point out that since

$$f_{v'\,v''} = \frac{8\,\pi^2\,\nu\,m}{3\,h\,e^2\,g_n}\cdot p_{v'\,v''} \tag{7.14}$$

the total emission coefficient of a band can be represented by

$$\varepsilon_{v''\,v'} = \frac{2\,\pi\,e^2\,h\,\nu^3\,g_n}{c^3\,m\,g_m}\cdot f_{v'\,v''}\,n_{v'}\,. \tag{7.15}$$

The introduction of the oscillator strength has the advantage that this is dimensionless.

Similarly, the emission coefficient of a band can also be described by

$$\varepsilon_{v''\,v'} = \frac{A_{v''\,v'}}{4\,\pi}\cdot n_v'\,h\,\nu \tag{7.16}$$

Finally, we give the expressions for the occupation numbers $n_{v'\,J'}$ and $n_{v'}$ as functions of the temperature T and the molecular particle density n_0 (assuming thermal equilibrium):

$$n_{v'\,J'} = n_0\,\frac{(2\,J' + 1)\,g_m\,\Phi_p'}{Q_i}\,\mathrm{e}^{-U_{m\,v'\,J'}/kT}\,, \tag{7.17}$$

$$n_{v'} = n_0\,\frac{g_m\,Q_r(m\,v')\,\bar{\Phi}_p}{Q_i}\,\mathrm{e}^{-U_{m\,v'}/kT}\,. \tag{7.18}$$

Q_i is the inner partition function of the molecule (see section 4.2.1) and $Q_r(m, v')$ the rotational component of the inner partition function for a given electron level m and given vibrational quantum number v'. Approximately, we have (see [7.11], p. 125)

$$Q_r = \frac{kT}{h\,c\,B_v}\,, \tag{7.19}$$

where B_v is taken from the corresponding tables for molecular data (e.g., in [7.11]).

The factors $\Phi_{p'}$ and $\bar{\Phi}_p$ allow for nuclear spin degeneracy. For heteronuclear molecules, $\Phi_{p'} = \bar{\Phi}_p = 1$ throughout, while for homonuclear molecules Φ_p is given by formula (3.13) and $\bar{\Phi}_p = 1/2$.

For more details about intensity formulas for diatomic molecular spectra see [7.38g].

7.2. Plasma Data from Absolute Band Intensities

7.2.1. Determination of the Number Density n_0 for a Certain Kind of Molecules from the Absolute Emission Coefficient, or Integral Absorption Coefficient, respectively, of a Molecular Band when the Absolute Oscillator Strength and Temperature are known

From emission measurements, we obtain, allowing for (7.15) and (7.18) the following expression for the emission coefficient of a band

$$\varepsilon_{v''\,v'} = \frac{2\,\pi\,e^2\,h\,\nu^3\,g_n\,Q_r(m, v')}{m\,c^3\,Q_i}\,\bar{\Phi}_p\,e^{-U_{m\,v'}/kT} f_{v'\,v''}\,n_0\,. \tag{7.20}$$

n_0 is subsequently obtained from the measured absolute emission coefficients $\varepsilon_{v''\,v'}$.

Let us now consider absorption measurements. In the case where induced emission can be neglected ($h\,\nu/kT > 3$), we have

$$\int\limits_{\text{(band)}} \varkappa_\nu\,d\nu = \frac{\pi\,e^2}{m\,c}\cdot n_{n\,v''}\,f_{v'\,v''}\,, \tag{7.21}$$

and since

$$n_{n\,v''} = n_0\cdot\frac{g_n\,Q_r(n, v'')\,\bar{\Phi}_p}{Q_i}\,e^{-U_{n\,v''}/kT} \tag{7.22}$$

we obtain

$$\int\limits_{\text{(band)}} \varkappa_\nu\,d\nu = \frac{\pi\,e^2}{m\,c}\,\frac{f_{v'\,v''}\,g_n\,Q_r(n\,v'')}{Q_i}\,\bar{\Phi}_p\,n_0\,e^{-U_{n\,v''}/kT} \tag{7.23}$$

as the equation of definition for n_0.

7.2.2. Temperature T from the Absolute Intensity and Total Absorption of a Molecular Band when the Pressure, Material Composition, and Absolute Oscillator Strength $f_{v' v''}$ are known

When emission measurements are available the emission coefficient $\varepsilon_{v'' v'}$ is calculated as a function of the temperature T for a given pressure p and material composition in accordance with (7.20) and with the help of the formulas given in section 4.2 for the plasma composition of chemically reacting gases (from these n_0 is calculated). Once the function $\varepsilon_{v'' v'}(T)$ is known we can state the value of T for each measured value of $\varepsilon_{v'' v'}$.

In the case of absorption measurements, $\int\limits_{(\text{band})} \varkappa_\nu \, d\nu$ is first measured and then calculated from (7.23) and the particle density $n_0(T)$ for a given pressure p (section 4.2). By comparison the sought value of T is finally obtained.

In conclusion, it should be mentioned that conversely the absolute oscillator strengths can be determined via expressions (7.20) and (7.23), provided the plasma composition, plasma temperature, and pressure are known.

7.3. Rotational Temperature

The rotational temperature is determined from the relative intensities of the rotational lines within a molecular band.

If not otherwise stated, the various formulas refer to the case of Boltzmann equilibrium between the rotational levels of the upper electronic state. This is only true in the optical thin case when the frequency of heavy particle collisions with an excited molecule is at least by a factor of 5 greater than the optical transition probability (which is of the order $10^6 - 10^8 \, s^{-1}$).

Taking into account the strong coupling between translational and rotational energy states [7.22], the rotational temperature derived from experiment is equal to the gas temperature, or in the case of thermal equilibrium, equal to the plasma temperature.

For pressures below 30 Torr, the Boltzmann equilibrium between rotational energy levels holds with certainty only for the basic, or, if exist, the metastable electronic state.

In the low pressure region when the transition probability exceeds the collision frequency between an excited molecule and the heavy particles, the formulas for evaluating rotational temperatures from relative line intensities within a molecular band are based on the assumption that during the excitation processes the rotational quantum numbers change according to the selection rules for optical transitions [7.21g].

These formulas, especially evaluated for the 1. negative band system of N_2^+, have been applied for the determination of gas temperature in aurorae [7.11y], [7.11z] or in low density gas flows containing nitrogen which is excited by electron beams [7.1g], [7.20i], [7.21g], [7.34e].

If the rotational temperature in the low pressure region is derived by means of the usual formulas valid in the case of Boltzmann distributed upper rotational levels, then a relative error with respect to the temperature ranging up to 20% for $T_{rot} > 1000$ °K must be expected. For lower temperatures, the error can be still greater.

In medium pressure discharges, deviations of the rotational energy distribution from a Boltzmann distribution may happen for some bands, due to collisions of the second kind, as has been shown by Sergeyenkova et al. [7.35f] for a nitrogen discharge and for the band $N_2(2+)$ at 4059 Å. For the band $N_2(2+)$ 3998 Å, no anomalies of this kind has been found.

The following statements refer to the case of Boltzmann equilibrium between rotational energy levels of the upper electronic state.

From (7.6), (7.7), (7.17), and (3.4) disregarding the nuclear spin degeneracy factor Φ_p' and the factor ν^4, it follows that the relative intensities of the rotational lines within a band are given by

$$\varepsilon_{J''\,J'} = \text{const.}\; S_{J''\,\Lambda''}^{J'\,\Lambda'}\, e^{-U_r/kT} \tag{7.24}$$

where

$$U_r \approx B_v\, h\, c\, J'\,(J'+1)\,. \tag{7.25}$$

A more accurate value for U_r can be obtained from (3.10). From (7.24) and (7.25) we have after taking logarithms

$$\lg \frac{\varepsilon_{J''\,J'}}{S_{J''\,\Lambda''}^{J'\,\Lambda'}} = \text{const} - \frac{B_v\, h\, c\, J'\,(J'+1)}{2.3026\, kT}\,. \tag{7.26}$$

If we plot $\lg \frac{\varepsilon_{J''\,J'}}{S_{J''\,\Lambda''}^{J'\,\Lambda'}}$ as a function of $J'\,(J'+1)$, we can determine the temperature from the slope of the straight line passing through the measured points. From (7.8) ($^1\Sigma$-$^1\Sigma$-transitions) we get for the P-branch $S = J' + 1$ and for the R-branch $S = J'$. Disregarding the accents by substituting J for J'', equation (7.26) becomes

P-branch: $$\lg(\varepsilon_J/J) = \text{const} - \frac{B_v\, h\, c}{2.3026\, kT}\, J\,(J-1)\,, \tag{7.26a}$$

R-branch: $$\lg(\varepsilon_J/J+1) = \text{const} - \frac{B_v\, h\, c}{2.3026\, kT}\,(J+1)\,(J+2)\,. \tag{7.26b}$$

For $^2\Sigma - {}^2\Sigma$-transitions, the quantum number J must be substituted by the quantum number K (Hund's coupling case b, see [7.11], p. 221). The rotational levels split into the doublet components $J = K \pm 1/2$. For the single band lines, formulas (7.26a) and (7.26b) hold again when J is substituted by K, with the exception that for homonuclear molecules the intensity of rotational lines alternate with quantum number K (see [7.11], p. 250). In the case of the $N_2^+\; {}^2\Sigma - {}^2\Sigma$ bands, the corresponding intensity factor is 2 (see [7.26a]).

Note, that in this text the definition $K = K''$ is used throughout while in some other publications one find the definition $K = K''$ for the R-branch and $K = K'$ for the P-branch.

As an illustration let us consider the measurements of Ornstein and Brinkman [7.26]. The rotational temperature of a carbon arc in air was determined from the relative intensities of the rotational lines of the P- and R-branch of the violet $B^2\,\Sigma - X^2\,\Sigma$ CN-band at 3884 Å. The radiation was not radially resolved, i.e., the emission coefficient ε was assumed proportional to the measured intensity I. With the definitions $p = K + 1$ for R-lines and $p = K$ for P-lines introduced by Ornstein and Brinkman, we have after replacing the emission coefficients in (7.26) by intensities I

$$\lg \frac{I}{p} = \text{const} - \frac{B_v\,h\,c}{2.3026\,kT}\,p\,(p \pm 1)\,, \qquad (7.27)$$

$+$ for R-lines, $-$ for P-lines.

Figure 7.1 gives a corresponding logarithmic representation (curve I refers to the P-branch, curve II to the R-branch). The rotational temperature was found to be 7000 °K from the slope of the straight lines.

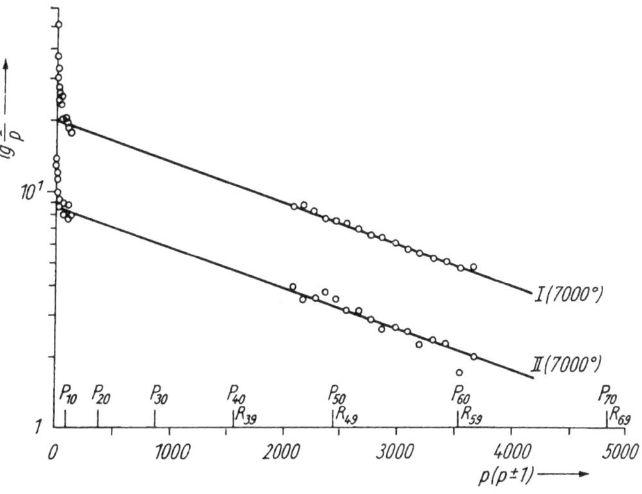

Figure 7.1. Plot for determining the rotational temperature in a carbon arc in air from the relative intensities I of the rotational lines of the CN band $^2\Sigma - {}^2\Sigma\;0 - 0$. p is the running variable for the rotational lines after (Ornstein and Brinkman [7.26])

Curve I refers to the P-branch, curve II to the R-branch

In the same work the rotational temperature for two AlO-bands was deter mined. Since the excitation of these bands reaches a maximum at severa' 1000 °K, it follows that when the radiation is not radially resolved the major portion of the radiation originates from a region, which possesses this low temperature. Consequently, the rotational temperature, which is measured on the basis of the radiation of these bands, is also of the order of this optimum temperature (norm temperature) (2950 °K for the $0 \rightarrow 0$ band and 3450 °K for the $1 \rightarrow 0$ band; the numbers 0 and 1 refer to the vibrational quantum number v).

From hence it follows that in the case of radially non resolved intensity measurements, the rotational temperature derived thereof is a measure for the temperature in the center of the plasma only in that case, when this temperature does not exceed the norm temperature (see section 5.2.6). Due to the temperature inhomogenity, the diagram functions $\lg (I/S) = f\left(J'\,(J'+1)\right)$ differ from straight lines [7.11y]. Deviations from straight lines can also be caused by self absorption [7.29i].

Due to the multiplet splitting of band lines which is a function of rotational quantum number, the apparent line profile varies in general from line to line. To obtain relative integral line intensities, it is then necessary to integrate over the whole line profile [7.26]. Substitution of the integral line profile by the peak intensity as measure for the relative line intensity is allowed only in those cases, when the apparatus profile is broader than the multiplet separation.

Another source of error affecting the determination of rotational temperatures is effected by the rotation — vibration interaction which is accompanied by a dependence of transition probability on rotational quantum number. This effect was observed for the molecules OH [7.19b], [7.20n], [7.41b], AgH [7.19c], CO [7.6o], O_2, H_2 and D_2 [7.9p]. If rotational temperatures are derived without allowance of this effect, there will be an error up to 3,5% [7.41b].

Considerable deviations between rotational temperature and gaskinetic temperature with $T_r < T_g$ are found in the reaction zone of hydrocarbon flames and in H_2-F_2-flames, when as indicator for the rotational temperature the radial OH is used [7.5b], [7.12b], [7.27a]. These deviations can be attributed to an overexcitation of rotational levels by chemical processes.

To illustrate the range of application for the determination of the rotational temperature we have listed in Table 7.1 some of the relevant experimental data.

7.4. Vibrational Temperature

The vibrational temperature can be determined on the basis of measured relative band intensities.

Unless indicated we assume Boltzmann equilibrium between the vibrational energy levels of the upper electronic state. This is true so long as the total frequency of inelastic collisions (collisions which change vibrational energy levels) is in excess of the optical transition probability for electronic transitions.

In a plasma in thermal equilibrium, the vibrational temperature is simply the plasma temperature.

In a non thermal plasma with $T_g < T_e$, the vibrational temperature T_v is intermediate between the electron temperature T_e and the gas kinetic temperature T_g, because the molecular vibrational energy levels are both influenced by heavy particle collisions and by electronic collisions. To the latter problem see [7.11b] and [7.11c]. The strong interaction between electronic translation and molecular vibration causes also a lowering of the difference between T_e and T_g in atomic gases when molecular gases are added.

Table 7.1. Determination of rotational temperature T_r. Abbreviations: d. – discharge, hc. – hollow cathode, hf. – high frequency, hp. – high pressure, lp. – low pressure, mp. – medium pressure

Molecule	Band position [Å]	Type of plasma	T_r [10^3 °K]	Reference[1])
C_2	5129	Air heated in carbon tube	2.9 – 3.0	33
	5165	mp. d., Ne, Ar, Kr, Xe	2.0 – 3.0	20k
		hydrocarbon arc	6.0	17
CH	3900	$C_2H_2-O_2$-flame	3.75	20j
	4180 – 4240	hydrocarbon – O_2-flame	2.0 – 3.6	5a
CN	3883	hc. d., Ar, He + traces C and N_2	1.0 – 1.3	15a
		mp. d., Ne, Ar, Kr, Xe	2.0 – 3.0	20k
		carbon arc in air, $p \leq 1$ atm	4.3 – 8.7	20b, 33m.
	4216[2])	arc in air	4.1 – 5.6	34
	4216 + 4197[3])	shock wave, CO + N_2	3.8 – 5.5	18a
CO[4])	(3 – 1)-band ≈25000	hf. d., CO, 3 – 4 Torr	3.2	6o
H_2	≈1220	electron beam excited H_2, 0.5 Torr	3.1	34a
HF[5])	(3 – 0)transition	H_2-F_2-flame	4.0	11k
N_2	3300	plasma jet, Ar + trace N_2, 100 Torr	4.3 – 6.0	41c
	3371	rf. torch, N_2, air	2.0 – 6.3	6, 6c, 6p, 21
	3805	mp. d., Ne, Ar, Kr, Xe	2.0 – 3.0	20k
	3998	mp. d., N_2	0.4 – 2.0	9q
N_2^+	3914	electron beam excited lp. N_2 flow	0.08 – 0.4	1g, 20i, 21g, 34e.
		Aurora	0.2 – 2.2	11y, 11z. 35h.
		low voltage arc, hc. d., mp. d. in rare gases with traces of N_2 or in pure N_2	0.4 – 3.0	6m, 9q, 15a, 20k, 26a.
	4278	lp. d., N_2	0.4 – 0.9	6l
NO	2370	hf. d., N_2 + trace O	1.0 – 5.7[4])	11v
OH[6])	2811	H_2-flame in O_2 or Air	2.3 – 3.0	44i
OH[6])	2811	detonation (2 H_2 + O_2)	4.5 – 4.8	12a
	3064	rf torch, air	2.2 – 4.0	6
		flame	2.3 – 3.4	20j, 44i
		shock wave, Ar + O_2 + CH_4	1.7 – 2.5	20n
	3139	flame	2.6 – 3.9	9f

[1]) References are given in short symbols (20 n means [7.20n]).
[5]) Non resolved band, see section 7.5.1.
[3]) Absorption measurement, see section 7.6.
[4]) Vibration band.
[2]) Radial temperature profile.
[6]) Corrected against vibration-rotation-interaction.

In a low pressure plasma the situation may arise that the frequency of inelastic collisions with a vibrational excited level is lower than the optical transition probability. While for the upper electronic state the vibrational levels deviate from a Boltzmann distribution, such a distribution can be found for the vibrational levels of the electronic ground state.

To find the corresponding distribution temperature T_v from emission measurements, formulas are needed for the population of the vibrational levels of the upper electronic state in terms of the population of the vibrational levels of the ground electronic state, depending from the excitation mechanism from the ground electronic state.

On this basis, Muntz [7.21g] has evaluated formulas for the determination of vibrational temperatures in low density nitrogen gas streams excited by electron beams from relative intensities of molecular bands belonging to a given electronic transition. For the electronic excitation processes, a changing of vibrational quantum numbers according to the Franck-Condon principle is assumed.

In the following, we refer only to the case of thermal equilibrium.

From (7.12) and (7.18) we get for the intensity of a band

$$\varepsilon_{v'' v'} = \frac{16\,\pi^3\,\nu^4}{3\,c^3}\,\frac{n_0\,p_{v'' v'}}{Q_i}\,Q_r(m, v')\,\mathrm{e}^{-U_{m\,v'}/kT}\,\overline{\Phi}_p\,. \tag{7.28}$$

If band strengths $p_{v''\,v'}$ are known for two bands with different excitation energies $U_{m\,v'}$, then, according to (7.28), from their relative intensities the vibrational temperature T_v can be derived. Considering the emission coefficients $\varepsilon_{v''\,v'}$ of several bands, which correspond to the same electron transition, and assuming the rotational partition functions $Q_r(m\,v')$ to be independent of the vibrational quantum number v', we obtain the following expression for the relative emission coefficient of the band

$$\frac{\varepsilon_{v''\,v'}}{\nu^4} \sim p_{v''\,v'} \cdot \mathrm{e}^{-U_{v'}/kT}\,. \tag{7.29}$$

(See section 3.2.2 for the calculation of the vibrational contribution U_v' to the internal energy).

From (7.13) we again have

$$\sum_{v''} \frac{\varepsilon_{v''\,v'}}{\nu^4} \sim \mathrm{e}^{-U_{v'}/kT} \tag{7.30}$$

or, taking logarithms

$$\lg \sum_{v''} \frac{\varepsilon_{v''\,v'}}{\nu^4} = \mathrm{const} - \frac{U_{v'}}{2.3026\,kT}\,. \tag{7.31}$$

The vibrational temperature can be found with the help of (7.31) by taking the expressions on the left-hand side of (7.31) from the measurements and plotting them as a function of $U_{v'}$ (see Figure 7.2: the intensity was taken instead of the emission coefficient). The temperature results from the slope of the straight line through the measured points.

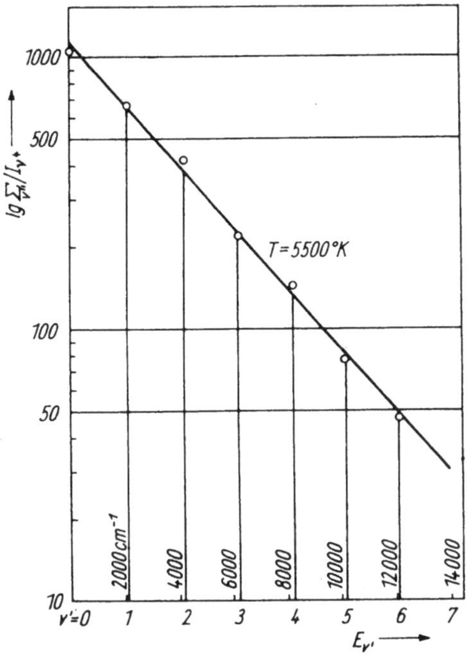

Figure 7.2. Plot for determining the vibrational temperature in a carbon arc in air from the relative band intensities. $\lg \sum_{v''} I/\nu^4$ is linearly dependent on U'_v, the vibrational contribution to the internal energy of the upper level. The temperature follows from the slope of the straight line (after Ornstein and Brinkman [7.25])

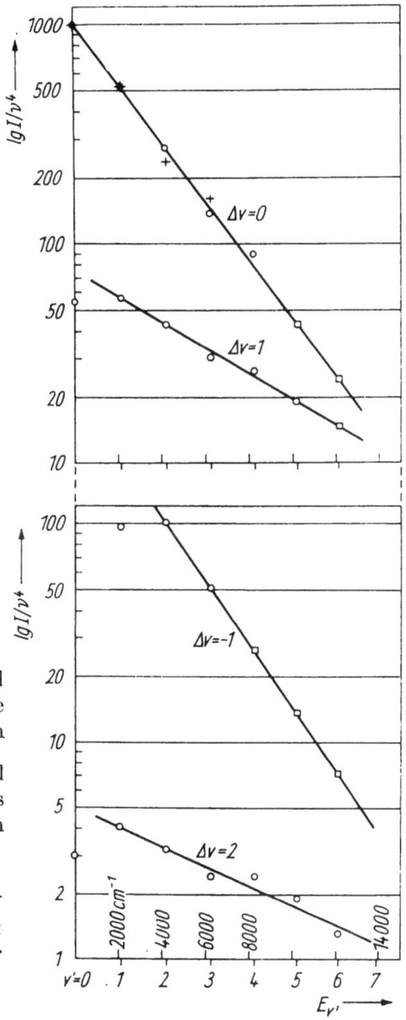

Figure 7.3. $I/(\nu^4)$ (I = band intensity) as a function of the vibrational energy $U_{v'}$, for different band series, whereby $\Delta v = v' - v''$ = const (after Ornstein and Brinkman [7.25])

Specifically (see [7.25]), temperature measurements in a carbon arc in air were dealt with on the basis of the relative band intensities of the $B^2\,\Sigma - X^2\,\Sigma$ band system of CN, and a temperature of 5500 °K was thus obtained.

Sometimes it can happen that the intensity cannot be measured for all bands; overlapping, for example, can have a disturbing influence on the measurement. It was found that bands with $v' - v''$ = constant are described by

$$\frac{I}{\nu^4} \sim e^{-\alpha\, U_{v'}/kT}\,, \qquad \alpha = f(v' - v'')\,. \tag{7.32}$$

Equation (7.32) can be regarded as an interpolation formula to find the missing intensities from incomplete band intensity measurements.

Figure 7.3 serves to illustrate the functional dependence of lg I/ν^4 on $U_{v'}$, from which the relationships become evident. The measured points are plotted as circles and crosses, while the extrapolated values are indicated by squares

7.5. Temperature Determination from Molecular Spectra for Nonresolved Rotational Structure

7.5.1. Rotational Temperature for Nonresolved Rotational Structure

It can happen that the rotational lines within a molecular band can no longer be observed separately due to blurring and overlapping. This is, for example, the case when a prism spectrograph is used, which in general possesses a fairly low dispersion. Furthermore, the possibility exists that the rotational lines are so strongly broadened due to physical causes that they can no longer be observed separately.

We shall attempt to find out whether it is possible to determine the rotational temperature from the spectral energy distribution within the band.

For this it is necessary to calculate the spectral intensity distribution for several temperatures. The method becomes more effective the stronger the profiles for different temperatures differ from one another.

To calculate the blurred intensity we start with the equation for the spectrally nonblurred energy distribution within a band, whereby the rotational lines can be approximated by δ-functions

$$\varepsilon(\nu) = \sum_{J''J'} \varepsilon_{J''J'}\, \delta\left(\nu - \frac{U_{J'} - U_{J''}}{h}\right). \tag{7.33}$$

$\varepsilon_{J''J'}$ is given by (7.24) and depends on temperature.

To obtain the blurred distribution $\bar{\varepsilon}_\nu$, equ. (7.33) is joined either with the true slit function (if necessary, the line profile function), with a rectangular function or with a triangular function $R(\nu)$ (half-width b and height 1):

$$\bar{\varepsilon}_\nu = \int \sum_{J''J'} \varepsilon_{J''J'}\, \delta\left(\nu' - \frac{U_{J'} - U_{J''}}{h}\right) R(\nu - \nu')\, \mathrm{d}\nu'$$

or

$$\bar{\varepsilon}_\nu = \sum_{J''J'} \varepsilon_{J''J'}\, R\left(\nu - \frac{U_{J'} - U_{J''}}{h}\right). \tag{7.34}$$

The temperature then results by comparing the measured profile with the profiles calculated under the assumption of different temperatures. The temperature corresponding to the best matched profile is designated as the rotational temperature.

Figure 7.4 shows several rotational profiles obtained by Phillips [7.33] for the 0—0 bands of the TiO α-system for the $^3\Pi - {}^3\Pi$ transition. It follows from the figure that in this particular case the rotational profile depends only slightly on temperature. This means that a determination of the rotational temperature may prove difficult when the bands are not resolved.

As a quantitative measure for the rotational temperature, Spier and Smit-Miessen [7.38] as well as Spier and Smit [7.37] proposed the linear portion of

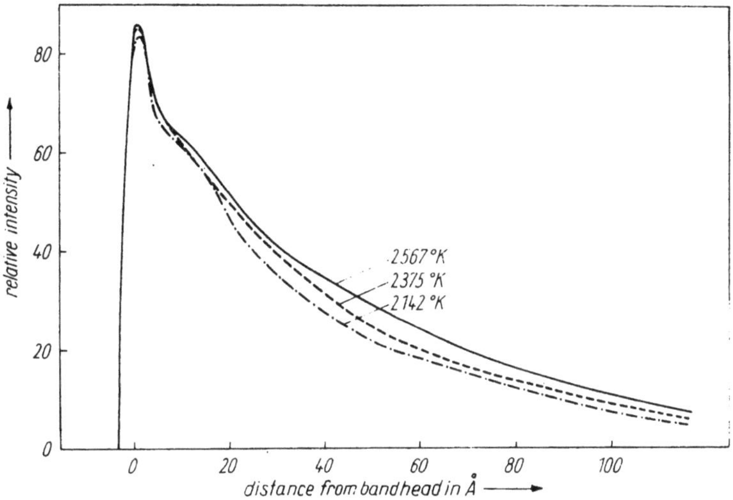

Figure 7.4. Blurred profile of the 0—0 bands of the TiO α-system (transition $C^3\Pi - X^3\Pi$) for different temperatures (after Phillips [7.33])

the logarithmically represented rotational profile, or else the difference of the ordinates in the linear portion of that profile for 2 given wavelengths (see Figure 7.5 and [7.36]).

Prokofiev et al. [7.34] took as a measure for the rotational temperature the intensity ratio within the nonresolved CN-band (0,1) for the wavelength 4216 (band head) and for a wavelength falling in the tail of the band. The experimental results of these authors are listed in Table 7.1.

7.5.2. Vibrational Temperature Determination when the Bands are not resolved and Partially Overlap

This problem was attacked, among others, by Spier, Smit, and Smit-Miessen ([7.36], [7.37], and [7.38]).

Assuming that the relative transition probabilities $p_{v'' v'}$ are known, the blurred spectral intensity distribution, which depends on the width of the spectrograph slit, can be calculated for 2 or more bands belonging to the same electron transition. To do this we go back to (7.6), (7.7), and (7.17) and obtain the following expression for the relative intensity of a molecular line

$$\varepsilon_{v'' J'' v' J'} \sim \nu^4 \, S_{J'' \Lambda''}^{J' \Lambda'} \, p_{v'' v'} \, \Phi_{p'} e^{-(U_{v'} + U_{J'})/(kT)} \, . \qquad (7.35)$$

Now we again calculate, as in section 7.5.1, the blurred intensity distribution:

$$\bar{\varepsilon}_\nu = \int \sum_{v'' J'' v' J'} \varepsilon_{v'' J'' v' J'}\, \delta\left(\nu' - \frac{U_{v'} + U_{J'} - U_{v''} - U_{J''}}{h}\right) R(\nu - \nu')\, d\nu'$$

or

$$\bar{\varepsilon}_\nu = \sum_{v'' J'' v' J'} \varepsilon_{v'' J'' v' J'}\, R\left(\nu - \frac{U_{v'} + U_{J'} - U_{v''} - U_{J''}}{h}\right). \tag{7.36}$$

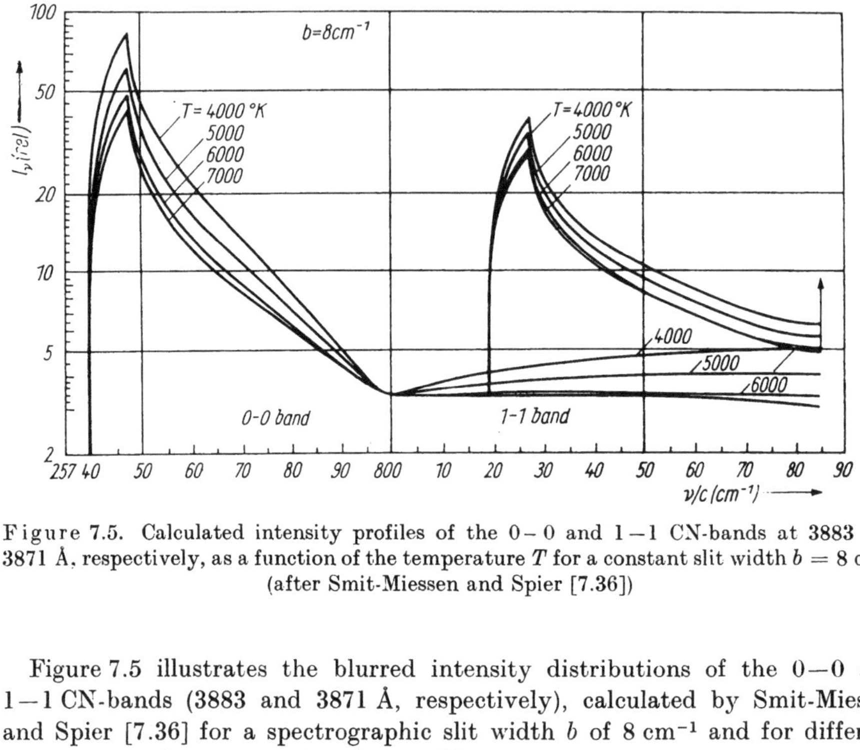

Figure 7.5. Calculated intensity profiles of the 0 – 0 and 1 – 1 CN-bands at 3883 and 3871 Å, respectively, as a function of the temperature T for a constant slit width $b = 8\ \mathrm{cm}^{-1}$ (after Smit-Miessen and Spier [7.36])

Figure 7.5 illustrates the blurred intensity distributions of the 0—0 and 1—1 CN-bands (3883 and 3871 Å, respectively), calculated by Smit-Miessen and Spier [7.36] for a spectrographic slit width b of $8\ \mathrm{cm}^{-1}$ and for different temperatures between 4000 and 7000 °K.

The temperature can now be determined:

a) from the ratio of the areas under the profile curves when the bands do not overlap ([7.37] and [7.38]) and from the ratio of portions under the profile curves when the bands partly overlap (see Figure 7 in [7.33]),

b) from the ratio of the maximum intensities at the band heads ([7.37] and [7.38]).

For this the ratio of the areas or that of the partial areas (case a), or the maximum intensity ratio (case b) must be calculated as a function of temperature. Figure 7.6 shows the relevant plots for the ratio of the areas under the profile curves of the 0—0 and 1—1 CN-bands (3883 and 3871 Å) for different slit widths b.

The vibrational temperature is finally obtained from the measured values and the graphs.

The situation becomes more simple when the rotational profile depends only slightly on temperature. Then the maximum spectral band intensity measured at the band head is proportional to the intensity integrated over the band.

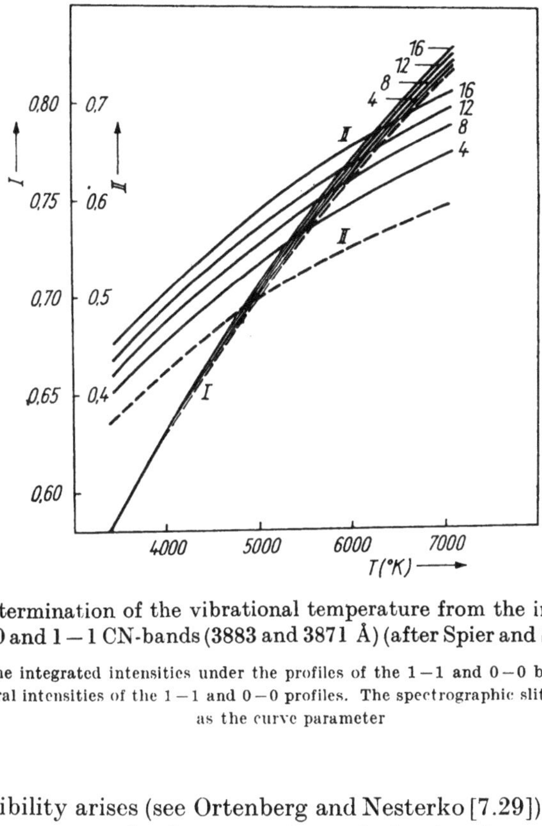

Figure 7.6. Determination of the vibrational temperature from the intensity ratio of the nonresolved 0—0 and 1—1 CN-bands (3883 and 3871 Å) (after Spier and Smit-Miessen [7.38])

Curves *I*: ratio of the integrated intensities under the profiles of the 1—1 and 0—0 bands; Curves II: ratio of the maximum spectral intensities of the 1—1 and 0—0 profiles. The spectrographic slit width *b* is given in cm^{-1} as the curve parameter

Thus, the possibility arises (see Ortenberg and Nesterko [7.29]), to determine the vibrational temperature for known values of the band strengths $p_{v''v'}$ from the expression (compare (7.29))

$$\lg \frac{I_{v''v'}}{\nu^4 p_{v''v'}} = \text{const} - \frac{U_{v'}}{2.3026\, kT}. \tag{7.37}$$

$I_{v''v'}$ here signifies the relative spectral intensity of the $v'' - v'$ band in the intensity maximum. The T-determination follows by plotting the measured values on a $\lg I_{v''v'}/\nu^4 p_{v''v'}$ versus $U_{v'}$ graph. T is then obtained from the slope of the straight line through the measured points.

To illustrate the range of application for the determination of vibrational temperature we have listed in Table 7.2 some of the relevant experimental data.

Table 7.2. Determination of the vibrational temperature T_v

Molecule	Transition	Plasma	T_v-range [10^3 °K]	Reference[1])
AlO	$A^2\Sigma - X^2\Sigma$	arc in air, Al-electrodes	5.0–7.6	38j
BaO	$A^1\Sigma - X^1\Sigma$	flame	2.2–2.3	29
BeO	$B^1\Sigma^+ - X^1\Sigma^+$	arc in air	4.2–4.3	27a
C_2	$A^3\Pi_g - X^3\Pi_u$	flame	2.2–2.3	29
		hydrocarbon arc	3.0–12.8	17, 18.
CH	$A^2\Delta - X^2\Pi$	hydrocarbon flame	2.0–3.4	5a
CN	$B^2\Sigma - X^2\Sigma$	arc in air	4.0–6.2	12g, 25, 33m, 34
		shock wave, Ar + traces C, N	7.0	35b.
N_2	$C^3\Pi_u - B^3\Pi_g$	low pressure disch. in N_2	2.0–6.4	33m, 39c.
N_2^+	$B^2\Sigma_u^+ - X^2\Sigma_g^+$	electron beam excited low-pressure gas flow	1.0[2])	21g
		Aurora	2.0	11z
		shock wave, N_2	1.4–6.6	1
OH	$A^2\Sigma - X^2\Pi$	flame	2.6–4.2	9f, 41a.
		shock wave, Ar + H_2O	3.3–4.3	40

[1]) The reference is given in short symbols (for example, 29 stands for [7.29]).
[2]) Low-density corrected values.

7.5.3. Electron Temperature for Partly Overlapping Bands

The electron temperature can be determined via the absolute intensity of the molecular spectrum at a given ν. When the bands overlap the spectral energy distribution must be found by means of (7.36) and (7.20) for an electron transition: this procedure involves extensive summations. The quantities $f_{v'v''}$ occurring in the calculation must be known beforehand. The procedure become even more laborious when bands of several molecular species overlap one another.

An example involving a spectrum consisting of overlapping bands is shown in Figure 7.7, where the absorption coefficient $\varkappa_\nu$ (which is linked to ε_ν via the expression $\varepsilon_\nu = \varkappa_\nu B_\nu$) is given as a function of frequency for the molecular radiation of nitrogen at a temperature $T = 6630$ °K, for a density ratio $\varrho/\varrho_0 = 0.88$ (ϱ_0 = density under normal conditions) (Wurster [7.44]).

In practice, the spectral distributions must be calculated for different temperatures, after which the experimental values for the spectral distribution (designated by hollow points) are fitted into the calculated curves).

A less time consuming procedure consists of carrying out measurements for one wavelength only and to determine the temperature from the measured intensity with the help of previously calculated or experimentally determined graphs $I_\nu = n_0 f(T)$, whree n_0 is the number of molecules per cm³. Figure 7.8 shows such a plot for N_2-radiation for the wavelengths 776 and 877 nm.

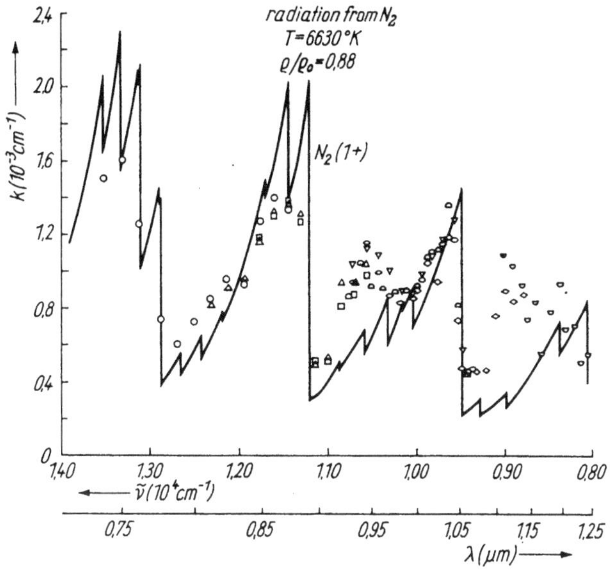

Figure 7.7. Spectral distribution of the absorption coefficients for a nitrogen plasma with $T = 6630\,°K$, $\varrho/\varrho_0 = 0.88$, in the range from 0.75 to 1.25 μm (after Wurster [7.44])

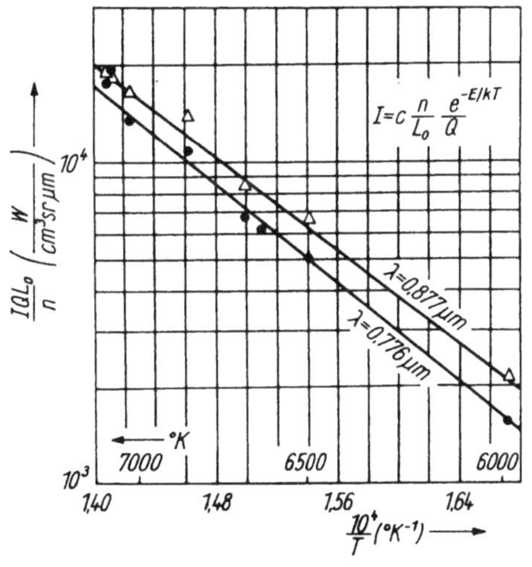

Figure 7.8. Molecular spectral emission coefficient of the radiation of an N_2-plasma as a function of temperature for the wavelengths $\lambda = 0.776$ and 0.877 μm (after Wurster [7.44])

7.6. Determination of Plasma Parameters from Absorption

These methods are based on equations (7.6), (7.7), (7.17), and Kirchhoff's law

$$\int_{\text{(line)}} \varepsilon_\nu \, d\nu = B_\nu \int_{\text{(line)}} \varkappa_\nu \, d\nu \tag{7.38}$$

where

$$B_\nu = \frac{2\,h\,\nu^3}{c^2} \frac{1}{e^{h\nu/kT} - 1}\,. \tag{7.39}$$

Then the total absorption coefficient is given by

$$\int_{\text{(line)}} \varkappa_\nu \, d\nu = \frac{8\,\pi^3\,\nu}{3\,h\,c\,Q_i} \cdot n_0 S_{J''\,\Lambda''}^{J'\,\Lambda'} p_{v''\,v'}\, \Phi_{p'} (1 - e^{-h\nu/kT})\, e^{-U_{n v'' J''}}\,. \tag{7.40}$$

$S_{J''\,\Lambda''}^{J'\,\Lambda'}$ follows from equations (7.8)–(7.10).

Equation (7.40) can be used to determine the electron temperature as well as the rotational and vibrational temperature. Kostkowski and Broida [7.15] applied equation (7.40) to calculate the rotational temperature in an acetylene-oxygen flame. They employed a chopped-light system, in which a continuous background source was used, and measured the total absorption $\int \varkappa_\nu \, d\nu$ for different rotational lines of the R-branch of the 0–0 band of the $^2\Pi - {}^2\Sigma$ transition in the OH molecular spectrum between 3067 and 3090 Å. In accordance with the relationship

$$\lg \frac{\int_{(\text{line } J''\,J')} \varkappa_\nu \, d\nu}{S_{J''\,\Lambda''}^{J'\,\Lambda'} \Phi_{p'}} = \text{const} - \frac{U_r(J'')}{2.3026\,kT} \tag{7.41}$$

the temperature was found to be $T = 2400-2800$ °K.

7.7. Transition Probabilities, Oscillator Strengths and Band Strengths $p_{v''\,v'}$

The experimental determination of the transition probability, oscillator strength, and band strengths $p_{v''\,v'}$, is usually effected by means of intensity measurements on one or several molecular bands, or also by means of absorption measurements, whereby in both cases the plasma temperature and composition must be known or else determined in some other way. The desired quantities then result from equations which are given in sections 7.1 and 7.2.1, so that we need not repeat them here. It should only be mentioned, that when applying methods which presuppose a small optical depth, the plasma must be checked for slight self-absorption (see [7.16]).

For details on the determination of oscillator strengths from lifetimes, see Bennett and Dalby [7.2].

When absolute quantities are measured, they are usually specified in the form of oscillator strength $f_{v' v''}$; relative values are usually given in terms of the band strength $p_{v'' v'}$ (rel.).

The band strength can also be theoretically calculated. It is given by the expression

$$p_{v'' v'} = \sum_{i,j} |\Psi_{v'} \boldsymbol{R}^e_{ij} \Psi_{v''} \, \mathrm{d}r|^2 \tag{7.42}$$

(see Nicholls and Stewart [7.24]). The matrix elements are summed over the degenerate upper level i and lower level j.

Expression (7.42) can be broken up into two factors, one of which essentially depends only on the electron transition and the other on the vibrational transition:

$$p_{v'' v'} = q_{v'' v'} \sum_{i,j} |\boldsymbol{R}^e_{ij}|^2 . \tag{7.43}$$

$q_{v'' v'}$ Franck-Condon-factor is given by [7.24]

$$q_{v'' v'} = [\int \Psi_{v' v''} \, \mathrm{d}r]^2 , \tag{7.44}$$

where Ψ_v is the solution to the wave equation

$$\frac{\mathrm{d}^2\psi_v}{\mathrm{d}r^2} + \frac{2\mu}{\hbar^2} \{ U_v - U(r) \} \psi_v = 0 . \tag{7.45}$$

μ is the reduced mass of the two nuclei of the molecule. $U(r)$ can be described by

$$U(r) = D_e \{1 - \mathrm{e}^{-\beta(r - r_e)}\}^2 \tag{7.46}$$

(Morse-potential) where D_e is the energy of dissociation and r_e is the equilibrium internuclear distance.

The $q_{v'' v'}$, of course, determine first of all the distribution of the radiated energy over the different vibrational transitions $v' \to v''$ occurring in one electron transition. However, for a given electron transition the quantity $\sum_{i,j} |\boldsymbol{R}^e_{ij}|^2$ is dependent on the internuclear distance as well as on the vibrational state. Nevertheless, this dependence is weaker than that of $q_{v'' v'}$ on v'' and v' and can be expressed as

$$\sum_{ij} |\boldsymbol{R}^e_{ij}|^2 = R^{e^2}_{nm}(\overline{r_{v' v''}}) , \tag{7.47}$$

in which the electronic transition moment R^{nm}_e is an analytically expressible function of the "R-centroid" defined by

$$\overline{r_{v' v''}} = \frac{\int \psi_{v'} \psi_{v''} \, r \, \mathrm{d}r}{\int \psi_{v'} \psi_{v''} \, \mathrm{d}r} . \tag{7.48}$$

Like $q_{v' v''}$, the quantity $\overline{r_{v' v''}}$ can be evaluated immediately from the potential curve $U(r)$ [7.23c].

For more details on the calculation of molecular transition probabilities and equivalent quantities, we refer the reader to Nicholls and Stewart [7.24], Ortenberg and Antropov [7.28a], Spindler [7.38a] and Nicholls [7.23q].

On the determination of oscillator strengths from lifetimes, see Bennett and Dalby [7.2].

In conclusion, we give a survey of the relevant literature on transition probabilities, oscillator strengths, and band strengths of molecular bands (see Table 7.3) and of Franck-Condon factors (Table 7.4).

Table 7.3. Bibliographic references on transition probabilities, oscillator strengths and band strengths $p_{v''v'}$ of molecular bands. The symbols left hand of the reference numbers characterize the methods of determination of the respective data and have the following meanings: A – absorption measurements, Calc. – calculated data, Comm. – comments, E – emission measurements, L – life time measurements, Oth. – other methods, rel – relative values. References are given in short form (for example, 10a instead of [7.10a])

Molecule	Reference
AlO	*E*: 38i. *E*, *rel*: 10a, 39a.
B_2	*Calc.*, *rel*: 23u
BH	*Calc.*, *rel*: 23u
BO	*Calc.*, *rel*: 23u. *E*: 35a.
BaO	*E*: 27b
BeO	*Calc.*, *rel*: 38l. *E*: 6*i*, 27b, 38k.
C_2	*A*: 38d, 38e. *Calc.*; 6a, 35d. *Calc.*, *rel*: 11g, 23u. *E*: 1d, 7b, 9r, 11x, 20l, 23, 38f.
CF:	*E*: 9r
CH	*L*: 1m
CN	*A*: 9h, 18a. *A*, *rel* + *Comm.*: 18c. *Calc.*: 35d. *Calc.*, *rel*: 23u. *Comm.*: 9g. *E*: 6k, 7a, 23, 33h, 34c. *L*: 21a, 42a. *Oth.*: 29f, 33c.
CO:	*A*: 34d. *E*: 20m, 23b, 35, 38l, 38m. *E*, *rel*: 35. *L*: 21a.
CO^+:	*Calc.x*: 11h, 35d. *E*: 12, 19a, 23f, 35a. *L*: 1l, 12, 19a, 35e. *Oth.*: 11d.
CaO	*E*: 27b
Cl_2	*Calc.*, 21f
D_2	*Calc.*: 11j
F_2	*Calc.*: 21f. *E*: 20d + *Correction* 20f.
H_2	*Calc.*: 11j, 29h.
H_2^+	*Calc.*: 1h, 1i, 1j, 20a.
H_2^-	*Calc.*: 9l
HeH^{++}	*Calc.*: 1e, 1f, 6d.
Li_2	*Calc.*: 11e
LiH	*Calc.*: 12c
MgH	*E*: 20d + *Corr.* 20f.
MgO	*E*: 20d + *Correction* 20f, 21n, 27b.
N_2	*A*: 5f. *Calc.*: 11m, 12d, 35d. *Calc.*, *rel*: 11i, 11l. *Comm.*: 6f. *E*: 5c, 6j, 6p, 12f, 18b, 34b, 39e, 42c. *E*, *rel*: 5d. *L*: 6b, 7g, 23i, 24b.
N_2^+	*Calc.*: 12d, 35d. *E*: 1, 6p, 6q, 12f, 14a, 18b, 34b, 35j, 39e, 42d. *L*: 7g, 23i, 24b.
NH	*E*: 9s. *L*: 1m, 7g.
NO	*A*: 1a, 1b, 29c, 44d. *A*, *E*: 20h. *Calc.*: 12d, 12f, 35d. *Calc.*, *rel*: 23u. *Comm.*: 11a. *E*: 42b. *E*, *rel.*: 35.
Na_2	*Calc.*: 6e
O_2	*A*: 7c. *A*, *E*: 20g. *Calc.*: 12d. *Calc.*, *rel.*: 23u. *E*: 12e, 12f.
O_2^+	*E*, *rel.*: 35. *L*: 11w.
OH	*A*: 9i. *Calc.*: 6g. *Calc.*, *rel*: 23u. *Comm.*: 6n. *E*: 2, 9b, 23. *L*: 2a.
PN	*E*, *rel*: 35c
PbO	*Calc.*, *rel*: 9q.
SiN	*E*, *rel*: 38c
SiO	*A*: 20e
TiO	*E*: 27b
VO	*E*: 27b

Table 7.4. Bibliographic references on calculated Franck-Condon-factors. References are given in short symbols (for example **23**v is used instead of [7.23v])

Molecule	Reference
AlO	23h, 35g, 38h, 38i.
B_2	23u
BH	23u
BN	23u
BO	23u, 23v.
BaO	27b, 28b, 39b.
BeO	23v, 27b, 38k, 38l.
C_2	9c, 9e, 11g, 23s, 23u, 28, 38a.
CH	5e
CH^+	23v
CN	9a, 9c, 11q, 23k, 23u, 33e, 38a, 44b.
CO	9a, 9o, 11p, 11s, 23g, 23v, 28, 33b, 33e, 33g, 35m, 35n, 39d.
CO^+	9d, 11h, 11s, 23f, 23g, 33e.
$CO \rightarrow CO^+$	21b, 21d, 23t.
CS	7f, 33f.
CaH	27b
CaO	28b, 35i, 44h.
CrO	44a. Comments: 21k.
D_2	9p
H_2	9o, 9p, 21c, 23n, 29g.
$H_2 \rightarrow H_2^+$	23t
He_2	35q
HgCl	38p
I_2	23j, 38b, 44e, 44f.
LaO	21h, 28b.
Li_2	21e
MgH	27b
MgH^+	33a
MgO	21m, 21n, 23h, 27b, 28b, 33k, 33l.
N_2	5e, 9b, 9d, 9n, 9o, 11m, 11r, 11s, 11t, 23d, 23g, 23p, 35k, 44g.
N_2^+	9c, 9j, 9k, 9n, 11r, 23d, 23e, 23g, 35k.
$N_2 \rightarrow N_2^+$	23g
NO	5e, 7h, 7i, 9d, 9n, 11x, 11s, 14b, 14c, 23h, 23l, 23u, 23v, 28, 29a, 29c, 29d, 39d.
NO^+	23t
$NO \rightarrow NO^+$	23t
Na_2	9o, 21b, 21d, 44e.
O_2	1c, 9c, 9d, 9m, 9n, 9o, 9p, 11f, 11n, 11o, 11s, 11u, 12f, 21b, 23a, 23e, 23g, 23r, 23u, 23v, 29b, 29c, 29e, 39d, 44c.
O_2^+	9n, 11r, 11s, 23c, 23v, 38o.
$O_2 \rightarrow O_2^-$	23t
OD	7d
OH	2, 7d, 7e, 11s, 23m, 23u, 38n.
PN	35c
PbO	23u
RbH	33d, 44e.
SD	23v
SH	23v
SO	35p
ScO	28b
SiN	38c
SiO	20c, 23h.
SrH	35o
SrO	23h
TiO	9c, 21i, 27b, 33j.
VO	21l, 27b, 33i, 33j.
YO	21j, 27b, 28b.

CHAPTER 8

Determination of Particle Densities in Plasmas by Means of the Refractive Index. Laser Diagnostics

8.1. Relationship between the Optical Refractivity and Particle Number Density

Inferences on the particle densities in plasmas can be drawn on the basis of the optical refractivity. The following relationship exists (see [8.5]):

$$(n-1)_{\text{plasma}} = (n-1)_{\text{atoms}} + (n-1)_{\text{ions}} + (n-1)_{\text{electrons}}\,. \tag{8.1}$$

For a given frequency $(n-1)_{\text{plasma}}$ is a linear function of the particle densities (in the case of a simple plasma with at the most singly-ionized particles, $(n-1)$ is a linear function of N_0, N_i, and N_e, when these respectively denote the number of neutral particles, ions, and electrons per cm^3*.

The refractive index of the neutral gas components is given by the expression (see [8.3])

$$(n-1)_{\text{atoms}} = \frac{2\pi e^2 N_0}{m} \sum_k \frac{f_k}{\omega_k^2 - \omega^2}\,, \tag{8.2}$$

where the summation is to extend over all resonance lines with cyclic frequency ω_k and oscillator strength f_k. If the resonance lines fall into the ultraviolet region (this applies for example in the case of Argon), then it is allowed to use the formula of Cauchy (see [8.3])

$$(n-1)_{\text{atoms}} = (A + B/\lambda^2)\,\frac{N_0}{N_0\,(0\,°\text{C},\ 760\ \text{Torr})}\,, \tag{8.3}$$

where N_0 (0 °C, 760 Torr) $= 2.69 \cdot 10^{19}\ \text{cm}^{-3}$ is the particle density under normal conditions. Values for A and B, taken from [8.1b], [8.3], and [8.26], are listed in Table 8.1 for several gases. As an illustration the tables also lists the index of refraction for the relevant gas under normal conditions for a wavelength in the range of the Na D-lines at 589 nm.

Further data about refractive indices are available for the species Ar [8.9b], [8.15l]; B^+, Be, Be^{++}, C^{++} [8.15h]; CH_4, C_2H_6, CO_2 ([8.20d] p. 390): Cu [8.12a]; F^- [8.15h]; H [8.8g]; He [8.15h]; Kr [8.15l]; Li^+ [8.15h]; N [8.2a]; N^{+++}, Na^+, Ne [8.15h]; O [8.2a]; Xe [8.15l]. In addition, the refractive index in a thermal Argon plasma can be found in [8.14o] under the condition $p = 1.1$ atm, $T = 5000-20{,}000$ °K, $\lambda = 6328$ Å, or in [8.14p], for the same pressure, for $T = 10{,}000-22{,}000$ °K and $\lambda = 6328$ Å respectively $\lambda = 1.15\,\mu$m (He—Ne-Laser wavelength).

* To avoid confusion, we have designated the particle densities by upper-case letters.

Table 8.1. Constants A and B for determining the refractive index of gases in accordance with (8.3)

Atom or molecule	A	B $[10^{-14}\ \text{cm}^2]$	$n - 1$ (589 nm)
He	$3.48 \cdot 10^{-5}$	0.08	$3.50 \cdot 10^{-5}$
Ne	$6.66 \cdot 10^{-5}$	0.16	$6.71 \cdot 10^{-5}$
Ar	$27.92 \cdot 10^{-5}$	1.56	$28.37 \cdot 10^{-5}$
Kr	$41.89 \cdot 10^{-5}$	9.92	$42.73 \cdot 10^{-5}$
X	$68.23 \cdot 10^{-5}$	6.92	$70.2 \cdot 10^{-5}$
Hg	$87.8 \cdot 10^{-5}$	19.9	$93.5 \cdot 10^{-5}$
Air	$28.71 \cdot 10^{-5}$	1.63	$29.18 \cdot 10^{-5}$
H_2	$13.58 \cdot 10^{-5}$	1.02	$13.84 \cdot 10^{-5}$
O_2	$26.63 \cdot 10^{-5}$	1.35	$27.2 \cdot 10^{-5}$
N_2	$29.06 \cdot 10^{-5}$	2.24	$29.7 \cdot 10^{-5}$
CO	$32.7 \cdot 10^{-5}$	2.65	$33.4 \cdot 10^{-5}$
NO	$28.9 \cdot 10^{-5}$	2.02	$29.7 \cdot 10^{-5}$
OH			$25.0 \pm 3 \cdot 10^{-5}$

For the refractive index of the ion components we assume, insofar no other information is given, the same constants A and B as for the neutral gas components:

$$(n - 1)_{\text{ions}} \approx \left(A + \frac{B}{\lambda^2}\right) \cdot \frac{N_i}{2.69 \cdot 10^{19}\ \text{cm}^{-3}}\,. \tag{8.4}$$

If a higher precision is needed, $(n - 1)_{\text{ions}}$ must be calculated according to a formula of type (8.2) in which N_0 is substituted by N_i.
The index of refraction of the electron component is given by

$$(n - 1)_{\text{electrons}} = -\frac{2\pi e^2 N_e}{m\,\omega^2}\,. \tag{8.5}$$

Formula (8.5) needs to be corrected when the electronic collision frequency comes into the order of the frequency of light, as may be the case for a far infrared test beam [8.9], [8.14n].

Combining we obtain the expression

$$(n - 1)_{\text{plasma}} = \frac{A + B/(\lambda^2)}{2.69 \cdot 10^{19}\ \text{cm}^{-3}}\,(N_0 + N_i) - \frac{2\pi e^2 N_e}{m\,\omega^2}\,. \tag{8.6}$$

To gain some idea on the order of magnitude of the influence of the neutral and electron gas on the index of refraction, we give the following formulative relationship in the case of argon for the wavelength 546 nm:

$$(n - 1)_{\text{Ar I}} = +\,1.06 \cdot 10^{-23}\,N_{\text{Ar I}}\,,$$

$$(n - 1)_{\text{electrons}} = -\,13.33 \cdot 10^{-23}\,N_e\,.$$

From this it is evident that at this wavelength the electrons have a 10-fold greater influence on the refractive index than the neutral particles. Furthermore, the sign is reversed, i.e., the electrons tend to decrease the index of refraction.

Since from (8.6) it follows that for a given frequency $(n - 1)_{\text{plasma}}$ is a linear function of N_0 as well as of N_e $(= N_i)$, the quantities N_0 and N_e cannot be separately found on the basis of one measurement of $n - 1$. If the degree of ionization is small (smaller than 1%), a choice of sufficiently short wavelengths and a monochromatic measurement of n will enable the determination of the neutral gas particle density N_0 alone, since in this case the influence of the electrons on the refractive index is practically negligible.

Both N_0 and N_e can be calculated with the help of (8.6) when the index of refraction n is measured for at least 2 different wavelengths, since the respective influences of the neutral gas and the electrons are dependent on wavelength to different degrees (the refractive index of the atoms varies in contrast to the refractive index of the electrons only slightly with wavelength).

In a gas mixture composed of two components with number densities N_1 and N_2, the quantity $(n - 1)$ is a function of both number densities. In consequence of this, it is not possible to determine the quantities N_1 and N_2 from a measurement of $(n - 1)$ alone. To get N_1 and N_2, besides of the quantity $(n - 1)$, further quantities respectively sets of quantities with different behavior with respect to a variation of N_1 and N_2 have to be measured (for example, the mass density, or the temperature together with the pressure). On this basis, Cary [8.9a] has determined the dissociation degree and the dissociation rate in shock heated N_2. The neutral components in this experiment are N and N_2.

Vice versa, starting from the resultant refractivity $(n - 1)$ of a partially dissociated two-atomic molecular gas in the equilibrium zone of a shock wave with known degree of dissociation, it is possible to obtain values for the atomic contribution to the refractivity index $\Delta (n - 1)/(\Delta N_{\text{atom}})$, as has been shown by Alpher and White [8.2a] for O and N and by Brinkschulte and Muntenbruch [8.8g] for H.

There are several techniques for determining the refractive index. Essentially these are: 1. the interferometric method, 2. the Schlieren technique, and 3. the shadowgraph technique. These will be treated below.

8.2. Interferometric Determination of the Refractive Index *n*

8.2.1. Interference Fringes

We shall consider a Mach-Zehnder interferometer system as shown in Figure 8.1 The light source S, which we assume to be monochromatic, is focussed on an aperture or slit A. The lens B renders the light bundle parallel, after which the latter is split by C_1 into two mutually interfering bundles. These two bundles are reunited behind the beam splitter C_2. A phase shift between the bundles occurs when a measurement object, whose refractive index differs from that of the surrounding medium, is placed at G_1. The object G_1 is imaged via the lens E on the film F. When the plates C_1 and C_2 and the mirrors D_1

and D_2 are all parallel to one another, no interference fringes are observed when the object is not located in the light path or when the measurement object is homogeneous and possesses a constant length along the light path.

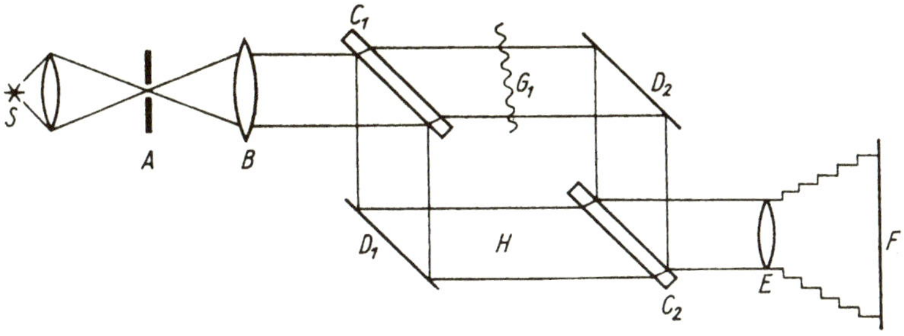

Figure 8.1. Mach-Zehnder interferometric system for producing interference fringes (after Weinberg [8.25], p. 207)

S — light source, A — aperture, C_1 and C_2 — beam spitters, D_1 and D_2 — mirrors, G_1 — measurement object, E — camera lens for focussing G_1 on the film F

The more or less uniformly illuminated image created in F is dark or more bright depending on the extent of the phase shift. Figure 8.2a shows an example of this in the case of an interference photograph of a carbon arc, which is not operating. The photograph shows nothing besides the carbon electrodes. When the carbon arc is actuated, the object in the ray path becomes optically inhomogeneous and interference fringes appear on the photograph (Figure 8.2b).

Interference fringes can also be produced when the object is not located in the light path by turning C_1 with respect to C_2 by an angle α_1 and D_1 with respect to D_2 by an angle α_2. Then the wavefronts which interfere are at an angle of $2(\alpha_1 + \alpha_2)$ to one another. The separation q between two neighboring fringes, relative to the plane G_1 (which is photographed on F) where the interference takes place, is given for the case when the angles α_1 and α_2 are small by the expression

$$q = \frac{\lambda}{2(\alpha_1 + \alpha_2)}. \tag{8.7}$$

If a homogeneous object of length L is placed lengthwise along the light path a shift in the interference results which can be expressed in units of the fringe interval as follows

$$s = \frac{\Delta(n-1) \cdot L}{\lambda}, \tag{8.8}$$

where $\Delta(n-1)$ is the difference between the refractive index of the object and that of the surrounding medium. For more details on interferometric techniques see Ladenburg et al. [8.17] and Kinder [8.15k]. For constructional details see, for example, [8.15f].

As an example we give an interference photograph obtained by Ascoli-Bartoli and Rasetti [8.7] for a longitudinally homogeneous object (Figure 8.3). The left portion of the photograph refers to the case in which a 32 cm long tube filled with argon (0.5 Torr) is placed in the light path. The right portion of the photograph displays a shift in the interference fringes, which results when an HF-discharge is triggered in the tube. In the above case a shift of $s = -0.06$ was obtained. The change in the refractive index $\Delta(n-1)$ is found from s with the help of equation (8.8). If $\Delta(n-1)$ is measured for 2 wavelengths, the neutral particle density and the electron density can be determined in accordance with the considerations in section 8.1.

To get two interference pictures of the same object and for the same time at two different wavelengths, the interferometric system shown in Figure 8.1 can be completed by addition of a second section E' F' placed below the half transparent mirror C_2, and by addition of two suitably chosen monochromatic filters placed between C_2 and E respectively C_2 and E' [8.2b].

While in the optical spectral region the interference pattern is recorded mainly photographically, in the IR region it is obligatory to use photoelectric recording [8.2a], [8.17b] (Only for very high intensive infrared radiation sources, photographically recording can be effected by means of liquid crystal screens (see section 2.3.1.).

8.2.2. Monochromatic Interferometric Technique for Axially Symmetric Inhomogeneous Plasmas

We shall treat an example of the photographic and computational procedure, and shall base our arguments on a study performed by Schmitz [8.5]. Figure 8.4b illustrates a corresponding interferometer photograph of a vertically burning carbon arc, which is located in the interferometer light path at G_1 (see Figure 1) and is irradiated at right angles to the axis of the arc. For comparison Figure 8.4a gives an interferometer photograph of the arc arrangement when the arc is not burning.

The computational procedure is as follows: for a given plane normal to the axis of the arc (indicated by an arrow in Figure 8.4b) the shift s of the fringes is measured as a function of the transverse coordinate x of the arc ($x = 0$ corresponds to the center of the arc). s is taken, as shown already above, with respect to the fringe interval. Figure 8.5 gives a corresponding $s(x)$-diagram.

Instead of (8.8) we now have the relationship

$$s = \frac{1}{\lambda} \int_{-y_0}^{+y_0} \Delta(n-1)\, dy\,, \tag{8.9}$$

where $\Delta(n-1)$ is the difference between the refractive index of the object and that of the surrounding medium: y is the coordinate along the line of

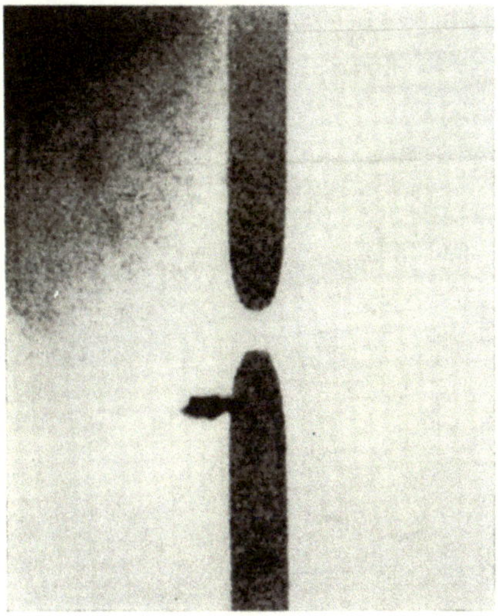

a)

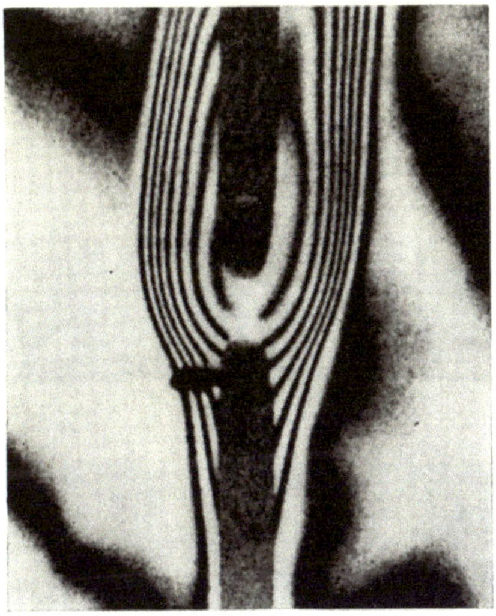

b)

Figure 8.2. Interferometer photographs for the case when the plates and mirrors are parallel (after Schmitz [8.21])

a) when arc is not burning

b) burning homogeneous carbon arc

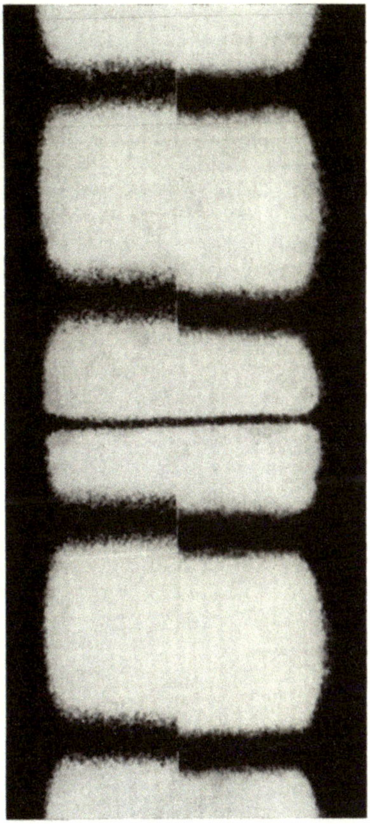

Figure 8.3. Interference photograph for a 32 cm long tube, filled with argon; initial pressure $p_0 = 0.5$ Torr. left: without, right: with triggered HF-discharge. Wavelength = 546 nm, (after Ascoli-Bartoli and Rasetti ([8.6] and [8.7]), Laboratorio Gas-Ionizzati, EURATOM-CNEN, Frascati, Rome)

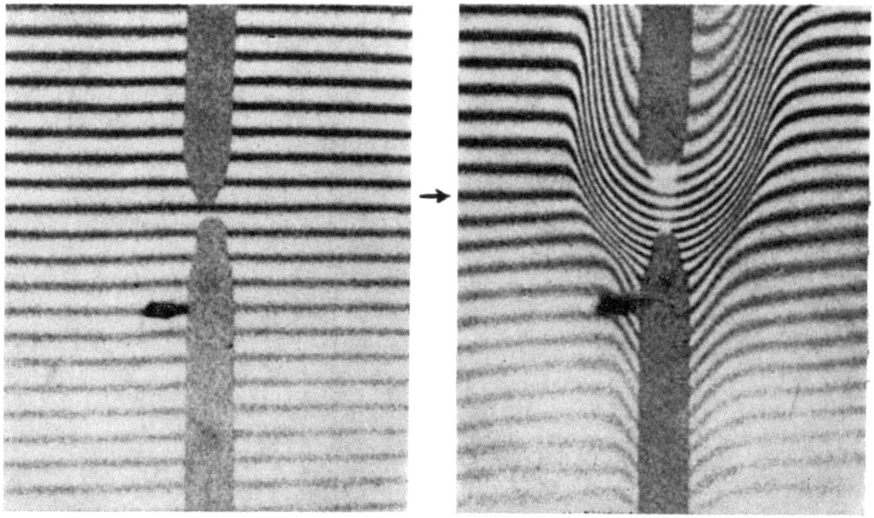

Figure 8.4. Interferometer photograph of a vertically burning carbon arc (b), compared to that for the arc when it was not burning (a) (after Schmitz [8.21])

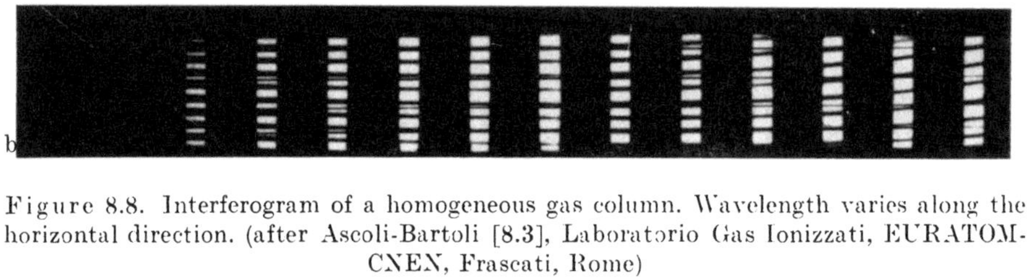

Figure 8.8. Interferogram of a homogeneous gas column. Wavelength varies along the horizontal direction. (after Ascoli-Bartoli [8.3], Laboratorio Gas Ionizzati, EURATOM-CNEN, Frascati, Rome)

a) with compensator (straight interference lines) b) without compensator (slanted interference lines)

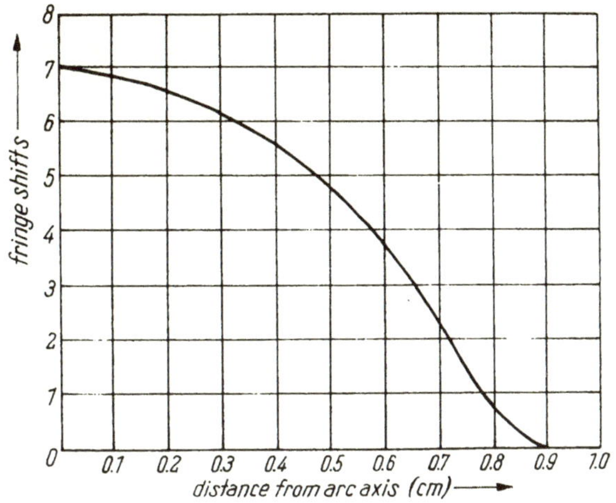

Figure 8.5. Fringe shift as a function of the transverse coordinate x of the arc (after Schmitz [8.21])

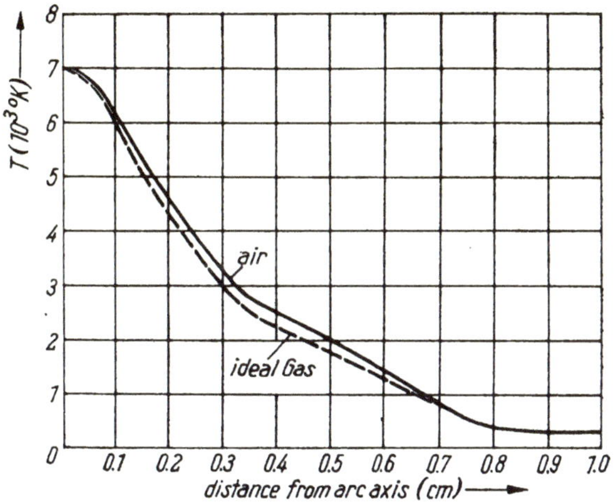

Figure 8.6. Radial temperature distribution in a homogeneous carbon arc, derived from the fringe shift (solid curve). The dashed curve refers to the case in which the molecular dissociation is not allowed for in the interpretation of the measurements (after Schmitz [8.21])

sight, and $-y_0$ and $+y_0$ are the coordinates along this same direction which form the outer limits of the arc.

Owing to the axial symmetry the radial distribution $\Delta (n - 1) (r)$ (compare section 2.10.1) follows from the transverse distribution $s(x)$ in accordance with

$$\Delta (n - 1) (r) = - \frac{\lambda}{\pi} \int_r^R \frac{s'(x) \, dx}{\sqrt{x^2 - r^2}} . \qquad (8.10)$$

From $\Delta (n - 1) (r)$ we finally obtain the quantity $\varrho(r) - \varrho$ (295 °K) (assuming the temperature of the surrounding medium is 295 °K) (via the relationship $\varrho/(n - 1) =$ const. With the known value for ϱ (295 °K) we then arrive at the density ϱ as a function of the radius r. From this we obtain the radial temperature distribution $T(r)$ (Figure 8.6) via the relationship $\varrho = f(T)$, $p = 1$ atm, whereby it should be taken into account that the arc plasma contains CO_2 and $(CN)_2$, and that additional dissociation products are present.

8.2.3. Application of the Interferometric Technique to Density Determination in Homogeneous Plasmas by Utilizing a Larger Spectral Range

In contrast to interferometric photographs of inhomogeneous plasmas (which are surrounded by the free atmosphere), from which the positions of the undisturbed interference lines can be found, in photographs of homogeneous plasmas the undisturbed lines cannot be obtained on the same photograph. Thus, in order to determine the fringe shift at least two photographs are required. The choice of a concrete measuring method depends on the strength of this shift.

8.2.3.1. Shift of the interference fringes by a fraction of the fringe interval

Ascoli-Bartoli et al. [8.3] considered this case. The plasma was produced by an HF-discharge in an approximately 30 cm long tube with plane-parallel end plates filled with argon or xenon under a slight pressure of several Torr (see ray path 1 in Figure 8.7). An open tube filled with air is placed in ray path 2. This serves to eliminate streaks which otherwise would occur in the vicinity of the discharge thus disturbing the measurements. The light source employed was a maximum-pressure xenon lamp XBO 162, which emits intense continuous radiation. The object to be investigated is imaged on the spectrograph slit in such a way that the interference fringes are directed along the dispersion direction (i.e., horizontally) and come to lie one above the other.

To obtain reference fringes on the photographic plate in order to facilitate the measurement of the fringe shift, horizontally stretched 0.01 mm thick tungsten filaments were placed in front of the spectrograph slit, (see Figure 8.8) which appear as sharp dark lines on the interferograms.

In order to determine fringe shifts it is necessary to obtain two interferograms in direct juxtaposition, i.e., one for the case when the discharge is triggered and one when the gas in tube 1 is not excited. For this purpose a screen with 18 consecutive rectangular openings was placed before the photographic plate.

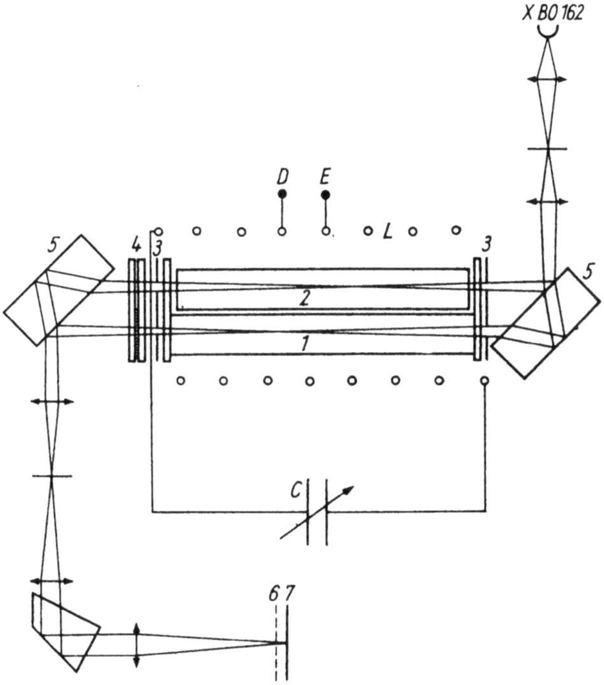

Figure 8.7. Layout employed by Ascoli-Bartoli et al. [8.3] for determining the particle densities in a tube filled with argon, triggered by an HF-discharge. (Laboratorio Gas Ionizzati, EURATOM-CNEN, Frascati, Rome)

1 — discharge tube with optical windows, *2* — open, air-filled tube, *3* — diaphragms, *4* — water-filled compensator, *5* — interferometer plates, *6* — screen with 18 windows, *7* — photographic plate. The circuit marked *CDE* serves to trigger the *HF*-discharge

First an interferogram was taken before the discharge was triggered (see Figure 8.8). The fact that certain wavelength ranges are blocked by the screen has no further disturbing effect. Then the photographic plate was horicontally shifted. A second interferogram was then taken after the discharge was triggered. In this fashion we obtain in each case 18 interferogram sections in direct juxtaposition which refer to the state with discharge and that without. Figure 8.3 shows a section of such an interferogram. In this particular case the fringe shift, expressed in units of the fringe interval, amounted to $s = 0.06$. The sensitivity of this system is so large, that a fringe shift of about 0.01 can still be observed.

When the fringe shift is measured for two suitably separated wavelength ranges, the neutral gas density N_0 as well as the electron density can be determined in accordance with the relationships given in section 8.1. The electron density in this case was found to be $7.6 \cdot 10^{14}$ cm^{-3}. The sensitivity of the device is sufficient to measure 10^{14} electrons per cm^{-3} or about 10^{15} argon atoms/cm^3 for a plasma length of 30 cm.

In conclusion it should be mentioned that due to the differing dependence of the refractive index on wavelength along the interferometer pathes 1 and 2 (Figure 8.7), the interference fringes are somewhat slanted (Figure 8.8b). This effect can be practically eliminated by placing a water-filled compensator in ray path 1 (4 in Figure 8.7). The refractive index of water and that of air depend on wavelength to nearly the same degree. Figure 8.8a illustrates that the interference fringes are in fact straightened out again due to the compensator.

8.2.3.2. Fringe shift over several fringe intervals

Shukhtin [8.22] developed a modified technique for the interferometry of plasmas enclosed in tubes. Here, the undisturbed interference fringe system and that disturbed by the discharge in the tube are directly superposed. We shall consider the case in which the interference pattern shifts over several fringe intervals as well as the case in which the shift amounts to only a fraction of the fringe interval.

We shall first treat the former case.

The interference pattern obtained when there is no discharge burning in the tube will be designated as the unshifted pattern, while that resulting as a consequence of the discharge will be referred to as the shifted pattern.

When the unshifted and shifted patterns are photographed one on top of the other, spectro-interferograms, such as shown in Figure 8.9, are obtained.

Assuming that the electrons do not contribute to the refractive index n (weak ionization), n and the gas density ϱ are then connected by

$$n - 1 = A\,\varrho\,. \tag{8.11}$$

The change in the optical path length is $L\,\Delta\,(n-1)$ or $L\,A\,\Delta\varrho$, where L denotes the path length through the plasma. If $\Delta\varrho$ is sufficiently large, then for certain wavelengths the change in the optical path length is a whole multiple of the light wavelength λ. At these points the interference fringes become more intense. At other wavelengths, for which the change in the optical path length is only an odd multiple of half the wavelength, the shifted fringes fall into the spaces of the unshifted pattern. In this case, either places displaying blurring or places with twice as many fringes can be seen on the spectro-interferogram.

The positions where blurring occurs are given by the expressions

$$\left.\begin{aligned} L\,A\,\Delta\varrho &= \frac{2\,k_1+1}{2}\,\lambda_1\,,\\ L\,A\,\Delta\varrho &= \frac{2\,k_2+1}{2}\,\lambda_2\,,\\ k_2 &= k_1+i\,, \end{aligned}\right\} \tag{8.12}$$

whereby the three unknowns k_1, k_2, and $\Delta\varrho$ can be calculated. In (8.12) k_1 and k_2 are whole numbers, while i is the number of places where a sharp pattern occurs between two blurred portions.

When the refractive index is determined by the electrons, we have the following relationship (see (8.5))

$$n-1=-\frac{N_e\,e^2}{2\,\pi\,m\,c^2}\,\lambda^2\,. \tag{8.13}$$

The interpretation of the corresponding interferogram follows as above, except that instead of (8.12) we have to solve the system of equations

$$\left.\begin{aligned} \frac{L\,N_e\,e^2}{2\,\pi\,m\,c^2}\,\lambda_1 &= \frac{2\,k_1+1}{2}\,,\\ \frac{L\,N_e\,e^2}{2\,\pi\,m\,c^2}\,\lambda_2 &= \frac{2\,k_2+1}{2}\,,\\ k_2 = k_1 &= i\,, \end{aligned}\right\} \tag{8.14}$$

where k_1 and k_2 are whole numbers and i has the same meaning.

When the electrons and the neutral gas simultaneously play an important role in determining the refractive index, the measuring technique remains the same. Here again an unshifted and a shifted interference pattern are superimposed on one another. In this case 3 places are observed displaying blurring of the interference pattern. The positions of these places are given by the equations

$$\left.\begin{aligned} L\,A\,\Delta\varrho+\frac{L\,N_e\,e^2}{2\,\pi\,m\,c^2}\,\lambda_1^2 &= \frac{2\,k_1+1}{2}\,\lambda_1\,,\\ L\,A\,\Delta\varrho+\frac{L\,N_e\,e^2}{2\,\pi\,m\,c^2}\,\lambda_2^2 &= \frac{2\,k_2+1}{2}\,\lambda_2\,,\\ L\,A\,\Delta\varrho+\frac{L\,N_e\,e^2}{2\,\pi\,m\,c^2}\,\lambda_3^2 &= \frac{2\,k_3+1}{2}\,\lambda_3\,,\\ k_2=k_1+i_1\,,\qquad k_3 &= k_1+i_2\,, \end{aligned}\right\} \tag{8.15}$$

where i_1 denotes the number of sharp regions between the blurred portions at wavelengths λ_1 and λ_2 and i_2 denotes the number of sharp spots beween the wavelengths λ_3 and λ_1. k_1, k_2, k_3, $\Delta\varrho$, and N can be obtained from the above system of equations.

The signs of i_1 and i_2 are as yet unknown, i.e., the direction in which the interference pattern shifts cannot be immediately deduced.

To determine the direction in which the pattern shifts a compensator which can be filled to different pressures, is inserted into the second ray path of the interferometer system.

First the interference pattern is photographed when the gas in the tube and in the compensator is not excited. Then the gas density in the compensator is changed by a given amount, the discharge is triggered in the tube, and the second interference pattern is superposed on the first.

When the sign of the change in the gas density inside the compensator agrees with that in the test volume, the amount by which the pattern shifts equals the difference between the shifts separately produced by the change in the compensator gas density and by that in the test volume. When the signs are opposite to one another the shift becomes equal to the sum of the individual shifts.

8.2.3.3. Shift in the interference pattern by a small fraction of a fringe width [8.22]

Here the method described in section 8.2.3.2 cannot be applied in its original form, since the variations in intensity of the superposed and the nonsuperposed pattern cannot be seen in the visible spectral range.

Blurred fringes are obtained in the following way:

A. An interferogram is taken when the gas in the compensator is maintained at a certain pressure and when the gas in the discharge tube is not excited. This interferogram is then superposed by another, which is taken when the gas pressure in the compensator is lowered by an amount $\Delta\varrho_0$. The position λ_{0k} of the k-th order blurring is given by

$$\frac{L_0 A_0 \Delta\varrho_0}{\lambda_{0k}} = \frac{2k+1}{2} . \tag{8.16}$$

B. An interferogram is taken for a given initial pressure in the compensator when the gas in the discharge tube is not excited. Then a second photograph is taken on the same film when the density in the compensator is decreased by a measurable (via the pressure) amount $\Delta\varrho_0$ and the test volume is excited by the discharge. The position of the blurring is now shifted, and is given by

$$\frac{2k+1}{2}\frac{\lambda_{0k}}{\lambda_k} \pm \frac{L A \Delta\varrho_x}{\lambda_k} - \frac{L N_e e^2}{2\pi m c^2}\lambda_k = \frac{2k+1}{2}, \tag{8.17}$$

where k is the order of the blurring and λ_k corresponds to the position where the blurring occurs. k is obtained from the calibration interferogram. We are now left with two unknowns $\Delta\varrho_x$ and N_e in (8.17), which refer to the excited gas. To determine these we must write two equations, which correspond to the blurrings of order k_1 and k_2:

$$\left.\begin{aligned} \frac{2k_1+1}{2}(\lambda_{0k_1} - \lambda_{k_1}) &= L A \Delta\varrho_x + \frac{L N_e e^2}{2\pi m c^2}\lambda_{k_1}, \\ \frac{2k_2+1}{2}(\lambda_{0k_2} - \lambda_{k_2}) &= L A \Delta\varrho_x + \frac{L N_e e^2}{2\pi m c^2}\lambda_{k_2}. \end{aligned}\right\} \tag{8.18}$$

This system of equations can be solved for N_e and $\Delta\varrho_x$.

Figure 8.9. Spectro-interferograms obtained when two interference patterns are superposed (after Shukhtin [8.22])

From top to bottom: a) superposition of two unshifted patterns; b) for wavelengths corresponding to the middle portion of the pattern the shift amounts to half a fringe width; c) the shift occurs over several fringe widths; d) the pattern are shifted by more than 100 fringes

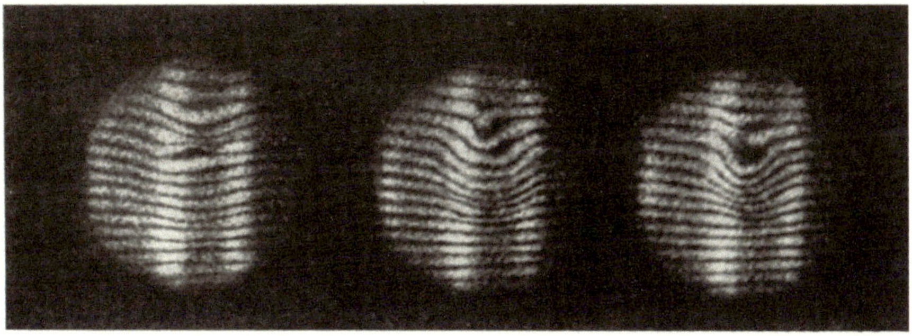

Figure 8.10. Photographs of the interference pattern of a theta-pinch at time intervals of 400 ns with exposure times of 10 ns, taken by means of an image converter. The spatial electron density distribution can be determined for different instances of time from the interference patterns (after Küpper [8.16])

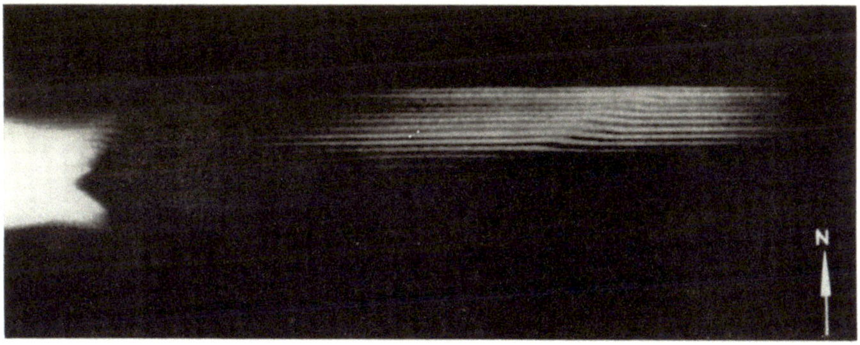

Figure 8.12. Time-resolved interferometer photograph along the axis of an electrodeless deuterium discharge in a cylindrical chamber, initial pressure 0.5 Torr. Slit along the radius. Time axis from left to right (after Medford et al. [8.20])

By this method fringe shifts as small as 1/500 of a fringe width can be still observed. For a tube length of 1 m an electron density of 10^{13} cm^{-3} or a neutral particle density of 10^{14} cm^{-3} can be measured. This differs somewhat according to the particular gas.

8.2.4. High-speed Interferometry

This is applied in those cases where neutral gas or electron densities are to be measured in plasmas which change rapidly with time (spark discharges, shock waves).

If one is satisfied with a single photograph of short exposure time, the usual interferometer arrangement can be used when a synchronizable short-time pulse discharge [8.8] or a Q-switched ruby laser [8.1a], [8.15c] is employed as the light source.

Several high-speed photographs of this type, taken at short time intervals from a test object rapidly changing with time, can be obtained by means of an image converter (see Küpper [8.16]). In this case the interference pattern is projected upon the photocathode of the image converter. By means of a suitable deflecting system it is ensured that consecutive images appear at different locations on the screen and can be photographed. The illumination time of the pulse lamp used to illuminate the system must be correspondingly great. Figure 8.10 illustrates 3 interferograms of a theta-pinch taken at time intervals of 400 ns with a 10 ns exposure time. From these photographs the spatial distribution of the electron density can be determined for different instances of time.

Monochromatic interferometer photographs with continuous time resolution can be taken by using a combination of an interferometer arrangement and a rotating-mirror camera. Such a system was employed by Medford et al. [8.20], [8.26c] to determine the electron density in an electrodeless discharge in a cylindrical chamber (see Figure 8.11). The discharge was irradiated in the axial direction. A pulse lamp served as the light source.

An example of a time-resolved interferogram is shown in Figure 8.12. The electron density can be derived from the deflection of the interference fringes; deflection in the direction of the arrow indicates an increase in the electron density. On the left-hand portion of the photograph the effect of the background luminosity of the discharge itself, which has a shock-wave structure, can be recognized. When the photograph was taken the initial pressure of the deuterium gas was 0.5 Torr.

The sensitivity of this system is sufficient to record a shift in the fringe pattern by 1/5 the interfringe separation, i.e., for optical path lengths of 1 cm the method is suitable for measuring electron densities of the order of 10^{17} cm^{-3}.

When the shift of the interference lines amounts to more than one order, then it is practical to detect this shift photoelectrically [8.9f]. Using a double slit detection system, together with the amount of the interference line shift also its direction can be recorded [8.91].

As light sources for time-resolved interferometry not only pulse discharge tubes (for constructional details see, for example, [8.12b]) can be used but also continuously working He-Ne-Laser [8.9f], [8.26d]. In [8.9f], a time resolution of 10 ns was obtained.

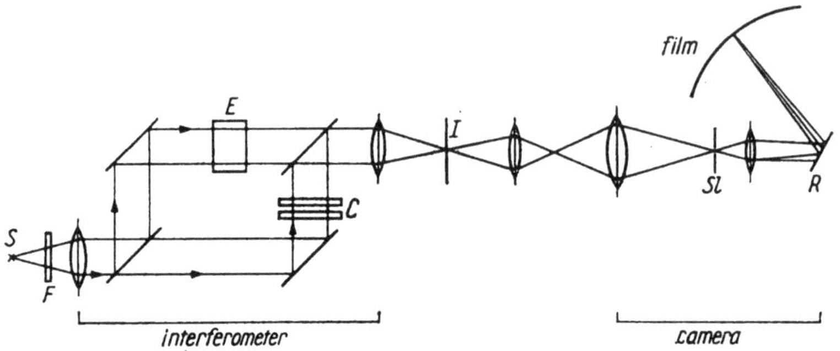

Figure 8.11. Interferometer layout with rotating-mirror camera for photographing time-resolved interferograms (after Medford et. al. [8.20])

S – source, *F* – filter, transmission range 4670 ± 35 Å, *C* – compensating plates, *E* – electrodeless discharge *I* – iris diaphragm, *Sl* – Camera Slit, *R* – rotating mirror

8.2.5. Other Types of Two-beam Interferometers

Brown, Bekefi, Llewellyn-Jones and Whitney [8.9], [8.17b] and also Weinberg ([8.25] p. 213) have described an interferometer, in which a diffraction grating was used as the light dispersion element. For details we refer the reader to the relevant literature.

A double slit interferometer illuminated by a He-Ne-Laser was used by German and Schreiber [8.14c] to evaluate the radial electron density profile of a high current argon arc.

In the Twyman-Greene-type interferometer used among others by Besse and Kelly [8.8d], the test beam crosses the plasma two times. Thus the sensitivity is doubled.

In the common path interferometer used by Dyson et al. [8.11], [8.11a], [8.12a] as well as in the differential interferometer used among others by Zimmermann [8.26h], the light splitting needed to get interferences is effected by means of crystal optical elements. In the differential interferometer, both the test beam and the reference beam cross the plasma with a slight lateral shift between these beams, thus the shift of interference fringes is proportional to $L \cdot \mathrm{d}n/\mathrm{d}r$, where L denotes the effective plasma length and $\mathrm{d}n/\mathrm{d}r$ the gradient of the refractive index perpendicular to the direction of the test beam.

For optical probing of a laser produced plasma, David et al. [8.9f] have used a Mach-Zehnder interferometer, adjusted to give a concentric interference pattern. The electron density was obtained from the phase shift of intensity within the central part of the Airy disk, which is detected photoelectrically.

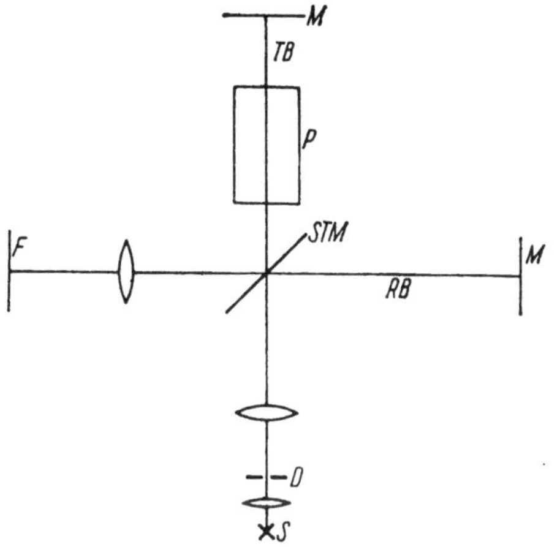

Figure 8.13. Twyman-Greene-type interferometer used in [8.8d] (schematic)
M – mirror, STM – semitransparent mirror, TB – test beam, RB – reference beam, P – plasma, S – radiation source, F – photographic film or photoelectric detector assembly

8.2.6. Laser Interferometer

The He-Ne-laser interferometer in the form used by Ashby, Jephcott et al. [8.7a] is schematically shown in Figure 8.14.

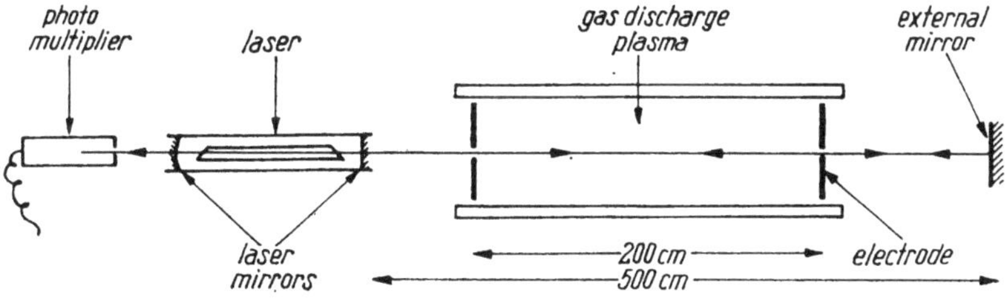

Figure 8.14. He-Ne-laser interferometer (after Jephcott et al. [8.7a])

The discharge plasma is optically coupled to the laser resonator by means of an external mirror. Changes of the refractive index of the plasma are indicated by the modulation of the laser beam intensity which is recorded by means of a photomultiplier on the left hand side from the laser. The number of intensity modulation periods corresponds to the number of fringe changes $2\,\Delta n \cdot L$, where L is the effective length of the plasma.

The probing wavelength, either 0.63 μm or 3.4 μm, can be isolated by means of a filter.

Due to the coupling of the laser output intensity at both working wavelengths, the shift of interference fringes can be recorded either via the intensity alteration of the visible radiation or the infrared radiation, independent from the probing wavelength.

The interferometer in this form is useful for the measurement of plasma densities which produces a shift of more than one interference fringes. With the infrared test beam, the minimal detectable electron density is $(3.10^{16}/L)\text{cm}^{-2}$. With an indicator wavelength of $\lambda = 0.63$ μm, the time resolution is relative low (1–10 μs), due to the inertia of laser mechanism. This limitation can be overcome by a direct recording of the test radiation at 3.4 μm which is monitored from a hole in the external mirror [8.14a], [8.15d]. Then a time resolution of 20–30 ns is obtainable.

The sensitivity of the laser interferometer can be raised by at least three methods:

1. Substitution of the plane external mirror by a spherical mirror of suitably chosen focal length [8.15d].
2. As indicator for the change in plasma refractivity the variation of laser oscillation frequency is used [8.14b], [8.15e].
3. Using as indicator for the change in plasma refractivity Δn the change of the laser output respectively the change of the corresponding photocurrent ΔI, then the sensitivity $\Delta I/\Delta n$ can be raised by installation of an optical feedback amplifier in the resonator volume [8.9k]. With a plasma length of 50 cm, the minimal detectable electron density was found to be $10^{12}\ \text{cm}^{-3}$.

The lower limit for the particle number density to be detected is determined by mechanical vibrations.

A laser interferometer system including a vibrating external mirror has been described by Freund [8.13a]. With this system, densities as low as $10^{15}\ \text{cm}^{-3}$ with a time resolution of 0.5 μs can be detected.

As probing wavelength of the He-Ne-laser interferometer, besides $\lambda = 0.63$ and 3.4 μm also the wavelength $\lambda = 1.15$ μm can be used [8.14b].

Relative long probing wavelengths (for example 10.6 μm) are obtainable by means of the CO_2-laser interferometer [8.17e], [8.20c].

Due to the small divergence angle of laser radiation, a relative high spatial resolution (0.3 mm) is obtainable [8.9k]. For more detail about modern interferometers see [8.20q].

8.2.7. Multiple Beam Interferometer

The experimental layout is such that the parallel monochromatic light of the laser lamp traverses the Fabry-Perot interferometer (gaseous discharge tube with 2 end plates, which serve as beam splitters, see section 2.1.3) and is focussed by a converging lens in a plane, on which the interference fringes appear. If the plasma excitation mechanism is alternately switched on and off in the tube, shifts in the interference pattern will occur, which constitute a measure for the change Δn in the refractive index in the tube. The actual relative shift of the fringes (= absolute shift of the interference pattern: separation between two fringes) amounts to

$$s = \frac{2t\,\Delta n}{\lambda}\,. \tag{8.19}$$

The absolute effectiveness of the method depends on the minimum value of s that can be determined. This is determined by the relative fringe width Δs, and thus depends on 1. the line width of the laser radiation, and 2. the resolution (resonance width) of the Fabry-Perot interferometer. Assuming that the resolving power of the interferometer limits the sensitivity, we obtain for the relative fringe width (again expressed in units of the interfringe separation)

$$\Delta s = \frac{1-r}{\pi\sqrt{r}}\,, \tag{8.20}$$

where r is the reflectivity of the interferometer plates.

Dougal [8.10] estimated the sensitivity of the technique when measuring electron densities, while assuming for the plates the very high reflectivity 0.98, and obtained a sensitivity of $\Delta n_e = 10^{13}\ \mathrm{cm}^{-3}$ for the wavelength 6328 Å.

8.2.8. Holographic Interferometry

Three dimensional interferometric charts from a plasma or a hot gas volume can be obtained using the holographic single or double exposure technique (see, for example, [8.14j]). Referring to the latter, at one of these exposures the plasma is switched on, while at the other exposure the plasma is not excited. Both exposure are superposed on the same photographic plate (by each of these exposures, the photographic plate is illuminated at the same time interval a) by direct laser light, b) by laser light passing through the test volume).

When the developed photographic plate (interference hologram) is illuminated by coherent light (for example, by light from a He-Ne-laser), regular interference pictures of the test volume can be photographed from this plate. When these photographs are taken for different visiting directions, a number of interference pictures (interferograms) are obtained. These interferograms correspond to different test beam directions. From these interferograms, the spatial distribution of particle number density in the plasma can be evaluated.

One of the advantages of holographic double exposure technique for obtaining interferograms consists in the elimination of disturbances due to plasma enclosing walls of low optical quality.

Table 8.2. Application of classical interferometry for the determination of temperature and particle number density in plasmas. The symbols in the head of the table have the following meanings: Q — quantity to be measured, V — numerical value, L — length of the test beam in the plasma, λ — wavelength of the test beam, Δt — time resolution. The symbols used in the tabulation are: *d.* — discharge, *hf.* — high frequency, *hp.* — high pressure, *lp.* — low pressure, *mp.* — medium pressure, *Or.* — Order of magnitude, *sd.* — spark or pulse discharge, *sh.* — shock wave, *td.* — toroidal discharge. References are given in short form (for example, 20b stands for [8.20b])

Case a) interferometry at one wavelength

Q	V	Plasma	L [cm]	Light-source	λ [nm]	Δt [µs]	Reference
T_g [°K]	360–5290	Xe, *mp. d.*	26	Hg, *hp. d.*	546		20b
$T(r)$ [°K]	300–3000	Ar, plasmajet	*Or.* 1	Hg, *hp. d.*	436	$1.6 \cdot 10^4$	26a
$T(r)$ [°K]	300–7000	Air, *hp. d.*	2	Hg, *hp. d.*	577/79		21
$T(r)$ [°K]	400–20000	H_2, *hp. d.*	*Or.* 1	Hg, *hp. d.*	546		14g
T_e [10^3 °K]	20–200	Laser induced plasma	*Or.* 1	He-Ne-Las.	633	*Or.* 0.01	9f
T_e [10^3 °K]	50–167	D_2, *sh.*	4	Ar *sd.*	467		20, 26c.
$\varrho(r, t)/\varrho_0$	0.004–5.3	Air, *sd.*	*Or.* 1	Spark	560		18, 19.
N_e [10^{15} cm^{-3}] -	$(1-6) \cdot 10^{-3}$	Ar, *lp. d.*	40	Hg, *lp. d.*	$3 \cdot 10^5$		9
N_e [10^{15} cm^{-3}]	0.1–3.8	H_2, *td.*	30	Xe, *sd.*	620	30	12, 12a.
N_e [10^{15} cm^{-3}]	0.1–20	He plasma, afterglow	20	He-Ne-Laser	633	*Or.* 1	26d
N_e [10^{15} cm^{-3}]	0.76	Ar, *lp. hf. d.*	30	Hg, *hp. d.*			6, 7
N_e [10^{15} cm^{-3}]	2–11	H_2, Θ-Pinch	25	He-Ne-Las.	633	*Or.* 0.1	9j
N_e [10^{15} cm^{-3}]	25–70	D_2, Θ-Pinch	70–120	Ruby Las.	694	*Or.* 0.1	14k, 15c.
N_e [10^{15} cm^{-3}]	45	H_2, *sh.*	0.3	Xe, *sd.*	553		201
N_e [10^{15} cm^{-3}]	50–5000	air, surface disch. channel	1.2	*sd.*	580	0.1	9e
N_e [10^{15} cm^{-3}]	54–110	H_2, *sh.*	*Or.* 5	expl. wire	546	0.1	15m
N_e [10^{15} cm^{-3})	84	Ar, *sh.*	8.2	*sd.*	560	0.1	2
N_e [10^{15} cm^{-1}]	*Or.* 100	H_2O, Θ-Pinch	16	Hg, *hp. d.*	550		16
N_e [10^{15} cm^{-3}]	$(3-20)\ 10^3$	Laser induced plasma	*Or.* 1	He-Ne-Las.	633	*Or.* 0.01	9f
N_e [10^{15} cm^{-3})	$\lesssim 10^4$	Laser induced plasma	0.1–0.4	Ruby Las.	694	0.001	1a
N_0 [10^{18} cm^{-3}]	1	gas burst into vacuum H_2, N_2	*Or.* 1	Ruby Las.	694	0.1	8b

Table 8.2. (Continuation)

Case b) interferometry at more than one wavelengths

N_e [10^{15} cm^{-3}]	N_0 [10^{15} cm^{-3}]	Plasma	L [cm]	Light-source	λ [nm]	Δt [μs]	Reference
0.1 – 2.2[1])	6 – 10[1])	Xe, *hf. d.*	30	Xe, *hp. d.*	420 – 650		3
0.5 – 4[1])	17 – 50[1])	Ar, *hf. d.*	30	Xe, *hp. d.*	420 – 650		3
5.6 – 120[2])		Ar, *hp. d.*	*Or* 1	He-Ne-Laser	633, 1150, 3390		14 c
10 – 150	240 – 1030	Hg, Xe *sh.*	5	Hg, *sd.*	405, 436, 546	1	24 b
14 – 96	1300	Ar, shock layer	2	Explod. wire	450 – 589	<1	20 e
220 – 280	460 – 1180	H_2, *sh.*	5	Xe + H_2, *sd.*	533, 454	8	17 d
1500 – 20 000	$6 \cdot 10^4 - 1.5 \cdot 10^5$	Laser focus	≤ 1	Ar^+-Laser	488, 514	<1	9 g

[1]) The radial profile of N_e and N_0 has been determined.
[2]) In addition to N_e, the values for T_e (9000 – 13000° K) are presented.

Table 8.3. Application of laser interferometry for plasma diagnostics
When not otherwise stated, the light source is a He – Ne-Laser. The abbreviations used are the same as in table 8.2

N_e-range [cm^{-3}]	Plasma	L [cm]	λ [nm]	Δt [μs]	References
$(5-400) \cdot 10^{10}$	Ar sd.	24	0.63 1.15	10	24e
$(3-1000) \cdot 10^{12}$	He, sd.	50	0.63	<10	9k
$(8-200) \cdot 10^{12}$	Ar, PIG d.	15	0.63 1.15 3.39	Or5	14l, 14m
$(1-24) \cdot 10^{13}$ 1)	Ar d., afterglow	47.5	0.63 1.15	5–10	14b
$(1-30) \cdot 10^{14}$	H_2, Z-Pinch, afterglow	148	3.39	Or 1	20j
$(1.2\text{-}2.8) \cdot 10^{14}$	He sd. 2)	22.5	0.63	0.01	15p
$8.5-10^{14}$	air, sd.	30	10.6 (CO_2)	Or 5	20c
$(1-10) \cdot 10^{15}$	Ar, sd. in magn. field	100	0.63		26g
$(1.6-32) \cdot 10^{15}$	Ar, sd., afterglow	56	0.63	2	23a
$(2-10) \cdot 10^{15}$	H_2, Θ	20	0.63	0.02	14a
$(2.9-3.9) \cdot 10^{15}$	Ar arc, modulated	19.3	3.39	Or 1	12c
$(1.7-7) \cdot 10^{16}$	Ar arc, 1 atm		0.63 + 3.39	1	14h
$(4-70) \cdot 10^{16}$	Xe sd.	Or 0.4	0.63 + 3.39	Or 10	20p
$(3-5) \cdot 10^{17}$	Kr. sd. afterglow	86	0.63	2	23a
$\lesssim 6 \cdot 10^{17}$	Al, laser spark	≈ 0.2	0.63, 1.15	Or 1	9c

1) In addition, the quantity N_0 has been given to $(9-15) \cdot 10^{15}$ cm^{-3}.
2) The plasma is enclosed in a F. P. Interferometer.

Mainly, holographic interferometry will be applied for the investigation of non cylindersymmetric short time plasmas.

Two-colour-interference holograms are obtainable when as light source a ruby laser in combination with a KH_2PO_4(KDP)-crystal is used. The latter transforms part of the 6943 Å radiation into the second harmonic radiation at $\lambda = 3472$ Å [8.15o].

While with the usual interference holographic systems only high density plasma can be investigated which produce more than one interference fringe shift, the investigation of plasmas which produce only fractional fringe shifts is possible by superposing the small shifts on an arbitrarily shaped background frings pattern, as has been shown by Jahoda et al. [8.15b].

With the single exposure technique, it is possible to obtain time sequences of holographic pictures in the ns range by using a multiple reflected laser test beam with crosses the plasma volume in different directions at different moments [8.26f].

The holographic interferometry has been applied to determine the electron density in laser sparks in air, using as light source a ruby laser [8.8f], [8.15n], [8.15o]. A time resolution of 10–15 ns was obtained. The electron density was in the range from 10^{18} to 7.10^{19} cm^{-3}, the plasma diameter was of the order 1–3 mm.

8.2.9. Plasmadiagnostic. Application of Interferometry

To illustrate the possibilities of interferometry for the determination of electron density and heavy particle density in plasmas, some of the relevant experimental data are presented in the following tables.

It must be noted that in turbulent plasmas like high-speed free streaming plasma jets, the interference fringes are smeared out when the time of exposure is longer than the characteristic time for the turbulent motion [8.26a].

8.3. The Schlieren and Shadowgraph Technique for Measuring the Refractive Index and the Particle Density in a Plasma

These methods are based on the light-ray deviations in regions where the refractive index changes.

To obtain Schlieren photographs for the quantitative measurements of axially symmetric plasmas, Sperling [8.23] employed the Schlieren system shown in Figure 8.15. A carbon arc illuminates a horizontal slit L. This is focused approximately in the ratio 1:1 by a Schlieren lens with a long focal distance on a second horizontal slit, which can be slid vertically up and down.

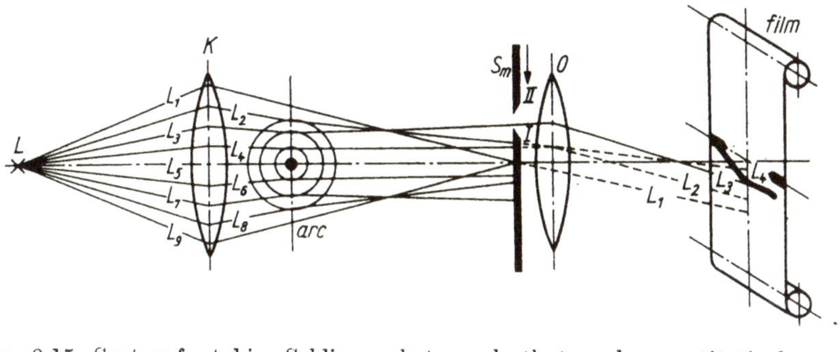

Figure 8.15. System for taking Schlieren photographs that can be quantitatively analyzed (after Sperling [8.23])

L – illuminated slit, K – Schlieren lens, $f = 2$ m, S_m – movable slit, O – camera objective, $f = 59$ cm

If the second slit is moved in such a way that the image of L is obscured in the plane of the second slit, no radiation will reach the film. When an optically inhomogeneous object (electric arc) is situated behind the Schlieren lens, the radiation from the former will be deviated. If the arc, which is located a short distance behind the Schlieren lens, is focused via a second objective lens O on the film, a sharp luminous line will appear on the latter. This line connects the points of constant light-ray deviation in the vertically burning arc, which is observed at right angles to the axis of the arc. We shall call this bright line an isocampt (curve connecting points of equal light-ray deviation).

Since several isocampts with different angle deviations are required for a quantitative analysis, the image of the arc is photographed with a framing camera. At the same time the second slit S_f is moved by means of an attached flexible shaft. In this manner within 1 sec all the required isocampts are photographed on 25 picture frames. For the geometrical orientation of the pictures a coordinate network is photographed at the same time.

Figure 8.16 illustrates film sections, which were taken for a narrow slit width (left) and a wide slit (right) and from which the isocampts can be obtained (in the case of the photographs taken with the wide slit each edge of the bright areas on the photograph represents an isocampt).

If we draw a straight line through each photograph normal to the axis of the arc (the line, for example, may pass through the center of the arc), then this line will intersect the different isocampts at a certain point x (transverse coordinate of the arc). Since each isocampt corresponds to a certain light-ray deviation of the slit image L at the point where the Schlieren lens is located (slit deviation), we can thus obtain points, correlating the slit deviation and the transverse coordinate x of the arc, which we can plot on a graph. Figure 8.17

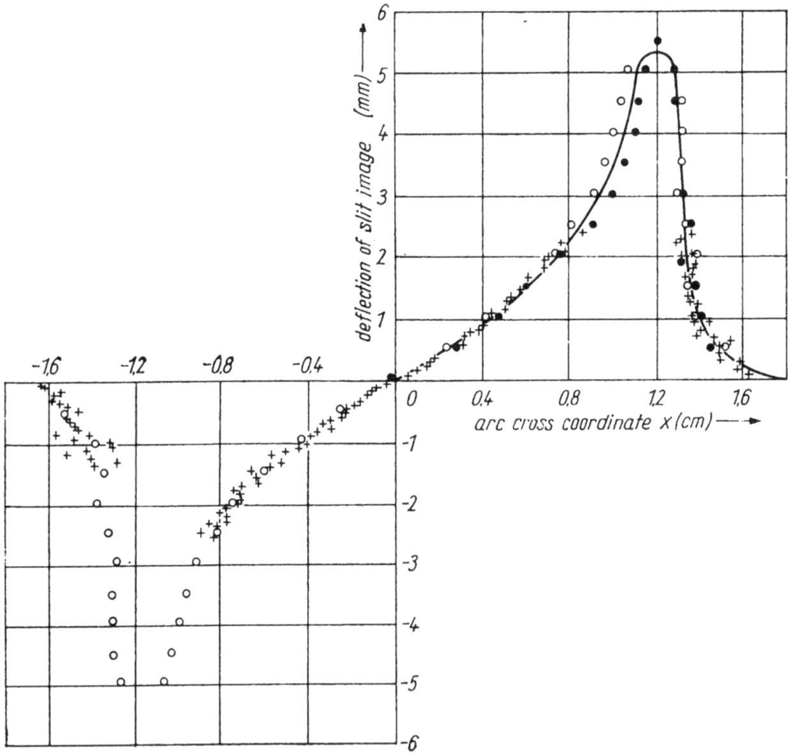

Figure 8.17. Light-ray deviation curve. Slit deviation as a function of the transverse coordinate x of the arc for the plane through the center of a 10 A carbon arc (after Sperling [8.23])

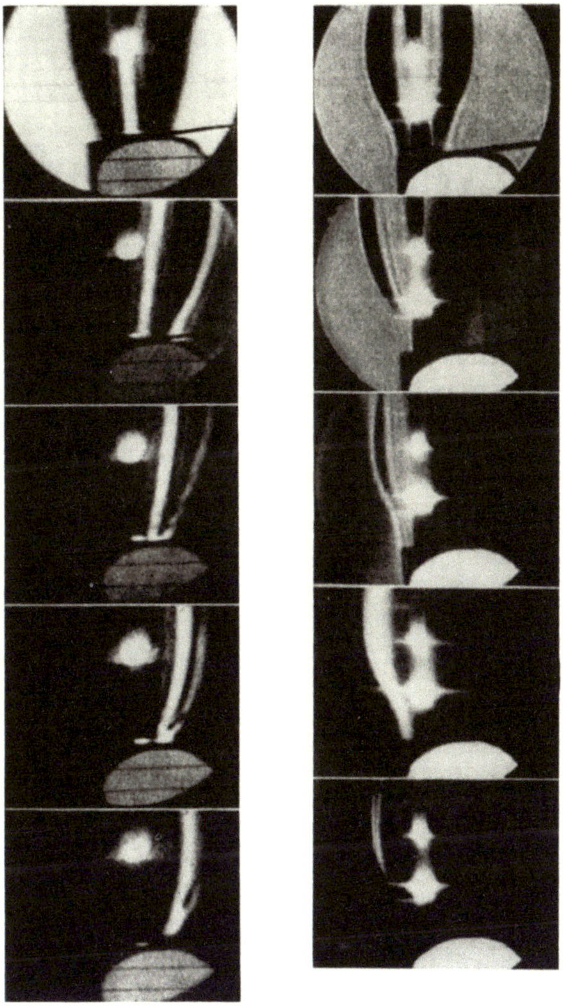

Figure 8.16. Film section of Schlieren photographs for different angular deviations. left: narrow slit, right: wide slit (after Sperling [8.23])

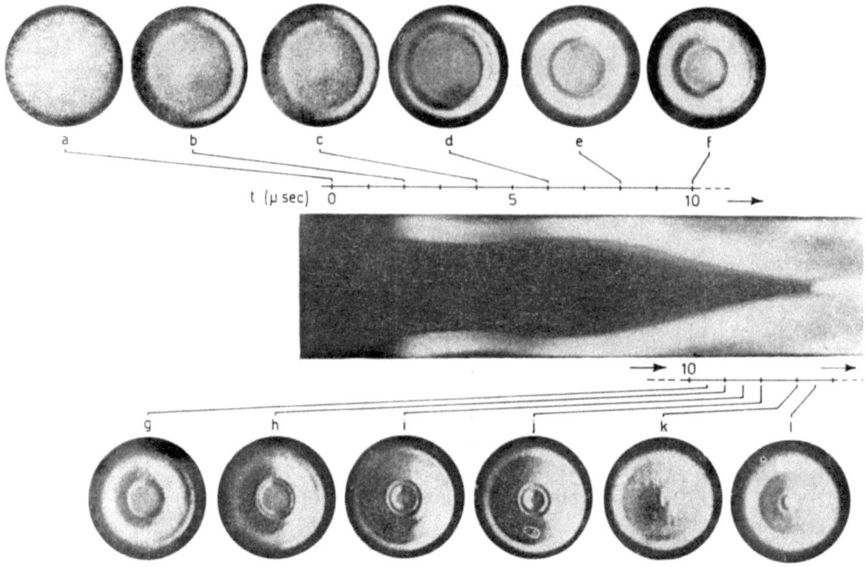

Figure 8.20. Schlieren photographs of an orthogonal pinch arrangement, whereby the plasma is produced by a condenser discharge via an inductive winding. Argon, initial pressure 10 Torr (after Ascoli-Bartoli and Martellucci [8.5], Laboratorio Gas Ionizzati, EURATOM-CNEN, Frascati, Rome)

top and bottom: individual photographs, middle: continuous, time-resolved streak photograph

gives an example of such a curve, which illustrates the relevant relationships in the plane through the center of the arc.

From the slit deviation, after multiplication with a constant factor, we finally obtain the angular deviation ε of the light ray, which is now given as a function of the transverse coordinate. Since the plasma, through which the radiation passes, can be regarded as axially symmetric, we get the following integral expression (see [8.23]) for $\varepsilon(x)$

$$\varepsilon(x) = \frac{2}{n_0} \int_x^{r_0} \frac{dn}{dr} \cdot \frac{x\, dr}{\sqrt{r^2 - x^2}}, \tag{8.21}$$

where the integration is taken along the line of sight transversing the radial range from x to r_0. The solution to the equation is

$$\frac{x}{r} \frac{1}{n} \frac{dn}{dr} = -\frac{1}{\pi} \int_r^{r_0} \frac{\varepsilon'(x)\, dx}{\sqrt{x^2 - r^2}}, \tag{8.22}$$

with the boundary condition $n(r_0) = n_0$. With regard to the numerical integration of the integral on the right-hand side, see section 2.10.1. Taking into account that the factor $1/n$ on the left-hand side of (8.22) can be assumed constant and equal to $1/n_0$, equation (8.22) can be readily integrated, giving n as a function of r. The temperature is now calculated from n and $(n - 1)$, respectively. To do this the plasma composition must be found as a function of temperature. From the plasma composition follows in accordance with the considerations in section 8.1 the refractive index, which is also given as a function of temperature and enables a correlation between the measured n-values and the temperature.

Figure 8.18 gives the thus obtained temperature field of a 10 A carbon arc. The isotherms are represented only up to a maximum temperature of 4400 °K. The value for the center of the arc were not included due to lack of accuracy.

Ascoli-Bartoli and Martellucci [8.4], [8.5] employed the Schlieren technique to determine the electron density in induced linear pinch discharges, whereby the plasma is irradiated along the axial direction. The layout for the photography is similar in principle to that shown in Figure 8.15, except for the fact that the light does not originate from a narrow illuminated slit L but from a tubular pulse lamp of finite width (S in Figure 8.19), and that instead of the movable slit S_f a knife edge K is used, which attenuates the light in accordance with the position of the radiation.

The deviation of the light ray upon transversing the plasma in the axial direction is dependent only on the distance of the light ray from the symmetry axis and is equal to

$$\varepsilon = \frac{L}{n} \frac{dn}{dr} \tag{8.23}$$

where the deviation is in the radial direction.

In the formula for the brightness of an image point in the plane F, where the image of the test object is formed, only the deviation normal to the side K occurs:

$$\varepsilon_{\perp} = \frac{L}{n} \frac{\mathrm{d}n}{\mathrm{d}r} \cos \varphi \,. \tag{8.24}$$

Here φ is the angle between the normal to the knife edge K and the direction of the gradient of the refractive index n, and L is the length of the plasma

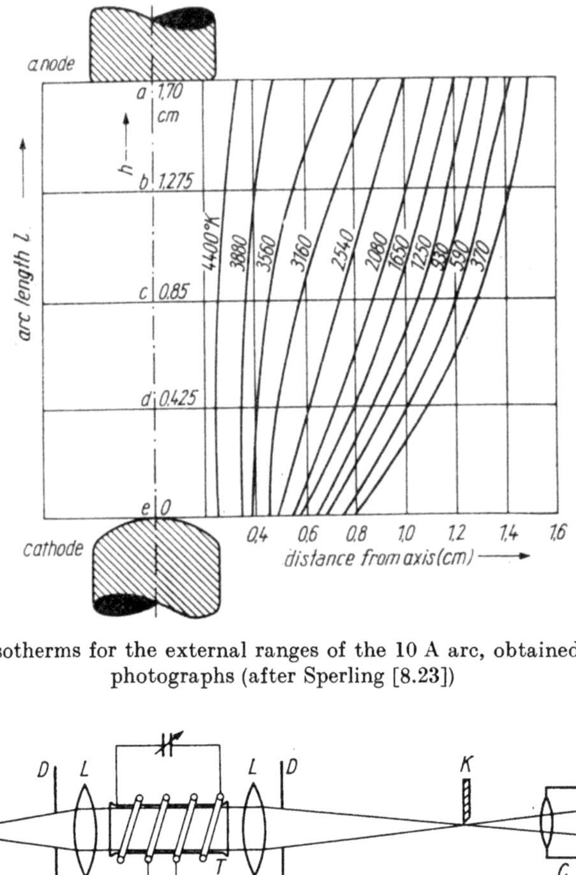

Figure 8.18. Isotherms for the external ranges of the 10 A arc, obtained from Schlieren photographs (after Sperling [8.23])

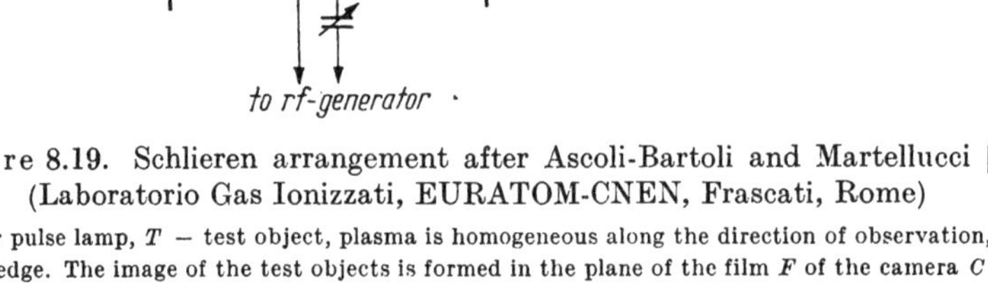

Figure 8.19. Schlieren arrangement after Ascoli-Bartoli and Martellucci [8.4] (Laboratorio Gas Ionizzati, EURATOM-CNEN, Frascati, Rome)

S – tubular pulse lamp, T – test object, plasma is homogeneous along the direction of observation, K – knife edge. The image of the test objects is formed in the plane of the film F of the camera C

traversed by the light. The brightness of the image point is a linear function of expression (8.24).

Single photographs can be taken in short time intervals one after the other (e.g., by means of image converter photography) as well as continuous time-resolved photographs [8.5]. On the basis of all these photographs we can determine the refractive index n for each radial coordinate and, consequently, the electron density N_e, which in a strongly ionized plasma contributes the major portion to the refractive index.

Figure 8.20 illustrates corresponding photographs of an orthogonally induced pinch discharge. Photometric recording of the photographs gives the electron density distribution.

If both the neutral gas component as well as the electron component of the refractive index are to be allowed for, one half of the image can be covered by a red filter, the other half by a blue filter, thus making it possible to simultaneously take Schlieren photographs for 2 different wavelengths [8.4].

Temple [8.24] developed a Schlieren technique, whereby additional diffracting elements are introduced to produce interference. These measures render the method more effective than is the case for the usual Schlieren technique.

A doubling in sensitivity is obtained with an arrangement of Lovberg [8.17c] including mirror optics, in which the test beam crosses the plasma two times.

In [8.8a], [8.15g] a Schlieren system is described which includes instead of one slit and one knife edge a series of illuminated slits and knife edges. On this way the plasma is illuminated by convergent-divergent light bundles, and thus local density gradients corresponding to the focus point can be determined.

In [8.2c] a direct reading Schlieren system is described in which density gradients are photographically recorded as function of a space coordinate. This system can be applied for the investigation of cylindrical symmetric plasma, the axis of which is parallel to the test beam.

For Schlieren systems with photoelectric detection see [8.15j], [8.24c], [8.24d].

For special Schlieren systems useful for the determination of radial density profiles in the case of plasmas with high density gradients (for example, laser produced plasmas), see [8.20u], [8.26b].

In certain aspects the shadowgraph technique is similar to the Schlieren technique. In one particular arrangement (see Weinberg [8.25], p. 150) a point-like diaphragm C, illuminated by the light source A, is placed before a collimator lens and the radiation transmitted via C is rendered parallel and reaches the screen H after transversing the test zone F (see Figure 8.21).

For an optically homogeneous test object the screen is uniformly illuminated. However, if in the test object there are components of the gradient of the refractive index normal to the line of sight, the previously parallel rays are deviated in the test object, whereby the deviation depends on the location of the line of sight which transverses the test zone. For the angular deviation ε

we have

$$\varepsilon = \int_{y_1}^{y_2} \frac{1}{n} \operatorname{grad}_\perp n \, dy \,, \tag{8.25}$$

where y is the coordinate along the line of sight, which intersects the plasma at y_1 and y_2. Due to this deviation certain patterns appear on the screen. For details on the computation of the intensity distribution of these shadowgraphs, see Weinberg [8.25], and Ascoli-Bartoli [8.2c].

Vasilev et al. [8.24f] have applied a photoelectric shadowgraph technique for the determination of density profiles in shock waves.

By placing a special diaphragm with parallel slit openings in the parallel light bundle in front of the test object (see Figure 8.21), bright fringes with kinks will appear on the screen when the test object is inhomogeneous (see Figure 8.22). For further details on the analytical procedure the reader is referred to Weinberg ([8.25], p. 173) as well as Fox and Weinberg [8.13]. Here other experimental methods of a similar nature are treated.

Some of the ways in which the Schlieren technique can be applied to the determination of plasma parameters are listed in Table 8.4, from which the corresponding experimental data can be taken.

For further references see [8.4], [8.5], [8.8].

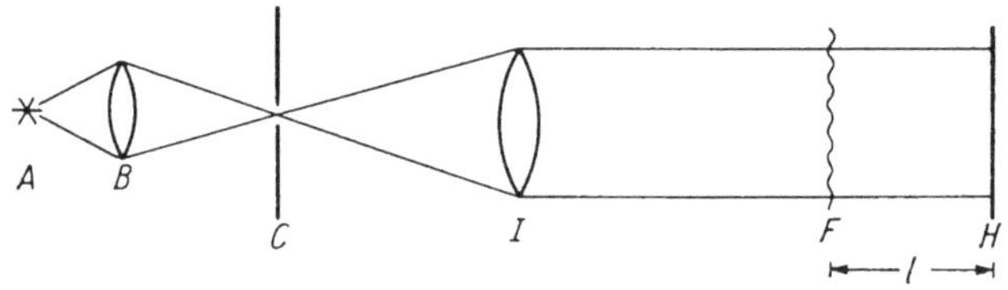

Figure 8.21. Experimental layout in the shadowgraph technique (after Weinberg [8.25], p. 150)

A – source, B – condenser, C – diaphragm, I – collimator lens, F – test zone, H – screen or film

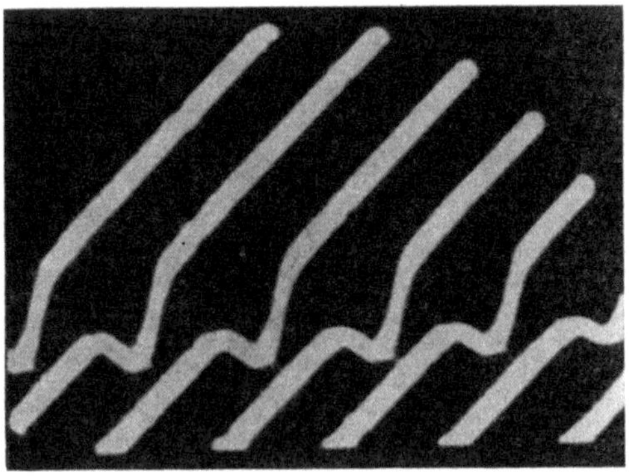

Figure 8.22. Investigation of flame structure by the fringe deviation method. Photograph of a fringe pattern of a laminar flame (after Fox and Weinberg [8.13])

Table 8.4. Application of the Schlieren technique for plasma investigations. The meanings of the symbols used are the same as in table 8.2

Q	V	Plasma	L [cm]	Light source	λ [nm]	Δt [ns]	Reference
$T(r)$ [°K]	370–4400	C-arc in air	≈ 3	arc lamp	red		23
	300–4000	C-arc in air	2	Hg, hp. d.			24d
	300–4300	N_2 plasma jet	1	Hg, hp. d.			24c
N_e [10^{16} cm^{-3}]	≤ 2	H_2, Θ-pinch	25	Ruby Laser	694	20	20s
	≤ 3	H_2, plasma accelerator	10	N_2, capill. discharge	500	50	17c

8.4. Laser Diagnostics

The scope of this section is restricted to the application of laser as light source in Faraday rotation and light scattering experiments. For the full scale of laser application in plasma diagnostics see the references [8.10b], [8.15r].

8.4.1. Determination of Electron Density from Faraday Rotation of Polarized Laser Beams in a Magnetized Plasma

When a polarized light ray passes through a plasma of length L in the direction of the magnetic lines of force, the plane of polarization is rotated by an angle [8.10]

$$\Theta_R \approx \frac{L}{2} \frac{\omega_B}{c} \frac{\omega_p^2}{\omega^2} \tag{8.26}$$

where

$$\omega_p^2 = \frac{4 \pi N_e e^2}{m} \tag{8.27}$$

denotes the square of the plasma frequency and

$$\omega_B = \frac{e B}{m c} \tag{8.28}$$

denotes the cyclotron frequency of the electrons. Combining equations (8.26) to (8.28) gives

$$\Theta_R \approx \frac{2 \pi e^3 L N_e B}{m^2 c^2 \omega^2}. \tag{8.29}$$

From this equation it follows that the angle of rotation for a given frequency ω is proportional to the electron density N_e, the plasma length L, and the magnetic induction B.

According to Graßmann and Wulff [8.14e], the minimal rotation angle which can be detected is 0.1°. With a ruby laser test beam of wavelength $\lambda = 6943$ Å, this sensitivity is sufficient for example to measure an electron density of $N_e = 5 \cdot 10^{16}$ cm^{-3} of a 20 cm long plasma placed in a magnetic field of 40 kG field strength ($\Theta_R = 0.3°$).

This estimate presupposes that the technique possesses a high sensitivity in electron density measurements for a test beam in the visible spectral region only in plasmas subjected to a high magnetic field, i.e., the method is primarily suited for N_e-determination in magnetically contained plasmas.

When the electron density N_e is known, then from a measurement of the rotation angle Θ_R, it is possible to determine the actual magnetic field strength B. This method is useful for the investigation of instationary magnetic confined plasmas where the magnetic field is influenced by plasma motion [8.12c], [8.14f].

The following tabulation contains some characteristic data about Faraday-rotation experiments, in which the quantity which had to be measured is heavy typed.

Plasma	Radiation source	wave-length	H [kG]	N_e [10^{15} cm^{-3}]	L [cm]	Δt	Reference
Θ-pinch, H_2	He-Ne-Laser	3.39 μm	50	**6–100**	20	1 μs	10a
pulse discharge, He	Ruby-Laser	6943 Å	40	**50**	20	<1 m	14e
Θ-pinch, H_2	He-Ne-Laser	6328 Å	**≤12**	60	24		12c
Θ-pinch, D_2	He-Ne-Laser	6328 Å	**20–80**	2–60	100	50 ns	14f

8.4.2. Non Resonant Scattering of Laser Radiation

From experiments on non resonant scattering of laser radiation, the electron density, electron temperature and, in principle, also the ion temperature can be determined.

Since the non resonant scattering of light radiation produces only a slight effect, for its detection a radiation source of high intensity, usually a Q-switched laser, is needed, in order to overcome the background radiation emitted by the plasma itself.

The laser radiation causes some additional heating of the plasma [8.15r], [8.26e] which leads to serious errors in temperature measurement if the irradiation intensity exceeds a value of 10^8 (T_e/°K) W/cm^2.

The scattering is effected by free and bound electrons.

In the region of single particle scattering (see section 8.4.2.1), the contribution of bound electrons on the scattered intensity can be eliminated when recording the spectral distribution of the scattered radiation, taking into account that free electrons cause a strong thermal Doppler broadening of the scattered radiation while the radiation scattered by bound electrons (Rayleigh-scattering, see [8.20r]) is restricted to a small spectral interval.

In the region of collective or coherent scattering (see section 8.4.2.2) the effect of bound electrons is negligible for full ionized high temperature plasmas composed from light elements, in which the number of bound electrons is only a small fraction of the total number of electrons.

8.4.2.1. Single particle scattering of laser radiation by free electrons (Thomson-scattering)

The case of single particle scattering of free electrons is realized when the characteristic parameter α defined by

$$\alpha = \frac{\lambda_0}{4\,\pi\,\lambda_D \sin\left(\frac{1}{2}\Theta\right)}, \tag{8.30}$$

is a small quantity against 1 [8.15r]. In this formula, λ_0 denotes the wavelength of incident radiation, Θ the angle between the direction of the incident radiation and the scattered radiation, λ_D the Debye-length defined by

$$\lambda_D = \frac{kT_e}{(4\,\pi\,N_e\,e^2)^{1/2}}\,.$$

α can be interpreted as the ratio between the length in the plasma over which the scattered radiation adds coherently, and the Debye length [8.14d].

For $\Theta = 90°$ and $\lambda_0 = 6943$ Å, the condition of Thomson scattering is $1.13 \cdot 10^{-6}\,(Ne/T)^{1/2} \ll 1$.

The total Thomson scattering cross section of a single electron is given by

$$\sigma_{Th} = \frac{8\,\pi}{3}\,r_0^2 = 0.665 \cdot 10^{-24}\ \mathrm{cm}^2\,, \tag{8.31}$$

in which r_0 denotes the classical electron radius (For comparison, the non resonant Rayleigh scattering cross section of unexcited atoms is lower by a factor $(\lambda_R/\lambda_0)^4$, where λ_R denotes the wavelength of the resonance line and $\lambda_0 > \lambda_R$ the wavelength of the incident radiation).

For the differential scattering cross section (scattering cross section per unit solid angle) one obtains

$$\frac{\mathrm{d}\sigma}{\mathrm{d}\Omega} = \frac{1}{4\pi}\,\sigma_{Th} \cdot \frac{3}{4}\,(1 + \cos^2\Theta)\,. \tag{8.32}$$

In practice, the fraction of the scattered radiation arriving at the radiation detector relative to the incident radiation is of the order 10^{-12} [8.20k]. The scattered radiation is polarized and has a profile due to the Doppler effect which depends on the electron velocity distribution. In the case of a Maxwellian velocity distribution, the scattered radiation has a Gaussian profile with total half width

$$\Delta\lambda_{1/2} = 4\,\lambda_0 \sin\left(\frac{1}{2}\,\Theta\right)\sqrt{\frac{2\,kT_e}{m\,c^2}\ln 2}\,. \tag{8.33}$$

From the measured half width and from the absolute intensity of scattered radiation, the electron temperature and the electron density can be evaluated.

Figure 8.23 shows an experimental arrangement used by Fünfer [8.14], in which the polarized radiation of a high-power ruby laser (20 Joule light energy for maximum light power of 100 kW, flash time 1 ms, wavelength 6936 Å) is directed along the axis of a theta-pinch. The scattered radiation emitted from the center of the tube is measured side-on. The radiation trans-

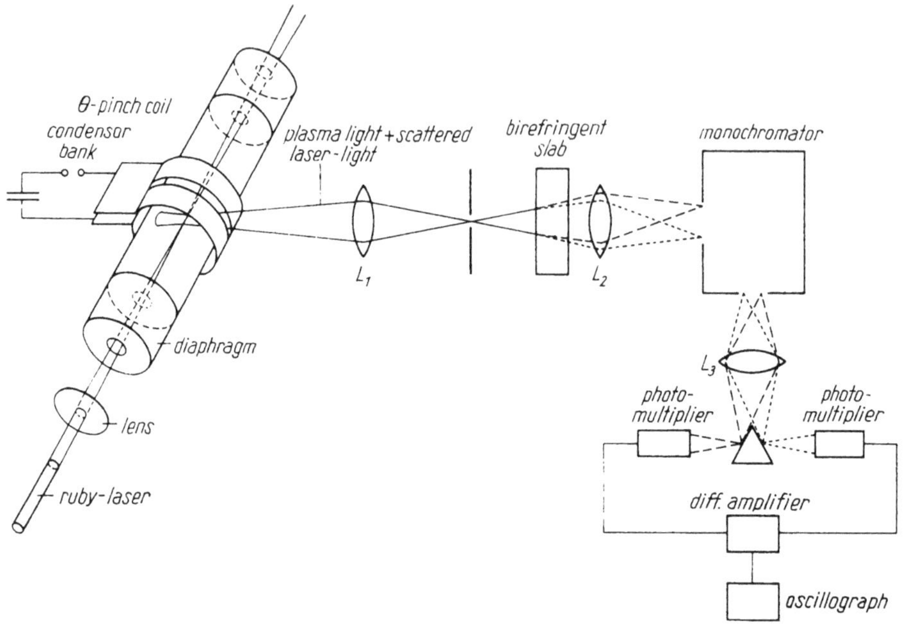

Figure 8.23. Experimental layout for the measurement of the scattering of a laser ray on electrons on a theta-pinch plasma. Explanation in text (after Fünfer et al. [8.14])

mitted through the plasma is also measured. In this way the ratio of the scattered radiative power and the total radiative power can be determined, thus providing a measure for the electron density in the plasma.

In order to ensure that the laser radiation scattered by the plasma will not be influenced by 1. the background radiation of the plasma itself, and 2. the laser radiation scattered by the container walls, etc., which is very intensive, the following preventive measures must be taken.

1. The first measure is based on the fact that the scattered laser radiation is polarized, while the radiation of the plasma itself possesses a vanishingly small polarization. The two radiation components can be separated as follows.

The scattered radiation with the plasma radiation is first separated after passage through a doubly refracting plate in accordance with the two polarization directions (one is parallel to the direction of polarization of the incident laser radiation) and is then led to two different points of the entrance slit of a monochromator. Behind the exit slit of the monochromator are located two photomultipliers, which receive both signals. By means of a difference amplifier the difference between both signals is recorded. This difference is a measure of the scattered radiation only.

2. To eliminate the laser radiation scattered at the container walls, which is about 100 times as strong as in the case of the background radiation of the plasma itself, the following fact is considered. While the radiation scattered at the solid surfaces is strictly monochromatic, the radiation scattered on electrons, whose mean thermal energy can be taken to be 100 eV, displays a strong Doppler broadening. Now we need only adjust the exit slit of the monochromator, whose transmission range is 165 Å, so far from the middle of the line profile, that all parts of the slit are separated by at least 5—15 Å from the center of the line. Then the radiation scattered by the solid surfaces will no longer be observed.

Absolute values for the electron density were not given in the above mentioned work.

More details about Thomson scattering experiments can be found in [8.15r]. For diagnostical applications see table 8.5 at the end of the next section.

8.4.2.2. Collective scattering of laser radiation ($\alpha \gg 1$)

For small enough scattering angles Θ (forward scattering), the characteristic parameter α becomes large against 1. Then the spectral distribution of the scattered radiation is determined by collective effects, and consists of a center line ("ionic line") with half width corresponding to the ionic Doppler width and two satellite lines ("plasma lines") deplaced from the wavelength of the incident beam by $\pm \Delta\lambda$, where $\Delta\lambda$ is given by

$$\Delta\lambda = \frac{\omega_p}{2\pi c} \lambda_0^2 \left(1 + \frac{3}{\alpha^2}\right)^{1/2}, \tag{8.34}$$

$$\omega_p = \left(\frac{4\pi N_e e^2}{m_e}\right)^{1/2} = \text{plasma frequency} \tag{8.35}$$

[8.20k]. From $\Delta\lambda$, the electron density is found.

For more details concerning the absolute intensity distribution which may be affected by strong magnetic fields or by collision effects, see, for example, [8.15r]. For disturbing effects due to scattering of laser radiation at strong density gradients see [8.1].

8.4.2.3. Intermediate region of non resonant scattering of laser radiation

In the case $\alpha \approx 1$, the spectral energy distribution of scattered radiation is obtained by superposition of the spectral energy distribution due to single particle scattering and that due to collective scattering. The ratio of both distributions, integrated over the wavelength, depends from α. For details, see, for example, [8.14d], [8.15i], [8.15r], [8.20k], [8.20t].

Table 8.5. Application of laser scattering experiments to plasma diagnostics. Θ = angle of scattering, Δt = time resolution

Element	Θ [°]	Range of kT_e [eV]	Reference	Range of N_e [10^{15} cm^{-3}]	Reference	Range of kT_i [eV]	Reference	Δt [ns]	Reference
Case a) low temperature plasma									
Ar	90	1.0–1.4	15a, 20g, 24a.	0.01–103	9i, 15a, 20g, 24a.				
H	13.5; 90	1	20k	2.4	20k			50	15a
H	90	2.3–4.5	8e[1]), 9d, 9h.	5–15	9d, 9h.	3.2–4.5	20m	30–100	8e, 9h.
H	170	2.2	22a						
Case b) high temperature plasma									
H	90	4.6–260	8c, 15q, 15s, 15t, 15v, 15w, 20a, 20h, 20i, 22b.	0.8–160	15q, 15s, 15t, 15v, 15w, 20h, 22b.			10–30	8c, 15q, 15s. 20i.
D	7; 90					175–650	20n	*Or.* 100	20n
D[2])	90	40–120	7b	4000	7b	390–470[3])	7b		

[1]) In addition, an axial velocity of 10^8 cm/s has been detected.
[2]) Plasma-focus-experiment.
[3]) In addition, $v_e = 1\ 7 \cdot 10^7$ cm/s has been measured.

CHAPTER 9

Determination of Plasma Parameters by Means of Continuous Radiation

Within the frame of this chapter we do not consider continua of molecular origin which have been handled in the review article of Finkelnburg and Peters [9.21] and in the articles of Mnatsakanyan [9.50d] and Wilkins [9.75b]. Information about molecular photodissociation and photoionization cross sections useful for the determination of particle number densities in relative cold gases in connection with absorption measurements can be found in table 6.6 of chapter 6.

Further, we exclude from our considerations two-photon emission continua resulting from transitions between two discrete electronic energy levels following the schema $h\nu_1 + h\nu_2 = U_2 - U_1$ which plays a role only in high temperature plasmas containing ions with a high charge number Z (see [9.20e]).

Finally, we exclude fictive continua given by the far line wings [9.61j], [9.75a] which can give in some cases a non vanishing contribution to continuous background intensity (10% in the case of a hydrogen plasma for wavelengths $\lambda < 1500$ Å, see [9.61j]).

First of all we have to consider the origin of the most important types of continuous radiation from plasma and we have to find out the quantitative relations for the dependence of radiation intensity upon particle number densities, electron temperature and wavelength or frequency.

If this problem is solved, then, vice-versa, starting from experimental values for the continuous radiation intensity it is possible in a great number of cases to determine particle number densities and the electron temperature. If not otherwise stated, the radiating system is considered to be a plasma in thermal equilibrium. But most of the formulas presented in this chapter are invariable against deviations from thermal equilibrium so long as there is a Maxwellian distribution of electron velocity and a Boltzmann distribution of electronic energy states while the gas- and ion temperature may deviate from the electron temperature. The reason for this invariability is that continuous emission is only little influenced by the motion of the heavy particles.

9.1. The Various Kinds of Continuous Radiation

We consider only continuous radiation originating from the motion of electrons because the intensity of radiation stemming from the motion of heavy particles is significant only in plasmas in which the heavy particle temperature is in excess of some million °K.

Radiation is generated when free electrons are accelerated or decelerated by the long or short ranging potential fields of other particles (ions, atoms and electrons). The generation of radiation of this kind can be interpreted classically and also by means of quantum mechanics. Following classical electrodynamics, the power radiated by an electron with acceleration $\mathrm{d}v/\mathrm{d}t$ is given by

$$R = \frac{2\,e^2}{3\,c^2}\,(\mathrm{d}v/\mathrm{d}t)^2\,. \tag{9.1}$$

Starting from this formula, the continuous emission coefficient of a plasma can be calculated leading to relations which are not much different from those derived quantummechanically so long as pure Coulomb fields are considered.

In quantum mechanics, the generation of continuous radiation is interpreted as consequence of transitions from upper continuous electronic energy states (corresponding to free electrons) to lower continuous (free) or discrete (bound) energy states. In the first case, the radiation generated is named free-free radiation, in the second case it is named free-bound radiation, inasmuch as distinction between these two mechanisms is wanted. In the case of free-bound radiation resulting from the interaction between free electrons and neutral atoms, the lower state of the corresponding transitions is that of a negative ion.

In quantum mechanics, the intensity radiated by an electron when making a transition from an upper level m to a lower level n is given by the formulas

$$I = \frac{64\,\pi^4\,e^2}{3\,c^3}\,\nu_{m\,n}^4\,|M|^2_{m\,n}\,, \tag{9.2}$$

$$M_{m\,n} = \int \Psi_m\,\boldsymbol{r}\,\Psi_n^*\,\mathrm{d}V\,, \qquad h\,\nu = U_m - U_n$$

in which the upper level (which is in any case continuous) is characterized by the initial velocity v_1 of the radiating electron respectively by the corresponding wave number $k_1 = m\,v_1/\hbar$. The total power of continuous radiation per unit volume and unit range of frequency is obtained by summing up respectively integrating the expression (9.2) over all possible lower energy states with frequency ν held constant, folding with the electron velocity distribution and multiplying with the product $N_e\,N_1$ (N_1 = number density of collision partners). After division through the total space angle $4\,\pi$, the spectral emission coefficient is obtained.

For the diverse kinds of continua originating from 1) acceleration of electrons in the field of ions, 2) acceleration of electrons in the field of neutral atoms and 3) acceleration of electrons in the field of electrons, we introduce the denotations electron-ion continuum, electron-atom continuum and electron-electron continuum. The intensities of these kinds of continua show differences with respect to their dependence from the number densities of the different kind of plasma particles which can be expressed by the formulas

$$\varepsilon_{\nu_{ei}} \sim N_e\,N_i \tag{9.3}$$

($\varepsilon_{\nu_{ei}}$ = emission coefficient of the electron-ion continuum = radiated power of $e - i$ continuum per unit volume, unit frequency and steradian),

$$\varepsilon_{\nu_{e0}} \sim N_e N_0 \tag{9.4}$$

and

$$\varepsilon_{\nu_{ee}} \sim N_e^2 . \tag{9.5}$$

Assuming that continuous radiation is generated only as result of two-particle collision processes, then it follows that the proportionality factors in relations (9.3)—(9.5) can depend only 1) from the electron temperature, 2) from the kind of the target ions (via the effective charge number Z and 3) from the kind of the target atom, and there can be no other dependences.

When applying the relations (9.3)—(9.5) to a given plasma showing continuous emission, it is possible to analyze the nature of the continuum if the particle number densities in this plasma can be changed independently by means of pressure variation while the temperature is held at a constant level. While it is possible by this means to distinguish between electron-atom and electron-ion continuum, it is in general not possible to differentiate between electron-ion and electron-electron continuum. This differentiation is meaningless for plasmas which contain only nonrelativistic electrons ($T_e < 10^8$ °K) because as result of electron-electron collisions only quadrupole radiation is emitted [9.64g], [9.76] which is of no importance relative to the dipole radiation originating from electron-ion collisions under the condition of non relativistic electron velocities.

Subsequent to these general statements, in the following sections we shall treat in detail the relations govering the emission of the diverse kinds of continua mentioned above, while our final statements are concerned with the plasmadiagnostical applications of these relations.

9.2. Electron-ion Continuum

We differentiate between the free-free continuum originating from electronic transitions between free continuous energy states to other free states (see the level diagram after Maecker and Peters [9.45] reproduced in figure 9.1) and the free-bound continuum originating from transitions between continuous and discrete energy states.

9.2.1. Free-free Radiation

We consider collision processes in which electrons are accelerated in the field of Z-fold charged ions. The calculation of emission intensity is performed taking into account the following 4 conditions:

1. The non relativistic approximation is used. This is correct for plasma temperatures below 10^8 °K.

2. Only collision processes involving two collision partners are considered. This is correct so long as the shortest wavelength of emitted bremsradiation is markedly greater than those distances between electron-ion pairs for which most of the radiation is generated.

3. The motion of ions is neglected.

4. As interaction potential the Coulomb potential is used. That means, the ions are considered as point charges. Screening effects are neglected.

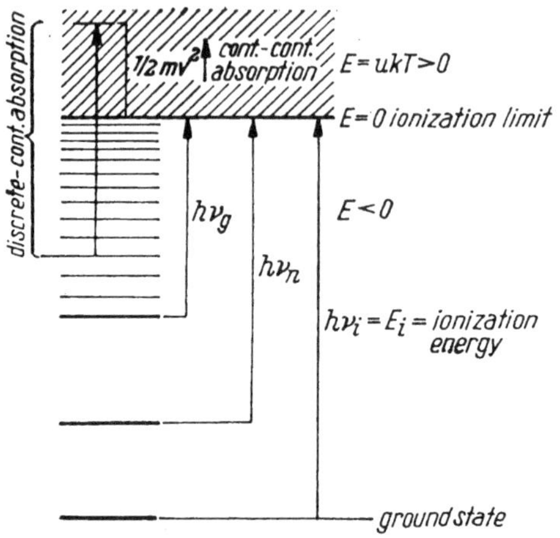

Figure 9.1. Level diagram, schematic. (After Maecker and Peters [9.45])

Corresponding calculations have been performed by Sommerfeld (see Hettner [9.30]) leading for the radiated power per unit frequency for a homogeneous electron stream of unit intensity (1 electron per second and cm^2) after integration over all angles of electron deflection and all directions of emission to the system of equations

$$W_\nu = C\,\eta^2\,L(\eta,\eta') \tag{9.6}$$

with

$$\left.\begin{aligned} C &= \frac{32\,\pi}{3}\,\frac{\hbar^2 e^2}{c^3 m^2} = 3.84\cdot 10^{-49}\ \text{erg cm}^2\ \text{s}\,, \\ \eta &= \frac{Z e^2}{\hbar v}\,, \qquad \eta' = \frac{Z e^2}{\hbar v'}\,, \qquad \frac{m}{2}\,(v^2 - v'^2) = h\,\nu\,, \end{aligned}\right\} \tag{9.7}$$

$$L(\eta,\eta') = \frac{\pi^2 x}{(e^{2\pi\eta}-1)(1-e^{-2\pi\eta'})}\,\frac{d}{dx}\,|F(i\,\eta, i\,\eta', 1, x)|^2\,, \tag{9.8}$$

$$x = -\frac{4\,\eta\,\eta'}{(\eta'-\eta)^2} = -\frac{4\,v\,v'}{(v-v')^2}\,. \tag{9.9}$$

The meanings of the symbols used are: v and v' — initial and final velocity of the electron under consideration, ν — frequency of emitted radiation. Instead of v the quantity $E = m\,v^2/2$ describing the initial kinetic energy of the electron with velocity v can be introduced. F is the hypergeometric function of 4 arguments as described for example by Magnus-Oberhettinger [9.46]. Then L gets a function of E, ν and Z.

The dependence of the quantity L from electron energy E and from frequency ν is shown graphically in figure 9.2. For simplicity, $Z = 1$ is assumed. As ordinate instead of L the quantity $W_\nu E$ is used which differs from L only by a constant factor ($W_\nu E = 8.40 \cdot 10^{-60} L$), while the frequency is given in

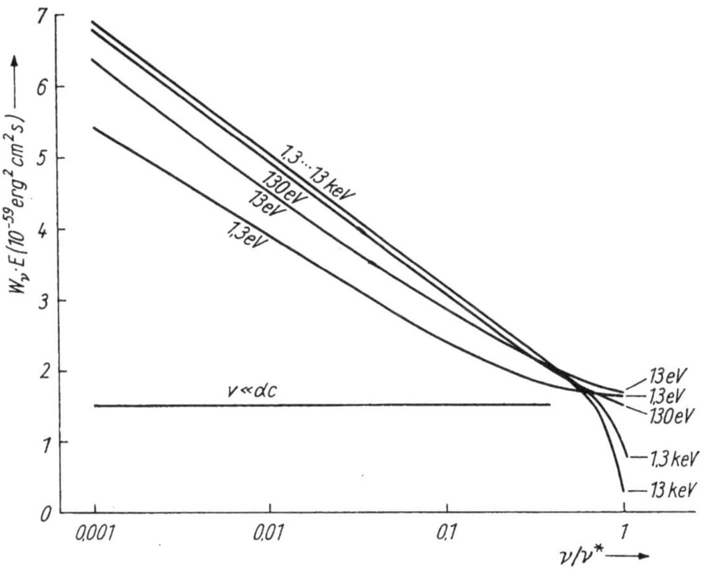

Figure 9.2. The quantity $W_\nu E = 8.40 \cdot 10^{-60} L$ as function of frequency and energy. (After Hettner [9.30])

dimensionless form ν/ν^* with $\nu^* = E/h$. As can be seen from figure 9.2., for $E \to 0$ the quantity $W_\nu E$ (respectively L) approaches a constant value corresponding to $L = \pi/\sqrt{3}$.

Figure 9.2 shows also that for low values of E the difference between the true quantity L and its limiting value rises with decreasing frequency ν. Hence it follows that in the region of infrared radiation it is not allowed to neglect the variation of L with ν and E.

A precise graphical representation for the quantity L is given by Schirmer [9.63].

For a Maxwellian distribution of electron velocity, the power emitted by free-free radiation per unit frequency, unit volume in solid angle 4π is given by

$$4\pi\,\varepsilon_\nu = \frac{32\sqrt{2\pi}}{3}\,\frac{Z^2 e^6 N_i N_e}{c^3 (m\,kT)^{3/2}} \int_{h\nu}^{\infty} L\, e^{-E/(kT)}\, dE\,. \tag{9.10}$$

In this formula, with N_e, N_i and Z are denoted the number densities of electrons and ions and the ionic charge number.

Due to the weak dependence of the quantity L from E, according to [9.30] a good approximation for (9.10) is obtained when L is taken before the integral while the quantity E entering into L is substituted by

$$E = 1/2\,(h\,\nu + kT) \quad \text{if} \quad h\,\nu < kT\,,$$

$$E = h\,\nu \quad \text{if} \quad h\,\nu \geq kT\,.$$

Then L converts into $\bar{L}(Z, \nu, T)$ leading to

$$4\,\pi\,\varepsilon_\nu = \frac{32\sqrt{2\,\pi}}{3}\;\frac{Z^2\,e^6\,N_i\,N_e}{c^3\,m^{3/2}\,(kT)^{1/2}}\;\bar{L}(Z, \nu, T)\,\mathrm{e}^{-h\nu/(kT)}\,. \tag{9.11}$$

If only low electron velocities are considered satisfying the relation $v \ll Z\,\alpha\,c$ in which α is the fine structure constant $= 1/137$, then in the frame of the "weak approximation" one obtains $L = \pi/\sqrt{3}$, and the continuous emission coefficient (radiated power per unit volume, unit frequency and unit solid angle) is described by the formula

$$\varepsilon_\nu \approx \frac{8(2\,\pi)^{1/2}}{3^{3/2}}\;\frac{Z^2\,e^6\,N_i\,N_e\,\mathrm{e}^{-h\nu/(kT)}}{c^3\,m^{3/2}\,(kT)^{1/2}} = 6.36\cdot 10^{-47}\,Z^2\,\frac{N_e\,N_i}{(kT)^{1/2}}\,\mathrm{e}^{-h\nu/(kT)} \tag{9.12}$$

$$[\mathrm{erg\;cm^{-3}\;sr^{-1}}]\,.$$

This formula has also been given by Maecker and Peters [9.45]. If higher precision is required, then instead of (9.12) the more precise formula (9.11) must be used.

For low frequencies, the formulation of Sommerfeld for the continuous emission coefficient leads to serious divergence problems which can be solved by replacing the pure Coulomb potential around the ions for example by a Debye-screened Coulomb potential [9.20f], [9.38b]. In classical theory, the divergence problem is solved by suitable termination of the collision parameter. Choosing as upper limit for it the mean distance between ions, Finkelnburg and Peters [9.21] obtained for the far infrared continuous emission coefficient the formula

$$h\,\nu \ll kT\,, \quad 2\,\pi\,\nu > \omega_p:$$

$$\varepsilon_\nu = 6.36\cdot 10^{-47}\,\frac{\sqrt{3}}{\pi}\,Z^2\,\frac{N_e\,N_i}{(kT)^{1/2}}\ln\left(\frac{kT}{1.44\cdot Z\,e^2\,N_i^{1/3}}\right),$$

in which Z denotes the charge number of the target ions, while $\omega_p = 5.64 \times 10^4\,(N_e/\mathrm{cm^{-3}})^{1/2}$ radian/s denotes the electron plasma frequency. For more literature in this field see [9.3a] p. 82, [9.8a], [9.18e], [9.18j], [9.26b] p. 122, [9.39], [9.60c]. The corresponding theories have been experimentally proved down to a circular frequency of twice the plasma frequency ω_p [9.8a]. (For a typical plasma with $N_e \approx 10^{16}\,\mathrm{cm^{-3}}$ we have $\omega_p \approx 10^{13}\,\mathrm{s^{-1}}$, the corresponding wave length is $\lambda_p \approx 0.3$ mm.)

For problems concerned with *IR* radiation in the neighbourhood of the plasma frequency see for example [9.18b], [9.18o] and [9.20b]. The special case of *IR* radiation in the region near the electron cyclotron frequency has been treated in [9.24d].

In general, the variableness of L can be taken into account by multiplying to the right hand side of equation (9.12) the corrective factor

$\bar{g} = \frac{\sqrt{3}}{\pi} L$ (Gaunt-factor):

$$\varepsilon_\nu = 6.36 \cdot 10^{-47} Z^2 \frac{N_e N_i}{(kT)} \mathrm{e}^{-h\nu/(kT)} \cdot \bar{g} . \tag{9.13}$$

The dependence of $\bar{g}$ upon temperature and frequency for $Z = 1$ is shown in figure 9.3. For the abscissa the dimensionless scale $u = h\nu/(kT)$ is used.

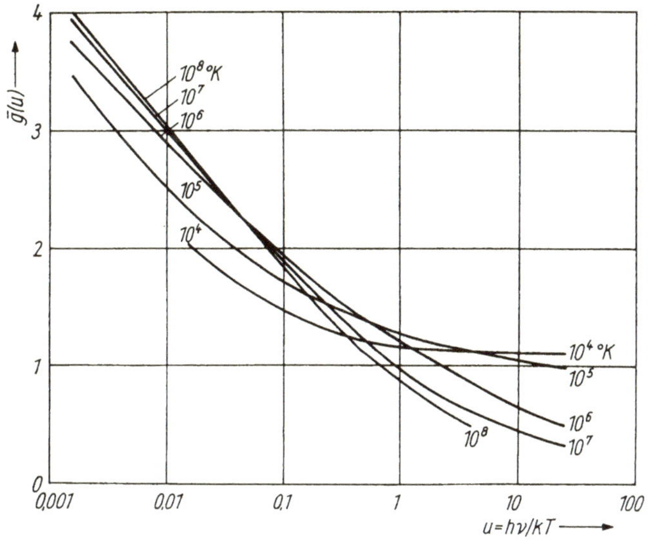

Figure 9.3. Mean value $\bar{g}$ as function of u ($u = h\nu/(kT)$) and T for $Z = 1$. (After Schirmer [9.63])

From the diagram in figure 9.3 it is also possible to take out $\bar{g}$ — values in the case $Z \neq 1$ if the parameter T written beside the curves is now interpreted as parameter T/Z^2.

Integrating (9.11) over the whole frequency range, we get the expression for the total free-free emission coefficient

$$S = \frac{32\pi\sqrt{2\pi}}{3\sqrt{3}} \frac{Z^2 e^6 N_i N_e (kT)^{1/2}}{c^3 m^{3/2} h} \bar{\bar{g}}$$

or

$$S = 1.41 \cdot 10^{-27} Z^2 N_i N_e T^{1/2} \bar{\bar{g}} \, [\mathrm{erg\ cm^3\ s^{-1}\ {}^\circ K^{-1/2}}] . \tag{9.14}$$

In the frame of the weak approximation $\bar{g}$ is equal to 1. In reality, $\bar{g}$ depends from T as shown in table 9.1. For laboratory plasmas we have $\bar{\bar{g}} \approx 1.4$.

Table 9.1. The factor $\bar{\bar{g}}$ as function of temperature (after Schirmer [9.63])

T [°K]	10^4	$4 \cdot 10^4$	10^5	$4 \cdot 10^5$	10^6	$4 \cdot 10^6$	10^7	$4 \cdot 10^7$	10^8
$\bar{\bar{g}}$	1.30	1.38	1.43	1.44	1.40	1.29	1.21	1.16	1.15

The expression for the total emission coefficient of bremsradiation S is of much interest for the estimation of radiative energy losses from high temperature plasmas used in thermonuclear fusion experiments. For illustration, S is plotted in figure 9.4 as function of T for the temperature range 10^4—10^8 °K assuming $N_e = N_i = 10^{15}$ cm^{-3} and $Z = 1$ (hydrogen plasma). The upper curve is the correct one while the lower curve is constructed under the assumption $\bar{\bar{g}} = 1$.

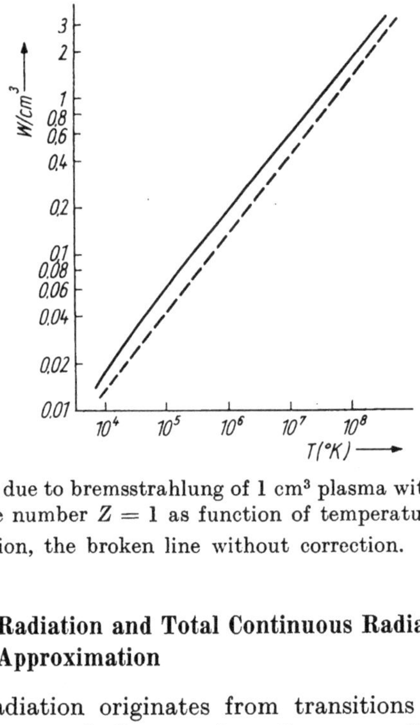

Figure 9.4. Power loss due to bremsstrahlung of 1 cm³ plasma with $N_e = N_i = 10^{15}$ cm^{-3} and effective ion charge number $Z = 1$ as function of temperature. The full line is calculated with $\bar{\bar{g}}$-correction, the broken line without correction. (After Hettner [9.30])

9.2.2. Free-bound Radiation and Total Continuous Radiation, Hydrogenic Approximation

The free-bound radiation originates from transitions between continuous electronic energy states and discrete (bound) states. When calculating the corresponding emission coefficient, transitions to all possible discrete energy states must be considered. Following Unsöld [9.69] and Maecker and Peters [9.45], the emission coefficient of free-bound radiation can be calculated in analogy to the emission coefficient of free-free radiation starting from formula (9.10) and using suitable chosen integration limits.

In first instance, the transitions to those discrete energy levels forming a level sequence with short interlevel distances down to a level energy of $h\,\nu_g$ below the ionization limit (see figure 9.1) are considered. Their contribution to the total continuous emission coefficient is taken into account either by lowering the lower integration limit in formula (9.10) by the amount $h\,\nu_g$ (for $\nu > \nu_g$) or by setting the lower integration limit equal to 0 (for $\nu \leq \nu_g$). Substituting L by $(\pi/3)\,\bar{g}$ where $\bar{g}$ is a numerical factor of order 1 which must not be specified in this stage of calculation, we obtain for the continuous emission coefficient the expressions

$$\nu \leq \nu_g: \quad \varepsilon_\nu = \frac{32\,\pi^2}{3\sqrt{3}}\,\frac{e^6\,Z^2}{c^3(2\,\pi\,m)^{3/2}}\,\frac{N_e\,N_i\,\bar{g}}{(kT)^{1/2}} = 6.36\cdot 10^{-47}\,Z^2\,\frac{N_e\,N_i\,\bar{g}}{(kT)^{1/2}}\,, \tag{9.15}$$

$$\begin{array}{l}\nu > \nu_g: \\ (\nu < \nu_n)\end{array} \quad \varepsilon_\nu = 6.36\cdot 10^{-47}\,Z^2\,\frac{N_e\,N_i}{(kT)^{1/2}}\,\mathrm{e}^{h(\nu_g-\nu)/(kT)}\,\bar{g} \tag{9.16}$$

which include the contribution of free-free radiation to the resulting continuum. To get the contribution of free-bound transitions alone, from (9.15) respectively (9.16) the expression (9.13) must be subtracted. The result is

$$\varepsilon_{\nu(f-b)} = 6.36\cdot 10^{-47}\,Z^2\,\frac{N_e\,N_i}{(kT)^{1/2}}\,\bar{g}\,(1-\mathrm{e}^{-h\nu/(kT)})\,, \quad \nu \leq \nu_g \tag{9.17}$$

respectively

$$\varepsilon_{\nu(f-b)} = 6.36\cdot 10^{-47}\,Z^2\,\frac{N_e\,N_i}{(kT)^{1/2}}\,\bar{g}\,\mathrm{e}^{-h\nu/(kT)}\,(\mathrm{e}^{h\nu_g/(kT)}-1)\,, \qquad \nu > \nu_g\,, \quad \nu < \nu_n\,. \tag{9.18}$$

A comparison between (9.15)—(9.18) shows that in the high frequency region (defined by $h\,\nu/(kT) \gg 1$ for $\nu \leq \nu_g$ respectively $\nu > \nu_g$ for $h\,\nu/(kT) \ll 1$) the free-free radiation is relative unimportant while in the low frequency region (defined by $h\,\nu/(kT) \ll 1$) the free-free radiation is dominant.

Formula (9.16) holds for frequencies up to the next of the threshold frequencies ν_n of the isolated recombination continua. For $\nu > \nu_n$, the contribution of the latter to the total continuous emission coefficient has to be taken into account separately (see section 9.2.3). Using the hydrogenic approximation we obtain [9.45]*

$$\nu > \nu_n: \qquad \varepsilon_{\nu_n} = \frac{128\,\pi^4\,e^{10}}{3\sqrt{3}\,(2\,\pi)^{3/2}\,h^2\,m^{1/2}\,c^3}\,\frac{Z^4\,G\,N_e\,N_i}{n^3\,(kT)^{3/2}}\,\mathrm{e}^{h(\nu_n-\nu)/(kT)} \tag{9.19}$$

in which n means the principal quantum number. The "Gaunt" factor G which in most cases differs from 1 by no more than 20% can be taken from tabulations [9.9]—[9.11], [9.36c], [9.38b], [9.60d].

A characteristic feature of the isolated recombination continua is the sudden onset of intensity at frequency $\nu = \nu_n$ followed by an exponentially decrease of intensity in the direction of increasing frequency with a steepness inversely proportional to the temperature T.

* In the original text [9.45], the term $(Z + s)^4$ stands instead of the term $Z^4\,G$.

The continua corresponding to recombination into the discrete lower electronic energy levels n are superimposed one to eachother, thus the resulting continuum becomes saw tooth shaped as indicated in figure 9.5.

For completion if must be noticed that the spectral emission coefficient $\varepsilon_{\nu, n}$ of an individual recombination continuum is connected with the photoionization cross section $\sigma_{\nu, n}$ and the absorption coefficient $\varkappa_{\nu, n}$ by the formulas

$$\varepsilon_{\nu, n} = \varkappa_{\nu, n} \cdot B_\nu(T) = N_n \, \sigma_{\nu, n} \cdot B_\nu(T)$$

in which $B_\nu(T)$ is the Planck function (see formula (2.19a)), while N_n is the number density of atoms in level n.

Difficulties arise when the formulas (9.15) and (9.16) are applied to real plasmas. The reasons for this are: 1) It is difficult to define $\bar{g}$ in the case of recombination radiation. 2) The electric field at place of a bound electron is no longer a pure Coulomb field given by the ionic charge number Z.

To overcome these difficulties, the factor $\bar{g}\, Z^2$ is substituted by a new factor $\overline{(Z + s)^2}$ [9.63] which depends (like $\bar{g}$) from frequency and temperature.

The factor $\overline{(Z + s)^2}$ can be derived by calibrating experiments as demonstrated by Schirmer [9.63] for the continuum emitted by a high pressure xenon lamp. For shorter wavelengths for which the transitions to lower energy states belonging to the zone of close spaced levels give the main contribution to the continuous intensity, a value of 2.5 was found for $\overline{(Z + s)^2}$, while in the direction of longer wavelengths (for frequencies $\nu < \nu_g$) a decrease of the factor $\overline{(Z + s)^2}$ was observed which approaches to the value 1.

On the other hand, the factor $\overline{(Z + s)^2}$ can be found theoretically by comparing the quantummechanically derived continuous emission coefficient with the expressions (9.17) and (9.18).

In general, the exact calculation of continuous intensity is very laborious (see [9.47b] chapter 3). A fairly good approximation can be obtained by means of the quantum defect method (*QDM*) to which reference is made in the next section.

9.2.3. Free-bound and Total Continuous Emission Coefficient in the Nonhydrogenic Case, evaluated by Means of the Quantum Defect Method (QDM)

For nonhydrogenic atoms (ions), the contribution of free-bound continuum $\varepsilon_{\nu, fb}$ to the total continuous emission coefficient ε_ν is found by summation over the discrete recombination continua:

$$\varepsilon_{\nu, fb} = \sum_{n, l} \varepsilon_{\nu, n l}\,, \qquad \frac{U_i - U_{nl}}{h} < \nu\,. \tag{9.20}$$

The sum is to take over all principal and angular quantum numbers for which relation $(U_i - U_{nl})/h < \nu$ is satisfied in which U_{nl} is the lower level energy of the recombining transition.

Under condition of local thermodynamic equilibrium (*LTE*), $\varepsilon_{\nu,\,nl}$ is of the form

$$\varepsilon_{\nu,\,n\,l} = \frac{B_\nu(T)\,N_0}{Q_0}\,g_{n\,l}\,\mathrm{e}^{-U_{n\,l}/(kT)}\cdot\sigma_{\nu,\,n\,l} \tag{9.21}$$

containing the photoionization cross section given by

$$\sigma_{\nu,\,n\,l} = \frac{4\,\pi\,\alpha\,a_0^2\,(U_{i,\,n\,l}+k^2)}{3\,g_{n\,l}}\,|M_{m\,n}|^2\,. \tag{9.22}$$

α is the fine structure constant ($= 1/137$), a_0 is the first Bohr radius and $g_{n\,l}$ is the statistical weight of level $n\,l$. The quantities $U_{i,\,n\,l}$ (ionization energy of level $n\,l$) and k^2 (kinetic energy of ejected electron in the corresponding photoionizing transition) are given in Rydberg units (13.6 eV). The quantity k^2 can be expressed in terms of ν by means of the relation $U_i + k^2 = h\,\nu/(13.6\ \mathrm{eV})$.

The main task is to find the matrix elements $M_{m\,n}$ which must be calculated from suitable chosen wave functions which depend upon the potential energy curve of the electron in the field of the residual ion. After Burgess and Seaton [9.13c], the electron is assumed to move in a modified Coulomb potential characterized by the quantum defect function $\mu(\varepsilon)$ (ε = energy of electron in question) which is defined for negative (discrete) energy terms $n\,l$ by the relation

$$\mu = n - n_l^* \tag{9.23}$$

in which n_l^* is the effective principal quantum number given by

$$\varepsilon_{n\,l} = -\frac{Z}{n_{l*}^2}\,, \tag{9.24}$$

and Z is the charge number of the residual ion. $\mu(\varepsilon)$ can be extrapolated to positive energy values for a given angular quantum number l.

Knowing the function $\mu(\varepsilon)$, then σ_{nl} is found by means of fairly simple formulas and tabulated data presented in [9.13b] and [9.13c] and reproduced for example in [9.26b].

In the high energy range (that means in the range of relative high quantum numbers) the tabulations are of reduced preciseness as shown for example in [9.1m], [9.60f] and [9.64b]. In [9.57a], [9.60f] and [9.60g], modified formulas are presented leading to more precise values for the continuous emission coefficient in the high energy range than the original quantum defect formulas.

In the region of higher quantum numbers n and for $l > 2$ or 3, the quantum defect μ converges to zero, and the photoionization cross section is then described by the hydrogenic formulas ([47b], p. 34):

$$\sigma_{\nu n\,(\mathrm{hydrogen})} = \frac{G\cdot 64\,\pi^4\,e^{10}\,m\,Z^4}{3\sqrt{3}\,c\,h^6\,\nu^3\,g_n\,n^3}\,,\qquad \nu > \nu_n\,. \tag{9.25}$$

The corresponding emission coefficient is given by relation (9.19), that means, the emission sets in at a threshold frequency $\nu = \nu_n$ and decreases exponentially with growing frequency ($\varepsilon_\nu \sim \mathrm{e}^{-h\nu/(kT)}$).

With these results, the calculation of electron-ion continuum becomes solely a summation problem.

Calculations of continuous emission coefficient on the basis of *QDM* with improved accuracy have been performed by Schlüter [9.64b] for the rare gases Ne, Ar, Kr and Xe for temperatures between 8000 and 16,000 °K and for wavelengths between 2000 and 7000 Å.

Considerable simplifications in the numerical procedure for evaluating the total continuous emission coefficients are possible on the basis of investigation performed by Biberman, Norman and Ulyanov [9.5b], [9.5c], [9.5d], [9.6a]. They write in analogy to the Kramers-Unsöld-Maecker-Peters-formulas ((9.17) and (9.18)):

$$\left.\begin{aligned} \varepsilon_\nu(T) &= 6.36 \cdot 10^{-47}\, N_e\, N_i\, (kT)^{-1/2}\, Z^2\, \xi(\nu, T)\,, && \nu \leq \nu_g\,. \\ \varepsilon_\nu(T) &= 6.36 \cdot 10^{-47}\, N_e\, N_i\, (kT)^{-1/2}\, Z^2\, \xi(\nu, T)\, e^{h(\nu_g-\nu)/(kT)}\,, && \nu \geq \nu_g\,. \end{aligned}\right\} \tag{9.26}$$

Formula (9.26) holds for frequencies up to the threshold frequency ν_n for the isolated recombination continua (see figure 9.1). The factor $\xi(\nu, T)$ is of the order 0.2—3 and depends from the quantum defect function $\mu(\varepsilon)$ for the lower l-values. If the quantum defect (for a given l) is independent from energy, then $\xi(\nu, T) = \xi(\nu)$ is independent from temperature T and is given by the expression

$$\xi(\nu) = 1 + \sum_{l=0}^{l_1} K_{l\,l'}\, \varepsilon^{m_{l\,l'}} \left\{ \cos^2 (\pi\, [\chi_{l\,l'}(\varepsilon) + \mu_{l'} - \mu_l]) - \cos^2 \left(\pi\, \chi_{l\,l'}(\varepsilon)\right) \right\}. \tag{9.27}$$

The symbols used in (9.27) are: l_1 = maximum angular quantum number for which the quantum defect is appreciable (in practice, $l_1 = 2$ or 3), $K_{01} = 0.75$, $K_{12} = 2.3$, $K_{23} = 6$; $m_{01} = m_{12} = 0.82$, $m_{23} = 0.90$; $l' = l + 1$; $\chi_{l\,l'}(\varepsilon) = a_{l\,l'} + b_{l\,l'}\sqrt{\varepsilon} + c_{l\,l'}\,\varepsilon$ if $\nu < \nu_g$ and $\chi_{l\,l'}(\varepsilon) = \chi_{l\,l'}(n_g^*, \varepsilon)$ if $\nu > \nu_g$. The coefficients $a_{l\,l'}, b_{l\,l'}$ and $c_{l\,l'}$ are given in [9.13b], [9.13c]. n_g^* is the effective quantum number corresponding to the threshold frequency ν_g; μ_l and $\mu_{l'}$ are the quantum defects corresponding to the angular quantum numbers l and l'. $\varepsilon = h\,\nu/Z^2$ scales in Rydberg units (= 13.6 eV). $\chi_{l\,l'}(n_g^*, \varepsilon)$ can be calculated by means of a formula given in [9.13b].

Analogous formulas for $\xi(\nu, T)$ in the case when $\mu(\varepsilon)$ varies linearely with ε are presented in [9.5c]. $\xi(\nu, T)$ varies relatively slowly with temperature T. Graphical representations of the factor $\xi(\nu)$ respectively $\xi(\nu, T)$ for a number of atoms respectively ions can be found for example in [9.5c].

Based on experiments with an argon plasma containing atoms, electrons and single charged ions, Gündel and Neumann [9.27b], [9.27c] have found the following simple half analytical expression for the continuous emission coefficient useful for plasmadiagnostical applications:

$$\varepsilon_\nu = A(\nu) \cdot N_0\, kT\, e^{-U_a(\nu)/(kT)}\,. \tag{9.28}$$

N_0 is the number density of neutral argon atoms and U_a is interpreted as "continuous excitation energy". The advantage of formula (9.28) over (9.26) for plasmadiagnostical application in the special case of argon is that in the whole visible-near ultraviolet spectral region the nonanalytical coefficients A and U_a in (9.28) are only functions of a single argument (ν) while the non-analytical quantity ξ in (9.26) and (9.27) would become a function of two variables (ν and T). The values of $A(\nu)$ and $U_a(\nu)$ in the case of argon are tabulated in [9.27c] for frequencies between $4.4 \cdot 10^{14}\ \mathrm{s}^{-1}$ and $1.0 \cdot 10^{15}\ \mathrm{s}^{-1}$.

9.2.4. Effect of Lowering of Ionization Energy on the Spectral Distribution of Continuous Emission Coefficient. Comparison between calculated and measured Spectral Distributions

Only in a high diluted plasma is the ionization limit, up to which discrete electronic levels corresponding to single-electron excitation exist, a well defined quantity which can be found in corresponding tables. Due to the Coulomb interaction of charged particles in real plasmas, the ionization energy is suppressed by a quantity ΔU_i which depends from plasma temperature and plasma density as shown in section 4.1.1. and figure 4.1. Due to this suppression, the formulas for the continuous emission coefficient are changed in the following way:

Formula (9.15) is formally retained, but its range of validity is shifted along the frequency scale by the quantity $\Delta U_i/h$, that means formula (9.15) holds now for frequencies $\nu \leq \nu_g - \Delta U_i/h$.

Formula (9.16) has to be interchanged with

$$\nu > \nu_g - \Delta U_i/h: \qquad \varepsilon_\nu = 6.36 \cdot 10^{-47}\, Z^2 \frac{N_e N_i}{(kT)^{1/2}}\, \mathrm{e}^{\frac{h(\nu_g-\nu)-\Delta U_i}{kT}}\, \bar{g}\,. \tag{9.16a}$$

For the isolated photoionization continua described originally by formula (9.19), the positions of the photoionization edges at which continuous emission sets in ar e also shifted by the quantity $\Delta\nu = \Delta U_i/h$. Then we get

$$\varepsilon_{\nu n} = \frac{128\,\pi^4\, e^{10}\, Z^4\, G}{3\sqrt{3}\,(2\,\pi)^{3/2}\, h^2\, m^{1/2}\, c^3\, n^3\, (kT)^{3/2}}\, N_e\, N_i\, \mathrm{e}^{+\frac{h(\nu_n-\nu)-\Delta U_i}{kT}}\,. \tag{9.19a}$$

A formula for ΔU_i has been derived by Armstrong [9.1o] under the assumption that the apparent shift of ionization edge is primarely caused by the effect of line merging at the series limit due to electron collision broadening of spectrum lines. For a simple plasma ($N_e = N_i$) he obtains

$$\Delta U_i = (\Delta U_n) = \frac{4.707 \cdot 10^{-6}\, N_e^{2/7}}{(kT)^{1/7}}\ [\mathrm{eV}]\,.$$

For comments on the problem of edge shift see also [9.75c] and sections 10.8.1 and 10.8.2.

For a model plasma, the resulting distribution of continuous intensity as function of frequency and the shift of the photoionization edges $\Delta U_i/h$ are schematically shown in figure 9.5.

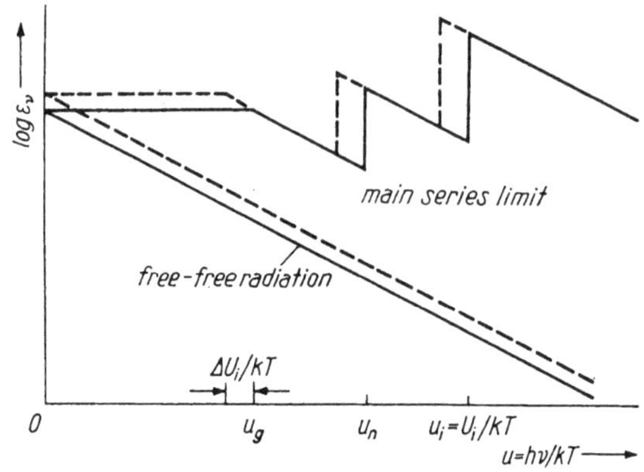

Figure 9.5. Schematic representation of the continuous spectrum emitted by a thermal plasma. One may notice in the figure the diverse kinds of continuous radiation (free-free radiation and radiation due to recombination of electrons into the group of close spaced levels and into isolated levels) and the shift of the emission edges by the amount $\Delta U_i(/kT)$ where ΔU_i is the lowering of ionization energy. (After Finkelnburg and Peters [9.21])

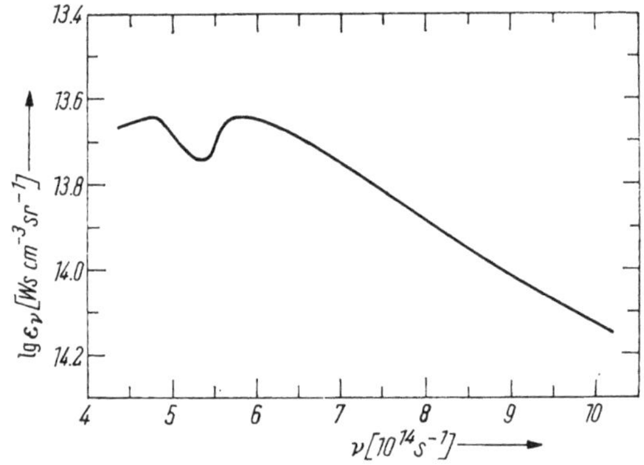

Figure 9.6. Continuous emission coefficient ε_ν of an argon plasma at $p = 1$ atm and $T = 12{,}000\,°K$ as function of frequency ν. (After Gündel and Neumann [9.27c])

This idealized intensity distribution will be in some cases a good approximation of the real one. This has been shown by Maecker and Peters [9.45] for a plasma containing the elements C, N and O (see also figure 9.9).

In contrast to this idealized intensity distribution we show in figure 9.6 the intensity distribution of the continuum emitted from an argon plasma for a pressure of 1 atm at $T = 12{,}000\,°K$, as observed by Gündel and Neumann. The data are taken from [9.27c]. The plasma was produced in a cascade arc.

The spectral intensity distribution shows principal deviations from that of the Maecker-Peters model:

At the expected limiting frequency $\nu_g = 4.6 \cdot 10^{14}\ \mathrm{s}^{-1}$ (corresponding to the position of level 3 d_6 down to which the electronic energy levels lie fairly close one to another, see figure 9.7), no marked edge is observed from the beginning

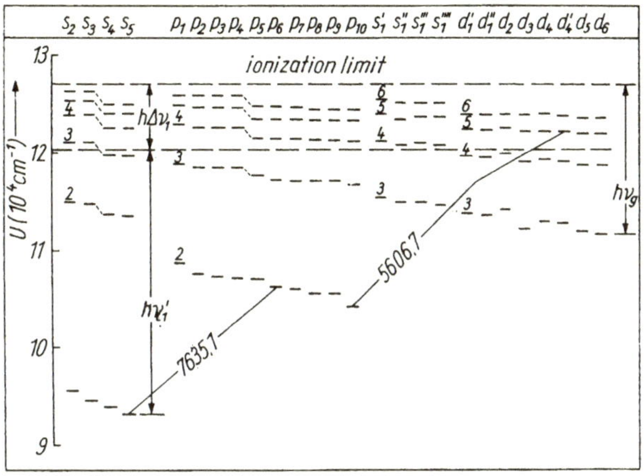

Figure 9.7. Level diagram of Argon I. (After H. N. Olsen [9.60])

of which the continuous emission coefficient in the model plasma will show an exponential decrease according to the law $\varepsilon_\nu \sim e^{-h\nu/(kT)}$. The observed decay begins at $\nu = 6 \cdot 10^{14}\ \mathrm{s}^{-1}$ and is much weaker than expected.

Furthermore, there are expected some other edges in the continuous intensity distribution in the visible-near ultraviolet spectral region, among them an edge at $\nu = 1.01 \cdot 10^{15}\ \mathrm{s}^{-1}$ corresponding to the position of the level 1 s_5. In reality, no edge is found in the neighborhood of this frequency (see also figure 5 in the work of Wende [9.74d).*

As Gündel and Neumann [9.27b] have shown, the nonvisibility of photo-ionization-photorecombination edges is a consequence of the statistical character of lowering of ionization potential, which causes not only a shift but also a smearing out of ionization edges along the frequency scale.

It must be noted that these conclusions are based on observations in the visible-near ultraviolet spectral region in which the photoionization edges of Ar are relative close spaced and they must not necessarily be valid in the vacuum ultraviolet region.

* The edges noticed in an older work of Olsen [9.60] are only fictitious, and they can be identified with Ar I spectral lines, as shown in [9.27b].

9.3. Electron-atom Continuum

The theoretical description of the continuous radiation originating from collisions between electrons and neutral atoms (e—a continuum) is more complex than that of the electron-ion continuum arising from a pure Coulomb interaction, where the spectral intensity distribution can be fairly accurately estimated from classical calculations.

The e—a continuum has been observed with considerable intensity in those plasmas containing atoms which have a high elastic cross section Q relative to electron collisions [9.55]. The existence of a correlation between intensity of e—a continuum and cross section Q has been confirmed indirectly by Taylor and Kivel [9.68] showing that the quantum mechanical formulas for the collision cross section and for the continuous radiation intensity include analogous terms. As a further argument for this correlation the authors refer to measurements of the continuous radiation intensity $I_\nu(T)$ in air and nitrogen in the parameter range $T = 6000-9000\ ^\circ\mathrm{K}$ and $\lambda = 2 - 8\ \mu\mathrm{m}$ performed by Taylor [9.66], [9.67] using as radiation source a reflected shock wave. In the wavelength region mentioned the continuous radiation stems almost exclusively from free-free transitions.

Considering only that part of radiation stemming from e—a collisions, the emission coefficient due to free-free radiation can be described formally by an expression analogous to (9.13)

$$\varepsilon_\nu = 6.36 \cdot 10^{-47} \frac{N_e \mathrm{e}^{-h\nu/(kT)}}{(kT)^{1/2}} \sum Z_l^2 N_l \tag{9.29}$$

in which the symbol l stands for the diverse kinds of neutral particles (N, N_2, O, O_2). The quantities Z_l^2 which have in first instance only formally meaning have been obtained on a pure experimental basis by applying formula (9.29) to emission coefficients obtained for variable plasma conditions (nitrogen plasma, air plasma, variable temperature). The quantities Z_l^2 have irrespective of their dimensionless character the meaning of inelastic cross sections relative to continuous emission. In the following table (taken from Taylor and Kivel) the values for Z_l^2 are compared with the corresponding values for the elastic cross section Q relative to electron collisions.

Table 9.2

Element	Z_l^2	Q [10^{-15} cm^2]
O	$(0.2 \pm 0.3)10^{-2}$	0.2
N	$(0.9 \pm 0.4)\ 10^{-2}$	(0.5)
N_2	$(2.2 \pm 0.3)\ 10^{-2}$	2

The data for Q are taken from Lin [9.41], [9.42] and from Neynaber et al. [9.56]. Irrespective of the fact that the contributions of the diverse kinds of neutral particles to the resulting free-free continuum has been derived with only low precision, a marked conformity between Z_l^2 and Q can be noticed.

While the atoms of mercury and hydrogen have evidently great cross sections relative to elastic electron collisions ($Q_{\mathrm{Hg}} = 1.2 \cdot 10^{-14}\,\mathrm{cm}^2$ after Brode [9.13], $Q_{\mathrm{H}} = 2.24 \cdot 10^{-14}\,\mathrm{cm}^2$ after Drawin [9.19]), the corresponding cross sections of the rare gases are relatively low (of the order $10^{-16}\,\mathrm{cm}^2$). Accordingly, the e—a continua can be generally well observed in plasmas containing the elements Hg and H in a considerably amount (see [9.7], [9.43] and [9.55]), while in heavy rare gas plasmas these continua are of importance only for low degrees of ionization (such continua have been observed in low pressure discharges [9.61a], [9.61b], [9.61c], [9.61o]).

Intensity formula for the e—a bremsstrahlung continuum derived on a quantum mechanical basis applicable both to atoms and to molecules for which the elastic collision cross section Q relative to electrons is only a slowly varying function of electron energy can be found in a number of papers (see for example [9.1b], [9.1f], [9.1r], [9.3a], [9.5d], [9.21a], [9.24f], [9.36d]). They can be brought to the form of equation (9.29), in which the effective charge number squared is asymptotically given by the expressions (see [9.5d])

$$\left.\begin{aligned} Z^2 &= \frac{2\sqrt{3}}{\pi^2 e^4}\,\overline{Q(T)}\,(kT)^2 && \text{if} \quad h\nu/(2\,kT) \ll 1 \\ Z^2 &= \frac{\sqrt{3}}{4\,\pi\,e^4\sqrt{\pi}}\,\overline{Q(T)}\,(h\,\nu)^{3/2}\,(kT)^{1/2} && \text{if} \quad \frac{h\,\nu}{2\,kT} \gg 1\,. \end{aligned}\right\} \tag{9.30}$$

By simple interpolation one obtains

$$Z^2 \approx \frac{2\sqrt{3}\,\overline{Q(T)}\,(kT)^{1/2}}{e^4\,\pi^2}\left[(kT)^{3/2} + \frac{\sqrt{\pi}}{8}\,(h\,\nu)^{3/2}\right].$$

If Q is a strongly depending function of ε, then the continuous emission coefficient of e—a bremsstrahlung can be evaluated from the formula (see [9.36a])

$$\varepsilon_\nu = \frac{5\cdot 10^{-3}\,h}{m^{3/2}\,c^2}\,N_e\,N_a\,(kT)^{-3/2}\int_{h\nu}^{\infty}\sqrt{\varepsilon\,(\varepsilon - h\,\nu)}\,[\varepsilon\cdot Q\,(\varepsilon - h\,\nu) +$$

$$+ (\varepsilon - h\,\nu)\cdot Q(\varepsilon)]\cdot \mathrm{e}^{-\varepsilon/(kT)}\,\mathrm{d}\varepsilon \tag{9.31}$$

in which T means the electron temperature.

As in the case of e—i continua, also in the case of e—a continua one distinguishes between free-free radiation and free-bound radiation. While the former appears in every case, the appearance of the latter presumes the existence of absolutely or relative stable negative ions the lifetime of which must exceed markedly the collision time. Such ions are formed for example in hydrogen, oxygen, mercury and in the halogenes.

The intensities of free-bound continua can be theoretically evaluated by means of the matrix elements*

$$|\int \psi_m (\boldsymbol{r}_1 + \boldsymbol{r}_2) \psi_n^* \, dV|^2 .$$

The quantities $\boldsymbol{r}_1$ and $\boldsymbol{r}_2$ are the radius vectors of the two outer electrons of the "two-electron system" substituted by the colliding system "atom + electron". Instead of the position operator $(\boldsymbol{r}_1 + \boldsymbol{r}_2)$ it is also possible to use the pulse- or acceleration operator. We do not go into detail of calculation but refer to literature [9.15], [9.16], [9.18], [9.23] and [9.59].

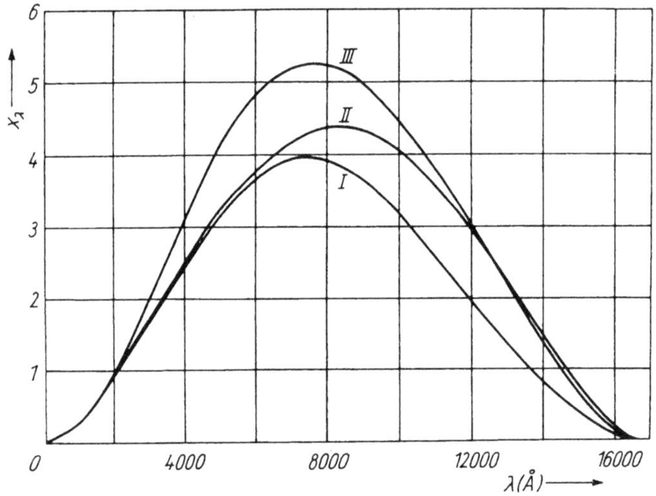

Figure 9.8. Absorption coefficient $\varkappa_\lambda$ in units of [10^{-17} cm^2 per H^- ion] calculated by means of the position matrix (*I*), momentum matrix (*II*) and acceleration matrix (*III*). (After Chandrasekhar [9.16])

A typical result of corresponding calculations is presented in figure 9.8 showing the continuous absorption coefficient for optical transitions between the ground state of the H^- ion and continuous energy states. The continuum (free-bound continuum) ranges from short wavelength up to the limiting wavelength at 16,550 Å corresponding to the ionization energy of the H^- ion = = 0.75 eV.

A tabulation of the total absorption coefficient of the e—a continuum of hydrogen as function of wavelength for the parameter range $\lambda = 4050$ to

* In general, the matrix elements are considered for allowed transitions only. As has been shown by Weinberg and Berry [9.74b], this restriction is not justified for example in the case of the *ea*-continuum of hydrogen for electron densities in excess of 10^{15} cm^{-3}, because the polarization of H^- by H^+ yields new possibilities of electron detachment. Thus in addition to the allowed continuum, a forbidden continuum is observed. See also [9.50t].

Table 9.3. Continuous absorption of the negative hydrogen ions due to free-free and bound-free transitions of electrons. Tabulated is the quantity $-\lg(\varkappa_\lambda/p_e)$ as function of $\Theta = 5040/T$ where $\varkappa_\lambda$ is the continuous absorption coefficient per unit number density of hydrogen atoms (1 H atom per cm³) and p_e is the electron pressure in [dyn/cm²]. The absorption coefficient includes a negative contribution from stimulated emission. (After Unsöld [9.70])

Δk^2	λ (Å)	$\Theta = 0.5$	0.6	0.7	0.8	0.9	1.0	1.2	1.4	1.6	1.8	2.0
[1]	—	28.454	28.338	28.242	28,160	28,089	28.026	27.920	27.831	27.756	27.690	27.633
0.005	182,300	23.932	23.816	23.720	23.638	23.567	23.504	23.397	23.308	23.233	23.168	23.110
0.010	91,130	24.533	24.416	24.319	24.237	24.166	24.102	23.995	23.906	23.829	23.763	23.705
0.015	60,750	24.882	24.364	24.667	24.585	24.513	24.449	24.341	24.251	24.175	24.108	24.050
0.020	45,560	25.128	25.010	24.913	24.829	24.757	24.693	24.584	24.494	24.417	24.350	24.292
0.025	36,450	25.318	25.199	25.101	25.017	24.945	24.880	24.771	24.680	24.603	24.537	24.479
0.030	30,380	25.472	25.353	25.254	25.170	25.096	25.032	24.922	24.832	24.755	24.688	24.630
0.040	22,780	25.712	25.592	25.492	25.407	25.334	25.269	25.159	25.068	24.992	24.926	24.868
0.050	18,230	25.896	25.775	25.675	25.590	25.516	25.451	25.341	25.251	25.174	25.108	25.051
0.060	15,190	26.012	25.872	25.749	25.634	25.523	25.415	25.196	24.970	24.737	24.500	24.261
0.070	13,020	26.026	25.841	25.667	25.500	25.337	25.178	24.871	24.576	24.294	24.024	23.764
0.080	11,390	26.018	25.801	25.599	25.408	25.225	25.051	24.721	24.413	24.124	23.849	23.587
0.090	10,130	26.013	25.779	25.562	25.360	25.169	24.989	24.652	24.341	24.049	23.774	23.511
0.100	9,113	26.016	25.771	25.547	25.340	25.146	24.963	24.623	24.310	24.018	23.742	23.479
0.120	7,594	26.040	25.785	25.555	25.344	25.147	24.963	24.622	24.308	24.016	23.740	23.477
0.140	6,509	26.079	25.820	25.588	25.376	25.179	24.995	24.655	24.340	24.048	23.772	23.508
0.160	5,695	26.123	25.864	25.631	25.419	25.222	25.038	24.696	24.383	24.091	23.815	23.551
0.180	5,063	26.169	25.910	25.678	25.465	25.269	25.084	24.743	24.430	24.137	23.860	23.597
0.200	4,556	26.215	25.956	25.724	25.512	25.315	25.131	24.789	24.476	24.183	23.906	23.643
0.225	4,050	26.272	26.013	25.782	25.569	25.372	25.187	24.845	24.531	24.239	23.962	23.698

[1] This line minus $2 \cdot \lg(\lambda/1000\ \text{Å})$ yields the quantity $-\lg(\varkappa_\lambda/p_e)$ for $\lambda > 180{,}000$ Å and the corresponding Θ

182,300 Å and $T = 2500-10{,}000$ °K is given by Unsöld ([9.70] p. 176). The tabulated data are given in a reduced form referring to the absorption of 1 neutral hydrogen atom and to an electron pressure of 1 dyn/cm². We reproduce these data in table 9.3.

At this place we remark that from the position of absorption edges of the free-bound electron-atom continua precise values for the electron affinity can be obtained. Using this method, the authors of [9.5a] have determined the electron affinities of the halogen atoms (see table 9.4).

Table 9.4. Determination of electron affinity from the position of the edges of the free-bound electron-atom continua

Element	Type of measurement	Method of gas heating	Electron affinity [eV]	Reference
Br	absorption	shock wave	3.363	[9.5a]
	emission	arc	3.359 – 3.366	[9.21b], [9.61g].
Cl	absorption	shock wave	3.613	[9.5a]
	emission	arc	3.614 – 3.616	[9.21b], [9.61g].
F	absorption	shock wave	3.448	[9.5a]
	emission	arc	3.400	[9.21b], [9.61e, f, g].
I	absorption	shock wave	3.063	[9.5a]
	emission	arc	3.061	[9.61g]

The e—a free-bound continua can also be observed in emission, as has been shown by Popp [9.61e] for the element F. Unlike the case of absorption continua, the long wavelength edges of the emission continua are shifted by about 50 Å in the direction of longer wavelengths. A possible explanation for this effect is the lowering of electron detachment energy by the microfield of the emitting plasma.

From the absolute emission coefficient of the e—a f—b continuum (electron affinity continuum), it is in principal possible to determine the concentration of negative ions.

Within the affinity continuum of F, Cl, Br and I, there have been observed maxima of the emission coefficient with a halfwidths of about 50—100 Å [9.61g]. In the case of Cl, the corresponding maximum has been identified as a molecular continuum [9.50s], whereas in an earlier work [9.61g] it was suggested that the corresponding maximum is due to an autoionizing line.

9.4. Separation of Electron-atom Continuum from Electron-ion Continuum

One method for separating the two main components of continuous radiation intensity is based upon their different behavior to changes of particle number densities. For the e-i continuum emitted by an one-element plasma, the emission coefficient is given by the formula

$$\varepsilon_{ei}(\nu, T) = S_{ei}(\nu, T)\, N_e\, N_i \text{ [erg cm}^{-3}\text{ s r}^{-1}] \tag{9.32}$$

with

$$S_{ei}(\nu, T) = 6.36 \cdot 10^{-47}\, \overline{(Z + s)^2}\, \frac{1}{(kT)^{1/2}}$$

while for the e-a continuum the analogous expression is

$$\varepsilon_{ea}(\nu, T) = S_{ea}(\nu, T)\, N_e\, N_a \tag{9.33}$$

in which $S_{ea}(\nu, T)$ is a more complex function of ν and T than $S_{ei}(\nu, T)$. Neglecting the dependence of the quantities S_{ei} and S_{ea} upon electron density, then one obtains for the total continuous emission coefficient the expression

$$\varepsilon_\nu = S_{ei}(\nu, T)\, N_e\, N_i + S_{ea}(\nu, T)\, N_e\, N_0 \tag{9.34}$$

where N_e, N_i and N_0 denote the number densities of electrons, ions and neutral particles. As seen from (9.34), a separation of continua can be realized by carrying out isothermal changes of the plasma state (by means of pressure variation) as a result of which different values for the quantity N_e/N_0 are obtained. Denoting for a given frequency ν and temperature T the continuous emission coefficient at pressure p_1 with ε_1 and at pressure p_2 with ε_2, then under the assumption that the particle number densities are known quantities, the quantities S_{ei} and S_{ea} can be found by solving the system of equations

$$\left.\begin{aligned} \varepsilon_1 &= S_{ei}\, N_{e1}\, N_{i1} + S_{ea}\, N_{e1}\, N_{o1}\,, \\ \varepsilon_2 &= S_{ei}\, X_{e2}\, N_{i2} + S_{ea}\, N_{e2}\, N_{o2}\,. \end{aligned}\right\} \tag{9.35}$$

Knowing S_{ei} and S_{ea} as function of frequency ν for some values of the temperature, the problem of separation of continuous spectra is solved.

Another method of separating the *e–a* continuum from the compound continuum has been applied by Boldt [9.9] who used as object of investigation a wall stabilized oxygen arc. The intensity of the *e–i* free-free continuum was calculated and subtracted from the total continuous intensity. The difference is due to the sum of the *e–i* free-bound continuum and of the *e–a* continuum. The separation of the latter is possible by making use of the markant (exponential) frequency dependence of *e–i* free-bound continuum at the series limits. A more detailed description of the method of separation can be found in the work [9.9] from which figure 9.9 is reproduced showing the wavelength dependence of the total continuous intensity for various temperatures.

The relative intensities I_{ei} and I_{ea} of the separated continua at wavelength $\lambda = 5000$ Å normalized to $I_{ei} + I_{ea} = 100\%$ are shown for various plasma temperatures in table 9.5. According to formula (9.35), the ratio $I_{ei}:I_{ea}$ should be proportional to N_i/N_0 if the temperature dependence of the quantities S_{ei} and S_{ea} could be neglected. It is suggested that this neglecting is justified for wavelengths not to close to the photoabsorption edges at which continuous intensity depends exponentially upon frequency (and also upon temperature) according to the law $I_\nu \sim K_1 + K_2\, e^{-h\nu/(kT)}$.

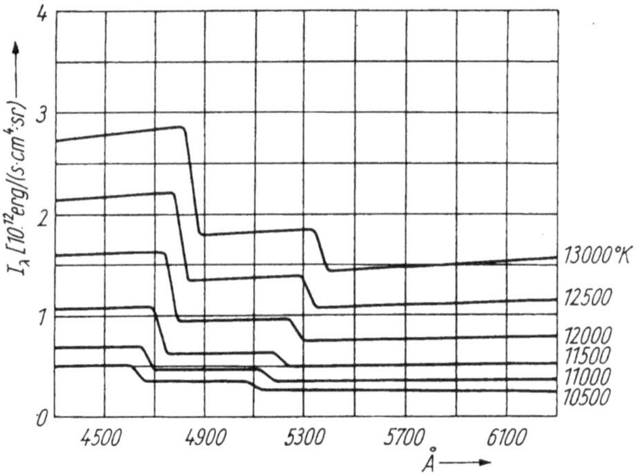

Figure 9.9. Spectral representation of continuous emission from an oxygen plasma at $p = 1$ atm for diverse temperatures. (After Boldt [9.9])

Table 9.5. Particle number densities and relative contributions I_{ei} and I_{ea} of $e-i$ and $e-a$ continuum to the total continuous intensity for an oxygen plasma at pressure $p=1$ atm. The experimental data are taken from Boldt [9.9]

T [°K]	N_i [10^{16} cm^{-3}]	N_0 [10^{17} cm^{-3}]	N_i/N_0	I_{ei}	I_{ea}	I_{ei}/I_{ea}	$\frac{I_{ei}}{I_{ea}} : \frac{N_i}{N_0}$
10,500	2.36	6.51	0.036	33	67	0.49	13.7
11,000	3,33	6.00	0.056	39	61	0.64	11.5
11,500	4.54	5.47	0.083	45	55	0.82	9.9
12,000	5.96	4.92	0.121	59	41	1.44	11.9
12,500	7.68	4.33	0.177	68	32	2.12	12.0
13,000	9.40	3.76	0.250	75	25	3.00	12.0

Table 9.5 shows that according to our suggestion the ratio $I_{ei}:I_{ea}$ is indeed approximately proportional to N_i/N_0. The mean proportionality factor is 11.8. Under special conditions, the separation of e–a and e–i continuum is possible on the basis of the different temperature dependence of the corresponding emission coefficients, see for example [9.1e].

9.5. Plasma Diagnosis by Means of Continuous Radiation

9.5.1. Determination of Plasma Parameters from Absolute and Relative Continuous Emission Coefficients

In first instance we consider a plasma which emits practically the electron-ion continuum only. Then in the case of a simple plasma (that means a one-element plasma containing only single charged ions) and for frequencies below the limiting frequency ν_g, the continuous emission coefficient is given by the formula

$$\varepsilon_\nu = 6.36 \cdot 10^{-47} N_e N_i (kT)^{-1/2} Z^2 \xi(\nu, T) . \tag{9.26a}$$

If the plasma contains more than one kind of ions, equation (9.26a) can be expanded to the form

$$\varepsilon_\nu = 6.36 \cdot 10^{-47} \frac{N_e}{(kT)^{1/2}} \sum_z z^2 N_z \xi_z(\nu, T) \tag{9.36}$$

in which N_z is the number density of ions with charge number z while $\xi(\nu, T)$ is a correction factor.

In the following we consider only the case of the simple plasma. To derive plasma parameters from measured values of ε_ν, equation (9.26a) can be used which contains the 4 unknowns T, N_e, N_i and ξ. Taking into account that $N_e = N_i$ and that the factor $\xi(\nu, T) \approx \xi(\nu)$ can be found from calibrating experiments (for example by application of the norm temperature method as will be shown later), then the number of unknowns (T, N_e) reduces to 2.

In order to derive the quantities T, N_e from only one measured value for ε_ν, then in addition to ε_ν some other quantities characterizing the plasma state must be known or must be determined by additional experiments.

If for example the pressure is known, then the quantity N_e can be expressed as a function of T so that the right hand side of (9.26a) is a function of T only. Thus, a measurement of ε_ν yields the value for T directly, while N_e is obtained subsequently by means of the Saha-equation.

If T is known and ε_ν has been measured, then from (9.26a) the electron density is obtained.

Formula (9.26a) can also be used for the determination of approximate values for the electron density N_e in non *LTE* plasmas in the absence of a Maxwellian distribution of electron velocity, if T_e is interpreted in terms of the mean kinetic electron energy.

The construction of the function $\varepsilon_\nu(T)$ for an one-element plasma at constant pressure will be explained by means of the following example referring to an argon plasma at $p = 1$ atm and to the wavelength $\lambda = 546$ nm. After Olsen [9.60], the quantity ξ in (9.26a) can be set equal to 2. The quantity $N_e = N_i$ can be calculated as function of temperature T using the formulas presented in chapter 4. Then by application of formula (9.26a) the wanted function $\varepsilon_\nu(T)$ is obtained which is shown in the diagram of figure 9.10. By means of this diagram, measured values for ε_ν can be transformed directly into temperature values. As seen from figure 9.10, the accuracy of temperature measurement is low for $T > 14{,}000$ °K because in this temperature range the quantity ε_ν is almost insensitive to changes in T.

From relative measurements of the continuous emission coefficient ε_ν it is also possible to determine the temperature of a plasma if the norm temperature method (see section 5.2.6.) is used provided that the pressure of the plasma is known. Furthermore, the emission coefficient ε_ν must be given either as function of a spatial coordinate (for example as function of radius) or for a given spatial coordinate (in the case of a transient plasma) as function of time. Further, for some values of the spatial respectively temporal coordinate falling into the range of measurement, the plasma temperature must attain the norm temperature T^*

at which $\varepsilon_\nu(T)$ reaches its maximum value $\varepsilon_\nu(T^*)$. Then it is possible to transform the measured relative values for ε_ν into the normalized values $\varepsilon_\nu^* = \varepsilon_\nu/\varepsilon_\nu(T^*)$ On the other hand, a diagram $\varepsilon_\nu^* = f(T)$ can be constructed by means of formula (9.26a) taking into account the relation between N_e and T derived from the Saha-equation. A knowledge of the quantity ξ is not needed because ξ cancels out in the corresponding calculations. By introducing the experimental value for ε_ν^* into the diagram $\varepsilon_\nu^* = f(T)$, the wanted value of the plasma temperature is obtained.

The practical procedure of deriving T from ε_ν^* is demonstrated in figure 9.11. The diagram on the left shows the calculated curve $\varepsilon_\nu^* = f(T)$, the diagram

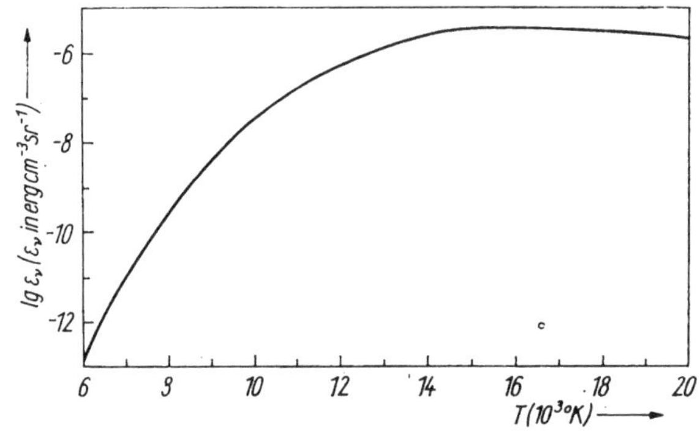

Figure 9.10. lg ε_ν as function of T for an argon plasma at $p = 1$ atm and $\lambda = 546$ nm, calculated with a value $\xi = 2$ taken from Olsen [9.60]

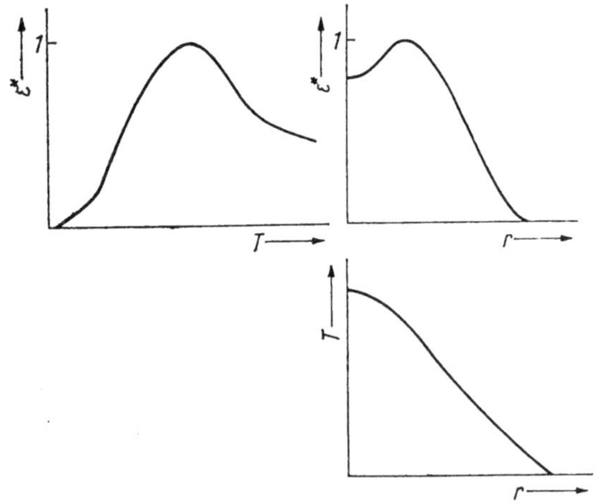

Figure 9.11. Evaluation of the radial temperature distribution $T(r)$ from the measured radial distribution of continuous emission coefficient $\varepsilon_\nu(r)$ by means of the norm temperature method (schematic)

beside it shows the experimental curve $\varepsilon_\nu^* = f(r)$ while the diagram below shows the wanted curve $T(r)$ resulting from comparison of coordinates.

In this connection we mention that Larenz was probably the first who applied the norm temperature method to a laboratory plasma (a water vapour plasma, see [9.40]).

In combination with absolute measurements of the continuous emission coefficient, the norm temperature method can be used for the determination of ξ. The latter quantity is obtained by dividing the measured value for $\varepsilon_\nu(T^*)$ through the value ε_ν/ξ at temperature T^* calculated by means of formula (9.26a).

When using the norm temperature method for the determination of temperature it is profitable but by no means necessary to perform the measurements for frequencies below the limiting frequency ν_g. For quickly finding the latter, the quantity ν_g is given for some elements in table 9.6. About the restricted sense of this limiting frequency see section 9.2.4.

Table 9.6. Limiting frequency ν_g for some elements

Element	ν_g [s^{-1}]	References
Ar	$(6.0-8.1) \cdot 10^{14}$	[9.1s], [9.14], [9.18g], [9.27c], [9.49a], [9.60], [9.64c], [9.74d].
C	$8.4 \cdot 10^{14}$	[9.45]
Fe	$1.36 \cdot 10^{15}$	[9.2]
H	$(6.4 \text{ and } 37) \cdot 10^{13}$	[9.2], [9.75a].
Hg	$1.3 \cdot 10^{15}$	[9.45]
N	$9.5 \cdot 10^{14}$	[9.45]
O	$1.0 \cdot 10^{15}$	[9.45]
Xe	$(5.0-5.5) \cdot 10^{14}$	[9.36h], [9.62].

If plasma parameters are determined from the continuous emission coefficient under the condition that a non vanishing contribution to the radiation stems from electron-atom collisions, then this contribution has to be taken into account. As in the case of the pure *e-i* continuum, the plasma parameters are found from a comparison between experimental and calculated emission coefficients. To derive the latter, it is time saving to use published data for the emission-respectively absorption coefficient of the *e-i* and *e-a* continuum, provided that such datas for the elements under consideration do exist (see the corresponding tables at the end of this chapter).

Finally, we notice the possibility of temperature determination from the ratio of continuous emission coefficient to the emission coefficient of an atomic spectral line. This method bases on the fact that (at least for a simple plasma) the ratio of the continuous emission coefficient, given for example by (9.28) to the emission coefficient of a neutral line, given for example by (5.33), is in first approximation a function of temperature only, provided that continuous emission is due to electron-ion interaction alone. The dependence of this ratio upon electron density N_e is only weak. This is demonstrated in figure 9.12 showing the temperature dependence of the intensity ratio H_β/H_{cont} and H_α/H_{cont} with N_e as parameter. The curves are taken from Griem [9.26] and refer to continuous wavelength bands of widths 100 Å positioned in the neighbourhood of the lines H_β respectively H_α.

Some examples for the application of plasmadiagnostical methods referred in this section are listed in the tables 9.9—9.11 at the end of this chapter.

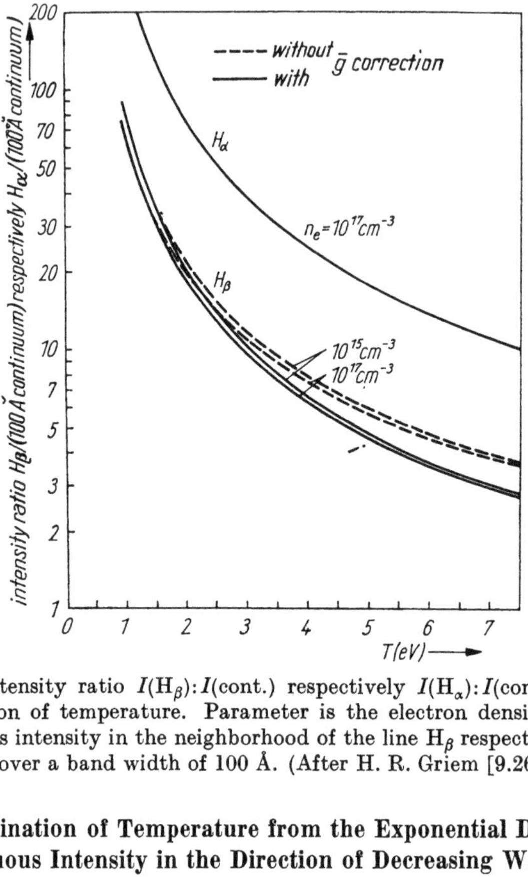

Figure 9.12. Intensity ratio $I(H_\beta):I(\text{cont.})$ respectively $I(H_\alpha):I(\text{cont.})$ for a hydrogen plasma as function of temperature. Parameter is the electron density. $I(\text{cont.})$ stands for the continuous intensity in the neighborhood of the line H_β respectively H_α integrated over a band width of 100 Å. (After H. R. Griem [9.26])

9.5.2. Determination of Temperature from the Exponential Decay of Continuous Intensity in the Direction of Decreasing Wavelength

This method makes use of the fact that behind the well isolated photoabsorption edges the continuous emission coefficient of the electron-ion continuum decays with increasing frequency according to the relation $\varepsilon_\nu \sim e^{-h\nu/(kT)}$ (see equations (9.16a) and (9.19a) derived for a hydrogen like plasma). For non hydrogenic plasmas, this relation is correct only approximately. The approximation becomes the better the higher the charge number Z of the plasma ion [9.24h] (that means, the exponential law can be well applied in the case of high temperature plasmas). Plotting ε_ν as function of ν semi-logarithmically, the corresponding diagram points should follow a straight line given by the relation

$$\lg \varepsilon_\nu = \text{const} - \frac{h\,\nu}{2.303\,kT}\,. \qquad (9.37)$$

In figure 9.13 is shown the continuous spectral intensity distribution in the region of soft X-rays as measured by Bearden et al. [9.3]. The continuum was

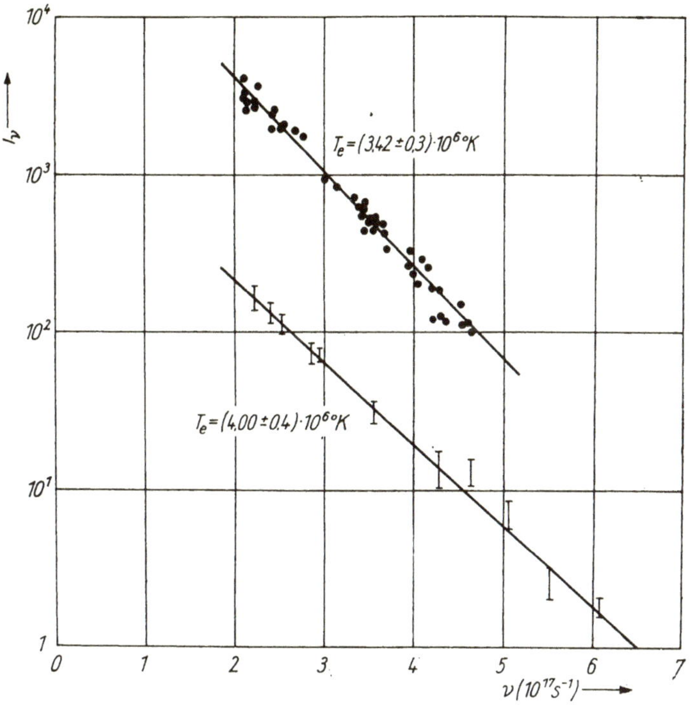

Figure 9.13. Decay of continuous intensity with increasing frequency in the soft X-ray region (after Bearden et al. [9.3])

The upper curve refers to a deuterium plasma with 6% admixture of O_2, the lower curve refers to a pure deuterium plasma

emitted from a deuterium plasma heated in a theta-pinch device of type Scylla. From the steepness of the exponential decay of continuous intensity toward short wavelengths the electron temperature was derived yielding the values $(3.4 \pm 0.3) \cdot 10^6$ °K for a deuterium plasma containing 6% O_2 and $(4.0 \pm 0.4) \times 10^6$ °K for the pure deuterium plasma.

Further examples for temperature determination from the exponential decay of continuous intensity toward short wavelength are listed in table 9.10 at the end of this chapter.

For technical details concerning soft X-ray diagnostics see [9.1g], [9.8] and [9.26a].

Table 9.7. References concerning theoretically calculated continuous emission-or absorption coefficients. Meaning of symbols: *ea* = electron-atom continuum, *ei* = electron-ion continuum, *ff* = free-free continuum, *fb* = free-bound continuum, *tot.* = total radiation (continuous radiation integrated over all frequencies)

System	Kind of continuum	Overall parameter range		Reference
		λ [μm]	T [10^3 °K]	
Air	*ea* + *ei*	0.14 – 1.12	6 – 13	1p
	ea	0.16 – 10	6 – 9	12a
	ea(ff)	0.2 – 4.9	3 – 5	36g, 50b.
Al	*ei*	0.5 – → ∞	4 – 12	5c
	ei(ff)	0.3 – 1.8	4 – 13	60e
Ar	*ei*	0.1 – → ∞	8 – 19	5c, 5d, 64a, 64b, 65a.
	ea(ff)	0.7		1u[1])
Br	*ea(fb)*			61k[2])
C	*ei*	0.062 – → ∞	3 – 30	1m, 5c, 5d, 13a, 60g, 75c.
	ei(fb)	0.275 – 0.99		1n[3]), 18m.
	ei(ff)	0.3 – 1.8	4 – 13	60e
	ea(fb)	0 – 1.1		11, 61k[2]).
C^+	*ei*	0.062 – 6	3 – 30	13a, 75c.
Ca^+	*ei*	0.09 – 0.41	7 – 12	60g
	ei(ff)	0.075 – 0.45	4 – 13	60e
Cl	*ei*	0.085 – 0.57	10.5 – 13	60g
	ei(ff)	0.3 – 1.8	4 – 13	60e
	ea(fb)	0.0275 – 0.51	10.5 – 12	18m, 36b, 61k[2]).
	ea(ff)	0.38 – 4.56	2.1 – 12.6	36b
Cs	*ei*	0.3 – → ∞		5d
	ei(fb)		1 – 4	1, 57a.
D	*ei*	0.50	11.6 – 1160	28a
F	*ei*	0.6328	30	13a
	ea(fb)	0.0275 – 0.35		18m, 61k[2]).
F^+	*ei*	0.6328	30	13a
H	*ei*	0.092 – → ∞	3.5 – 100	1m, 13a, 20d, 59a, 72, 73.
	ei, tot.		10 – 3500	37, 73.
	ei(fb)	0.2 – 0.65		1n
	ei(ff)	$3\cdot10^{-5}$ – 800	4 – 10^5	5, 39, 60e.
	ea	0.15 – 18.23	2.5 – 17.5	2, 34, 63.
	ea(fb)	0 – 1.65		3b, 15, 17, 18a, 18s, 18u, 23, 58a, 68b, 73[4])
	ea(fb)[5])	0.70 – 1.64		74a, 74b.
	ea(ff)	0.4 – → ∞	1.8 – 12.6	18t, 23a, 35a, 58, 59.
	ea + *ei*	0.13 – 0.4	5.8 – 35	20d, 26, 48, 71j[6]).
He	*ei*	0.06 – → ∞	8 – 1000	5c, 5d, 38a, 50e, 60d, 60i, 64e.
	ei(fb)	0.2 – 0.65		1n[3])
He^+	*ei(ff)*	0.075 – 0.45	4 – 13	60e
Hg	*ei*	0.2 – → ∞		5c, 5d.
	ea	0.26 – 1.00	6.0 – 8.2	7
I	*ea(fb)*			61k[2])
K	*ei*	0.3 – → ∞		5d
	ei(ff)	0.3 – 1.8	4 – 13	60e
r	*ei*	0.2 → ∞	8 – 16	5c, 5d, 64a, 64b.

Table 9.7. (Continuation)

System	Kind of continuum	Overall parameter range λ [μm]	T [10^3 °K]	Reference
Li	ei	0.18 — → ∞	2.5 — 20	5c, 5d, 38a, 50e, 76a.
	$ei(ff)$	0.3 — 1.8	4 — 13	60e
	$ea(fb)$	0 — 3.23		23, 68b.
	$ea(ff)$	0.3 — 2	5 — 20	76b
Mg	ei	0.1 — → ∞		5c, 5d, 60d, 60g.
	$ei(ff)$	0.3 — 1.8	4 — 13	60e
Mg^+	ei	0.075 — 0.39	8 — 13	60g
	$ei(ff)$	0.075 — 0.45	4 — 13	60e
N	ei	0.0012 — → ∞	3 — 230	1m, 5b, 5c, 5d, 60d, 60g, 60h, 65a, 75c.
	$ei(fb)$	0.2 — 0.65		1n³)
	$ei(ff)$	0.3 — 1200	3.5 — 29	60e, 72b.
	ea	0.4 — 1.0		36f
	$ea(ff)$	0.247 — → ∞	3.5 — 29	12, 50c, 72b.
	$ei + ea$	0.4955	9 — 18	50n
N^+	ei	0.06 — 5	3 — 24	75c
Na	ei	0.3 — → ∞		5d
	$ei(ff)$	0.3 — 1.8	4 — 13	60e
Ne	ei	0.2 — → ∞	8 — 16	5c, 5d, 64b.
O	ei	0.06 — → ∞	3 — 30	1m, 5b, 5c, 5d, 13a, 60d, 60g, 75c.
	$ei(fb)$	0.2 — 0.65		1n³)
	$ei(ff)$	0.3 — 1.8	4 — 13	60e
	ea	0.04 — 0.5		36f
	$ea(fb)$	0.0275 — 0.86		11a, 18i, 18m, 61k.
	$ea(ff)$	0.274 — → ∞	4 — 8	11b, 12, 50c.
	$ei + ea$	0.4348	9 — 18	50n
O^+	ei	0.06 — → ∞	3 — 30	5c, 5d, 13a, 75c.
O^{++} } O^{6+}	ei	0.3 — → ∞		5c, 5d
Rb	ei	0.3 — → ∞		5d
S	$ea(fb)$			61k²)
Si	ei	0.13 — → ∞	4 — 9	5c, 5d, 60g.
	$ei(ff)$	0.3 — 1.8	4 — 13	60e
	$ea(fb)$			61k²)
Xe	ei	0.2 — → ∞		5c, 5d, 35, 61q, 64a, 64b.

¹) Given are cross sections for electron energies between $U_e = 0.14$ and $U_e = 13.6$ eV.
²) Given are detachment cross sections for the electron energy range 0.07 — 12.1 eV above threshold.
³) Given are Gaunt factors for individual recombination continua.
⁴) Data are given for $p = 1000$ atm.
⁵) Forbidden and allowed continuum for $N_e = 4 \cdot 10^{13}$ cm^{-3}.
⁶) Included is the line wing contribution to the apparent continuum.

Table 9.8. Bibliography about experimental data for the continuous emission coefficient ε_ν respectively absorption coefficient $\varkappa_\nu$ as function of temperature and frequency. The meanings of the symbols used in column 3 are: *ea* = electron-atom continuum, *ei* = electron-ion continuum, *p. d.* = pulse discharge, *sh* = shock wave, *pl* = plasma jet, Θ = Theta pinch

Species	Kind of continuum	Method of heating	Overall parameter range				Reference
			λ [μm]	T [$10^3 \cdot K$]	p [atm]	N_e [10^{16} cm^{-3}]	
Air	*ei*	*p. d.*	0.11 – 0.48	20 – 36		0.11 – 50	25b, 56a.
	ea(ff)	*sh*	2 – 8	6 – 8	[1])		1i, 67a.
	ea + ei	*sh*	0.50 – 0.51	9.65	[2])		1k
Ar	*ei*	*sh*	0.24 – 0.60	5.8 – 17.5	0.5 – 3		1l, 4b, 50a.
	ei	*pl*	0.32 – 330	10 – 12.3	1	6.5 – 10	18d, 49a, 61.
	ei	*arc*	0.15 – 11	9.5 – 19	1 – 5	0.6 – 19	1s, 20l, 27b, 27c, 60, 63a, 64, 64c, 65a, 74c, 74d.
	ei	*p. d.*	0.25 – 0.40	34	6		18h
	ei(ff)	*pl*	50 – 170	13 – 14	1	11 – 15	18c
	ea(ff)	*arc*	0.35 – 0.75	5.4		3.4 10^{-3}	2a
	ea + ei	*sh*	2 – 5	8 – 11			67b
	ea + ei	*arc*	0.07 – 3.7	9 – 13.6	1 – 2		50p
Br	*ea(fb)*	*sh*	0.3 – 0.4	6 – 6.5			61m
		arc	0.263 – 0.37		1		21b
C	*ea(fb) + ei*	*arc*	0.3 – 0.58	12.5 – 14.5	1		29a
C, H, O[3])	*ei*	*p. d.*	0.2 – 0.6	39			57b
C, N, O[4])	*ei*	*arc*	0.22 – 0.67	10.9	1		45
Cl	*ea*	*arc*	0.24 – 0.46	6.6 – 8.3	1	0.06 – 0.7	50s
	ea(fb)	*sh*	0.3 – 0.4	6 – 6.5			61m
	ea(fb)	*arc*	0.26 – 0.34		1		21b
	ea + ei	*arc*	0.28 – 0.54	10.5 – 12.0	1		29
Cs	*ei(fb)*	*oven*	0.4 – 0.5				1a
F	*ea(fb)*	*arc*	0.26 – 0.36	7.5 – 8.8	1		21b, 61e, 61f.
H	*ei(ff)*	Θ	0.25 – 25	10^3	≈30		8a
	ea	*sh*	0.4 – 0.62	8.2		3	8c
	ea + ei	*sh*	0.39 – 0.57	8.6	29	4.8	73a
	ea + ei	*arc*	0.3 – 0.6	9.4 – 12.8	0.4 – 1		43, 75a.[5])

Table 9.8. (Continuation)

Species	Kind of continuum	Method of heating	Overall parameter range				Reference
			λ [µm]	T [$10^3 \cdot$ K]	p [atm]	N_e [10^{16} cm^{-3}]	
H + Kr	*ei* + *ea*(*H*)	*sh*	0.39 – 0.57	8.2 – 10.5	≈30		74
He	*ei*	*sh*	0.28 – 0.54	41.4 – 46.2	≈ 1	11 – 77	44
Hg	*ei*	*arc*	0.29 – 0.66	7.8	≈60		24
I	*ea*(*fb*)	*sh*	0.3 – 0.4	6 – 6.5			61 m
Kr	*ei*	*sh*	0.24 – 0.66	5.8 – 17.5		2 – 13	11, 4 b, 20, 50 a, 61 h.
Li^{++}	*ei*(*fb*)	*p. d.*	0.004 – 0.01	230		500	18 j
Mg	*ei*(*fb*)	*arc*	0.24 – 0.26			≈ 1	10 c
N	*ei*	*arc*	0.3 – 0.9	11 – 13	1	3.4 – 6.5	44 a, 50 h
		p. d.	0.11 – 0.60	20 – 25		3 – 50	25 a, 25 b
	ei(*fb*)	*sh*	0.07 – 10	9.65			1 k
		arc	0.09 – 0.117	13.2	1		50 k
	ea	*sh*	2 – 5.4	8.4 – 9.1			67 c
	ea(*fb*)	*sh*	0.5	9.25	[6]		1 j
	ea(*ff*)	*sh*	2 – 6	6.2 – 9.0	25 – 38	[7]	1 i, 61 d, 67 a
	ea + *ei*	*sh*	0.3 – 0.8	11 – 18		3.1 – 11	27 d
		arc	0.07 – 3.7	9 – 16	1 – 2		1 t, 10, 50 i, 50 j, 50 n, 50 p, 50 q
Na	*ei*(*fb*)	*sh*	0.3 – 0.41	5.7 – 10			61 l
Ne	*ei*	*arc*	0.26 – 0.65	14.7	1.5	3.8	64 c
	ea + *ei*	*sh*	1.8 – 5.4	9 – 15			67 b
O	*ei*	*p. d.*	0.11 – 0.6	20 – 25		3 – 50	25 a, 25 b.
	ea	*sh*	2 – 5.4	9.7 – 9.8			67 c
	ei + *ea*	*arc*	0.3 – 1	9 – 14	1		9, 50 n, 50 o.
SF_6	*ea*	*arc*	0.27 – 0.68	7.5 – 8.8	1		61 f
Xe	*ei*	*sh*	0.29 – 0.66	5.8 – 17.5		0.02 – 17	11, 4 b, 20, 50 a.
		arc	0.2 – 1.8	7.7 – 11.0	16 – 24		36 h, 62.
		p. d.	0.25 – 1.0	10.5 – 14.5	0.6		1 v, 18 h.
	ea + *ei*	*sh*	2 – 5	7.6 – 9.1			67 b

[1]) $\varrho/\varrho_0 = 0.9$.
[2]) $\varrho/\varrho_0 = 10^{-3} - 0.4$.
[3]) Capillary discharge through textolite-vapour = 47% H + 37% C + 16% O.
[4]) Carbon arc in air.
[5]) Line wing contribution to continuum is considered.
[6]) $\varrho/\varrho_0 = (2-20) \cdot 10^{-3}$.
[7]) $\varrho/\varrho_0 = 0.5 - 1$.

Table 9.9. Determination of electron temperature and electron density from absolute values of continuous emission coefficient (if not otherwise stated, the $e-i$ continuum is considered). Meanings of abbreviations: *dec.* = decaying plasma, *l. f.* = laser focus, *p. d.* = pulse discharge, *pl* = plasma jet, *rf* = *rf*-discharge, *sh* = shock wave, Θ = Theta-pinch, ζ = Zeta design

Medium	Method of heating	Quantity to be measured	Overall parameter range λ [μm]	T [10^3 °K]	p [atm]	N_e [10^{16} cm^{-3}]	Reference
Ar	*pl*	T_e	0.40 – 0.70	7 – 14	0.01 – 1	0.02 – 19	10a, 19a, 43a, 50r, 52, 54, 54a.
		N_e	0.45	3 – 5¹)	< 1	0.002 – 1	24c
	rf	T_e, N_e	0.45 – 0.57	8.7 – 9.7	0.2 – 1	0.3 – 1.2	24b, 61n.
	arc	$T_e(N_e)$	0.40 – 0.554	7 – 19	1 – 5	0.03 – 21	2b, 10b, 24a, 53, 60a, 61i, 77a.
	arc	T_e	2.5 – 4.85	12 – 28	1 – 15		27a
	arc	N_e	0.42 – 0.80	≤ 8.3	1	0.05 – 0.3	1d
C²)	*l. f.*	N_e	0.002 – 0.6	52 – 1050		7 – 1000	8e
Cs	diode	N_e	0.40 – 0.52	1.8 – 8	< 1	0.0007 – 0.36	18r, 20a, 50g.
D	Θ	N_e	0.35 – 0.65	9 – 29000		0.008 – 40	1a, 27, 28a, 32, 38.
	ζ	N_e	100 – 2000	20 – 100		0.01 – 0.5	28
H	*dec.*	N_e	0.5320			0.15 – 0.35	181
	sh	N_e	0.545	28		20 –	75
	p. d.	N_e	0.2 – 0.6	8 – 80		0.2 – 2500	8d, 23b, 77c.
		T_e³)	0.226 – 0.347	13 – 20		5 – 10	77
	Θ	N_e	0.49 – 0.54	810 – 1630		2 – 40	8b, 20c, 23b.
H_2O	*arc*	T_e³)	0.30 – 0.56	12 – 26	1	≈8	24c, 56a.
He	*sh*	N_e	0.39 – 0.59	20 – 49		0.8 – 30	1q, 4, 31.
	p. d.	N_e	0.30 – 0.50	≈ 60		120 – 1000	47
	Θ	N_e	0.32 – 0.65	10 – 115		10 – 2800	4c, 21c, 77b.
		T_e³)	0.35 – 0.65	115		2800	77b
LiH	*p. d.*	N_e	0.01 – 3.2	230 – 490		280 – 1500	18j, 18k.
N	*arc*	$T_e(N_e)$	0.496	8 – 14	1		77a
	arc	$T_e(N_e)$⁴)	0.515	6 – 6.5	1	0.01 – 0.04	24a
SF_6	*arc*	N_e	red	6 – 11	1	1.5 – 3	44b
Xe	*sh*	N_e	0.405 – 0.604			6.6 – 22	18q
	rf	$T_e(N_e)$	0.496 – 0.530	7 – 7.8	0.31 – 1	0.5 – 2.0	61n
	arc	N_e	0.444 – 0.516	7.4 – 10.8	≈10	4.3 – 83.1	51

¹) Recombining plasma; T_e is lower than the equilibrium value $T_e(N_e)$

²) Laser focus on polyethylene; see also table 9.10.

³) The norm temperature method applied to continuous radiation is used. See section 5.2.6.

⁴) The total continuous radiation ($ea + ei$ continuum) is considered.

9.5.3. Determination of Electron Density from the Long Wavelength Cutoff of Infrared Bremsstrahlung

As has been shown by Kimmitt and Niblett [9.36e], it is possible to determine peak values of electron density in a cylindrical plasma column from the long wavelength cut off of continuous infrared radiation. When plotted as function of frequency ν, the spectral intensity I_ν rises up in the direction from low to higher frequency values and becomes practical constant for $\nu \gtrsim \nu_p$. From the position of the corresponding edge, the plasma frequency ν_p $(\approx 9 \cdot 10^3 (N_e/\mathrm{cm}^{-3})^{1/2}\,\mathrm{s}^{-1})$ in the center of the plasma and hence the electron density N_e ($r = 0$) is obtained.

As object of investigation the authors used a deuterium plasma produced in a theta pinch device. The peak values for electron density and electron temperature are in the range of $N_e = (2.8-5.0) \cdot 10^{16}$ cm^{-3} and $T_e = (1.1-2.3) \times 10^6$ °K.

9.5.4. Determination of Electrical Conductivity from the Emission Coefficient of Continuous Infrared Radiation

Excluding continuous molecular emission, then in the far infrared region with $h\nu \ll kT$, the main part of continuous radiation is due to bremsstrahlung of the free electrons. The corresponding absorption coefficient can be obtained from the dispersion theory. At first, we write down the formula for the complex refractive index (see [9.18c], [9.18d])

$$n^2 = 1 - \left(1 - j\,\frac{\alpha}{\omega}\right) \cdot \frac{\omega_p^2}{\omega^2 + \alpha^2} \tag{9.38}$$

in which ω_p is the electron plasma frequency ($= 5{\cdot}64 \cdot 10^4 (N_e/\mathrm{cm}^{-3})^{1/2}$ radian/s) and α is the electron collision frequency given in s^{-1}.

In the case $\omega > \omega_p > \alpha$, the damping of radiation is described by the relation

$$I \approx I_0\,\mathrm{e}^{-\frac{\alpha}{c}(\omega_p/\omega)^2 x} = I_0\,\mathrm{e}^{-\varkappa x}$$

giving for the absorption coefficient $\varkappa$ the expression

$$\varkappa = \frac{\alpha}{c}\,(\omega_p/\omega)^2 = \frac{4\,\pi\,e^2\,\alpha\,N_e}{c\,m_e\,\omega^2}\,.$$

Taking into account Kirchhoff's radiation law in the form

$$\varepsilon_\nu = \varkappa_\nu\,\frac{2\,\nu^2}{c^2}\,kT$$

we obtain for the spectral emission coefficient

$$\varepsilon_\nu = \frac{2}{\pi}\,\frac{e^2\,\alpha\,N_e\,kT}{m_e\,c^3}\,. \tag{9.39}$$

Substituting into this formula the expression for the electrical conductivity

$$\sigma = \frac{e^2\,N_e}{m_e\,\alpha} \tag{9.40}$$

we get the resulting formula

$$\varepsilon_\nu = \frac{2}{\pi}\,\frac{e^4}{m_e^2\,c^3}\,\frac{N_e^2}{\sigma}\,(kT_e) \tag{9.41}$$

which is useful for the evaluation of the electrical conductivity σ of a plasma for which the infrared continuous emission coefficient ε_ν has been measured.

For comments about the reliability of the relations used above see L. Oster [9.60c].

For further details concerning the correlation between *IR* emission and electrical conductivity see also [9.20f], [9.68a].

9.5.5. Reference Tables Useful for the Determination of Plasma Parameters from Continuous Radiation

The references listed in table 9.7 contain theoretically evaluated data for the continuous emission coefficient ε_ν respectively absorption coefficient $\varkappa_\nu$, while the papers cited in table 9.8 inform about experimental values for ε_ν respectively $\varkappa_\nu$. References about photoionization cross sections are in general not included in these tables but can be found in table 6.5 of chapter 6.

Table 9.9 informs about the determination of plasma parameters from absolute emission coefficients while tables 9.10 and 9.11 are concerned with the determination of electron temperature from relative continuous emission coefficients respectively from the reatio of line to continuous intensity. To save space, the literature is given in short symbols, for example 1 p instead of [9.1 p].

Table 9.10. Determination of electron temperature from relative continuous intensity, especially from the decay of intensity toward short wavelengths behind the photoionization edges. Abbreviations: see head of table 9.9 and: ex. wi. = exploding wire, *l. sp.* = laser spark, *prom.* = prominence

Medium	Method of heating	Overall parameter range			Reference
		λ [Å]	T [10^3 °K]	N_e [10^{16} cm^{-3}]	
Air	*l. sp.*	0.4– 15	600– 2030	≈200	1c, 47a.
Al	*l. sp.*	15– 50	290– 380		64d
C[1])	*l. sp.*	20– 40	52– 1160	7–1000	8e, 8f.
Cs	diode	4000–5200	1.8– 4.6	(4–110) 10^{-4}	1h, 20a, 24e, 50g.
Cu	*ex. wi.*	0.6– 12	21000		72a
D	Θ	3.8– 15	2800–29000	5–7.2	3, 27, 32, 33, 61p, 65.
D+ trace O	Θ	6– 30	1460– 3400	≈5	3, 24h, 33, 61p, 65.
H	*dec.*	3200–5300	5– 30		181
	prom.	3646 ± $\Delta\lambda$	≈6		60b
	p. d.	2300–4300	8– 23	0.2–10	23b, 57b.
	Θ	5– 14	810– 1630	11 –13	8b
	[2])	3550–5005	12– 200		64f
H_2O	arc	3000–3800	13.3	[3])	36
	p. d.	3800–4400	20– 160	[4])	38c
Ne	*l. sp.*	4– 15	900– 1070		1c

[1]) Polyethylene plasma.
[2]) Tabulation of continuous intensity ratio $i(3550):i(5005)$ across the balmer series limit.
[3]) $p = 1$ atm.
[4]) $p = 4 - 65$ atm.

Table 9.11. Determination of electron temperature T_e from line: continuum intensity ratio
Abbreviations: see head of table 9.9

Medium	Method of heating	Wavelength in [Å] of line	Wavelength in [Å] of continuum	Overall parameter range T [10^3 °K]	Overall parameter range N_e [10^{16} cm^{-3}]	Reference
D	*p. d.*	4861	4970–5000	12.7– 58	0.13 – 0.9	1a, 28a.
	Θ	4861	4970	9 – 40	0.08 – 0.5	1a
	[1]	4861	5000	11.6–116		28a
H	*dec.*	4861	5305–5325	5 – 30		181
	dec.	6563	adjacent	15 –200	10 –1000	42a
	sh	4861	adjacent	14 – 25	8 – 25.4	4, 50f.
	arc	4861	5200–6500	8.2– 14.5	0.02 –340	18p, 57.
	p. d.	4861	3550–4256	8 – 23	0.2 – 1	23b
	[1] [2]	4861	3550; 4526	5.8–116	0.03 –300	20d
	[1] [2]	6563	3550; 4526	5.8–116	0.03 –300	20d
He	*sh*	I 4471, II 4686	5015; 5876	48	13 – 20	18n
	sh	I 3889, 4471, 4713, 5016	adjacent	20 – 49	12.2 – 16.5	4
	[1]	II 4686	adjacent	27 –100	5 – 30	4a

[1]) Given is a tabulation of the ratio line: continuum intensity as function of temperature.

[2]) A numerical procedure for the evaluation of T and N_e from the intensity ratio is also given in [9.43b].

CHAPTER 10

Determination of Plasma Parameters from Line Profiles. Fading of Spectral Lines at the Series Limit and the Problem of the Reduction in the Ionization Energy

Below we shall describe measurement techniques, by means of which translational temperatures, flow velocities, particle densities (in particular, electron densities), as well as magnetic and electric field strengths can be determined.

The translational temperature as well as the directional velocity of the plasma components emitting the relevant spectral lines are obtained from the Doppler broadening and Doppler shift. The broadening and shift caused by particle interaction provide a measure for finding the particle densities in the plasma, and in particular the electron density.

As a rule both types of line broadenings and line shifts occur simultaneously in addition to natural broadening due to radiation damping.

Upon analyzing the measurement data the different factors causing the broadening must be separated from one another. The Doppler profile can be readily separated from the dispersion profile, since the former is concentrated about the line center (see equation (2.51)), whereas the dispersion profile is predominant in the wings. Separation, as a rule, is effected through approximation of the observed profile by a Voigt profile (see section 2.9.2).

Line broadenings which possess similar profiles cannot be resolved. For example, it is impossible to directly resolve the Doppler broadening into a turbulent component and a thermal component. Further more, without additional information the natural line width cannot be distinguished from the broadening caused by the interaction of the charge carriers or the neutral particles.

More accurate determinations of plasma parameters from the line profile are, as a rule, only possible when the type of line broadening, which is to be measured, is the predominant component of the total line profile.

From the above it is evident that when analyzing the experimentally obtained line profiles several factors causing the broadening must be simultaneously allowed for. In the favorable case this consists of establishing the smallness of one broadening factor with regard to another.

Thus, in addition to the Doppler broadening and Doppler shift as well as the broadening and shift due to particle interaction, all of which are of special interest in plasma diagnostics, we must still consider the natural line broadening, which has only indirect significance in plasma diagnostics, since in isolated cases it enables transition probabilities to be determined.

The technique for recording the line profile and the method for resolving the profile were already treated in Chapter 2; we thus need only consider at the most a few cases of specific measuring systems for determining the line profile.

Finally, we shall discuss the fading of spectral lines at the series limits. This is of definite significance for the determination of charge carrier densities in rarefied plasmas. In connection with this we shall touch upon the problem of the reduction in the ionization energy.

10.1. Doppler Broadening

When the emitting atoms or ions move along the line of sight with disorderly velocities the spectral line is broadened by the Doppler effect. We assume a Maxwell distribution of the velocities of the emitting particles in the direction of observation (x-direction) (see Wulff [10.180]) and obtain for the relative portion of atoms or ions (in what follows we shall use for the sake of simplicity only the term "atom") $\mathrm{d}N/N$ in the velocity range v_x

$$\frac{\mathrm{d}N}{N} = \frac{1}{\sqrt{\pi}} \exp\left(-\frac{1/2\, M\, v_x^2}{kT}\right) \sqrt{\frac{M}{kT}}\, \mathrm{d}v_x \,. \qquad (10.1)$$

To obtain the Doppler profile we make use of the expression for the Doppler shift

$$\frac{\Delta\lambda}{\lambda} = \frac{\Delta\nu}{\nu} = \frac{v_x}{c}, \qquad (10.2)$$

where c denotes the light speed and $\Delta\lambda$ or $\Delta\nu$ denotes the line shift due to the movement of the atoms having velocity v_x. With

$$\frac{\Delta\lambda_{\mathrm{D}}}{\lambda} = \frac{\Delta\nu_{\mathrm{D}}}{\nu} = \frac{1}{c}\sqrt{\frac{2\,kT}{M}} = \frac{1}{c}\, v_w\,, \qquad (10.3)$$

where v_w is the most probable velocity and M is the atomic mass, we have the following expression for the Doppler profile

$$\frac{\mathrm{d}I}{I} = \frac{1}{\sqrt{\pi}}\, \mathrm{e}^{-(\Delta\lambda/\Delta\lambda_{\mathrm{D}})^2}\, \mathrm{d}\lambda/(\Delta\lambda_{\mathrm{D}}) = \frac{1}{\sqrt{\pi}}\, \mathrm{e}^{-(\Delta\nu/\Delta\nu_{\mathrm{D}})^2 \mathrm{d}\nu/\Delta\nu_{\mathrm{D}}}\,. \qquad (10.4)$$

The total half-width of the line displaying Doppler-broadening is given by

$$2\,\Delta'\lambda = 2\sqrt{\ln 2}\,\Delta\lambda_{\mathrm{D}} = 1.67\,\Delta\lambda_{\mathrm{D}} \quad \text{respectively} \quad 2\,\Delta'\nu = 1.67\,\Delta\nu_{\mathrm{D}} \qquad (10.5)$$

or

$$2\,\Delta'\lambda = 7.17 \cdot 10^{-7}\,\lambda\sqrt{\frac{T}{M^*}} \qquad (10.6)$$

(M^* = molecular weight, $2\,\Delta'\lambda$ and λ in Å; T in °K) .

From (10.6) it follows that the Doppler width increases with increasing temperature T and with decreasing atomic weight of the radiating atom.

It must be mentioned that formula (10.4) becomes incorrect when $\Delta\omega_D$ is much lower than the gaskinetic collision frequency ν_c, or again when the wavelength λ is much in excess of $2\pi l$, where l is the gaskinetic mean free path. In this case, the actual profile is narrower than that indicated by formula (10.6) [(10.166a] p. 73). In actual plasmas, deviations from (10.4) are only expected for wavelengths in the microwave region. More information on this problem can be found in [10.39b] and [10.138i].

Depending on the magnitude of the Doppler broadening different recording techniques are used. For broadenings below 1/10 Å in the optical spectral range in general interferometric devices are employed for measuring the line profile (see section 2.1.3). These devices can also be used for determining widths of several tenths of Å. For line profiles above $\sim$0.2 Å a grating spectrograph can be employed to resolve the profile, and line widths above 1 Å can be measured by means of prism spectrographs with a higher resolution.

There are also in use indirect methods of line profile determination. Patel [10.134a] for example has determined the Doppler profile of the term inverted transition Xe I $\lambda = 2.026\,\mu$m originating from a cylindrical Xe-He-plasma, by recording the optical gain as function of wavelength. The probing line which was emitted by a tunable Xe-He-laser source, was much narrower than the line under investigation.

Burhorn [10.21a] employed a Fabry-Perot interferometer to determine the profile of selected spectral lines of iron, which are emitted in a high-pressure arc at a temperature of $\sim$6000 °K and are about $3-4 \cdot 10^{-2}$ Å wide. The fact that the Doppler width is not overlapped by the broadening caused by the particle interaction, as is usually the case with high-pressure plasmas, is due to the fact that the relevant spectral lines originate from electron transitions between levels in an inner shell, which are shielded extensively by the outer electrons against disturbances produced by electron collisions.

Here it should be mentioned that such lines not only occur in iron-like atoms. The oxygen ion O II also emits lines (O II 4649 Å and 4705 Å, see Durand [10.27]), which up to an electron density of $3 \cdot 10^{16}$ cm^{-3} possess a Gaussian profile.

In short-lived plasmas a high-speed recording of the line profile can be effected by means of an oscillating Fabry-Perot interferometer (see section 2.5.4.4). For moderate registration speeds a Fabry-Perot interferometer coupled with a rotating glass square can also be used for time resolving [10.39c].

When the plasma to be investigated is inhomogeneous with respect to the line of observation, then the resulting profile of the spectral line displaying Doppler-broadening will be a superposition of Doppler profiles with different half widths. In cases of rotational symmetry, the original profiles belonging to the various radial coordinates within the plasma can be found by applying the Abel inversion technique, see [10.103a] and also section 2.10.1.

For the measurement of the Doppler width in high-temperature plasmas a grating spectrograph or grating spectrometer is suitable. In the case of stationary plasmas it is advantageous to record the line profile photoelectrically in conjunction with a high-precision continuous line-recording device. Stationary plasmas occur, for example, in a DCX-system (magnetically contained low-pressure high-current arc) and were investigated by McNally and Skidmore [10.107], who indicated the existence of a Gaussian line profile.

More accurate measurements gave the result that the temperature measured via the Doppler width is larger for observations taken normal to the direction of the magnetic field than for those taken along the direction of the magnetic field lines.

In high-temperature plasmas occurs the phenomenon that heavily charged ions possess a higher ion temperature than weakly charged ones (see, e.g., [10.107] and [10.184]). This can be explained by the fact that the electrostatic fields present in the plasma accelerate the more heavily charged ions more

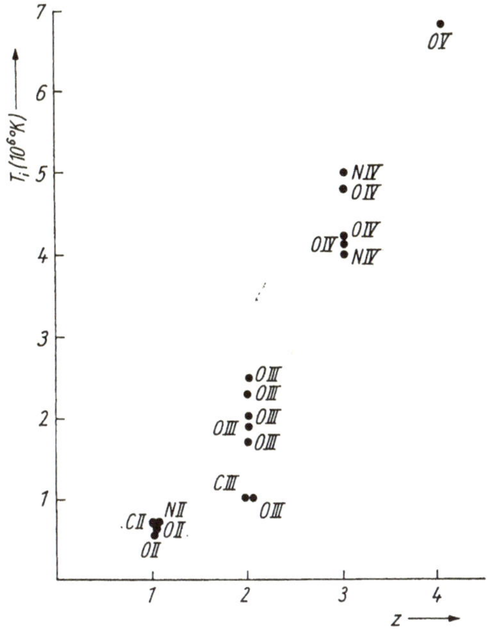

Figure 10.1. Dependence of the ion temperature T_i on the ionic charge z for a high-temperature plasma in the "alpha" toroidal-pinch system (after Zaidel' et al. [10.184])

strongly. Figure 10.1 shows the dependence of the ion temperature T_i on the ion charge number Z for a plasma produced in a pulsed toroidal pinch. The lowest temperature is exhibited by the O II ions with $T_i \approx 0.5 \cdot 10^6$ °K, while the O V ions have a temperature of about $7 \cdot 10^6$ °K.

The Doppler shift is not only observed in the visible spectral range but also in the range of soft X-ray radiation. Sawyer et al. [10.146] recorded the spectrum of a theta pinch in the range from 15 to 25 Å with a crystal spectrometer. The system was filled with deuterium. The maximum electron density amounted to $5 \cdot 10^{16}$ cm^{-3}. The profiles of the impurity lines of O VII, O VIII, and Ne IX were recorded continuously by revolving the crystal. A scintillator with multiplier served as the radiation detector.

Values between 1.08 and $1.15 \cdot 10^8$ °K were obtained for the Doppler temperature. A dependence of the Doppler temperature on the ion species was also established, whereby in addition reference was made to the Doppler temperature of C V and the nonspectroscopically determined Doppler temperature of D II in the ultraviolett spectral range. In fact, there results a linear relationship between the Doppler temperature and the ionic mass in the sense that all ions have the same mean velocity of $3 \cdot 10^7$ cm/s.

In magnetically contained plasmas it should be remembered that the Doppler effect is overlapped by the Zeeman effect. If the observations are taken normal to the magnetic field direction, then in the case of the normal Zeeman effect there result an unshifted π-component of the spectral line and 2 shifted σ-components. By employing a polarizer, which only transmits the π-direction, McNally and Skidmore [10.107] succeeded in freeing the line width from the influence of the Zeeman effect.

In astrophysics the Doppler broadening is utilized for measuring the corona temperature. On the basis of measurements carried out by different authors, especially on the red corona line Fe X 6374 Å, Unsöld [(10.170], p. 651) stated the mean corona temperature as $2.1 \cdot 10^6$ °K.

Unsöld ([10.170], p. 286) pointed out that the occurrence of turbulence can produce a Gaussian line profile as well, which together with the profile originating from the thermal Doppler effect again gives a Gaussian profile with a half-width equal to the square root of the sum of the square of the individual half-widths (see section 2.9.2). The two effects cannot be separated without taking additional measurements.

Another source of Doppler broadening is found in decaying plasmas when the decay is effected by dissociative recombination, as has been shown by Rogers and Biondi [10.139c] for a helium-plasma. The relevant recombination process is $He_2^+ + e \rightarrow He^* + He +$ recoil energy E_D. The excited atoms stemming from this process have in addition to their thermal velocity an velocity component of value $v_1 = (E_D/M)^{1/2}$ which causes an additional profile $F_1(v_z) = \frac{1}{2\,v_1}$ (z is the direction of observation). The result is an increased line width, from the study of which some inside can be obtained into the mechanism of recombination. Similar investigations were carried out for a neon afterglow [10.23c].

Doppler broadening is also observed in laser scattering experiments where it is used as means for the determination of electron and ion temperature

Table 10.1. Determination of temperature from the Doppler width of spectral lines. Abbreviations:

Column 1) *d.* = discharge, *h. c.* = hollow cathode, *h. p.* = high pressure, *l. p.* = low pressure, *m. c. arc* = magnetically confined arc, *pl. j.* = plasma jet, *r. f.* = radio frequency, *stat.* = stationary, *t. d.* = toroidal discharge.

Column 3) $\Delta\lambda$ = full half width.

Column 4) F. P. = Fabry-Perot interferometer, G. S. = grating spectrograph or grating spectrometer, respectively, I. C. = image converter, P. S. = prism spectrograph.

References are given in short symbols, for example, "44z" is given instead of [10.44z].

a) low temperature plasma ($T < \approx 10^5$ °K)

Type of plasma	Spectral lines λ [Å]	$\Delta\lambda$ [Å]	Spectral device	T [10^3 °K]	N_e [10^{13} cm^{-3}]	Reference
l. p. d., Ar	Ar I 4158–7147	0.01–0.03	F. P.	0.35–3.4	0.04–7	36, 44f, 75, 79a, 131d.
l. p. d., Ar	Ar II 4401–5009	0.01–0.04	F. P.	0.57–6.0	0.006–20	36, 44f, 79a, 185.
l. p. d., He	He I 5016	0.04–0.06	F. P.	0.47–1.28		74b.
l. p. d., Kr	Kr I 6056	0.002–0.003	F. P.	0.06–0.13		8f.
l. p. d., Air + Li + Cu	Li I 4972	0.18[1])	F. P.	5	120	44z.
l. p. d., Ne	Ne I 5400–11500	0.01–0.04	F. P.[2])	0.30–1.50	0.04–9	74a, 75, 104, 131d, 171f.
l. p. d., Xe + He	Xe I 20260	0.03	[3])	0.52		134a.
r. f. d., H	H_β 4861	0.15	G. S.	1.9	1.3	147
h. c. d. Fe + He + H_2	Fe I 4832–5307	0.02	F. P.	1.4–2.2	1–4	155a
Pl. j., *l. p.*, Ar	Ar I 4272	0.02–0.03	F. P.	2–3.3	0.003–3	17c
arc, *h. p.*, Fe + air	Fe I 4302–5307	0.03–0.04	F. P.	4–6.3	100–3000	21a
puls. d., *l. p.*, H_2O	OH 3092.39	0.03	G. S.[4])	2.85		14b
puls. d., *l. p.*, Ne	Ne I 5852	0.07	F. P.	6	1000	131
puls. d., *l. p.* He + O	O II 4649–4705	0.2	G. S.	33.5	$\leq$2000[5])	27
puls. d., *l. p.*, air	O II 4189	0.15	F. P.	38	2200–11000	131n[6])
m. c. arc, *stat.*, Ar	Ar II 4806	Order 0.1	F. P.	4.6–11.0	0.1–5	63j
t. d., *l. p.*, He	He I 4921	0.09–0.26	F. P.	2.3–21	0.2–6	63g
ion source, He + H_2 + N_2	He I 4922, 5016	0.04–0.07	F. P.	0.5–1.4	9–29	139d
	N II 4630	0.13–0.15	F. P.	20–30	9–29	139d

Table 10.1. (Continuation)

b) high temperature plasma ($T > \approx 10^5$ °K)

Type of plasma	Spectral lines λ [Å]	Δλ [Å]	Spectral device	T [10^6 °K]	Ne/ [10^{15} cm^{-3}]	Reference
m. c. arc, stat. (H + trace C)	C III 5696	0.2 – 0.4	F. P. + filter	0.02 – 0.14		103a
m. c. arc, stat.	C III 4647 – 5696	0.4 – 2.0	G. S.[7]	0.15 – 4.0	0.01 – 0.1	54, 107, 107a.
m. c. arc, stat.	He II 4686	0.2 – 0.6	F. P.	0.013 – 0.116	0.02 – 0.06	39c, 40.
m. c. arc, stat.	{Ar III, C II, III He II, N II – IV, O II – III}	<1	G. S.[7]	0.19 – 2.6	0.01 – 0.2	44, 107.
plasma cluster	C II 4267	0.3 – 0.6	P. S.	0.09 – 0.29	Order 0.2	165a, 165b.
rf plasmoid	O I 4368	0.07 – 0.12	G. S.	0.008 – 0.022		166b
rf plasmoid	O II 4649	0.2 – 0.3	G. S.	0.08 – 0.140		166b
Plasma source	N II 3995	0.25	P. S.	0.1		156b
magn. trap, He	He II 4686	0.3 – 0.7	G. S.[7]	0.035 – 0.15	Order 0.1	128c
t. d., D + impurity	{C III 4647	0.3 – 0.5	F. P.	0.13 – 0.28		{88a
	D_β 4861	0.4 – 0.6	F. P.	0.027 – 0.064		
	He II 4686}	0.6 – 0.8	F. P.	0.14 – 0.26		}
t. d., H or He	He II 4686	Order 1	G. S.; F. P.[8]	0.1 – 1.6	≤34[5]	56, 63f, 133, 136f.
	He I 5016		F. P.[8]	0.014 – 0.070		136f
	H_β 4861		F. P.[8]	0.006 – 0.174	0.001 – 0.017	63g, 136f.
	O V 2781 – 4915	Order 0.4	G. S.; F. P.[8]	0.2 – 1.7	0.002 – 0.030	63f, 136f.
Θ-Pinch	{C II – IV, N II – IV, O II – VI, Si III – IV range 350 – 5000}	Order 0.1 – 1	G. S.	0.3 – 13		181, 184.
Θ-Pinch, D	O VII – VIII, Ne IX 15 – 25	Order 0.4	crystal spectrometer	110	50	146
Θ-Pinch, D, H	C V 2271	1.4 – 2.0	G. S.	1.5 – 17.4	2 – 30	17b, 156a.

Table 10.1. (Continuation)

Type of plasma	Spectral lines λ [Å]	$\Delta\lambda$ [Å]	Spectral device	T [10^6 °K]	N_e [10^{15} cm^{-3}]	Reference
Θ-Pinch, H	C V 2271	0.5	F. P. multi-channel	1	100 – 150	27a
	O VI 3811	0.3 – 0.6	F. P.	0.22 – 0.67	100 – 150	27a
Corona	Fe X 6374, Ni XV 6702	0.9	G. S.	2.0		25a, 170.
	Fe XIII 10747/98, Fe XIV 5303, Ni XV 6702	1 – 1.5	G. S. + I. C.	1.9 – 2.1		75a, 170.

1) The Doppler width is eliminated from total line width by Voigt profile analysis.
2) In [10.171f], the line width is derived from absorption profile obtained by using a wavelength scanned laser as light source.
3) The line width is derived from optical gain analysis.
4) A high resolution grating is used.
5) For $N_e > 10^{16}$ cm^{-3}, the observed line width is influenced by Stark effect broadening.
6) A time resolution of 10 ns is obtained.
7) An Echelle grating is used in [10.44], [10.54] and [10.128c].
8) The line profile is obtained from multichannel recording of the Fabry-Perot interference pattern.

preferably in high temperature hydrogen and hydrogenic plasmas. The corresponding methods have been dealt with already in section 8.4.2.

To illustrate the range of application of the measurement technique discussed in this section the relevant experimental data have been listed in Table 10.1.

10.2. Doppler Shift

If the atoms or ions carry out a drift movement, the spectral lines will be shifted in accordance with equation (10.2), where v_x is the drift velocity of the particles along the line of sight. When the particles move in the direction of the observation apparatus the spectral lines are shifted toward the short-wave or otherwise toward the long-wave portion of the spectrum.

Conversely, the components of the drift velocity along the line of sight can be determined from the Doppler shift.

To measure the shift the same spectroscopic devices are used as in the determination of the Doppler broadening. To establish the actual shift an unshifted line, emitted by a steady plasma, must be recorded together with the shifted line.

Shifts below 1/10 Å can be measured by means of a Fabry-Perot interferometer. In the experimental arrangement employed by Frish and Kagan [10.35] the interferometer is placed between the collimator objective and the prism of a spectrograph (see Figure 2.8), which instead of the usual symmetrical slit has a slit with two independently movable slit fixtures (see Figure 10.2). First, a photograph is taken when the second slit fixture is centered, while the other slit half is opened. This produces a certain interference pattern on the photographic plate (right-hand side of lower figure). Then the left slit half is covered and the right one opened and a second spectral photograph is taken, which gives the second pattern (left-hand side of lower figure). If during the two exposures the velocity components of the light-emitting particles along the line of sight are different in magnitude, the two patterns appearing on the spectroscopic plates are shifted relative to one another.

Frish and Kagan employed an arrangement for the determination of ion velocities and ion mobilities in a low-pressure argon arc (see Figure 10.3), which burns through a capillary AB. First the radiation leaving the capillary from the left is measured and then that from the right. As a result two interference patterns are obtained which are shifted in opposite directions to one another. The opposing shift corresponds to the double Doppler shift.

The two photographs can also be taken for the same direction of observation but different flow velocities (effected by changing to polarity). This does not influence the shift of the two interference patterns.

Frish and Kagan [10.36] pointed out that an ion drift not only exists in the axial direction but also in the transverse direction. This movement is a result of the considerable radial electric field strengths in gaseous discharges. When observing along a direction normal to the axis of the arc, where cylindrical symmetry of the discharge tube is assumed, the spectral line is not shifted but merely more strongly broadened, since the radiation coming from ranges in which the ions move toward the observer reaches the detection device in the same way as the radiation associated with ions moving away from the observer. This broadening is overlapped by the thermal Doppler broadening. In order to determine the temperature from the line profile obtained normal to the axis a corresponding correction must be introduced.

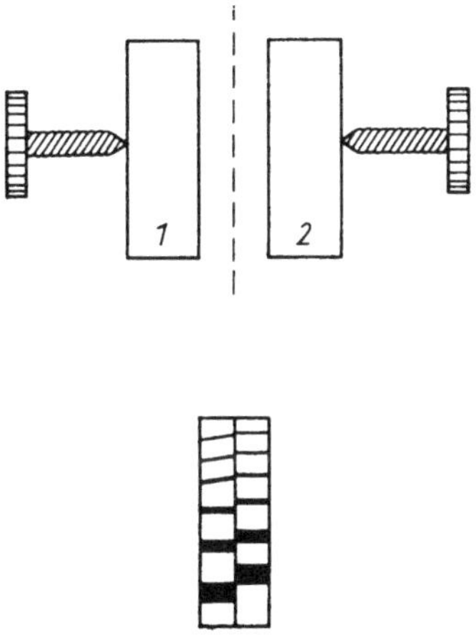

Figure 10.2. Slit system after Frish and Kagan [10.35] for determining Doppler shifts (explanation in text)

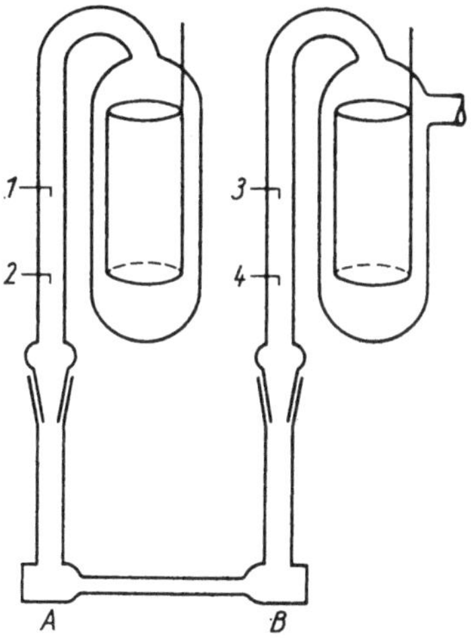

Figure 10.3. Discharge tube with measuring probes *1* to *4* determining the ion drift velocity in the capillary *AB* (after Frish and Kagan [10.35])

Zaidel' et al. [10.183] investigated the ion drift in the direction of the electric field of a toroidal discharge while observing along the direction of the circular discharge axis. Spectral resolution was effected by a combination of a Fabry-Perot interferometer and a prism spectrograph. Figure 10.4 shows the optical ray path before the spectrograph slit *sl*. From this figure it follows that the slit receives radiation from ranges with different velocity components v_x in the observation direction and that therefore the effective measured velocity is smaller than the true flow velocity along the toroid axis. The procedure consists of taking two photographs for different electric field directions. The technique differs from that of Frish and Kagan described above [10.35] only in that instead of the slit with the two movable halves an ordinary slit is employed, which, however, is still movable as a whole.

Larger line shifts ($\delta\lambda > 0.1$ Å) were appropriately determined with the help of a grating spectrograph (see, e.g., Zaidel' et al. [10.182]) or an Echelle grating with preresolution (Bovey and Gabriel [10.18]).

A direct recording of the line shift was effected by Zaidel' et al. [10.182] with the use of the following arrangement. In the plane of the exit slit of the

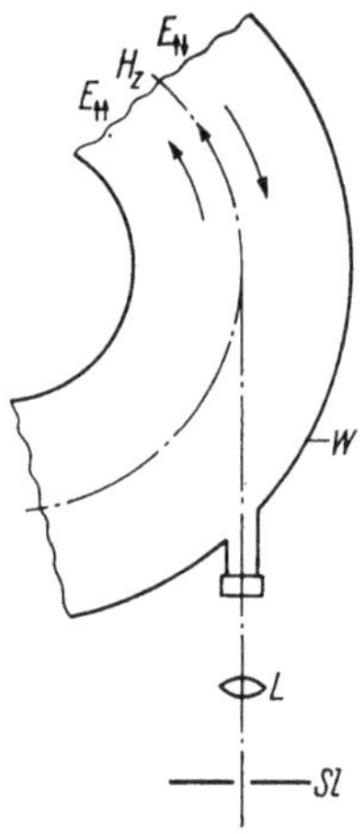

Figure 10.4. Layout for forming the image of a toroidal discharge on a spectrograph slit in order to determine the drift velocity along the discharge axis from the Doppler effect (after Zaidel' et al. [10.183])

W — chamber wall, *L* — lens, *SL* — spectrograph slit, *Hz* — direction of the axial magnetic field $E\uparrow\uparrow$ and $E\uparrow\downarrow$ — electric field direction parallel and antiparallel to the magnetic field, respectively

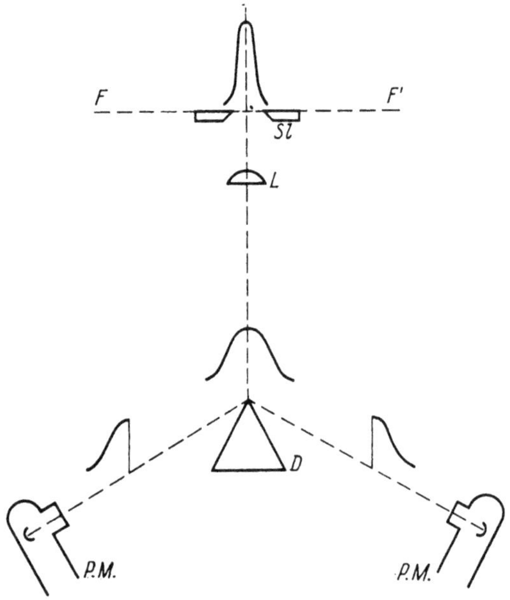

Figure 10.5. Arrangement of the beam splitter for determining line shifts (after Zaidel' et al. [10.182])

Sl — exit slit of the spectrograph *DSF*-8, *F*—*F'* — focal plane, *L* — cylindrical lens, *D* — prism beam splitter

grating spectrograph is placed an additional exit slit, which completely lets through the radiation to be studied. The latter is imaged and magnified (10×) by a lens (cylindrical) on a prism beam splitter (see Figure 10.5). The dissected radiation reaches two photomultipliers, whose photocurrents are recorded on a split-beam oscillograph. The beam splitter is adjusted in such a way that both multipliers give the same current reading when the spectral line displays no shift. A shift in the spectral line now becomes evident in that the photomultipliers indicate different currents. The method is not independent of the width of the shifted line.

This disadvantage is eliminated in the measuring system developed by Drobyshevskii [10.26]. The experimental layout differs from that shown in Figure 10.5 in that the prism beam splitter is revolved through an angle α about the optical axis of the incident ray. Figure 10.6 illustrates the position of the unshifted line and that of the line shifted by $d \cdot \delta\lambda$ as well as the dividing line which occurs due to the presence of the prism beam splitter. d again is the linear dispersion.

If the spectral line, which is considered to be symmetric, is not shifted, the emission to the left and right of the dividing line is equally intense, Let us denote in this case the strength of the spectral line on both sides of the dividing line by S_{10}. If the line is shifted by $d \cdot \delta\lambda$, the intensities S_{11} and S_{21} on both sides of the dividing line will differ. Ensuring that α does not exceed a minimum value $\alpha_{\min}$, which is given by

$$\tan\alpha_{\min} = (\Delta\lambda_{\max} + 1/2 \cdot b_{\max})\, d/l\,, \tag{10.7}$$

where $b_{\max}$ is the maximum line width and $2\,l$ is the height of the spectrum, the line shift $\delta\lambda$ can be obtained from the measured values S_{11} and S_{21} in accordance with

$$\delta\lambda = \frac{S_{21} - S_{11}}{S_{21} + S_{11}} \cdot \frac{l \tan\alpha}{d}\,. \tag{10.8}$$

In deriving the above formula a continuous background was not allowed for. If this is present, we have instead of (10.8)

$$\delta\lambda = \frac{S_{21} - S_{11}}{S_{21} + S_{11} - 2\,S_k}\,, \tag{10.9}$$

where S_k is the continuous intensity, which is measured inclusively by each multiplier. S_k is determined by sliding the exit slit with respect to the spectrum in such a way that the line radiation no longer passes through.

When applying the above described measuring technique it should be ensured that the spectrograph slit is illuminated uniformly along its whole length.

Drobyshevskii used this method to determine the drift velocities in a toroidal discharge, whereby the resolution of an autocollimating prism spectrograph was sufficient to enable velocities of 2 km/s and higher to be measured.

For further special systems for determination of line shifts see, for example, [10.1 a] and [10.44 e].

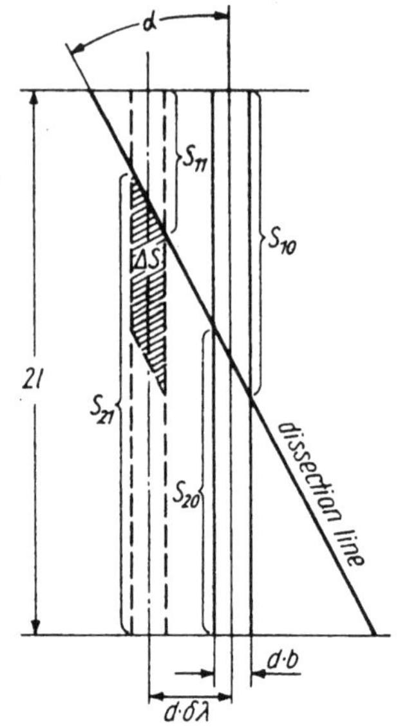

Figure 10.6. Schematic diagram for determination of line shifts when employing a slanting prism beam splitter (after Drobyshevskii [10.26])

α — slope angle of the dividing line with respect to the spectral line, $S_{10} = S_{20}$ — intensity of the half-spectral line in relative measure (expressed as a multiplier current); S_{11} — intensity component of the shifted line on the right of the dividing line, S_{21} — intensity component of the line on the left of the dividing line, $d\delta\lambda$ — shift and $d \cdot b$ — width of the spectral line, d — linear dispersion, $2l$ — height of spectrum

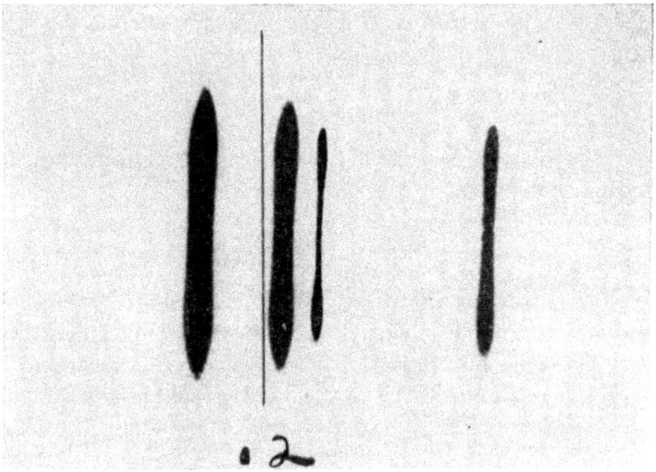

Figure 10.7. Echelle spectrogram of a DCX-discharge at right angles to the discharge axis (after Griffin et al. [10.54])

Spectral lines on the left: 479th order C III 4647, 4650, and 4651 Å as well as a sharp line of the 511th order of Hg 4358 for wavelength calibration; right: 480th order 4647 A

The angular velocity of a rotating plasma can also be measured via the Doppler shift. This rotation occurs in all high-temperature plasmas with stabilizing axial magnetic fields because of the $\boldsymbol{E}\times\boldsymbol{B}$-drift, provided a radial electric field is present, or in the so-called unipolar arrangements, in which the rotation is caused by a force density $\boldsymbol{j}\times\boldsymbol{B}$.

Griffin et al. [10.54] examined the plasma rotation of the first named type in a DCX-arrangement. They formed the image of a narrow plasma streak on the entrance slit of the spectroscopic device at right angles to the discharge axis.

The device contained a high-resolution Echelle grating. Figure 10.7 shows a spectrogram, which clearly illustrates that the spectra lines emitted by the DCX-plasma are inclined. To find the connection between the inclination and the angular velocity ω of the rotating plasma let us consider Figure 10.8

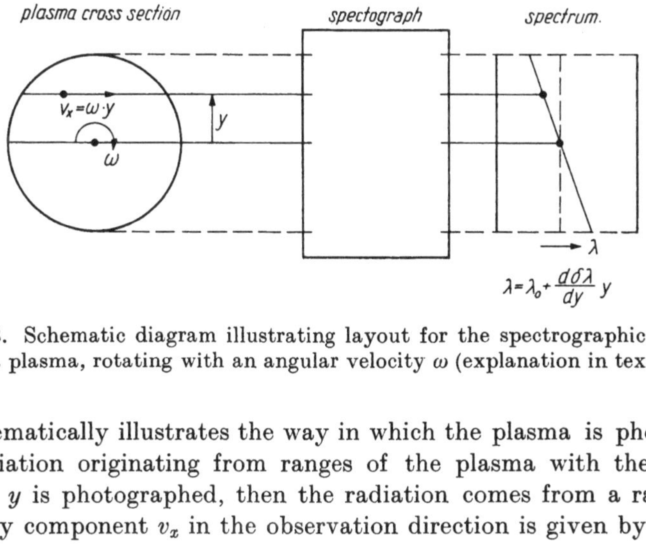

Figure 10.8. Schematic diagram illustrating layout for the spectrographic recording of a plasma, rotating with an angular velocity ω (explanation in text)

which schematically illustrates the way in which the plasma is photographed. If the radiation originating from ranges of the plasma with the transverse coordinate y is photographed, then the radiation comes from a range, where the velocity component v_x in the observation direction is given by

$$v_x = y\,\omega\,. \tag{10.10}$$

This velocity component leads to a shift in the wavelength by $\delta\lambda = \lambda_0 \cdot v_x/c$ (equation (10.2)). The quantity $\mathrm{d}(\delta\lambda)/\mathrm{d}y$ can be determined from the slope of the spectral line on the basis of the linear dispersion d and the image scale of the optical system. We then obtain

$$\omega = \frac{\mathrm{d}v_x}{\mathrm{d}y} = \frac{c}{\lambda_0}\,\frac{\mathrm{d}\delta\lambda}{\mathrm{d}y}\,. \tag{10.11}$$

The experimental results gave rotational frequencies of $\nu = \omega/(2\,\pi)$ up to nearly 200 kHz.

Doppler shift measurements are also useful for the determination of the radial velocity profile in plasma jets when the direction of observation is inclined with respect to the direction of flow. For the evaluation of local velocity data, Abel inversion technique must be applied. For details see [10.17c], [10.130b].

Abel inversion technique is also needed for the determination of angular (respectively radial) velocities in axialsymmetric plasmas from line shift measurements, when the plasma is observed from the side and when the angular (respectively radial) velocity depends from the radius. For details see [10.13a] and [10.180i], respectively.

Examples of the application of the Doppler effect for determining plasma velocities are listed in Table 10.2 along with certain experimental data.

10.3. Determination of Magnetic Field Strengths from the Zeeman Effect

As we know the spectral lines are split when the emitting atoms or ions are present in a magnetic field. The splitting is given by

$$\delta\lambda = g \, \frac{e\,\lambda^2}{4\,\pi\,m\,c^2} \cdot B\,. \tag{10.12}$$

With regard to the type of splitting we shall distinguish between several cases:

a) Normal Zeeman effect. If the radiation direction is parallel to the magnetic field direction, a right and left-circularly polarized component result (r and l), which are shifted by $\pm\delta\lambda$ with respect to the undisturbed wavelength λ_0, where

$$\delta\lambda = \frac{e\,\lambda^2}{4\,\pi\,m\,c^2} \cdot B = 4.66 \cdot 10^{-13}\,\lambda^2\,B \tag{10.13}$$

$$(\lambda \text{ in Å},\ B \text{ in Gauss})\,.$$

Both components have the same intensity. When the radiation is emitted at right angles to the magnetic field direction, an unshifted π-component and two σ-components shifted by $\pm\delta\lambda$ result. The symbols π and σ denote that the relevant components are linearly polarized respectively parallel and perpendicular to the magnetic field direction. Both σ-components each have half the intensity of the unshifted π-component. The normal Zeeman effect is observed for transitions between levels with $S = 0$.

b) Anomalous Zeeman effect. Here the spectral lines can split into more than 2 or 3 components in a magnetic field; the factor g in (10.12) is in most cases not equal to 1 and depends on the L- and J-values of the upper and lower levels. If the magnetic field is so large that the subsequent splitting is larger than the natural fine-structure splitting, the anomalous Zeeman effect goes over into the normal Zeeman effect (Paschen-Back effect).

To measure magnetic fields in plasmas Babcock [10.5], Jahoda et al. [10.70], as well as Hübner [10.64], [10.64a] employed the longitudinal Zeeman effect, whereby only those lines were selected which possessed only two Zeeman components.

Table 10.2. Determination of the velocity of plasma components from the Doppler shift of spectral lines. The abbreviations have the same meanings as in table 10.1. Additional abbreviation used in column 4: F. P. mult. = Fabry-Perot interferometer in combination with a multichannel recording system. References are given in short symbols as indicated in head of table 10.1

a) and b): axial and radial velocity

Plasma	Spectral lines λ in [Å]	Shift in [Å]	Spectral device	v [km/s]	Reference
		a) axial velocity			
l. p. d., Ar	Ar II 4401 – 4933	0.0004 – 0.004	F. P.	0.04 – 0.23	35, 44f, 79a, 185.
shock wave, H_2	H_β 4861	0.02	G. S.	1.5	55b
pl. j., *l. p.*, Ar	Ar I 4158 – 4300	0.015 – 0.04	F. P.	1 – 3	17c
pl. j., *l. p.*, N_2	N II 3995	0.04 – 0.06	F. P.[1])	3.4 – 5.3	44e
t. d., H_2	{C III 2297, N IV, O III – V:2530 – 3580}	0.01 – 0.1	F. P.[1])	1 – 10	11, 181, 182, 183.
plasma gun, C, H, O	C II 4267	Order 0.2	P. S.	13 – 30	165a, 165b.
		b) radial velocity			
Θ-pinch, H_2 + impurity	N V	[2])	F. P. mult.	11	27a
	O VI	[2])	F. P. mult.	8 – 16	27a

[1]) Special recording arrangement.

[2]) v_{radial} is evaluated from the different line widths obtained when the plasma is observed end-on and side-on.

Table 10.2. (Continuation)

c) azimuthal velocity

Type of plasma	Spectral lines used, λ [Å]	Shift in [Å]	Spectral decive	v [km/s]	$\frac{\omega}{2\pi}$ [kHz]	Reference
m. c. arc, Ar	Ar I 4044+4300	0.002	F. P.	≤ 0.15		186
m. c. arc, Ar	Ar II 4806		F. P.		10–20	63 j
m. c. arc, Ar	Ar II 4348		G. S.	3–6	100–900	16 a
m. c. arc, H_2	Hβ 4861	≤ 0.23	Echelle-G. S.	0–14	200–1200	18, 25 d.
m. c. arc, (DCX)	C III 4647–4651	0.1	Echelle-G. S.		90–185	54
Homopolar device, Ar	Ar II 4013–4806	$\lesssim 0.016$	F. P.	0.25–2.0		132 d, e.
Homopolar device, Ar	Ar I 4158–4300		F. P.	0.3–1.2		132 d
Homopolar device, Air	N I 4358	0.03–0.6	P. S.[1])	2–45		26
pl. j., *l. p.*, rotating	Ar II 4348	0–0.06	F. P.	0–4.5	50–230	13 a
Θ-pinch	C II 4267	0.1–0.4	F. P. mult.		53–95	75 b
Θ-pinch	C V 2271	0.03	G. S.		1000–2000	75 b

[1]) Special recording arrangement.

The measurement of the splitting $\delta\lambda$, from which the magnetic field B can be calculated via equation (10.13), is rendered more difficult due to the fact that the Zeeman effect in the most favorable case is overlapped by the Doppler effect. This overlapping occurs as shown in Figure 10.9. The two Doppler profiles shifted by $2\,\delta\lambda$ give a resultant profile, which can hardly be distinguished from a Doppler profile.

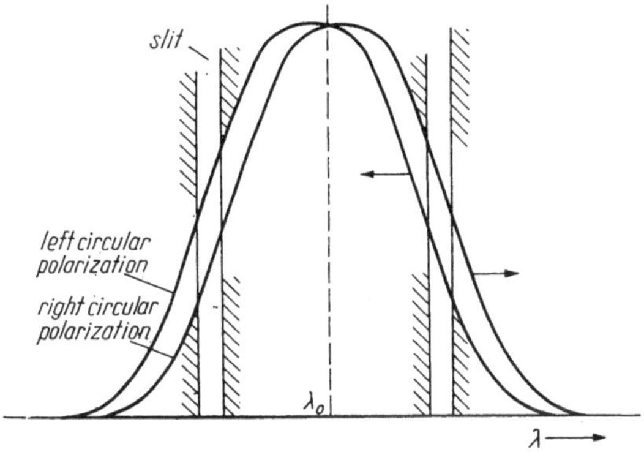

Figure 10.9. Overlapping of two oppositely circularly polarized Doppler-broadened Zeeman components (after Jahoda et al. [10.70])

The Zeeman splitting is measured with the help of a polarization system by first reproducing only the Doppler-broadened r-component and then the Doppler-broadened l-component. The shift $2\,\delta\lambda$ of the two profiles in principal is measured by delimiting the field to a portion of the line emission in the flank of the Doppler profile with the spectrograph exit slit (see Figure 10.9). When the profile shifts by $2\,\delta\lambda$, the intensity of the radiation limited by the slit changes according to

$$\frac{\Delta I}{I} = 2\,k \cdot \delta\lambda\,, \tag{10.14}$$

where k is the relative steepness $\frac{1}{I_\lambda} \cdot \frac{dI_\lambda}{d\lambda}$, which can be determined, for example, by sliding the slit a measurable distance along the line flank at the point of interest, at the same time recording the intensity I_λ.

Figure 10.10 schematically illustrates the experimental arrangement employed by Hübner. The radiation of the plasma — we are dealing with a nitrogen

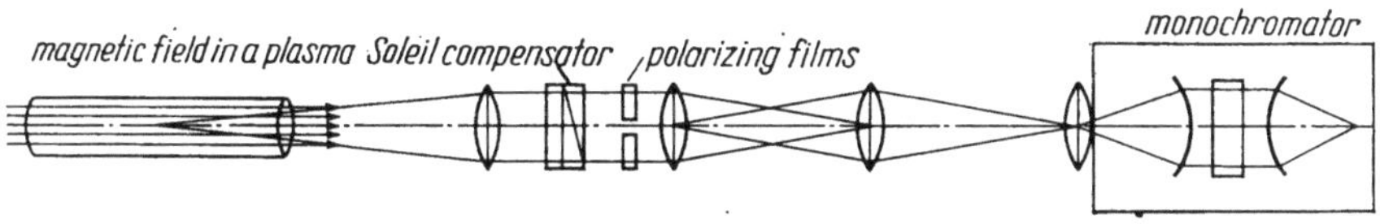

Figure 10.10. Layout for measuring the Zeeman shift in Doppler-broadened spectral lines (after Hübner [10.64])

discharge in a longitudinal magnetic field — passes through a lens and reaches a Soleil compensator, which works like a $\lambda/4$-plate and converts circularly polarized light of one sense into linearly polarized light. Thus, the r- and l-waves are converted into two mutually perpendicular linearly polarized waves. The radiation then passes through two polarization foils, which are arranged in such a manner that one transmits the original r-component and the other the original l-component, whereby the same radiation intensity passes through

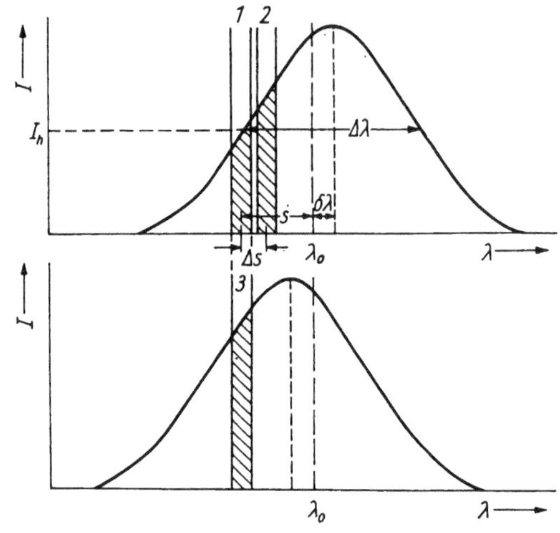

Figure 10.11. Intensity distribution of both polarization components and position of the slit in the exit plane of the monochromator (after Hübner [10.64])

both foils. In this way the two slit halves of the monochromator receive oppositely polarized radiation.

Accordingly, in the plane of the exit slit of the monochromator the two line profiles are displayed one upon the other in the manner shown in Figure 10.11. To measure the line shift $\delta\lambda$ the exit slits 1, 2, and 3 are used, which are connected via light conductors fabricated from 50 µm thick glass fibres to 3 multipliers of equal sensitivity. The spectral intensities I_1, I_2, and I_3 in the center of each of the three exit slits are respectively given by

$$\left.\begin{aligned} I_1 &= A\left(-s-\delta\lambda-d+\frac{\Delta\lambda}{2}\right)+I_h+I_k\,,\\ I_2 &= A\left(-s+\Delta s-\delta\lambda-d+\frac{\Delta\lambda}{2}\right)+I_h+I_k\,,\\ I_3 &= A\left(-s+\delta\lambda-d+\frac{\Delta\lambda}{2}\right)+I_h+I_k\,. \end{aligned}\right\} \qquad (10.15)$$

Here it has been assumed that the slits are located along the linear portion of the line flank. A is the flank slope $\mathrm{d}I_\lambda/\mathrm{d}\lambda$, s the distance between slits 1 respectively 3 and the line center λ_0 of the unshifted line, $\delta\lambda$ the shift to be measured, d the line shift due to the Doppler effect, $\Delta\lambda$ the full half-width, and Δs the distance between slit 1 and 2. I_h is half the maximum spectral intensity and I_k is the intensity of the continuum.

The value of $\delta\lambda$ follows in accordance with (10.15) from the expression

$$\delta\lambda = \frac{1}{2}\,\frac{I_3 - I_1}{I_2 - I_1}\,\Delta s\,,$$

and with (10.12) we have

$$B = 1.07 \cdot 10^{12}\,\frac{\lambda_0^{-2}\,\Delta s\,(I_3 - I_1)}{g\,(I_2 - I_1)}\,, \qquad (10.16)$$

B in Gauss, λ_0 and Δs in Å.

With this technique the time dependence of the magnetic field can be obtained.

Since these measurements are carried out electronically, disturbances due to noise occur which limit the sensitivity, depending on the band width of the detection device. For a band width of 40 kHz Hübner obtained a limiting sensitivity of 400 G.

A spatial resolution cannot be directly effected by this method. At the most this can be achieved by successively delimiting different parts of the plasma, whose radiation is under investigation.

Jahoda et al. [10.70] developed a similar experimental arrangement for measuring magnetic field strengths via the Zeeman effect (Figure 10.12). The radiation from a theta-pinch (Scylla) is rendered incident upon the entrance slit of the spectroscopic device, passes via a collimator lens through a Babinet-

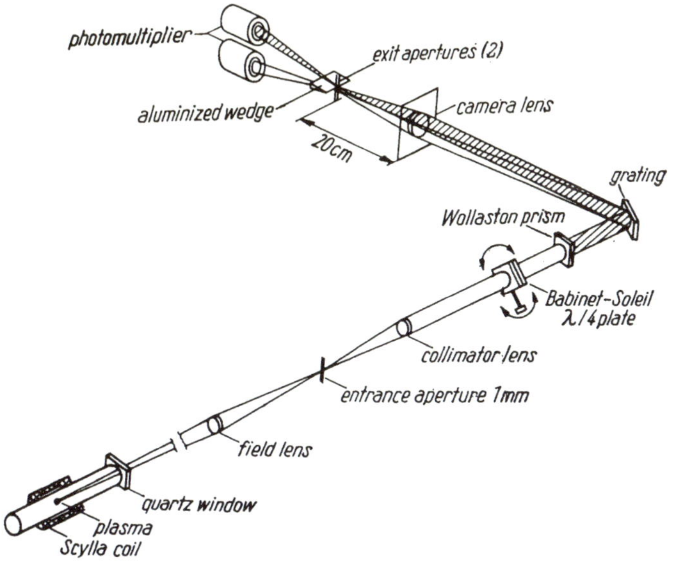

Figure 10.12. Experimental layout for measuring the Zeeman effect of Doppler-broadened lines (after Jahoda et al. [10.70])

Soleil $\lambda/4$-plate, in which the circularly polarized light is converted to linearly polarized light. Upon passing through the Wollaston prism the light bundle is split into two parts, which distinguish themselves in accordance with their polarization direction. The splitting amounts to 15 angle minutes in a vertical plane parallel to the grating rulings, so that both light bundles are incident upon the grating at the same horizontal angle and after spectral resolution in the exit plane of the spectrograph will be displayed one upon the other. Via a flat prism beam splitter the radiation then reaches the two secondary electron multipliers. The slits in the exit plane of the spectrograph lie one above the other.

The intensities let through by both slits are denoted by I_2 and I_1, respectively. To obtain the absolute value of B, a second spectrogram (in contrast to the experimental arrangement of Hübner) at another time point is required, whereby the radiation reaching the photographic plate must be emitted from a volume with known (strong) magnetic field strength. The respective intensities let through the two slit can be denoted by I_2^0 and I_1^0. Then the magnetic field strength follows from the relationship

$$\frac{B}{B_0} = \frac{I_2 - I_1}{I_2^0 - I_1^0}. \tag{10.17}$$

The detection accuracy limit is about 1 kG in this case.

Babcock [10.5] employed a plane grating with a camera objective lens of 23 m focal length to measure the Zeeman splitting of solar radiation. The dispersion in the 5th order was 11 mm/Å. Since there was sufficient time to carry out the observations, the Zeeman shift was measured in a somewhat different way than that described above.

In front of the entrance slit of the spectrograph was placed an electro-optical waveplate made of a crystal of ammonium dihydrophosphate (ADP) cut with respect to the Z axis. To this were attached transparent electrodes, to which an alternating voltage of 9 kV amplitude at a frequency of 120 Hz was applied. This had the effect that the wave plate produced a periodic shift of $\approx\lambda/4$, thus converting the circularly polarized light into linearly polarized light. When a Nicol prism is then inserted into the ray path the experimental arrangement operates as an analyzer for right- and left-circularly polarized radiation, which is periodically and alternately let through. If the exit slit is centered on the flank of the profile of the Doppler-broadened line, the Zeeman shift results from the intensity difference of the r- and l-radiation let through by the exit slit, and the magnetic field strength then follows from (10.12). To raise the accuracy two slits, arranged symmetrically with respect to the line center, and a difference amplifier are employed. For details on the electronic circuitry we refer the reader to a paper of Babcock [10.5]. The precision of the system is so large that magnetic field strengths above 1 G can be accurately measured. The limiting sensitivity is 0.1 G.

Finally, we list below some of the most important experimental data on Zeeman splitting.

Table 10.3. Determination of magnetic field strengths from Zeeman effect

Reference	Type of plasma	Spectral line λ in [Å]	g-factor	$\Delta\lambda$ [Å] [1]	H [kG]	Threshold value in [G]	Threshold value in [Å]
Babcock [10.5]	Sun	Fe I 5250.2	1	0.1	0.001 – 0.02	0.1	$2 \cdot 10^{-5}$
Hübner [10.64], [10.64a]	Z-pinch, N_2	N II 4447	1		0.4 – 20	400	0.004
Eberhagen et al. [10.27a]	Θ-pinch, H_2 + impurity	N V 4603	1	$\approx$0.5	13 – 18		
		O VI 3811	1	0.3 – 0.6	15 – 27		
Jahoda et al. [10.70]	Θ-pinch	C V 2277.3, 2277.9	1.8	$\approx$3	3 – 50	$\approx$1000	0.004
		C III 2296.9	1		3 – 50	$\approx$1000	0.004

[1]) $\Delta\lambda$ = full half width of the line under consideration.

10.4. Natural Line Broadening

The natural line width has no direct significance in plasma diagnostics. However, this is of interest for two reasons: 1. under favorable conditions the transition probability of one or several spectral lines can be determined from the natural line width (see sections 5.3.7 and 5.3.10), 2. the natural line width must often be allowed for when temperatures and particle densities are to be determined from a line profile. This essentially applies to narrow lines. For lines, whose width is large compared to the natural line width, corrective measures need not be taken.

Due to the small width of the natural line profile, for its determination it is profitable to use indirect methods, see, for example, [10.36a].

At this stage it will suffice to indicate the calculation of the natural line profile. This is achieved by considering the emission process, which can be represented as a damped oscillation of an oscillator (see Shpolskii [10.153]):

$$t \leq 0 : f(t) = 0 ,$$

$$t > 0 : \quad f(t) = a\, e^{-\frac{\gamma}{2} t} \cos(2\pi \nu_0 t + \delta) . \tag{10.18}$$

γ is the damping constant, ν_0 the fundamental frequency of the damped oscillation, and δ a phase angle. Fourier expansion gives

$$f(t) = \int_{-\infty}^{+\infty} a(\nu)\, e^{i 2\pi \nu t}\, d\nu , \qquad a(\nu) = \int_{-\infty}^{+\infty} f(t)\, e^{-i 2\pi \nu t}\, dt , \tag{10.19}$$

or

$$a(\nu) = \frac{a}{\gamma/2 - i\, 2\pi(\nu_0 - \nu)} . \tag{10.20}$$

The distribution of the emitted energy of the damped wave over the frequency results from the square of $a(\nu)$:

$$I_\nu = \frac{\text{const}}{4\pi^2(\nu_0 - \nu)^2 + \frac{\gamma^2}{4}} . \tag{10.21}$$

The constant is determined in such a way that integrating over the frequency gives the value I. We thus obtain

$$I_\nu = \frac{\gamma I}{4\pi^2(\nu_0 - \nu)^2 + \frac{\gamma^2}{4}} . \tag{10.22}$$

The profile represented by (10.22) is designated as a dispersion profile. Upon normalizing it has been assumed that the full half-width $\Delta\nu_n$ in frequency units

$$\Delta\nu_n = \frac{\gamma}{2\pi} \tag{10.23}$$

is small compared to the fundamental frequency ν_0.

The damping constant γ depends on the lifetimes of the upper and lower level in accordance with

$$\gamma = \frac{1}{\tau_m} + \frac{1}{\tau_n} . \tag{10.24}$$

The lifetime of the upper level is given by

$$\frac{1}{\tau_m} = \sum_{n'<m} A_{n'm} , \tag{10.25}$$

whereby one must sum over all lower levels n', to which optical transitions from level m exist.

In similar fashion we obtain

$$\frac{1}{\tau_n} = \sum_{n''<n} A_{n''n} . \tag{10.26}$$

In the special case that the lower level is the ground level or is metastable $(1/\tau_n = 0)$ and only the transition from m to n exists, we have

$$\gamma = A_{nm} . \tag{10.27}$$

The natural line width is of the order of magnitude of 10^{-4} Å (see Traving [10.166], p. 105) and is in most cases negligible compared to the thermal Doppler width. An exception to this is, for example, an atomic or ionic beam, excited to luminescence and observed side-on.

About the enhancement of natural line broadening due to induced emission and absorption processes, see [10.132g].

10.5. Line Broadening due to Particle Interaction

In a plasma of finite density the electron levels of the atoms and ions are subject to perturbations changing with time which are caused by other particles. These include electrons, atoms, and ions in the ground state and excited state.

As a consequence of the interaction the emitted spectral lines are broadened and shifted. The type of broadening and shift depends to an important extent on whether the perturbing particles move at high or low velocities. The influence of fast particles (in general electrons) on the line profile is described by the collision theory of line broadening, the profile alteration due to slow particles (usually ions or atoms) is described by the statistical theory. The limits of the range of validity of both theories will be discussed below.

Further, we shall distinguish between the adiabatic and nonadiabatic theory of line broadening. In the framework of the adiabatic theory it is assumed that the perturbations themselves do not lead to any quantum jumps. This, as a rule, is the case when the frequency shift of the line due to the perturbing particle is smaller than the frequency interval between the upper electron level and the closest neighboring level. For nondegenerate levels (i.e., not for hydrogen), the adiabatic condition is

$$\tau_s \ll \frac{\hbar}{U_m - U'_m}, \quad \text{or} \quad \tau_s \gg \frac{\hbar}{U_m - U'_m}, \tag{10.28}$$

(see Sadjian [10.145]). Here τ_s is the disturbance time and is given by

$$\tau_s \approx 1.59\, \varrho/v\,, \tag{10.29}$$

(Unsöld [10.170], Chapter 11), where ϱ is the distance of closest approach of the perturbing particle upon passing by the emitting atom and v is the relative velocity. U_m is the excitation energy of the upper level, U'_m is that of another level to which transitions can occur. When the level is degenerate the adiabatic condition becomes

$$\Delta\omega\, \tau_s \gg 1 \tag{10.30}$$

(see Traving [10.166]), where $\Delta\omega$ denotes the shift in the circular frequency of the emitting atom produced by the perturbing particle.

While the adiabatic perturbations can be treated classically, only quantum mechanical considerations can be applied to nonadiabatic perturbations.

In this chapter we shall discuss the most important types of line broadening due to particle interaction. We shall not go into great detail, for which we refer the reader to the relevant literature ([10.6], [10.19], [10.20], [10.20a], [10.23f], ([10.47a] p. 63), [10.50], [10.52], [10.114], [10.145], [10.157a], [10.166], [10.166a], and [10.170]).

10.5.1. Classical Collision Theory

10.5.1.1. The Lorentz theory

In the collision theory of Lorentz [10.102] and Weisskopf [10.174] the emitting atom is considered as a classical oscillator with circular frequency ω_0, which due to the effect of the perturbing particles is knocked out of phase after statistical time intervals, which group themselves around a mean value of τ (average time between two collisions). Let us assume that between these collisions the oscillator will emit the wave with the unperturbed circular frequency ω_0. Further, it is assumed that no correlation exists between the phase before and after the collision. Fourier expansion of the chopped-off wave train and averaging over all relevant time intervals t between two collisions and over all collision parameters ϱ gives the line profile (dispersion profile)

$$I(\omega) = \frac{1}{\pi\tau} \frac{1}{(\omega_0 - \omega)^2 + (1/\tau)^2} \tag{10.31}$$

with the normalization

$$\int_{-\infty}^{+\infty} I(\omega)\, d\omega = 1\,. \tag{10.32}$$

Let us introduce the full half-width γ in circular frequency units:

$$\gamma = \frac{2}{\tau} = 2\,Z\,. \tag{10.33}$$

Z is the optical collision frequency. From (10.31) it thus follows that

$$I(\omega) = \frac{\gamma}{2\,\pi}\,\frac{1}{(\omega_0 - \omega)^2 + (\gamma/2)^2}\,. \tag{10.34}$$

The theory yields a line broadening but no line shift (in reality, line broadening without line shift is characteristic for inelastic collisions).

The half-width γ can be connected with the density N of the perturbing particles as follows

$$\frac{\gamma}{2} = Z = \pi\,\varrho^2\,N\,v\,, \tag{10.35}$$

where ϱ signifies a suitable optical radius.

10.5.1.2. The Lenz and Weisskopf theory

To find the radius ϱ, Lenz [10.97] and Weisskopf [10.174] took the following relationship for the shift $\Delta\omega$ in the circular frequency of the radiating atom produced by a perturbing particle passing at a distance r

$$\Delta\omega = \frac{C_p}{r^p}\,, \tag{10.36}$$

where the exponent p depends on the type of perturbation, and C_p must be determined by quantum mechanics or empirically.

In the approximation it is assumed that the perturbing particle flies past the atom rectilinearly. With v = relative velocity of the perturbing particle and ϱ = collision parameter, we obtain for $r(t)$

$$r^2 = \varrho^2 + v^2\,t^2\,. \tag{10.37}$$

When the particle flies past, the oscillation suffers a total phase shift of

$$\eta(\varrho) = \int_{-\infty}^{+\infty} \Delta\omega\,\mathrm{d}t = \int_{-\infty}^{+\infty} \frac{C_p\,\mathrm{d}t}{(\varrho^2 + v^2\,t^2)^{p/2}} = \frac{\alpha_p\,C_p}{v\,\varrho^{p-1}} \tag{10.38}$$

where

$$\alpha_p = \sqrt{\pi}\,\frac{\Gamma\left(\frac{p-1}{2}\right)}{\Gamma\left(\frac{p}{2}\right)}\,. \tag{10.39}$$

p depends on the type of perturbation:

$p = 2$, $\alpha_2 = \pi$: linear Stark effect, such as occurs in the case of the hydrogen atom;

$p = 3$, $\alpha_3 = 2$: resonance broadening, resonance of similar particles, resonance line broadening;*

$p = 4$, $\alpha_4 = \pi/2$: quadratic Stark effect;

$p = 6$, $\alpha_6 = 3\pi/8$: van der Waals broadening.

In the framework of this approximation only those perturbations are considered as collisions for which the total phase change η exceeds a value η_0 of the order of magnitude 1. If we set $\eta_0 = 1$, the corrsponding collision parameter ϱ_W (= Weisskopf radius) is given by

$$\varrho_W = \left(\frac{\alpha_p C_p}{v}\right)^{\frac{1}{p-1}}. \tag{10.40}$$

The line profile then follows from (10.31) with

$$\gamma = \frac{2}{\tau} = 2\pi \varrho_W^2 N v. \tag{10.41}$$

This representation still has some shortcomings, insofar as 1. perturbations with collision parameter $\varrho > \varrho_W$ are neglected, 2. the phase correlation before and after the collisions is not considered, and 3. the radius ϱ_W, up to which the perturbations are reckoned as collisions, is arbitrary.

10.5.1.3. Lindholm theory of collision broadening

This arbitrariness does not appear in the Lindholm theory of collision broadening ([10.99], [10.100], [10.101]), which was subsequently improved from the mathematical standpoint by Sobel'man [10.157]. The oscillator is described by

$$f(t) = e^{i\int_0^t \omega(t)\,dt} = e^{i\omega_0 t + i\eta(t)}. \tag{10.42}$$

Here

$$\eta(t) = \int_0^t \left(\omega(t) - \omega_0\right) dt = \int_0^t \Delta\omega(t)\,dt \tag{10.43}$$

denotes the sum of all phase shifts over the time interval from 0 to t. The process of the phase shift itself is assumed to take place in an infinitesimally short time. The wave train can be imagined somewhat in the fashion shown in Figure 10.13. The magnitudes of the phase jumps are statistically distributed

* According to Grechikhin [10.44g], a broadening of type $p = 3$ can also be effected by the gradient of the electric microfield, but its application for plasmadiagnostical purpose is restricted to optical transitions between spectral levels of low excitation energy.

in accordance with the probability of encountering a certain collision parameter ϱ.

The spectral intensity of this oscillation, which is obtained from the Fourier coefficients found via the correlation calculation, is given by

$$I(\omega) = \frac{\gamma}{2\pi} \frac{1}{(\omega - \omega_0 - \Delta)^2 + (\gamma/2)^2}. \tag{10.44}$$

Figure 10.13. Wave train $f(t)$ with phase jumps (after Breene [10.20], Figure 2.3)

From (10.44) it follows that the spectral line is not only broadened (full half-width in circular frequency units $= \gamma$) but also shifted by the amount Δ.

For γ and Δ follow different expressions, depending on the exponent p.

The quantities γ, Δ, γ/Δ for $p = 3$, 4, and 6 are listed in Table 10.4, which also contains the value for η_0 occurring in the Weisskopf theory of line broadening.

Table 10.4. Line-broadening parameters after Lindholm

p	γ	η_0	Δ	γ/Δ
3	$2\pi^2 C N$	0.64	0	
4	$11.37 \cdot C^{2/3} v^{1/3} N$	0.64	$9.85 \cdot C^{2/3} v^{1/3} N$	1.16
6	$8.08 \cdot C^{2/5} v^{3/5} N$	0.61	$2.94 \cdot C^{2/5} v^{3/5} N$	2.75

The relative velocity v between the perturbing particle and the radiating atom is not constant but is subject to a Maxwell distribution. It suffices, however, to take a constant value for v (Unsöld [10.170] Ch. 11), given by

$$v = \sqrt{\frac{8\,kT}{\pi}\left(\frac{1}{m_1} + \frac{1}{m_2}\right)}, \tag{10.45}$$

where m_1 and m_2 are the respective masses of the atom and the perturbing particle.

The case $p = 2$ (linear Stark effect) must be handled separately because the interaction energy between radiating atom (dipole) and disturber particle (monopole) depends not only from the internuclear distance but also from the direction of dipole relative to the internuclear axis. Then, $\eta(\varrho)$ is the product of expression (10.38) (for $p = 2$) with $\cos\Theta$, when Θ is the angle between the dipole direction and a line drawn through the nucleus of the radiating atom which is normal to the path of the perturber ([10.166a] p. 78). To avoid divergences when calculating the line width, the range of ϱ which has to be considered must be cut off at the collision parameter $\varrho = \varrho_{max}$. For ϱ_{max} we can take either the Debye length D or the mean distance r_0 between the perturbing particles. Which of the two values is to be taken depends on the magnitude of δ, which gives the number of perturbing particles in the Debye sphere. For this we have

$$\delta = \frac{4\pi}{3} D^3 N , \tag{10.46}$$

where

$$D = \left(\frac{kT}{4\pi e^2 \left(N_e + \sum_i N_i z_i^2\right)}\right)^{1/2} \tag{10.47}$$

or

$$\delta = 490 \cdot T^{3/2} N^{-1/2} \quad \text{for} \quad N_e = N_i = N \quad \text{and} \quad z = 1 . \tag{10.48}$$

If $\delta < 1$, i.e., if the number of particles in the Debye sphere is less than 1, ϱ must be integrated up to r_0. On the other hand, if $\delta > 1$, ϱ is integrated up to $\varrho = D$, since only those charge carriers which are further than a distance D from the atom can be considered as shielded.

For r_0 we have the relationship

$$\frac{4}{3} \pi r_0^3 N = 1 \qquad \text{or} \qquad r_0 = \left(\frac{3}{4\pi N}\right)^{1/3} . \tag{10.49}$$

As result, we get for the line width the formula

$$\gamma_2 = \frac{32}{3} \frac{C_2^2 N}{v} \left(2.566 - \ln \frac{2 C_2}{v \varrho_{\max}}\right). \tag{10.50}$$

10.5.1.4. Range of application of the collision theory of line broadening

The derivation of the collision theory essentially consisted of carrying out a Fourier expansion for the radiation during the time interval τ between two collisions, whereby those perturbations were considered as collisions for which a phase change of $\eta \geq 0.6$ was produced.

In the collision theory the radiation processes, which occur during the collision time τ_s itself while the perturbing particle flies past within a distance ϱ_0, were neglected. These processes give rise to a line with a half-width of the order of $1/\tau_s$ and a shift $\Delta\omega(\varrho)$ in accordance with (10.36) for $r = \varrho$ and $\tau_s \approx 1.5\,\varrho/v$. While this additional broadening is only significant for the line intensity for distances $\geq \Delta\omega(\varrho_0)$ from the line center, it does not influence the line profile in the neighborhood of the line center.

The exact value for $\Delta\omega$, up to which the collision theory holds for the line profile, is according to Holstein (see Traving [10.166]).

$$\Delta\omega_H = \frac{v^{p/(-1)}}{C_p^{1/(p-1)}} \cdot \left(\frac{p \cdot 2^{p/2}}{9}\right)^{p/(p-1)} . \tag{10.51}$$

This limiting frequency corresponds more or less to the frequency shift, which occurs when the perturbing particle is at a distance equal to the Weisskopf radius from the radiating atom.

The numerical factor on (10.51) varies from 0.2 ($p = 2$) to 7.4 ($p = 6$). It follows from expression (10.51) that the limiting frequency lies increasingly further away from the line middle the faster the perturbing particles. Further, the limiting frequency is independent of the particle density. From this it follows that in the majority of cases the principal portion of the line profile determined by electron collisions can be described by the collision theory, whereas the perturbations caused by ions cannot be described by this theory due to their lower velocities. Exceptions to the above occur in 1. very high-density plamas, where essentially the electron perturbations can also no longer be described by the collision theory, 2. rarefied high-temperature plasmas, in which the ionic perturbations can be described by the collision theory.

If the value $\Delta\omega = \Delta\omega_H$ is exceeded, the perturbations can be described by the statistical theory, which will be discussed in the following section.

10.5.2. Statistical Theory of Line Broadening

10.5.2.1. One-particle approximation

In statistical theories the perturbations are considered as quasistationary. The frequency shift $\Delta\omega$ is again assumed to be a function of the distance r of the perturbing particle from the radiating atom in accordance with equation (10.36).

In the framework of the one-particle approximation ([10.166], p. 17) we only consider the effect of a single perturbing particle, whose whereabouts in arbitrary points of space is assumed as equally probable.

The line profile follows from the probability that there is a certain frequency shift $\Delta\omega$, which again in accordance with equation (10.36) follows from the probability $W(r)$ that the closest particle is located between r and $r + dr$ and that inside the range r there occur no perturbing particles:

$$W(r) = 4\pi r^2 N \mathrm{e}^{-\frac{4\pi}{3} r^3 N} . \tag{10.52}$$

The line profile is given by

$$I(\omega) = \frac{3}{p} \left(\frac{\Delta\omega_0}{\Delta\omega}\right)^{(p+3)/p} \mathrm{e}^{-\left(\frac{\Delta\omega_0}{\Delta\omega}\right)^{3/p}} \cdot \frac{1}{\Delta\omega_0} \tag{10.53}$$

where

$$\int_0^\infty I(\omega)\, \mathrm{d}\omega = 1$$

and the normal shift

$$\Delta\omega_0 = \frac{C_p}{r_0^p} = C_p \left(\frac{4\pi}{3} N\right)^{p/3} . \tag{10.54}$$

Equation (10.53) represents an assymmetric profile which, depending on the sign of C_p, displays a trend toward the short-wave or long-wave side of the unperturbed circular frequency ω_0 (toward that side to which a shift in the line would result in the Lindholm theory). If the line splits into several components, such as in the case for hydrogen, there will be a corresponding superposition of the statistical profiles associated with these components. When the splitting is symmetric, the components, will again give symmetric profiles.

Here it is essential that the velocity of the perturbing particles no longer enters into consideration.

The statistical theory must be further refined. The one-particle approximation is only valid for the wings, which corresponds to small distances between the perturbing particle and the atom. Toward the line center the perturbing influences of a greater number of particles must be allowed for.

10.5.2.2. Many-particle theory, Holtsmark distribution

If several particles simultaneously exert a perturbing influence, the perturbations will be superimposed on one another in some way. For the case $p = 2$ and 4 (linear and quadratic Stark effect) the perturbations are superimposed vectorially, for $p = 6$ in scalar fashion, while for $p = 3$ no simple superposition law can be stated.

For $p = 6$ (van der Waals interaction) the distribution function is given by

$$I(\omega) = \frac{1}{2}\left(\frac{\Delta\omega_0}{\Delta\omega}\right)^{3/2} \mathrm{e}^{-\frac{\pi}{4}\frac{\Delta\omega_0}{\Delta\omega}} \cdot \frac{1}{\Delta\omega_0} . \tag{10.55}$$

The asymptotic behavior is again in agreement with that according to (10.53). The half-width and the shift are given by

$$\Delta\omega_{1/2} = 0.822 \cdot \pi^3\, C_6\, N^2 , \tag{10.56}$$

$$\omega_{\max} - \omega_0 = \frac{8}{27}\, \pi^3\, C_6\, N^2 . \tag{10.57}$$

The ratio of the broadening to the shift is 2.78 in agreement with the collision theory, which gives 2.75.

For a general broadening theory in the case $p = 6$ see Traving ([10.166], p. 23).

We shall now consider the case in which the perturbations add up vectorially ($p = 2$ and 4). Then the line profile is given by the Holtsmark distribution, which represents the probability distribution $W(\beta)$ of the reduced microfield strength β in the region of the radiating atom:

$$W(\beta) = \frac{2}{\pi\,\beta}\int\limits_0^\infty v\,\sin(v)\,\mathrm{e}^{-(v/\beta)^{3/p}}\,\mathrm{d}v \tag{10.58}$$

where

$$\beta = E/E_0 , \qquad E_0 = e\,\gamma_p\,N^{p/3} , \qquad E = e/(r^2) . \tag{10.59}$$

Here E is the perturbing field strength produced by a particle and E_0 the "normal field strength" (frequently this is denoted by F_0 in the literature).

The quantities γ_p assume the following values:

$$p = 2\,, \quad \gamma_p = 2.6031\,,$$
$$p = 3\,, \quad \gamma_p = 3.2899\,,$$
$$p = 4\,, \quad \gamma_p = 4.9548\,.$$

For $p = 2$ the asymptotic expressions for $W(\beta)$ are

$$W(\beta) = \frac{4}{3\,\pi}\,\beta^2\,(1 - 0.4628\,\beta^2 + 0.1227\,\beta^4 - 0.02325\,\beta^6 \ldots)\,, \quad \beta \ll 1\,, \tag{10.60}$$

$$W(\beta) = \frac{1.496}{\beta^{5/2}}\left(1 + \frac{5.107}{\beta^{3/2}} + \frac{14.43}{\beta^3} + \cdots\right) \quad \text{for} \quad \beta \gg 1\,. \tag{10.61}$$

In the neighborhood of 3.5, where the series converge badly, expression (10.58) must be numerically integrated. The values for $W(\beta)$ are listed in Table 10.5.

From $W(\beta)$ the line profile follows in accordance with

$$I(\omega)\,\mathrm{d}\omega = W(\beta)\,\mathrm{d}\beta\,. \tag{10.62}$$

Table 10.5. Holtsmark distribution $W(\beta)$ (after Sadjian [10.145])

$\beta = \alpha^{1/2}$	$W(\beta)$	$\beta = \alpha^{1/2}$	$W(\beta)$
0.000	0.000	2.80	0.206
0.10	0.004	2.90	0.191
0.20	0.017	3.00	0.176
0.30	0.037	3.10	0.163
0.40	0.063	3.20	0.150
0.50	0.095	3.30	0.140
0.60	0.130	3.40	0.128
0.70	0.167	3.50	0.119
0.80	0.204	3.60	0.109
0.90	0.239	3.70	0.101
1.00	0.272	3.80	0.093
1.10	0.300	3.90	0.087
1.20	0.324	4.00	0.081
1.30	0.343	4.20	0.070
1.40	0.356	4.40	0.061
1.50	0.364	4.60	0.053
1.60	0.367	4.80	0.047
1.70	0.365	5.00	0.041
1.80	0.359	5.20	0.037
1.90	0.350	5.40	0.032
2.00	0.337	5.60	0.029
2.10	0.323	5.80	0.026
2.20	0.307	6.00	0.024
2.30	0.290	6.50	0.016
2.40	0.273	7.00	0.011
2.50	0.256	7.50	0.009
2.60	0.239	8.00	0.007
2.70	0.222		

For the case of the quadratic Stark effect ($p = 4$) we have

$$\left(\frac{\Delta\omega}{\Delta\omega_0}\right)^{1/2} = \frac{E}{E_0} = \beta\,. \tag{10.63}$$

With

$$\alpha = \frac{\Delta\omega}{\Delta\omega_0} \tag{10.64}$$

we obtain

$$I(\omega)\,\mathrm{d}\omega = \frac{W(\sqrt{\alpha})}{2\sqrt{\alpha}\,\Delta\omega_0}\,, \tag{10.65}$$

where $\Delta\omega_0$ is given by

$$\Delta\omega_0 = C_4\,E_0^2 \tag{10.66}$$

and E_0 is defined by (10.59).

10.5.2.3. Charge carrier interaction among the perturbing particles

Hitherto it was assumed that the perturbing particles are independent of one another. If the perturbing particles are ions, their positions in a dense plasma are then no longer independent of one another, since the probability of one ion being in the neighborhood of another decreases due to electrostatic repulsion. Consequently, deviations appear in the Holtsmark distribution $W(\beta)$ (see Ecker [10.29], and also Mozer and Baranger [10.130f]). The corrected distribution function $W_r(\beta)$ is characterized by the parameter $r = r_0/(\sqrt{2}\,D)$, with r_0 and D given by equations (10.49) and (10.47) (Note that $\sqrt{2}\,D$ in our notation corresponds to the Debye length in the notation of [10.130f] which does not include shielding by ions). $W_r(\beta)$ is tabulated for atomic and single charged ionic emitters for the parameter range $\beta = 0.1 - 10.0$ and $r = 0.0 - 0.8$ in [10.130f] and ([10.47a] p. 442), respectively, from which the data of table 10.6 are taken (For $r = 0$ we have the Holtsmark distribution already tabulated in table 10.5).

10.5.2.4. Range of application of the statistical theory of line profiles

The range of application of the statistical theory of line profiles begins approximately for those values of $\Delta\omega$, for which the range of application of the collision theory ends, i.e., for $\Delta\omega > \Delta\omega_H$, where $\Delta\omega_H$ is the Holstein limiting frequency defined in equation (10.51). Further details can be found in section 10.5.1.4. For the special case of Van der Waals broadening ($p = 6$) at high gas pressures, see the comments given in section 10.5.4.3.

10.5.3. Determination of the Stark Constants C_p

The Stark constants C_p can be found by means of a quantum mechanical perturbation calculation or by experiment with the use of strong stationary electric fields. The calculation procedures, the results of which will be briefly discussed, are distinguished according to the nature of the Stark effect.

Table 10.6. Electric microfield probability $W_r(\beta)$ for neutral and single charged emitters. Parameter is $r = r_0/(2\sqrt{D})$, where r_0 and D are given by (10.49) and (10.47). The data are taken from Mozer and Baranger [10.130f] (the last digit has been left out)

	Neutral emitters				Single charged emitters			
β \ r	0.2	0.4	0.6	0.8	0.2	0.4	0.6	0.8
0.2	0.0285	0.0510	0.0876	0.1387	0.0295	0.0554	0.1009	0.1695
0.4	0.1040	0.1748	0.2753	0.3924	0.1075	0.1888	0.3133	0.4675
0.6	0.2020	0.3091	0.4327	0.5452	0.2084	0.3313	0.4838	0.6295
0.8	0.2947	0.4039	0.4990	0.5624	0.3031	0.4286	0.5468	0.6284
1.0	0.3620	0.4432	0.4885	0.5033	0.3710	0.4652	0.5242	0.5443
1.2	0.3959	0.4364	0.4380	0.4212	0.4042	0.4529	0.4602	0.4418
1.4	0.3987	0.4016	0.3754	0.3429	0.4053	0.4121	0.3864	0.3498
1.6	0.3785	0.3544	0.3152	0.2772	0.3831	0.3598	0.3180	0.2757
1.8	0.3446	0.3053	0.2625	0.2252	0.3475	0.3069	0.2600	0.2186
2.0	0.3048	0.2595	0.2185	0.1848	0.3058	0.2585	0.2129	0.1751
2.2	0.2644	0.2194	0.1824	0.1534	0.2643	0.2167	0.1752	0.1419
2.4	0.2268	0.1852	0.1532	0.1290	0.2258	0.1816	0.1452	0.1164
2.6	0.1934	0.1567	0.1296	0.1094	0.1918	0.1526	0.1212	0.0966
2.8	0.1645	0.1331	0.1103	0.0936	0.1626	0.1288	0.1020	0.0809
3.0	0.1401	0.1137	0.0946	0.0806	0.1381	0.1093	0.0866	0.0685
3.5	0.0951	0.0784	0.0662	0.0567	0.0931	0.0744	0.0592	0.0467
4.0	0.0666	0.0561	0.0481	0.0416	0.0648	0.0526	0.0422	0.0333
4.5	0.0482	0.0415	0.0361	0.0319	0.0467	0.0385	0.0311	0.0246
5.0	0.0360	0.0315	0.0278	0.0248	0.0347	0.0290	0.0236	0.0186
5.5	0.0276	0.0246	0.0220	0.0196	0.0265	0.0224	0.0183	0.0145
6.0	0.0216	0.0195	0.0177	0.0159	0.0208	0.0177	0.0145	0.0114
7.0	0.0141	0.0130	0.0120	0.0110	0.0135	0.0116	0.0096	0.0075
8.0	0.0098	0.0092	0.0085	0.0079	0.0093	0.0081	0.0066	0.0052
9.0	0.0071	0.0067	0.0063	0.0059	0.0068	0.0059	0.0048	0.0038
10.0	0.0054	0.0051	0.0048	0.0046	0.0051	0.0044	0.0036	0.0028

10.5.3.1. Linear Stark effect, $p = 2$

In the case of hydrogen or a hydrogenic ion ($z_1 < 1$) with l-degeneracy the levels split into several components. The splitting is given by (see Unsöld [10.170], Ch. 11)

$$\Delta\omega_k = \frac{e\,a_0}{\hbar\,z_1}\,\frac{3}{2}\cdot n_k\,E\,, \tag{10.67}$$

where n_k is a whole number, characteristic of one of the components, a_0 is the Bohr radius, and z_1 is the ionic charge of the atomic hull in whose field the emission electron moves.

Because of the definition $\Delta\omega_k = C_k/(r^k)$ (see equation (10.36)), and since $E = e\,z_2/(r^2)$ and $a_0 = \hbar^2/(m\,e^2)$ (z_2 = ionic charge of the perturbing ion), we have

$$C_k = \frac{3}{4\pi}\,\frac{h\,z_2}{m_e\,z_1}\,n_k\,. \tag{10.68}$$

From this we can define a mean value for the Stark constant

$$C = \frac{3\,h\,z_2}{4\,\pi\,m_e\,z_1}\,\bar{n}_k = 1.738\,\frac{z_2\,\bar{n}_k}{z_1} \tag{10.69}$$

where

$$\bar{n}_k = \frac{\Sigma\,n_k\,I_k}{\Sigma\,I_k}. \tag{10.70}$$

Here the I_k denote the relative intensities of the Stark components, which, for example, result from the wave mechanical calculations of Schrödinger [10.154]. For the Balmer lines we have the following (see Unsöld [10.170], Ch. 11)

Line	H_α	H_β	H_γ	H_δ	H_n
$\bar{n}_k$	2.24	5.96	11.8	15.9	$\approx \frac{1}{2}\,n\,(n-1)$

10.5.3.2. Quadratic Stark effect, $p = 4$

The term shift in the case of the one-particle approximation, taking hydrogen eigenfunctions and the correct energy values, is found after a perturbation calculation (see Sadjian [10.145] and Bethe and Salpeter [10.14]) to equal

$$\Delta U_{n,\,l,\,m} = \\ = \frac{E^2\,9\,n^2\,e^2\,a_0^2}{4\,Z^2\,(2\,l+1)}\cdot\left[\frac{(n^2-(l+1)^2)\,((l+1)^2-m^2)}{(2\,l+3)\,(U_{n,\,l}-U_{n,\,l+1})}+\frac{(n^2-l^2)\,(l^2-m^2)}{(2\,l-1)\,(U_{n,\,l}-U_{n,\,l-1})}\right] \tag{10.71}$$

ΔU, U in erg.

In equation (10.71) $U_{n,\,l}$ and $U_{n,\,l+1}$ denote the unperturbed energies, where n and l respectively denote the principal and orbital quantum number of the upper level. $\Delta U_{n,\,l,\,m}$ is the term shift proportional to E^2, corresponding to the line component with magnetic quantum number m. Z is the effective charge of the atomic hull (Z characterizes the field in which the emission electron is moving). $a_0 = 5.2917 \cdot 10^{-9}$ cm is the first Bohr radius; $e = 4.8029 \cdot 10^{-10}$ e · stat. cgs units.

Assuming that the disturbance of the upper spectral level is given by (10.71) while the disturbance of the lower level can be neglected, we obtain with $\Delta\omega_{n,\,l,\,m} = \frac{\Delta U_{n,\,l,\,m}}{\hbar} = \frac{C_{4\,m}}{r^4}$ and $E = \frac{e}{r^2}$, for the Stark constant the expression

$$C_{4\,m} = \frac{\Delta\omega_{n,\,l,\,m}\,e^2}{E^2}. \tag{10.72}$$

Now C_4 results by taking a suitable mean over the different m-components.

In the general (non hydrogenic) case, the Stark effect constant is given by

$$C_4 = C_{4,i} - C_{4,f},$$

when i and f respectively denotes the initial and final spectral energy level. $C_{4,i}$ is given by ([10.166a] p. 116)

$$C_{4,i} = \frac{e^2}{2\hbar} \bar{\alpha}_i = \frac{e^4}{3\hbar^2} \sum_j \frac{|\mathbf{R}_{ij}|^2}{\omega_{ij}} = \frac{e^2}{3\hbar^2 g_i} \sum \frac{S_{ij}}{\omega_{ij}}, \tag{10.73}$$

when the sum is to be taken over all levels j in the neighbourhood of level i for which optical transitions exist. In terms of oscillator strength this formula changes into (see equation (5.86))

$$C_{4,i} = 7.8 \cdot 10^{-7} \sum_j f_{ij} \lambda_{ij}^2, \tag{10.74}$$

(λ_{ij} in cm; negative f_{ij} for emission, compare ([10.166a] p. 116)). An analogous expression holds for $C_{4,f}$.

Here it should be mentioned that in the presence of large electric field strengths the quadratic Stark effect can go over into the linear Stark effect. This phenomenon was observed, for example, in helium (see Unsöld [10.168] and [10.170], Ch. 11).

Data on Stark constants can be obtained, from the works of Joos and Saur [10.72] and Kopfermann [10.90] as well as from the references listed in the appendix at the end of section 10.5.

10.5.3.3. Resonance broadening ($p = 3$)

This broadening takes place upon the interaction of like excited and non-excited atoms. The broadening involves resonance levels and thus those spectral lines originating from transitions between the resonance levels and the ground level as well as from transitions from higher levels to a resonance level. When the perturbations take place the light quanta are transferred from one atom to another. The resonance broadening was calculated, quantum mechanically among others, by Fursov and Vlasov [10.39]. Their results are in approximate agreement with those of the statistical theory (see [10.166], p. 231). In each case the Stark constant was found to be

$$C_3 = \frac{e^2 f_{nm}}{2\, m\, \omega_0}, \tag{10.75}$$

where f_{nm} is the oscillator strength of the resonance line, originating from a transition from level m with the unperturbed circular frequency ω_0.

The line profile is given in accordance with [10.39] as

$$I(\omega) = \frac{1}{2\pi^2} \cdot \frac{1 - e^{-\left(\frac{\omega - \omega_0}{0{,}394\,\delta}\right)}}{(\omega - \omega_0)^2}. \tag{10.76}$$

δ is the full half-width of the profile, and is given by

$$\delta = \frac{6 f e^2 N}{m\, \omega_0} = 12\, C_3 N. \tag{10.77}$$

Other authors have found other numerical factors, which all together do not deviate more than 60% from one another.

The formula for the line width derived by Ali and Griem [10.1h] contains an additional factor $(g_a/g_e)^{1/2}$:

$$\delta_{\text{Ali, Griem}} = 3.84\,(g_a/g_e)^{1/2}\,N\,\frac{e^2 f}{m\,\omega_0}\,.$$

In this formula, g_a and g_e denote the statistical weights of the ground ("absorbing") level and the resonance ("emitting") level.

The case when the resonance broadening and the broadening by dispersion forces are of the same order has been treated in [10.97a].

A survey of different calculations can be found in [10.20], p. 231 (Breene).

In addition to its importance in particle density measurements, resonance broadening is also significant in determining oscillator strengths. This was discussed previously in section 5.3.8.

We accentuate that the validity of the theory of resonance broadening as quoted in this section has been verified for a greater number of elements including Hg [10.43], Cd [10.128], Ar [10.115], He [10.180d] and In [10.134d]. An unusual high broadening leading to apparent line widths much in excess of those following from formula (10.77) was observed for the resonance lines of the elements Ca, Sr, Mg [10,31f], Na, K, Rb, Cs [10.138k], [10.151], Li [10.180d].

In the case of alkali resonance lines, Ya'akobi [10.180d] and Cayless [10.21f] were able to show that the excess broadening is merely a result of the simultaneous effect of customary resonance broadening and self absorption. The degree of self absorption in the center of the resonance lines in cases where the resonance broadening is dominant, is given by the formula (see (10.77) and (5.68) with the substitution $\gamma = \delta$)

$$\varkappa_0 = \frac{4\,\pi^2}{3\,\lambda} \approx \frac{13}{\lambda}\,,$$

leading to values for $\varkappa_0$ of the order 10^5 cm^{-1}.

Thus, the undisturbed profile of a resonance line in the case when $\gamma \approx \delta$ can only be observed in the region of the line wings [10.134d]. To evaluate the damping constant ($\gamma \approx \delta$) in the region of the line wings, special techniques are available, for example the hook method described in [10.134d] (see section 5.3.7).

From absorption measurements on Ca vapour heated up to a temperature of about 1000 °C, Espenhain et al. [10.31f] found at the place of the resonance line a broad absorption band. From the partial density dependence of the corresponding absorption coefficient ($\varkappa_\nu \sim N_{\text{Ca}}^2$) as measured in the wings of the absorption band, the authors conclude that the absorption is due to Ca_2-quasimolecules.

In the case of the resonance lines of Sr and Mg [10.31f], [10.175a], the nature of the broadening effect is still an open question.

10.5.3.4. Van der Waals broadening, $p = 6$

This type of broadening is observed when an excited atom is perturbed by a particle of another species. The same type of broadening is observed for like broadeners if the broadened line is not a resonance line nor a line ending on a resonance level. The profile can be calculated from the Lindholm theory of collision broadening with the restriction that in some cases the ratio between the line width and the line shift deviates from the theoretical value 2.75.

Assuming that the perturbation energy ΔU is like that of induced dipoles (see [10.170]) p. 331 and [10.166a] p. 121), we obtain in the case of LS coupling

$$\Delta U = -\frac{1}{6}\frac{e^2 a_0^2}{r^6}\alpha_p \frac{n_*^2}{Z^2}\{5 n_*^2 + 1 - 3 l (l + 1)\} . \tag{10.78}$$

The meanings of the symbols are: α_p = polarizability of the perturber, n_* = effective quantum number of the radiating atom (see equation (5.100)), Z the charge number of the radiator when the emission electron is removed, l the angular quantum number. Neglecting the perturbation energy of lower level of the radiating atom or ion, respectively, we get for the interaction constant C_6 the expression

$$C_6 = \frac{\Delta U r^6}{\hbar} = 1.01 \cdot 10^{-9}\alpha_p \frac{n_*^2}{Z^2}\{5 n_*^2 + 1 - 3 l (l + 1)\} . \tag{10.79}$$

(When the perturbation of the lower level can not be neglected, then the resulting quantity C_6 is given by the difference of two expressions (10.79) taken for the upper and lower spectral level).

When the perturbation is effected by hydrogen atoms, then we obtain with $\alpha_p = 0.663 \cdot 10^{-24}$ cm^3 under the assumption $n_* \gg 1$

$$C_6 = -1.01 \cdot 10^{-32}\left(\frac{Z \cdot 13.6 \text{ eV}}{U_i - U_r}\right) \tag{10.80}$$

in which U_i and U_r denotes the ionization energy and excitation energy, respectively, of the radiating system (atom or ion).

Numerous experimental studies on the van der Waals broadening have been carried out. These mainly involve the resonance lines of alkali and other readily vaporizable metals, which are imbedded in a principal gas (inert gas, hydrogen, nitrogen, flame gas, CO_2, and others), whereby the latter is maintained at a pressure of 1/100 up to about 1,000 atm. In these studies an apparent change in the oscillator strength of the spectral line was observed for high pressures of the admixture, which later, however, proved to be wrong. For more details see section 5.3.11.

Here we refer the reader to a comprehensive survey (1957) of studies on van der Waals broadening compiled by Ch'en and Takeo [10.23]. Further, we refer to [10.138y]. The order of magnitude of the half-width caused by admixtures amounts to about 0.1 Å per atm. pressure.

10.5.4. Special Pressure Broadening Phenomena; Molecular Line Broadening

10.5.4.1. Pressure broadening and -shift of high series member spectral lines

This phenomenon is observed only in absorption in relatively cold gaseous systems (in plasmas with appreciable electron density, high series terms disappear due to the ionization of the upper spectral levels).

We assume that the broadening of a high series member line is determined solely by the broadening of the upper level. The volume of the sphere within the orbiting electron which corresponds to the high excited level is occupied by a greater number of neutral atoms. These atoms are polarized by the electric field between the orbiting electron and the central ion, thus causing a polarization line shift and broadening. Further line broadening and line shift is effected by the scattering interaction between the (very slowly moving) orbiting electron and the perturbers. If the perturber particles have vanishing electron affinity (this applies for perturbing rare gas atoms), then from the line shift as well as from the line width conclusions can be drawn on the elastic scattering cross section of slow electrons interacting with the perturber atoms (respectively perturber molecules). For the theoretical background see, for example, [10.1 e], [10.23 a], [10.128 b].

If the perturbers are electronegative atoms (for example alkali atoms), the main effect of level broadening is given by the formation of negative ions constituted from the orbiting electron and the perturbing atom [10.138 f]. In this case, from the line width no conclusion can be drawn on the scattering cross section.

Not affected by negative ion formation is the line shift, from the study of which data for elastic electron scattering cross sections can be derived [10.138 f], [10.117 a].

10.5.4.2. Satellite bands

Satellite bands in the form of narrow diffuse bands are observed in the neighbourhood of spectral lines of metal atoms when the spectral levels are disturbed by foreign atoms or sometimes foreign molecules [10.23], [10.31 a], [10.70 c], [10.70 e], [10.70 g], [10.138 y]. These bands are observed mainly in absorption but appear also in emission [10.23], [10.31 p], [10.70 b], [10.70 f], [10.91 a], [10.91 b], [10.139 e], [10.139 f], [10.154 a]. In a first approximation, the position of these bands depends only upon the potential curve of the excited atom-perturber pair for which is characteristic the existence of a flat potential minimum [10.76 a]. Like the original line, the satellite bands show also the effect of pressure broadening and -shift [10.44 i].

A band like structure was also observed in the wings of the self-broadened resonance lines of Mg I, $\lambda = 2852$ Å [10.132 e].

Because the satellite band phenomenon is of restricted value for plasma diagnostics, we got not into further detail but refer to the literature.

10.5.4.3. Pressure broadening at high perturber densities

While at low perturber densities the pressure broadening can be well interpreted on the base of the Van der Waals interacting potential between perturbing and perturbed particle $U(r) = -\hbar C_6/r^6$ with the corresponding line shift C_6/r^6 (see formula (10.36)), this is no longer correct for high perturber densities where the mean particle distance is so low that repulsive forces must be allowed for. This is best done by replacing the van der Waals potential by the Lennard Jones potential $U(r) = -\hbar C_6/r^6 + \hbar C_{12}/r^{12}$. As a consequence, the pressure- and temperature dependence of line width and line shift is modified, while the ratio between width and shift is no longer constant ([10.9a], [10.63a], [10.63b]).

Other high density effects are the suppression of multiplet structure of spectral lines [10.70d] and the fusion of regular lines with neighbouring satellite lines ([10.23] p. 64).

10.5.4.4. Molecular line broadening

Like atomic spectral lines, molecular band lines show also the effect of natural line broadening [10.42a], Doppler broadening [10.14b], Stark broadening and pressure broadening.

The information about Stark broadening is very sparse. Breene [10.20c] has calculated Stark widths and -shifts for the Schumann-Runge system of O_2 (see also Herman and Breene [10.59a]), for the 1. and 2. positive band system of N_2 and for the β- and γ-system of NO. The calculations are performed for air at a relative density $\varrho/\varrho_0 = 1$ and for a temperature ranging from 3000 to 18,000 °K. At 14,000 °K, the Stark width is of the order 1 cm^{-1} (about 0.2 Å in the visible spectral region). Buttrey et al. [10.21e] have investigated the broadening of the $P = 26-33$ lines of the N_2^+ band at 3914 Å for the parameter range $N_i = 3 \cdot 10^{16} - 4 \cdot 10^{17}$ cm^{-3}, finding for the line width a proportionality to $(N_i)^{1/2}$. For $N_i = 2 \cdot 10^{17}$ cm^{-3}. a width of 2 cm^{-1} (about 0.3 Å) was found.

On the field of pressure broadening, the majority of information is concerned with the broadening of vibration-rotational lines, see the survey articles ([10.20a] p. 57) and [10.23].

Investigations about pressure broadening of molecular band lines involving electronic transitions have been made for Br_2 [10.163b], for N_2 and N_2^+ [10.20b], for O_2 [10.20b], [10.21d], for NO [10.20b], [10.165f], and for the OH band at 3064 Å [10.31d], [10.42a], [10.63h], [10.88b] and [10.131c]. The authors of [10.63h] used for the determination of the line profile the Zeeman scanning technique (this technique is also useful for the determination of narrow atomic line profiles [10.154b]).

The broadening constant (line width: pressure) depends upon the nature of the perturber gas [10.31d] and is of the order 0.1 cm^{-1}/atm or $\approx$ 0.01 Å/atm. When the perturber gas is H_2O [10.31d] or a flame gas [10.131e], the broadening constant depends from the rotational quantum number too.

For the quadruple lines of H_2 molecules, the effect of collision narrowing of the Doppler profile has been observed [10.138g].

10.5.5. Quantum Theory of Collision Broadening by the Stark Effect

Classical theory ceases to be valid when the de Broglie wavelength $\Lambda = h/m\,v$ is no longer small compared to the distance between the perturber and the emitting atom.

Since the de Broglie wavelength for ions of the order of 1/10 Å is small compared to the characteristic distances between the plasma particles themselves, perturbations due to ions can be treated classically. Electrons possess a de Broglie wavelength of the order of 10 Å, and hence the validity of the classical theories is questionable, at least in plasmas under high pressure.

In practice only the electron collision broadening need be treated from the standpoint of quantum mechanics. The quantum mechanical calculation of the line broadening due to electrons is based on the time dependent equations of perturbation theory, whereby the perturbed eigenfunctions are to be expanded in terms of the eigenfunctions of the unperturbed Hamiltonian operator H_0. The perturbed Hamiltonian operator contains the component of the unperturbed atom, the radiation field, the perturbing particle, the interaction between the emitting atom and the perturber, and the interaction between the emitting atom and the radiation field. The electrons are assumed to be plane waves.

When the broadening is small compared to the level separations (this is not the case for the hydrogen atom with degenerate levels) the following expression is obtained for the line profile (see Kivel [10.84] and Sadjian [10.145]):

$$I(\omega) \sim \frac{1}{\left(\frac{\gamma_2}{2}\right)^2 + (\omega - \omega_0 + \delta_1 - \delta_2)^2}. \tag{10.81}$$

Here, δ_1 and δ_2 denote the term shift of the upper and lower state, respectively, and γ_2 (neglecting γ_1) is the full half-width. The shift of the upper level 2 is given by

$$\delta_2 = -\left(\frac{2\,N_e\,v_e\,\sigma\,\pi}{3}\right) \sum_n{}' \left(\frac{\boldsymbol{r}_{2\,n}}{a_0}\right) \frac{(\omega_{2\,n} - \delta_2)}{|\omega_{2\,n} - \delta_2|} = \sum_n{}' \delta_{2\,n}\,, \tag{10.82}$$

where the sum is to be taken over all levels n, which can be excited from level 2 by collisions with electrons of velocity v_e. Further

$$\sigma = \frac{\pi}{k^2} \quad \text{where} \quad k = m\,v_e/\hbar\,. \tag{10.83}$$

$\boldsymbol{r}_{2\,n}$ is the dipole moment of the transition from level 2 to level n. $\hbar\,\omega_{2\,n}$ is the energy difference between levels 2 and n. Since we assume $\delta_2 < \omega_{2\,n}$, the sign of the line shifts determined by the different levels depends on the sign of $\omega_{2\,n}$. This means that when the perturbing electron causes a transition to a higher level, energy is released and a red-shifted photon is emitted.

Assuming the electron velocities correspond to a Maxwell distribution, we have instead of (10.82)

$$\delta_2 = -\left(\frac{N_e\,h}{6\,m}\,\Lambda\right) \left\{ \sum_{\hbar\omega_n > \hbar\omega_2} \left(\frac{\boldsymbol{r}}{a_0}\right)^2 e^{-\hbar\omega_{2\,n}/(kT)} - \sum_{\hbar\omega_n < \hbar\omega_2} \left(\frac{\boldsymbol{r}_{2\,n}}{a_0}\right)^2 \right\}. \tag{10.84}$$

Λ is the de Broglie wavelength $h/m\,v$.

There exists a corresponding expression for δ_1.

The full half-width γ is given by

$$\gamma_{nl} = \frac{N_e h}{\pi\, 3\, m} \Lambda \sum_{l'=l,\, l+1} \left(\frac{\mathbf{r}(n, l'; n, l'-1)}{a_0}\right)^2 \cdot 2 \ln \left| \frac{4\, \varepsilon}{U_{n\, l'} - U_{n,\, l'-1}} \right| . \quad (10.85)$$

where summation is taken over the closest neighbors of the state (n, l), from which the transition originated (n is assumed as constant). ε denotes the kinetic energy of the electrons.

The broadening thus produced is symmetrical with respect to the shifted line center.

The line broadening is smaller than that given by the Lindholm theory. This is due to the fact that in (10.84) positive and negative terms occur, part of which compensate one another.

10.5.6. Nonadiabatic Theory of Line Broadening due to the Quadratic Stark Effect (Vainshtein and Sobelman)

The results that will be discussed here have the advantage that they can be represented in closed form. As Vainshtein and Sobel'man [10.170a] and Mandel'shtam and Mazing [10.112] have shown, the Lindholm collision theory fails in the case of line broadening in a plasma of high temperature. Mandel'shtam and Mazing [10.112] supported this statement by means of the line profiles of Ar II and He I lines in flash plasmas. The selected lines experience the quadratic Stark effect. The reason for the deviation of the electron collision broadening from that expected according to the Lindholm theory is due to the fact that at higher electron velocities the Weisskopf radius or the radius r_0 (see section 10.5.1.) can no longer be regarded as constant, but decreases in proportion to $1/v$ in the sense of the Born scattering cross section with increasing electron velocity v. Consequently, for the line width we have $\gamma \sim v^{-1}$ instead of $\gamma \sim v^{1/3}$ in the case of the quadratic Stark effect, which is here exclusively considered. Therefore, inelastic collisions gain in significance and for sufficiently large electron velocities the inelastic cross section is larger than the elastic cross section. Since inelastic collisions do not produce a shift but only a broadening, so that the shift is only determined by elastic collisions, the ratio of line width to line shift, which should equal 1.16 according to the Lindholm theory, becomes increasingly larger with increasing electron velocity. Due to the decreasing influence of the electrons on the shift, the shifting influence of the ions becomes more significant with increasing electron velocity (increasing temperature), while at the same time the broadening influence of the ions reaches the order of magnitude of electron collision broadening. Thus, with increasing temperature the ratio of line width to line shift approaches again to value 1.16.

From [10.170a] the broadening and the shift are given by

$$\gamma = \gamma_0\, J'(\beta)\,, \qquad \Delta = \Delta_0\, J''(\beta)\,. \quad (10.86)$$

γ_0 and Δ_0 are respectively, the full half-width and the shift in circular frequency units according to the Lindholm theory. For the case when only one of the two levels of the spectral transition is perturbed, the parameter β is given by the expression

$$\beta = \frac{Z\,\mu}{m}\,\frac{\pi\,\Delta U}{8\,kT}\left(\frac{S_{12}}{3\,g_1\,a_0^2\,e^2}\right)^{1/2} = \frac{Z\,\mu}{m}\,\frac{\pi\,\Delta U}{8\,kT}\left(f_{21}\,\frac{13.6\text{ eV}}{\Delta U}\right)^{1/2}. \tag{10.87}$$

Here Z denotes the charge and μ the mass of the perturber, ΔU is the energy difference between the level U_1, in question and the closest neighboring level U_2 (perturbation level), to which transitions can take place during the perturbation. It is assumed that only one perturbation level exists*; S is the line strength of the transition from U_1 to U_2, f_{21} the corresponding oscillator strength, g_1 is the statistical weight of the level in question. The parameter β essentially depends on temperature and differs strongly for electrons and ions due to their different masses. The ratio β is known when, besides the temperature, the quantities ΔU and S are also known. Instead of S the Stark constant C_4 can also be known (e.g., by experiment), which is related to S and ΔU via the expression

$$C_4 = \frac{Z^2\,e^2\,S}{3\,\hbar\,\Delta U\cdot g}. \tag{10.88}$$

J' and J'' are tabulated in [10.170a] as functions of β. The functional dependence is shown in Figure 10.14. For $\beta > 1$ the Lindholm theory is valid.

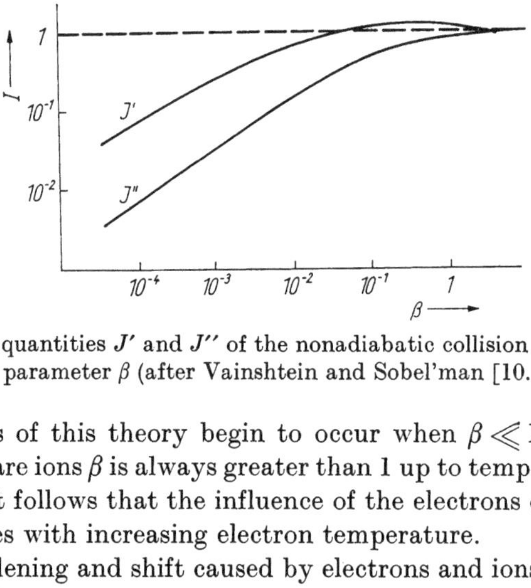

Figure 10.14. The quantities J' and J'' of the nonadiabatic collision theory as a function of the parameter β (after Vainshtein and Sobel'man [10.170a])

Strong deviations of this theory begin to occur when $\beta \ll 1$. When the perturbing particles are ions β is always greater than 1 up to temperatures of 10^6 °K. From the figure it follows that the influence of the electrons on the broadening and shift decreases with increasing electron temperature.

The total broadening and shift caused by electrons and ions can be expressed by the relationships

$$\gamma = \gamma_0^e\left\{J'(\beta_e) + \left(\frac{Z^2\,m}{M}\right)^{1/6} J'(\beta_i)\right\}, \tag{10.89}$$

$$\Delta = \Delta_0^e\left\{J''(\beta_e) + \left(\frac{Z^2\,m}{M}\right)^{1/6} J''(\beta_i)\right\}. \tag{10.90}$$

* The case in which several perturbation levels exist is described in [10.157a], page 546.

The ratio γ/Δ is a unique function of β_e. Figure 10.15 illustrates the experimentally determined dependence of γ/Δ as a function of β_e for an argon and helium plasma.

For the quantities γ and Δ as a function of β_e there exists good agreement between theory and experiment (see Figure 10.16).

For a given spectral line, the ratio $\gamma/\Delta = f(\beta_e)$ is a function of electron temperature T_e alone. But this dependence is too weak as to allow a determination of T_e from measured γ/Δ-values [10.138r].

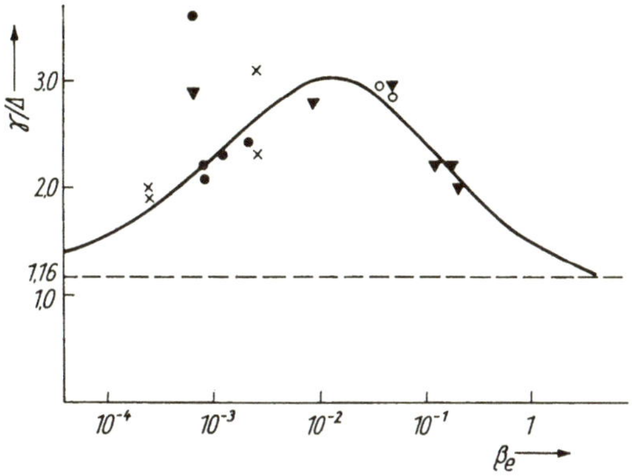

Figure 10.15. Ratio γ/Δ as a function of the parameter β_e. The dashed line corresponds to the Lindholm theory. The different measurement points refer to: × – Ar II lines, ● – He I lines (after measurements of Mandl'shtam and Masing), ▲ – He I lines (after measurements of Wulff [10.179]) (after Mandel'shtam and Mazing [10.112])

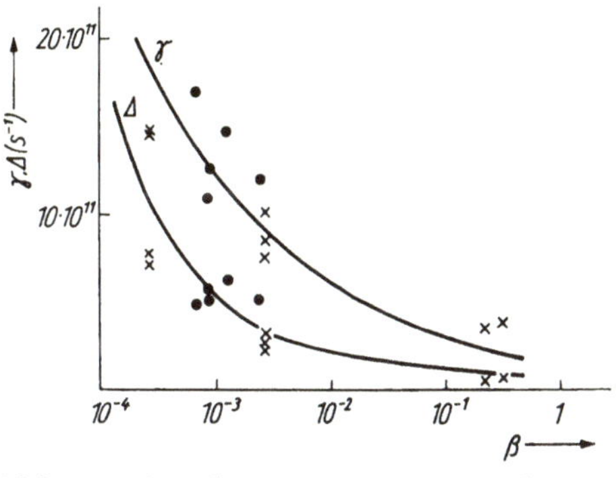

Figure 10.16. Line width γ and line shift Δ as a function of the parameter β_e. Solid curve: calculated values; × – measurement values, for known ΔU, ● – measurement values for calculated ΔU (after Mandel'shtam and Mazing [10.112])

10.5.7. Interaction of Broadening Caused by Electrons and Ions

In the simplest case the line broadening can be described by the collision theory, which gives a dispersion profile. Then the half-widths and the shifts, determined separately by the electrons and ions, add linearly:

$$\gamma = \gamma_e + \gamma_i\,, \qquad \Delta = \Delta_e + \Delta_i\,. \tag{10.91}$$

For the special case of the quadratic Stark effect, we have

$$\gamma = \gamma_0^e \left(1 + \left(\frac{Z_2^2\, m}{M}\right)^{1/6}\right), \qquad \Delta = \Delta_0^e \left(1 + \left(\frac{Z_2^2\, m}{M}\right)^{1/6}\right). \tag{10.92}$$

In the nonadiabatic case this expression goes over into expressions (10.89) and (10.90).

The situation becomes more complicated when the electron collision broadening must be described by the collision theory and the ionic broadening by the statistical theory, such as in the case for the broadening of the hydrogen lines, lines of hydrogenic ions, and certain neutral helium lines, displaying the linear Stark effect, and also in some cases of broadening by the quadratic Stark effect.

The total profile is then obtained from the relationship (Griem and Shen [10.53])

$$I(\omega) = \int W(E)\, I_e(\omega, E)\, \mathrm{d}E\,, \tag{10.93}$$

where $W(E)$ is the field strength distribution, determined by the ions. $I_e(\omega, E)$ is the electron collision profile, which occurs when the atom is disturbed quasi-statically by an ion in such a way that during the rapidly occurring electronic perturbation the additional field strength perturbation E due to the ion can be regarded as constant in time. $I_e(\omega, E)$ must be determined quantum mechanically. The total profile is obtained by averaging over all possible ionic perturbation field strengths E, which is effected in accordance with (10.93) by multiplying $I_e(\omega, E)$ with the weight function $W(E)$.

The computational procedures in the case of linear Stark effect are discussed in detail in [10.46], [10.47], [10.49], [10.51], [10.53], and [10.88] by Griem, Kolb, and Shen. Explicit data on line profiles for different temperatures and pressures can be found for hydrogenic lines in [10.50], [10.52] and ([10.47a] p. 445) (see also appendix at the end of section 10.5).

In the tables of Griem et al., the line profile is given in the reduced form $S(\alpha)$, where α denotes the reduced wavelength separation from the line center and is given by

$$\alpha = \frac{\Delta\lambda}{E_0}\,. \tag{10.94}$$

E_0 denotes the normal field strength defined by

$$E_0 = \left(\frac{4\,\pi\,N}{3}\right)^{2/3} \cdot e \tag{10.95}$$

for the case of the linear Stark effect. The tabulated data refer to the parameter range $T = (0.5-8) \cdot 10^4\,{}^\circ\mathrm{K}$, $N_e = 10^{16} - 10^{19}\,\mathrm{cm}^{-3}$. For the applicability of approximate solutions in a specific case, see section 10.5.8.2a.

While in the case of linear Stark effect the line profile is symmteric, the profile of spectral lines displaying quadratic Stark effect is asymmetric. In terms of the reduced wavelength

$$x = \frac{\lambda - \lambda_0 - d\lambda}{w_\lambda} \tag{10.96}$$

the line profile $j(x)$ is given by ([10.47a] p. 90)

$$j(x) = \frac{1}{\pi} \int_0^\infty \frac{W_r(\beta)\, d\beta}{1 + (x - \alpha^{4/3} \beta^2)^2}. \tag{10.97}$$

Here d_λ and w_λ denote the shift and half-half width due to electron impact, expressed in wavelength units, while the ion broadening parameter (or asymmetry parameter) α (not to be confused with the reduced wavelength α in the case of linear Stark effect) is given by

$$\alpha = \left(\frac{C_4\, E_0^2}{e^2\, w}\right)^{3/4} \tag{10.98}$$

w in formula (10.98) denotes the electron impact half-half width in angular frequency units.

The profile $j(x)$ is tabulated in [10.47a] for various values of the ion broadening parameter α and Debye shielding parameter r (for definition see head of table 10.6). $W_r(\beta)$ is given in table 10.6.

A tabulation of the quantities w_λ, d_λ/w_λ and α for various spectral lines emitted by a number of elements (He through Ca and neutral Cs) covering the temperature range 2500–80,000 °K is given in ([10.47a] p. 454). The tabulated values refer to an electron density of $N_e' = 10^{16}$ cm^{-3}. To get the corresponding values for other electron densities N_e, the quantities w_λ and α have to be multiplied with N_e/N_e' and $(N_e/N_e')^{1/4}$, respectively.

The resulting half-half width and Stark shift for neutral emitters is given by the formulas

$$w_{\lambda,\,\text{result}} \approx w_\lambda(N_e/N_e')\,[1 + 1.75\,\alpha(N_e/N_e')^{1/4}\,(1 - 0.75\,r)]\,, \tag{10.99}$$

$$d_{\lambda,\,\text{result}} \approx w_\lambda(N_e/N_e')\left[\frac{d_\lambda}{w_\lambda} \pm 2\,\alpha(N_e/N_e')^{1/4}\,(1 - 0.75\,r)\right] \tag{10.100}$$

The formulas (10.99) and (10.100) are valid for $\alpha \leq 0.5$ and $r \leq 0.8$. For greater values of α and r, the line width and line shift must be derived from the full profile $j(x)$.

Note, that deviations between measured and calculated line broadening parameters are due to incorrect oscillator strengths respectively matrix elements used for the calculation of these parameters [10.159], [10.173].

For single ionized emitters, the calculations of Griem [10.47a] based on classical straight path approximation, neglecting the lower state broadening, give too low values for the line width as compared with experimental values (see, for example, [10.20f]). For improved calculations giving better agreement see [10.1f], [10.1g], [10.23j], [10.47c], [10.47f], [10.138s]. Simplified formulas for the widths of higher charged ion lines has been worked out by Davies and Roberts [10.23k]. They can be used approximately also for describing the widths of single charged ion lines.

10.5.8. Line Profiles due to Superposition of Zeeman Effect and Linear Stark Effect

When the Zeeman-splitting (see section 10.3) comes into the same order of magnitude as the line width due to linear Stark effect, then the resultant profile cannot be obtained by a simple folding procedure but must be found by detailed calculations, as has been done for the profiles of the hydrogen lines Ly_α, Ly_β and H_α [10.131i], [10.131j], [10.131k], [10.131l], [10.131m] for the parameter range $N_e = 10^{15}-10^{18}$ cm^{-3}, $H \leq 1.2 \cdot 10^5$ G, $T = 10^4-4 \cdot 10^4$ °K. Calibrating experiments giving the profile of the lines $H_\alpha - H_\delta$ as function of N_e in the range $(3-10) \cdot 10^{15}$ cm^{-3} at $H = 69$ kG has been performed by Drawin et al. [10.25e], showing a sufficient agreement with theory.

Similar calculations for the profiles of some hydrogen lines of He I are in preparation (see [10.23p], [10.23q]).

In practice, the corrections of the Stark profile due to the superposition of the Zeeman effect are of importance only if the magnetic field strength is in excess of 1—10 kG.

10.5.9. Determination of Plasma Parameters from Line Broadening due to Particle Interaction

10.5.9.1. Separation of different factors causing line broadening

Line broadening due to particle interaction is used to determine the plasma parameters and in particular the particle densities. It should be ensured that the broadening caused by interacting particles is not confused by the Doppler effect, but that the latter is so small that it can be either neglected or eliminated. The elimination of thermal Doppler broadening is not critical when the gas temperature is known and when the Doppler width does not exceed the width related to particle interaction. In cases of unknown gas temperature, the Doppler width can be eliminated by decomposing the total line profile into a Gaussian profile and a dispersion profile (see section 2.9.2 and [10.119a]), but the accuracy of this procedure is relative low.

For emission lines displaying the Stark effect and the Doppler effect only, Kling and Beth [10.84c] applied a separating procedure yielding the electron density N_e and the Doppler temperature T directly. First, for a number of N lines with known Stark constants $C_1, \ldots, C_N$ differing one from each other, the full half widths $\Delta\lambda_n$, $n = 1 \ldots N$ has to be measured. Subsequent, for each line a curve $N_e = f^{(n)}(T)$ has to be constructed for which $\Delta\lambda = \Delta\lambda_n$. From the intersectional point of all curves, the actual value for N_e and T can be derived.

Based on a similar principle, de Vries [10.171h] was able to determine the electron density ($N_e = 10^{15}-10^{17}$ cm^{-3}) and the ion temperature ($T_i = 10^4-10^5$ °K) in a He-plasma from the widths of the lines He II 4686 and 3203 Å.

In the following, we base our discussion on a profile determined only by the particle interaction. Here we should take into account that several factors causing broadening can overlap one another. If we are dealing with spectral lines which display a linear Stark effect upon interaction with charge carriers, such as in the case for hydrogen and hydrogenic ions, the line width is so strongly influenced by this effect that as a rule the influence of the broadening produced by the neutral particles can be neglected. On the other hand, for lines which exhibit the quadratic Stark effect due to charge carrier interaction, the broadening caused by interaction of the emitting atoms and the neutral perturbers must be allowed for. Resonance broadening plays a large role as an additional broadening produced by neutral particles (see section 10.5.3.3); this takes effect when the upper or the lower level of the spectral line is a resonance level. Rompe and Schulz [10.141] succeeded in separating the influence of the charge carrier broadening from that due to resonance broadening in the total broadening of certain mercury lines emitted by a maximum-pressure mercury arc.

They investigated the half-widths of the lines 5770 Å (transition $2\,{}^1P_1 - 3\,{}^3D_2$) and 10,140 Å ($2\,{}^1P_1 - 2\,{}^1S_0$) (see transition scheme in Figure 10.17) as a function

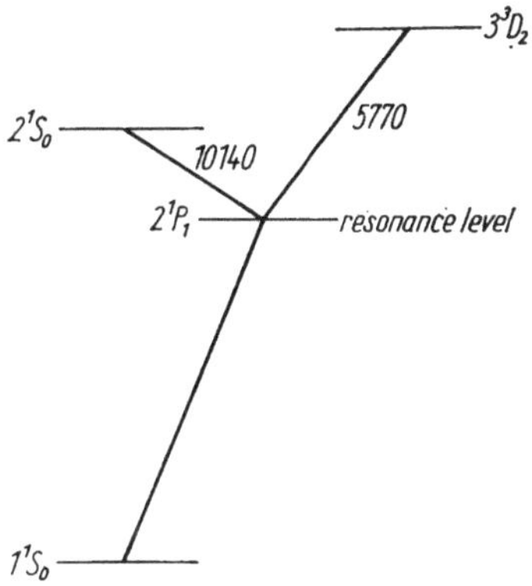

Figure 10.17. Transition scheme for the spectral lines Hg I 5770 and 10,140 Å

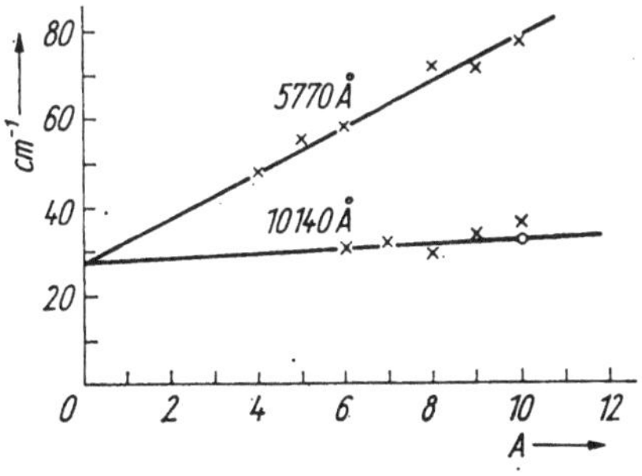

Figure 10.18. Dependence of the full half-width of the lines 5770 and 10,140 Å on the arc current intensity (after Rompe and Schulz, Z. Physik **110** (1938) 223)

of current intensity. Both lines correspond to transitions to a resonance level, and are therefore broadened by charge carriers and resonance broadening.

Figure 10.18 gives the full half-width of these two lines as a function of the current intensity I ($I \sim N_e$) (as a measure of the wave number; multiplication by $2\pi c$ gives the full half-width γ in circular frequency units).

Since both lines correspond to transitions to the same level, the difference of the widths is given only by the difference between the levels $2\,^1S_0$ and $3\,^3D_2$, for which only the quadratic Stark effect is active. The fact that the width of the line 10,140 hardly varies with increasing current intensity (i.e., with increasing electron density), lies herein that the quadratic Stark effect of the term $2\,^1S_0$, which lies relatively deep below the ionization limit, is vanishingly small. Then the difference of the line widths applies practically only to the broadening of the $3\,^3D_2$ level. For this Rompe and Schulz [10.141] gave a Stark constant of $C = 2.57 \cdot 10^{-13}$ cgs. Allowing for the broadening due to electrons and ions $\Big($the total broadening is equal to the broadening produced by the electrons multiplied by the factor $\left\{1 + \left(\frac{2\,m_e}{m_i}\right)^{1/6}\right\}\Big)$ we obtain the following expression for the width of the $3\,^3D_2$ level:

$$\gamma\,(3\,^3D_2) = 4.41 \cdot 10^{-6} \cdot N_e\, T^{1/6} \quad \text{rad} \cdot \text{s}^{-1}\,. \tag{10.101}$$

10.5.9.2. Applicability limits of the statistical theory, the collision theory, the adiabatic, and nonadiabatic theory. Examples

a) *Linear Stark effect*

The frequency up to which the collision theory holds is equal to the Holstein limiting frequency, defined by equation (10.51), and in the case of the linear Stark effect is given by

$$\Delta\omega_{\mathrm{H}}\,(p = 2) = \frac{0.2\,v^2}{C_2}\,. \tag{10.102}$$

For the Balmer lines of hydrogen one obtains the limiting values $\Delta\lambda_{\mathrm{H}}$, listed separately in Table 10.7 for electrons and ions as the perturbing particles.

Table 10.7. Limiting values $\Delta\lambda_H$ for the validity of the collision theory of line broadening for the case of electrons and protons as the perturbing particles (after Traving [10.166], p. 74)

T		25,000°	10,000°	5000°	3000 °K
H_α	electrons	1160 Å	460	220	140
	protons	1.26 Å	0.50	0.24	0.16
H_β	electrons	240	96	48	28
	protons	0.26	0.10	0.06	0.04
H_γ	electrons	96	38	18	12
	protons	0.10	0.04	0.02	0.012
H_δ	electrons	64	26	12	8
	protons	0.06	0.02	0.014	0.008

Since different theories are applied to the line broadening in the case of the hydrogen atom, the following schematization must be considered (see Traving [10.166], p. 74):

1. Below a proton concentration, which is subject to the condition

$$N_1 \frac{4\pi}{3} \varrho_W^3 = 1$$

(ϱ_W is given by (10.40)) or

$$N_1 = 3{,}6 \cdot 10^{11} \frac{T^{3/2}}{(n(n-1))^3} \qquad (10.103)$$

the Lindholm collision theory applies to the broadening caused by the protons (small perturber density). n is the principal quantum number of the upper level. The collision broadening due to electrons need not be allowed for in this range.

2. In the range $N_1 \leq N < N_2$, where N_2 via

$$\frac{1}{\gamma_e} = \frac{r_0}{v_i}$$

with r_0 from (10.49) is determined by

$$N_2 = 5{,}5 \cdot 10^{12} \frac{T^{3/2}}{(n(n-1))^3} \qquad (10.104)$$

the statistical theory gives a valid description of the line broadening due to ions (sections 10.5.5. and 10.5.6). The broadening by electrons is slight and in most cases can be neglected.

3. We finally consider the range $N_2 \leqq N < N_3$.
Let N_3 be defined by

$$\gamma_e = \Gamma$$

where Γ is the full half-width of the statistically broadened line. This amounts to $\Gamma = 2 \cdot 2.9\, \Delta\omega_0$ for symmetrical splitting into line components, where $\Delta\omega_0$ is given by (10.54). For hydrogen lines we obtain

$$N_3 = 5.6 \cdot 10^{15} \frac{T^{3/2}}{(n(n-1))^3}. \qquad (10.105)$$

In this range the ion broadening is described by the statistical theory, the electron broadening by the collision theory, and both broadenings overlap one another independently. According to Griem [10.45d] the total profile is approximately given by

$$I(\Delta\lambda) = \frac{\Delta\lambda_n}{\pi} \frac{I_0}{\Delta\lambda^2 + \Delta\lambda_n^2} + \frac{\Delta\lambda_n}{\pi} \int\limits_{-\infty}^{+\infty} \frac{S\left(\frac{x}{E_0}\right)}{(\Delta\lambda - x)^2 + \Delta\lambda_n^2} \, \mathrm{d}\left(\frac{x}{E_0}\right) \qquad (10.106)$$

where

$$\Delta\lambda_n = \gamma \frac{\lambda^2}{4\pi c}. \qquad (10.107)$$

S is the statistical profile resulting from the superposition of several Holtsmark distributions corresponding to the different line components (see Traving [10.166], p. 77).

γ is obtained from expression (10.50) where C_2 is to be replaced by the mean value of the Stark constants for the Stark components (see section 10.5.3.1). For sufficiently high ion densities the one-particle approximation (section 10.5.2.1) can be used instead of the Holtsmark distribution (see Figure 10.19).

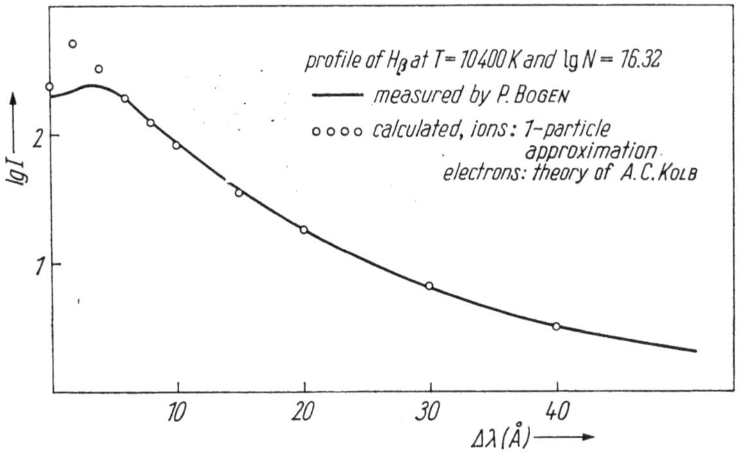

Figure 10.19. Comparison of H_β-line profile measured by Bogen [10.17] with the theoretical profile calculated in accordance with the one-particle approximation for the ion broadening and the Kolb theory for the electron collision broadening (after Traving [10.166], p. 86)

The figure contains the line profile of H_β measured by Bogen [10.17]. The line was emitted by a water vapor plasma at $T = 10{,}400$ °K and an ion density $\lg N = 16.32$ (solid curve). The figure further gives the theoretical profile, whereby the one-particle approximation was employed for the statistical ion broadening and the theory developed by Kolb [10.87] was taken for the electron collision broadening. (Kolb used for the quantity γ an expression somewhat different from (10.50).) The theoretical profile is in good agreement with the experimental one.

If higher demands are to be imposed on accuracy then $1/v$ in the above equations must be averaged over the Maxwell velocity distribution.

For transitions with low quantum numbers in the above stated particle density range it is advisable to refer back to the line profile calculated by Griem, Kolb, and Shen (see section 10.5.7) on the basis of the nonadiabatic theory which is tabulated for example in [10.47a]. Note, that these profiles are symmetric, while in reality the hydrogen lines show a slight red-blue asymmetry [10.40c], [10.60d], [10.131g], [10.155b], [10.155c], [10.159a], [10.174b] which is of no significance in those cases when the electron density has to be derived from the full half width.

4. Above the density N_3 the half-width determined by electron collision is larger than that determined by ionic collision. The resulting profile follows from a superposition of the statistical ionic profile upon the electron collision profile. So far this case has practically not been realized.

5. Above the density N_4, which is given by an expression analogous to (10.103)

$$N_4 = 2.8 \cdot 10^{16} \frac{T^{3/2}}{(n\,(n-1))^3} \qquad (10.108)$$

the electron broadening must also be statistically calculated.

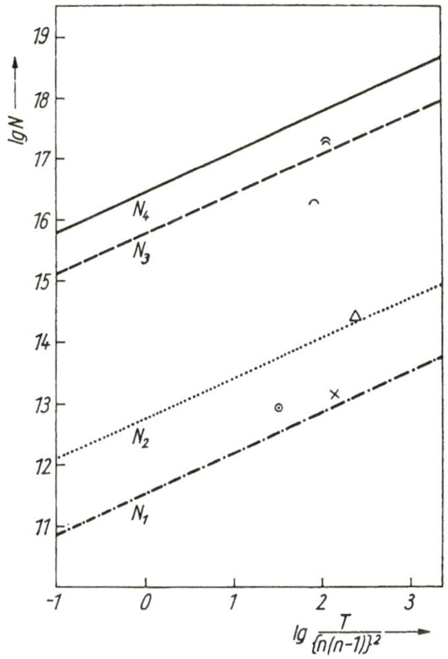

Figure 10.20. Limiting particle densities $N_1 - N_4$ (see text) as a function of $\lg \frac{T}{(n\,(n-1))^2}$. The symbols in the figure refer to the H_β line, which was emitted by the following light sources: ⌢⌢ water-stabilized arc 250 A, ⌢ water-stabilized arc 15 A, ⊙ solar atmosphere, △ O 8.5 main-sequence star (10 Lac), × B3 supergiant (55 Cyg). (after Traving [10.166], p. 76)

Figure 10.20 illustrates the limiting densities $N_1 - N_4$ plotted as functions of $\lg \frac{T}{(n\,(n-1))^2}$. The symbols in the figure refer to the H_β line, which was emitted by different sources.

Finally, we mention the case of a moderately rarefied high-temperature plasma ($N_i = 10^{13}$–10^{15} cm^{-3}, $T = 10^4 - 10^6$ °K), whereby according to Kogan [10.86] deviations of the Holtsmark statistical profile occur due to the thermal motion of the ions.

b) *Quadratic Stark effect*

As an example we shall consider the broadening of a spectral line with the Stark constant $C = 2.57 \cdot 10^{-13}$ cgs (this corresponds to the broadening of the Hg I 3 $^3D_2$ level, see section 10.5.8.1). The Holstein limiting frequency in the case of the quadratic Stark effect is

$$\Delta\omega_H = 2.15 \cdot \frac{v^{4/3}}{C_4^{1/3}} .$$

The limit up to which the collision theory can be applied is given when $\gamma/\Delta\omega_{\mathrm{H}} > 1$. With $\gamma = 11.37\, C_4^{2/3}\, v_4^{1/3}\, N$ the condition for the applicability of the collision theory in the case of electron broadening is

$$\frac{2.19 \cdot 10^{-18}\, N_e}{T^{1/2}} < 1 . \qquad (10.109)$$

For $T = 10^4$ °K it follows

$$N_e < 4.6 \cdot 10^{19}\ \mathrm{cm}^{-3} . \qquad (10.110)$$

In practice ($T \approx 10{,}000$ °K, $p < 100$ atm) we can therefore reckon throughout with electron collision broadening.

The condition for applying the collision theory to the ion broadening is in the case of perturbing particles, which have the mass of the mercury atom (= 200.6:16 oxygen mass), given by

$$\frac{9.43 \cdot 10^{-16}\, N_i}{T^{1/2}} < 1 . \qquad (10.111)$$

With $T = 10^4$ °K it follows that

$$N_i < 1.06 \cdot 10^{17}\ \mathrm{cm}^{-3}. \qquad (10.112)$$

Now the broadening by electrons in the validity range of the collision theory is about 8 times larger than the broadening produced by ions, so that the latter represents only a corrective factor. Thus, we can also describe the ion broadening in approximate fashion by the collision theory for ion densities of several 10^{17} cm^{-3} and obtain for the total width

$$\gamma = \gamma_e \left[1 + \left(\frac{2\, m_e}{m_i}\right)^{1/6}\right], \qquad (10.113)$$

where γ_e is the half-width determined by electron collision broadening alone.

For higher electron temperatures (according to Mandel'shtam and Masing [10.112] the following is evident for temperatures significantly exceeding 10,000 °K), the adiabatic theory no longer holds for electron collision broadening. With regards to the nonadiabatic theory see section 10.5.6. The transition from adiabatic to nonadiabatic conditions is manifested herein that the ratio of line broadening to line shift exceeds the value 1.16 obtained in the Lindholm theory.

c) *Transition from quadratic to linear Stark effect*

For some spectral lines, particularly in the helium I spectrum, a transition from the quadratic to the linear Stark effect is observed in the presence of high electric field strengths. At the same time "forbidden" spectral transitions take place which contradict the selection rule $\Delta l = \pm 1$. For example, besides the allowable transitions $2\,^1P - 4\,^1D$ of He I there occur the forbidden transitions $2\,^1P - 4\,^1P$ and $2\,^1P - 4\,^1F$, as soon as the field strength E exceeds 15 kV/cm (see Foster [10.34], and also Bethe and Salpeter [10.14] sect. 57). In this case the transition probabilities of the allowable lines are altered.

The way in which the broadening is to be calculated depends on the magnitude of the charge carrier density. If the latter is so low that

$$\gamma \ll 2\pi \frac{\Delta E}{h} \tag{10.114}$$

applies, wherein ΔE is the energy difference between the upper spectral level of the broadened line and the nearest perturbing level [10.31c], [10.44u], then the broadening can be described in terms of the quadratic Stark effect with some correction. This case can be realized when the forbidden components of the spectral line under consideration have a greater distance from the allowed one.

In the region of higher electron densities, the broadening is strongly influenced by linear Stark effect. This case is complicated and the computation has only been carried out in a few cases, such as for example, by Unsöld [10.168] for the lines He I 4471 and 4026 Å.

In the case when the quadratic Stark effect dominates, the line broadening can be described in a first approximation by the collision broadening theory of Lindholm. Then for the broadening due to one kind of charged perturber particles we have the formula (see table 10.4 in section 10.5.1.3)

$$\gamma = 11.37 \cdot C_4^{2/3}\, v^{1/3}\, N\,. \tag{10.115}$$

In order to allow for the influence of the linear Stark effect, Traving ([10.166], p. 88) derived for the frequency shift as a function of the perturber distance the relation

$$\Delta\omega = \frac{C_2}{r^2}\left(\frac{\varrho_0^2}{\varrho_0^2 + r^2}\right), \tag{10.116}$$

which for $r \ll \varrho_0$ goes over into $\Delta\omega = C_2/r^2$ (linear Stark effect) and for $r \gg \varrho_0$ into $\Delta\omega = C_2\,\varrho_0^2/r^4 = C_4/r^4$. This statement is now introduced into the Lindholm collision theory. The resulting line broadening γ and shift Δ are given by

$$\gamma = \frac{2\pi^3 N C_2^2}{v}\int_0^\infty \sin^2\left(u - \frac{u\,u_0}{\sqrt{u^2 + u_0^2}}\right)\frac{du}{u^3}, \tag{10.117}$$

$$\Delta = \frac{\pi^3 N C_2^2}{2v}\int_0^\infty \sin 2\left(u - \frac{u\,u_0}{\sqrt{u^2 - u_0^2}}\right)\frac{du}{u^3}. \tag{10.118}$$

The parameter

$$u_0 = \frac{\pi\, C_4}{2\, \varrho_0^3} \cdot \frac{1}{\bar{v}} = \frac{2\, C_4}{\varrho_0^3} \cdot \frac{1}{\bar{v}} \tag{10.119}$$

is a measure for the decreasing influence of the linear Stark effect with increasing u_0.

The values $\gamma(u_0):\gamma\,(u_0 = \infty)$ and $\Delta(u_0):\Delta\,(u_0 = \infty)$ are tabulated in [10.166], p. 89. Here, instead of the quantities γ and Δ the quantities σ_r and σ_i have been listed, which are connected to γ and Δ via the respective relationships $\gamma = 2\, N\, v\, \sigma_r$ and $\Delta = N\, v\, \sigma_i$.

The parameter u_0 is different for electrons and ions. This, however, does not change the fact that the γ- and Δ-values, respectively determined by the electrons and ions, add linearly.

It should be mentioned that for high electron temperatures the deviation from classical conditions must be taken into account. In particular, from the measurement results of Wulff [10.179] on helium lines for $T = 30{,}000$ °K and $N_e = 3 \cdot 10^{16}\ \mathrm{cm}^{-3}$ if follows that for lines, which display a predominant quadratic Stark effect, the line shift is approximately twice as small as in the case where the classical theory is valid, while the broadening can be described by the Lindholm theory. This agrees with the arguments put forward in section 10.5.6.

10.5.9.3. Different methods for determining plasma parameters from line profiles

Plasma parameters, in particular particle densities, can be determined from the half-width of spectral lines, the line shift, and the total line profile.

If particle densities are to be found from the half-width, it must be clarified which theory is compatible with this half-width. This can be done, for example, by recording the total line profile in a few exemplary cases and comparing this to the theoretical line profile. If the line profiles are in agreement, the theory can be considered as valid. In most cases it suffices to use classical theories.

It is more difficult to determine the particle densities from the line shift. When the perturbing particles are electrons, the classical theories in most cases no longer apply and the shift is determined along with the temperature.

General guidelines can only be established when the plasma parameters are to be determined from the half-width.

As long as the collision theory applies the line-width γ is proportional to the perturber density N_2 which includes different perturbing particles with different weight factors. The latter, among other things, depend on charge and mass and are weakly temperature dependent via the particle velocity. We shall briefly indicate the relationship between half-width, temperature, and density of the perturbers.

For $p = 2$ (linear Stark effect, singly charged perturbers):

$$\gamma_i \sim N_i\, T^{-1/2}\,, \tag{10.120}$$

the exact relationship follows from (10.50) and (10.45).

For $p = 3$ (resonance broadening):

$$\gamma = 2\pi^2 C_3 N . \tag{10.121}$$

For $p = 4$ (quadratic Stark effect and singly charged ions):

$$\gamma = 9.70 \cdot 10^2 C_4^{3/2} N_e T^{1/6} \left[1 + \left(\frac{2 m_e}{m_i}\right)\right]^{1/6} . \tag{10.122}$$

For $p = 6$ (van der Waals broadening):

$$\gamma = 8.08 \cdot C_6^{2/5} \left(\frac{8 k}{\pi \mu}\right)^{3/10} N T^{3/10} , \tag{10.123}$$

where N is the concentration of perturbing particles and μ is the reduced mass of the emitting atom and the perturber.

Equations (10.120)–(10.123) can be brought on the form

$$\gamma = A \cdot T^b N . \tag{10.124}$$

Formula (10.124) is of most practical use as interpolation formula for the determination of particle density N from measured line width γ if the quotient γ/N is known from calibrating measurements and if the calibrating temperature differs from the temperature of the plasma under consideration not more than by a factor 2. Within this temperature limit, formula (10.124) with the exponent b given by (10.120)–(10.123) can also be used if the line broadening phenomenon deviates from the classical behavior accepted in the Lindholm theory.

Due to the weak dependence of γ/N from T, it is not possible to determine the particle temperature from collision broadening measurements directly (an indirect determination of T_e from γ is possible if T_e is a unique function of N_e, as it is the case, for example, in a simple thermal plasma of known pressure). On the other hand, the weak temperature dependence enables to determine particle number densities from γ when the temperature is only known approximately.

In the range of the statistical broadening there exists a somewhat different relationship between the half-width and the particle density. On the basis of the Holtsmark distribution in the case of the linear Stark effect we have for the full half-width for symmetrically broadened lines

$$\Delta\omega_{1/2} = 7.8 C_2 \cdot N_i^{2/3} \tag{10.125}$$

(see Margenau [10.113]).

In most cases it happens that in addition to the statistical broadening of the ions the collision broadening due to electrons must be allowed for. For the most important Balmer lines and certain helium lines the half-widths as function of ion density and temperature can be derived from the line broadening data given by Griem, Kolb, and Shen [10.50], [10.52], which are also reproduced in [10.47a].

A direct tabulation of the full half width (respectively fractional intensity width) for some hydrogen lines based on the profiles calculated by Griem, Kolb and Shen has been given by Hill [10.60b], [10.60c] for electron temperatures between 10,000 and 40,000 °K and for electron densities of $10^{14}-10^{19}$ cm^{-3}. For the purpose of interpolation the formula

$$N_e = C(N_e, T_e)\, \Delta\lambda^{3/2} \tag{10.126}$$

($\Delta\lambda$ = full half width in Å, while C is a slowly varying function of N_e and T_e) is of much use which can be brought on the form

$$\frac{N_e}{(\Delta\lambda)^{3/2}} = \sum_{j=0}^{3} C_j (\ln \Delta\lambda)^j\,. \tag{10.127}$$

The temperature dependent coefficients $C_0 \cdots C_3$ for the lines $H_\alpha \cdots H_\delta$ can be taken from [10.60c]. For the line H_β widely used in electron density determination, we reproduce the coefficients $C_0 \cdots C_3$ from the work of Greig et al. [10.45b] (note, that $\Delta\lambda$ is the full half width in Å, while N_e is given in cm^{-3}).

Coefficients $C_0 \cdots C_3$ for the line H_β

T [10^3 °K]	C_0	C_1	C_2	C_3
10	$3.684 \cdot 10^{14}$	$-1.430 \cdot 10^{13}$	$-1.335 \cdot 10^{12}$	$+8.878 \cdot 10^{10}$
20	$3.663 \cdot 10^{14}$	$-1.284 \cdot 10^{13}$	$-6.095 \cdot 10^{12}$	$+1.340 \cdot 10^{12}$
30	$3.627 \cdot 10^{14}$	$-1.223 \cdot 10^{13}$	$-2.808 \cdot 10^{12}$	$+4.119 \cdot 10^{11}$
40	$3.611 \cdot 10^{14}$	$-1.419 \cdot 10^{13}$	$-2.269 \cdot 10^{12}$	$-8.960 \cdot 10^{11}$

For nonhydrogen lines showing linear Stark effect, formula (10.126) can be used for extrapolation purposes, with $C(N_e, T_e) \approx$ const at least for a temperature range 1:2 and for an electron density range 1:10.

In the case of spectral lines, which exhibit an intensity drop in the center and an intensity maximum on both sides of the line center, the charge carrier density can also be measured from the distances at which these maxima appear. However, first calibration measurements between the distance at which the maxima occur and the half-width of the line must be performed [10.16b], [10.82]. The method employed by Kitaeva [10.79] for determining the charge carrier density from the H_β line of a plasma containing hydrogen and by Hermel [10.59b] on the line He I 4471 Å for the determination of electron density in a pulsed plasma containing He is very convenient insofar as no intensity calibration of the spectrogram has to be carried out. This method yields results which are only slightly less accurate than those obtained by the method of determining the charge carrier density from the half-width.

The electron density can be determined also from the intensity ratio at two properly selected wavelength positions within a broadened spectrum line, as shown by Powell [10.138b] for the H_β line, or, alternatively, from the ratio of the center intensities, obtained for two different spectrographic slit widths [10.170b].

Table 10.8. Experimental studies on spectral line broadening due to Stark effect and plasmadiagnostical applications. Abbreviation: Qr = Order of magnitude; $\Delta\lambda$ = full half width. References are given in short symbols (for example, "23o" in used instead of "[10.23o]")

Emitter	Z	Main elements in plasma	Parameter range T [10^3 °K]	N_e [10^{15} cm^{-3}]	$\Delta\lambda$ [Å]	Reference
Al I	13	N, O	15	700	3 – 4[1])	23o
Al II	13		Or 20	100–400	0.3 – 2	110
Al III	13		30 –40	100–600	0.4 – 6	110, 119, 112a.
Ar I	18	Ar	9 –21	5–400	0.1 – 15	21g, 31g[2]), 63i, 66a[2]), 66b, 115, 116, 123c[2]), 131c, 136, 138a, 138l, 154c, 158a.
Ar II	18	Ar, H	10 –40	2–560	0.2 – 10	8d, 21g, 21h, 66b, 66c, 70a, 88c, 110, 111, 112, 112a, 114c, 117, 118, 131c, 138a, 138n.
C I	6	C, H	10 –14	20–500	1 –>8	57a, 93a[1]), 123c[2]).
C II	6	C, He, H, O	12 – 25	40–900	1.4 –14	44t, 93a[1]), 138t.
Ca I	20		6	1	0.07–0.09	115, 116.
Ca II	20	C, H, He, O	25	235	0.24	138t
Cd I	48	C, Cd, H	1 –30	70–1700	0.04– 55	8[3]), 93d, 128[3]).
Cd II	48	C, Cd, H	10 –12	70–200	0.4 – 4	93d[1])
Cl I	17	Ne	12	100	2 – 7	123c[2])
Cs I	55	Al, Ar, Cl, Cs, N, O	2 –8	0.03–10	0.2 – 24	1, 25b, 31k, 44c, 44u, 45c, 108a[2]), 162.
Cu I	29	N, O	5 –7	3–38	0.07– 1.4	5c, 119a[4]).
D_β	1	D	13 – 35	1.2–3.8	2 – 4	56a
D_β	1	D	12 –19	74–120	[5])	134b
$H_\alpha - H_\gamma$	1	H	2 – 70	0.013–4300	0.2 –430	12, 13, 14c, 23d, 31, 41a, 52, 58, 60d, 79, 101b, 105, 106, 106a, 119b, 123d, 132, 134c, 148, 172, 177, 178, 180g.
H_δ	1	H	21 –26	13–30	22 – 40	60d
$H_\varepsilon = H_7 - H_{16}$	1	H	1.8[6])–8	0.013–5	0.3 – 12	13, 57, 149, 171a, 31o.
$Pa_{\alpha,\beta}$	1	H		15	20 – 40	58
Pa_{6-13}	1	H	2	0.013	1.2 – 5.3	171b, 171c.
$H_{\alpha-\delta}$	1	H, Ar	7.3 –17	0.2–200	1 – 70	1c, 1i, 2, 3a, 20e, 44b, 69, 78, 79, 81, 82, 130d, 134, 135a, 156, 174a, 174b.
		H, C, He, O	25	235	16	138t

Table 10.8. (Continuation)

Emitter	Z	Main elements in plasma	Parameter range T [10^3 °K]	N_e[10^{15} cm^{-3}]	$\Delta\lambda$ [Å]	Reference
$H_{\alpha-\delta}$		H, O	7–16	20 –200	5 – 80	5c, 17, 24, 25, 45d, 73, 89, 156, 178a.
		H, He	>10–>23	0.2–10000	0.7–250	31j, 78, 79, 82, 98c.
		H, Kr	*Or* 10	95	9	81
		H, N	12.1	0.82	40	156
		H, Ne	6–20	0.015–290	0.1–100	10a, 95, 104, 131, 134.
		H, Ti	20	26	4	25c
H, Ly_α	1	H, Ar	12	84	[5])	17d
He I	2	C, H, He, N Na, O	1.8[6])–>49	0.02–7000	0.1–280	1d, 16b, 23g, 23n[7]), 31j, 50, 59c[8]), 70a, 97c, 98c, 112a, 114c, 118, 138t, 139b[5]), 145, 155, 171a, 172, 178d, 179.
He II	2	He	43–200	0.08–4000	0.2–40	2a[5]), 11a, 12, 13, 23g, 36b, 59d[1]), [9]), 92, 109, 138e, 138h, 178d, 179.
Hg I	80	Hg, Xe	6–30	3–1200	0.08–38[3])	8, 43, 85, 135, 138c[10]), 140, 142.
K IX	19			980000	0.26	23k.
Kr I	36	Kr	9.5–10.6	68–259	2 –7	158a
Kr II	36		30–40	100	0.4 –1.8	110, 112a, 118.
Li I	3		5–30	0.7–800	0.2–>160	44y, 45, 60a, 101d, 180b, 180e[11])
Mg I	12		8–30	100–1200	0.1–1.3	45, 91, 96, 145, 180a.
Mg II	12		8–15	50–152	1.5–12[2])	101a
N I	7	N	12–14	10–160	0.001–14.3	130c[9]), 130e, 159.
N II	7	He, N	20–60	15–3000	0.3–10	12a, 12b, 23n, 112b, 138o, 165c.
N III	7	He, N	40–45	110–190	0.2–0.4	31e
Na I	11	He, N, Na	5–20	0.5–70	0.01–1.5	44r, 77a, 80, 91, 132h, 158.
Ne I	10	Ne	6–12	10–100	0.08–50	44f[2]), 131
Ne II	10	Ne	30–40	100	0.1–1.9	110, 112a, 118.
O I	8	H, O	12–14	10–129	0.002–71	74, 123c[2]), 130c[9]), 130e[9]), 176, 178a.
O II	8	N, O	34–36	50–156	0.15–0.7	54a, 131n.
O VI	8			980000	0.5–1.1	23k
P I	15		*Or* 12	100	12	123c
S I	16	H, Ne, O, S	11–12	5–100	1.5–32	20d, 20f, 123c.
S II	16		11–12	63–75	0.4–0.7	20d, 20f.
Si I	14	Ne	12	100	6–14	123c
Si III	14		20–40	100–600	0.2–2.6	110, 112a, 119.

Table 10.8. (Continuation)

Emitter	Z	Main elements in plasma	Parameter range T [10^3 °K]	N_e[10^{15} cm^{-3}]	$\Delta\lambda$ [Å]	Reference
Si IV	14		30 –40	100– 600	0.4 – 1.4	112a, 119.
Sn IV	50		20 –40	100– 600	0.3 – 2.8	110, 112a, 119.
Ti I	22			30– 60	[1])	119c
Ti II	22			30 – 60	[1])	119c
Tl I	81		6	0.6	0.05– 0.1	80
Xe I	54	Xe	7.3– 9.0	37– 196	4 –12	138d, 158a.
Zn I	30	C, H, Zn	8 –40	50–2000	0.04–18	8, 93c, 165c.
Zn II	30	C, H, Zn	10 –12	100– 700	0.6 –18	93c[1])

[1]) Values for γ/N_e are given.
[2]) Line shift data are given.
[3]) Line width is influenced by resonance broadening.
[4]) Line width is influenced by Doppler-broadening.
[5]) Line wing investigation.
[6]) This temperature value corresponds to electron energy distribution in the low energy range.
[7]) Comment.
[8]) Data for the separation of the two line maxima are given.
[9]) Vacuum UV lines.
[10]) Corrected for van der Waals broadening.
[11]) Corrected for self absorption.

When determining the plasma parameters from the line shift, simple relationships can be obtained only under the assumption that the adiabatic theory of Lindholm is applicable:

$$p = 4 \text{ (quadratic Stark effect)}: \qquad \Delta = \frac{1}{1.16}\,\gamma\,, \tag{10.128}$$

where γ follows from (10.122):

$$p = 6 \text{ (van der Waals broadening)}: \quad \Delta = \frac{1}{2.75}\,\gamma\,, \tag{10.129}$$

where γ is found from (10.123).

In other cases, appropriate formulas must be derived quantum mechanically. For interpolation, the formula

$$\frac{\Delta}{N} = B \cdot T^b \tag{10.130}$$

can be used, where the coefficient B must be derived from calibrating experiments or published data, if any.

10.5.10. **Table Appendix**

In concluding section 10.5 we have listed a number of references on experimental and theoretical investigations of the line broadening produced by particle interaction, whereby we have also cited those works, in which plasma parameter are determined from the line broadening.

Table 10.8 contains experimental data on the line broadening, produced by the linear and quadratic Stark effect and their application to plasma parameter

Table 10.9. Calculated data for profiles of spectral lines broadened by Stark effect

a) Tabulation of statistical line profiles

Reference	Spectral lines
Schmaljohann [10.150]	$H_\alpha - H_\gamma$
Verwej [10.171]	$H_\alpha - H_\gamma$
Underhill and Waddell [10.167]	H I, Lyman-Balmer-Paschen- and Brackett-series lines up to $n = 18$

b) Tabulated profiles of hydrogen and helium lines, taking into account electron collision broadening

Reference	Spectral line λ in [Å]	T-range [10^3 °K]	N_e-range [10^{15} cm^{-3}]
Griem et al. [10.47a], [10.50]	H_α	10–40	10– 1000
	H_β, H_γ	10–40	1– 100
	H_δ	10–40	0.1– 10
	Ly_α	10–40	100– 10000
	Ly_β	10–40	10– 1000
Griem et al. [10.47a], [10.52]	He II 4686	5–80	1– 10000
	He II 3203	10–80	1– 1000
Griem et al. [10.47a], [10.52]	He I 3965	20	300
	He I 4471	20	300
Griem et al. [10.48]	He I 522–46950	5–80	10–100000

c) Other line broadening data. Literature is given in short symbols (see head of table 10.8)

Emitter	Z	Parameter range T_e [10^3 °K]	Parameter range N_e [10^{15} cm^{-3}]	Reference
Al I	13	2.5–80		23o, 47a.
Al II	13	2.5–80		47a
Ar I	18	2.5–80		47a
Ar II	18	2.5–200		1f, 1g, 18a, 18b, 23e, 23l, 47a, 47c
B II	5	2.5–80		47a
Ba I	56		1.3 –7000	44q.[1])
Ba II	56		100	23e
Be I	4	2.5–80		47a
Be II	4	2.5–80		47a
C I	6	2.5–80		47a, 178b.
C II	6	2.5–80		18b, 18d, 23e, 47a, 47c, 178b.

Table 10.9. (Continuation)

Emitter	Z	Parameter range T_e [10^3 °K]	N_e [10^{15} cm^{-3}]	Reference
Ca I	20	2.5–80		47a
Ca II	20	2.5–80		18b, 18c, 23e, 47a, 170e.
Cd I	48		9.3 –6700	44q¹)
Cl I	17			47a, 138q.
Cl II	17	2.5–80		47a
Cs I	55	2.5–80		44p, 44q¹), 44u, 47a.
Cu I	29	5	0.5 –200	44p
F I	9	2.5–80		47a
F II	9	2.5–80		47a
$H_\alpha - H_\delta$	1	5–50	0.01–10000	5a, 5b, 75c.
$H_6 - H_{14}$	1	1.8	0.013²)	136a, 136b.
Ly_α	1	10–20	84–1000	47b²), 131h, 156c, 156d, 156e⁵), 171g⁴).
He I	2	1.8–80		8b³), 8c³), 18c, 23f, 44a, 47a, 47d, 47e³), 112a, 132f, 136a²), 136c²), 136d²), 138p.
K I	19	2.5–80		44p, 47a.
Li I	3	2.5–80		44r, 44z, 47a.
Li II	3	2.5–80		47a
Mg I	12	2.5–80		44q¹), 47a.
Mg II	12	2.5–80		18d, 47a.
N I	7	2.5–80		47a, 178b.
N II	7	2.5–80		18b, 23e, 23k, 23n, 47a, 47c, 178b.
Na I	11	2.5–80		18c, 44r, 47a.
Na II	11	2.5–80		47a
Ne I	10	2.5–80		47a
Ne II	10	2.5–80		47a
O I	8	2.5–80		47a, 178b.
O II	8	2.5–80		47a, 178b.
P I	15	2.5–80		47a
P II	15	2.5–80		47a
S I	16	2.5–80		47a
S II	16	2.5–80		18d, 23e, 23j, 47a, 47c, 138s.
S III	16	10–20		18d
Si I	14	2.5–80		47a
Si II	14	2.5–80		18d, 47a.
Si III	14	10–20		18d
Zn I	30		5.0 –940	44q¹)

¹) The effect of broadening in inhomogeneous electric microfield is considered.
²) Line wing investigation.
³) Forbidden components are included in calculations.
⁴) Calculated with many body formalism.
⁵) Incomplete presentation.

Table 10.10. Reference list about van der Waals type broadening. References are given in short symbols, see head of table 10.8

a) Broadening by rare gases

Emitter or Absorber	Perturbing gas				
	Ar	He	Kr	Ne	Xe
Ag I	23b	23b			
Ar I	63b, c, d, 91e, 94a, 158a, 158b.				
Ba I	134e	134e.	134e	134e	
Ba II	54g	54g			
Ca I	63a, b, 134e, 138j.	61, 62, 63, 63a, b, 134e, 138h.	63a, b, 134e, 138j.	63a, b, 134e. 138j.	138j
Cd I	8a[1]), 138j, 166c.	8a[1]), 138j.	8a[1]), 138j.	8a[1]), 138j.	8a[1]), 138j.
Cs I	9a, 22, 21l, 31h, 36d, 138z.	22, 31h, 36d, 40a.	21n	9a, 36c, 36d, 40b.	31h, 97d.
Cu I	94	94			
Fe II	60				
Ga I	134d	134d		134d	
H I		123			
He I	63b	63b	63b		
Hg I	3b, 37a, 54d, 93e, 121, 121a, 138w, z, 144, 165d.	31l, 44h, 44l, 120, 138v, 138z, 165e.	54d	31l, 121, 121a.	54d, 138c.
In I	134d	134d		134d	
K I	37b, 63e, 65a, 138u, 138z.	36e, 37b.	63b	37b	
Kr I	128b	128b	8f, 63b, 158a, 170e.	128b	
Li I	63b, 94.	63b, 94.	63b.	63b.	
Mg I[2])					
Na I	9, 10, 12c, 37c, 63b, 84a, 84b, 114a, 138z, 139.	10			
Ne I		54f, 109a, 123a, 175b.		63b, 91d, 91e, 123a.	
Ni I	96a	96a			
Rb I	21j, m, 44j, l, 54b, 84a, 132b, c, 138x, z.	21j, 54b, 132b, c, 138x, z.	44i, 44l[3]), 54b.	44k, 44l, 132b, c.	
Si I	23m, 31n.				
Sr I[2])	10, 134e.	10, 134e.	134e.	134e	
Ti I	23h, 93b.				
Ti II	93b.				
Tl I	31i	31i			
Xe I					158a

Table 10.10 (Continuation)

b) Broadening by D_2, H_2 and N_2

Emitter or Absorber	Perturbing gas D_2	H_2	N_2
Ca I		93	114c
Cs I		36d	36c, 36d, 138z.
Fe I		93	76b
Hg I	21b, c, 44l.	21b, 21c[4]), 37a, 44l, 138w, z.	37a, 138w, z, 144.
K I		37b	37b, 65a, 173b.
Na I		37c, 114a.	9, 10, 37c, 77a, 114a.
Rb I	44g, 44l.	21j, 21c, 44g, 44l.	21c, 138x, 138z.
Sr I			10
Tl I		31i	31i, 77a.

c) Broadening by other gases

Emitter or Absorber	Perturbing gas: Air	CO_2	CS	H	H_2O	Hydro-carbon-flame	O	O_2	Rb
Ca I						158			
Cu I				5c			5c		
Fe I				93					
Hg I	80	37a 144			37a	158		21b, 37a, 144.	
K I									21k
Li I						158			
Na I	80	9 10	37			65 158			
Sr I		10							
Ti I				119c					
Ti II				119c					
Zn I						63h			

1) Broadening cross section is determined by Hanle- and modulated light double magnetic resonance technique.
2) Reduced broadening parameters scaling per unit polarizability of perturbing gas are given in [10.173].
3) Satellite band broadening is included in the investigation.
4) Also considered as broadener is *HD* and T_2.

determination, while Table 10.9 refers to calculated Stark profiles and Stark widths, respectively. Table 10.10 lists a number of characteristic studies, in which the van der Waals broadening and shift, respectively, is investigated, while table 10.11 is concerned with resonance broadening of spectral lines. Finally, in Table 10.12 references are listed in which Stark constants are given. A survey of their determination can be found, for example, in [10.17e].

Not included in table 10.8 are a great number of such investigations concerned with the determination of charged particle number density under the condition

Table 10.11. Bibliographic references on resonance broadening

Emitter or Absorber	Z	Reference
Ag I	47	14a
Ar I	18	63d, 94a, 136e, 158b.
Ba I	56	134f
Ca I	20	134f
Cs I	55	45a, 180h.
Cd I	48	14a
He I	2	23i, 36a, 92b, 97a, 136e, 170d.
Hg I	80	43
In I	49	134d
K I	19	97b, 101c, 165h, 180h.
Na I	11	165g, 173a, 180h.
Ne I	10	91d, 92a.
Rb I	37	21i, 165i, 180h.
Sr I	38	134f

$N_e = N_i$ in plasmas containing hydrogen as additive element, from the half width $\Delta\lambda$ of the spectral lines H_α and H_β, if such communications contain no new information about the dependence of $\Delta\lambda$ from N_e and T_e which in these works have been assumed to follow the well approved theory of Griem, Kolb and Shen.

10.6. Limits of the Ranges in which the Line Broadening is Predominantly Determined from the Doppler Effect and Charge Carrier Interaction

In many cases it is desired to obtain quick insight into the parameter ranges for which a predominant Doppler effect or predominant charge carrier broadening is expected. Margenau [10.113] took this task upon himself and compiles his results in the form of three representations, which we give in Figure 10.21 a—c.

In the figures N denotes the density of perturbing particles and T the temperature. Figure 10.21 a refers to the linear Stark effect (for a hydrogen plasma with Stark constant $C_2 = 10$ this corresponds approximately to the case in which the H_β line is broadened). Figure 10.21 b applies to the quadratic Stark-effect for light, emitting elements, and is based on an atomic weight of 1 and a Stark constant of $C_4 = 10^{-12}$. Figure 10.21 c is connected with the quadratic Stark effect in the case of heavy emitting particles, whereby the atomic weight of the latter is taken to be 50.

Below the dashed line the line broadening due to the Doppler effect is predominant, while above the line broadening caused by particle interaction predominates.

Table 10.12. Bibliographic references on Stark-constants[1])

System	Z	Reference
Ag I	47	38
Al I	13	23 o, 67.
Ar I	18	72, 126, 131 c.
Ar II	18	72, 108, 114 c, 118, 125, 131 b, 131 c.
Au I	79	38
Ba I	56	44 q[2]), 90, 124, 137, 138.
Ba II	56	4, 72.
Bi I	83	59
Br I	35	3
C I	6	67, 68 a.
C II	6	44 t
Ca I	20	41, 72, 90.
Ca II	20	131 a
Cd I	48	1 b, 6 a, 7, 8, 37 d, 44 q[2]), 75 e, 139 a.
Cl I	17	3
Cr I	24	90
Cs I	55	8 e, 17 e, 44 p, 44 q[2]), 44 u, 55 a, 72, 75 d, e, 114 b, 131 a, 161 a, 180 f.
Cu I	29	5 c, 38, 44 p.
F I	9	98
Fe I	26	72, 76 b.
H I	1	72, 131 f.
He I	2	34, 72, 114 c, 118, 127, 136 c, 145, 163 a.
He II	2	72, 160, 161.
Hg I	80	6 a, 7, 8, 15, 15 a, 15 b, 68, 72, 75 e, 139 a.
I I	53	3
K I	19	17 e, 44 p, 54 c, 72, 131 a, 156 f, 175.
Kr I	36	72, 158 a.
Kr II	36	118
Li I	3	31 b, 44 r, 44 z, 66 d, 72, 131 a, 145.
Mg I	12	6 a, 7, 8, 28, 44 q[2]), 72, 135 a.
Mg II	12	28, 72.
N I	7	152 c, d.
N II	7	59, 152, 165 c.
Na I	11	39 a, 44 r, 44 w, x, 72, 77, 77 a, 89 a, 90, 131 a.
Ne I	10	72
Ne II	10	72, 118.
O I	8	76, 83, 134 g, 134 h, 165.
O II	8	134 g
Rb I	37	8 e, 72, 75 e, 114 b, 131 a.
Si I	14	173
Sr I	38	72, 90, 178 c.
Xe I	54	72, 158 a.
Zn I	30	6 a, 7, 8, 37 d, 44 q[2]), 93 c, 130 g, 138 m, 139 a, 164, 165 c.

[1]) A survey of the determination of Stark-constants is given in [10.17 e].

[2]) The effect of energy shift due to electric field inhomogeneities is included (the Stark constant is C_3).

In addition the diagrams include the division of the $N - T$ region into the regions I, II, and III, in which the broadening due to particle interaction must be described by different theories (see also section 10.5.8.2):

Region I: collision theory for ions and electrons,
Region II: statistical theory for ions and collision theory for electrons,
Region III: statistical theory for ions and electrons.

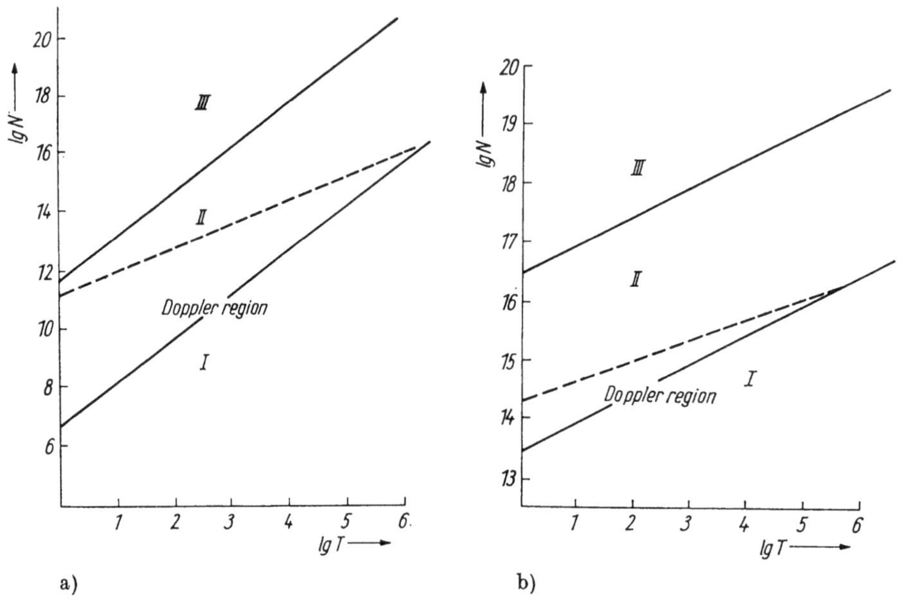

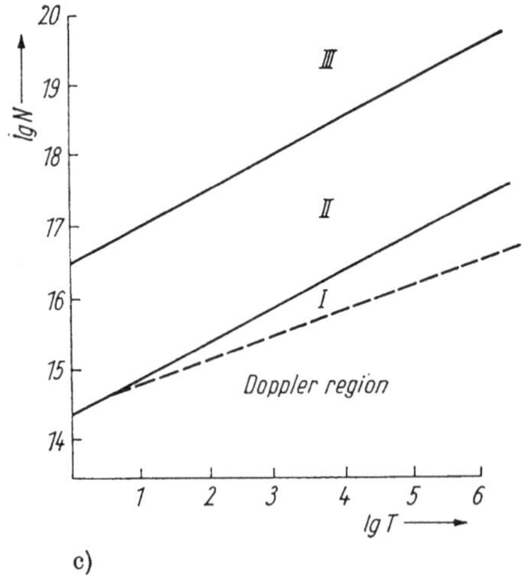

Figure 10.21a–c. $N - T$ diagrams with boundary lines (N = perturber density). Below the dashed line the Doppler effect predominates, while above this line the line broadening due to particle interaction is predominant. The solid lines effect a separation of the regions *I*, *II*, and *III* (see text), in which the broadening produced by electrons and ions, respectively, must be described by different theories (after Margenau [10.113])

The individual figures respectively refer to
a) linear Stark effect, $C_2 = 10$, at. wt. 1;
b) quadratic Stark effect, $C_4 = 10^{-12}$, at. wt. 1;
c) quadratic Stark effect, $C_4 = 10^{-12}$, at. wt. 50

10.7. Determination of Electron Density from the Emission Coefficient of Forbidden Lines

The electric microfield in a plasma not only causes a broadening of allowed spectral lines but also gives rise to the appearance of forbidden spectral lines in the neighbourhood of allowed lines. Thus, forbidden/allowed line pairs are formed. The forbidden and allowed component of such a line pair in general have a common lower level. Then the upper levels are neighbored, and in a great number of cases they are coupled one to another by allowed optical transitions.

The forbidden transitions can be classified into two groups:

Group a. Transitions between odd levels or between even levels, following the selection rules

$$\Delta J = 0, \pm 1, \pm 2,$$

$$\Delta M = 0, \pm 1, \pm 2.$$

The transition probability is a function of the homogeneous part of the electric microfield ($A \sim E^2$ if the allowed line component displays quadratic Stark effect) and can be derived from perturbation calculations (see, for example, [10.44n], [10.44o], [10.123b], [10.123d]).

Group b. Transitions between odd and even levels and vice versa, following the selection rules

$$\Delta J = 0, \pm 1, \pm 2, \pm 3 \text{ (excepted } J = 0 \to J = 0),$$

$$\Delta M = 0, \pm 1, \pm 2, \pm 3.$$

The transition probability in this case is a function of the inhomogeneous part of the electric microfield and can be found, for example, by perturbation calculations [10.44m], [10.123b].

For plasmadiagnostical purpose, practicable formulas for the intensity ratio of forbidden/allowed line components of a line pair taking into account level perturbation not only by the ion field component but also by electron collisions, are given for some elements by Grechikhin and coworkers. They are reproduced by part in the following table 10.13. Further data can be found in [10.44n], [10.180c] (for Li), in [10.44x] (for Na), in [10.44s] for Cs and in [10.171a] (for He).

It must be noticed that in the case of highly turbulent plasmas the forbidden lines can be surrounded by satellite lines stemming from non thermal plasma oscillations, as has been shown by Kunze et al. [10.92c] in experiments with a He plasma produced in a Theta pinch discharge.

Tabel 10.13. Formulas for the ratio of forbidden/allowed emission coefficient as function of electron density N. References are given in short symbols, see head of table 10.8

Emitter	Forbidden transition		Allowed transition		Ratio of forbidden:allowed emission coefficient. N in [cm^{-3}], T in [°K]	Validity range of N [10^{15} cm^{-3}]	Reference
	λ [Å]	transition	λ [Å]	transition			
Cs I	3320	$6S_{1/2}-10F$	3289.3	$6^2S_{1/2}-14^2P_{1/2}$	$2.2 \cdot 10^{-17}\,N$	2–40	44m
	3328	$6S_{1/2}-9F$	3314.1	$6^2S_{1/2}-13^2P_{1/2}$	$1.0 \cdot 10^{-17}\,N$	5–100	44m
	3369	$6S_{1/2}-8F$	3348.8	$6^2S_{1/2}-12^2P_{1/2}$	$4.2 \cdot 10^{-18}\,N$	10–200	44m
	4016		4023		$1.3 \cdot 10^{-26} \exp(-2100/T) \times N^{4/3}$	400–5100	44p
	4111		2889		$5.0 \cdot 10^{-26} \exp(-2000/T) \times N^{4/3}$	150–1300	44p
	7219	5^2D-6^2G	7229	$5^2D_{3/2}-6^2F_{5/2}$	$9.6 \cdot 10^{-22}\,N^{4/3}$	0.6–6	44p, 44s.
	7271		7280	$5^2D_{5/2}-6^2F_{5/2,\,7/2}$	$7.1 \cdot 10^{-22}\,N^{4/3}$	0.8–8	44p
Cu I	3852.8	$4P_{1/2}-5F$	3654	$4^2P_{1/2}-6^2D_{3/2}$	$6.1 \cdot 10^{-21}\,N^{4/3}$	0.2–1.5	5d
	4015.8	$4P_{1/2}-4F$	4022	$4^2P_{1/2}-5^2D_{3/2}$	$1.6 \cdot 10^{-22}\,N^{4/3} + 7.0 \cdot 10^{-19}\,N$	3–20	5d, 44m, 44p.
	4056.8	$4P_{3/2}-4F$	4063	$4^2P_{3/2}-5^2D_{3/2}$	$1.7 \cdot 10^{-22}\,N^{4/3}$	3–20	5d, 44p, 44v.
Li I	3922.5	$2P-5P$	4132	2^2P-5^2D	$5.1 \cdot 10^{-23}\,N^{4/3}$	4–40	44m, 44n, 44s.
	4601.4	$2P-4F$	4602.9	2^2P-4^2D	$4.7 \cdot 10^{-21}\,N^{4/3} + 1.6 \cdot 10^{-16}\,N$	0.1–2	44n, 44s.
	4635.6	$2P-4P$	4602.9	2^2P-4^2D	$4.9 \cdot 10^{-24}\,N^{4/3}$	10–200	44m, 44n, 44s.
	6239.8	$2P-3P$	6103.6	2^2P-3^2D	$2.1 \cdot 10^{-25}\,N^{4/3}$	100–3000	44m, 44n, 44s.
Na I	4973.4	$3P_{1/2}-5F$	4978.6	$3P_{1/2}-5^2D_{3/2}$	$1.83 \cdot 10^{-21}\,N^{4/3}$	0.6–3	44x
	4977.6	$3P_{3/2}-5F$	4978.6	$3P_{1/2}-5^2D_{3/2}$	$1.9 \cdot 10^{-21}\,N^{4/3}$	0.6–3	44x
	5528.2	$3P_{1/2}-5P$	5682.7	$3P_{1/2}-4^2D_{3/2}$	$2.66 \cdot 10^{-25}\,N^{4/3} + 9.8 \cdot 10^{-22}\,N$	400–1000	44x
	5669.8	$3^2P_{1/2}-4^2F_{5/2,\,7/2}$	5682.6	$3P_{1/2}-4^2D_{3/2}$	$1.38 \cdot 10^{-22}\,N^{4/3}$	4–20	44x
	5675.3	$3^2P_{3/2}-4^2F_{5/2,\,7/2}$	5688.2	$3^2P_{3/2}-4^2D_{3/2}$	$1.43 \cdot 10^{-22}\,N^{4/3}$	4–20	44x
	7519.9	$3P_{3/2}-4P_{1/2}$	8183.3	$3^2P_{1/2}-3^2D_{3/2}$	$(2.97 \cdot 10^{-26}\,N^{4/3} + 4.72 \cdot 10^{-22}\,N) \times \exp(-1600/T)$	3000–10000	44x

Figure 10.22. Fusion of Balmer lines at the series limit. Last line observed: $n_m = 20$ (after Schlüter [10.147])

10.8. Fusion of Lines at the Series Limit and the Problem of the Reduction in the Ionization Energy

10.8.1. Fusion of Lines at the Series Limit (after Inglis and Teller)

As is well known the higher levels become increasingly broader as the charge carrier density increases. When the charge carrier density is kept constant the level broadens as the principal quantum number n increases until the broadened level corresponding to n_m extends right up to the next higher level. This level fuses in the statistical mean with the next higher level. Each individual photon, which is emitted during a transition from such a blurred level, possesses of course a relatively sharp frequency, determined by the instantaneous perturbation, provided this can be regarded as quasistationary (in the sense of the statistical theory of line broadening), but which appears blurred due to averaging over many emission events.

To obtain a quantitative estimate for n_m corresponding to the highest level which can be still observed as discrete, Inglis and Teller [10.66] introduced the following simplifications:

1. Considering the high quantum numbers n_m to be expected the emitting atom will be regarded as hydrogenic.

2. Only the influence of the statistical line width on the smearing is taken into account; the electron broadening is neglected; electrons will only be allowed for when the broadening caused by them can be described by the statistical theory.

After introducing a correction into the numerical factor (see Traving [10.166], p. 99, and Schlüter [10.147]) they obtained

$$\lg N_i = 23.14 - 7.5 \lg n_m + 4.5 \lg Z \quad \text{for } T > \frac{2.3 \cdot 10^5 Z}{n_m} \tag{10.131}$$

and

$$\lg (N_e + Z^{3/2} N_i) = 23.14 - 7.5 \lg n_m + 4.5 \lg Z \quad \text{for } T < \frac{2.3 \cdot 10^5}{n_m} . \tag{10.132}$$

In this formula, Z denotes the charge number of the emitting system without the emission electron (the factor $Z^{3/2}$ in formula (10.132) is taken over from [10.17a]).

For nonhydrogenic atoms we must take instead of the actual principal quantum number n_m the effective quantum number n_m^* as the basis for equations (10.131) and (10.132). n_m^* is defined by the expression

$$n_m^{*2} = \frac{13.60 \text{ eV}}{U_i - U_{n_m}} \cdot Z^2 , \tag{10.133}$$

where U_i is the ionization energy, and U_{n_m} is the excitation energy of the last discrete level.

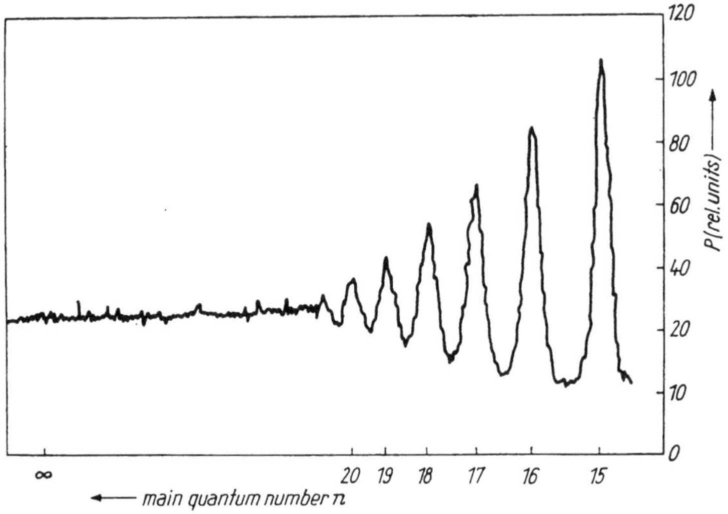

Figure 10.23. Photometer curve of the Balmer lines in the neighborhood of the series limit. P – photometer reading (after H. Schlüter [10.147])

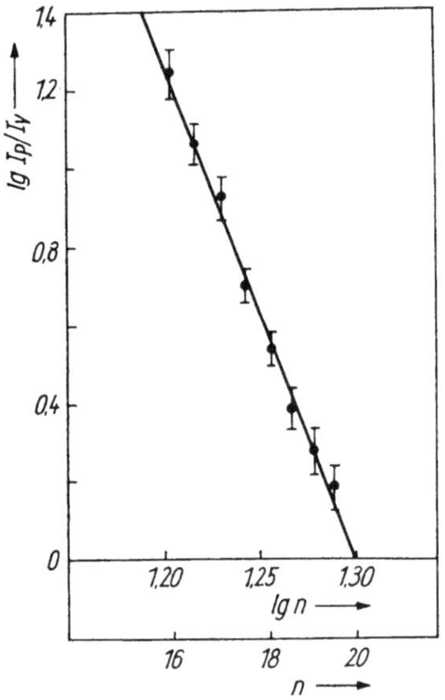

Figure 10.24. Ratio of the maximum and minimum intensity (I_p and I_v) of the Balmer lines in the vicinity of the series limit as a function of the principal quantum number n (after Schlüter [10.147])

Expressions (10.131) and (10.132) are used to determine the electron density in a plasma from the fusion of the spectral line at the series limit.

The number n_m can be determined by counting the number of visible lines of a spectral series on a spectrogram (see Figure 10.22) or on a photometer recording (Figure 10.23). To exclude subjective errors in interpreting the last line, it is advisable to plot the ratio of the maximum intensity I_P of the line and the minimum intensity I_v as a function of lg n (Mohler [10.130] and Schlüter [10.147]). Figure 10.24 illustrates such a diagram, which is based upon the photometer recording in Figure 10.23. In all the techniques used it was found that $n_m = 20$. The corresponding electron density is given by (10.132) and amounts to $N_e = 1.3 \cdot 10^{13}$ cm^{-3}.

The electron density determined by Inglis and Teller still involves an uncertainty factor of about 50%.

A higher accuracy is obtainable, when in the statistical line broadening mechanism responsible for the line merging, the influence of Debye shielding on the microfield distribution is taken into account, as has been shown by Vidal [10.171d], [10.171e] for a He- and H-plasma, respectively. The wanted N_e-values are obtained from the position of the $I_P/I_v = f(n)$-curve in the corresponding diagram.

Lastly we have to mention that Armstrong [10.2b] has derived a formula for the last observable line as function of electron density on the basis of electron collision broadening theory. When T_e is of the order 10^4 °K, the result is practically the same as that derived from the theory of Inglis and Teller.

Concerning the application of the Inglis-Teller method we give the following table.

Table 10.14. Application of the Inglis-Teller-method for the determination of electron density

Emitter	Medium	T_e [10^3 °K]	N_e [10^{15} cm^{-3}]	n_m	Reference
C V	C, H	400	15,000	8	17a
Cs I	Cs	2.1 – 8	0.002 – 12	8 – 22	25b, 130.
H I	H	1.8[1]) – 70	0.012 – 90	7 – 23	55b, 147, 171a, 171d.
H I	H_2O	12.7	84	7	73
He I	He	1.8[1]) – 30	0.026 – 32	7 – 20	171a, 171d, 178d,179.

[1]) The relative low temperature of T_e = 1800 °K corresponds to the low energy part of free electron energy distribution.

10.8.2. Weakening of High Series Member Lines by Microfield Ionization of the Upper Levels. Termination of the Partition Function

As shown by Vujnović [10.171i] for a hydrogen plasma and by Gündel [10.54e] for the general case, the emission coefficient of a spectral line originating from an upper level near the ionization limit is weakened by the effect of level ionization due to the microfield in the plasma. According to the argument of Gündel, the main influence on a certain atom in an excited level m with excitation energy U_m is effected by the nearest ion at a distance r. Considering this ion alone, the condition that the level m is ionized is

$$U_i - U_m = \chi_m < \frac{e^2}{r}\left(4\, Z_{\mathrm{eff}}\left(\frac{r}{2}\right) - 1\right), \tag{10.134}$$

where Z_{eff} denotes the effective charge number of the atom relative to a bound electron orbiting at a medium distance $\frac{r}{2}$ from the atomic nucleus, and χ_m is the nominal ionization energy of level m. Because r is a quantity statistical distributed around the mean value r_0 given by

$$r_0 = \left(3/(4\,\pi\, N_e)\right)^{1/3} \tag{10.135}$$

there exist for any given quantity m and N_e a certain positive probability (depending from m and N_e) that the level m is ionized or not ionized, respectively. This is equivalent to a reduction of the statistical weight of the discrete level m. Then for the actual (reduced) statistical weight g_m of the level m we can write (see [10.54e])

$$g_m = g_{m_0} \cdot p(\chi_m, N_e)\,. \tag{10.136}$$

g_{m_0} is the nominal statistical weight, and $p < 1$ is the probability that the level m is not ionized. Taking into account that the ionization is effected by the interaction of more than one ion simultaneously as well as by inelastic electron collisions, Gündel [10.54e] direved for $p(\chi_m, N_e)$ in the case of neutral emitters the formula

$$p(\chi_m, N_e) \approx \exp\left(-f(\chi_m)^3\left[\left(\frac{1.4\,\chi_{i\,0}}{-\,\varepsilon_m}\right)^3 + \left(\frac{\chi_{e\,0}}{-\,\varepsilon_m}\right)^{7/2}\right]\right). \tag{10.137}$$

The factor

$$f(\chi_m) = 1 + \frac{4}{3}\,\frac{\gamma_0}{(2\cdot R\,y)^{1/2}}\,(\chi_m)^{1/2} \tag{10.138}$$

allows for the nonhydrogenic behavior of the radiation atoms (for hydrogen atoms, we have $f = 1$). γ_0 is the quantum defect related to the angular quantum number $l = 0$ which is defined by the equation

$$\chi_{n,\,0} = \frac{R\,y}{(n - \gamma_0)} \tag{10.139}$$

with $R\,y = 13.6$ eV. γ_0 depends from the term structure of the excited atom and can be derived from term tables (for example, from [10.130a]). The other quantities are given by

$$\chi_{i\,0} = \frac{3\,e^2}{r_0}, \tag{10.140}$$

$$\chi_{e\,0} = (2\,\pi^{3/2}\,e^6\,N_e\sqrt{kT})^{2/7}\,. \tag{10.141}$$

The quantity $\chi_{i\,0}$ describes the reduction of ionization potential due to one disturbing ion in the case of hydrogen emitters, and the quantity χ_{i0} describes the reduction of ionization potential due to inelastic electron collisions.

Because the temperature dependence in expression (10.137) is very weak, it can be practically neglected.

Formula (10.137) can be utilized for the determination of electron density, taking as a measure for N_e the ratio of emission coefficients of two spectral lines 1 and 2 with markedly different excitation energies U_1 and U_2, one of which must be in a close position to the ionization energy. Then we have for the ratio of emission coefficients the relation

$$\frac{\varepsilon_1}{\varepsilon_2} = \frac{A_1 g_{1,0} \lambda_2}{A_2 g_{2,0} \lambda_1} \cdot \exp\left(\frac{U_2 - U_1}{kT}\right) \cdot \frac{p_1(\chi_1, N_e)}{p_2(\chi_2, N_e)}, \qquad (10.142)$$

from which N_e can be derived if $\varepsilon_1/\varepsilon_2$ has been measured and if the quantities A_1, A_2 and T (electron temperature) are known. When the difference $(U_1 - U_2)$ does not exceed 1 eV, then it is sufficient for the calculation to have a rough value for T.

To illustrate the microfield influence on spectral line intensities, the ratio p (Ar I 5739 Å):p (Ar I 4272 Å) is presented in Figure 10.25 as function of $N_e^{1/3}$. The two lines have excitation energies which are lower by 1.23 and 0.45 eV, respectively, than the ionization limit.

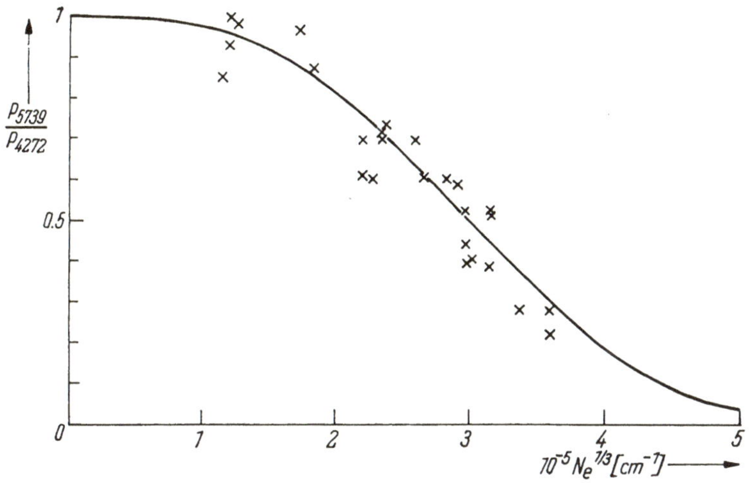

Figure 10.25. The ratio p_{5739}/p_{4272} for two Ar I spectral lines as function of electron density. After Gündel [10.54e]

Formerly, plasma parameters has been determined from line intensity measurements (see chapter 5) without allowing for the microfield influence on line intensities. This can lead to serious errors for the plasma parameters to be determined in those cases where the electron density exceeds 10^{15} cm^{-3} and when the lines used for diagnostics have excitation energies differing by not more than 1 eV from the ionization limit. In this case, the methods subjected to the greatest errors are the methods of temperature determination from relative line intensities (section 5.2.2—5.2.4), and the norm temperature method (section 5.2.6). According to Gündel [10.54e], the application of the latter method without correction will lead to an overestimation of the electron temperature.

A second consequence of the reduction of the statistical weight of an excited level caused by the microfield is the termination of the partition function

$$Q = \sum_{m=0}^{\infty} g_m \exp\left(-U_m/(kT)\right) \tag{10.143}$$

where g_m is given by (10.136). Expression (10.143) can be brought to the form

$$Q = \sum_{U_m=0}^{U_m=U_i-\Delta U_Q} g_{m_0} \exp\left(-U_m/(kT)\right) . \tag{10.144}$$

The quantity ΔU_Q which can be interpreted as the "lowering of ionization energy responsible for the termination of partition function", is a function of N_e and T_e and depends like g_m from the term structure of the atom characterized by the term defect γ_0.

A graphical representation of ΔU_Q taken from [10.54e] is given in Figure 10.26. For a rough estimate it is allowed to terminate the summation at $U_m = U_i - 1.2\,(N_e \cdot \text{cm}^3/10^{18})^{1/3}$ eV.

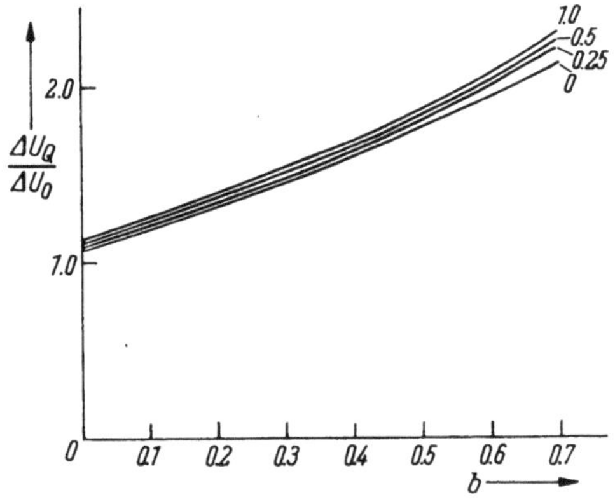

Figure 10.26. $\Delta U_Q/\Delta U_0$ as function of the quantity $b = \frac{4}{3}\gamma_0 \sqrt{\frac{\Delta U_0}{2\,Ry}}$ with $\Delta U_0 = (N_e \cdot \text{cm}^3/10^{18})^{1/3}$ eV. Parameter of the curves is the quantity $\Delta U_0/(kT_e)$. After Gündel [10.54e]

The microfield ionization of high excited atomic levels influences also the true continuous spectral energy distribution, resulting in a red shift and a diffusion of the photorecombination edges.

Due to the appearance of pseudo continua in form of narrow bands (composed from spectral lines at the series limits blurred out by the effect of line merging described in section 10.8.1 and positioned at the red flank of the photorecombination edges), an additional edge shift in the observed continuum results. The position of the resultant edge $U_n - \Delta U_n$ is given by line merging theory (see section 9.2.4 and 10.8.1).

10.8.3. Reduction in the Ionization Energy

It is of great importance, for example, when calculating the particle densities as a function of temperature and pressure with the use of the Saha equation, that the ionization energy is accurately known. For atoms and ions in a plasma this energy is lower than for isolated atoms, since the former are influenced by the interaction of the charge carriers.

The reduction in the ionization potential is brought about by the Stark effect, i.e., it is due to the interaction of the electric field E and the emitting atom, whose potential curve $U(r)$ is given a downward trend due to the additional term $e|E|\,z$ (see, e.g., Blokhintsev [10.16], p. 364).

The decisive electric field strengths E are those produced by the electric microfields. According to Ecker and Kröll [10.29a], [10.30] only the stationary (statistical mean) component is significant, since the time-dependent microfield disturbances originating in electron and ion collisions in most cases have a time constant which is smaller than the period of the electron in the last existing orbit.

Since we have already given the results of the corresponding theory in section 4.1.1., at this stage we shall merely indicate the expressions for the case of a simple plasma ($N_e = N_i$):

$$\Delta U_i = \begin{cases} \varkappa e^2 & \text{for} \quad N_e \le N_{e,cr}\,, \\ \dfrac{C e^2}{r_0} & \text{for} \quad N_e \le N_{e,cr}\,. \end{cases} \tag{10.145}$$

Here $\varkappa$ denotes the Debye length, C is a numerical constant, r_0 is the mean distance between the charge carriers, $N_{e,cr}$ is equal to those electron densities, for which just one charged particle is located inside the Debye sphere. Numerically, equation (10.145) becomes

$$\left.\begin{aligned} \Delta U_i &= 2.95 \cdot 10^{-8} (N_e/T)^{1/2} && \text{for} \quad N_e \le 2.57 \cdot 10^7\, T^3 = N_{e,cr}\,, \\ \Delta U_i &= 4.9 \cdot 10^{-7} N_e^{1/3} && \text{for} \quad N_e \ge 2.57 \cdot 10^7\, T^3 \\ & (\Delta U_i \text{ in eV})\,. \end{aligned}\right\} \tag{10.146}$$

From (10.146) it follows that for small electron densities the reduction in the ionization energy is dominated by the "polarization interaction", which finds expression in the term $\varkappa e^2$. For high densities the reduction due to "electrostatic interaction" predominates. This is expressed by the term $C e^2/r_0$, where r_0 is the mean distance between the charge carriers.

Recently, Ecker and Kröll [10.30a] have shown that in the range of low electron densities satisfying the relation $N_e \le N_{e,cr}$, the lowering of ionization energy ΔU_i is greater by a factor 1.4 than indicated in the upper part of formula (10.145) and (10.146), respectively, when the actual structure of the microfield is taken into account.

Of the same order of magnitude is the correction factor

$$(1 + \xi(\Delta U_p/(kT))^{1/2}$$

given by Gündel [10.54e], where ξ varies between 0.6 and 0.7, and ΔU_p stands for $\varkappa e^2$.

Besides these statistical theories, there exist further theories on the reduction of U_i. Some of these are mentioned in [10.30].

Unsöld [10.169] considered the potential energy of an electron in the field of two nuclei, each having charge Z, which are separated by a distance $r_m = \left(\frac{4\pi}{3} N_e\right)^{-1/3}$ (corresponding to the mean distance between an atom and the nearest ion). The electron together with one of the nuclei is taken to represent an atom, whose ionization energy is the subject under discussion.

The potential curve displays a maximum in the middle which lies below the zero axis by the amount

$$\Delta U_{i\,\text{Unsöld}} = Z^{2/3} \cdot 0.7\,(N_e/10^{18})^{1/3}\ \text{eV}\,. \tag{10.147}$$

Unsöld identified this energy difference with the lowering of ionization energy.

According to Ecker and Kröll this derivation is valid only in the case of a quasistationary interaction between the perturbing particles and the atom, whose ionization energy is studied. Quasistationarity in the sense described by Ecker and Kröll occurs when the duration of the microfield influence is larger than the period of an electron around an orbit corresponding to the last existing level. For the validity of the Unsöld derivation this leads to the relationship

$$N_e \gg 1.11 \cdot 10^5 \left(\frac{T}{M^*}\right)^3, \tag{10.148}$$

where M^* is the molecular weight of the ion. This condition can be satisfied in a whole series of plasmas. For heavier plasmas the condition is more readily satisfied than in lighter ones. For example, taking $M^* = 10$, the condition becomes for $T = 10^4\ ^\circ\text{K}$: $N_e \gg 1.11 \cdot 10^{14}\ \text{cm}^{-3}$.

Now it must be established that in a relative wide parameter range, the reduction in the ionization energy given by Unsöld is greater than that found by Ecker and Kröll [10.30], so that the question concerning which theory is actually correct bears some significance.

We still have to discuss the theory developed by Rother [10.143], which is based on the following considerations.

A level ceases to exist when the ionization frequency Z_i of this level becomes larger than the rotational frequency f_u of an electron, which is located at this level. For the last existing level n_m it is assumed that Z_i and f_u are equal. This leads to the equation

$$\sigma_{n_m} = \sqrt{\frac{2\,kT}{m}}\, N_e\, e^{-U_{n_m}/(kT)} = \frac{U_{n_m} - U_{n_m+1}}{h}\,, \tag{10.149}$$

where $\sigma_{n_m} = 4\,\pi\, a_0^2\, n_m^4$ is the ionization cross section, which can be assumed as constant for electrons, whose kinetic energy exceeds the value U_{n_m} (this is taken as positive and is reckoned from the ionization potential). The exponential factor can be set equal to 1. Finally, we have the following expression for the maximum principal quantum number n_m:

$$n_m^7 \approx \frac{3.4 \cdot 10^{25}}{N_e \sqrt{T}}\,. \tag{10.150}$$

Since $U_{n_m} = R\,h\,c/(n_m^2)$ it follows from (10.150) that the reduction in the ionization energy is given by

$$\Delta U_{i\,\mathrm{Rother}} = 7 \cdot 10^{-7}\, n_e^{2/7}\, T^{1/7}\ \mathrm{eV}\,. \tag{10.151}$$

Rother believed he had verified his theory experimentally after finding an agreement between the theoretically calculated limiting quantum number n_m and the quantum number of the last visible line at the series limit. However, as we saw in section 10.8.1, this is no proof, since a line which can no longer be observed by optical means may still exist physically when it is statistically blurred over a greater wavelength range.

Which theory is closest to reality must be decided by experiment, whereby one must rely upon the equation of state of the plasma $p = f(T, N_e)$. As long as no decisive clarification is reached, we suggest that the theory of Ecker and Kröll [10.30] should be given preference, since on the average it gives the smallest correction values, as can be seen from the following table.

Table 10.15. Lowering of ionization energy for a simple plasma ($N_e = N_i$) according to Ecker and Kröll [10.30], Unsöld [10,169] and Rother [10.143]. ΔU_i is given in eV

N_e	$T = 10^3\,°$K			$T = 10^4\,°$K			$T = 10^5\,°$K			$T = 10^6\,°$K		
[cm^{-3}]	E. K.	U.	R.	E. K.	U.	R.	E. K.	U.	R.	E. K.	U.	R.
10^{12}	0.00	0.01	0.00	0.00	0.01	0.01	0.00	0.01	0.01	0.00	0.01	0.01
10^{13}	0.00	0.02	0.01	0.00	0.02	0.01	0.00	0.02	0.02	0.00	0.02	0.03
10^{14}	0.01	0.03	0.02	0.00	0.03	0.03	0.00	0.03	0.04	0.00	0.03	0.05
10^{15}	0.03	0.07	0.04	0.01	0.07	0.05	0.00	0.07	0.07	0.00	0.07	0.10
10^{16}	0.09	0.15	0.07	0.03	0.15	0.10	0,09	0.15	0.14	0.00	0.15	0.19
10^{17}	0.23	0.33	0.13	0.09	0.33	0.18	0.03	0.33	0.26	0.01	0.33	0.36
10^{18}	0.49	0.70	0.26	0.30	0.70	0.36	0.09	0.70	0.50	0.03	0.70	0.70
10^{19}	1.06	1,51	0.50	0.93	1.51	0.70	0.30	1.51	0.98	0.09	1.51	1.36
10^{20}	2.28	3.26	0.97	2.28	3.26	1.34	0.93	3.26	1.88	0.30	3.26	2.60

10.9. Determination of Macroscopic Electric Field Strengths via the Stark Effect

Macroscopic field strengths E in a plasma can be determined via the splitting and shifting of spectral lines when E is not shielded by the microfield strength E_0 (see section 10.5.2.2). In the case of the linear Stark effect this leads to the condition

$$E \geq 3.8 \cdot 10^{-7} N_e^{2/3} \text{ V/cm}, \tag{10.152}$$

For the quadratic Stark effect the numerical factor is somewhat different. For $N_e = 10^{15}$ cm^{-3} and 10^{18} cm^{-3} the lower perceptibility limit thus becomes $E = 3{,}800$ V/cm and 380,000 V/cm, respectively.

Of course, in order to determine E the Stark constants and the splitting pattern for the linear Stark effect must be known.

In [10.21], [10.32], [10.33], and [10.103] the electric field strengths produced by glow discharges before the cathode were measured via the stationary Stark effect. The field strengths were of the order of 10^4 V/cm.

In [10.84d], the radial electric field strength in a Penning discharge at low pressure is investigated. It was found, that the observed broadening of the Stark components is due to electrical field fluctuations.

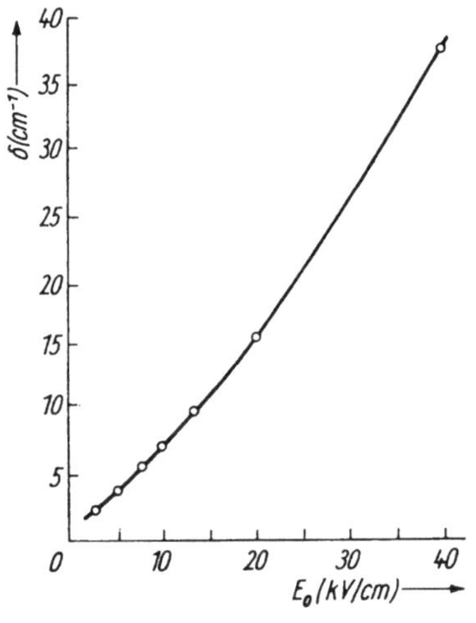

Figure 10.27. Dependence of the half-width δ of the H_β line as a function of the alternating electric field amplitude E_0 for a discharge triggered by 3.2 cm waves (after Mizuk [10.129])

On the basis of the Stark effect on certain Hg I lines in the spectral range from 2536 to 5802 Å St. John and Winans [10.71] tried to determine the electric field strength before the cathode of an arc burning in mercury vapor and found that the values must exceed $6 \cdot 10^6$ V/cm. The data on the Stark effect were taken from a work of Ishida and Hiyama [10.68].

The Stark effect can also be used to measure strong rf-fields. Instead of a line shift a time-averaged line broadening is obtained. Mizuk [10.129] considered the special case of the broadening of the H_β line upon interaction between an electromagnetic wave of 9400 MHz ($\lambda = 3.2$ cm) and a hydrogen plasma in a wave guide.

From the results a plot of the half-width δ of the H_β line was obtained as a function of the field strength amplitude E_0 (see Figure 10.27). For field strengths up to 8 kV/cm there exists a linear relationship between δ and E_0

$$\delta = \delta_0 + 0.625 \cdot E \, \frac{\text{cm}^{-1}}{\text{kV/cm}}, \tag{10.153}$$

the half-width δ_0 for a vanishingly small applied alternating electric field is dependent upon the line broadening produced by the plasma itself. In the special case of a capillary discharge for a hydrogen filling at 0.5 Torr the half-width produced by the plasma itself was $\delta_0 = 0.75 \text{ cm}^{-1}$. This corresponds to a half-width of 0.18 Å for a wavelength of 4861 Å.

For analogeous investigations in which the H_β- and H_γ-line, respectively, is used as r-field indicator, see [10.98b] and [10.98a], respectively.

CHAPTER 11

Validity Limits of the Methods of Thermal Spectroscopy discussed in Chapters 5 to 10. Some Special Methods of Nonthermal Spectroscopy

11.1. Validity Limits of the Spectroscopic Methods treated in Chapters 5 to 10

The above described methods for determining the plasma parameters were interconnected with a series of assumptions which imposed certain requirements on the plasma. The most important of these assumptions are listed below:

a) the electron velocities possess a Maxwell distribution,
b) the velocity distributions of the atoms, molecules, and ions are all Maxwell distributions,
c) the electron levels are occupied in accordance with the Boltzmann distribution,
d) the degree of ionization satisfies the Saha equation,
e) the occupation of the rotational and vibrational levels corresponds to a Boltzmann distribution, and
f) the electron temperature as well as the temperature of the atoms, molecules, and ions are all equal to one another.

It so happens that in most cases for a given measurement method not all the above stated conditions have to be fulfilled.

Further details are given in Table 11.1, in which the most important spectroscopic techniques and their corresponding prerequisites are listed.

Conditions a) to f) stated in the table impose certain requirements on the plasma under investigation. We must now enquire into the properties the plasma must have in order to satisfy these requirements.

In the following we shall discuss the above assumptions in connection with the series of experimental methods listed in Table 11.1. For this we state the technique, for which the named condition must be fulfilled.

Furthermore, we must consider alterations in the computational procedure which are demanded when some of the prerequisites are not satisfied.

The condition for the validity of Kirchhoff's law, which is the basis of the temperature determination from absorptivity and emissivity measurements, is not explicitly stated in Table 11.1. This condition as well should be clarified in connection with the other prerequisites.

Finally, we must consider more closely the quasistationarity condition as well as the condition of local Boltzmann equilibrium in the case of spatially inhomogeneous plasmas.

Our representation involves a number of simplifications. A more detailed study of the problems concerned with plasma equilibrium is given by Drawin [11.8b].

11.1.1. Condition a): The Electrons Possess a Maxwell Velocity Distribution

This condition is required when temperatures are determined from line intensities, the intensity of continuous radiation, or via particle densities. Also for the measurement of temperature from the Doppler broadening of laser radiation scattered by free electrons it is desirable that the electron velocities correspond to a Maxwell distribution. If this distribution does not occur, then only an equivalent temperature corresponding to the mean kinetic energy of the electrons along the direction of observation can be obtained.

When the electrons do not have a Maxwell distribution of velocities, the electron density n_e can be determined only approximately from the absolute intensity of continuous radiation; the temperature cannot be determined; the electron density can be only approximately found via the line profile determined by the linear and quadratic Stark effect, since the temperature also plays a role in the profile, though not as strongly as the electron density.

We now ask ourselves how deviations from the Maxwell distribution are brought about.

In plasmas heated by electric fields these deviations can occur when the energy acquired by an electron per mean free path is no longer small compared to the mean thermal energy of the electron.

On the other hand, the Maxwell distribution can be disrupted due to inelastic collisions of the electrons with atoms or ions, whereby only those electrons take part whose energy is sufficient for the excitation of levels. Because of these collisions the electron gas is depleted of fast electrons. This depletion can be compensated once the interaction between the electrons themselves becomes so strong that the Maxwell velocity distribution of the electrons is re-established in spite of energy losses due to inelastic collisions. This leads to a condition for the minimum degree of ionization.

If compensation is not achieved, especially for plasmas with a small degree of ionization, the energy loss suffered by the electrons due to inelastic collisions can be compensated when collisions of the second type occur nearly as frequently as the inelastic collisions. This leads to a condition for a minimum electron concentration.

Thus the conditions for the existence of a Maxwell velocity distribution of the electrons are:

1. The energy acquired by the electrons per mean free path λ_e must be small compared to the mean thermal electron energy $3/2 \cdot kT_e$.

2. (1): The number of electron-electron collisions ν_{ee} per unit volume and time must be large compared to the number of inelastic collisions ν_{eu} per unit volume and time. This is a prerequisite for a plasma with a larger degree of ionization. According to Rother [11.38] we must have

$$\frac{n_e}{n_0} \geq q_i \, U_i \, U_e^2 \cdot 10^{-4} \,. \tag{11.1}$$

Table 11.1.

Method	a) Electrons: Maxwell distribution	b) Atoms, molecu- les, ions: Maxwell distribution	c) Electron states: Boltzmann distribution
a. Intensity of atomic spectral lines, optically thin layer	×		×
2. Intensity of ionic spectral lines, optically thin layer	×		×
3. Emissivity and absorptivity	×		×
4. Absorption of resonance lines			
5. Absol. intensity molec. lines + bands	×	×	×
6. Rel. intensity molec. lines + bands		×	
7. Meas. of refractive index at one or more wavelengths			
8. Faraday rotation of light			
9. Thomson scattering of a laser ray on plasma electrons	desirable		
10. Intensity of the electron-ion continuum	×		
11. Doppler profile of spectral lines			
		×	
12. Zeeman effect			
13. Linear and quadratic Stark effect	desirable		
14. Resonance broadening			
15. Van der Waals broadening		desirable	
16. Fusion of lines at the series limit			
17. Stationary Stark effect			
18. Intensity ratio forbidden: allowed line			

Spectroscopic methods of plasma diagnostics

d) Degree of ionization: Saha equation	e) Rot. and vibr. levels: Boltzmann distribution	f) $T_e = T_i = T_0 = T_M = T$	Type of measurement	Determinable parameter	Dependence of measured values in thermal equil. on:
		desirable	Absol. intensity	T_e, n_0	T, P
			Rel. intensity	T_e	T
desirable		desirable	Absol. intensity	T_e, n_i	T, P
			Rel. intensity	T_e	T
×			Absol. measurement	T_e	T
			Integral absorption	n_0	T, P
	×	×	Absol. measurement	T, n_M	T, P
	×		Rel. measurement	T_{rot}, T_{vibr}	T
				n_0, n_e	T, p
				n_e, H	T, p, H
			Absol. scattering intens.	n_e	T, p
			Doppler profile Scattered radiation	T_e	T
desirable		desirable	Absol. intensity	T_e, n_e	T, p
			Intensity drop toward short wavelengths	T_e	T
			Total profile	$n_0(v_x)$	$n_0(v_x)$
			Total profile	$n_i(v_x)$	$n_i(v_x)$
			Half-width	T_0, T_i	T
			Shift	$\bar{v}_{0x}$, $\bar{v}_{ix}$	$\bar{v}_{0x}$, $\bar{v}_{ix}$
			Shift	H	H
desirable		desirable	Total profile, Line width, Line shift (})	n_e, (T_e)	T, p
			Half-width	n_0	T, p
			Half-width	n Admixture (gas)	T, p
			Determ. of last visible line	n_e	T, p
			Shift of lines	E	E
			relative	n_e	T, p

In this approximation formula U_i denotes the ionization energy in eV, $U_e = kT/e$ in eV, q_i is a numerical factor of the order of 1 and characterizes the ionization cross section (argon: $q_i = 0.24$, neon: $q_i = 0.06$). For argon with $U_i = 15.7$ eV we obtain, for example, for $U_e = 1$ eV minimum degree of ionization of $n_e/n_0 = 4 \cdot 10^{-4}$.
Instead of condition 2. (1) we can have condition 2. (2) for the existence of the Maxwell distribution.

2. (2) The number of inelastic collisions must be almost equal to the number of collisions of the second type. However, for plasmas not produced by radiation this is identical with the condition for the existence of a Boltzmann distribution of the electrons, which we shall come back to later on.

11.1.2. Condition b): The Atoms, Molecules, and Ions have a Maxwell Velocity Distribution

This is a prerequisite for determining temperature and particle densities from the absolute and relative intensity of molecular lines and bands, the Doppler profile of spectral lines, and the van der Waals broadening.

If the Maxwell distribution is disrupted, information on the plasma parameters can no longer be derived from the intensity in the molecular spectrum. On the other hand, however, we can ascertain an equivalent temperature from the Doppler profile of the spectral lines, which is obtained, for example, via the mean square velocity of the particles in the direction of observation:

$$\frac{m}{2} \overline{v_x^2} = \frac{1}{2} kT_{\text{equ.}} \,. \tag{11.2}$$

The plasma satisfies condition b) firstly, when the free pathlengths of the relevant plasma components are small compared to the geometrical plasma dimensions, and secondly, when the energy $Z\, e\, E\, \lambda_i$ acquired by the ions per mean free pathlength in an electric field is small compared to the average thermal energy $\frac{3}{2} kT_i$.

11.1.3. Condition c): The Occupation of the Electron Levels up to the Limiting Quantum Number n_m is given by a Boltzmann Distribution. Condition d): The Degree of Ionization Satisfies the Saha Equation

The existence of a Boltzmann distribution of the electrons over the levels is a prerequisite for the determination of temperatures and particle densities via spectral line intensities, as well as for the determination of the temperature from the emissivity and absorptivity, since only in this case is Kirchhoff's law valid.

The Saha equation must apply when using Kirchhoff's law for the determination of the temperature from the emissivity and absorptivity, provided continuous radiation can be measured as well. Furthermore, it is employed to obtain the electron temperature via the electron concentration, which, for example, can be approximately determined from the intensity of ionic spectral lines, the intensity of the continuous spectrum, or via the Stark broadening of spectral lines even when the Saha equation does not hold.

Griem [11.18] and Biberman and Ulyanov [11.2] have shown that due to the strong interaction via electron collisions between the higher excited levels and the continuous electron states, which correspond to the free electrons, the condition for the validity of the Saha equation is identical to that for the existence of a Boltzmann electron distribution, whereby we need only discuss the condition for the latter. This follows from the considerations below.

As is well known, for a plasma with not too low degree of ionization, a change in the occupation number of a level m is effected by electron collision or by radiation. In an optically dense homogeneous plasma (the plasma must be optically dense for all wavelengths) we can assume a detailed equilibrium for the level changes in the sense that the radiation processes, which contribute to the filling of the level m, occur just as frequently as the radiation processes, which lead to a depletion of the electrons in level m. The same applies to the electron collision processes.

In optically thin plasmas the radiation equilibrium is disturbed, insofar as for equal values of the electron temperature and the particle densities the emission processes remain approximately as frequent as before, while the number of excitation and ionization processes determined by radiation absorption is strongly reduced. The occupied levels are described by a Boltzmann distribution only when the corresponding electron collision processes, which lead to an alteration in the occupation number of the level m, are much more frequent than the processes of radiation emission from level m (see Finkelnburg and Maecker [11.13], p. 309). Then the principle of the detailed equilibrium can be applied to the collision processes only, which is a sufficient condition for the existence of a Boltzmann distribution of the electrons among the levels. The condition for this is

$$\bar{A}_{nm} \ll n_e \bar{q}_n^m v_e \left(\frac{e U_m}{kT} + 1\right), \tag{11.3}$$

where $\bar{A}_{nm}$ and $\bar{q}_n^m$ are suitable mean values for the transition probability and the excitation cross section, respectively. Taking $\bar{A}_{nm} = 10^7 \text{ s}^{-1}$, $v_e = 10^8 \text{ cm s}^{-1}$, $\bar{q}_n^m = 10^{-16} \text{ cm}^2$ (see Finkelnburg and Maecker) and the excitation energy as 10 eV, the inequality is satisfied above $n_e = 10^{14} \text{ cm}^{-3}$ with a quality factor of 1 and for $n_e + 10^{15} \text{ cm}^{-3}$ with a quality factor of 10.

In actual cases, relation (11.3) must be applied to the transition corresponding to the greatest energy gap in the level sequence of the atom (or ion) under consideration.

If this gap corresponds to a resonance transition, $\bar{A}_{nm}$ must be reduced by a factor $g \gg 1$ following from the theory of resonance radiation trapping (see, for example, Holstein [11.22]). In addition, relation (11.3) must be applied to the transition corresponding to the greatest but one energy gap. That means, $n_{e,\min}$ (minimal electron density above which a Boltzmann distribution of electronic levels will exist) must be determined in a great number of cases from two inequalities of form (11.3) which must apply simultaneously. For details of calculation see [11.8b].

Explicit formulas for $n_{e,\min}$ in the case of hydrogen atoms or hydrogen like term systems can be found in [11.2a], [11.8b] and [11.18].

It must be noticed that estimations about the existence or nonexistence of Boltzmann equilibrium cannot be made for such levels corresponding to electron excitation in inner shells (such levels are observed for example for iron atoms, see Burhorn [11.4]). These inner shell electrons are screened by the outer electrons, and it is to be expected that they have a relative low excitation cross section for which no data are available. Thus, from the existence of Boltzmann equilibrium between the outer shell electronic levels, no conclusion can be drawn on the existence of Boltzmann equilibrium between the inner shell electronic levels.

To get an impression about the order of magnitude of $n_{e,\min}$ in actual plasmas, we present the results of some relevant investigations.

For this we could refer to those studies, in which the electron temperature the gas density and the electron density are determined by independent methods, or to those studies whereby the electron temperature is first determined via relative spectral line intensities and then via absolute intensities. The equality of the two temperature thus determined would confirm the agreement between the partial Boltzmann distribution and the total Boltzmann distribution.

Semenova and Kokhanenko [11.42] investigated the pressure dependence in the range from 10 to 500 Torr of the intensity ratio of the lines Sr I 4892 and 5156 Å with excitation energies of 4.8 and 4.7 eV, respectively, as well as for the lines Zn I 3346 and 6362 Å ($U_m = 7.71$ and 7.75 eV, respectively), emitted by a carbon arc burning in air at a current intensity of 6 A. Above a pressure of $p = 100$ Torr the constancy of the intensity ratios of these lines, which have different excitation functions, indicates the existence of a Boltzmann distribution. Unfortunately, in this work no data are given on the electron densities.

Experiments of McNally [11.31] on a magnetically contained low-pressure high-current arc, which burned in carbon vapor, exhibited a temperature of about 24,000 °K, and had an electron density of 10^{14} cm^{-3}, showed that for this relatively small electron density the electron levels of the ions C II to C IV, which emit radiation in the optical range, are occupied in accordance with a Boltzmann distribution. Mass-spectroscopic measurements, which were carried out for the same electron density and at a temperature of 40,000 °K indicated that C IV and C V occur with equal frequency. Thus, the validity of the Saha equation for this plasma is established for the above given temperature and electron density.

Neumann [11.34a] investigated a medium pressure discharge in Argon in the pressure range 30—120 Torr and finds at an electron temperature of about 10^4 °K $n_{e,\min} = 3.10^{14}$ cm^{-3}.

For a Helium plasma at atmospheric pressure with an electron temperature of about 13,000 °K, Bott [11.3a] finds that a Boltzmann equilibrium will exist for $n_e = 1.5 \cdot 10^{15}$ cm^{-3}, while Hattenburg and Kostkowski [11.20] as well as Kolesnikov and Sobolev [11.28] have found for $n_{e,\min}$ values between 6.10^{15} cm^{-3} and 10^{16} cm^{-3}.

For O II at 24,000 °K, a Boltzmann equilibrium was found to exist with security when $n_e = 10^{16}$ cm^{-3} [11.1e]. The actual value for $n_{e,\min}$ was not found because the investigations were not extended to electron densities below 10^{16} cm^{-3}.

For Cs I diluted in Argon, a Boltzmann equilibrium was found to exist for $n_e = 10^{13}$ cm^{-3} [11.17b].

11.1.3.1. Condition for partial Boltzmann distribution

By a partial Boltzmann distribution is understood an electron distribution, in which the occupation numbers of the levels with principal quantum number n and upwards correspond to a Boltzmann distribution of electrons and can be described by the electron temperature T_e.

In his search to establish the existence of a partial Boltzmann distribution from the level n upwards Griem [11.18] started by considering the excitation frequency of the level n to all higher levels n' due to electron collisions:

$$\sum_{n'>n} R^c_{n'n} = \sum_{n'>n} n_e \langle v \, \sigma_{n'n} \rangle \, . \tag{11.4}$$

Here the expression in the brackets denotes the mean value of the product of the electron velocity and the effective cross section for electron collision excitation from level n to level n'. The frequency with which this occurs should be ten times greater than that with which the occupation number of the level n is reduced due to quantum transitions to a lower level n'. In the case of a one-electron system, we have

$$\sum_{n'<n} R^{r,z}_{n'n} \frac{\alpha^4 \, c \, z^4}{a_0 \, n^{9/2}} \, , \tag{11.5}$$

where α is the Sommerfeld fine-structure constant ($\alpha = e^2/(\hbar \, c)$, z is the effective charge number of the rest ion as seen by the emission electron, and a_0 is the first Bohr radius.

The condition for the validity of Boltzmann statistics from a level n upwards is now given for a one-electron system by

$$n_e \geqq 7.4 \cdot 10^{18} \frac{z^7}{n^{17/2}} \left(\frac{kT_e}{U^z_H}\right)^{1/2} [\text{cm}^{-3}] \, , \qquad U^z_H = z^2 \, U_H \, , \tag{11.6}$$

(U_H = ionization energy of the hydrogen atom). This relationship holds for the level $n = 2$ (first excited level) and upwards. Of course, we can expect that when (11.6) is valid the occupation level corresponds, subject to a deviation smaller than 10%, to a Boltzmann distribution.

For the derivation of (11.6) see Griem [11.18].

For hydrogen at a temperature of $kT_e = 1$ eV it follows from (11.6) that thermal equilibrium occurs above the principal quantum number $n = 2$ when the electron density exceeds $n_e = 6 \cdot 10^{15}$ cm^{-3}.

Expression (11.6) naturally applies only to hydrogen atoms and hydrogenic ions such as He II or Li III. But for higher quantum numbers n, relation (11.6) can be extended to other term systems by identifying the quantity U_H^z with the ionization energy U_i of the term system under consideration and introducing an effective charge number z via the relation

$$z^2 = \frac{U_i}{13.6 \text{ eV}} . \tag{11.7}$$

For lower quantum numbers n, information about the existence of partial Boltzmann distribution can be obtained from the formulas given in [11.8b].

If the Boltzmann distribution is only partially valid, say for levels upwards from n, the Saha equation of course no longer applies in its original form, but instead we have the expression, also called the Saha-Boltzmann equation (for the case of simple ionization)

$$\frac{n_e n_i}{n_m} = \frac{g_e Q_i}{g_m} \frac{(2\pi m_e kT_e)^{3/2}}{h^3} \mathrm{e}^{-U_m^*/(kT)} , \tag{11.8}$$

where

$$Q_i \sim g_i$$

(see F. Robben [11.37]). Here $n_e = n_i$ are the electron and ion concentrations respectively, n_m is the occupation number per cm³ of the level m with the statistical weight g_m, U_m^* is the ionization energy of the level m, where $m \geq n$, Q_i is the partition function of the ion, and g_i is the statistical weight of the ionic ground level.

It is often practical to use the following relationship instead of (11.8) (see Biberman and Ulyanov [11.2]):

$$y_e^2 = y_m , \tag{11.9}$$

where $y_e = n_e/(n_e^0)$ and $y_m = n_m/(n_m^0)$ are respectively the relative electron concentration and the relative occupation of the level m, referred to the electron concentration and the level occupation for the case when the Saha equation is valid.

Expressions (11.8) and (11.9) enable us to calculate the electron concentration from the occupation number of a higher excited term which can be determined via the absolute intensity of one of the lines emitted from this term [11.37].

In this connection it should be mentioned that when the partial Boltzmann distribution is valid the electron temperature can be determined from the relative intensity of spectral lines with different excitation energies in accordance with the techniques described in sections 5.2.1 to 5.2.3, if the following conditions are satisfied:

a) The upper levels of the spectral lines used for diagnostics must be in a not too close proximity to the ionization limit (see section 10.8.2). Otherwise, the spectral lines are weakened by ionization of the upper levels due to the microfield in the plasma.

b) The degree of ionization must not be too low. If $x \ll 10^{-2}$, the distribution of excited levels forming a level sequence with energy separation between neighboured levels of $\Delta E < kT_g$ (this applies in general to the higher excited levels) may be strongly influenced by collisions with heavy particles. Then, the level distribution is determined rather by the gas temperature than by the electron temperature in the case of $T_e \neq T_g$. For experimental indication of this effect see [11.16a], [11.17b], [11.27b] p. 103, 126, [11.28b], [11.28c]. For theoretical details concerning the influence of inelastic heavy particle collisions on level distribution see [11.8b]. About the influence of these collisions on temperature determination after the line reversal method see the begin of chapter 6.

11.1.4. Condition e): The Rotational and Vibrational Levels of Molecules possess a Boltzmann Distribution

This prerequisite must be fulfilled when temperatures and molecular densities are to be determined via the intensity of molecular lines. The additional requirement that the rotational and vibrational temperature are equal to the temperature of distribution of the translational degree of freedom of the heavy particles presupposes in the case of nonthermal equilibrium that the coupling between the rotational and vibrational levels on the one hand and translational degrees of freedom of the heavy particles on the other hand dominate all other couplings.

In order to have a Boltzmann distribution between rotational and vibrational levels, respectively, the relaxation time of these levels, determined by particle collision frequencies (see sections 11.1.7.1 and 11.1.7.2) must be clearly shorter than the radiative life time which is of the order of 10^{-8} s.

The additional requirement of equal rotational-vibrational and gas kinetic temperature seems to be fulfilled for pressures above 1 atm*.

As a proof for the strong coupling between rotational and translational degrees of freedom it can be said that the rotational degrees of freedom are already in equilibrium with the translational degrees of freedom after a small number of molecular collisions (in the case of nitrogen the number of collisions is five, see Greene and Toennis [11.17]). However, the interaction between the translational and vibrational degrees of freedom of the molecules is less strong and can be of the same order of magnitude as the interaction between molecular vibrations and electron translation (see following section). For low degrees of

* This is not correct, in general, for the reaction zone of flames, where the Boltzmann equilibrium can be disturbed due to chemiexcitation [11.17a].

ionization the last process plays no role in the establishment of vibrational equilibrium. Proof that for pressures above 1 atm the coupling between vibrational and translational degrees of freedom of the molecules is decisive can be deduced from relaxation-time measurements for the vibrational degrees of freedom [11.35], which agree with the theoretically expected relaxation times, assuming that the vibrational degrees of freedom enter into equilibrium with the translational degrees of freedom of the molecules (see [11.34]).

At the present time there is no proof of the existence of a Boltzmann distribution of the vibrational states for lower pressures (for rotational energy levels, Boltzmann distribution is observed at least for pressures above some ten Torr [11.34a].

11.1.5. Condition f): The Translational Temperatures of the Electrons, Ions, and Neutral Gas Atoms or Molecules are equal to one another

This prerequisite is required when the plasma temperature is determined from the intensity of molecular lines, since the occupation of the electron levels is determined by the electron temperature whereas the occupation of the rotational and vibrational levels is determined by the molecular temperature. When the two temperatures do not agree, the calculation of the occupation numbers of the energy states becomes very complicated.

The equality of the electron temperature, ion temperature, and neutral gas temperature is in general employed whendetermining the plasma parameters via the absolute intensities of spectral lines. If the prerequisite is not fulfilled, the pressure balance no longer applies in the form $p = (n_e + n_i + n_0)\,kT$ (in the case of a simple plasma), but

$$p = k\,(n_e\,T_e + n_i\,T_i + n_0\,T_0) \qquad (11.10)$$

with up to three different temperatures. Equations (5.1) to (5.3), which give the emission coefficient of a spectral lines as a function of the particle densities n_z^A and the electron temperature T_e, remain valid provided only the Boltzmann distribution applies. The Saha equation in the form

$$\frac{n_e\,n_i}{n_0} = \frac{2\,(2\,\pi\,m\,kT_e)^{3/2}}{h^3}\,\frac{Q_i}{Q_0}\cdot \mathrm{e}^{-\frac{e\,U_{i,\,\mathrm{eff}}}{kT_e}}, \qquad (11.11)$$

remains valid as before. As a rule the ion temperature agrees with the neutral gas temperature, since the interaction between these heavy particles is large. To establish the plasma properties with regard to a thermal plasma only one more parameter need be determined (T_e and $T_0 = T_i$ instead of $T_e = T_i = = T_0 = T$).

It remains to be investigated what differences between electron and gas temperature are to be expected in the case of a high-pressure plasma.

We must start from the fact that in a plasma, which is produced by ohmic heating, the electrons acquire nearly all the energy going into the plasma from the electric field due to their large mobility. This energy is again liberated via several processes (see Gurevich and Podmoshenski [11.19]):

1. By elastic collisions with heavy particles. In a weakly ionized plasma ($n_e/n_0 < 0.01$) which contains no rare gases only elastic collisions with the neutral gas need be taken into account. By means of these collision processes the heavy particle component obtains its energy, whereby a temperature T_g (gas temperature) is maintained.

2. In the case that the molecular gas exists in pure form or as an admixture of at least 1% in an atomic gas, the energy loss of the electron gas due to excitation of molecular vibrations must be allowed for. This type of energy transfer is extremely effective and exceeds all other energy transfer processes. Via the strong coupling between the vibrational and translational degrees of freedom the latter preserve a considerable energy supply, and a temperature T_v (vibrational temperature) $\approx T_g$ (gas temperature) is maintained. This differs considerably less from the electron temperature than in case 1.

3. Inelastic collisions, whereby electron levels are excited, must be allowed for. Equivalently, the energy loss due to radiation can be taken into account. This process is significant in the case of arcs with high power concentrations of the order of 100 W/cm³ and higher, whereby a high percentage of the arc's power is radiated (see Neumann [11.33]).

4. The energy loss of the electron gas due to electronic heat conduction, including heat conduction by ambipolar diffusion, must be considered.

Gurevich et al. [11.1a], [11.19] set up the following system of equations for the power balance per cm³ arc plasma for the case of a low degree of ionization:

$$L = 4\,\frac{m}{M}\,k\,(T_e - T_g)\,n_e\,\bar{v}\,(n_0\,Q_e^0 + n_i\,Q_e^i) + n_e\,n_M\,k\,(T_e - T_v)\,\bar{v}\,Q_v$$
$$+ \frac{n_e\,n_0\,Q_{\text{exc.}}\,(U_{\text{exc.}} + kT_e)}{\sqrt{2\pi\,m\,kT_e}} \cdot e^{U_{\text{exs.}}/(kT)}\,k\,(T_e - T_{\text{exc.}}) + \frac{1}{r}\,\frac{\mathrm{d}}{\mathrm{d}r}\left(r\,\varkappa_e\,\frac{\mathrm{d}T_e}{\mathrm{d}r}\right). \quad (11.12)$$

Here m denotes the electron mass, M the mass of the heavy particles, T_e the electron temperature, T_g the gas temperature, n_0 the concentration of heavy particles, n_e the electron density, n_i the ion density, v the mean velocity of the electrons, Q_e^0 and Q_e^i the elastic effective cross sections for collisions of electrons with neutral atoms and ions, respectively, n_m the molecular concentration, T_v the vibrational temperature, Q_v the inelastic energy exchange cross section between kinetic energy of electrons and vibrational energy, $Q_{\text{exc.}}$ the excitation cross section for electron collision, $U_{\text{exc.}}$ the excitation energy, $T_{\text{exc.}}$ the excitation temperature, and $\varkappa_e$ the electronic conductivity.

From this equation we obtain in a relatively simple fashion the difference $T_e - T_g$, when neglecting (insofar as this is permissible) the third term, which gives the energy loss of the electrons due to inelastic collisions, identifying the vibrational temperature T_v with the gas temperature T_g, and taking suitable values for the cross sections.

Gurevich and Podmoshenskii applied the system of equations (11.12) to arcs, which burned in air, in argon with approximately 1% admixture of a molecular gas, and in mercury. The arcs in air and argon burned between carbon electrodes so that these plasma contained certain admixtures of C.

Of interest is the method with which Gurevich and Podmoshenskii determined the difference between the electron and the gas temperature by carrying out an experimental check on (11.12) (this method has already been used by Witte [11.46]).

When the arc is suddenly turned off by short circuiting, the intensity of a spectral line emitted by the arc is reduced from I_0 to I_1, within a time of about 10^{-7} s. The further decrease in the intensity occurs slowly within a time of about 10^{-3} s. The rapid intensity drop lies herein that after the power supply to the electrons via the electric field is switched off, the electron temperature T_e drops very rapidly to the value of the gas temperature T_g, or to a lower value, which can deviate from T_g by a value up to $T_e - T_g$. The temperature jump $\Delta T_{\text{meas.}}$ of the electron gas follows from I_0, I_1, the electron temperature identified with the excitation temperature T_0 just before the arc is switched off, and the excitation energy U_m of the spectral line:

$$\Delta T_{\text{meas.}} = \frac{kT_0^2}{U_m} \ln \frac{I_0}{I_1} . \tag{11.13}$$

The method is so sensitive that a temperature jump of about 30 °K can still be established.

In later experiments with arcs in rare gases and in hydrogen, Aleksandrov et al. [11.1] found that under some experimental conditions instead of a sudden decrease of line intensity after arc interruption a sudden rise of intensity will happen with rise time of the order 10^{-7} s, followed by a relative slow decrease within a time interval of 50–100 µs.

The sudden rise of intensity in the period of rapid cooling of the electron gas from T_e to T_g indicates that the upper spectral level of the corresponding transition is more strongly coupled to the continuous electronic energy states (free electrons) than with the ground state. Assuming constant electron density n_e, the temperature jump $\Delta T_{\text{meas.}}$ can be found from the Saha-Boltzmann formula (11.8) leading to the expression

$$\Delta T_{\text{meas.}} = \sqrt{\frac{2}{3} \frac{kT^3}{U_m^*} \ln \left(\frac{I_1}{I_0}\right)} \quad \text{for} \quad \Delta T \ll T . \tag{11.13a}$$

For spectral lines of metal atoms, only the effect of a sudden intensity decrease was observed.

Figure 11.1 shows an oscillogram of the intensity of the line Zn I 4811 Å, which was emitted by a decaying arc in air at 100 Torr.

The measurement results for different arcs are listed in Table 11.2 along with some data partly calculated in accordance with (11.12) including the difference $\Delta T_e'$ when considering elastic collisions only and ΔT_e when also allowing for the other factors.

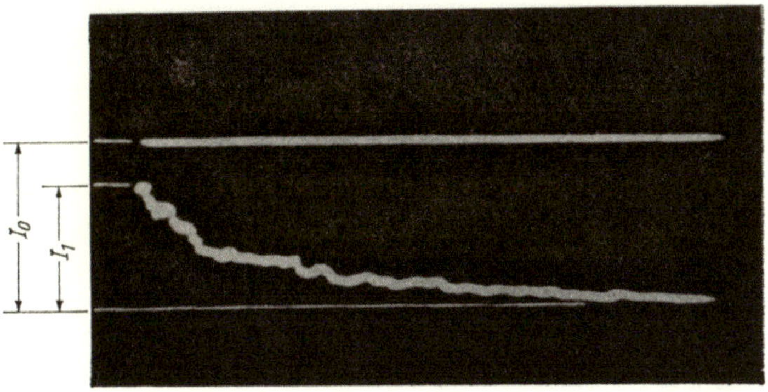

Figure 11.1. Time behavior of the emission of the spectral line Zn I 4811 Å, emitted by a decaying air arc produced at a pressure of 100 Torr. The intensity jump from I_0 to I_1 at the onset of decay is clearly seen (after Gurevich and Podmoshenskii [11.19])

As can be seen from Table 11.2 the measured temperature difference $\Delta T_{\text{meas.}}$ corresponds approximately to the temperature difference ΔT_e calculated from equation (11.12). The difference $\Delta T_e'$, however, has values which are too large.

For an argon arc at $p = 760$ Torr an unlikely high temperature difference $T_e - T_g$ of 20,000 °K is obtained when considering the elastic collisions only. This value is reduced to 10,000 °K when allowing for the energy loss of the electrons due to electronic heat conduction (in the case of the pure argon plasma). When taking into account a 1% molecular gas content a difference $T_e - T_g$ of only 400° (measured 120°) is obtained because of the transfer of electronic energy towards exciting molecular vibrations.

For a pure argon plasma the calculation of $T_e - T_g$ also yields very high values. This is due to the fact that the effective elastic cross section for electron-argon collisions is very small ($\approx 10^{-16}$ cm^2 for slow electrons). Similar relationships are expected for the other heavy inert gases. For all other gases it can be expected that for pressures starting at $p = 1$ atm the difference $T_e - T_g$ is too insignificant to be allowed for when analyzing the measurement data, since it essentially only enters into the calculation of the pressure balance [11.18].

It remains to mention that Gurevich and Podmoshenskii confirmed that $T_e - T_g$ in the case of a mercury arc depends explicity only on the gas pressure and not on the current intensity (and thus the electron density) explicitely.

Finally, we should mention the results of some other authors who measured the gas temperature and electron temperature by independent techniques. The results refer to the following plasmas:

1. Hydrogen HF-discharge. $n_e = 1.3 \cdot 10^{13}$ cm^{-3}, $T_e = 2000$ °K*, $T_g = 1900$ °K (Schlüter [11.39]).

* The relative low value of the electron temperature $T_e = 2000$ °K is derived from the low energy part of the electron velocity distribution function.

Table 11.2. Difference between electron and gas temperature for different arcs. L is the power of the arc/cm³, $\Delta T_{\text{meas.}} = a\,(T_e - T_g)$, $1 \leq a \leq 2$; $\Delta' T_e = T_e - T_g$ is calculated accordance with (11.12), assuming that only elastic collisions are active in the energy transfer, $\Delta T_e = T_e - T_g$ is calculated from (11.12), neglecting only the energy loss due to excitation of the electron levels (after Gurevich and Podmoshenskii [11.19])

Arc gas	p [Torr]	L [ergs^{-1} cm^{-3}]	T [°K]	n_e [cm^{-3}]	$\Delta T_{\text{meas.}}$ [°K]	Energy loss of electrons in [erg s^{-1} cm^{-3}] in accordance with (11.12), neglecting the excitation of the electron levels			$\Delta' T_e$ [°K]	ΔT_e [°K]
						Elastic collisions	Excitation of vibrations	Electronic heat conduction		
Air	760	$1 \cdot 10^{10}$	6000	$2 \cdot 10^{15}$	<30	$4 \cdot 10^6 (T_e - T_g)$	$2.0 \cdot 10^8 (T_e - T_g)$	$5 \cdot 10^7$	2500	50
	100	$5 \cdot 10^9$	5800	$8 \cdot 10^{14}$	250	$2.8 \cdot 10^5 (T_e - T_g)$	$10^7 (T_e - T_g)$	$15 \cdot 10^7$	7000	500
Argon + 1% molecular gas	760	$2 \cdot 10^9$	7500	10^{15}	120	$1 \cdot 10^5 (T_e - T_g)$	$2.5 \cdot 10^6 (T_e - T_g)$	10^9	20000	400
	100	$1 \cdot 10^9$			600					
Mercury	~1000	$1.2 \cdot 10^9$	6500	$5 \cdot 10^{15}$	30	$5 \cdot 10^6 (T_e - T_g)$		$2 \cdot 10^8$	200	120
	~ 100	$2 \cdot 10^8$	5800	$4 \cdot 10^{14}$	800	$8 \cdot 10^4 (T_e - T_g)$		$6 \cdot 10^7$	2500	1000

2. Helium arc. Equality of T_e and T_g within the limits of the measurement accuracy for $T = 3-4 \cdot 10^4$ °K and for electron densities around $2 \cdot 10^{16}$ cm^{-3} (Durand [11.9]).
3. Iron arc. Equality of T_e and T_g within the limits of the measurements accuracy for $p = 1$ atm, $T = 4000-6300$ °K, $n_e = 10^{15} - 3 \cdot 10^{16}$ cm^{-3} (Burhorn [11.4]).

11.1.6. Validity of Kirchhoff's Law

This law must hold when the plasma temperature is determined via the emissivity and absorptivity. This is then identified with the electron temperature, since electron collisions are the primary cause of radiation processes.

As Mandel'shtam and Sukhodrev [11.31 b] as well as Bekefi et al. [11.1 d] have shown, the existence of Kirchhoff's law is connected with the following three requirements:

a) The electrons must possess a Maxwell distribution of velocities.
c) The occupation number of the electron levels must correspond to a Boltzmann distribution.
d) The degree of ionization must be such that it can be described by the Saha equation.

If these prerequisites are not satisfied, the temperature formally determined from Kirchhoff's law is in general greater or smaller than that temperature, which can be defined on the basis of the mean kinetic energy of the electrons.

The conditions for prerequisites a), c), and d) were already stated above.

For flame gases of not too low density seeded with alkali, it has been found that a Boltzmann equilibrium between spectral levels of the alkali atoms can be established due to inelastic collisions with molecules. In this case, the application of Kirchhoff's law to an atomic transition for example in the form of the line reversal method, yields instead of the electron temperature the gas temperature (see begin of chapter 6).

11.1.7. Condition of Quasistationarity for Plasmas Changing with Time

The spectrocopic plasma diagnostic methods mentioned in Chapters 5 to 10 presuppose, insofar as plasma temperatures are to be determined, the quasistationarity of certain plasma properties. That is, either the plasma must be constant in time or the temporal changes of state must take place so slowly that the plasma can be considered as passing through equilibrium states.

If τ is the relaxation time for the establishment of an equilibrium distribution of the energy states within one degree of freedom, on the basis of which the relevant temperature is to be determined, the condition of quasistationarity becomes (see Griem [11.18])

$$\left|\frac{T(t+\tau) - T(t)}{T(t)}\right| = \delta \ll 1 , \qquad (11.14)$$

i.e., the temperature may only vary by a small fraction during the relaxation time τ.

In order to estimate whether we can count on quasistationarity for a specific plasma which is variable with time, we must estimate the relaxation times of the different degrees of freedom.

11.1.7.1. Relaxation time of the translational and rotational degrees of freedom

As known, the translational temperature is obtained via the Doppler effect from the energy distribution within the translational degrees of freedom.

The relaxation times of the translational degrees of freedom are given by

$$\frac{1}{\tau_e} = Z_e = \sqrt{2}\, n_e\, v_e\, q_e^e + n_i\, v_e\, q_i^e + n_0\, v_e\, q_0^e\,, \tag{11.15}$$

$$\frac{1}{\tau_i} = Z_i = \frac{M}{m}\, n_e\, v_e\, q_e^i + \sqrt{2}\, n_i\, v_i\, q_i^i + n_0 \sqrt{v_0^2 + v_i^2}\, q_0^i\,, \tag{11.16}$$

$$\frac{1}{\tau_0} = Z_0 = \frac{m}{M}\, n_e\, v_e\, q_e^0 + n_i \sqrt{v_0^2 + v_i^0}\, q_i^0 + \sqrt{2}\, n_0\, v_0\, q_0^0\,. \tag{11.17}$$

For $n_e = n_i = n_0 = 10^{15}$ cm^{-3}, $v_e = 10^8$ cm s^{-1}, $m/M = 10^{-4}$, $q_e^e = q_i^e = q = 10^{-15}$ cm^2 we have, $\tau_e = 3 \cdot 10^{-9}$ s, $\tau_i = \tau_0 = 3 \cdot 10^{-7}$ s.

The rotational degrees of freedom, from which the rotational temperatures are determined, enter into equilibrium with the translational degrees of freedom of the molecules after a relatively small number of collisions. In the case of nitrogen 5 collisions are required [11.17]. We thus have

$$\frac{1}{\tau_R} \approx \frac{1}{5}\, Z_0\,; \tag{11.18}$$

with the above given numerical values for the cross sections etc., the relaxation time of the rotational degrees of freedom amounts to $\tau_R = 2 \cdot 10^{-6}$ s.

In practice, at pressures of 1 atm and higher the relaxation time need no longer be allowed for due to its smallness.

11.1.7.2. Relaxation time of the vibrational degrees of freedom

As is well known the vibrational temperature of the molecules is determined via the distribution of the vibrational energy states. The relaxation time of the vibrational degree of freedom is given by (see also [11.34])

$$\frac{1}{\tau_v} = P\, Z_0\,, \tag{11.19}$$

where

$$P = C\, \mathrm{e}^{-t_c/t_v}\,, \tag{11.20}$$

P is the probability that translational energy is transferred to vibrational energy per collision. C is a numerical factor, t_c the collision time, which is inversely proportional to the velocity of the molecule, and t_v is the period of a molecular vibration. At tower temperatures P increases in an approximately exponential fashion with temperature.

For temperatures between room temperature and 8000 °K the values obtained for P lie between 10^{-7} and 10^{-1} [11.29]. For $T = 6000$ °K, values for P lie has values lying between $3 \cdot 10^{-2}$ (O_2) and $3 \cdot 10^{-4}$ (N_2), i.e., for relaxation of the molecular vibrational degrees of freedom 30 (in the case of oxygen) up to 3000 (for nitrogen) collisions are required. Intermediate values apply to other molecules.

Given $n_M = 10^{18}$, $T = 4000$ °K, $v_M = 10^5$ cm/s, $q = 10^{-15}$ cm² (high-pressure plasma), relaxation times from $\tau = 3 \cdot 10^{-7}$ s (O_2) to $3 \cdot 10^{-5}$ s (N_2) are obtained. The relaxation time of the molecular vibrations in an air plasma at atmospheric pressure and a temperature of 4000 °K is approximately 30 μs [11.35], in conformity with the above data.

11.1.7.3. Relaxation time of the dissociation equilibrium

The relaxation time of the dissociation equilibrium is given by (see [11.29])

$$\frac{1}{\tau_{\text{diss.}}} = P_D \, Z \, e^{-D/kT} , \tag{11.21}$$

where D is the dissociation constant. P_D assumes values between 1 and 1000.

11.1.7.4. Relaxation time of the Boltzmann distribution and the ionization equilibrium

The relaxation time is of the order of magnitude of the lifetime of the ground state with regard to collisions producing ionization and excitation. In general, only those excitation collisions need be considered which cause transitions from the ground level (1) to the first excited level (2), since these excitations occur most frequently. From level 2 numerous higher levels are excited or else ionizing collisions of this level take place. Since the frequency at which these latter processes occur by far exceeds that of the excitation events from the ground level, these processes can be neglected upon estimating of the relaxation time.

According to Griem [11.18] the time τ can be estimated by multiplying the reciprocal excitation rate of the atoms in the ground state with the fraction of atoms, which are to be excited or ionized. Griem replaced this fraction by the ratio $n_+/(n_+ + n_0)$, where n_+ is the concentration of "ions" and n_0 is the concentration of "atoms" (in the case of higher ionization stages "ions" and "atoms" respectively refer to the particles in the upper and in the lower ionization stage). Then for hydrogen or hydrogenic ions the relaxation time is given by

$$\tau = \frac{1.15 \cdot 10^7 \, z^3 \, n_+}{f_{21} \, n_e \, (n_+ + n_0)} \left(\frac{U_2 - U_1}{U_H^z}\right) \left(\frac{kT}{U_H^z}\right)^{1/2} \exp\left(\frac{U_2 - U_1}{kT}\right) [\mathrm{s}] . \tag{11.22}$$

Given hydrogen with $n_e = 10^{16}$ cm^{-3} and $kT_e = 1$ eV the relaxation time is about 3 μs, for ionized helium with $n_e = 10^{18}$ cm^{-3} and $kT = 4$ eV this is approximately 0.3 μs.

The relaxation time for the partial Boltzmann distribution is shorter. Griem [11.18] gives for it the expression

$$\tau_n = \frac{4.6 \cdot 10^7 z^3}{n^4 n_e} \left(\frac{kT}{U_H^z}\right)^{1/2} \exp\left(\frac{U_{n+1} - U_n}{kT}\right) [\mathrm{s}] . \tag{11.23}$$

Here the relaxation times for the above given data on temperature and electron density are less than 10^{-3} µs. Short relaxation times are of interest insofar they enable the determination of the electron temperature via the relative occupation of the higher levels.

A more exact and detailed treatment of the problem of Boltzmann equilibrium in transient plasmas can be found in [11.8b], [11.8c].

11.1.8. Condition for Local Boltzmann Equilibrium in Inhomogeneous Plasmas

The condition for local Boltzmann equilibrium in inhomogeneous plasmas is as follows: the temperature, insofar as possible, does not vary by more than 1% over a distance which is equal to the relaxation distance d or expressed analytically

$$\left|\frac{T(r) - T(r+d)}{T(r)}\right| = \varepsilon \ll 1 . \tag{11.24}$$

The relaxation distance d is equal to those distance over which the excited atoms diffuse during the relaxation time τ:

$$d = \lambda \left(\frac{\tau}{\tau_0}\right)^{1/2} , \tag{11.25}$$

where $\tau_0 = \lambda/v$ is the mean free flight duration of the electrons. After a series of computations Griem [11.18] obtained

$$d = \frac{7 \cdot 10^{14}}{A^{1/4}} \left(\frac{kT}{U_H^1}\right)^{1/2} \exp\left(\frac{U_2 - U_1}{2\,kT}\right) \left(\frac{U_2 - U_1}{f_{21}\, U_H^1}\right)^{1/2} [(n_0 + 10\, n_+)(n_0 + n_+)]^{-1/2} \,[\mathrm{cm}] . \tag{11.26}$$

A here denotes the atomic weight. For hydrogen with $kT_e = 1$ eV, $n_0 = 10^{17}$ cm^{-3}, $n_e = 10^{16}$ cm^{-3} we obtain a distance d of about 0.2 cm, over which the plasma must be homogeneous, so that the condition of local thermodynamic equilibrium can be regarded as fulfilled. In real plasmas this condition is often not satisfied, especially in the outer zones of discharge plasmas. For more details concerning Boltzmann equilibrium in inhomogeneous plasmas see [11.8b].

11.1.9. Compilation of the Conditions Required for the Application of the Spectroscopic Plasma Diagnostic Techniques described in Chapter 5 to 10

We shall tabulate the most important requirements imposed upon the plasma and the necessary properties, which the plasma must possess in order to satisfy these requirements. The requirements of quasistationarity and local homogeneity are not specifically mentioned.

Table 11.3. Most important prerequisites for the application of spectroscopic measuring techniques

Requirements of the plasma	Necessary properties of the plasma
a) The electrons possess a Maxwell distribution of velocities	1. $e E \lambda_e \ll \frac{3}{2} kT$ 2.1. either $\nu_{ee} \gg \nu_{eu}$ (i.e. $n_e/n_0 > 10^4 \dots 10^{-3}$) or 2.2. level occupation after Boltzmann
b) The plasma components K (atoms, molecules, ions) possess a Maxwell distribution of velocities	1. $\lambda_K \ll$ plasma dimension 2. $Z e E \lambda_i \ll \frac{3}{2} k T_i$
c) Electron levels are occupied in accordance with a Boltzmann distribution $n_m = \frac{g_m}{Q} n_0 e^{-U_m/k T_e}$ d) Ionization according to Saha equation	1.1. plasma is optically dense or 1.2. $n_e \gg \frac{\bar{A}_{nm}}{\bar{q}_{nm} v_e \left(\frac{e U_m}{kT_e} + 1\right)} \approx 10^{14}$ to 10^{16} cm^{-3}
e) Occupation of rotational and vibrational levels – Boltzmann distribution	requirement certainly fulfilled when $p \geq 1$ atm
f) $T_e = T_g$	in general fulfilled for $p \geq 1$ atm; for pure, heavy inert gases either $p \geq 10$ atm or $p \geq 1$ atm and $n_e/n_0 > 0.1$ must hold

11.2. Special Techniques of Nonthermal Spectroscopy

Whereas in thermal spectroscopy the emission coefficient of a spectral line is uniquely determined by temperature and pressure as well as by the gas composition, in the case of nonthermal, plasma data are required on the electron temperature (provided a Maxwell distribution exists) or on the electron velocity distribution in order to determine these quantities. Further, the number of different particle species per cm^3 or the partial pressures of the different particle species and their temperatures must be known. In those cases where stepwise excitations from resonance levels play a role, the re-absorption of this resonance emission must be taken into account; then the emission coefficient of a spectral line still depends on the geometry of the emitting plasma (this applies not only to spectral lines corresponding to transitions from higher excited levels but also to resonance lines; see [11.32a] for the emission of the Ba II resonance lines emitted by a quiescent Barium plasma).

To evaluate the radiative properties of a plasma which is not in thermal equilibrium one requires data on the effective cross sections for inelastic collisions of the first and second kind as well as information on the transition probabilities for the transitions connected with the radiation. An estimate of the degree of self-absorption of the plasma requires further data on the line width, which essentially consists of a Doppler broadening, a Stark broadening, and a resonance broadening (insofar as broadening of the resonance levels is concerned).

The emission coefficient $\varepsilon_{n\,m}$ of a spectral line is determined as in a thermal plasma by means of the expression (see equation (5.1))

$$\varepsilon_{n\,m} = \frac{1}{4\,\pi}\, n_m\, A_{n\,m}\, h\, \nu_{n\,m}\,. \tag{11.27}$$

The occupation number n_m for a stationary nonthermal optically thin plasma follows from a system of equations (see Biberman and Ulyanov [11.2]), in which the occupation numbers of all others levels also appear:

$$y_n\, n_n^0 \left[y_e\, n_e^0\, j_{e\,n} + \sum_{k=1}^{n_0} (A_{k\,n} + y_e\, n_e^0\, j_{k\,n}) \right] - \sum_{k=1}^{n_0} y_k\, n_k^0\, (A_{n\,k} + y_e\, n_e^0\, j_{n\,k})$$
$$- y_e^2\, (n_e^0)^2\, (\alpha_{n\,e} + y_e\, n_e^0\, j_{n\,e}) = 0\,. \tag{11.28}$$

Equation (11.28) is a balance for the depletion and refilling of the level n due to radiation and electron collision processes, whereby transitions between different discrete levels as well as transitions to continuous states (free electrons) are considered. Inelastic collisions by heavy particles are neglected.

The individual variables signify:

$y_n = \frac{n_n}{n_n^0}$	relative concentration of the excited atoms in level n, referred to the equilibrium concentration in the case of Boltzmann equilibrium,
$y_e = \frac{n_e}{n_e^0}$	relative electron concentration, referred to the equilibrium concentration,
$j_{e\,n}$	collision integral for ionizing collisions,
$A_{k\,n}$	transition probability for $n \to k$ transitions; $A_{k\,n} = 0$ for $n \leq k$,
$j_{k\,n}$	collision integral for electron collision excitations of the first and second type, which lead to the transitions $n \to k$,
$\alpha_{n\,e}$	recombination coefficient for two-body recombinations, which contribute to the filling of level n,
$j_{n\,e}$	coefficient for three-body recombinations, which contribute to the filling of level n.

Biberman and Ulyanov introduced simplifications by considering the case in which the excited levels above n, including the continuous levels (free electrons), possess a partial Boltzmann distribution of electrons. Then it follows that $y_n = y_{n+1} \cdots = y_e^2 = \text{const}$. Now we are left with only a few unknowns (for $n = 2$ there are only the unknowns y_1 and y_2), and the system of equations can be relatively easily solved when allowing for the particle balance $\Sigma n_n + 2 n_e = n$, provided of course, that the collision integrals, transition probabilities, and recombination coefficients are known, which is the case for hydrogen.

The problem of finding the level distribution becomes more complicated if the time dependence and the local inhomogeneity of plasma parameters cannot be neglected [11.2b]. In the case of optically thick resonance radiation, the excitation of resonance levels by photoabsorption processes must be considered too.

We shall not go any further into the problems of the treatment of non-thermal emission, since it is too laborious and has not yet gained significance in spectroscopic plasma diagnostics.

The following is established:

1. The partial Boltzmann distribution of electrons applies to the higher levels. The modifications of the methods of thermal spectroscopy, which in this case must be effected, were already described in 11.1.3.1.

3. The excitation of the lower levels is primarily due to electron collision excitation from the ground level or cascade excitations, in particular from the first excited level. Corresponding experiments, in which successive excitations in a low-pressure discharge were recorded, were performed by different authors, such as for example, Kagan et al. [11.23] and Frish and Revald [11.16] for neon, Razumovskaya [11.36] and Kagan et al. [11.24] for argon, and Vilenskaya [11.42c] for mercury.

The case, in which only direct excitations from the ground state play a role for the spectral intensity, is discussed in more detail below.

11.2.1. Direct Excitation of Relatively Low Levels

From (11.27) it follows that the emission coefficient of a spectral line is directly proportional to the occupation number n_m of the upper level.

For direct excitation from the ground state (this applies to relatively small electron concentrations) the line emission is described by

$$I_{nm} \sim n_1 n_e \int_{v_{1k}}^{\infty} Q_{m1}(v) F(v) v \, dv , \qquad (11.29)$$

where n_1 is equal to the number of atoms per cm^3 in the ground state (this is approximately equal to the total number of atoms per cm^3), and Q_{m1} is the excitation cross section of the level m for transitions from the ground state. v_{1k} here corresponds to the minimum electron velocity at which excitations are possible.

The following expression specified by Drawin [11.7] and stemming from Seaton [11.41] gives the excitation cross section as a function of the electron temperature, or what amounts to the same, as a function of the mean kinetic electron energy U for the transitions $j \to n$.

$$Q_{nj} = \frac{8\pi}{\sqrt{3}} \pi a_0^2 \frac{(13.6 \text{ eV})^2}{U U_{nj}} f_{nj} \bar{g}\left(\frac{U}{U_{nj}}\right), \tag{11.30}$$

where U_{nj} denotes the excitation energy and f_{nj} the oscillator strength for the transition $j \to n$. The factor $\bar{g}$ follows from the following table:

$U/(U_{nj})$	1.0	1.2	1.5	2.0	3	5	10	30	100
$\bar{g}$ (neutral parti)	0.00	0.032	0.064	0.11	0.21	0.33	0.56	0.98	1.33
$\bar{g}$ (ions)	0.20	0.20	0.20	0.20	0.24	0.33	0.56	0.98	1.33

For the specific case of excitation from the ground state ($j = 1$) to the state $n = m$, equation (11.30) becomes

$$Q_{m1} = 1.28 \cdot 10^{-15} \frac{(13.6 \text{ eV})^2}{U U_m} f_{m1} \bar{g}\left(\frac{U}{U_m}\right). \tag{11.31}$$

Here U_m is the excitation energy of the level m and f_{m1} the associated oscillator strength for the transition $1 \to m$.

When a Maxwell distribution applies, we can integrate over the velocities in (11.29), giving

$$I_{nm} \sim n_1 n_e j_{m1} \,. \tag{11.32}$$

where n_1 is the concentration of the atoms in the ground state and j_{m1} is the collision integral obtained from (11.29) and depends on the electron temperature T_e.

Empirical formulas yielding j_{m1} directly can be found in [11.8a] and [11.8b].

As an example we give the quantity $I_{m1}/n_1 n_e$, which is proportional to the collision integral j_{m1}, as a function of temperature in the case of resonance radiation of a low-pressure discharge for pressures between 2 and $6 \cdot 10^{-3}$ Torr (Figure 11.2).

With expression (11.32) plasma parameters can be found from the measured emission coefficient of a spectral line, whose upper level is essentially excited by direct electron collision excitation. In particular, the gas density $n \approx n_1$ can be determined for known values of T_e and n_e, which, for example, are known from probe measurements.

Whether direct excitation occurs or whether cascade excitation must also be considered, can be deduced in the case of electrically produced plasmas from the dependence of the line emission on current intensity at constant pressure. When direct excitation dominates, it follows from (11.32) that I_{nm} and the current intensity I (the latter in the case of a low-pressure discharge is proportional to the electron density n_e) are linearly related (see Figure 11.3). However, in the range of higher electron densities, where cascade excitations must be allowed for, the following relationship is obtained for constant electron temperature and given atom number density

$$I_{nm} \sim A\, n_e + B\, n_e^2 \tag{11.33}$$

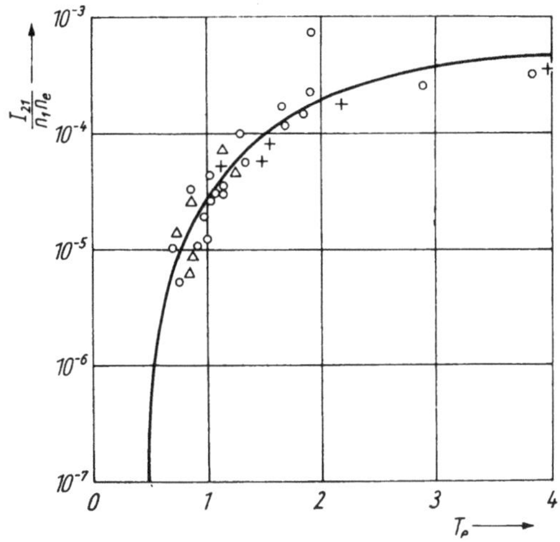

Figure 11.2. The quantity $I_{21}/(n_1 n_e)$, proportional to j_{21}, as a function of temperature for the case of resonance emission of a sodium low-pressure discharge at pressures of $2 - 6 \cdot 10^{-3}$ Torr (after Kaptsov [11.25])

or

$$I_{nm} \sim a I + b I^2 . \qquad (11.34)$$

Figure 11.4 illustrates as an example the intensity of the spectral line Hg II 3983 Å as a function of the current intensity for a Hg low-pressure discharge at $p = 3 \cdot 10^{-3}$ Torr in the current intensity range from 0.4 to 2.4 A.

From the curved behavior of the line intensity as a function of the electron density excitation cross sections, in particular for collisions of the second type, can be determined, provided there are sufficient data, especially on the line emission of different spectral lines.

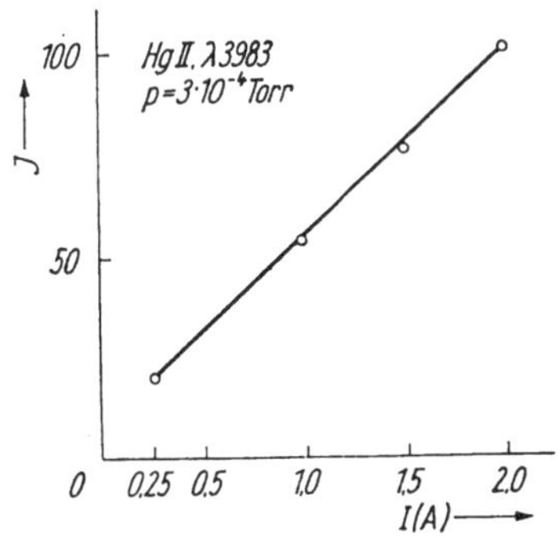

Figure 11.3. Dependence of the line intensity of the line Hg II 3983 Å on the current intensity. Hg low-pressure discharge at $p = 3 \cdot 10^{-4}$ Torr. Direct excitation predominates (after Frish [11.15])

For more details on the calculation and experimental determination of effective inelastic cross sections for excitation and collisions of the second type see Massey [11.32], Frish [11.15], Seaton [11.41], Fite [11.14], Kagan et al. [11.23] and [11.24], Frish and Revald [11.16], and Drawin [11.8a].

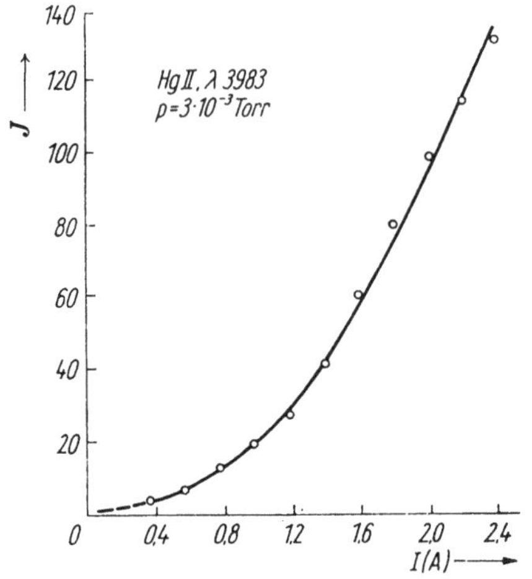

Figure 11.4. Dependence of the intensity of the spectral line Hg II 3983 Å on the current intensity. Hg low-pressnre discharge at $p = 3 \cdot 10^{-3}$ Torr. Cascade processes are predominant in the excitation of spectral lines (after Frish [11.15])

11.2.2. Determination of Electron Temperature T_e in Rarefied Plasmas from the Emission Coefficient of Spectral Lines with Low Excitation Quantum Numbers

11.2.2.1. Determination of electron temperature from the absolute emission coefficient of spectral lines emitted by helium-like ions in rarefied high-temperature plasmas

The fundamentals of this method were already discussed in section 11.2.1. Kaufman and Williams [11.26] employed this method to determine the electron temperature in a deuterium plasma, produced in a Sceptre III A machine, from the absolute emission coefficients of spectral lines of the helium-like ions C V 2270.9 Å and B IV 2821.7 Å, both corresponding to the transitions $2\,{}^3P_2 \to 2\,{}^3S_1$. The elements C and B were added in controllable amounts in the form of diborane to deuterium.

The spectral line emission coefficient follows from the expression derived from (11.29), assuming that the effective cross section for the excitation of the level $2\,{}^3P_2$ from the ground state $1\,{}^1S_0$ is constant above the threshold energy of the excited electrons and can be set equal to Q.

$$\varepsilon = 2\left(\frac{2\,kT_e}{\pi\,m_e}\right)^{1/2} n_1\,n_e\,Q\,h\,\nu\left(1 + \frac{U_m}{kT_e}\right)\exp\left(-\frac{U_m}{kT_e}\right). \qquad (11.35)$$

Here n_1 is the number density of helium-like ions in the ground state. For Q a value of $0.2\,\pi\,a_0^2$ (a_0 is the first Bohr radius), estimated by Seaton, was assumed for both ions.

n_e was calculated, first assuming a totally ionized deuterium plasma, and secondly for an impurity admixture of 10% with a mean ionic charge of 5. Further, a central plasma concentration was assumed, whose value was found on the basis of the toroidal magnetic field intensification, which was determined by magnetic probes.

In determining the contributions of the elements B and C it was assumed that during the discharge these are just as great as before the discharge. Further it was assumed that the major portion of carbon in the center of the discharge tube was in the ground state of the ion C V and the major portion of boron in the ground state of the ion B IV. This was confirmed by the fact the spatially resolved intensity of the B IV line in the center of the discharge tube far exceeded the intensity of all lines emitted by other boron ions (see, e.g., Figure 11.5).

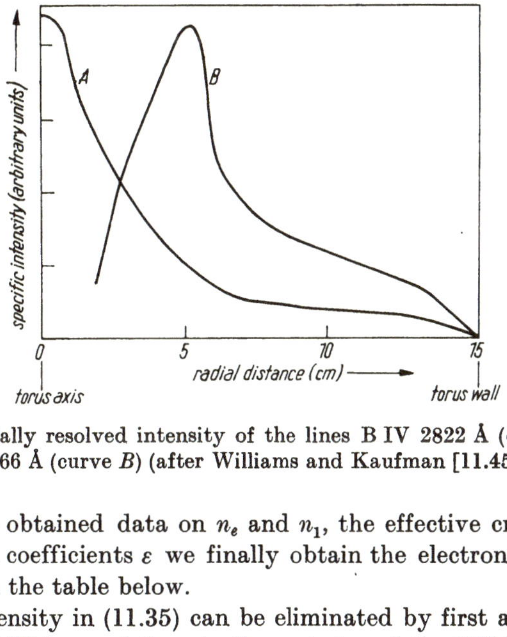

Figure 11.5. Radially resolved intensity of the lines B IV 2822 Å (curve A) and B III 2066 Å (curve B) (after Williams and Kaufman [11.45])

From the thus obtained data on n_e and n_1, the effective cross sections, and the line emission coefficients ε we finally obtain the electron temperature T_e, which is listed in the table below.

The electron density in (11.35) can be eliminated by first applying it to the emission of the B IV ion and then to the emission of the C V ion. The electron density then cancels out and we are left with the equation ([11.26], [11.45])

$$\frac{\varepsilon'}{\varepsilon} = \frac{n_1'}{n_1}\,\frac{Q'\,\nu'}{Q\,\nu}\,\frac{1 + U_m'/(kT_e)}{1 + U_m/(kT_e)}\,\mathrm{e}^{\frac{U_m - U_{m'}}{kT_e}}\,. \tag{11.36}$$

The accented quantities correspond to the B IV ion, and the remaining ones correspond to the C V ion. Via equation (11.36) we also obtain an electron temperature, which is listed in the following table

Ion	U_m [eV]	p [Torr]	n_1 [cm^{-3}]	ε [erg cm^{-3} s^{-1}]	T_e [°K]
C V	304	$1.1 \cdot 10^{-4}$	$1 \cdot 10^{13}$	$9 \cdot 10^{2}$	$2{,}2 \cdot 10^{5}$
B IV	203	$2.1 \cdot 10^{-4}$	$6 \cdot 10^{13}$	$1 \cdot 10^{5}$	$2{,}0 \cdot 10^{5}$
T_e from the relative B IV/(CV)					$3{,}0 \cdot 10^{5}$

p in the tables denotes the partial pressure of the admixture (methane or diborane)

The deviations in the measured temperatures are connected with inaccuracies in the data employed or with possible deviations of the electron velocity from the Maxwell distribution.

11.2.2.2. Determination of electron temperature in rarefied plasmas from relative emission coefficients of spectral lines belonging to the same level system

Assuming that the upper levels of the spectral lines in question are filled in chief by direct electron excitation from the ground state, then the absolute emission coefficient of a spectral line belonging to the transition $m \to n$ is given by the formula (see [11.34b], p. 44)

$$\varepsilon_{n\,m} = h\,\nu\,n_e\,n_1\,j_{m\,1}(T_e)\,\frac{A_{n\,m}}{\sum\limits_{k=1}^{m-1} A_{k\,m}}\,, \tag{11.37}$$

where n_1 denotes the ground state density and $j_{m\,1}$ the collision integral defined in section 11.2.1. A like relation holds for the emission coefficient of a second spectral line belonging to the transition $m' \to n'$. Taking the ratio of both relations, the quantities n_e and n_1 cancel, thus we get

$$\frac{\varepsilon_{n\,m}}{\varepsilon_{n'\,m'}} = \frac{\nu}{\nu'}\,\frac{j_{m\,1}(T_e)}{j_{m'\,1}(T_e)}\,\frac{A_{n\,m}}{A_{n'\,m'}}\,\frac{\sum\limits_{k=1}^{m'-1} A_{k\,m'}}{\sum\limits_{k=1}^{m=1} A_{k\,m}} = f_{n\,m}(T_e)\,. \tag{11.38}$$

Relation (11.38) can be used for the determination of electron temperature from the ratio of measured emission coefficients of such spectral lines which have collision integrals with different temperature dependence. That means in practice, the excitation energies of the two spectral lines must be quite different.

It is not profitable for the purpose of temperature determination to use spectral lines with different temperature dependence of the excitation functions when the excitation energies of these lines are nearly the same. Then for not too low particle number densities, the transition rate between the two near resonant upper levels due to inelastic particle collisions becomes so high that the intensity ratio of the two spectral lines becomes a function of particle number density, while the temperature dependence of the line intensity ratio is weakened. The significance of this effect has been demonstrated by Drawin

Table 11.4. Determination of electron temperature in rarefied plasmas from the ratio of emission coefficients of spectral lines emitted by the same ion

Abbreviations in column 2: m. c. arc = magnetically confined arc; l. p. d. = low pressure discharge. hy. pl. = hypothetical plasma; in column 7: N_1 = ground level ion density

System	Type of plasma	Line pair λ [Å]	Upper level	T-range [10^3 °K]	Actual (A) respectively limiting (L) parameter range			Reference
					n_e [cm^{-3}]	n_1 [cm^{-3}]	p [mTorr]	
He I	m. c. arc, He	4922/4713 4437/4121 5047/4713	$4^1D/4^3S$ $5^1S/5^3S$ $4^1S/4^3S$	100–200	$3 \cdot 10^{11}$ (A, L)			11.4a
He I	l. p. d., He	5047/4472 5016/4472	$4^1S/4^3D$ $2^1S/4^3D$	190			1 (A, L)	11.28
He I	hy. pl., He	5876/5015	$3^3D/3^1P$	400–10000			1–3 (A)	11.43
N V	Zeta, D + trace N	1239/209	$2P/3P$	140–300	$2 \cdot 10^{14}$ (A) 10^{16} (L)	$\frac{4 \cdot 10^{13}/D}{(L)}$ [1]	0.5–2.5 (A)	11.20a
NV	hy. pl., N	1239/209	$2P/3P$	50–300	10^{16} (L)			11.42b
O IV	Stellarator, H + trace O	790/554 790/610 610–554	$2s\,2p^2{:}\,^2D/^2P$ $2s\,2p^2{:}\,^2D/^2S$ $2s\,2p^2{:}\,^2S/^2P$	150–300	(6–17) · 10^{12} (A)			11.21
Li-like ions, C IV – Si XII	hy. pl.	$2s-2p/2s$ $-3p$	$2P/3P$		$10^{11}\,U_i^{3/2}\,(kT_e)^2$ (L)[2]	$\frac{20\,\Delta\nu_D}{f\,D}$ (L)[3]		11.20a

[1]) D = geometrical depths of the plasma in the line of sight, in [cm].
[2]) U_i = ionization energy of the considered ion in [eV]; kT_e in [eV].
[3]) $\Delta\nu_D$ = Doppler half width in frequency units [s^{-1}]. f = oscillator strength. D = geometrical depths of the plasma in the line of sight, in [cm].

and Henning [11.8d] for the triplet-singulet line pair He I 4713 Å ($4\,^3S - 2\,^3P$) and 4921 Å ($4\,^1D - 2\,^1P$) for particle densities in excess of 10^{11} cm^{-3}.

Relation (11.38) has been applied for the purpose of temperature determination to a number of plasmas; a few examples are given in table 11.4. The table contains also, if known, values for the maximal electron density above which step excitation of spectral levels cannot be ignored.

11.2.3. Determination of the Ionization Temperature for Low Particle Densities

In a highly rarefied plasma, as in the case of a plasma in thermal equilibrium the electron temperature can be determined, provided the electrons possess a Maxwell velocity distribution, from the intensity ratio of two spectral lines emitted by differently charged ions of the same element. This was the subject of discussion in section 5.2.5.1 for a thermal plasma. The advantage of this method is that the geometrical dimensions of the test object need not be accurately known in the direction of observation. Thus, such a method is well suited for the determination of electron temperatures of astrophysical objects.

Let us assume that the spectral lines are excited by direct electron collisions. Then we obtain an expression similar to (11.36) for the ratio of the spectral intensities I' and I^*, associated with the differently charged ions. To obtain the electron temperature on the basis of the measured value for I'/I via (11.36), an additional relationship between the ratio of the respective numbers n_1' and n_1 of the differently charged ions and the electron temperature is required. Such a relationship is designated as an ionization equation.

In the complete ionization (or charge carrier balance) equation, the ionizing collisions and photoionization of all excited states must be allowed for. Further, recombination by electron collisions and by three-body collisions must also be taken into account along with all inelastic collisions and excitation processes, since the latter produce an alteration in the occupation of the discrete levels leading to a change in the ionization and recombination conditions of the plasma. Thus we arrive at the system of equations (11.28).

According to Elwert ([11.11], [11.12]) a simplification can be introduced in the case of highly rarefied plasmas, whereby collisions of the excitation type or of the second type are not taken into account at all, photoionizations are neglected due to the low radiation density, while recombination by three-body collision can be neglected with respect to photo recombination because of the small particle density.

The balance equation for ionization and recombination processes for the $i \rightleftarrows i + 1$ ionization equilibrium is given by

$$n_i\, n_e\, S_{12} = n_{i+1}\, n_e\, Q_{21}\,, \tag{11.39}$$

where

$n_i\, n_e\, S_{12}$ is the number of ionization events per cm^3 and s,

$n_{i+1}\, n_e\, Q_{21}$ is the number of photorecombinations per cm^3 and s.

* In the homogeneous plasma approximation used in astrophysical applications, the ratio of emission coefficients ε'/ε appearing in formula (11.36) can be substituted by the intensity ratio I'/I.

S_{12} and Q_{21} can be regarded as collision integrals, which are connected with the ionization cross section and with the recombination cross section. When calculating S_{12} and Q_{21} it should be remembered that not only ionizations originating from the ground state but also from excited states can play a role. The significance of the excited levels in a rarefied plasma, however, is regarded as minor. Further, it should be remembered that photo-recombinations can occur not only in the lowest state n but also in higher states.

For a rarefied plasma Elwert obtained the following ionization formula, also called the corona formula:

$$\frac{n_{i+1}}{n_i} = 8.3 \cdot 10^5 \frac{\zeta_{(n)}}{g\,n} \left(\frac{U_{i,H}}{U_i}\right)^2 \frac{e^{-U_i(kT_e)}}{U_i/(kT_e)} . \tag{11.40}$$

Here two uncertainty factors of the order of 1, which were stated by Elwert, have been left out. $U_{i,H}$ is the ionization energy of hydrogen (= 13.6 eV), U_i the ionization energy of the ion, n the principal quantum number of the ground state of the ion i, and $\zeta_{(n)}$ the number of electrons in the ground level of the ion of species i. g is a factor between 1.4 and 4, which allows for the recombination in higher levels. Elwert [11.11] plotted this factor as a function of the ionization energy U_i, the electron temperature T_i, and the principal quantum number n of the lowest level. The ratio of the ion concentrations for 2 different ionization stages with ionic charges differing by more than 1 from one another is given as a product of several factors of the type (11.40).

The above relationship were used to determine the electron temperature via the intensity ratio of the spectral lines Fe X 6374 Å and Fe XIV 5303 Å emitted by the solar corona. Neglecting again the uncertainty factors the ratio of the ion densities is given by (see Elwert [11.11])

$$\frac{n_{\text{Fe XIV}}}{n_{\text{Fe X}}} = \left(620 \frac{f_2}{f_1} \frac{e^{-x}}{x}\right)^4 \quad \text{where} \quad x = \frac{320}{(kT_e)_{\text{eV}}} . \tag{11.41}$$

The intensity ratio I_{5303}/I_{6374} is given by Edlén [11.10] for the relevant electron temperature range as

$$\frac{I_{5303}}{I_{6374}} = 2.4 \cdot \frac{n_{\text{Fe XIV}}}{n_{\text{Fe X}}} . \tag{11.42}$$

The connection between the intensity ratio, which according to Waldmeier [11.43] deviates above and below from the value 1 by one order of magnitude, and the associated corona temperature follows from (11.41) and (11.42). Below are given some numerical values stemming from the work of Elwert.

I_{5303}/I_{6374}	0.1	1	10
$T_e/[10^5\,°K]$	6.4	7.1	7.8

Accordingly, the corona temperature varies between approximately 600,000 and 800,000 °K.

Blaha [11.3] considered the possibility that the result can be erroneous because of the fact that the lines measured have maximum intensities which fall into regions of variable temperature as a consequence of the different excitation conditions. Since in the measurements radiation is simultaneously observed from regions of variable temperature, it was disputable whether the Fe XIV radiation was mainly emitted from a region with a temperature of about $1.2 \cdot 10^6$ °K and the Fe X line predominantly from a region of lower temperature (e.g., 500,000 °K). In such a case the above method becomes useless, and it is more practical for the temperature determination to use lines of ions corresponding to successive ionization stages, for which the excitation conditions are not so different.

Accordingly, Lyot [11.30] measured the intensity ratio of the lines Fe XIV 5303 Å and Fe XV 7060 Å.

$$\frac{I_{5303}}{I_{7060}} = \frac{100}{2.2} .$$

From this Blaha [11.3] obtained a corona temperature of 10^6 °K on the basis of more accurate calculations of the ionization equilibrium and some effective inelastic cross sections.

For further discussions about the problem of corona temperature see [11.45a].

Measurements of the ionization temperature have also been carried out for high temperature laboratory plasmas.

Zaidel et al. [11.46a] determined the electron temperature T_e (20,000 to 150,000 °K) in a hydrogen plasma generated in the toroidal plasma device "Alpha" from the intensity ratios of spectral lines emitted by impurity ions (Al III, IV, Ar I, II, III, C I, II, III, IV, N II, III, IV and O II, III, IV, V) in two successive ionization stages.

Averin et al. [11.1b] determined the electron temperature T_e (140,000 °K to 500,000 °K) in a toroidal discharge device of type Beta via the intensity ratio of the line pair O V 2781 Å/O VI 1032 Å.

In cases where the plasma parameters vary strongly with time, the evaluation of the electron temperature from line emission data is more complex than in the stationary case, because instead of the time independent rate equations describing the density of excited levels, a system of coupled time dependent rate equations must be solved. Such a procedure has been applied by Goldman and Kilb [11.16b] to a deuterium plasma in a theta pinch device containing the elements C and O as impurity. In this case, the electron temperature as function of time $T_e(t)$ was derived from the time dependent emission data of spectral lines originating from the ions C III, IV, V and O III, IV, V, VI, VII. A time resolution of 0.1 μs was obtained. For T_e, values between 60,000 and 3,500,000 °K has been found.

In conclusion we refer the reader to some references, in which statements are made on the ionization cross section and recombination coefficients:

Massey [11.32], Seaton [11.40], Bates and Dalgarno [11.1c], Seaton [11.41], Fite [11.14], Drawin [11.7], [11.8], [11.8a], and Lotz [11.29a], [11.29b].

For other versions of the ionization equation see Dewan [11.5], [11.6], and Mahn [11.31a].

11.2.4. Determination of Electron Density in Rarefied Plasmas from Line Emission

11.2.4.1. The method of Suckewer

Suckewer [11.42a] considers two spectral lines 1 and 2 originating from two different upper levels m_1 and m_2. Assuming that these levels are filled by electronic step excitation processes whereby the number of steps is n_1 and n_2, respectively, with $n_2 \neq n_1$, then the line intensities are proportional to $N_e^{n_1}$ and $N_e^{n_2}$, respectively, and the quotient of the line intensities is proportional to $N_e^{n_1-n_2}$. For suitably selected spectral lines and within a limited range of electron density, the line intensity ratio $I_{1/2}$ shows a negligible temperature dependence, so that from I_1/I_2 the quantity N_e can be derived directly. Explicit calculations has been performed for the intensity ratio of the lines N V 247 Å and 209 Å, showing the applicability of this method for electron densities between 10^{14} cm^{-3} and 10^{16} cm^{-3}.

11.2.4.2. Determination of electron density from the intensity of autoionizing lines

For a number of metal atoms, spectral lines have been observed originating from upper spectral levels lying above the ionization limit which are filled by excitation of inner shell electrons. Some of these levels are highly unstable against autoionizing transitions to isoenergetic continuous energy states, while some of these levels are relative stable.

The character of stability of these levels is governed by selection rules. In the range of high electron densities ($n_e \gg 10^{16}$ cm^{-3}), a Boltzmann equilibrium among the "displaced" levels is established via inelastic electron collisions, the rate of which must be higher than the probability of autoionizing transitions. For low electron densities, the population of autoionizing levels is markedly lowered relative to the population of nonautoionizing levels. Accordingly, in the case of two spectral lines having nearly the same excitation energy one of which is autoionizing while the other is not, the intensity ratio is a sensitive function of electron density [11.27a]. This fact can be used for the determination of electron densities below 10^{16} cm^{-3}. About the derivation of corresponding formulas describing the intensity ratio of such lines as function of electron density, see Grechikhin [11.16d].

For copper, suitable line pairs for the determination of electron density are Cu I 4378.20 Å (with the autoionizing upper level $e\,^4D_{5/2}$) and 4248.96 Å (with the non autoionizing upper level $e\,^4D_{1/2}$ [11.27]). Using these lines, Grankova and Kiselevskii [11.16c] measured the electron density (0.9—3.5$10^{15}$ cm^{-3}) in an alternating current arc as function of time.

Bibliography

[2.1] Aarts, J. F. M. and F. J. de Heer, J. Opt. Soc. Amer. **58** (1968) 1666.

[2.1a] Anacker, F. and R. Mannkopff, Z. Physik **155** (1959) 1.

[2.2] Anacker, F. and R. Mannkopff, Z. Physik **155** (1959) 16.

[2.2a] De Angelis, A., S. Martellucci and G. Tondello, Proc. 6. internat. Conf. Ion. Phenom. Gases, Paris 1963, Vol 4, p. 41.

[2.3] Angerer, E. von and G. Joos, Wissenschaftliche Photographie. Akad. Verlagsges. Geest and Portig KG, Leipzig 1959.

[2.3a] Anonymous, Elektron. Rundschau **21** (1967) 107.

[2.3b] Artemyev, V. V., Optika i Spektroskopiya **25** (1968) 635.

[2.4] Astoin, N., B. Vodar and J. Romand, J. Phys. Radium **16** (1965) 491.

[2.4a] Avellone, J. C., P. L. Byard and F. L. Damm, Rev. sci. Instrum. **38** (1967) 1808.

[2.5] Balashov, I. F., M. P. Vanyukov, V. R. Muratov und Y. V. Nilov, Optika i Spektroskopiya **9** (1960) 790.

[2.6] Bardócz, Á., Rev. Univ. Mines **15** (1959) 344.

[2.7] Bardócz, Á., Spectrochim. Acta **16** (1960) 1173.

[2.8] Bardócz, Á., Acta Technica Ac. Sci. Hung. **40** (1962) 179.

[2.9] Bardócz, Á. and F. Varsanyi, Z. Naturforsch. **10a** (1955) 1031.

[2.9a] Barr, W. L., J. Opt. Soc. Amer. **52** (1962) 885.

[2.9b] Bashilov, V. A., Pribory Tekhn. Eksp. No. 2 (1965) 183.

[2.10] Bauer, A., Proc. 4. Internat. Conf. Ionization Phenomena in Gases, Vol. 2, North Holland Publ., Co. Amsterdam 1960, p. 817.

[2.11] Bayer-Helms, F., Z. angew. Phys. **15** (1963) 330.

[2.11a] Bayer-Helms, F., PTB-Mitt. (Phys.-Techn. Bundesanstalt, Braunschweig, Germany) **74** (1964) 201; 312.

[2.12] Beams, J. W., Phys. Rev. **35** (1930) 24.

[2.13] Bearden, A. J., F. L. Ribe, G. A. Sawyer and T. F. Stratton, Phys. Rev. Letters **6** (1961) 257.

[2.13a] Becker, L. and H.-W. Drawin, Z. Instrum.kde **72** (1964) 251.

[2.13b] Becker, L. and H.-W. Drawin, Z. Instrum.kde **73** (1965) 155.

[2.14] Bergstedt, K., Z. Physik **155** (1959) 23.

[2.14a] Bernstein, M. J., Appl. Optics **5** (1966) 174.

[2.14b] Besshaposhnikov, A. A., I. K. Kuchuberiya, T. S. Melnikova and N. V. Simonova, Pribory Tekhn. Eksp. No. 4 (1965) 169.

[2.14c] Besshaposhnikov, A. A., A. E. Voloshin and K. Kuchuberiya, Pribory Tekhn. Eksp. No. 5 (1965) 204.

[2.15] Biondi, M. A., Rev. sci. Instrum. **27** (1956) 36.

[2.15a] Bird, P. F. and G. L. Schott, J. quant. Spectrosc. Radiat. Transfer **5** (1965) 783.

[2.15b] Birkeland, J. W. and J. P. Oss, Appl. Optics **7** (1968) 1635.

[2.16] Blau, H. H., W. S. Martin and E. Chaffee, in: Temperature, its Measurement and Control in Science and Industry, Vol. 3, 2, Reinhold Publ. Corp. New York 1962, p. 1035.

[2.17] Bötticher, W. and K. J. Repenning, Z. angew. Physik **14** (1962) 536.
[2.17a] Bötticher, W., O. Roder and K. Wobig, Z. Physik **175** (1963) 480.
[2.17b] Bogen, P., H. Conrads, G. Gatti and W. Kohlhaas, J. Opt. Soc. Amer. **58** (1968) 203.
[2.17c] Bogen, P. and D. Rusbüldt, Phys. Fluids **11** (1968) 2022.
[2.18] Boldt, G., Proc. 5. internat. Conf. Ionization Phenomena in Gases, Vol. 1, North Holland Publ. Co., Amsterdam 1962, p. 925.
[2.19] Bondarev, A. F., Optika i Spektroskopiya **12** (1962) 510.
[2.20] Bortfeldt, J., Proc. 5. internat. Conf. Ionization Phenomena in Gases, Vol. 2, North Holland Publ. Co., Amsterdam 1962, p. 1931.
[2.21] Bovey, L. and A. H. Gabriel, Spektrochim. Acta **18** (1962) 631.
[2.22] Bradley, D. J., Brit. J. appl. Phys. **13** (1962) 83.
[2.22a] Burgess, D. D. and J. Cooper, Proc. 7. Intern. Conf. Phenomena in Ionized Gases (Belgrade 1965) Vol. 3, p. 158.
[2.22b] Brömer, H. H. and J. Hesse, Z. Instrum.Kde **74** (1966) 126.
[2.22c] Buser, R. G. and J. J. Kainz, J. Opt. Soc. Amer. **55** (1965) 12.
[2.23] Butslov, M. M., A. G. Plakhov and V. V. Shankin, Optika i Spektroskopiya **12** (1962) 419.
[2.24] Calker, J. van and E. Tacke, Z. Naturforsch. **4a** (1949) 573.
[2.25] Camm, J. C., Rev. sci. Instrum. **31** (1960) 278.
[2.26] Candler, C., Modern Interferometers. Hilger & Watts Ltd., London 1951.
[2.27] Castle, J., W. Woodbury and W. A. Shelton, Proc. 3. Internat. Congress on High-Speed Photography, Butterworths Sci. Publ., London 1957.
[2.27a] Church, C. H., R. G. Schlecht and I. Liberman, J. quant. Spectrosc. Radiat. Transfer **8** (1968) 403.
[2.27b] Clayton, J. O., Rev. sci. Instrum. **34** (1963) 1391.
[2.27c] Conrads, H., Z. Physik **200** (1967) 444.
[2.27d] Cook, A.H., Monthly Notices Roy. Astron. Soc. **139** (1968) 141.
[2.28] Cooper, J. and J. R. Greig, J. sci. Instrum. **40** (1963) 433.
[2.28a] Cooper, J. and J. R. Greig, A rapid scanning Fabry-Perot-Spectrometer. Rep. Imperial Coll. Sci. Technol., Univ. of London, 1965.
[2.28b] Cremers, C. J. and R. C. Birkebak, Appl. Optics **5** (1966) 1057.
[2.29] Curzon, F. L. and J. R. Greig, J. sci. Instrum. **38** (1961) 239.
[2.29a] Dahm, U., PTB-Mitt. (Phys.-Techn. Bundesanstalt, Braunschweig, Germany) **76** (1966) 107.
[2.30] Dean, D. E., Amer. J. Phys. **24** (1956) 66.
[2.30a] Drawin, H. W., Z. Physik **172** (1963) 181.
[2.30b] Dike, R. S. and E. L. Kremp, Rev. sci. Instrum. **36** (1965) 1256.
[2.30c] Dobbins, R. A., L. Crocco and I. Glassmann, Rev. sci. Instrum. **34** (1963) 162.
[2.31] Dunn, G. H., R. Geballe and D. Pretzer, Phys. Rev. **128** (1962) 2200.
[2.32] Durand, J., Z. Naturforsch. **18a** (1963) 281.
[2.33] Dutton, W. L. and R. C. Hirt, Appl. Spectrosc. **13** (1959) 120.
[2.34] Eastman Kodak C., Photographs of Heat Rays. Amer. Pat. 2643538 Publ. 16. 6. 1953.
[2.34a] Eberhagen, A., Z. Physik **143** (1955) 312.
[2.34b] Ebeling, D., F. Früngel and J. F. Suarez, Optik **25** (1967) 359.
[2.34c] Eberhagen, A., M. J. Bernstein und H. Hermansdorfer, Rep. Inst. Plasmaphys., Garching near Munich, Germany, No. 1/41, Aug. 1965.
[2.34d] Eck, J. van and F. J. de Heer, Proc. 6. internat. Conf. Ion. Phenom. Gases, Paris 1963, Vol. 4, p. 11.
[2.35] Edlén, B., Nova Acta Reg. Soc. Sci. Ups. IV, **9**, No. 6 (1934).
[2.35a] Edlén, B., Rep. Progr. Phys. **26** (1963) 181.
[2.35b] Elder, P., T. Jerrick and J. W. Birkeland, Appl. Optics **4** (1965) 589.

[2.35c] ELISEEV, V. V., V. D. RUSANOV, A. R. STRIGANOV and D. A. SHEGLOV, Zh. prikl. Spektrosk. 5 (1966) 396.

[2.36] ELSTE, G., Z. Astrophys. **33** (1953) 39.

[2.37] EULER, J., S.-B. Heidelberger Akad. Wiss., math.-naturwiss. Kl. (1956/57) 413.

[2.38] EULER, J. and R. LUDWIG, Arbeitsmethoden der optischen Pyrometrie. Verlag G. Braun, Karlsruhe 1960.

[2.38a] FABRE, E. and L. HERMAN, Rev. Opt. No. **41** (1962) 339.

[2.39] FAWCETT, B. C., A. H. GABRIEL, W. G. GRIFFIN, B. B. JONES and R. WILSON, Nature **200** (1963) 1303.

[2.40] FEATES, F. S., B. KNIGHT and E. W. T. RICHARDS, Spectrochim. Acta **18** (1962) 485.

[2.40a] FORREST, M. J., J. sci. Instrum. **44** (1967) 26.

[2.41] FREEMAN, M. P. and S. KATZ, J. Opt. Soc. Amer. **50** (1960) 826.

[2.42] FRICKE, K., Bestimmung der Halbwertsbreite von Spektrallinien mit dem Perot-Fabry-Interferometer. Diss. T. H. Braunschweig 1957.

[2.42a] FROMMHOLD, L. and W. A. FEIBELMAN, J. sci. Instrum. **44** (1967) 182.

[2.43] GABRIEL, A. H., J. sci. Instrum. **37** (1960) 50.

[2.44] GABRIEL, A. H. and W. A. WALLER, J. sci. Instrum. **40** (1963) 10.

[2.44a] GABRIEL, A. H., J. R. SWAIN and W. A. WALLER, J. sci. Instrum. **42** (1965) 94.

[2.45] GENKIN, Y. Y. and I. A. RUMYANTSEV, Izvest. Akad. Nauk SSSR Ser. Fiz. **24** (1960) 384.

[2.46] GOODERUM, P. B. and G. P. WOOD, NACA Technical Note 2173, Aug. 1950.

[2.46a] GORENFLO, R., Z. angew. Math. Mech. **45** (1965) T33—T35 (Sonderheft).

[2.46b] GÖRLICH, P. and H.-J. POHL, Optik **25** (1967) 25.

[2.47] GREINER, H., Z. Naturforsch. **12a** (1957) 735 and **13a** (1958) 61 (Correction).

[2.48] GRIFFIN, P. M., J. R. MCNALLY and G. K. WERNER, Temperature, its Measurement and Control in Sci. and Ind. Vol. 3,1, Reinhold Publ. Corp., New York 1962, p. 651.

[2.49] GRIFFIN, W. G. and R. W. P. MCWHIRTER, Proc. Conf. Optical Instrum. and Techniques, London 1961.

[2.49a] GRODSKI, J. J. and B. W. SCHUMACHER, Rev. sci. Instrum. **39** (1968) 702.

[2.50] GROTH, W. E., Z. phys. Chem. **37** (1937) 307.

[2.50a] GÜNTHER, K., Beitr. Plasmaphys. **10** (1970) 469.

[2.51] HAMAKER, H. C., Diss. Utrecht 1934 and Physica **3** (1936) 561.

[2.52] HARRISON, G. R., R. C. LORD and J. R. LOOFBOUROW, Practical Spectroscopy, Prentice-Hall Inc., Englewood-Cliffs 1959.

[2.53] HARTMANN, W. and F. BERNHARD, Fotovervielfacher und ihre Anwendung in der Kernphysik. Akademie-Verlag Berlin 1957.

[2.53a] HARTWIG, M., W. ESCHKE and D. ZIRWER, Monatsber. Dtsch. Akad. Wiss. Berlin 7 (1965) 375.

[2.53b] HATTENBURG, A. T., Appl. Optics **6** (1967) 95.

[2.54] HEARN, A. G., B. B. JONES and S. A. RAMSDEN, Plasma Phys. **4** (1962) 22.

[2.55] HENNING, F., Temperaturmessung. Johann Ambrosius Barth, Leipzig 1951.

[2.56] HEROUX, L. and H. E. HINTEREGGER, Rev. sci. Instrum. **31** (1960) 280.

[2.56a] HILL, R. A., Appl. Optics **4** (1965) 1593.

[2.57] HOFFMANN, F. and C. TINGWALDT, Optische Pyrometrie. Verlag Vieweg, Braunschweig, 1938, S. 130.

[2.58] HOLM, R. and A. LOTZ, Wiss. Veröff. Siemens-Werken **13** (1934) 87.

[2.59] HONERJÄGER-SOHM, M. and H. KAISER, Spectrochimica Acta **2** (1944) 396.

[2.59a] HOPPMANN, K. and E. WUTTKE, Tech.-wiss. Abh. Osram **9** (1967) 137.

[2.59b] IPATKI, I. S., B. P. BULATOV and E. A. ANTONOV, Pribory Tekhn. Eksp. No. 3 (1966) 66.

[2.60] Jahoda, F. C., E. M. Little, W. E. Quinn, O. A. Sawyer and T. F. Stratton, Phys. Rev. **119** (1960) 843.

[2.61] Jahoda, F. C., F. L. Ribe, G. A. Sawyer and T. F. Stratton, Proc. 5. internat. Conf. Ion. Phenomena in Gases Vol. 2, North Holland Publ. Co., Amsterdam 1962, p. 1987.

[2.61a] Jayroe, R. R. and R. G. Fowler, J. Opt. Soc. Amer. **57** (1967) 513.

[2.61b] Jensen, V. O., RISÖ-Rep., Denmark, No. 74, 1964.

[2.61c] Johnson, F. S., K. Watanabe and R. Towsey, J. Opt. Soc. Amer. **41** (1951) 702.

[2.62] Jucker, H. and E. K. Rideal, J. chem. Soc. (1957) 1058.

[2.63] Jürgens, G., Z. Physik **134** (1952) 21.

[2.64] Kaporski, L. N., F. S. Pedos, N. S. Swentitski and S. I. Shlepkova, Izv. Akad. Nauk SSSR Ser. Fiz. **26** (1962) 968.

[2.64a] Keilmann, F., Appl. Optics **9** (1970) 1319.

[2.65] Kistiakowsky, G. B., Rev. sci. Instrum. **2** (1931) 549.

[2.66] Knopp, C. F., C. F. Gottschlich and A. B. Cambel, J. quant. Spectrosc. Radiat. Transfer **2** (1962) 297.

[2.66a] Knopp, C. F. and A. B. Cambel, Rev. sci. Instrum. **37** (1966) 332.

[2.67] Kohlrausch, F., Praktische Physik. B. G. Teubner Verlags.ges. Leipzig 1950.

[2.68] Körber, W. G. and H. E. Röllig, in: Optik u. Spektrosk. aller Wellenlängen. Akademie-Verlag, Berlin 1962, p. 441.

[2.69] Koltsov, V. V., Optika i Spektroskopiya 8 (1960) 582.

[2.70] Korolev, F. A., High-Resolution Spectroscopy, Gos. Izd. tekhn. — teor. Literatury, Moscow 1953.

[2.71] Kostkowski, H. J. and R. D. Lee, in: Temperature, its Measurement and Control in Sci. and Ind., Vol. 3,1, Reinhold Publ. Corp., New York 1962, p. 449.

[2.72] Kovaleva, T. A., A. E. Melamid, A. N. Pertsev and A. N. Pisarevski, Pribory Tekhn. Eksp. No. 5 (1966) 5.

[2.73] Krause, H. and K. Krebs, Optik **20** (1963) 471.

[2.74] Krefft, H., F. Rössler and Rüttenauer, Z. techn. Phys. **18** (1937) 20.

[2.75] Lang, K., Mikrochim. Acta (1955) 278.

[2.76] Larenz, F. O. and F. G. Dunnington, Phys. Rev. **35** (1930) 396.

[2.77] Larrabee, R. D., The spectral emissivity and optical properties of tungsten. Techn. Rep. 328 Res. Lab. Electronics Mass. Inst. Technol., April 1957.

[2.78] Larrabee, R. D., J. Opt. Soc. Amer. **49** (1959) 619.

[2.79] Lau, E., E.-G. Kind and G. Roose, Optik **15** (1958) 162.

[2.80] Lawrence, E. O. and N. E. Edlefson, Rev. sci. Instrum. **1** (1930) 45.

[2.81] Lebedev, S. V., S. L. Mandelshtam and G. M. Rodin, Zh. eksp. teoret. Fiz. **37** (1959) 349; German Transl.: Phys. Abh. SU **2** (1960) 120.

[2.81a] Lincke, R., in: Plasmadiagnostics, North Holland Publ. Co. Amsterdam 1968, p. 347.

[2.81b] Liu, I. D. and M. Steinberg, J. quant. Spectrosc. Radiat. Transfer 8 (1968) 161.

[2.81c] Low, M. J. D., Appl. Spectrosc. **18** (1964) 187.

[2.82] Löwe, F., Optische Messungen des Chemikers und des Mediziners. Theodor Steinkopff, Dresden und Leipzig 1949.

[2.83] Lukirski, A. P. and I. A. Karpovich, Optika i Spektroskopiya **6** (1959) 685.

[2.83a] Lush, H. J., J. sci. Instrum. **42** (1965) 597.

[2.83b] Magdeburg, H., Z. Naturforsch. **20a** (1965) 980.

[2.83c] Magdeburg, H. und U. Schley, Z. angew. Physik **20** (1966) 465.

[2.84] Nobbs, J. Mck., Spectrochim. Acta **15** (1959) 939.

[2.84a] Maisenhälder, F., Thesis Stuttgart 1962.

[2.85] Malyshev, G. M., G. T. Razdobarin and L. V. Sokolova, Doklady Akad. Nauk SSSR **145** (1962) 768.

[2.86] MALYSHEV, G. M., G. T. RAZDOBARIN and L. V. SOKOLOVA, Zh. tekhn. Fiz. **33** (1963) 191.
[2.86a] MANDEL, L., Progr. Opt. **2** (1963) 181.
[2.86b] MANTHEY, H. and W. BÖTTICHER, Z. angew. Phys. **25** (1968) 77.
[2.87] MARSHAK, I. S., Usp. Fiz. Nauk **77** (1962) 229.
[2.88] MEHLTRETTER, J. P., Z. Astrophys. **51** (1960) 32.
[2.88a] MEWE, R. and R. F. DE VRIES, Plasma Physics **6** (1964) 591.
[2.89] MEYER, A. E. H. and E. O. SEITZ, Ultraviolette Strahlen. Verl. Walter de Gruyter and Co., Berlin 1949.
[2.90] MIELENZ, K. D., Optik **13** (1956) 437.
[2.91] MILNE, G., W. STAUDENMAIER and T. PUTNAM, J. Opt. Soc. Amer. **47** (1957) 117.
[2.91a] MIYAMOTO, K., M. OTSUKA, J. FUJITA, S. KAWASAKI, N. INCUE, Y. SUZUKI and T. UCHIDA, Japan. J. appl. Phys. **5** (1966) 970.
[2.91b] MORGAN, F. J., A. H. GABRIEL and M. J. BARTON, J. sci. Instrum. **2** (1968) 998.
[2.92] MOSER, H., Optik **12** (1955) 362.
[2.93] MOSER, H., S.-B. math. naturwiss. Kl. Bayer. Akad. Wiss., Munich 1955, 53.
[2.93a] NÄSSL, W., Z. angew. Phys. **19** (1965) 238.
[2.94] NAGIBINA, I. M. and V. K. PROKOFYEV, Spektralnye pribory i tekhnika spektroskopii, Mashgiz, Moskow-Leningrad 1963.
[2.95] NESTOR, O. H. and H. N. OLSEN, SIAM-Rev. **2** (1960) 200.
[2.96] NEUMANN, W., Beitr. Plasmaphys. **2** (1962) 80.
[2.97] NEUMANN, W., Beitr. Plasmaphys. **2** (1962) 252.
[2.97a] NEUMANN, W., in: Ergebn. Plasmaphys. Gaselektronik Vol. 1, Akad. Verl. Berlin 1967, p. 3.
[2.97b] NEUMANN, W. and F. SERICK, Beitr. Plasmaphys. **7** (1967) 203.
[2.98] NEWMAN, P. A. and R. BINDER, Rev. sci. Instrum. **32** (1961) 351.
[2.99] NULL, M. R. and W. W. LOZIER, J. Opt. Soc. Amer. **52** (1962) 1156.
[2.100] OHKAWA, T., H. K. FORSEN, A. A. SCHUPP and D. W. KERST, Phys. Fluids **6** (1963) 846.
[2.101] OLSEN, H. N., J. quant. Spectrosc. Radiat. Transfer **3** (1963) 59.
[2.102] OLSEN, H. N., J. quant. Spectrosc. Radiat. Transfer **3** (1963) 305.
[2.103] PAETZOLD, H. K., Z. Naturforsch. **2a** (1947) 219.
[2.103a] PAQUETTE, D. R. and W. L. WIESE, Appl. Optics **3** (1964) 291.
[2.104] PATRICK, R. M. and M. CAMAC, in: Plasma Hydromagnetics. Stanford Univ. Press, Stanford 1962, p. 97.
[2.104a] PEACOCK, N. J., J. COOPER and J. R. GREIG, Proc. Phys. Soc. **83** (1964) 803.
[2.105] PEARCE, W. J., in: Optical spectrometric measurements of high temperatures, Univ. of Chicago Press, Chicago 1961.
[2.105a] PENSELIN, S. and A. STEUDEL, Z. Physik **142** (1955) 21.
[2.105b] PIRRONI, J. S. and R. R. STEVENS, Rev. sci. Instrum. **38** (1967) 382.
[2.106] PISMENNY, W. D., I. M. PODGORNY and S. SUKEWER, Zh. eksp. teoret. Fiz. **43** (1962) 2008.
[2.106a] RATH, J., W. MANTHEY and W. BÖTTICHER, Z. angew. Phys. **22** (1967) 350.
[2.106b] QUINN, T. J. and C. R. BARBER, Metrologia **3** (1967) 19.
[2.107] RAUTIAN, S. G., Usp. Fiz. Nauk **66** (1958) 475.
[2.108] REICH, H. J., Z. angew. Phys. **16** (1963) 299.
[2.109] RÖSSLER, F., Ann. Physik **10** (1952) 177.
[2.110] RÖSSLER, F., Ann. Physik **1** (1958) 122.
[2.111] ROLLETT, J. S. and L. A. HIGGS, Proc. Phys. Soc. Lond. **79** (1962) 87.
[2.112] RUSIN, A. D. and V. M. TATEVSKI, Pribory Tekhn. Eksp. No. 4 (1958) 101.
[2.113] SAIDEL, A. N., G. M. MALYSHEV, A. B. BERESIN and G. T. RAZDOBARIN, Zh. tekhn. Fiz. **30** (1960) 1437.
[2.114] SAMSON, J. A. R., J. Opt. Soc. Amer. **52** (1962) 525.

[2.114a] Samson, J. A. R. and W. Liebl, Rev. sci. Instrum. **33** (1962) 1340.
[2.114b] Savis, F. W., B. J. F. Luyken, F. J. de Heer and J. Aarts, in: Sympos. Calibrat. Methods in UV and X-ray regions of the spectrum, Munich 1968, S. 67.
[2.114c] Schanda, J., Acta Imeko III., Stockholm 1964, 24-HU-135, p. 9.
[2.114d] Schmeidler, W., Integralgleichungen mit Anwendungen in Physik und Technik. I. Lineare Integralgleichungen. Akad. Verlagsges. Geest & Portig KG, Leipzig 1955, p. 213.
[2.115] Schneider, R., Proc. 5. internat. Conf. Ionization Phenomena in Gases, Vol. 2, North Holland Publ. Co., Amsterdam 1962, 1945.
[2.115a] Schneider, R. T., Proc. 6. internat. Conf. Ion. Phenom. Gases, Paris 1963, Vol. 4, p. 23.
[2.115b] Schneider, R. T., Engng. aspects of Magnetohydrodynamics, Gordon & Breach, N. Y., London 1964, 153.
[2.116] Schneider, R. and M. Mailänder, Z. angew. Phys. **12** (1960) 521.
[2.117] Schneider, R., F. Maisenhälder and M. Kling, Z. angew. Phys. **13** (1961) 211.
[2.117a] Schurer, K., Appl. Optics **7** (1968) 461.
[2.118] Schwartz, F. R. and B. J. Pernick, Rev. sci. Instrum. **33** (1962) 765.
[2.119] Semenov, R. I., E. Y. Fradkin and M. P. Chaika, Optika i Spektroskopiya **7** (1959) 785.
[2.119a] Shannon, P. E. V., Rev. sci. Instrum. **34** (1963) 1281.
[2.119b] Shumaker, J. B. and C. R. Yokley, Appl. Optics **3** (1964) 83.
[2.119c] Smith-Saville, R. J. and S. Ness, J. sci. Instrum. **44** (1967) 631.
[2.120] Spillman, G. R., W. S. Cooper and J. M. Wilcox, Appl. Optics **2** (1963) 205.
[2.120a] Steudel, A. and S. Stotz, Z. Physik **151** (1958) 233.
[2.121] Stewart, J. E., Appl. Optics **1** (1962) 75.
[2.121a] Stojanoff, C. G., AIAA-J. **4** (1966) 1766.
[2.121b] Stokes, A. D., J. Opt. Soc. Amer. **57** (1967) 1100.
[2.121c] Streissle, H. P., Doct. Diss. Techn. Hochsch. Stuttgart 1965.
[2.122] Tanaka, Y., R. E. Huffmann and J. C. Larrabee, J. quant. Spectrosc. Radiat. Transfer **2** (1962) 451.
[2.123] Thackeray, D. P. C., J. sci. Instrum. **35** (1958) 248.
[2.124] Tikhodeyev, G. M., Izv. Akad. Nauk SSSR Ser. tekhn. Nauk. **8** (1957) 78.
[2.125] Török, T., Optik u. Spektroskopie aller Wellenlängen, Akad. Verl. Berlin 1962, p. 573.
[2.125a] Turner, E. B., in: Plasma diagnostic Techniques, Acad. Press, N. Y. London 1965, p. 319.
[2.125b] Tyrén, F., Z. Physik **111** (1938) 314.
[2.125c] Tyrén, F., Nova Acta Soc. Sci. Uppsal. IV **12**, No. 1, 1940.
[2.126] Unsöld, A., Physik der Sternatmosphären. Springer Verlag Berlin-Göttingen-Heidelberg 1955.
[2.127] Uvarova, V. M., N. K. Sukhodrev, A. A. Pankova, M. P. Shpolski and A. N. Kovanova, Izv. Akad. Nauk SSSR Ser. Fiz. **26** (1962) 967.
[2.127a] Vanhauwermeiren, R., Proc. 8. internat. Conf. Phenom. Ionized Gases, Contributed Pap., IAEA Vienna 1967, p. 504.
[2.127b] Vidal, C.-R., Z. Instrum.-Kde **74** (1966) 316.
[2.127c] Vilesov, F. I., M. E. Akopyan and V. I. Kleimenov, Pribory Tekhn. Eksp. No. 6 (1963) 150.
[2.127d] Volkov, Ya. F., V. G. Dyatlov and N. I. Mitina, Teplofizika **3** (1965) 659.
[2.128] Vos, J. C. de, The emissivity of tungsten ribbon. Diss. Univ. Amsterdam 1953
[2.129] Vos, J. C. de, Physica **20** (1954) 690.
[2.130] Weber, P. E., Optik **15** (1958) 592.

[2.131] Weizel, W. and G. Meister, Spektralanalyse durch Messung des Interferenzkontrastes. Forsch.-Ber. Landes Nordrhein-Westfalen Nr. 334, Westdeutscher Verlag, Köln and Opladen 1956.
[2.131a] Wende, B., Z. angew. Phys. **20** (1966) 473.
[2.131b] Wende, B., Optik **25** (1967) 284.
[2.131c] Whitney, W. T., Rev. sci. Instrum. **36** (1965) 1656.
[2.132] Wiese, W. L., Rev. sci. Instrum. **31** (1960) 943.
[2.133] Wiese, W. L. and J. B. Shumaker, J. Opt. Soc. Amer. **51** (1961) 937.
[2.134] Willenberg, H., ATM No. 334 (1963) 249.
[2.135] Wurster, W. H., Rev. sci. Instrum. **28** (1957) 1093.
[2.136] Yokley, C. R. and J. B. Shumaker, Rev. sci. Instrum. **34** (1963) 551.
[2.136a] Young, A. T., Rev. sci. Instrum. **38** (1967) 1336.
[2.136b] Zajdel, A. N. and E. Ya. Shreider, Spektroskopiya vakuumnogo ultrafioleta, Izd. Nauka, Moscow 1967.
[2.136c] Zschaeck, H., VIII. internat. Coll. Techn. Hochschul. Ilmenau 1963, Pt. 4, p. 521.
[2.137] Zschaeck, H., Exp. Techn. Phys. **12** (1964) 373.
[3.1] Allen, C. W., Astrophysical Quantities, Athlone Press London 1963.
[3.2] Finkelnburg, W., Einführung in die Atomphysik. Springer-Verlag, Berlin-Göttingen-Heidelberg 1962.
[3.3] Herzberg, G., Spectra of Diatomic Molecules. D. van Nostrand Co., Princeton, Toronto, N. Y. und London 1961.
[4.1] Altmann, R. M., J. chem. Phys. **32** (1960) 2.
[4.1a] Artmann, J., Z. Physik **183** (1965) 65.
[4.2] Baumann, G., Z. phys. Chem. **14** (1958) 173.
[4.2a] Blaise, J. and R. Vetter, Compt. Rend. Acad. Sci. **256** (1963) 630.
[4.2b] Bögershausen, W. and H. Krempl, Z. angew. Phys. **22** (1967) 165.
[4.2c] Bögershausen, W., J. Hingsammer and H. Krempl, Ber. Bunsenges. phys. Chem. **7** (1967) 64.
[4.3] Boldt, G., Z. Physik **154** (1959) 319.
[4.3a] Bötticher, W., O. Roder and K. H. Wobig, Z. Physik **175** (1963) 480.
[4.3b] Boumans, P. W. J. M., Theory of spectrochemical excitation, Hilger & Watts Ltd., London 1966.
[4.4] Bredt, I., Z. Naturforsch. **6a** (1951) 103.
[4.5] Burhorn, F., H. Maecker and T. Peters, Z. Physik **131** (1951) 18.
[4.6] Burhorn, F. and R. Wienecke, Z. phys. Chem. **212** (1959) 105.
[4.7] Burhorn, F. and R. Wienecke, Z. phys. Chem. **213** (1960) 37.
[4.8] Burhorn, F. and R. Wienecke, Z. phys. Chem. **215** (1960) 269.
[4.9] Burhorn ,F. and R. Wienecke, Z. phys. Chem. **215** (1960) 285.
[4.9a] Chowdhury, S. S., J. quant. Spectrosc. Radiat. Transfer **9** (1969) 125.
[4.9b] Clementi, E., Astrophys. J. **133** (1961) 303.
[4.9c] Czernichowski, A., Teplofizika **6** (1968) 809.
[4.9d] Czernichowski, A., Teplofizika **6** (1968) 989.
[4.10] D'Ans, J. and E. Lax, Taschenbuch für Chemiker und Physiker. Springer-Verlag, Berlin 1949.
[4.10a] Döbele, H. F., R. Wienecke and S. Witkowski, Z. Naturforsch. **21a** (1966) 390.
[4.10b] Drawin, H.-W. and P. Felenbok, Data for plasmas in local thermodynamic equilibrium, Gauthier-Villars Paris 1965.
[4.10c] Drellishak, K. S., D. P. Aeschliman and A. B. Cambel, Phys. Fluids **8** (1965) 1590; Erratum: Phys. Fluids **8** (1965) 2304.
[4.10d] Drellishak, K. S., C. F. Knopp and A. B. Cambel, Phys. Fluids **6** (1963) 1280.

[4.10e] EBERHAGEN, A. and W. LÜNOW, Tabellen zur Auswertung von Intensitätsmessungen am Wasserstoffplasma. Inst. Plasmaphys. Garching near Munich, Laboratory Report IPP 1/23 and IPP 6/20, June 1964.

[4.11] EBERT, H., Physikalisches Taschenbuch, VEB Deutscher Verlag der Wissenschaften, Berlin 1962.

[4.12] ECKER, G. and W. KRÖLL, Phys. Fluids **6** (1963) 62.

[4.12a] FABRE, E. and L. HERMAN, Revue d'Optique **41** (1962) 339.

[4.12b] FRIE, W., Z. Physik **201** (1967) 269.

[4.13] FRIE, W. and H. MAECKER, Z. Physik **168** (1962) 206.

[4.13a] FRIEDRICH, J., Techn. wiss. Abh. Osram **9** (1967) 15.

[4.13b] GALAN, L. DE, R. SMITH and J. D. WINEFORDNER, Spectrochim. Acta **23 B** (1968) 521.

[4.13c] GARSTANG, R. H., Internat. Astron. Union, Sympos. No. 26 (1966) 57.

[4.14] GAYDON, A. G., Dissociation energies and spectra of diatomic molecules, Chapman and Hall, London 1953.

[4.15] GODNEW, I. N., Berechnung thermodynamischer Funktionen aus Moleküldaten, VEB Dt. Verl. Wissenschaften, Berlin 1963 (Translated from Russian).

[4.15a] GOLDMAN, K., Physica **29** (1963) 1024.

[4.16] GROSS, R. A. and C. L. EISEN, Dynamics of conducting Gases. Northwestern Univ. Press, Evanston, Illinois 1960, p. 15.

[4.16a] HERMEL, A., Spektroskopische Untersuchung eines Helium-Wasserstoff-Mischplasmas in einer quasistationären Impulsentladung. Dissertation Humboldt-Univ. Berlin 1967; Beitr. Plasmaphys. **9** (1969) 261.

[4.17] HERZBERG, G., Spectra of Diatomic Molecules. D. van Nostrand Co., Inc., Princeton, Toronto, London and New York 1961.

[4.18] HIMPAN, J., Z. Naturforsch. **13a** (1958) 239.

[4.18a] HOCHSTIM, A. R., Theoretical calculation of thermodynamic properties of air, in: Combustion and Propulsion, 5. AGARD Coll. on High Temperature Phenomena, Pergamon Press Oxford 1963, p. 3.

[4.19] HODGMAN, C. D., Handbook of Chemistry and Physics, Chemical Rubber Publ. Co., Cleveland, Ohio 1961.

[4.19a] IRMER, J. and M. WORM, Z. phys. Chem. **231** (1966) 215.

[4.19b] IRMER, J. and M. WORM, Beitr. Plasmaphys. **8** (1968) 217.

[4.19c] HUNGMAN, V. S., L. V. GURVICH and N. P. RTISHCHEVA, Electricity from MHD Vol. 3, IAEA Wien 1966, p. 79.

[4.19d] JUNGMAN, V. S., L. V. GURVICH and N. P. RTISHCHEVA, Teplofizika **4** (1966) 507.

[4.19e] KLOSS, H. G., Mon. Ber. Dtsch. Akad. Wiss. **6** (1964) 93.

[4.19f] KLÜBER, O., Z. Naturforsch. **22a** (1967) 1599.

[4.19g] KRINBERG, L. A., Teplofizika **5** (1967) 936.

[4.19h] KROEPELIN, H., Combustion and Propulsion, High Temperature Phenomena. Pergamon Press, Oxford 1963, p. 45.

[4.19i] KROEPELIN, H. and K.-U. HOFFMANN, Optik **22** (1965) 599.

[4.19j] KROEPELIN, H., D. E. KIPPING and H. PIETRUCK, Equilibrium partial pressures in the systems $C + H_2 + N_2$ and $C + H_2 + NH_3$ at temperatures from 1000 to 12,000 °K. In: Progr. internat. Res. on Thermodynamic and Transport Properties, Academic Press N. Y. London 1962, p. 626.

[4.19k] KROEPELIN, H. and K. K. NEUMANN, Thermodynamics and Transport Properties of Gases, Liquids and Solids, McGraw Hill Book Co., Inc. 1959, v. 453.

[4.19l] KROEPELIN, H. and E. WINTER, Thermodynamics and Transport Properties of Gases, Liquids and Solids, McGraw Hill Book Co., Inc. 1959, p. 438.

[4.20] KÜMMEL, F., Beitr. Plasmaphys. **1** (1961) 94.

[4.20a] KUTHE, R. and K.-K. NEUMANN, Z. Flugwiss. **13** (1965) 411.

[4.20b] KUTHE, R. and K. NEUMANN, Nizkotemperaturnaya Plazma, Izd. Mir Moscow 1967, p. 138.

[4.21] LANDOLT-BÖRNSTEIN, Zahlenwerte und Funktionen ..., Vol. I, 1, Springer-Verlag, Berlin-Göttingen-Heidelberg 1950.

[4.22] LANDOLT-BÖRNSTEIN, Zahlenwerte und Funktionen ..., Vol. II,4, Springer-Verlag, Berlin-Göttingen-Heidelberg 1962.

[4.22a] LEWIS, C. H. and E. G. BURGESS, AIAA-J. **1** (1963) 1928.

[4.23] LEWIS, B. and VON ELBE, G., Combustion, Flames and Explosion of Gases, Academic Press, N. Y.-London 1961.

[4.23a] LICK, W. J. and H. W. EMMONS, Thermodynamic properties of Helium to 50,000 °K, Cambridge, Mass., Harvard Univ. Press 1962.

[4.23b] LOTZ, W., J. Opt. Soc. Amer. **57** (1967) 873.

[4.23c] LUDWIG, H. C., Weld. Res. Suppl. Weld. J. (1959) 296s – 300s.

[4.23d] MCCHESNEY, M. and N. R. JONES, Proc. Phys. Soc. Lond. **84** (1964) 983.

[4.23e] MAIN, R. P. and E. BAUER, J. quant. Spectrosc. Radiat. Transfer **7** (1967) 527.

[4.23f] MARTINEK, F., Thermodynamics and Transport Properties of Gases, Liquids and Solids, McGraw Hill Book Co., Inc., NewYork-Toronto-London 1959, p. 130.

[4.24] MARYNOWSKI, C. W., R. C. PHILLIPS, J. R. PHILLIPS and N. K. HIESTER, Ind. Engng. Chem. Fundamentals **1** (1962) 52.

[4.25] MOORE, C. E., Atomic Energy Levels. National Bureau of Standards Circular Nr. 467, Washington, Vol. 1 (1949), Vol. 2 (1952), Vol. 3 (1958).

[4.25a] MORGENROTH, H., Ann. Physik **3** (1959) 473.

[4.25b] MÜCK, G. and H. P. POPP, Z. Naturforsch. **23a** (1968) 1213.

[4.26] NEUMANN, K. K., Z. Elektrochem. **66** (1962) 551.

[4.27] NEUMANN, K. K., Ber. Bunsenges. phys. Chem. **67** (1963) 377.

[4.28] OLSEN, H. N., Phys. Fluids **2** (1959) 614.

[4.29] OLSEN, H. N., Temperature, its Measurement and Control in Science and Industry, Bd. III, 1, Reinhold Publ. Co., New York 1962, p. 593.

[4.30] OLSEN, H. N., J. quant. Spectrosc. Radiat. Transfer **3** (1963) 59.

[4.30a] OLSEN, H. N., in: Physico-Chemical Diagnostics of Plasmas, Northwestern Univ. Press, Evanston 1963, p. 47.

[4.30b] OLSEN, H. N., J. quant. Spectrosc. Radiat. Transfer **3** (1963) 305.

[4.30c] OPPENHEIM, I. and D. R. HAFFMANN, J. chem. Phys. **39** (1963) 101.

[4.31] PENG, T. and A. L. PINDROH, Magnetohydrodynamics, Northwestern Univ. Press, Evanston 1962, p. 67.

[4.31a] PIETSCH, G., Z. phys. Chem. Neue Folge **81** (1972) 69.

[4.32] PLESHANOV, A. S., Fizicheskaya gazodinamika, teploobmen i termodinamika gazov vysokich temperatur. Izd. Akad. Nauk SSSR, Moscow 1962, p. 36.

[4.33] PLESHANOV, A. S. and S. G. SAITSEV, Fizicheskaya gazodinamika, teploobmen i termodinamika gazov vysokich temperatur, Izd. Akad. Nauk SSSR, Moscow 1962, p. 15.

[4.33a] POPENOE, C. H. and J. B. SHUMAKER, J. Res. Nat. Bur. Stand. **69A** (1965) 495.

[4.34] PREDVODITELEV, A. S., E. P. STUPOCHENKO et al., Tables of thermodynamic functions of air for the temperature range 6000 – 12,000 °K and pressure range 0.001 – 1000 atm. Cleaver-Hume Press Ltd., London 1958.

[4.35] PREDVODITELEV, A. S. et al., Tablitsy termodinamicheskich funktsii vozducha dlya temperatur ot 12,000 – 20,000 °K i davlenii ot 0.001 – 1000 atm, Izd. Akad. Nauk SSSR, Moscow 1959.

[4.36] RAYMOND, J. L., J. chem. Engng. Data **7** (1962) 190.

[4.36a] REED, T. B., Recent developments in plasma generation, Rep. MIT, Lincoln Lab., MS-904, ca 1963.

[4.36b] RICHARDS, P. H., C. F. GOZNA and J. G. STEEL, Proc. 8. internat. Conf. Phenom. Ionized Gases, Contributed Pap., IAEA Vienna 1967, p. 235.

[4.37] ROZHDESTVENSKII, I. B. and Y. V. SAMUILOV, Fizicheskaya gazodinamika, teploobmen i termodinamika gazov vysokich temperatur. Izd. Akad. Nauk SSSR, Moscow 1962, p. 103.
[4.38] RUTNER, E., W. H. MCLAIN and K. SCHELLER, J. chem. Phys. **24** (1956) 173.
[4.39] SAIDEL, A. N., W. K. PROKOFIEV and S. M. RAISKI, Spektraltabellen. VEB Verlag Technik, Berlin 1961 (translated from Russian).
[4.39a] SAMUILOV, E. V. and K. K. OLEVINSKI, in: Fizicheskaya Gazodinamika, Izd. Nauka, Moscow 1964, p. 10.
[4.39b] SHERMAN, A. and F. MARTINEK, Planet. Space Sci. **3** (1961) 271.
[4.40] STUPOCHENKO, E. V., I. P. STAKHANOV, E. V. SAMUILOV, A. S. PLESHANOV and I. B. ROSHDESTVENSKII, in: Physical gas dynamics, Pergamon Press, Oxford-London-New York-Paris 1961, p. 1.
[4.41] UNSÖLD, A., Physik der Sternatmosphären, Springer Verlag Berlin-Göttingen-Heidelberg 1955, p. 79.
[4.41a] VEDENEYEV, V. I., L. V. GURVICH, V. N. KONDRATYEV, V. A. MEDVEDEV and YE. L. FRANKEVICH, Bond energies, ionization potentials and electron affinities, Edward Arnold Publ. Ltd. London 1966 (translated from Russian).
[4.42] WRAY, K. L., Hypersonic flow research. Acad. Press, N. Y.-London 1962, p. 181.
[5.1] AARTS, J. and G. BOSCH, Physica **30** (1964) 1673.
[5.1a] AARTS, J., D. HARTING and C. J. BAKKER, Physica **20** (1954) 1250.
[5.1b] ABDRAZAKOV, A., Z. ZHEYENBAYEV, F. G. KAVICH, R. I. KONAVKO, M. A. SAMSONOV and V. S. ENGELSHT, Issledovanie elektricheskoi dugi i plazmatrona, Izd. Ilim, Frunze 1968, p. 48.
[5.1c] ACKERMANN, H., Z. Physik **194** (1966) 253.
[5.1d] ADCOCK, B. D., J. quant. Spectrosc. Radiat. Transfer **7** (1967) 385.
[5.2] ADCOCK, B. D. and W. E. G. PLUMTREE, J. quant. Spectrosc. Radiative Transfer **4** (1964) 29.
[5.3] ADDINK, N. W. H., Spectrochim. Acta **15** (1959) 349.
[5.3a] ADDINK, N. W. H., Proc. 14. Coll. Spectrosc. internat., Debrecen, Hungary 1967, p. 201.
[5.3b] AGARBICEANU, I., I. KUKUREZIANU, I. POPESKU and V. VASILIU, Optika i Spektroskopiya **14** (1963) 18.
[5.3c] AGNEW, L. and C. SUMMERS, Proc. 7. internat. Conf. Phenom. Ionized Gases, Belgrade 1965, Vol. 2, p. 574.
[5.3d] AHLBORN, B., Z. Naturforsch. **20a** (1965) 466.
[5.4] ALEKSEEV, M. A., Trud. Sibir. Fiz.-tekhn. Inst. **32** (1953) 21; Diss. Tomsk 1951.
[5.4a] ALEKSEEV, M. A. and V. FILIPPOVA, Spektroskopiya, metodi i primenenie, Izd. Nauka, Moscow 1964, p. 39.
[5.4b] ALEKSEEV, M. A. and E. A. KOMAROVA, Spektroskopiya, metodi i primenenie, Izd. Nauka, Moscow 1964, p. 37.
[5.5] ALLEN, C. W., Astrophysical Quantities, Athlone Press, London 1955.
[5.6] ALLEN, C. W., Mon. Not. R. astr. Soc. **117** (1957) 622.
[5.7] ALLEN, C. W., Mon. Not. R. astr. Soc. **121** (1960) 299.
[5.8] ALLEN, C. W., Astrophysical Quantities, Athlone Press, London 1963.
[5.9] ALLEN, C. W. and A. S. ASSAD, Mon. Not. R. astr. Soc. **117** (1957) 36.
[5.9a] ALLEN, C. W. and C. H. CORLISS, Monthly Notices Roy. Astron. Soc. **126** (1963) 37.
[5.10] ALLEN, J. S. V. and C. E. HESTFAL, Phys. Rev. **47** (1935) 926.
[5.11] ALLER, L. H., Astrophys. J. **97** (1943) 135.
[5.12] ALLER, L. H., Astrophys. J. **109** (1949) 244.
[5.13] ALLER, L. H. and D. H. MENZEL, Astrophys. J. **102** (1945) 239.
[5.13a] ALTICK, P. L. and E. N. MOORE, Phys. Rev. Lett. **15** (1965) 100.
[5.13b] ALTMAN, E. L. and M. P. CHAIKA, Optika i Spektroskopiya **19** (1965) 968.

[5.13c] ANDERSON, D. K., Phys. Rev. **137** (1965) A 21.
[5.13d] ANDERSON, E. M., E. K. ANDERSON and V. F. TRUSOV, Optika i Spektroskopiya 21 (1966) 114.
[5.13e] ANDERSON, E. M., E. K. ANDERSON and V. F. TRUSOV, Optika i Spektroskopiya **22** (1967) 861.
[5.13f] ANDERSON, E. M., Z. A. BUSKA, R. O. GRINBERG and A. K. SAULGOZHA, Vestnik Leningrad Univ. Ser. Fiz. i Khim. No. 4 (1956) 27.
[5.14] ANDERSON, E. M. and V. A. TSILITIS, Optika i Spektroskopiya **16** (1964) 177.
[5.15] ANDERSON, E. M. and V. A. TSILITIS, Optika i Spektroskopiya **16** (1964) 382.
[5.16] ANDERSON, E. M., V. A. TSILITIS and E. A. SOROKINA, Optika i Spektroskopiya 23 (1967) 188.
[5.16a] ANDERSON, E. M., V. A. TSILITIS and E. S. SOROKINA, Optika i Spektroskopiya 23 (1967) 513.
[5.16b] ANKUDINOV, V. A., S. V. BOBASHOV and E. P. ANDREEV, Zh. eksp. teoret. Fiz. 48 (1965) 40.
[5.17] ARDENNE, M. VON, Zh. eksp. teoret. Fiz. **26** (1954) 362.
[5.17a] ARMSTRONG, J. F., B. J. ELLIOTT, D. W. GEORGE, H. K. MESSERLE and A. D. STOKES, Internat. Sympos. Magnetohydrodynamic Electric Power Generation, Paris 1964, Vol. 1, p. 149.
[5.17b] ARMSTRONG, B. H. and P. S. KELLY, Bull. Amer. Phys. Soc. **7** (1962) 643.
[5.17c] ARMSTRONG, L. and R. LIEBERMANN, J. quant. Spectrosc. Radiat. Transfer **9** (1969) 123.
[5.17d] ARMSTRONG, B. H. and K. L. PURDUM, Phys. Rev. **150** (1966) 51.
[5.18] AUSLÄNDER, J., Helv. phys. Acta **11** (1938) 562.
[5.18a] BAKANOVICH, G. I. and L. I. GRECHIKHIN, Teplofizika **3** (1965) 520.
[5.18b] BAKOS, J. and J. SZIGETI, Acta phys. hung. **21** (1966) 149.
[5.18c] BAKOS, J. and J. SZIGETI, Acta phys. hung. **23** (1967) 341.
[5.18d] BAKOS, J. and J. SZIGETI, Optika i Spektroskopiya **23** (1967) 478.
[5.18e] BAKOS, J., J. SZIGETI and L. VARGA, Phys. Letters **20** (1966) 503.
[5.18f] BARACH, J. P., J. appl. Phys. **39** (1968) 1771.
[5.18g] BARACH, J. P. and T. W. MAYES, Phys. Fluids **11** (1968) 89.
[5.18h] BARKER, J. R. and J. V. MICHAEL, J. Opt. Soc. Amer. **58** (1968) 1615.
[5.18i] BARRAT, J.-P., Thesis Paris 1959.
[5.18j] BARRAT, J.-P., J.-L. COJAN and Y. LECLUSE, Compt. Rend. Acad. Sci. **260** (1965) 1893.
[5.18k] BARRAT, J.-P., J.-L. COJAN and Y. LECLUSE, Compt. Rend. Acad. Sci. AB **262** (1966) B 609.
[5.18l] BASCHEK, B., W. H. KEGEL and G. TRAVING, Z. Astrophys. **56** (1963) 282.
[5.18m] BÄSTLEIN, C., G. BAUMGARTNER and B. BROSA, Z. Physik **218** (1969) 319.
[5.19] BATES, D. R., Proc. Roy. Soc. Lond. A **188** (1946) 359.
[5.20] BATES, D. R. and A. DAMGAARD, Astrophys. J. **107** (1948) 383.
[5.21] BATES, D. R. and A. DAMGAARD, Phil. Trans. A **242** (1949) 101.
[5.22] BATES, D. R. and H. S. W. MASSEY, Proc. Roy. Soc. Lond. A **177** (1941) 329.
[5.23] BAUDER, U., Z. Physik **205** (1967) 303.
[5.23a] BAUMANN, M. and G. WANDEL, Phys. Letters **22** (1966) 283.
[5.23b] Beam-Foil Spectroscopy, North Holland Publ. Co., Amsterdam 1970.
[5.24] BELL, G. D., M. H. DAVIS, R. B. KING and P. M. ROUTLY, Astrophys. J. **127** (1958) 755.
[5.25] BELL, G. D., M. H. DAVIS, R. B. KING and P. M. ROUTLY, Astrophys. J. **129** (1959) 437.
[5.26] BELL, G. D. and R. B. KING, Astrophys. J. **133** (1961) 781.
[5.26a] BELOKON, V. A., A. I. PETRUKHIN and V. A. PROSKURYAKOV, Zh. eksp. teoret. Fiz. 48 (1965) 50.

[5.26b] BELOUSOVA, I. M. and D. B. GUREVICH, Zh. tekhn. Fiz. **31** (1961) 1337.

[5.26c] BELYAKOV-BODIN, I. B. and S. L. MANDELSHTAM, Zh. tekhn. Fiz. **10** (1940) 332.

[5.26d] BENNETT, W. R., Proc. 2. internat. Conf. Quantum Electronics, Columbia Univ. Press New York 1961, p. 28.

[5.26e] BERG, H. F., Z. Physik **199** (1967) 445.

[5.26f] BERG, H. F., K. L. ECKERLE, R. W. BURRIS and W. L. WIESE, Astrophys. J. **139** (1964) 751.

[5.26g] BERG, H. F. and W. ERVENS, Z. Physik **206** (1967) 184.

[5.27] BERGSTEDT, K., Z. Physik **155** (1959) 23.

[5.27a] BERGSTRÖM, I., J. BROMANDER, R. BUCHTA, L. LUNDIN and I. MARTINSON, Phys. Letters **28A** (1969) 721.

[5.27b] BERKNER, K., W. S. COOPER, S. N. KAPLAN and R. V. PYLE, Phys. Letters **16** (1965) 35.

[5.27c] BERSUKER, I. B., Soviet Phys.-Doklady **2** (1957) 167.

[5.27d] BERTHELOT, C. A., J. quant. Spectrosc. Radiat. Transfer **5** (1965) 495.

[5.27e] BERTHELOT, C. A., Spectrochim. Acta **22** (1966) 829.

[5.27f] BETH, M. U., W. L. BOHN and G. NEDDER, Z. Naturforsch. **21a** (1966) 1203.

[5.28] BETHE, H., Handb. Phys. 24/1, Springer-Verlag, Berlin 1939, Chap. 3.

[5.29] BEUTEL, M., Ann. Physik **36** (1939) 533.

[5.29a] BIALAS-ZABAWA, A., M. KUCHARSKI, E. SKULSKA, J. URBACZKA and Z. WALACH, Acta phys. polon. **30** (1966) 897.

[5.29b] BIALAS-ZABAWA, A., F. PIECHA, E. SKULSKA, D. STREK, P. SZEWOZYK and Z. WALACH, Acta phys. polon. **29** (1966) 241.

[5.29c] BIALAS-ZABAWA, A., E. SKULSKA and Z. WALACH, Acta phys. polon. **26** (1964) 175.

[5.29d] BICKEL, W. S., Phys. Rev. **162** (1967) 7.

[5.29e] BICKEL, W. S. and S. BASHKIN, Phys. Letters **20** (1966) 488.

[5.29f] BICKEL, W. S. and A. S. GOODMAN, Phys. Rev. **148** (1966) 1.

[5.29g] BIENIEWSKI, T., Monthly Notices Roy. Astron. Soc. **127** (1964) 359.

[5.30] BIERMANN, L., Z. Physik **44** (1943) 232.

[5.31] BIERMANN, L., Z. Astrophysik **22** (1943) 157.

[5.32] BIERMANN, L., Nachr. Akad. Wiss. Göttingen Math. physk. Kl. Nr. 2 (1946) 116.

[5.33] BIERMANN, L., Veröff. Göttingen Nr. 86 (1947).

[5.34] BIERMANN, L., Veröff. Göttingen Nr. 87 (1947).

[5.35] BIERMANN, L., in: Landolt-Börnstein, Zahlenwerte und Funktionen Vol. I,1 Springer-Verlag Berlin 1950, p. 260.

[5.36] BIERMANN, L. and K. LÜBECK, Z. Astrophys. **25** (1948) 325.

[5.37] BIERMANN, L. and K. LÜBECK, Z. Astrophys. **26** (1948) 43.

[5.38] BIERMANN, L. and E. TREFFTZ, Z. Astrophys. **26** (1949) 213.

[5.38a] BIGEON, M. C., J. Physique **28** (1967) 51.

[5.39] BILLETER, W., Helv. phys. Acta **7** (1934) 505.

[5.40] BILLETER, W., Helv. phys. Acta **7** (1934) 841.

[5.40a] BOERSCH, H., J. GEIGER and B. SCHRÖDER, Abh. dtsch. Akad. Wiss. Berlin, Kl. Math., Phys., Techn. No. 1 (1967) 15.

[5.40b] BOERSCH, H., J. GEIGER and B. SCHRÖDER, Proc. 5. internat. Conf. Phys. Electronic Atomic Collisions, Nauka. Leningrad 1967, p. 481.

[5.40c] BOERSCH, H. and H. J. REICH, Optik **22** (1965) 289.

[5.40d] BÖGERSHAUSEN, W. and H. KREMPL, Z. angew. Phys. **18** (1965) 290.

[5.40e] BÖGERSHAUSEN, W. and H. KREMPL, Z. angew. Phys. **22** (1967) 165.

[5.40f] BOKOVA, N. A. and G. B. SUKHANOVA, Izvest. VUZ Fiz., Tomsk No. 5 (1964) 176.

[5.41] BOLD, H. J. VAN DEN, Diss. Utrecht 1945.

[5.42] BOLDT, G., Diss. Univ. Kiel 1957.

[5.43] Boldt, G., Z. Physik **150** (1958) 205.
[5.44] Boldt, G., Proc. 6. internat. Conf. Ionization Phenomena in Gases, Paris 1963, Vol. 3, p. 309.
[5.45] Boldt, G., Z. Naturforsch. **18a** (1963) 1107.
[5.45a] Boldt, G. and F. Labuhn, Rep. Max-Planck-Inst. MPI-PAE/Extraterr. 5/67, Munich 1967.
[5.45b] Boldt, G. and F. Labuhn, Z. Naturforsch. **22a** (1967) 1613.
[5.46] Bolotin, A. B., Diss. Vilna 1954.
[5.47] Bolotin, A. B. and A. P. Yutsis, Zh. eksp. teoret. Fiz. **24** (1953) 537.
[5.48] Bolotin, A. B., I. B. Levinson and L. I. Levin, Zh. eksp. teor. Fiz. **24** (1955) 449.
[5.48a] Bondar, V. A. and L. I. Kiselevski, Zh. prikl. Spektrosk. **5** (1966) 273.
[5.48b] Bondar, V. A. and L. I. Kiselevski, Teplofizika **5** (1967) 949.
[5.48c] Boni, A. A., J. quant. Spectrosc. Radiat. Transfer **8** (1968) 1385.
[5.48d] Bott, J. F., Phys. Fluids **9** (1966) 1540.
[5.48e] Bötticher, W., O. Roder and K. H. Wobig, Z. Physik **175** (1963) 480.
[5.49] Bouma, T., Diss. Utrecht 1930.
[5.49a] Boumans, P. W. J. M., Theory of Spectrochemical excitation, Hilger & Watts Ltd., London 1966, p. 20.
[5.50] Braddick, H. J. and R. W. Ditchburn, Proc. roy. Soc. Lond. A **143** (1934) 472.
[5.51] Braddick, H. J. and R. W. Ditchburn, Proc. roy. Soc. A **150** (1935) 478.
[5.52] Brannen, E., F. R. Hunt, R. H. Aldington and R. W. Nicholls, Nature **175** (1955) 810.
[5.53] Brannen, E., F. R. Hunt, R. H. Aldington and R. W. Nicholls, Phys. Rev. 99 (1955) 1658.
[5.54] Brehm, B., W. Demtröder and O. Osberghaus, Z. Naturforsch. **16a** (1961) 843.
[5.54a] Bridges, J. M., Diss. Univ. Maryland 1966.
[5.54b] Bridges, J. M. and W. L. Wiese, Phys. Rev. **159** (1967) 31.
[5.54c] Bridgett, K. A. and T. A. King, Proc. Phys. Soc. Lond. **92** (1967) 75.
[5.54d] Brochard, J., Thesis Paris 1950.
[5.54e] Brown, W. A., Phys. Fluids **9** (1966) 1273.
[5.55] Bruck, H., Ec. Norm. Supér. Paris, Lab. Phys. Nr. 1 (1942).
[5.55a] Brückner, R., Ber. dtsch. Keram. Ges. **40** (1963) 603.
[5.55b] Brule, M., J. Lagrue, J.-P. Barrat and J.-L. Cojan, Compt. Rend. Acad. Sci. **263** (1966) B1312.
[5.55c] Bucka, H., Coll. internat. Centre nat. rech. sci. No. 162 (1967) 131.
[5.55d] Bucka, H., B. Budick, R. J. Goshen and S. Marcus, Phys. Rev. **144** (1966) 96.
[5.55e] Bucka, H., B. Grosswendt and H. A. Schüssler, Z. Physik **194** (1966) 193.
[5.55f] Bucka, H. and J. Ney, Z. Physik **194** (1966) 208.
[5.55g] Bucka, H., J. Ney and C. Heppke, Z. angew. Phys. **20** (1966) 354.
[5.55h] Bucka, H. and G. Oppen, Ann. Physik **10** (1962) 119.
[5.55i] Budick, B., Phys. Rev. **168** (1968) 89.
[5.55j] Bues, I., H. J. Patt and J. Richter, Z. angew. Phys. **22** (1967) 345.
[5.56] Burger, H. C., J. B. van Milaan and L. S. Ornstein, Physica **4** (1937) 730.
[5.57] Burhorn, F., Z. Physik **140** (1955) 440.
[5.58] Burhorn, F. Z. Physik **155** (1959) 42.
[5.59] Burhorn, F., H. Maecker and T. Peters, Z. Physik **131** (1951) 28.
[5.59a] Busigin, E. P., V. G. Grigoryants, B. G. Zhukov and I. P. Yavor, Teplofizika 6 (1968) 181.
[5.59b] Byard, P. L., J. quant. Spectrosc. Radiat. Transfer 7 (1967) 559.
[5.59c] Byard, P. L., J. quant. Spectrosc. Radiat. Transfer 8 (1968) 1543.
[5.59d] Bychkova, T. V., V. G. Kirpilenko, S. G. Rautian and A. S. Khaikin, Optika i Spektroskopiya 22 (1967) 678.

[5.59c] CAMHY-VAL, C. and A.-M. DUMONT, Compt. Rend. Acad. Sci. 267 (1968) 689.
[5.60] CARST, A. and R. LADENBURG, Z. Physik 48 (1928) 192.
[5.61] CARTER, W. W., Phys. Rev. 76 (1949) 962.
[5.62] CHAMALAUN, F., Diss. Utrecht 1934.
[5.63] CHANDRASEKHAR, S., Astrophys. J. 102 (1945) 223.
[5.63a] CHASHINA, G. I. and E. YA. SHREIDER, Optika i Spektroskopiya 20 (1966) 511.
[5.63b] CHAMULEAU, F. J., Physica 1 (1934) 518.
[5.63c] CHARATIS, G., Thesis Univ. Michigan 1962.
[5.63d] CHARATIS, G. and T. D. WILKERSON, Phys. Fluids 5 (1963) 1661.
[5.63e] CHASHCHINA, G. I. and Y. Y. SHREIDER, Optika i Spektroskopiya 22 (1967) 519.
[5.63f] CHEN, C. J., Phys. Rev. 163 (1967) 1.
[5.64] CH'EN, S. Y. and A. SMITH, Physica 25 (1959) 1289.
[5.64a] CHENEVIER, M., J. DUFAYARD and J. C. PEBAY-PEYROULA, Phys. Letters A 25 (1967) 283.
[5.64b] CHIPLIS, I. V. and K. K. USHPALIS, Liet. fiz. Rink., Vilnius, USSR, B2, No. 25 (1961) 23; Engl. Transl.: Optical Transition Probabilities, Jerusalem 1963, p. 191.
[5.64c] CHIPLIS, I. V. and K. I. USHPALIS, Liet. fiz. Rink., Vilnius, USSR, B2, No. 25 (1961) 31. Engl. Translation: Optical Transition Probabilities, Jerusalem 1963, p. 197.
[5.64d] CHUPP, E. L., L. W. DOTCHIN and D. J. PEGG, Phys. Rev. 175 (1968) 44.
[5.65] CLAAS, W. J., Rech. astr. Obs. d'Utrecht 12, Pt. 1, 1951.
[5.65a] COATES, P. B. and A. G. GAYDON, Proc. Roy. Soc. Lond. A 293 (1966) 452.
[5.66] COENEN, P. A. and H. A. KRAMERS, Physica 3 (1936) 341.
[5.66a] COHEN, M. and P. S. KELLY, Canad. J. Phys. 45 (1967) 2079.
[5.67] COJAN, J. L. and M. THIBEAU, C. R. Acad. Sci., Paris 249 (1959) 1489.
[5.67a] CORLISS, C. H., J. Res. Nat. Bur. Stand. 66 A (1962) 5.
[5.67b] CORLISS, C. H., J. Res. Nat. Bur. Stand. 66 A (1962) 169.
[5.68] CORLISS, C. H., J. Res. Nat. Bur. Stand. A 66 (1962) 497.
[5.68a] CORLISS, C. H., J. Res. Nat. Bur. Stand. 69 A (1965) 87.
[5.69] CORLISS, C. H. and W. R. BOZMAN, Nat. Bur. Stand. Monogr. Nr. 53 (1962); see also [5.49a] p. 28.
[5.69a] CORLISS, C. H. and J. L. TECH, Nat. Bur. Stand. Monogr. 108 (1968).
[5.69b] CORLISS, C. H. and B. WARNER, Astrophys. J., Suppl. 8 (1964) 395.
[5.69c] CORLISS, C. H. and B. WARNER, J. Res. Nat. Bur. Stand. A 70 (1966) 325.
[5.70] CROSSWHITE, H. M., Spectrochim. Acta 4 (1950) 122.
[5.70a] CUNNINGHAM, P. T., J. Opt. Soc. Amer. 58 (1968) 1507.
[5.70b] CUNNINGHAM, P. T. and J. K. LINK, J. Opt. Soc. Amer. 57 (1967) 1000.
[5.70c] CURRY, R. H., J. quant. Spectrosc. Radiat. Transfer 6 (1966) 357.
[5.70d] CZYZAK, S. J. and T. K. KRUEGER, Monthly Not. Roy. Astron. Soc. 126 (1963) 177.
[5.71] DALGARNO, A. and N. LYNN, Proc. Phys. Soc. Lond. A 70 (1957) 802.
[5.72] DAVIS, M. H., P. M. ROUTLY and R. B. KING, Proc. Conf. Stellar Atmospheres. Indiana Univ. 1955, p. 47.
[5.72a] DAY, R. A. and H. R. GRIEM, Phys. Rev. 140 A (1965) 1129.
[5.73] DEMPSTER, R. R., Astrophys. J. 96 (1942) 295.
[5.74] DEMTRÖDER, W., Z. Physik 166 (1962) 42.
[5.74a] DEREVSHCHIKOV, V. A., Teplofizika 5 (1967) 423.
[5.74b] DEREVSHCHIKOV, V. A. and M. A. DERYABINA, Teplofizika 4 (1966) 20.
[5.74c] DESAI, S. V. and W. H. CORCORAN, J. quant. Spectrosc. Radiat. Transfer 8 (1968) 1721.
[5.75] DICKERMAN, P. J. and B. P. ALPINER, J. quant. Spectrosc. Radiat. Transfer 2 (1962) 305.

[5.75a] DICKERMAN, P. J. and R. W. DEUEL, J. quant. Spectrosc. Radiat. Transfer 4 (1964) 807.
[5.76] DICKERMAN, P. J. and J. C. MORRIS, Optical spectrometric measurements of high temperatures. Chicago Univ. Press, Chicago 1961, p. 170.
[5.76a] DIERMEIER, R. and H. KREMPL, Z. Physik **200** (1967) 239.
[5.76b] DIERMEIER, R. and H. KREMPL, Proc. 14. Coll. Spectrosc. internat. Debrecen, Hungary 1967, Vol. 2, p. 547.
[5.77] DITCHBURN, R. W., Proc. Roy. Soc. Lond. **A 117** (1927) 486.
[5.77a] DITCHBURN, R. W. and R. D. HUDSON, Proc. Roy. Soc. Lond. **A 256** (1960) 53.
[5.78] DITCHBURN, R. W. and P. H. JUTSUM, Nature, Lond. **165** (1950) 723.
[5.79] DITCHBURN, R. W. and G. V. MARR, Proc. Phys. Soc. Lond. **A 66** (1953) 655.
[5.80] DITCHBURN, R. W., J. TUNSTEAD and YATES, Proc. Roy. Soc. Lond. **A 181** (1943) 386.
[5.80a] DÖBELE, H. F., R. WIENECKE and S. WITKOWSKI, Z. Naturforsch. **21a** (1966) 390.
[5.80b] DOERFFEL, K. and J. LICHTNER, Wiss. Z. Tech. Hochsch. Chemie Leuna-Merseburg **8** (1966) 19.
[5.81] DOHERTY, L. R., Diss. Univ. Michigan 1961.
[5.81a] DONIN, V. I., Zh. prikl. Spektrosk. **5** (1966) 724.
[5.81b] DONIN, V. I. and V. P. SHCHEBOTAYEV, Optika i Spektroskopiya **20** (1966) 740.
[5.81c] DOUGLAS, A. S. and R. H. GARSTANG, Proc. Cambridge Phil. Soc. **58** (1962) 377.
[5.81d] DOW, J. D. and R. S. KNOX, Phys. Rev. **152** (1966) 50.
[5.82] DRAWIN, H. W., Z. Physik **146** (1956) 295.
[5.83] DRIEL, H. VAN, Diss. Utrecht 1935.
[5.84] DRONOV, A. P., A. G. SVIRIDOV and N. N. SOBOLEV, Optika i Spektroskopiya **5** (1958) 490.
[5.84a] DUBROVSKAYA, O. N., V. M. SINEGLASOV and B. E. MOSHKIN, Zh. prikl. Spektrosk. **2** (1965) 470.
[5.84b] DUBROVSKI, L. N. and YU. I. GALUSHKIN, Optika i Spektroskopiya **22** (1967) 876.
[5.85] DUPUY, G., Compt. Rend. Acad. Sci., Paris **222** (1946) 654.
[5.86] DURAND, J., Z. Naturforsch. **18a** (1963) 281.
[5.86a] DURAND, J., Rep. Inst. Plasmaphys. Lausanne, Switzerland, LRP 29/66, 1966.
[5.86b] EBERHAGEN, A. E., Z. Physik **143** (1955) 312.
[5.87] EBERHAGEN, A., Z. Physik **143** (1955) 392.
[5.88] EDDY, J. A., L. L. HOUSE and H. ZIRIN, Astrophys. J. **133** (1961) 299.
[5.88a] EGOROVA, K. A., Zh. prikl. Spektrosk. **6** (1967) 288.
[5.88b] EICHINGER, R. and H. KREMPL, Z. angew. Phys. **12** (1960) 461.
[5.89] EICKE, H. F., Z. Physik **168** (1962) 227.
[5.90] ELENBAAS, W., The high pressure mercury vapour discharge. North Holland Publ. Co., Amsterdam 1951, S. 15.
[5.90a] EL-KHALAFAWY, T., Rep. Inst. Plasmaphys. KFA Jülich, Germany, JÜL-12-PP, Juni 1961.
[5.91] ELLET, A., Phys. Rev. **33** (1929) 124.
[5.91a] ELLIS, E. and N. D. TWIDDY, Proc. 9. internat. Conf. Ioniz. Phenomena in Gases, Bukarest 1969, p. 17.
[5.92] ENGLER, H. D., Z. Physik **144** (1956) 343.
[5.93] ESTABROOK, F. B., Astrophys. J. **115** (1952) 571.
[5.94] ESTABROOK, F. B., Astrophys. J. **113** (1951) 684.
[5.94a] EVANS, D. L. and R. S. TANKIN, Phys. Fluids **10** (1967) 1137.
[5.94b] FAIRCHILD, C. E., Doct. Diss. Univ. Washington 1962.
[5.95] FAIRCHILD, C. E. and K. C. CLARK, Phys. Rev. Letters **9** (1962) 100.
[5.95a] FAWCETT, B. C., N. J. PEACOCK and R. D. COWAN, J. Physics **B 1** (1968) 295.

[5.95b] FEICHTNER, J. D., J. H. GALLAGHER and M. MIZUSHIMA, Phys. Rev. **164** (1967) 44.
[5.96] FILIPPOW, A., Z. Physik **69** (1931) 526.
[5.97] FILIPPOW, A., Phys. Z. SU **1** (1932) 289.
[5.98] FILIPPOV, A., Zh. eksp. teoret. Fiz. **2** (1932) 24.
[5.98a] FILIPPOW, A. N., Trudy Gosudarst. Opt. Inst. Leningrad 8, No. 85 (1932) 1. Engl. Translation: Optical Transition probabilities, Jerusalem 1963, p. 1.
[5.99] FILIPPOW, A. N., Zh. eksp. teoret. Fiz. **3** (1933) 520.
[5.100] FILIPPOW, A., Phys. Z. SU **5** (1934) 1.
[5.101] FILIPPOW, A. and N. KREMENEWSKI, Phys. Z. SU **1** (1932) 299.
[5.102] FILLIPOW, A. and W. K. PROKOFJEW, Z. Physik **56** (1929) 458.
[5.103] FILIPPOW, A. N. and W. K. PROKOFJEW, Z. Phys. **85** (1933) 647.
[5.103a] FISCHER, C. F., J. quant. Spectrosc. Radiat. Transfer **8** (1968) 755.
[5.103b] FLANNERY, M. R. and A. L. STEWART, Monthly Notices Roy. Astron. Soc. **126** (1963) 387.
[5.104] FOCK, V., Z. Physik **61** (1930) 126.
[5.105] FOCK, V. and M. PETRASHEN, Phys. Z. SU **8** (1935) 547.
[5.105a] FORBRICH, C. A., W. C. MARLOW and D. BERSHADER, Phys. Rev. **173** (1968) 150.
[5.106] FOSTER, E. W., Proc. phys. Soc. Lond. **80** (1962) 882.
[5.107] FOSTER, E. W., Proc. phys. Soc. Lond. **79** (1962) 94.
[5.107a] FOSTER, E. W., Rep. Progr. Phys. **27** (1964) 470.
[5.107b] FOSTER, E. W., Proc. phys. Soc. Lond. **90** (1967) 275.
[5.107c] FOWLER, R. G., T. M. HOLZBERLEIN and C. H. JACOBSON, Phys. Rev. **140** (1965) A 1050.
[5.107d] FOWLER, R. G., T. M. HOLZBERLEIN, C. H. JACOBSEN and S. J. B. CORRIGAN, Proc. phys. Soc. Lond. **84** (1964) 539.
[5.107e] FOWLER, R. H. and E. A. MILNE, Monthly Notices Roy. Astron. Soc. **83** (1923) 403; **84** (1924) 499.
[5.108] FRERICHS, R., Z. Physik **31** (1925) 305.
[5.109] FRERICHS, R., Ann. Physik **81** (1926) 807.
[5.110] FRIE, W., Z. Physik **172** (1963) 99.
[5.111] FRIE, W. and H. MAECKER, Z. Physik **162** (1961) 69.
[5.111a] FRIEDRICHS, H., Z. Astrophys. **60** (1964) 176.
[5.111b] FRISH, S. E., Izvest. Akad. Nauk SSSR Ser. Fiz. **14** (1950) 711.
[5.111c] FRISH, M. S., Optika i Spektroskopiya **22** (1967) 19.
[5.111d] FRISH, S. E. and O. P. BOCHKOVA, Vestnik Leningrad Univ. Ser. Fiz. i Khim. No 16 (1961) 40; Engl. Translation: Optical Transition Probabilities, Jerusalem 1963, p. 204; Addendum and Correction: Vestnik Leningrad Univ. Ser. Fiz. i Khim. No. 4 (1962) 73; Engl. Translation: Optical Transition Probabilities, Jerusalem 1963, p. 222.
[5.112] FRISH, S. E. and O. P. BOCHKOVA, in: Fizicheskie problemy spektroskopii Vol. 1, Izd. Akad. Nauk SSSR, Moskau 1962, p. 142.
[5.112a] FRISH, S. E. and A. P. MAKAROV, Optika i Spektroskopiya **24** (1968) 831.
[5.112b] FRISH, M. S. and G. P. STARTSEV, Optika i Spektroskopiya **16** (1964) 724.
[5.113] FÜCHTBAUER, C. and F. GÖSSLER, Z. Physik **87** (1933) 89.
[5.114] FÜCHTBAUER, C. and W. HOFMANN, Ann. Physik **43** (1914) 96.
[5.114a] FUCKS, W., W. L. BOHN, G. HEINRICH and P. PLATZ, Z. Physik **170** (1962) 409.
[5.114b] FUKUDA, F., K. ISHII, R. OKASAKA and T. FUJIMOTO, J. Phys. Soc. Japan **20** (1965) 2309.
[5.115] FURSSOW, W. S., M. N. OGANOW and A. R. STRIGANOW, Doklady Akad. Nauk SSSR **101** (1955) 453; German Translation: Phys. Abh. SU **10**, Gasentl. u. Plasmaphys. Folge **1** (1957) 97.
[5.115a] GABRIEL, A. H. and D. W. O. HEDDLE, Proc. Roy. Soc. Lond. **A 258** (1960) 124.

[5.115b] GAILLARD, M. and J. E. HESSER, Astrophys. J. **152** (1968) 695.
[5.115c] DE GALAN, L. and J. D. WINEFORDNER, J. quant. Spectrosc. Radiat. Transfer 7 (1967) 703.
[5.115d] GALLAGHER, A., Phys. Rev. **157** (1967) 24.
[5.115e] GALLAGHER, A. C., Colloqu. internat. Centre nat. rech. sci. Nr. 162 (1967) 217.
[5.115f] GALLAGHER, A. and A. LURIO, Phys. Rev. **136** (1964) A87.
[5.116] GARRET, P. H., Phys. Rev. **40** (1932) 779.
[5.117] GARSTANG, R. H., Monthly Notices Roy. Astron. Soc. **110** (1950) 612.
[5.118] GARSTANG, R. H., Monthly Notices Roy. Astron. Soc. **114** (1954) 118.
[5.119] GARSTANG, R. H., Vistas Astr. **1** (1955) 268.
[5.119a] GARSTANG, R. H., Monthly Notices Roy. Astron. Soc. **117** (1957) 393.
[5.119b] GARSTANG, R. H., Monthly Notices Roy. Astron. Soc. **118** (1958) 234.
[5.119c] GARSTANG, R. H., Monthly Notices Roy. Astron. Soc. **124** (1962) 321.
[5.119d] GARSTANG, R. H., Commun. Univ. Lond. Obs. No. 57 (1962) (Suppl. zu Monthly Notices Roy. Astron. Soc. **124** (1962) 321).
[5.119e] GARSTANG, R. H., The Observatory **82** (1962) 50.
[5.120] GARSTANG, R. H., Ann. Astrophys. **25** (1962) 109.
[5.121] GARSTANG, R. H., J. Opt. Soc. Amer. **52** (1962) 845.
[5.121a] GARSTANG, R. H., J. Res. Nat. Bur. Stand. **68A** (1964) 61.
[5.121b] GARSTANG, R. H., Publ. Astron. Soc. Pacific **78** (1966) 399.
[5.121c] GARSTANG, R. H., Astrophys. J. **148** (1967) 579.
[5.121d] GARSTANG, R. H. and J. VAN BLERKOM, J. Opt. Soc. Amer. **55** (1965) 1054.
[5.121e] GARSTANG, R. H. and J. A. WAWE, The Observatory **82** (1962) 210.
[5.121f] GARSTANG, R. H. and S. J. HILL, Publ. Astron. Soc. Pacific **78** (1966) 70.
[5.121g] GARSTANG, R. H. and L. J. SHAMEY, in: The magnetic and related stars, Mono Book Corp. Baltimore 1967.
[5.121h] GARSTANG, R. H. and L. J. SHAMEY, Astrophys. J. **148** (1967) 665.
[5.121i] GARTON, W. R. S., W. H. PARKINSON and E. M. REEVES, Astrophys. J. **140** (1964) 1269.
[5.121j] GEIGER, J., Z. Physik **177** (1964) 138.
[5.121k] GENEUX, E. and B. WANDERS-VINCENZ, Phys. Rev. Lett. **3** (1959) 422.
[5.121l] GENEUX, E. and B. WANDERS-VINCENZ, Helv. Phys. Acta **33** (1960) 185.
[5.121m] GEORGE, D. W., Brit. J. appl. Phys. **16** (1965) 1105.
[5.122] GERICKE, W. E., Z. Astrophys. **53** (1961) 68.
[5.122a] GETSLOV, KH. B. and A. V. IVANOVA, Optika i Spektroskopiya **25** (1968) 629.
[5.122b] GLADUSHCHAK, V. I. and E. YA. SHREIDER, Optika i Spektroskopiya **14** (1963) 815.
[5.123] GLEMBOTSKII, I. I., T. D. STROZKIT'E and A. P. YUTSIS, Trudy Akad. Nauk Lith. SSR, B **2** (1955) 15.
[5.124] GLENNON, B. M. and W. L. WIESE, Nat. Bur. Stand. Monogr. Nr. 50 (1962).
[5.125] GOLD, M. T., Monthly Notices Roy. astron. Soc. **109** (1949) 471.
[5.126] GOLD, A. and R. S. KNOX, Phys. Rev. **113** (1959) 834.
[5.127] GOLDBERG, L., Astrophys. J. **82** (1935) 1.
[5.128] GOLDBERG, L., Astrophys. J. **84** (1936) 11.
[5.129] GOLDBERG, L., Astrophys. J. **90** (1939) 414.
[5.130] GOLDBERG, G. I., Bulletin of the Pulkovo Observatory No. 156 (1956) 126.
[5.131] GOLDBERG, L., E. A. MÜLLER and L. H. ALLER, Astrophys. J. Suppl. Ser. **5**, Nr. 45 (1960) 1.
[5.131a] GOLDFARB, V. M. and S. V. DRESVIN, Teplofizika **3** (1965) 333.
[5.131b] GOLDFARB, V. M. and E. V. ILINA, Trudy Leningrad. Mekhan. Inst. No. 33 (1963) 142.
[5.131c] GOLDFARB, V. M., E. V. ILINA and A. I. KASABOV, Proc. 14. Coll. Spectrosc. internat., Debrecen, Hungary 1967, Vol. 2, p. 603.

[5.131d] GORDON, G. M., O. A. COTTER and E. R. PARKER, Welding J. **35** (1956) 109s.
[5.132] GORYACHEV, V. N., Trudy Sibir. phys. tech. Inst. **30** (1950) 61; Diss. Tomsk 1950.
[5.133] GOTTSCHALK, W. M., Astrophys. J. **108** (1948) 326.
[5.133a] GOUGH, W. and G. W. SERIES, Proc. Phys. Soc. Lond. **85** (1965) 469.
[5.133b] DRANKOVA, D. A. and L. I. KISELEVSKY, Zh. prikl. Spektrosk. **3** (1965) 168.
[5.133c] GRASHEW, P. F., Fizicheskie problemi Spektroskopii, Izd. Akad. Nauk SSSR, Moskau 1962, Vol. 1, p. 36.
[5.133d] GRECHIKHIN, L. I., Optika i Spektroskopiya **25** (1968) 655.
[5.134] GREEN, L. C., Astrophys. J. **109** (1949) 289.
[5.134a] GREEN, L. C., E. K. KOLCHIN and N. C. JOHNSON, in: Abundance Determinat. Stellar Spectra, Acad. Press London-N. Y. 1966, p. 77.
[5.135] GREEN, L. C., P. P. RUSH and C. D. CHANDLER, Astrophys. J. Suppl. Ser. **3**, Nr. 26 (1957) 37.
[5.136] GREEN, L. C. and N. E. WEBER, Astrophys. J. **111** (1950) 58.
[5.137] GREEN, L. C. and N. E. WEBER, Astrophys. J. **111** (1950) 587.
[5.138] GREEN, L. C. and N. E. WEBER and E. KRAWITZ, Astrophys. J. **113** (1951) 690.
[5.139] GRECHIKHIN, L. I. and V. D. SHIMANOVICH, Optika i Spektroskopiya **13** (1962) 626.
[5.139a] GRECHIKHIN, L. I. and Y. S. TYUNINA, Teplofizika **1** (1963) 399.
[5.139b] GRIDNEVA, S. M. and G. A. KASABOV, in: Electricity from MHD, IAEA Vienna 1966, Vol. 1, p. 73.
[5.139c] GRIDNEVA, S. M. and G. A. KASABOV, Proc. 7. internat. Conf. Phenom. Ionized Gases, Belgrade 1965, Vol. 2, p. 581.
[5.139d] GRIDNEVA, S. M. and G. A. KASABOV, Report Kurchatov Inst. Moscow ca 1967.
[5.139e] GRIEM, H. R., Naval Res. Lab. Rep. 6085, Washington 1964.
[5.139f] GRIEM, H. R., Plasma Spectroscopy, McGraw Hill Book Co., New York 1964.
[5.140] GRIEM, H. R., A. C. KOLB and K. Y. SHEN, Phys. Rev. **116** (1959) 4.
[5.141] GRIFFITS, J. H. E., Proc. Roy. Soc. Lond. A **143** (1934) 588.
[5.142] GRUZDEV, P. F., Fizicheskie problemy spektroskopii Vol. 1, Izd. Akad. Nauk USSR, Moscow 1962, p. 36.
[5.143] GRUSDEV, P. F., Optika i. Spektroskopiya **13** (1962) 302.
[5.144] GRUSDEV, P. F., Optika i Spektroskopiya **13** (1962) 451.
[5.144a] GRUSDEV, P. F., Optika i Spektroskopiya **22** (1967) 169.
[5.144b] GRUSDEV, P. F., Optika i Spektroskopiya **22** (1967) 313.
[5.144c] GRUSDEV, P. F., Optika i Spektroskopiya **25** (1968) 3.
[5.144d] GRUSDEV, P. F. and V. K. PROKOFYEV, Optika i Spektroskopiya **21** (1966) 255.
[5.144e] GÜNDEL, H. and W. NEUMANN, Beitr. Plasmaphys. **7** (1967) 221.
[5.144f] GÜNDEL, H. and W. NEUMANN, Monatsber. dtsch. Akad. Wiss. Berlin **9** (1967) 604.
[5.145] GURVICH, L. V., Optika i Spektroskopiya **5** (1958) 205.
[5.145a] GURVICH, L. V. and I. V. VEITS, Izvest. Akad. Nauk SSSR Ser. Fiz. **22** (1958) 673.
[5.145b] HAMEED, S., A. HERZENBERG and M. G. JAMES, J. Physics **B 1** (1968) 822.
[5.145c] HANLE, W., Z. Physik **30** (1924) 93.
[5.145d] HANLE, W., Ergebn. exakt. Naturwiss. **4** (1925) 214.
[5.145e] HANUS, W., Bull. acad. polon. sci. ser. sci. math. astron. phys. **9** (1961) 287.
[5.146] HARGREAVES, J., Proc. Camb. phil. Soc. **25** (1929) 75.
[5.147] HARRISON, G. R., J. Opt. Soc. Amer. **17** (1928) 389.
[5.147a] HARRISON, G. R., J. Opt. Soc. Amer. **19** (1929) 109.
[5.148] HARRISON, G. R. and H. ENGWICHT, J. Opt. Soc. Amer. **18** (1929) 287.
[5.149] HARTREE, D. R., The calculation of atomic structures. John Wiley & Sons Inc., New York 1956.

[5.150] Hartree, D. R. and W. Hartree, Proc. Roy. Soc. Lond. **A 150** (1935) 9.
[5.151] Hartree, D. R. and W. Hartree, Proc. Roy. Soc. Lond. **A 154** (1936) 588.
[5.152] Hartree, D. R. and W. Hartree, Proc. Roy. Soc. Lond. **A 164** (1938) 167.
[5.153] Hartree, D. R., W. Hartree and B. Swirles, Phil. Trans. **A 238** (1939) 229.
[5.154] Heard, J. F., Monthly Notices Roy. Astron. Soc. **94** (1934) 458.
[5.155] Hefferlin, R., J. Opt. Soc. Amer. **49** (1959) 680.
[5.156] Hefferlin, R., J. Opt. Soc. Amer. **49** (1959) 948.
[5.157] Hefferlin, R., J. Opt. Soc. Amer. **52** (1962) 338.
[5.158] Hefferlin, R., B. Cobb, D. Hall and C. Lehman, Astrophys. J. **132** (1960) 259.
[5.159] Hefferlin, R. and J. Gearhart, J. quant. Spectrosc. Radiat. Transfer **4** (1964) 9.
[5.159a] Heierman, J. H., Thesis Utrecht 1937.
[5.160] Held, E. F. M. van der, Diss. Utrecht 1932.
[5.161] Held, E. F. M. van der and J. H. Heierman, Physica **2** (1935) 71.
[5.162] Held, E. F. M. van der and J. H. Heierman, Physica **3** (1936) 31.
[5.163] Held, E. F. M. van der and L. S. Ornstein, Z. Physik 77 (1932) 459.
[5.163a] Helliwell, T. M., Phys. Rev. **135** (1964) A 325.
[5.164] van Hengstum, J. P. A. and J. A. Smit, Physica **22** (1956) 86.
[5.164a] Henning, H., Z. Astrophys. **62** (1965) 109.
[5.164b] Hermel, A., Diss. Humboldt Univ. Berlin 1967; see also Beitr. Plasmaphys. **9** (1969) 261.
[5.165] Heron, S., R. W. P. McWhirter and E. H. Rhoderick, Nature, Lond. **174** (1954) 564.
[5.166] Heron, S., R .W. P. McWhirter and E. H. Rhoderick, Proc. Roy. Soc. Lond. **A 234** (1956) 565.
[5.166a] Heroux, L., Proc. Phys. Soc. Lond. **83** (1964) 121.
[5.166b] Heroux, L., Phys. Rev. **153** (1967) 156.
[5.166c] Heroux, L., Phys. Rev. **161** (1967) 47.
[5.166d] Hertzberg, M. and R. Holland, J. quant. Spectrosc. Radiat. Transfer **5** (1965) 313.
[5.167] Herzfeld, K. F. and K. L. Wolf, Ann. Physik **76** (1925) 71.
[5.168] Herzfeld, K. F. and K. L. Wolf, Ann. Physik **76** (1925) 567.
[5.168a] Hese, A. and H.-P. Weise, Z. Physik **215** (1968) 95.
[5.168b] Hess, H., H.-G. Kloss, K. Rademacher and K. Seliger, Beitr. Plasmaphys. **2** (1962) 171.
[5.168c] Hesser, J. E., Phys. Rev. **174** (1968) 68.
[5.168d] Hesser, J. E. and B. L. Lutz, J. Opt. Soc. Amer. 58 (1968) 1513.
[5.169] Hey, P., Diss. Univ. Kiel 1959.
[5.170] Hey, P., Z. Physik **157** (1959) 79.
[5.171] Hill, W. E., Temperature, its measurement and control in Science and Industry, Vol. III, 1, Reinhold Publ. Co., New York 1962, p. 581.
[5.172] Hill, A. and A. B. King, J. Opt. Soc. Amer. **41** (1951) 315.
[5.172a] Hindmarsh, W. R. and K. A. Thomas, Proc. Phys. Soc. Lond. **A 77** (1961) 1193.
[5.172b] Hinnov, E., Report Princeton Univ., Plasmaphysics Lab., MATT 194, Mai 1963.
[5.173] Hirschfeld, G. and H. E. Weinschenk, Schweißtechnik **10** (1960) 438.
[5.173a] Hofman, W., Report Max-Planck-Inst., MPI-PAE/Extraterr. 11/67, Munich 1967.
[5.173b] Holstein, T., Phys. Rev. **73** (1947) 1212.
[5.173c] Holstein, T., Phys. Rev. **83** (1951) 1159.
[5.173d] Holzberlein, T. M., Doct. Diss. Univ. Oklahoma 1963.
[5.173e] Holzberlein, Rev. sci. Instrum. **35** (1964) 1041.

[5.173f] HOUSE, L. L., Astrophys. J. **149** (1967) 211.
[5.174] HOUZIAUX, L. and M. P. SADOINE, Bull. Soc. roy. Sci. Liége **30** (1961) 287.
[5.175] HUANG, S. S., Astrophys. J. **108** (1948) 354.
[5.175a] HUGHES, R. H., H. R. DAWSON and B. M. DOUGHTY, J. Opt. Soc. Amer. **56** (1966) 830.
[5.176] HULDT, L., Ark. Mat. Astr. Fys. **36A**, Nr. 3 (1948).
[5.177] HULDT, L. and A. LAGERQVIST, Ark. Fys. **5** (1952) 91.
[5.178] HULDT, L. and A. LAGERQVIST, J. Opt. Soc. Amer. **42** (1952) 142.
[5.178a] HULPKE, E., E. PAUL and W. PAUL, Z. Physik **177** (1964) 257.
[5.178b] HUTCHERSON, J. W., B. C. MUNDY and R. HEFFERLIN, Appl. Spectrosc. **20** (1966) 12.
[5.179] HYLLERAAS, E. A., Z. Physik **54** (1929) 347.
[5.180] HYLLERAAS, E. A., Z. Physik **106** (1937) 395.
[5.181] INGLIS, D. E. and E. TELLER, Astrophys. J. **90** (1939) 439.
[5.181a] IRWIN, J. and R. A. NODWELL, Canad. J. Phys. **44** (1966) 1781.
[5.182] ISLAMOV, I. I. and A. N. FILIPPOV, Zh. eksp. teor. Fiz. **3** (1933) 524.
[5.183] JAHN, R. E., Proc. 5. internat. Conf. Ionization Phenomena in Gases, Vol. I, North Holland Publ. Co., Amsterdam 1962, p. 955.
[5.184] JAHN, R. E., Brit. J. appl. Phys. **14** (1963) 585.
[5.184a] JAHN, R. E., Proc. 6. internat. Conf. Ion. Phenom. Gases, Paris 1963, Vol. 4, p. 15.
[5.184b] JAHN, E., Proc. 14. Coll. spectrosc. internat., Debrecen, Hungary 1967, Vol. 2, p. 741.
[5.185] JANOUCH, F., Vestnik Leningrad Univ. Ser. Matem. Fiz. i Khim. No. 1 (1955) 135.
[5.185a] JEAN, P., M. MARTIN, J.-P. BARRAT and J.-L. COJAN, Compt. Rend. Acad. Sci. Paris **264** (1967) B 1709.
[5.185b] JEAN, P., M. MARTIN and D. LECLER, Compt. Rend. Acad. Sci. Paris **264** (1967) B 1791.
[5.186] JONGERIUS, H. M., Diss. Univ. Utrecht 1961.
[5.186a] JORDAN, J. A., G. S. BAKKEN and R. E. YAGER, J. Opt. Soc. Amer. **57** (1967) 530.
[5.187] JÜRGENS, G., Z. Physik **134** (1952) 21.
[5.188] JÜRGENS, G., Z. Physik **138** (1954) 613.
[5.189] KARSTENSEN, F., Z. Physik **187** (1965) 165.
[5.190] KARSTENSEN, F. and J. SCHRAMM, Z. Physik **195** (1966) 370.
[5.190a] KARSTENSEN, F. and J. SCHRAMM, J. Opt. Soc. Amer. **57** (1967) 654.
[5.190b] KARSTENSEN, F. and J. SCHRAMM, Z. Astrophys. **68** (1968) 214.
[5.191] KAST, W., Z. Physik **79** (1932) 731.
[5.191a] KASTNER, S. O., K. OMIDVAR and J. H. UNTERWOOD, Astrophys. J. **148** (1967) 269.
[5.192] KATZ, E. and L. S. ORNSTEIN, Physica **4** (1937) 757.
[5.193] KELLY, P. S., J. quant. Spectrosc. Radiat. Transfer **4** (1964) 117.
[5.193a] KELLY, P. S., J. quant. Spectrosc. Radiat. Transfer **4** (1964) 426.
[5.193b] KELLY, P. S., Astrophys. J. **140** (1964) 1247.
[5.194] KELLY, P. S. and B. H. ARMSTRONG, Phys. Rev. Letters **9** (1962) 426.
[5.195] KERSTEN, J. A. H., Diss. Utrecht 1941.
[5.196] KERSTEN, J. A. H. and L. S. ORNSTEIN, Physica **8** (1941) 1124.
[5.197] KESSLER, R., Compt. Rend. Acad. Sci. Paris **236** (1953) 1335.
[5.198] KESSLER, W. and D. V. DOBENECK, Z. anal. Chem. **198** (1963) 35.
[5.199] KEY, J., Diss. Utrecht 1937.
[5.199a] KHOKHLOV, M. Z., Akad. Nauk SSSR, Krymskaia Astrofizicheskaya Observ., Izvestia **22** (1960) 128.
[5.199b] KIBBLE, B. D., Phys. Rev. **153** (1967) 9.

[5.199c] KIMURA, I. and A. KANZAWA, AIAA-J. 1 (1963) 310.
[5.199d] KINDLMANN, P. J. and W. R. BENNETT, Bull. Amer. Phys. Soc. 8 (1963) 87.
[5.200] KING, R. B., Astrophys. J 94 (1941) 27.
[5.201] KING, R. B., Astrophys. J. 95 (1941) 78.
[5.202] KING, R. B., Astrophys. J. 105 (1947) 376.
[5.203] KING, R. B., Astrophys. J. 108 (1948) 87.
[5.204] KING, R. B. and A. S. KING, Astrophys. J. 82 (1935) 377.
[5.205] KING, R. B. and A. S. KING, Astrophys. J. 87 (1938) 24.
[5.206] KING, R. B. and A. S. KING, Astrophys. J. 87 (1938) 30.
[5.207] KING, R. B., B. R. PARNES and M. H. DAVIS and K. H. OLSEN, J. Opt. Soc. Amer. 45 (1955) 350.
[5.208] KING, R. B. and D. C. STOCKBARGER, Astrophys. J. 91 (1940) 488.
[5.209] KING, R. B. and D. C. STOCKBARGER, Phys. Rev. 58 (1940) 187.
[5.210] KING, R. B. and D. C. STOCKBARGER, Astrophys. J. 95 (1942) 78.
[5.211] KING, G. W. and J. H. VAN VLECK, Phys. Rev. 56 (1939) 464.
[5.212] KINGSBURY, R. F., Phys. Rev. 99 (1955) 1846.
[5.212a] KIRSANOVA, N. N., Zh. prikl. Spektrosk. 6 (1967) 398.
[5.212b] KISELEVSKI, L. I. and V. D. SHIMANOVICH, Optika i Spektroskopiya 24 (1968) 506.
[5.212c] KISELEVSKI, L. I., V. D. SHIMANOVICH and A. K. SHIPAI, Teplofizika 6 (1968) 30.
[5.212d] KISELEVSKI, L. I. and E. P. TRUKHAN, Zh. prikl. Spektrosk. 9 (1968) 928.
[5.212e] KLOSE, J. Z., Phys. Rev. 141 (1966) 181.
[5.212f] KLOSE, J. Z., J. Opt. Soc. Amer. 58 (1968) 1509.
[5.212g] KNOPP, C. F., C. F. GOTTSCHLICH and A. B. CAMBEL, J. quant. Spectrosc. Radiat. Transfer 2 (1962) 297.
[5.213] KNOX, R. S., Phys. Rev. 110 (1958) 375.
[5.213a] KNOX, R. S. and L. D. DEXTER, Phys. Rev. 104 (1956) 1245.
[5.213b] KOCK, M. and J. RICHTER, Z. Astrophys. 69 (1968) 180.
[5.214] KOENIG, H. D. and A. ELLETT, Phys. Rev. 39 (1932) 576.
[5.214a] KOMAROVSKII, V. A., N. P. PENKIN and I. N. SHABANOVA, Optika i Spektroskopiya 25 (1968) 155.
[5.214b] KOOPMAN, D. W., J. Opt. Soc. Amer. 54 (1964) 1354.
[5.214c] KOROLEV, F. A. and V. I. ODINTSOV, Optika i Spektroskopiya 18 (1965) 968.
[5.214d] KOROLEV, F. A. and ZH. ZHEENBAYEV, Fizicheskie problemi spektroskopii, Izd. Akad. Nauk SSSR, Moskau 1962, Vol. 1, S. 107.
[5.215] KÖRWIEN, H., Z. Physik 91 (1934) 1.
[5.216] KOHN, H. and H. J. HÜBNER, Phys. Z. 34 (1933) 278.
[5.217] KOLESNIKOV, V. N. and V. V. BOGDANOVA, Optika i Spektroskopiya 1 (1956) 846.
[5.218] KOLESNIKOV, V. N. and L. V. LESKOV, Usp. Fiz. Nauk 65 (1958) 1.
[5.219] KOLESNIKOV, V. N. and N. S. SLEPCHENKO, in: Fizicheskie problemy spektroskopii Vol. 1, Izd. Akad. Nauk SSSR, Moskau 1962, p. 41.
[5.220] KOPFERMANN, H. and R. LADENBURG, Z. Physik 48 (1928) 51; 65 (1930) 167.
[5.221] KOPFERMANN, H. and W. TIETZE, Z. Physik 56 (1929) 604.
[5.222] KOPFERMANN, H. and G. WESSEL, Z. Physik 130 (1951) 100.
[5.223] KOROLEV, F. A., V. I. ODINTSOV and Y. V. FURSOVA, Optika i Spektroskopiya 16 (1964) 555.
[5.223a] KOSLOV, M. G. and G. P. STARTSEV, Optika i Spektroskopiya 24 (1968) 8.
[5.223b] KÖSTLIN, H., Z. Physik 178 (1964) 200.
[5.223c] KOWALSKI, J. and G. ZU PUTLITZ, Z. Physik 208 (1968) 459.
[5.223d] KRAUSS, L., J. quant. Spectrosc. Radiat. Transfer 7 (1967) 373.
[5.223e] KREISLER, J., Acta phys. polon. 4 (1935) 151.
[5.224] KREMPL, H., Z. Physik 167 (1962) 302.

[5.225] KREMPL, H., Z. anal. Chem. **198** (1963) 21.
[5.225a] KREMPL, H. and G. SCHMID, Spectrochim. Acta **23B** (1968) 819.
[5.225b] KRIBEL, R. E., Phys. Fluids **11** (1968) 1579.
[5.225c] KRIBCHENKOVA, V. S., Optika i Spektroskopiya **25** (1968) 966.
[5.225d] KROEPELIN, H. and K.-U. HOFFMANN, Optik **22** (1965) 599.
[5.225e] KRUEGER, T. K. and S. J. SZYZAK, Astrophys. J. **144** (1966) 1194.
[5.226] KRUITHOF, A. M., Physica **10** (1943) 493; Diss. Utrecht 1943.
[5.226a] KRUPENIKOVA, T. and M. CHAIKA, Optika i Spektroskopiya **20** (1966) 1088.
[5.227] KRYSMANSKI, K. H., Ann. Physik **6** (1960) 169.
[5.227a] KRYSMANSKI, K. H., Ann. Physik **15** (1965) 207.
[5.228] KUCK, C., Diss. Univ. Kiel 1957.
[5.229] KUHN, W., Naturwiss. **13** (1925) 724.
[5.230] KUHN, W., Danske Vidnsk. Selsk. Medd. 7, Nr. 12, 1926.
[5.231] KUHN, W., Naturwiss. **14** (1926) 48.
[5.231a] KUHN, H. G. and E. L. LEWIS, Proc. Roy. Soc. Lond. **A 299** (1967) 423.
[5.232] KUNISZ, M. D., Bull. Acad. polon. Sci. (math., astr., phys.) **1** (1953) 52.
[5.232a] KUNISZ, M. D., Acta phys. polon. **17** (1958) 455.
[5.232b] KUNISZ, M. D. and J. SEGUIR, J. Phys. Radium **21** (1960) 527.
[5.232c] KUSCH, H.-J. and E. R. MEWES, Z. Naturforsch. **22a** (1967) 676.
[5.233] KWATER, G. S., Zh. eksp. teor. Fiz. **11** (1941) 421.
[5.234] KWATER, G. S., J. Phys. USSR **6** (1942) 145.
[5.235] KWATER, G. S., Izvestiya Akad. Nauk USSR, phys. ser. **9** (1945) 236.
[5.236] KWATER, G. S., Vestnik Leningrad Univ. Ser. Fiz. i Khim. No. 2 (1947) 135.
[5.237] KWATER, G. S., Diss. Univ. Leningrad 1947.
[5.238] KWATER, G. S. and T. G. MEISTER, Vestnik Leningrad Univ. Ser. Fiz. i Khim. No. 9 (1952) 137.
[5.238a] LABUHN, F., Z. Naturforsch. **20a** (1965) 998.
[5.239] LADENBURG, R., Z. Elektrochem. **36** (1930) 631.
[5.240] LADENBURG, R. and S. LEVY, Z. Physik **65** (1930) 189.
[5.241] LADENBURG, R. and S. LEVY, Z. Physik **88** (1934) 461.
[5.242] LADENBURG, R. and E. THIELE, Z. Physik **72** (1931) 697.
[5.243] LADENBURG, R. and G. WOLFSOHN, Z. Physik **63** (1930) 616; **65** (1930) 207; **83** (1933) 234.
[5.243a] LANGE, W., J. LUTHER, A. STEUDEL and H. WALTHER, Phys. Letters **20** (1966) 166.
[5.244] LANGSTROTH, G. O. and D. R. MCRAE, Canad. J. Res. **A 16** (1938) 17.
[5.245] LARENZ, R. W., Z. Physik **129** (1951) 327.
[5.246] LARENZ, R. W., Z. Phys. **129** (1951) 343.
[5.246a] LAWRENCE, G. M., Astrophys. J. **148** (1967) 261.
[5.246b] LAWRENCE, G. M., Phys. Rev. **175** (1968) 40.
[5.246c] LAWRENCE, G. M. and B. D. SAVAGE, Phys. Rev. **141** (1966) 67.
[5.246d] LEBEDEV, S. V., S. L. MANDELSHTAM and G. M. RODIN, Zh. eksp. teoret. Fiz. **37** (1959) 349.
[5.247] LEBEDEVA, V. V., Fizicheskie problemy spektroskopii, Vol. 1, Izd. Akad. Nauk SSSR, Moskau 1962, p. 43.
[5.247a] LEBLANC, F. J., O. OLDENBERG and N. P. CARLETON, J. chem. Phys. **45** (1966) 2200.
[5.247b] LECLER, D., J. Physique **29** (1968) 611.
[5.247c] LES, Z. and H. NIEWODNICZANSKI, Acta phys. polon. **20** (1961) 701.
[5.248] LETFUS, V., Bull. astr. Inst. (Csl.) **12** (1961) 161.
[5.248a] LETFUS, V., J. Opt. Soc. Amer. **51** (1961) 1151.
[5.248b] LETFUS, V., Central Astron. Inst. of Czechosl., Bulletin **13** (1962) 98.
[5.249] LETFUS, W., Optika i Spektroskopiya **21** (1966) 666.

[5.250] LEVINSON, I. B., A. B. BOLOTIN and L. I. LEVIN, Trudy Univ. Vilna 5 (1956) 49.
[5.251] LEVINSON, I. B. and A. A. NIKITIN, Rukovodstvo po teoreticheskomy vychisleniyu intensivnostei linii v atomnych spektrach. Izd. Leningrad Univ. 1962.
[5.251a] LEVY, S., Z. Physik 72 (1931) 578.
[5.251b] LEWIS, E. L., Proc. Phys. Soc. Lond. 92 (1967) 817.
[5.251c] LEWIS, M. R., T. MARSHALL, E. H. CARNEVALE, F. S. ZIMNOCH and G. W. WARES, Phys. Rev. **164** (1967) 94.
[5.251d] LINCKE, R. and H. R. GRIEM, Phys. Rev. **143** (1966) 66.
[5.252] LINGEN, D. VAN, Physica **3** (1936) 977.
[5.253] LINGEN, D. VAN, Diss. Utrecht 1936.
[5.253a] LINK, J. K., J. Opt. Soc. Amer. **56** (1966) 1195.
[5.253b] LINK, J. K., J. Opt. Soc. Amer. **56** (1966) 1262.
[5.254] LOOKE, M., Untersuchung der Temperaturverteilung und der Strömungsgeschwindigkeit im stationären und impulsmodulierten Plasmastrahl. Diplomarbeit Humboldt-Universität Berlin 1963.
[5.254a] LOOKE, M., W. NEUMANN and K. RADEMACHER, Beitr. Plasmaphys. 4 (1964) 239.
[5.254b] LUBOWIECKA, Acta Phys. Polon. **25** (1964) 848.
[5.254c] LURIO, A., Phys. Rev. **126** (1962) 1768.
[5.254d] LURIO, A., Phys. Rev. **140** (1965) A 1505.
[5.254e] LURIO, A. and R. NOVICK, Phys. Rev. **134** (1964) A 608.
[5.254f] LURIO, A., R. L. DEZAFRA and R. J. GOSHEN, Phys. Rev. **134** (1964) A 1198.
[5.254g] LUTHER, J. and H. WALTHER, Phys. Letters **23** (1966) 551.
[5.254h] LVOV, B. V., Optika i Spektroskopiya **19** (1965) 507.
[5.254i] MA, I.-J., G. ZU PUTLITZ and G. SCHÜTTE, Z. Physik **208** (1968) 276.
[5.255] MCCARROLL, R. J. and P. G. WAKELY, The Airglow and the Aurorae. Pergamon Press, London & New York 1956, p. 337.
[5.255a] MCCONKEY, J. W., D. J. BURNS, K. A. MORAN and K. G. EMELEUS, Phys. Letters **22** (1966) 416.
[5.255b] MCCONKEY, J. W., D. J. BURNS, K. A. MORAN and J. A. KERNAHAN, Nature Lond. **217** (1968) 538.
[5.255c] MCCONKEY, J. W. and J. A. KERNAHAN, Nature, Lond. **220** (1968) 1017.
[5.255d] MCCONNELL, J. C. and B. L. MOISEIWITSCH, J. Physics **B 1** (1968) 406.
[5.255e] MCDOWELL, M. R. C. and A. D. STAUFFER, Phys. Letters **12** (1964) 207.
[5.255f] MCGREGOR, W. K., in: Physico-Chemical Diagnostics of Plasmas, Northwestern Univ. Press, Evanston 1964, p. 143.
[5.256] MCGREGOR, W. K. and M. T. DOOLEY, in: Temperature, its Measurement and Control in Science and Industry, Vol. III, 2, Reinhold Publ. Co., New York 1962, p. 467.
[5.257] MCLEAN, E. A., Temperature, its Measurement and Control in Science and Industry, Vol. III, 1, Reinhold Publ. Co., New York 1962, 607.
[5.258] MCLEAN, E. A., Proc. 6. internat. Conf. Ion. Phenom. Gases, Paris 1963, Vol. 3, p. 389.
[5.259] MCNALLY, J. R., in: Optical Spectrometric Measurements of high temperatures. Univ. of Chicago Press, Chicago 1961, S. 70.
[5.259a] MCNALLY, J. R., Plasma Physics **10** (1968) 903.
[5.260] MAECKER, H., Z. Physik **129** (1951) 108.
[5.261] MAECKER, H., Z. Physik **135** (1953) 13.
[5.262] MAECKER, H., Z. Physik **136** (1953) 119.
[5.262a] MAKAROV, YU. V. and A. M. MAKSIMOV, Zh. tekhn. Fiz. **35** (1965) 658.
[5.262b] MALAKHOV, V. P. and E. P. OSTAPCHENKO, Zh. prikl. Spektrosk. **2** (1965) 489.
[5.262c] MALAKHOV, V. P. and E. P. OSTAPCHENKO, Zh. prikl. Spektrosk. **3** (1965) 14.

[5.262d] MALAKHOV, V. P., E. P. OSTAPCHENKO, V. I. PUGNIN and V. A. STEPANOV, Proc. 7. internat. Conf. Phenom. Ionized Gases, Belgrad 1965, Vol. 1, p. 559.

[5.262e] MALONE, B. S. and W. H. CORCORAN, J. quant. Spectrosc. Radiat. Transfer **6** (1966) 443.

[5.262f] MARANTZ, H., R. I. RUDKO and C. L. TANG, Appl. Phys. Lett. **9** (1966) 409.

[5.262g] MARGOSHES, M. and B. F. SCRIBNER, J. Res. Nat. Bur. Stand. **67A** (1963) 561.

[5.263] MARR, G. V., Proc. Roy. Soc. **A 224** (1954) 83.

[5.263a] MARR, G. V. and R. HEPPINSTALL, Proc. Phys. Soc. Lond. **87** (1966) 547.

[5.264] MASON, R. C., Physica **5** (1938) 777.

[5.265] MASTRUP, F., Diss. Univ. Kiel 1957.

[5.266] MASTRUP, F. and F. WIESE, Z. Astrophys. **44** (1958) 259.

[5.266a] MATHIS, J. S., Astrophys. J. **125** (1962) 318.

[5.266b] MAY, R. M., Nuclear Fusion **4** (1964) 244.

[5.266c] MAZING, M. A. and N. P. PENKIN, Optika i Spektroskopiya **21** (1966) 749.

[5.267] MELCHENKO, V. S., Zh. eksp. teoret. Fiz. **28** (1955) 628.

[5.267a] MENDLOWITZ, H., Astrophys. J. **138** (1963) 1277.

[5.267b] MENDLOWITZ, H., J. Res. Nat. Bur. Stand. **71 A** (1967) 149.

[5.267c] MENDLOWITZ, H., Astrophys. J. **154** (1968) 1099.

[5.267d] MILES, B. M. and W. L. WIESE, Bibliography on Atomic Transition Probabilities, January 1916 through June 1969. Nat. Bur. Stand. Spec. Publ. 320, New York 1970.

[5.268] MENZEL, D. H. and L. H. ALLER, Astrophys. J. **94** (1941) 436.

[5.269] MENZEL, D. H. and E. GOLDBERG, Astrophys. J. **85** (1937) 40.

[5.270] MENZEL, D. H. and C. L. PEKERIS, Monthly Notices Roy. Astron. Soc. **96** (1935) 77.

[5.271] MICHELS, A., DE KLUIVER, H. and D. MIDDELKOOP, Physica **24** (1958) 543.

[5.271a] MILLER, W. F. and C. L. PERKINS, Spectrochim. Acta **8** (1956) 195.

[5.272] MILLER, W. F. and R. L. PLATZMAN, Spectrochim. Acta **8** (1956) 195.

[5.273] MINKOWSKI, R., Z. Physik **36** (1926) 839.

[5.274] MINKOWSKI, R. and W. MÜHLENBRUCH, Z. Physik **63** (1930) 198.

[5.275] MISHRA, B., Phys. Rev. **77** (1950) 153.

[5.276] MISHRA, B., Proc. Camb. phil. Soc. **48** (1952) 511.

[5.277] MISRA, Diss. London 1948.

[5.278] MITCHELL, A. C. G., Phys. Rev. **43** (1933) 887.

[5.279] MITCHELL, A. C. G. and E. J. MURPHY, Phys. Rev. **46** (1934) 53.

[5.280] MITCHELL, A. C. G. and M. ZEMANSKY, Resonance Radiation and Excited Atoms. Cambridge Univ. Press, Cambridge 1934, S. 151.

[5.281] MITROFANOVA, L. A., Izvestiia Pulkovo Astronomicheskaia observatoriia **149** (1952) 81.

[5.282] MITROFANOVA, L. A., Izvestiia Pulkovo Astronomicheskaia Observatoriia **151** (1953) 45.

[5.283] MITROFANOVA, L. A., Izvestiia Pulkovo Astronomicheskaia Observatoriia **153** (1955) 107.

[5.284] MITROFANOVA, L. A., Izvestiia Pulkovo Astronomicheskaia Observatoriia **20** (1958) 52.

[5.285] MITROFANOVA, L. A., Izvestiia Pulkovo Astronomicheskaia Observatoriia **21** (1960) 185.

[5.286] MOHLER, F. L. and C. BOECKNER, J. Res. Nat. Bur. Stand. **3** (1929) 303.

[5.287] MOORE, C. E., A multiplet table of astrophysical interest. Part 1: Table of multiplets; Part 2: Finding list of all lines in the table of multiplets. Nat. Bureau of Standards, Washington 1959.

[5.287a] MORACK, J. L. and C. E. FAIRCHILD, Phys. Rev. **163** (1967) 125.

[5.287b] MORGULIS, N. D., I. L. BABICH, YU. A. DYADYN and V. P. KOVALENKO, Optika i Spektroskopiya 18 (1965) 931.
[5.288] MOROSOVA, N. G. and G. P. STARTSEV, Izv. Akad. Nauk SSSR Ser. Fiz. 26 (1962) 929.
[5.288a] MOROSOVA, N. G. and G. P. STARTSEV, Optika i Spektroskopiya 17 (1964) 327.
[5.288b] MOROSOVA, N. G. and G. P. STARTSEV, Optika i Spektroskopiya 18 (1965) 899.
[5.289] MOROSOVA, N. G., G. P. STARTSEV and M. S. FRISH, Fizicheskie problemy spektroskopii Vol. 1, Izd. Akad. Nauk SSSR, Moskau 1962, p. 45.
[5.289a] MORSE, F. A. and F. KAUFMAN, J. chem. Phys. 42 (1965) 1785.
[5.289b] MOSKVIN, YU. V. and N. N. CHESNIKOVA, Teplofizika 3 (1965) 370.
[5.290] MOTSCHMANN, H., Z. Physik 143 (1955) 77.
[5.290a] MOTSCHMANN, H., Z. Physik 200 (1967) 93.
[5.290b] MOTSCHMANN, H., Z. Physik 214 (1968) 42.
[5.290c] MÜCK, G. and H.-P. POPP, Z. Naturforsch. 23a (1968) 1213.
[5.291] MÜLLER, F., Helv. phys. Acta 7 (1934) 488.
[5.292] MÜLLER, F., Helv. phys. Acta 7 (1934) 813.
[5.292a] MÜLLER, D., Z. Naturforsch. 23a (1968) 1707.
[5.292b] MURPHY, P. W., J. Opt. Soc. Amer. 58 (1968) 1200.
[5.292c] MURPHY, P. W., Astrophys. J. 153 (1968) 301.
[5.293] MUTÔ, T., Proc. phys. math. Soc. Japan 12 (1930) 93.
[5.294] NAGIBINA, I. M., V. K. PROKOFYEV and G. P. PETROV, Optika i Spektroskopiya 8 (1960) 376.
[5.294a] NAQVI, A. M., Proc. 7. internat. Conf. Phenom. Ionized Gases, Belgrad 1965, Vol. 2, p. 558.
[5.295] NAQVI, A. M. and S. P. TALWAR, Monthly Notices Roy. Astron. Soc. 117 (1957) 463.
[5.295a] NAUMENKO, P. A., Teplofizika 2 (1964) 176.
[5.295b] NEDOSPASOV, A. V. and I. YA. SHIPUK, Investigation of plasma conductivity in transversal magnetic field. Doklady na Parizhskuyu konferentsiyu po ionizatsionnim yavleniyam v gazach, Moscow 1964.
[5.295c] NEUMANN, K. K., Optik 14 (1957) 311.
[5.296] NEUMANN, W., Beitr. Plasmaphys. 2 (1962) 80.
[2.596a] NEUMANN, W. and U. SACKLOWSKI, Beitr. Plasmaphys. 8 (1968) 57.
[5.296b] NEUMANN, W. and F. SERICK, Beitr. Plasmaphys. 7 (1967) 203.
[5.296c] NEWSON, G. H., Astrophys. J. 154 (1968) 1095.
[5.297] NICHOLLS, R. W. and A. L. STEWART, Allowed transitions. In: Atomic and Molecular Processes, Academic Press, New York und London 1962, S. 47.
[5.297a] NICOLS, L. L. and W. E. WILSON, Appl. Optics 7 (1968) 167.
[5.297b] NIKITIN, A. A., Uchenye Zapiski Leningrad Gosudarst. Univ. 17, No. 190 (1957) 3 (Ser. Mat. Nauk No. 29); Engl. Übersetz.: Optical Transition Probabilities, Jerusalem 1963, S. 130.
[5.297c] NIKITIN, A. A., Vestnik Leningrad Univ. Ser. Mat. Mekhan. Astron. Ser. 3 No. 13 (1962) 113; Engl. Übersetz.: Optical transition Probabilities, Washington 1963, S. 235.
[5.298] NIKITIN, A. A. and G. V. GORDIENKO, Doklady Akad. Nauk Armen. SSR 20 (1955) 165.
[5.298a] NIKOLAEVSKIY, L. S., A. F. SIMONENKO and S. G. GRENISHIN, Zh. prikl. Spektrosk. 4 (1966) 485.
[5.299] NIKONOVA, J. I. and V. K. PROKOFYEV, Optika i Spektroskopiya 1 (1956) 290.
[5.300] NIKONOVA, J. I. and V. K. PROKOFYEV, Fiz. Sb. Lvovsk Univ. No. 4 (1958) 318.
[5.301] NIKONOVA, J. I. and V. K. PROKOFYEV, Fizicheskie problemy spektroskopii Vol. 1, Izd. Akad. Nauk SSSR, Moskau 1962, S. 51.

[5.301a] Nodwell, R. A., H. W. van Andel and A. M. Robinson, J. quant. Spectrosc. Radiat. Transfer 8 (1968) 859.
[5.301b] Nodwell, R. A. and A. M. Robinson, Proc. 7. internat. Conf. Phenom. Ionized Gases, Beograd 1965, Vol. 1, p. 368.
[5.302] Norman, G. E., Optika i Spektroskopiya **14** (1963) 593.
[5.302a] Norman, G. E., Optika i Spektroskopiya **19** (1965) 657.
[5.302b] Odintsov, A. I., Optika i Spektroskopiya **14** (1963) 322.
[5.302c] Odintsova, G. A. and E. P. Vaulin, Zh. prikl. Spektrosk. 7 (1967) 749.
[5.302d] Oertel, H., Shock Tubes, Springer Verlag Wien u. New York 1966, p. 745.
[5.303] Olsen, H. N., in: Temperature, its measurements and control in science and industry Vol. 3, 1, Reinhold Publ. Co., New York 1962, 593.
[5.304] Olsen, H. N., J. quant. Spectrosc. Radiat Transfer **3** (1963) 59.
[5.305] Olsen, H. N., J. quant. Spectrosc. Radiat. Transfer **3** (1963) 305.
[5.306] Olsen, K. H., P. M. Routly and R. B. King, Proc. Conf. Stellar Atmospheres. Indiana Univ. (1955) p. 46.
[5.307] Olsen, K. H., P. M. Routly and R. B. King, Astrophys. J. **130** (1959) 688.
[5.308] Ornstein, L. S. and T. Bouma, Phys. Rev. **36** (1930) 679.
[5.309] Ornstein, L. S., J. P. A. van Hengstrum and H. Brinkman, Physica **5** (1938) 145.
[5.310] Ornstein, L. S. and J. Key, Z. Physik **85** (1933) 565.
[5.311] Ornstein, L. S. and J. Key, Physica **1** (1934) 945.
[5.311a] Orville, R. E., M. A. Uman and A. M. Sletten, J. appl. Phys. **38** (1967) 895.
[5.312] Osberghaus, O. and K. Ziock, Z. Naturforsch. **11a** (1956) 762.
[5.312a] Osherovich, A. L. and G. M. Petelin, Doklady Akad. Nauk SSSR **129** (1959) 544.
[5.312b] Osherovich, A. L. and I. G. Savich, Optika i Spektroskopiya **4** (1958) 715.
[5.312c] Osherovich, A. L. and Ya. F. Verolainen, Doklady Akad. Nauk SSSR **164** (1965) 1022.
[5.312d] Osherovich, A. L. and Ya. F. Verolainen, Optika i Spektroskopiya **22** (1967) 329.
[5.312e] Osherovich, A. L. and Ya. F. Verolainen, Vestnik Leningrad Univ. No. 4 (1967) 140.
[5.312f] Osherovich, A. L. and Ya. F. Verolainen, Optika i Spektroskopiya **24** (1968) 162.
[5.312g] Ostroumenko, P. P. and V. S. Rossikhin, Optika i Spektroskopiya **19** (1965) 653.
[5.313] Ostrovskii, Y. I., Optika i Spektroskopiya **2** (1957) 673.
[5.314] Ostrovskii, Y. I., Optika i Spektroskopiya **3** (1957) 391.
[5.315] Ostrovskii, Y. I., Diss. Leningrad 1958.
[5.316] Ostrovskii, Y. I., G. F. Parchevski und N. P. Penkin, Optika i Spektroskopiya **1** (1956) 821.
[5.317] Ostrovskii, Y. I., G. F. Parchevski and N. P. Penkin, Fiz. Sb. Univ. Lvov No. 4 (1958) 316.
[5.318] Ostrovskii, Y. I. and N. P. Penkin, Optika i Spektroskopiya **3** (1957) 193.
[5.319] Ostrovskii, Y. I. and N. P. Penkin, Optika i Spektroskopiya **3** (1957) 391.
[5.320] Ostrovskii, Y. I. and N. P. Penkin, Optika i Spektroskopiya **4** (1958) 719.
[5.321] Ostrovskii, Y. I. and N. P. Penkin, Optika i Spektroskopiya **5** (1958) 345.
[5.322] Ostrovskii, Y. I. and N. P. Penkin, Optika i Spektroskopiya **9** (1960) 703.
[5.323] Ostrovskii, Y. I. and N. P. Penkin, Optika i Spektroskopiya **10** (1961) 8.
[5.324] Ostrovskii, Y. I. and N. P. Penkin, Optika i Spektroskopiya **10** (1961) 429.
[5.325] Ostrovskii, Y. I. and N. P. Penkin, Optika i Spektroskopiya **11** (1961) 3.
[5.326] Ostrovskii, Y. I. and N. P. Penkin, Optika i Spektroskopiya **11** (1961) 565.
[5.327] Ostrovskii, Y. I. and N. P. Penkin, Optika i Spektroskopiya **12** (1962) 669.

[5.328] Ostrovskii, Y. I., N. P. Penkin and L. N. Shabanova, Doklady Akad. Nauk SSSR **120** (1958) 66.
[5.329] Ostrovskii, Y. I., N. P. Penkin and L. N. Shabanova, Izv. Akad. Nauk SSSR Ser. Fiz. **22** (1958) 725.
[5.329a] Otten, E. W., Z. Physik **170** (1962) 336.
[5.330] Ottinger, C. and K. Ziock, Z. Naturforsch. **16a** (1961) 720.
[5.330a] Ovechkin, G. V., Sh. Bachtovarshoyev und L. E. Sandrigailo, Zh. prikl. Spektrosk. **7** (1967) 441.
[5.331] Parchevski, G. F. and N. P. Penkin, Vestnik Leningrad Univ. Ser. Fiz. i Khim. No. 11 (1954) 113.
[5.332] Parchevski, G. F. and N. P. Penkin, Zh. eksp. teoret. Fiz. **28** (1955) 379.
[5.333] Parchevski, G. F. and N. P. Penkin, Zh. eksp. teoret. Fiz. **28** (1955) 766.
[5.334] Parchevski, G. F. and N. P. Penkin, Izv. Akad. Nauk SSSR Ser. Fiz. **19** (1955) 8.
[5.334a] Pardies, J., Compt. Rend. Acad. Sci. Paris **266B** (1968) 1586.
[5.334b] Pardies, J., Compt. Rend. Acad. Sci. Paris **267A** (1968) 1144.
[5.334c] Parker, L. W. and J. R. Holmes, J. Opt. Soc. Amer. **43** (1953) 103.
[5.334d] Parkes, D. A., L. F. Keyser and F. Kaufman, Astrophys. J. **149** (1967) 217.
[5.335] Paul, W., Z. Physik **124** (1944) 121.
[5.336] Pearce, W. J., in: Optical spectrometric measurements of high Temperatures. Univ. of Chicago Press, Chicago 1961, p. 125.
[5.336a] Pebay-Peyroula, J. C., Thesis Paris 1959.
[5.337] Penkin, N. P., Zh. eksp. teoret. Fiz. **17** (1947) 355.
[5.338] Penkin, N. P., Zh. eksp. teoret. Fiz. **17** (1947) 1114.
[5.339] Penkin, N. P., Izv. Akad. Nauk SSSR Ser. Fiz. **11** (1947) 217.
[5.340] Penkin, N. P., Optika i Spektroskopiya **2** (1957) 545.
[5.341] Penkin, N. P., J. quant. Spectrosc. Radiat. Transfer **4** (1964) 41.
[5.342] Penkin, N. P. and T. P. Redko, Optika i Spektroskopiya **9** (1960) 680.
[5.343] Penkin, N. P. and L. N. Shabanova, Optika i Spektroskopiya **12** (1962) 3.
[5.343a] Penkin, N. P. and L. N. Shabanova, Optika i Spektroskopiya **12** (1962) 254.
[5.344] Penkin, N. P. and L. N. Shabanova, Optika i Spektroskopiya **14** (1963) 12.
[5.345] Penkin, N. P. and L. N. Shabanova, Optika i Spektroskopiya **14** (1963) 167.
[5.345a] Penkin, N. P. and L. N. Shabanova, Optika i Spektroskopiya **18** (1965) 896.
[5.345b] Penkin, N. P. and L. N. Shabanova, Optika i Spektroskopiya **23** (1967) 22.
[5.345c] Penkin, N. P. and I. Y. Slavenas, Optika i Spektroskopiya **15** (1963) 9.
[5.346] Penkin, N. P. and I. J. J. Slavenas, Optika i Spektroskopiya **15** (1963) 154.
[5.346a] Pery-Thorne, A. and J. E. Chamberlain, Proc. Phys. Soc. Lond. **82** (1963) 133.
[5.346b] Pery-Thorne, A. and W. R. S. Garton, Proc. Phys. Soc. Lond. **76** (1960) 833.
[5.347] Petrashen, M. I. and I. V. Abarenkov, Vestnik Leningrad Univ. Ser. Fiz. i Khim. No. 5 (1954) 141.
[5.348] Petrie, W., Canad. J. Res. **A 25** (1947) 42.
[5.348a] Petrova, G. P., Fizicheskaya gazodinamika i svoistva gazov pri visokich temperaturach, Izd. Nauka, Moskau 1964, S. 73.
[5.349] Pfender, E. and W. Bez, Proc. 5. internat. Conf. Ionization Phenomena in Gases, Vol. 1, North Holland Publ. Co. Amsterdam 1962, p. 987.
[5.349a] Pfennig, H., R. Stelle and E. Trefftz, J. quant. Spectrosc. Radiat. Transfer **5** (1965) 335.
[5.349b] Pickett, R. C. and R. Anderson, J. quant. Spectrosc. Radiat. Transfer **9** (1969) 697.
[5.349c] Popenoe, C. H. and J. B. Shumaker, J. Res. Nat. Bur. Stand. **A 69** (1965) 495.
[5.349d] Porter, C. E., Phys. Letters **2** (1962) 292.
[5.349e] Prag, A. B., Doct. Diss. Univ. Washington 1964.

[5.349f] PRAG, A. B., C. E. FAIRCHILD and K. C. CLARK, Phys. Rev. **137** (1965) A 1358.
[5.349g] PRILEZHAEVA, N. A. and V. V. KOKHANENKO, Zh. prikl. Spektrosk. 7 (1967) 522.
[5.350] PROKOFYEV, V. K., Phil. Mag. **311** (1927) 1010.
[5.351] PROKOFYEV, V. K., Z. Physik **50** (1928) 701.
[5.352] PROKOFYEV, V. K., Z. Physik **58** (1929) 255.
[5.353] PROKOFYEV, V. K., Zh. eksp. teoret. Fiz. **1** (1931) 123; Z. Physik **57** (1929) 387.
[5.354] PROKOFYEV, V. K., Zh. eksp. teoret. Fiz. **1** (1931) 111.
[5.355] PROKOFYEV, V. K., Zh. eksp. teoret. Fiz. **4** (1934) 347.
[5.356] PROKOFYEV, V. K. and A. N. FILIPPOV, Zh. eksp. teoret. Fiz. **4** (1934) 31; Z. Physik **85** (1933) 647.
[5.357] PROKOFYEV, V. K. and G. GAMOV, Z. Physik **44** (1927) 887.
[5.358] PROKOFYEV, V. K., I. M. NAGIBINA and G. P. PETROVA, Optika i Spektroskopiya **8** (1960) 376.
[5.359] PROKOFYEV, V. K. and G. SHTANDEL, Zh. eksp. teoret. Fiz. **4** (1934) 359.
[5.360] PROKOFYEV, V. K. and V. N. SOLOVYEV, Z. Physik **48** (1928) 276.
[5.361] PROKOFYEV, V. K. and V. N. SOLOVYEV, Trudy Gos. Opt. Inst. **4**, No. 41 (1928).
[5.361a] ZU PUTLITZ, G. and K. V. VENKATARAMU, Z. Physik **209** (1968) 470.
[5.362] RASUMOVSKAYA, L. P., Optika i Spektroskopiya **14** (1963) 189.
[5.363] REGEMORTER, H. VAN, J. Phys. Radium **20** (1959) 907.
[5.364] REINDL, H. P., Diss. Utrecht 1946.
[5.364a] REINKE, P., Z. Astrophys. **66** (1967) 234.
[5.365] RICHTER, J., Diss. Univ. Kiel 1957.
[5.366] RICHTER, J., Z. Physik **151** (1958) 114.
[5.367] RICHTER, J., Z. Astrophys. **51** (1961) 177.
[5.367a] RICHTER, J., Z. Astrophys. **53** (1961) 262.
[5.367b] RICHTER, J., Z. Astrophys. **61** (1965) 57.
[5.367c] RIEMANN, M., 11. Coll. Spectrosc. internat., Beograd 1963, A-16.
[5.367d] RIEMANN, M., Diss. Tech. Hochsch. Ilmenau 1963.
[5.367e] RIEMANN, M., Monatsber. dtsch. Akad. Wiss. **5** (1963) 316.
[5.367f] RIEMANN, M., Z. Physik **179** (1964) 38.
[5.368] RIGHINI, G., Z. Astrophys. **10** (1935) 344.
[5.368a] RITTER, G. J. and G. W. SERIES, Proc. Roy. Soc. Lond. A **238** (1956/57) 473.
[5.369] ROBBEN, F., W. B. KUNKEL and L. TALBOT, Phys. Rev. **132** (1963) 2363.
[5.369a] ROBERTS, J. R. and K. L. ECKERLE, Phys. Rev. **153** (1967) 87.
[5.369b] ROBERTS, T. G. and W. L. HALES, Rep. U. S. Army Missile Command. Redstone Arsenal, Alabama, RR-TR-62-8, 1962.
[5.370] RODER, O., Z. Astrophys. **55** (1962) 38.
[5.371] ROSCHDESTWENSKI, D. S., Ann. Physik **39** (1912) 307.
[5.372] ROSCHDESTWENSKI, D. S. and N. P. PENKIN, Izv. Akad. Nauk SSSR Ser. Fiz. **5** (1941) 635.
[5.373] ROSCHDESTWENSKI, D. S. and N. P. PENKIN, J. Phys. (USSR) **5** (1941) 319.
[5.374] RUBENSTEIN, P. J., Phys. Rev. **58** (1940) 1007.
[5.375] RUDKJØBING, M., Publ. Kopenhagen Obs. Nr. 124 (1940).
[5.376] RUDKJØBING, M., Publ. Kopenhagen Obs. Nr. 145 (1947).
[5.376a] RUDKO, R. I. and C. L. TANG, J. appl. Phys. **38** (1967) 4731.
[5.377] RUDNICK, P., Phys. Rev. **48** (1935) 807.
[5.377a] RUDTSIKAS, Z., J. VIZBARAITE and A. YUTSIS, Liet. fiz. Rink, Vilnius, USSR **5** (1965) 63.
[5.378] SAGORJANSKAJA, J. W., Zh. eksp. teor. Fiz. **19** (1949) 447.
[5.378a] SALOMAN, E. B., Phys. Rev. **152** (1966) 79.
[5.379] SAMBURSKY, Proc. nederl. Akad. Wetensch. **34** (1941) 339.
[5.380] SANDO, K. M. and S. T. EPSTEIN, J. chem. Physics **43** (1965) 1620.
[5.380a] SAVAGE, B. D. and W. ERVENS, Astrophys. J. **146** (1966) 940.

[5.380b] Savage, B. D. and G. M. Lawrence, Astrophys. J. **146** (1966) 940.
[5.380c] Schiff, B. and C. L. Pekeris, Phys. Rev. **134** (1964) A 638.
[5.380d] Schillbach, H., Ann. Physik **18** (1933) 721.
[5.380e] Schnapauff, R., Z. Astrophys. **68** (1968) 431.
[5.381] Schön, K., Ann. Physik **28** (1937) 649.
[5.382] Schnautz, H., Spectrochim. Acta **1** (1939) 173.
[5.383] Schouten, J. W., Diss. Utrecht 1941.
[5.384] Schouten, J. W. and J. A. Smith, Physica **10** (1943) 661.
[5.385] Schrödinger, E., Ann. Physik **80** (1926) 437.
[5.385a] Schulz-Gulde, E., J. quant. Spectrosc. Radiat. Transfer **9** (1969) 13.
[5.386] Schütz, W., Z. Physik **38** (1926) 864.
[5.387] Schütz, W., Z. Physik **64** (1930) 682.
[5.388] Schütz, W., Ann. Physik **18** (1933) 705.
[5.389] Schuttevaer, J. W., Diss. Utrecht 1937.
[5.390] Schuttevaer, J. W., M. J. de Bont and van den Broek, T. H., Physica **10** (1943) 544.
[5.391] Schuttevaer, J. W. and J. A. Smith, Physica **10** (1943) 502.
[5.392] Schwarz, K. H., Physica **7** (1940) 361.
[5.393] Seaton, M. J., Proc. roy. Soc. **A 208** (1951) 418.
[5.394] Seaton, M. J., Proc. Phys. Soc. Lond. **A 67** (1954) 927.
[5.394a] Seka, W. and F. L. Curzon, J. quant. Spectrosc. Radiat. Transfer **8** (1968) 1147.
[5.395] Semenova, O. P. and M. A. Levchenko, Doklady Akad. Nauk SSSR **151** (1963) 90.
[5.395a] Semenova, O. P. and M. A. Levchenko, Zh. prikl. Spektrosk. **1** (1964) 218.
[5.396] Seward, R. S., Phys. Rev. **37** (1931) 344.
[5.397] Sherebtsova, K. I., Zh. eksp. teor. Fiz. **9** (1939) 566.
[5.397a] Shabanova, L. N., Optika i Spektroskopiya **15** (1963) 828.
[5.397b] Shackleford, W. L., J. quant. Spectrosc. Radiat. Transfer **2** (1962) 245.
[5.397c] Shackleford, W. L., J. quant. Spectrosc. Radiat. Transfer **5** (1965) 303.
[5.398] Shortley, G. H., Phys. Rev. **47** (1935) 295.
[5.399] Shumaker, J. B., Rev. sci. Instrum. **32** (1961) 65.
[5.399a] Shumaker, J. B. and C. H. Popenoe, J. Opt. Soc. Amer. **57** (1967) 8.
[5.400] Shumaker, J. B. and W. L. Wiese, Temperature, its Measurement and Control in Science and Industry, Vol. III, 1, Reinhold Publ. Co., New York 1962, p. 575.
[5.400a] Shumaker, J. B. and C. R. Yokley, Appl. Optics **3** (1964) 83.
[5.400b] Simonenko, A. F., Zh. prikl. Spektrosk. **9** (1968) 332.
[5.400c] Sinclair, D. and H. W. Webb, Phys. Rev. **50** (1936) 440.
[5.401] Slavnas, I. J. J., Optika i Spektroskopiya **16** (1964) 390.
[5.401a] Slovetski, D. I., Teplofizika **5** (1967) 401.
[5.401b] Smårs, E. A., Ark. Fys. **29** (1965) 97.
[5.401c] Smith, W. W. and A. C. Gallagher, Joint Inst. Lab. Astrophys., Boulder, Colorado, USA, Rep. 52, 1965.
[5.401d] Smith, W. W. and A. C. Gallagher, Phys. Rev. **145** (1966) 26.
[5.402] Snoek, J. L., Z. Physik **52** (1928) 654.
[5.403] Snoek, J. L. and L. S. Arnstein, Z. Physik **50** (1928) 600.
[5.403a] Sobelman, I. I., Vvedenie v teoriyu atomnich spektrov. Fizmatgis, Moscow 1963.
[5.404] Sobolev, N. N., Zh. eksp. teor. Fiz. **13** (1943) 131.
[5.404a] Solarski, J. and W. L. Wiese, J. Opt. Soc. Amer. **52** (1962) 592.
[5.404b] Solarski, J. E. and W. L. Wiese, Phys. Rev. **135** (1964) A 1236.
[5.405] Soleillet, D., Compt. Rend. Acad. Sci., Paris **187** (1928) 723.

[5.405a] VON SPECHT, J., Z. Naturforsch. **23a** (1968) 1499.
[5.405b] VON SPECHT, J., Z. Naturforsch. **24a** (1969) 457.
[5.405c] SPITZER, M., Compt. Rend. Acad. Sci., Paris **260** (1965) 3907.
[5.405d] SPITZER, L. and J. L. GREENSTEIN, Astrophys. J. **114** (1951) 407.
[5.405e] STACEY, D. N. and J. M. VAUGHAN, Phys. Letters **11** (1964) 105.
[5.405f] STARTSEV, G. P. and M. G. KOZLOV, Proc. 5. internat. Conf. Phys. Electronic Collisions, Nauka, Leningrad, SSSR 1967, p. 616.
[5.405g] STATZ, H., F. A. HORRIGAN, S. H. KOOZEKANANI, C. L. TANG and G. F. KOSTER, J. appl. Phys. **36** (1965) 2278.
[5.405h] STEELE, R. and E. TREFFTZ, J. quant. Spectrosc. Radiat. Transfer **6** (1966) 833.
[5.406] STEINHÄUSER, A., Z. Physik **95** (1935) 669.
[5.407] STEINHÄUSER, A., Z. Physik **99** (1936) 300.
[5.408] STEKELENBURG, L. H. M. VAN, Diss. Utrecht 1943.
[5.409] STEKELENBURG, L. H. M. VAN, Physica **14** (1948) 189.
[5.410] STEPHENSON, G., Nature, Lond. **167** (1951) 156.
[5.411] STEPHENSON, G., Proc. Phys. Soc. Lond. **A 64** (1951) 458.
[5.412] STEPHENSON, G., Proc. Phys. Soc. Lond. **A 64** (1951) 666.
[5.413] STEVENSON, A. F., Proc. Roy. Soc. Lond. **A 137** (1932) 298.
[5.414] STEWART, Proc. Phys. Soc. Lond. **A 67** (1954) 917.
[5.414a] STOJANOFF, C. G., Proc. 8. internat. Conf. Phenom. Ionized Gases, Contributed Pap., IAEA Vienna 1967, p. 226.
[5.415] STONE, P. M., Phys. Rev. **127** (1962) 1151; Correction: Phys. Rev. **135AB** (1964) 2.
[5.416] STRIJLAND, J. C. and A. J. NANASSY, Physica **24** (1958) 935.
[5.416a] STROTSKITE, T. D. and A. P. YUTSIS, Trudy Akademii Nauk Litovskoi SSR, Ser. B, 2, Nr. 25 (1961) 93; Engl. Translation: Optical Transition Probabilities, Jerusalem 1963, p. 224.
[5.417] SUGIURA, Y., Phil. Mag. **4** (1927) 495.
[5.418] SUGIURA, Y., Sci. Pap. Inst. Phys. Chem. Res. Tokyo **11** (1929) 1.
[5.418a] SUKEWER, S., Teplofizika **3** (1965) 199.
[5.418b] SY, A. A., Rev. haut. Temp. Réfract. **2** (1965) 65.
[5.418c] SYNEK, M. and W. LICHODZIEJEWSKI, Phys. Letters **22** (1966) 578.
[5.418d] TAPPE, W. and J. VAN CALKER, Z. analyt. Chem. **198** (1963) 13.
[5.419] TATUM, J. B., Monthly Notices Roy. Astron. Soc. **122** (1961) 311.
[5.419a] TEPLOVA, S., M. CHAIKA and V. CHERENKOVSKY, Optika i Spektroskopiya **25** (1968) 346.
[5.420] TERPSTRA, J. and J. A. SMIT, Physica **24** (1958) 937.
[5.421] THIEL, E. and H. KREMPL, Z. angew. Physik **14** (1962) 341.
[5.421a] THORNTON, J. A. and A. B. CAMBEL, J. quant. Spectrosc. Radiat. Transfer **4** (1964) 539.
[5.421b] TIDWELL, E. D. and J. I. SHIPP, 8. internat. Conf. Phenom. Ionized Gases, Contrib. Pap., IAEA Vienna 1967, p. 453.
[5.422] TREFFTZ, E., Z. Astrophys. **26** (1949) 240.
[5.423] TREFFTZ, E., Z. Astrophys. **28** (1950) 67.
[5.424] TREFFTZ, E., Z. Astrophys. **29** (1951) 287.
[5.425] TREFFTZ, E., Probleme des Plasmas in Phys. u. Astronomie, Akademie-Verlag, Berlin 1958, p. 102.
[5.426] TREFFTZ, E. and J. BIERMANN, Z. Astrophys. **30** (1952) 275.
[5.427] TREFFTZ, E., A. SCHLÜTER, K. H. DETTMAR and K. JÖRGENS, Z. Astrophys. **44** (1957) 1.
[5.427a] TREFFTZ, E. and R. N. ZARE, J. quant. Spectrosc. Radiat. Transfer **9** (1969) 643.
[5.427b] TRUKHAN, P. and L. Y. KISELEVSKI, Doklady Akad. Nauk Beloruss. SSR, **11** (1967) 122.

[5.428] TRUMPY, B., Z. Physik **44** (1927) 575.
[5.429] TRUMPY, B., Z. Physik **47** (1928) 804.
[5.430] TRUMPY, B., Z. Physik **50** (1928) 228.
[5.431] TRUMPY, B., Z. Physik **54** (1929) 372.
[5.432] TRUMPY, B., Z. Physik **57** (1929) 787.
[5.433] TRUMPY, B., Z. Physik **61** (1930) 54.
[5.434] TRUMPY, B., Z. Physik **66** (1930) 720.
[5.435] TRUMPY, B., Z. Physik **71** (1931) 720.
[5.435a] TSILITIS, V. A., Optika i Spektroskopiya **25** (1968) 641.
[5.435b] TSIUNAITIS, G. K. and A. P. YUTSIS, Zh. eksp. teoret. Fiz. **28** (1955) 452.
[5.435c] TSUI-FANG and J. D. CRAGGS, J. Electronics and Control **4** (1958) 385.
[5.346] TUNSTEAD, Proc. Phys. Soc. Lond. **A 66** (1953) 304.
[5.436a] TURKO, M. N., Izv. Sibirsk. Otdel. Akad. Nauk SSSR Nr. 6 (1961) 14.
[5.436b] TURNER, R., Phys. Rev. **140 A** (1965) 426.
[5.437] UFFORD, C. W., Astrophys. J. **85** (1937) 249.
[5.437a] UNDERHILL, A. B., Astrophys. J. **151** (1968) 765.
[5.438] UNSÖLD, A., Physik der Sternatmosphären, Springer Verlag, Berlin-Göttingen-Heidelberg 1955.
[5.438a] VAINSHTEIN, L. A., Optika i Spektroskopiya **3** (1957) 313.
[5.438b] VAINSHTEIN, L. A., Diss. Moskau 1958.
[5.438c] VAINSHTEIN, L. A., Izv. Akad. Nauk SSSR, Ser. Fiz. **22** (1958) 671.
[5.438d] VAINSHTEIN, L. A., Transactions of the P. N. Lebedev-Physics Inst. **15**, Pt. 1 (1962) 1 (CBE-TR).
[5.438e] VAINSHTEIN, L. A. and L. A. MINAYEVA, Zh. prikl. Spektrosk. **9** (1968) 60.
[5.438f] VAINSHTEIN, L. A. and I. A. POLUEKTOV, Optika i Spektroskopiya **12** (1962) 460.
[5.438g] VAINSHTEIN, L. A. and V. M. YAVORSKI, Doklady Akad. Nauk SSSR **87** (1952) 919.
[5.438h] VAINSHTEIN, L. A. and V. M. YAVORSKI, Zh. eksp. teoret. Fiz. **27** (1954) 712; German Translation: Phys. Abh. SU **10** (1957) 64.
[5.438i] VARSAVSKY, C. M., Thesis Harvard Univ., Dep. Astronomy, Nov. 1958.
[5.438j] VASIN, V. L., L. V. LESKOV and F. A. SAVIN, Fizicheskie problemi Spektroskopii, Izd. Akad. Nauk SSSR, Moscow 1962, Vol. 1, S. 39.
[5.438k] VAUGHAN, J. M., Phys. Rev. **166** (1968) 13.
[5.438l] VEROLAINEN, YA. F. and A. L. OSHEROVICH, Optika i Spektroskopiya **20** (1966) 929.
[5.438m] VESELOV, M. G., Zh. eksp. teoret. Fiz. **19** (1949) 959.
[5.438n] VESELOV, M. G., Vestnik Leningrad Univ. Ser. Fiz. i Khim. Nr. 8 (1953) 181.
[5.438o] VETSHINKIN, S. I., Optika i Spektroskopiya **14** (1963) 317.
[5.438p] VIDAL, C.-R., Z. Naturforsch. **19a** (1964) 1018.
[5.438q] VIDALE, G. L., Technical Inf. Series Rep. R 60 SD 331, Gen. Electr. Co., Space Sci. Lab., 1960.
[5.438r] VEROLAINEN, YA. F. and A. L. OSHEROVICH, Optika i Spektroskopiya **25** (1968) 466.
[5.439] VILLARS, D., J. Opt. Soc. Amer. **42** (1952) 552.
[5.440] VINTI, J. P., Phys. Rev. **42** (1932) 632.
[5.440a] VISBARAITE, Y. I., A. I. KANTSERYAVICHUS and A. P. YUTSIS, Optika i Spektroskopiya **1** (1956) 9.
[5.441] VOORHOEVE, P. G., Diss. Utrecht 1946.
[5.441a] VÖRÖS, T. and A. PETKOV, Proc. 14. Coll. Spectrosc. Internat., Debrecen, Hungary 1967, Vol. 2, p. 583.
[5.442] WAIBEL, F., Z. Physik **53** (1929) 459.
[5.443] WALTERS, A. K., E. I. NIKONOVA and G. P. STARTSEV, Doklady mezhvuz. nauchn. konferentsii po spektroskopii i spektr. analizu, Tomsk Univ., Tomsk 1960, p. 15.

[5.444] WALTERS, A. K., E. I. NIKONOVA and G. P. STARTSEV, Optika i Spektroskopiya **16** (1964) 717.
[5.445] WALTERS, A. K. and G. P. STARTSEV, Optika i Spektroskopiya **17** (1964) 483.
[5.446] WARNER, B., Commun. Univ. Lond. Observ. No. 66 (1965).
[5.447] WARNER, B., Commun. Univ. Lond. Observ. No. 70 (1966).
[5.448] WARNER, B., Commun. Univ. Lond. Observ. No. 71 (1966).
[5.449] WARNER, B., Z. Astrophys. **69** (1968) 399.
[5.450] WARNER, B. and E. L. G. POWELL, Commun. Univ. Lond. Observ. Nr. 67 (1964).
[5.451] WEBB, H. W. and H. A. MESSENGER, Phys. Rev. **66** (1944) 77.
[5.452] WEILER, J., Ann. Physik **1** (1929) 361.
[5.453] WEINGEROFF. M., Z. Physik **67** (1931) 679.
[5.454] WEINZIERL, G., Elektrie **19** (1965) 359.
[5.455] WEISS, A. W., Astrophys. J. **138** (1963) 1262.
[5.455a] WEISS, A. W., Phys. Rev. **162** (1967) 71.
[5.455b] WEISS, A. W., J. Res. Nat. Bur. Stand. **71 A** (1967) 157.
[5.455c] WEISS, A. W., J. Res. Nat. Bur. Stand. **71 A** (1967) 163.
[5.455d] WENDE, B., Diss. Tech. Hochsch. Karlsruhe, Fak. Elektrotechn., 1966.
[5.455e] WENDE, B., Z. Physik **198** (1967) 1.
[5.455f] WENDE. B., Z. Physik **213** (1968) 341.
[5.456] WESSEL, G., Z. Physik **126** (1949) 440. Correction: Z. Physik **130** (1951) 100.
[5.456a] WESSLER, R., Compt. Rend. Acad. Sci. Paris **236** (1953) 1335.
[5.457] WHEELER, J. A., Phys. Rev. **43** (1933) 258.
[5.458] WIESE, W., Diss. Univ. Kiel 1957.
[5.459] WIESE, W. L., Proc. 10. Coll. Spectrosc. Internat., Spartan Books, Washington D. C. 1963, p. 37.
[5.459a] WIESE, W. L., Proc. 8. internat. Conf. Phenom. Ionized Gases, Contrib. Pap., IAEA Vienna 1967, p. 447.
[5.460] WIESE, W. L. and P. W. MURPHY, Phys. Rev. **131** (1963) 2108.
[5.461] WIESE, W. L. and J. B. SHUMAKER, J. Opt. Soc. Amer. **51** (1961) 937.
[5.461a] WIESE, W. L., M. W. SMITH and B. M. GLENNON, Atomic Transition Probabilities, Vol. 1, Nat. Bur. Stand. Washington 1966; WIESE, W. L., M. W. SMITH and B. M. MILES, Atomic Transition Probabilities, Vol. 2, Nat. Bur. Stand. Washington 1969.
[5.462] WILD, J. P., Astrophys. J. **115** (1952) 206.
[5.463] WILKERSON, T. D., Diss. Univ. Michigan 1961.
[5.464] WILKERSON, T. D., O. LAPORTE, U. S. Air Force Office of Scientific Research (AFOSR) Report Nr. 1115, 1961.
[5.465] WILKINSON, P. G., J. quant. Spectrosc. Radiat. Transfer **5** (1965) 503.
[5.465a] WILKINSON, P. G., J. quant. Spectrosc. Radiat. Transfer **6** (1966) 823.
[5.465b] WILKINSON, P. G., Canad. J. Phys. **45** (1967) 1709.
[5.465c] WILLIAMS, W. L. and E. S. FRY, Phys. Rev. Lett. **20** (1968) 1335.
[5.465d] WILSON, K. H. and W. E. NICOLET, J. quant. Spectrosc. Radiat. Transfer **7** (1967) 891.
[5.466] WOBIG, K. H., Z. Astrophys. **55** (1962) 100.
[5.467] WOLFE, H. C., Phys. Rev. **41** (1932) 443.
[5.468] WOLFSOHN, G., Z. Phys. **83** (1933) 234.
[5.468a] WOLNIK, S. J., R. O. BERTELS, G. S. LARSON and E. H. CARNEVALE, Phys. Fluids **11** (1968) 1002.
[5.468b] WU, T. Y., Phys. Rev. **58** (1940) 1114.
[5.469] YAMANOUCHI, T. and KOTANI, Proc. math. phys. Soc. Japan **22** (1940) 60 and 569; **24** (1942) 351.
[5.470] YILMAZ, H., Phys. Rev. **100** (1955) 1148.

[5.470a] Yos, J. M., AVCO, Wilmington, Mass., USA, Rep. RAD-TR-65-7, May 1965.
[5.470b] YURYEV, M. S., Optika i Spektroscopiya **25** (1968) 295.
[5.470c] YUTSIS, A. P., K. K. USHPALIS, V. I. KAVETSKIS and I. V. LEVINSON, Optika i Spektroskopiya **1** (1956) 1.
[5.470d] YUTSIS, A. P., YA. I. VISBARAITE and K. K. ERINGIS, Trudy Akad. Nauk. Litovskoi SSR Ser. B, **3** (1961) No 26, p. 99; Engl. Transl.: Optical Transition Probabilities, Jerusalem 1963, p. 229.
[5.470e] YUTSUM, P. H. and G. V. MARR, Proc. Roy. Soc. Lond. **A 219** (1953) 89.
[5.470f] DE ZAFRA, R., R. J. GOSHEN, A. LANDMAN and A. LURIO, Bull. Amer. Phys. Soc. **7** (1962) 433.
[5.470g] DE ZAFRA, R. L. and W. KIRK, Amer. J. Phys. **35** (1967) 573. Erratum: Amer. J. Phys. **36** (1968) 1020.
[5.470h] DE ZAFRA, R. L. and A. MARSHALL, Phys. Rev. **170** (1968) 28.
[5.470i] ZARE, R. N., J. chem. Phys. **47** (1967) 3561.
[5.471] ZEHDEN, W. and M. ZEMANSKY, Z. Phys. **72** (1931) 442.
[5.472] ZEMANSKY, M. N., Z. Phys. **72** (1931) 587.
[5.473] ZHEYENBAYEV, Z., I. P. NESMACHNAYA, A. I. POLOVIKOV and V. S. ENGELSHT, Issledovanie elektricheskoi dugi i plazmatrona, Izd. Ilim, Frunze 1968, p. 23.
[5.474] ZIOCK, K., Z. Physik **147** (1957) 99.
[5.475] ZWAAN, A., Naturwiss. **17** (1928) 121.
[5.476] ZWAAN, A., Diss. Utrecht 1929.
[6.1] ABOUD, A. A., J. P. CURTIS, R. MERCURE and W. A. RENSE, J. opt. Soc. Amer. **45** (1955) 767.
[6.1a] ALKEMADE, C. TH. J., Proc. 10. Coll. Spectrosc. internat., Washington 1963, p. 143.
[6.1b] ALTICK, P. L., Phys. Rev. **169** (1968) 21.
[6.2] ASTOIN, N., J. Rech. Nr. 38 (1957) 1.
[6.3] BARTELS, H., Z. Phys. **127** (1950) 243.
[6.4] BARTELS, H., Z. Phys. **128** (1950) 546.
[6.5] BARTELS, H., Probleme des Plasmas in Physik und Astronomie. Akademie-Verlag, Berlin 1958, p. 170.
[6.6] BARTELS, H. and R. BEUCHELT, Z. Phys. **149** (1957) 594.
[6.6a] BASOW, N. G., B. L. BOROWICH, V. S. ZUYEV and YU. YU. STOILOV, Th. eksp. teoret. Fiz. **28** (1968) 2079.
[6.7] BATES, D. R., Mon. Not. R. astr. Soc. **106** (1946) 423.
[6.8] BATES, D. R. and H. S. W. MASSEY, Proc. roy. Soc. **A 177** (1941) 281.
[6.9] BATES, D. R. and M. J. SEATON, Mon. Not. R. astr. Soc. **109** (1949) 698.
[6.9a] BAUDER, U. and U. PLANTIKOW, Proc. z. internat. Conf. Phenom. Ionized Gases, Belgrade 1965, Vol. 1, p. 664.
[6.10] BAUER, A., Proc. 4. internat. Conf. Ionization Phenomena in Gases, Bd. 2, North Holland Publ. Co., Amsterdam 1960, 817.
[6.11] BAUER, S. H., Hypersonic Flow Research. Academic Press, New York and London 1962, p. 143.
[6.12] BAUSERMAN, G. W., C. H. BRIEN and T. ZANDSTRA, Rev. sci. Instrum. **25** (1954) 640.
[6.12a] BEIJER, L. B., Proc. 7. internat. Conf. Phenom. Ionized Gases, Belgrade 1965, Vol. 3, p. 182.
[6.13] BEKEFI, G., J. L. HIRSHFIELD and S. C. BROWN, Phys. Fluids **4** (1961) 173.
[6.13a] BELL, K. L. and A. E. KINGSTON, Proc. Phys. Soc. Lond. **90** (1967) 337.
[6.13b] BERGE, O. E. and W. BÖTTICHER, Z. Naturforschg. **19a** (1964) 1460.
[6.13c] BEYNON, J. D. E., Nature **207** (1965) 405.
[6.13d] BEYNON, J. D. E., Proc. Phys. Soc. Lond. **89** (1966) 59.
[6.13e] BEYNON, J. D. E. and R. B. CAIRNS, J. sci. Instrum. **41** (1964) 111.

[6.13f] BIRKELAND, J. W. and W. G. BRAUN, Proc. 7. internat. Conf. Phenom. Ionized Gases, Belgrade 1965, Vol. 1, S. 697.

[6.13g] BLACKWELL, H. E., G. S. BAJWA, G. S. SHIPP and G. L. WEISSLER, J. quant. Spectrosc. Radiat. Transfer **4** (1964) 249; IEEE Trans. Nucl. Sci. **11** (1964) 199.

[6.13h] BLEEKRODE, R., J. appl. Phys. **38** (1967) 5062.

[6.13i] BLEEKRODE, R., Proc. 8. internat. Conf. Phenom. Ionized Gases, Contributed Pap., IAEA Wien 1967, p. 150.

[6.14] BOLDT, G., Proc. 5. internat. Conf. Ionization Phenomena in Gases, Bd. 1, North Holland Publ. Co., Amsterdam 1962, p. 925.

[6.15] BRADDICK, H. J. and R. W. DITCHBURN, Proc. roy. Soc. **A 143** (1934) 472.

[6.15a] BRAUN, W. G., Rev. sci. Instrum. **36** (1965) 802.

[6.15b] BRAUN, W. G. and J. P. OSS, Proc. 9. internat. Conf. Phenom. Ionized Gases, Bukarest 1969, p. 634.

[6.15c] BREDERLOW, G. W., W. RIEDMÜLLER and M. SALVAT, Z. Naturforsch. **23a** (1968) 731.

[6.16] BUCHELE, D., Temperature, its Measurement and Control in Science and Industry, Bd. 3, 2, Reinhold Publ. Co., New York 1962, p. 879.

[6.16a] BURAKOV, V. S., P. A. NAUMENKOV and V. P. IVANOV, Zh. prikl. Spektrosk. **8** (1968) 738.

[6.16b] BURGESS, A. and M. J. SEATON, Monthly Notices Roy. Astron. Soc. **120** (1960) 121.

[6.16c] BYARD, P. L. and R. E. ROLL, J. quant. Spectrosc. Radiat. Transfer **5** (1965) 715.

[6.16d] CAIRNS, R. B. and J. A. R. SAMSON, Phys. Rev. **139** (1965) A 1403.

[6.16e] CARNEVALE, E. H., S. WOLNIK, G. LARSON, C. CAREY and G. W. WARES, Phys. Fluids **10** (1967) 1459.

[6.16f] CENTER, R. E., Phys. Fluids **10** (1967) 1777.

[6.17] CLOUSTON, J. G., A. G. GAYDON and I. R. HURLE, Proc. roy. Soc. **A 252** (1959) 143.

[6.17a] COMES, F. J. and A. ELZER, Z. Naturforsch. **19a** (1964) 721.

[6.17b] COMES, F. J. and A. ELZER, Z. Naturforsch. **23a** (1968) 133.

[6.17c] COMES, F. J., F. SPEIER and A. ELZER, Z. Naturforsch. **23a** (1968) 125.

[6.17d] COOK, G. R. and B. K. CHIUG, J. chem. Physics **43** (1965) 1794.

[6.17e] COOK, G. R. and P. H. METZGER, Proc. 6. internat. Conf. Ion. Phenom. Gases, Paris 1963, Vol. 1, p. 149.

[6.17f] COOK, G. R. and P. H. METZGER, J. chem. Phys. **41** (1964) 321.

[6.17g] COOK, G. R. and M. OGAWA, Canad. J. Phys. **43** (1965) 256.

[6.17h] COWAN, R. and G. DIEKE, Rev. mod. Phys. **20** (1948) 418.

[6.18] DALGARNO, A., Proc. phys. Soc. Lond. **A 65** (1952) 663.

[6.19] DITCHBURN, R. W., Proc. roy. Soc. **A 236** (1956) 216.

[6.19a] DITCHBURN, R. W., J. quant. Spectrosc. Radiat. Transfer **2** (1962) 361.

[6.19b] DITCHBURN, R. W., J. TUNSTEAD and J. G. YATES, Proc. Roy. Soc. Lond. **181 A** (1943) 386.

[6.20] DITCHBURN, R. W., P. H. JUTSUM and G. V. MARR, Proc. roy. Soc. **A 219** (1953) 89.

[6.21] DITCHBURN, R. W. and G. V. MARR, Proc. phys. Soc. Lond. **A 66** (1953) 655.

[6.22] DITCHBURN, R. W. and U. ÖPIK, Atomic and Molecular Processes. Academic Press, New York and London 1962, p. 79.

[6.23] DITCHBURN, R. W., J. TUNSTEAD and J. G. YATES, Proc. roy. Soc. **A 181** (1943) 386.

[6.24] DITCHBURN, R. W. and P. A. YOUNG, J. atmos. terr. Phys. **24** (1962) 127.

[6.24a] EDERER, D. L. and D. H. TOMBOULIAN, Phys. Rev. **133** (1964) A 1525.

[6.25] EHLER, A. W. and G. L. WEISSLER, J. opt. Soc. Amer. **45** (1955) 1035.

[6.25a] English, P. E. and S. S. Hitchcock, J. sci. Instrum. 2 (1968) 984.
[6.26] Evans, R. A., J. appl. Phys. 28 (1957) 1005.
[6.26a] Faizullov, F. S., Trudy Fiz. Inst. P. N. Lebedeva, Akad. Nauk SSSR, Vol. 18 (1962) 105.
[6.27] Faizullov, F. S., N. N. Sobolev and Y. M. Kudryavtsev, Doklady Akad. Nauk SSSR 127 (1959) 541.
[6.28] Faizullov, F. S., N. N. Sobolev and Y. M. Kudryavtsev, Optika i Spektroskopiya 8 (1960) 761.
[6.28a] Ferrari, C. and J. H. Clarke, in: High temperatures in Aeronautics, Pergamon Press London 1964, p. 239.
[6.28b] Forman, P. R. and W. B. Kunkel, Phys. Fluids 11 (1968) 1528.
[6.28c] Garrett, W. R. and H. T. Jackson, Phys. Rev. 153 (1967) 28.
[6.28d] Garton, W. R. S., W. H. Parkinson and E. M. Reeves, Proc. Phys. Soc. Lond. 88 (1966) 771.
[6.28e] Gaydon, A. G., in: Douzième conseil de chimie, Interscience Publ. New York 1962.
[6.29] Gaydon, A. G., I. R. Hulre, Proc. roy. Soc. A 262 (1961) 38.
[6.30] Gaydon, A. G. and I. R. Hurle, The Shock Tube in high Temperature chemical Physics, Chapman and Hall Ltd., London 1963.
[6.30a] Gaydon, A. G., I. R. Hurle and G. H. Kimbell, Proc. Roy. Soc. Lond. A 273 (1963) 291.
[6.30b] Gezalov, Ch. B. and A. V. Ivanova, Zh. prikl. Spektrosk. 9 (1968) 173.
[6.30c] Gezalov, Ch. B. and A. V. Ivanova, Teplofizika 6 (1968) 416.
[6.31] Göing, W., Z. Phys. 131 (1952) 603.
[6.31a] Grechikhin, L. I., Teplofizika 1 (1963) 198.
[6.31b] Greig, J. R., Brit. J. appl. Phys. 16 (1965) 957.
[6.32] Hargreaves, J., Proc. Camb. phil. Soc. 35 (1929) 75.
[6.32a] Hashmi, M. A., J. van der Houven van Oordt and J.-G. Wegrowe, Nuclear Fusion 8 (1968) 197.
[6.32b] Henry, R. J., J. chem. Phys. 44 (1966) 4357.
[6.32d] Hernandez, J. P., Phys. Rev. 167 (1968) 108.
[6.32e] Hildalgo, M. B., Astrophys. J. 153 (1968) 981.
[6.32f] Hinnov, E., J. E. Hesser and L. C. Johnson, Rep. Plasma Phys. Lab. Princeton Univ., MATT-462, 1967.
[6.32g] Hinnov, E., J. G. Hirschberg, F. W. Hofman and N. Rynn, Phys. Fluids 6 (1963) 1779.
[6.32h] Hiramoto, T., S. Matsunaga, H. Shirakata and S. Yano, Electricity from MHD, IAEA Vienna, 1966, Vol. 1, S. 85.
[6.32i] Hofmann, F. W., Phys. Fluids 7 (1964) 532.
[6.32i-1] Hooymayers, H. P. and P. L. Lijnse, J. quant. Spectrosc. Radiat. Transfer 9 (1969) 995.
[6.32j] Hudson, R. D., Phys. Rev. 135 (1964) A 1212.
[6.32k] Hudson, R. D., J. chem. Phys. 43 (1965) 1790.
[6.32l] Hudson, R. D. and V. L. Carter, Phys. Rev. 139 (1965) A 1426.
[6.32m] Hudson, R. D. and V. L. Carter, J. Opt. Soc. Amer. 57 (1967) 651.
[6.32n] Huffmann, R. E., J. chem. Phys. 40 (1964) 356.
[6.32o] Huffmann, R. E., J. E. Larrabee and Y. Tanaka, J. chem. Phys. 40 (1964) 2261.
[6.32p] Huffmann, R. E., Y. Tanaka and J. C. Larrabee, J. chem. Phys. 39 (1963) 902.
[6.32q] Huffmann, R. E., Y. Tanaka and J. C. Larrabee, J. chem. Phys. 39 (1963) 910.
[6.32r] Huffmann, R. E., Y. Tanaka and J. C. Larrabee, Proc. 6. internat. Conf. Ion. Phenom. Gases, Paris 1963, Vol. 1, p. 145.

[6.32s] Hurle, I. R. and A. L. Russo, J. chem. Phys. **43** (1965) 4434.
[6.32t] Hurle, I. R., A. L. Russo and J. G. Hall, J. chem. Phys. **40** (1964) 2076.
[6.32u] Ivanova, A. V., Optika i Spektroskopiya **16** (1964) 925.
[6.32v] Jacobs, T. A. and R. D. Giedt, J. chem. Phys. **39** (1963) 749.
[6.32w] Jarmain, W. R. and R. W. Nicholls, Proc. Phys. Soc. Lond. **84** (1964) 417.
[6.32x] Johnston, P. D., J. Phys. **D 1** (1968) 479.
[6.32y] Jonston, R. R., B. H. Armstrong and O. R. Platas, J. quant. Spectrosc. Radiat. Transfer 5 (1965) 49.
[6.33] Jutsum, P. H., Proc. Phys. Soc. Lond. **A 67** (1954) 190.
[6.33a] Kadyshevich, A. Ye., Usp. Fiz. Nauk **76** (1962) 683.
[6.33b] Kandyba, V. V., Acta IMEKO T. 2 (1964) 359.
[6.33c] Khare, S. P., Proc. 5. internat. Conf. Phys. Electronic Atomic Collisions, Nauka, Leningrad 1967, p. 625.
[6.33d] Khare, S. P., Phys. Rev. **173** (1968) 43.
[6.33e] Kivel, B., J. quant. Spectrosc. Radiat. Transfer 2 (1962) 509.
[6.33f] Klein, L., Appl. Optics 7 (1968) 677.
[6.33g] Kobsev, G. A., G. E. Norman and K. I. Seryakov, Teplofizika 4 (1966) 473.
[6.33h] Konkov, A. A., A. P. Ryazin and V. S. Rudnev, J. quant. Spectrosc. Radiat. Transfer 7 (1967) 339.
[6.34] Kopec, U., Ann. Physik **12** (1963) 209.
[6.34a] Kreye, W. C., Proc. 8. internat. Conf. Phenom. Ionized Gases, Contributed Pap., IAEA Vienna 1967, p. 100.
[6.35] Kudryavtsev, Y. M., N. N. Sobolev, L. N. Tunitski and F. S. Faisullov, in: Fizicheskie problemy spektroskopii Vol. 1, Izd. Akad. Nauk SSSR, Moscow 1962, p. 111.
[6.35a] Kuhn, G. and R. S. Tankin, J. quant. Spectrosc. Radiat. Transfer 8 (1968) 1281.
[6.35b] Kuyatt, C. E. and J. A. Simpson, in: Atomic Collision Processes, North Holland Publ. Co., Amsterdam 1964, p. 191.
[6.35c] Lapworth, K. C., in: The high temperature aspects of hypersonic flow, Pergamon Press Oxford, London, New York, Paris 1964, p. 255.
[6.35d] Lee, P. and G. L. Weissler, Astrophys. J. **115** (1952) 570.
[6.36] Lee, P. and G. L. Weissler, Proc. Roy. Soc. Lond. **A 219** (1953) 71.
[6.37] Lee, P. and G. L. Weissler, Phys. Rev. **99** (1955) 540.
[6.38] Losev, S. A. and N. A. Generalov, in: Fizicheskie problemy spektroskopii Vol. 1, Izd. Akad. Nauk SSSR Moskau 1962, p. 114.
[6.38a] Lowry, J. F., D. H. Tomboulian and D. L. Ederer, Phys. Rev. **137** (1965) A 1054.
[6.38b] Macek, J., Proc. Phys. Soc. Lond. **92** (1967) 365.
[6.38c] McGuire, E. J., Phys. Rev. **161** (1967) 51.
[6.38d] McGuire, E. J., Phys. Rev. **175** (1968) 20.
[6.38e] Madden, R. P., D. L. Ederer and K. Codling, Phys. Rev. **177** (1969) 136.
[6.38f] Marr, G. V., Photoionization processes in gases, Academic Press, New York, London 1967.
[6.39] Makayama, T., M. Y. Katamura and K. Watanabe, J. chem. Phys. **30** (1959) 1180.
[6.40] Maleyev, Y. N. and D. S. Yermakov, Optika i Spektroskopiya **13** (1962) 598.
[6.41] Mandelshtam, S. L. and N. K. Suchodrev, Izv. Akad. Nauk SSSR, Ser. Fiz. **19** (1955) 11.
[6.42] Marr, G. V., Proc. Roy. Soc. **A 224** (1954) 83.
[6.42a] Marr, G. V., Proc. Phys. Soc. **A 67** (1954) 196.
[6.42b] Marr, G. V., Photoionization Processes in Gases, Academic Press New York, London 1967.

[6.42c] MATSUNAGA, F. M., R. S. JACKSON and K. W. WATANABE, J. quant. Spectrosc. Radiat. Transfer **5** (1965) 329.

[6.42d] METZGER, P. H. and G. R. COOK, J. chem. Phys. **41** (1964) 642.

[6.42e] METZGER, P. H. and G. R. COOK, J. Opt. Soc. Amer. **55** (1965) 516.

[6.42f] MITCHEL, A. and M. ZEMANSKY, Resonance Radiation and excited Atoms, Cambridge Univ. Press, Cambridge 1934.

[6.42g] MNATSAKANYAN, A. KH., Teplofizika **6** (1968) 236.

[6.43] MOHLER, F. L. and C. BOECKNER, J. Res. Nat. Bur. Stand. **3** (1929) 303.

[6.43a] MOORES, D. L., Proc. Phys. Soc. Lond. 88 (1966) 843.

[6.43b] MOSKVIN, YU. V., Teplofizika **3** (1965) 821.

[6.43c] MOTZ, R. O., Optical determination of barium plasma density. Phys. Fluids **10** (1967) 2282.

[6.43d] MUNTZ, E. P., Phys. Fluids **5** (1962) 80.

[6.43e] MUNTZ, E. P. and D. J. MARSDEN, in: Rarefied Gas Dynamics, Academic Press Inc., New York 1963, Vol. II, p. 495.

[6.43f] NORCROSS, D. W. and P. M. STONE, J. quant. Spectrosc. Radiat. Transfer **6** (1966) 277.

[6.43g] OSTROUMENKO, P. P., Zh. prikl. Spektrosk. 7 (1967) 752.

[6.43h] PATCH, R. W., J. quant. Spectrosc. Radiat. Transfer **5** (1965) 137.

[6.44] PAVLOVSKAYA, Y. N. and V. K. PROKOFYEV, in: Fizicheskie problemy spektroskopii, Vol. 1, Izd. Akad. Nauk SSSR, Moscow 1962, p. 138.

[6.44a] PEACH, G., Mem. Roy. Astron. Soc. **71** (1967) 13.

[6.45] PERY-THORNE, A. and W. R. S. GARTON, Proc. Phys. Soc. Lond. **76** (1960) 833.

[6.45a] PHILLIPS, M., Phys. Rev. **39** (1932) 905.

[6.46] PINGEL, H. and E. W. SIEKMEIER, Z. Physik **166** (1962) 176.

[6.46a] PRADERIE, F., Special Rep. Smithonian Instn. Astrophys. Obs. No. 167 (1964) 52.

[6.46b] PREOBRASHENSKI, N. G., Izvest. VUZ Fiz., Tomsk, No. 6 (1959) 31.

[6.46c] PREOBRASHENSKI, N. G., Optika i Spektroskopiya 7 (1959) 273.

[6.46d] RAUTENBERG, T. H. and P. D. JOHNSON, Appl. Optics **3** (1964) 487.

[6.46e] RIEDMÜLLER, W., Messung der Elektronentemperatur in einem Argon-Kalium-Plasma. Inst. Plasmaphys. Garching, Report 3/31, Okt. 1965.

[6.46f] ROBBEN, F. and L. TALBOT, Phys. Fluids **9** (1966) 633.

[6.46g] ROBINSON, E. J. and S. GELTMAN, Phys. Rev. **153** (1967) 4.

[4.46h] ROSS, K. J. and G. V. MARR, Proc. Phys. Soc. Lond. **85** (1965) 193.

[6.47] RÖSSLER, F., in: Temperature, its Measurement and control in Science and Industry, Vol. 3, 2, Reinhold Publ. Co., New York 1962, p. 509.

[6.47a] ROTHE, D. E., Phys. Fluids **9** (1966) 1643.

[6.47b] RUSTGI, O. P., J. Opt. Soc. Amer. **54** (1964) 464.

[6.47c] SAMSON, J. A. R., J. Opt. Soc. Amer. **54** (1964) 420.

[6.47d] SAMSON, J. A. R., J. Opt. Soc. Amer. **54** (1964) 842.

[6.47e] SAMSON, J. A. R., J. Opt. Soc. Amer. **54** (1964) 876.

[6.47f] SAMSON, J. A. R., J. Opt. Soc. Amer. **55** (1965) 935.

[6.47g] SASAKI, Y., Jap. J. appl. Phys. **5** (1966) 439.

[6.47h] SCHMIDT, K., Proc. 6. internat. Conf. Ion. Phenom. Gases, Paris 1963, Vol. 3, p. 323.

[6.47i] SCHOTT, G. L., J. chem. Phys. **32** (1966) 710.

[6.47j] SCHULZ, G. and H. J. GUTJAHR, Z. angew. Phys. **25** (1968) 269.

[6.47k] SCHUMACHER, B. W. and E. O. GADAMER, Canad. J. Phys. **36** (1958) 659.

[6.47l] SCHWENN, R., Rep. Inst. Plasmaphys., Garching near Munich, IPP 3/42, Okt. 1966.

[6.48] SEATON, M. J., Proc. Roy. Soc. Lond. **A 208** (1951) 408.

[6.49] SEATON, M. J., Proc. Roy. Soc. Lond. **A 208** (1951) 418.

[6.50] Seaton, M. J., Proc. Phys. Soc., Lond. **A 67** (1954) 927.
[6.50a] Sewell, K. G., Phys. Rev. **138** (1965) A 418.
[6.50b] Sewell, K. G., J. Opt. Soc. Amer. **57** (1967) 1058.
[6.50c] Shevelko, V. P., Report T-18619, phys. Instit. Akad. Nauk USSR, Moscow, 4. April 1970.
[6.51] Simon, L., Optik **19** (1962) 621.
[6.52] Sobolev, N. N., V. N. Alyamovski, A. F. Dronov, F. S. Faisullov, V. F. Kitayeva, Y. M. Kudryavtsev, and A. G. Sviridov, in: Proc. 5. internat. Conf. Ionization Phenomena in Gases, Vol. 2, North Holland Publ. Co., Amsterdam 1962, p. 2122.
[6.53] Sobolev, N. N., A. V. Potatov, V. F. Kitaeva, F. S. Faisullov, V. N. Alyamovski, E. T. Antropov and I. L. Isaev, Izv. Akad. Nauk SSSR Ser. Fiz. **22** (1958) 730.
[6.54] Soshnikov, V. N., Optika i Spektroskopiya **6** (1959) 315.
[6.55] Startsev, G. P. and M. S. Frish, Iz. Akad. Nauk SSSR, Ser. Fiz. **26** (1962) 927.
[6.56] Stewart, A. L., Proc. Phys. Soc. Lond. **A 67** (1954) 917.
[6.56a] Stewart, A. L. and T. G. Webb, Proc. Phys. Soc. Lond. **82** (1963) 532.
[6.57] Strong, H. M. and F. P. Bundy, J. appl. Phys. **25** (1954) 1527.
[6.58] Sun, H. and G. L. Weissler, J. chem. Phys. **23** (1955) 1372.
[6.59] Sun, H. and G. L. Weissler, J. chem. Phys. **23** (1955) 1625.
[6.60] Tanaka, Y., A. S. Jursa and F. Le Blanc, J. chem. Phys. **26** (1957) 862.
[6.61] Tappe, W. and J. van Calker, Z. anal. Chem. **198** (1963) 13.
[6.61a] Torvén, S., Ark. Fys. **29** (1965) 533.
[6.61b] Tourin, R. H., J. Heat Transfer No. 5 (1962) 164.
[6.62] Tourin, R. H., Temperature, its Measurement and Control in Science and Industry, Bd. 3, 2, Reinhold Publ. Co., New York 1962, p. 455.
[6.62a] Tourin, R. H. and B. Krakow, Appl. Optics **4** (1965) 237.
[6.63] Trumpy, B., Z. Phys. **71** (1931) 720.
[6.64] Tsuchiya, S., Aeronaut. Res. Inst., Univ. Tokyo, Rep. **27** (1961) 157 (Report No. 366).
[6.65] Tunstead, J., Proc. phys. Soc. Lond. **A 66** (1953) 304.
[6.65a] Unsöld, A., Physik der Sternatmosphären, Springer Verlag, Berlin-Göttingen-Heidelberg 1955.
[6.66] Vinti, J. P., Phys. Rev. **44** (1933) 524.
[6.66a] Vitchas, A. F., V. S. Golubev and V. K. Zhibritski, Electricity from MHD, IAEA, Vienna 1966, Vol. 1, p. 63.
[6.66b] Volk-Levanovich, M. V., Inzhenerno-fiz. Zhurnal (J. Engng. Phys.) **5**, No. 8 (1962) 87.
[6.66c] Volk-Levanovich, M. V., Inzhenerno-fiz. Zhurnal (J. Engng. Phys.) **7**, No. 10 (1964) 98.
[6.67] Wainfan, N., W. C. Walker and G. L. Weissler, J. appl. Phys. **24** (1953) 1318.
[6.68] Wainfan, N., W. C. Walker and G. L. Weissler, Phys. Rev. **99** (1955) 542.
[6.69] Walker, W. C. and G. L. Weissler, J. chem. Phys. **23** (1955) 1540.
[6.70] Walker, W. C. and G. L. Weissler, J. chem. Phys. **23** (1955) 1547.
[6.71] Walker, W. C. and G. L. Weissler, J. chem. Phys. **23** (1955) 1962.
[6.72] Watanabe, K., J. chem. Phys. **22** (1954) 1564.
[6.72a] Weissler, G. L., J. quant. Spectrosc. Radiat. Transfer **2** (1962) 383.
[6.73] Weissler, G. L. and P. Lee, J. opt. Soc. Amer. **42** (1952) 200.
[6.74] Weissler, G. L., P. Lee and E. I. Mohr, J. opt. Soc. Amer. **42** (1952) 84.
[6.74a] Werner, A., Aust. J. Phys. **7** (1954) 25.
[6.75] Wheeler, J. A., Phys. Rev. **43** (1933) 258.
[6.76] Wilesov, F. I., Uspekhi fiz. Nauk **81** (1963) 667.

[6.77] WINKLER, E. M., Physical measurements in Gas Dynamics and Combustion, Part 1 Section 1.5, Oxford Univ. Press, Oxford 1955.
[6.78] ZAREM, A. M., F. R. MARSHALL and S. M. HAUSER, Rev. sci. Instrum. **29** (1958) 1041.
[6.78a] ZATSENIN, YU. A., E. G. POPOV and M. A. ZIKULIN, Zh. eksp. teor. Fiz. **54** (1968) 112.
[6.78b] ZIEGLER, C. A., L. L. BIRD, K. H. OLSON, J. A. HULL and J. A. MORREAL, Rev. sci. Instrum. **35** (1964) 450.
[6.78c] ZWICKER, H. and U. SCHUMACHER, Proc. 6. internat. Conf. Ion. Phenom. Gases, Paris 1963, Vol. 2, p. 593.
[7.1] ALLEN, R. A., J. quant. Spectrosc. Radiat. Transfer **5** (1965) 511.
[7.1a] ANTROPOV, E. T., A. P. DRONOV and N. N. SOBOLEV, Optika i Spektroskopiya **17** (1964) 654.
[7.1b] ANTROPOV, E. T., A. P. DRONOV and N. N. SOBOLEV, Doklady Akad. Nauk SSSR **153** (1963) 67.
[7.1c] ANTROPOV, E. T., N. N. SOBOLEV and V. P. CHEREMISINOV, Optika i Spektroskopiya **16** (1964) 208.
[7.1d] ARNOLD, J. O., J. quant. Spectrosc. Radiat. Transfer **8** (1968) 1781.
[7.1e] ARUTRE, A. M., R. A. B. BOND and J. HYSLOP, Proc. Phys. Soc. Lond. **A 70** (1957) 617.
[7.1f] ARTURE, A. M. and J. HYSLOP, Proc. Phys. Soc. Lond. **A 70** (1957) 489.
[7.1g] ASHKENAS, H., Phys. Fluids **10** (1967) 2509.
[7.1h] BATES, D. R., J. chem. Phys. **19** (1951) 1122.
[7.1i] BATES, D. R., R. T. S. DARLING, S. C. HAWE and A. L. STEWART, Proc. Phys. Soc. Lond. **A 66** (1953) 1124.
[7.1j] BATES, D. R., R. T. S. DARLING, S. C. HAWE and A. L. STEWART, Proc. Phys. Soc. Lond. **A 67** (1954) 533.
[7.1k] BENNETT, R. G. and F. W. DALBY, J. chem. Phys. **31** (1959) 434.
[7.1l] BENNETT, R. G. and E. W. DALBY, J. chem. Phys. **32** (1960) 1111.
[7.1m] BENNETT, R. G. and E. W. DALBY, J. chem. Phys. **32** (1960) 1716.
[7.2] BENNETT, R. G. and F. W. DALBY, J. chem. Phys. **36** (1962) 399.
[7.2a] BENNETT, R. C. and F. W. DALBY, J. chem. Phys. **40** (1964) 1414.
[7.3] BETHKE, G. W., J. chem. Phys. **31** (1959) 662.
[7.4] BETHKE, G. W., J. chem. Phys. **31** (1959) 669.
[7.5] BRINKMAN, H., Diss. Utrecht 1937.
[7.5a] BROIDA, H. P., J. chem. Phys. **21** (1953) 340.
[7.5b] BROIDA, H. P., J. chem. Phys. **21** (1953) 1165.
[7.5c] CARLETON, N. P. and O. OLDENBERG, J. chem. Phys. **36** (1962) 3460.
[7.5d] CARLETON, N. P. and C. PAPALIOLIOS, J. quant. Spectrosc. Radiat. Transfer **2** (1962) 241.
[7.5e] CHILDS, D. R., J. quant. Spectrosc. Radiat. Transfer **4** (1964) 283.
[7.5f] CHING, B. K., G. R. COOK and R. A. BECKER, J. quant. Spectrosc. Radiat. Transfer **7** (1967) 323.
[7.6] CHOCHLOV, M. S., Zh. eksp. teor. Fiz. **29** (1955) 645.
[7.6a] COULSON, C. and G. LESTER, Trans. Faraday Soc. **51** (1955) 1605.
[7.6b] CUNIO, B. E. and R. E. W. JANSSON, J. quant. Spectrosc. Radiat. Transfer **8** (1968) 1763.
[7.6c] CRISTESCU, G. D. and R. GRIGOROVIOI, Bull. Soc. Roum. Phys. **42** (1941) 37.
[7.6d] DALGARNO, A., N. LYNN and E. J. A. WILLIAMS, Proc. Phys. Soc. Lond. **A 69** (1956) 610.
[7.6e] DAVIES, D., Trans. Faraday Soc. **54** (1958) 1429.
[7.6f] DESAI, D. D., K. S. KORGACKAR and B. B. LAUD, Proc. Phys. Soc. Lond. **A 65** (1952) 228.

[7.6g] DIEKE, G. H. and H. M. CROSSWHITE, J. quant. Spectrosc. Radiat. Transfer 2 (1962) 97.

[7.6h] DIXON, R. N. and R. W. NICHOLLS, Canad. J. Phys. **36** (1958) 127.

[7.6i] DRAKE, G. W. F., D. C. TYTE and R. W. NICHOLLS, J. quant. Spectrosc. Radiat. Transfer 7 (1967) 639.

[7.6j] DRONOV, A. P., N. N. SOBOLEV and F. S. FAIZULLOV, Optika i Spektroskopiya **21** (1966) 267, 538.

[7.6k] DRONOV, A. P., N. N. SOBOLEV, F. S. FAIZULLOV and V. A. BOIKO, Optika i Spektroskopiya **21** (1966) 727.

[7.6l] DUFFENDACK, O. S. and K.-T. CHAO, Phys. Rev. **56** (1939) 176.

[7.6m] DUFFENDACK, O. S., R. REVANE and A. S. ROY, Phys. Rev. **45** (1934) 807.

[7.6n] DYNE, P. J., J. chem. Phys. **28** (1958) 999.

[7.6o] EATON, D. R., Canad. J. Phys. **38** (1960) 390.

[7.6p] ECKERT, H. U., F. L. KELLY and H. N. OLSEN, J. appl. Phys. **39** (1968) 1846.

[7.6q] EGOROV, V. N., L. N. TUNIZKI and E. M. CHERKASOV, Zh. prikl. Spektrosk. 8 (1968) 479.

[7.7] ELLIOTT, A., Proc. Phys. Soc. Lond. **45** (1933) 627.

[7.7a] FAIRBAIRN, A. R., AIAA-J. **2** (1964) 1004.

[7.7b] FAIRBAIRN, A. R., J. quant. Spectrosc. Radiat. Transfer **6** (1966) 325.

[7.7c] DE FARMER, A. J., W. FABIAN, B. R. LEWIS, K. H. LOKAN and G. N. HADDAD, J. quant. Spectrosc. Radiat. Transfer **8** (1968) 1739.

[7.7d] FELENBOK, P., Ann. Astrophys. **26** (1963) 393.

[7.7e] FELENBOK, P., Compt. Rend. Acad. Sci. **256** (1963) 2334.

[7.7f] FELENBOK, P., Proc. Phys. Soc. **86** (1965) 676.

[7.7g] FINK, E. and K. H. WELGE, Z. Naturforsch. **19a** (1964) 1193.

[7.7h] FLINN, D. J., R. J. SPINDLER, S. FIFER and M. KELLY, J. quant. Spectrosc. Radiat. Transfer **4** (1964) 271.

[7.7i] FLINN, D. J., R. J. SPINDLER, S. FIFER and M. KELLY, J. quant. Spectrosc. Radiat. Transfer **4** (1964) 271.

[7.8] FLOYD, A. L. and R. B. KING, Proc. Conf. Stellar Atmospheres. Indiana Univ. (1955) 44.

[7.9] FLOYD, A. L. and R. B. KING, J. opt. Soc. Amer. **45** (1955) 249.

[7.9a] FRASER, P. A., Proc. Phys. Soc. Lond. **67** (1954) 939.

[7.9b] FRASER, P. A. and W. R. JARMAIN, Proc. Phys. Soc. Lond. **66** (1953) 1145.

[7.9c] FRASER, P. A., W. R. JARMAIN and R. W. NICHOLLS, Astrophys. J. **119** (1954) 286.

[7.9d] FRASER, P. A., W. R. JARMAIN and R. W. NICHOLLS, Astrophys. J. **122** (1955) 55.

[7.9e] FUMIO, K., Publ. Astron. Soc. Japan **12** (1960) 420.

[7.9f] DE GALAN, L. and J. D. WINEFORDNER, J. quant. Spectrosc. Radiat. Transfer 7 (1967) 703.

[7.9g] GIPPIUS, E. F., E. M. KUDRYAVTSEV, A. N. PECHENOV and N. N. SOBOLEV, Teplofizika **5** (1967) 32.

[7.9h] GIPPIUS, E. F., E. M. KUDRYAVTSEV, A. N. PECHENOV, N. N. SOBOLEV and V. P. FOKEYEV, Teplofizika **2** (1964) 181.

[7.9i] GOLDEN, D. M., F. P. DEL GRECO and F. KAUFMAN, J. chem. Phys. **39** (1963) 3034.

[7.9j] GRANDMONTAGNE, J. Phys. Radium **22** (1961) 842.

[7.9k] GRANDMONTAGNE, R., J. d'INCAN and J. JANIK, J. Phys. Radium **20** (1959) 559.

[7.9l] GUPTA, B. K., Physica **26** (1960) 335.

[7.9m] HALMANN, M. and T. LAULICHT, J. chem. Phys. **42** (1965) 137.

[7.9n] HALMANN, M. and T. LAULICHT, J. chem. Phys. **43** (1965) 438.

[7.9o] HALMANN, M. and T. LAULICHT, J. chem. Phys. **43** (1965) 1503.

[7.9p] HALMANN, M. and I. LAULICHT, J. quant. Spectrosc. Radiat. Transfer 8 (1968) 935.

[7.9q] HAMADA, T., Proc. K. Akad. Wetensch. Amsterdam **39** (1936) 50.

[7.9r] HARRINGTON, J. A., A. MODICA and D. R. LIBBY, J. chem. Phys. **44** (1966) 3380.

[7.9s] HARRINGTON, J. A., A. P. MODICA and D. R. LIBBY, J. quant. Spectrosc. Radiat. Transfer 6 (1966) 799.

[7.10] HÉBERT, G. R. and R. W. NICHOLLS, Proc. Phys. Soc. Lond. 78 (1961) 1024.

[7.10a] HÉBERT, G. R. and D. C. TYTE, Proc. Phys. Soc. Lond. **83** (1964) 629.

[7.11] HERZBERG, G., Spectra of diatomic Molecules, D. van Nostrand Co., Inc., Princeton, Toronto, London, New York 1961.

[7.11a] HUGHES, R. F., J. Opt. Soc. Amer. **50** (1960) 740.

[7.11b] HURLE, I. R., in: Nizkotemperaturnaya Plazma, Mir, Moscow 1967, p. 168.

[7.11c] HURLE, I. R. and A. L. RUSSO, J. chem. Phys. **43** (1965) 4434.

[7.11d] ISAACSON, L., E. P. MARRAM and T. WENTINK, J. quant. Spectrosc. Radiat. Transfer 7 (1967) 691.

[7.11e] ISHIGURU, E., K. KAYAMA, M. KOTANI and Y. MIZUNO, J. Phys. Soc. Japan **12** (1957) 1355.

[7.11f] IVANOV, V. N., Optika i Spektroskopiya 7 (1959) 873.

[7.11g] JAIN, D. C., J. quant. Spectrosc. Radiat. Transfer **4** (1964) 427.

[7.11h] JAIN, D. C. and R. C. SAHNI, J. quant. Spectrosc. Radiat. Transfer **6** (1966) 705.

[7.11i] JAIN, D. C. and R. C. SAHNI, J. quant. Spectrosc. Radiat. Transfer **7** (1967) 475.

[7.11j] JAMES, H. M. and A. S. COOLIDGE, Phys. Rev. **55** (1939) 184.

[7.11k] JANIN, J., A. BOUVIER and J. CHEVALEYRE, Compt. Rend. Acad. Sci. **263** (1966) B291.

[7.11l] JANSSON, R. E. W., Proc. Phys. Soc. Lond. 87 (1966) 851.

[7.11m] JANSSON, R. E. W. and B. E. CUNIO, J. quant. Spectrosc. Radiat. Transfer 8 (1968) 1747.

[7.11n] JARMAIN, W. R., Canad. J. Phys. **41** (1963) 414.

[7.11o] JARMAIN, W. R., Canad. J. Phys. **41** (1963) 1926.

[7.11p] JARMAIN, W. R., R. EBISUZAKI and R. W. NICHOLLS, Canad. J. Phys. **38** (1960) 510.

[7.11q] JARMAIN, W. R. and P. A. FRASER, Proc. Phys. Soc. Lond. **66** (1953) 1153.

[7.11r] JARMAIN, W. R., P. A. FRASER and R. W. NICHOLLS, Astrophys. J. **118** (1953) 228.

[7.11s] JARMAIN, W. R., P. A. FRASER and R. W. NICHOLLS, Astrophys. J. **122** (1955) 55.

[7.11t] JARMAIN, W. R. and R. W. NICHOLLS, Canad. J. Phys. **32** (1954) 201.

[7.11u] JARMAIN, W. R. and R. W. NICHOLLS, Proc. Phys. Soc. **84** (1964) 417.

[7.11v] JECHT, U. and W. KESSLER, Z. Phys. **178** (1964) 133.

[7.11w] JEUNEHOMME, M., J. chem. Phys. **44** (1966) 4253.

[7.11x] JEUNEHOMME, M. and R. P. SCHWENKER, J. chem. Phys. **42** (1965) 2406.

[7.11y] JONES, A. V. and A. W. HARRISON, J. atm. terr. Phys. **6** (1955) 336.

[7.11z] JONES, A. V. and D. M. HUNTEN, Canad. J. Phys. **38** (1960) 458.

[7.12] JOSHI, K. C., V. D. P. SASTRI and S. PARTHASARATHI, J. quant. Spectrosc. Radiat. Transfer **6** (1966) 215.

[7.12a] JUST, T. and H. G. WAGNER, Z. Elektrochem. **61** (1957) 678.

[7.12b] KANE, W. R. and H. P. BROIDA, J. chem. Phys. **21** (1953) 347.

[7.12c] KARO, A. M., J. chem. Phys. **32** (1960) 907.

[7.12d] KECK, J. C., R. A. ALLEN and R. L. TAYLOR, J. quant. Spectrosc. Radiat. Transfer **3** (1963) 335.

[7.12e] KECK, J. C., J. C. CAMM and B. KIVEL, J. chem. Phys. **28** (1958) 723.

[7.12f] KECK, J. C., J. C. CAMM, B. KIVEL and T. WENTINK, Ann. Physics **7** (1959) **1**.

[7.12g] KEY, J., Meting van overgangswaarschijnlijkheden in het spectrum de alkalien. Thesis Utrecht 1937.
[7.13] KING, R. B., Astrophys. J. **108** (1948) 87.
[7.14] KING, A. S., Astrophys. J. **108** (1948) 429.
[7.14a] KISELEVSKI, L. I. and V. D. SHIMANOVICH, Optika i Spektroskopiya **24** (1968) 506.
[7.14b] KIVEL, B., H. MAYER and H. BETHE, Ann. Physics **2** (1957) 1.
[7.14c] KIVEL, B., H. MAYER and H. BETHE, Ann. Physics **2** (1957) 57.
[7.15] KOSTKOWSKI, H. and H. P. BROIDA, J. opt. Soc. Amer. **46** (1956) 246.
[7.15a] KREMPL, H., J. MAIERHOFER and H. MEINEL, Z. angew. Phys. **22** (1967) 171.
[7.16] KRINDACH, N. I., N. N. SOBOLEV und L. N. TUNITSKI, Optik u. Spektroskopie **15** (1963) 298.
[7.17] KROEPELIN, H., Proc. 5. internat. Conf. Ionization Phenomena in Gases, Vol. 2, North Holland Publ. Co., Amsterdam 1962, p. 1830.
[7.17a] KROEPELIN, H. and D. E. KIPPING, in: Progr. internat. Res. Thermodyn. and Transport Properties, Acad. Press N. Y., Lond. 1962, p. 649.
[7.18] KROEPELIN, H. and R. KUTHE, Z. Naturf. **16a** (1961) 1092.
[7.18a] KUDRYAVTSEV, E. M., E. F. GIPPIUS, A. N. PECHENOV and N. N. SOBOLEV, Teplofizika **1** (1963) 73 and 218; dito and S. S. DERBENEVA, Teplofizika **1** (1963) 376.
[7.18b] KUPRIYANOVA, E. B., V. N. KOLESNIKOV and N. N. SOBOLEV, J. quant. Spectrosc. Radiat. Transfer **9** (1969) 1025.
[7.18c] LAMBERT, D. L., J. quant. Spectrosc. Radiat. Transfer **8** (1968) 1265.
[7.19] LAPP, M., J. quant. Spectrosc. Radiat. Transfer **1** (1961) 30.
[7.19a] LAWRENCE, G. M., J. quant. Spectrosc. Radiat. Transfer **5** (1965) 359.
[7.19b] LEARNER, R. C. M., Proc. Roy. Soc. Lond. **A 269** (1962) 311.
[7.19c] LEARNER, R. C. M., Proc. Roy. Soc. Lond. **A 269** (1962) 327.
[7.20] LESKOV, L. W. and L. P. VASILYEVA, Izv. Akad. Nauk. SSSR. Ser. Fiz. **22** (1958) 696.
[7.20a] LEWIS, J. T., M. R. C. MCCARROLL and B. L. MOISEIWITSCH, Proc. Phys. Soc. Lond. **A 68** (1955) 565.
[7.20b] LOCHTE-HOLTGREVEN, W. and H. MAECKER, Z. Physik **105** (1937) 1.
[7.20c] MCGREGOR, A. T., R. W. NICHOLLS and W. R. JARMAIN, Canad. J. Phys. **39** (1961) 1215.
[7.20d] MAIN, R. P., D. J. CARLSON and R. A. DUPUIS, J. quant. Spectrosc. Radiat. Transfer **7** (1967) 805.
[7.20e] MAIN, R. P., A. L. MORSELL and W. J. HOCKER, J. quant. Spectrosc. Radiat. Transfer **8** (1968) 1527.
[7.20f] MAIN, R. P. and A. SCHADE, J. quant. Spectrosc. Radiat. Transfer **9** (1969) 713.
[7.20g] MARR, G. V., Canad. J. Phys. **42** (1964) 382.
[7.20h] MARR, G. V., Proc. Phys. Soc. Lond. **83** (1964) 293.
[7.20i] MARRONS, P. V., Phys. Fluids **10** (1967) 521.
[7.20j] MASAHIRO, T. and A. TOMIO, Technol. Repts. Kansi Univ., No. 8 (1966) 23.
[7.20k] MASSEY, J. T. and S. M. CANNEN, J. appl. Phys. **36** (1965) 361.
[7.20l] MENTALL, J. E. and R. W. NICHOLLS, Proc. Phys. Soc. Lond. **86** (1965) 873.
[7.20m] MEYER, V. D., A. SKERBELE and E. N. LASSETTRE, J. chem. Phys. **43** (1965) 805.
[8.20n] MIYAMA, H. and T. TAKEYAMA, J. chem. Phys. **39** (1963) 851.
[7.21] MOLLWO, L., Ann. Phys. **2** (1958) 97.
[7.21a] MOORE, J. H. and D. W. ROBINSON, J. chem. Phys. **48** (1968) 4870.
[7.21b] MORTON, E. W., J. chem. Phys. **41** (1964) 930.
[7.21c] MORTON, E. W., J. Res. Nat. Bur. Stand. **68** (1964) 631.
[7.21d] MORTON, E. W. and F. MORRIS, J. chem. Phys. **35** (1961) 1902.

[7.21e] MOSKVIN, YU. V. and I. T. YAKUBOV, Izvest. VUZ Fiz., Tomsk, No. 4 (1961) 173.

[7.21f] MULLIKEN, R. and C. RICKE, Rep. Progr. Phys. 8 (1941) 1231.

[7.21g] MUNTZ, E. P., Phys. Fluids 5 (1962) 80.

[7.21h] MURTHY, N. S., Nature **190** (1961) 430.

[7.21i] MURTHY, N. S., Current Sci. **31** (1962) 414.

[7.21j] MURTHY, N. S. and B. N. MURTHY, Proc. Phys. Soc. Lond. **90** (1967) 881.

[7.21k] MURTHY, N. S. and S. NAGARAJ, Proc. Phys. Soc. Lond. **84** (1964) 827.

[7.21l] MURTHY, N. S., T. K. S. SETTY and M. K. V. SUMATHI, Indian J. Phys. **38** (1964) 428.

[7.21m] NAZIMOVA, N. A. and V. A. SOKOLOV, Izv. VUZ Fiz., Tomsk, No. 2 (1961) 143.

[7.21n] NAZIMOVA, N. A., Zh. prikl. Spektrosk. 7 (1967) 169.

[7.22] NEUMANN, W., Beitr. Plasmaphysik **3** (1963) 12.

[7.23] NICHOLLS, R. W., Proc. phys. Soc., Lond. **A 69** (1956) 741.

[7.23a] NICHOLLS, R. W., Canad. J. Phys. **38** (1960) 1705.

[7.23b] NICHOLLS, R. W., Nature **186** (1960) 958.

[7.23c] NICHOLLS, R. W., J. Res. Nat. Bur. Stand. **A 65** (1961) 45.

[7.23d] NICHOLLS, R. W., J. Res. Nat. Bur. Stand. **A 65** (1961) 451.

[7.23e] NICHOLLS, R. W., Canad. J. Phys. **40** (1962) 523.

[7.23f] NICHOLLS, R. W., Canad. J. Phys. **40** (1962) 1772.

[7.23g] NICHOLLS, R. W., J. quant. Spectrosc. Radiat. Transfer **2** (1962) 433.

[7.23h] NICHOLLS, R. W., J. Res. Nat. Bur. Stand. **A 66** (1962) 227.

[7.23i] NICHOLLS, R. W., J. atm. terr. Phys. **25** (1963) 218.

[7.23j] NICHOLLS, R. W., J. chem. Phys. **38** (1963) 1029.

[7.23k] NICHOLLS, R. W., J. Res. Nat. Bur. Stand. **A 68** (1964) 75.

[7.23l] NICHOLLS, R. W., J. Res. Nat. Bur. Stand. **A 68** (1964) 535.

[7.23m] NICHOLLS, R. W., in: Supersonic flow, chemical processes and Radiative Transfer, Pergamon Press, New York 1964, p. 413.

[7.23n] NICHOLLS, R. W., Astrophys. J. **141** (1965) 819.

[7.23o] NICHOLLS, R. W., Canad. J. Phys. **43** (1965) 1390.

[7.23p] NICHOLLS. R. W., J. chem. Phys. **42** (1965) 804.

[7.23q] NICHOLLS, R. W., J. quant. Spectrosc. Radiat. Transfer **5** (1965) 647.

[7.23r] NICHOLLS, R. W., J. Res. Nat. Bur. Stand. **A 69** (1965) 369.

[7.23s] NICHOLLS, R. W., J. Res. Nat. Bur. Stand. **A 69** (1965) 397.

[7.23t] NICHOLLS, R. W., J. Physics **B 1** (1968) 1192.

[7.23u] NICHOLLS, R. W., P. A. FRASER and W. R. JARMAIN, Combustion and Flame **3** (1959) 13.

[7.23v] NICHOLLS, R. W., P. A. FRASER, W. R. JARMAIN and P. P. MCEACHRAN, Astrophys. J. **131** (1960) 399.

[7.24] NICHOLLS, R. W. and A. L. STEWART, Atomic and molecular processes. Academic Press, New York and London 1962, p. 47.

[7.24a] NICHOLLS, R. W. and D. C. TYTE, Proc. Phys. Soc. Lond. **91** (1967) 489.

[7.24b] NICOLS, L. L. and W. E. WILSON, Appl. Optics 7 (1968) 167.

[7.25] ORNSTEIN, L. S. and H. BRINKMAN, Kon. Akad. Wetensch. Amsterdam, Proc. **34** (1931) 33.

[7.26] ORNSTEIN, L. S. and H. BRINKMAN, Kon. Akad. Wetensch. Amsterdam, Proc. **34** (1931) 498.

[7.26a] ORNSTEIN, L. S. and W. R. VAN WIJK, Z. Physik **49** (1928) 315.

[7.27] ORTENBERG, F. S., Optika i Spektroskopiya **9** (1960) 151.

[7.27a] ORTENBERG, F. S., Ukrain. Fiz. Zh., Kiev **5** (1960) 645.

[7.27b] ORTENBERG, F. S., Diss. Moskau 1961.

[7.28] ORTENBERG, F. S., Optika i Spektroskopiya **16** (1964) 729.

[7.28a] ORTENBERG, F. S. and E. T. ANTROPOV, Usp. Fiz. Nauk **90** (1966) 237.

[7.28b] ORTENBERG, F. S., V. B. GLASKO and A. I. DMITRIEV, Astron. Zh. **40** (1964) 332.
[7.29] ORTENBERG, F. S. and N. A. NESTERKO, Optika i Spektroskopiya **10** (1961) 270.
[7.29a] ORY, H. A., Astrophys. J. **139** (1964) 346.
[7.29b] ORY, H. A., Astrophys. J. **139** (1964) 357.
[7.29c] ORY, H. A., J. chem. Phys. **40** (1964) 562.
[7.29d] ORY, H. A., A. P. GITTLEMAN and J. P. MADDOX, Astrophys. J. **139** (1964) 346.
[7.29e] ORY, H. A. and P. GITTLEMAN, Astrophys. J. **139** (1964) 357.
[7.29f] PARTHASARATHI, S., V. D. P. SASTRI and K. C. JOSHI, J. quant. Spectrosc. Radiat. Transfer **6** (1966) 903.
[7.29g] PATCH, R. W., J. chem. Phys. **41** (1964) 1881.
[7.29h] PEEK, J. M. and E. N. LASSETTRE, J. chem. Phys. **38** (1963) 2392.
[7.29i] PENNER, S. S., J. chem. Phys. **21** (1953) 686.
[7.30] PENNER, S. S., J. Opt. Soc. Amer. **50** (1960) 627.
[7.31] PHILLIPS, J. G., Astrophys. J. **119** (1954) 274.
[7.32] PHILLIPS, J. G., Astrophys. J. **125** (1957) 153.
[7.33] PHILLIPS, J. G., in: Optical spectrometric Measurements of high Temperatures. Univ. of Chicago Press, Chicago 1961, p. 217.
[7.33a] PILLOW, M. E., Proc. Phys. Soc. Lond. **A 62** (1949) 237.
[7.33b] PILLOW, M. E. and A. L. ROWLATT, Proc. Phys. Soc. Lond. **A 75** (1960) 162.
[7.33c] POLETTO, G. and M. RIGUTTI, Z. Astrophys. **60** (1964) 199; Erratum: Z. Astrophys. **60** (1965) 289.
[7.33d] POOTS, G., Proc. Phys. Soc. Lond. **A 66** (1953) 1181.
[7.33e] POPOVA, T. N., Izvest. VUZ Fiz., Tomsk, No. 1 (1959).
[7.33f] POPOVA, T. N., Izvest. VUZ Fiz., Tomsk, No. 1 (1960) 124.
[7.33g] POPOVA, T. N. and N. M. SASHENKO, Optika i Spektroskopiya **12** (1962) 791.
[7.33h] PRASAD, S. S., Indian J. Phys. **34** (1960) 584.
[7.33i] PRASAD, S. S., Indian J. Phys. **37** (1963) 547.
[7.33j] PRASAD, S. S., Proc. Phys. Soc. Lond. **A 82** (1963) 419.
[7.33k] PRASAD, Proc. Phys. Soc. Lond. **85** (1965) 810.
[7.33l] PRASAD, S. S. and K. PRASAD, Proc. Phys. Soc. Lond. **A 80** (1962) 311.
[7.33m] PRILESHAYEVA, N. A. and V. V. KOKHANENKO, Zh. prikl. Spektrosk. **7** (1967) 522.
[7.34] PROKOFYEV, V. K., D. B. GUREVICH, I. M. BELOUSOVA and Y. A. SNIGIREV, Optika i Spektroskopiya **7** (1959) 14; German Translation: Phys. Abh. SU **1** (1959) 518.
[7.34a] REICH, H. J. and H. SCHMORANZER, Z. Physik **186** (1965) 305.
[7.34b] REIS, V. H., J. quant. Spectrosc. Radiat. Transfer **4** (1964) 783.
[7.34c] REIS, V. H., J. quant. Spectrosc. Radiat. Transfer **5** (1965) 585.
[7.34d] RICH, J. C., Astrophys. J. **153** (1968) 327.
[7.34e] ROBBEN, F. and L. TALBOT, Phys. Fluids **9** (1966) 644.
[7.35] ROBINSON, D. and R. W. NICHOLLS, Proc. Phys. Soc. Lond. **71** (1958) 957.
[7.35a] ROBINSON, D. and R. W. NICHOLLS, Proc. Phys. Soc. **75** (1960) 817.
[7.35b] ROSA, R. J., Shock wave spectroscopy and Engineering Magnetohydrodynamics, Thesis Cornell Univ. 1956.
[7.35c] SANKARANARAYANAN, S., Physica **29** (1963) 1403.
[7.35d] SCHADEE, A., J. quant. Spectrosc. Radiat. Transfer **7** (1967) 169.
[7.35e] SCHWENKER, R. P., J. chem. Phys. **42** (1965) 1895.
[7.35f] SERGEJENKOVA, E. A., V. S. SAVIN and S. S. VASILYEV, Zh. fiz. Khim. **40** (1966) 2573.
[7.35g] SHARMA, A., J. quant. Spectrosc. Radiat. Transfer **7** (1967) 289.
[7.35h] SHEPHERD, G. G. and D. M. HUNTEN, J. atm. terr. Phys. **6** (1955) 328.
[7.35i] SHIRNOV, N. I. and L. A. KRONROD, Optika i Spektroskopiya **19** (1965) 871.
[7.35j] SHUMAKER, J. B., Proc. 6. internat. Conf. Ion. Phenom. Gases, Paris 1963, Vol. 3, p. 311.

[7.35k] Shumaker, B., J. quant. Spectrosc. Radiat. Transfer **9** (1969) 153.
[7.35l] Silverman, S. M. and E. N. Lassettre, J. chem. Phys. **41** (1964) 3727.
[7.35m] Singh, N. L. and D. C. Jain, Proc. Phys. Soc. Lond. **78** (1961) 399.
[7.35n] Singh, N. L. and D. C. Jain, Proc. Phys. Soc. Lond. **A 77** (1961) 817.
[7.35o] Singh, P. D. and Y. P. Srivastava, J. quant. Spectrosc. Radiat. Transfer **8** (1968) 1443.
[7.35p] Smith, I. W. M., J. quant. Spectrosc. Radiat. Transfer **8** (1968) 1437.
[7.35q] Smith, A. L. and J. W. Meriwether, J. chem. Phys. **42** (1965) 2984.
[7.36] Smit-Miessen, M. M. and J. L. Spier, Physica **9** (1942) 193.
[7.37] Spier, J. L. and J. A. Smit, Physica **9** (1942) 597.
[7.38] Spier, J. L. and M. M. Smit-Miessen, Physica **9** (1942) 422.
[7.38a] Spindler, R. J., J. quant. Spectrosc. Radiat. Transfer **5** (1965) 165.
[7.38b] Steinfeld, J. I., R. N. Zare, L. Sones, M. Lesk and W. Klemperer, J. chem. Phys. **42** (1965) 25.
[7.38c] Stevens, A. E. and H. I. S. Ferguson, Canad. J. Phys. **41** (1963) 240.
[7.38d] Sviridov, A. G., N. N. Sobolev and N. S. Novgorodov, J. quant. Spectrosc. Radiat. Transfer **6** (1966) 337.
[7.38e] Sviridov, A. G., N. N. Sobolev, M. S. Novgorodov and G. A. Arutyunova, J. quant. Spectrosc. Radiat. Transfer **6** (1966) 875.
[7.38f] Sviridov, A. G., N. N. Sobolev and V. M. Sutovski, J. quant. Spectrosc. Radiat. Transfer **5** (1965) 525.
[7.38g] Tatum, J. B., Astrophys. J. Suppl. **14** (1967) 21.
[7.38h] Tawde, N. R. and V. M. Korwar, Proc. Phys. Soc. Lond. **80** (1962) 794.
[7.38i] Tawde, N. R. and V. M. Korwar, Proc. nat. Inst. Sci., India **A 29** (1963) 325.
[7.38j] Tawde, N. R. and V. M. Korwar, Indian J. pure appl. Phys. **3** (1965) 198.
[7.38k] Tawde, N. R. and N. S. Murthy, Proc. Indian Acad. Sci. **A 51** (1960) 219.
[7.38l] Tawde, N. R. and N. S. Murthy, J. Opt. Soc. Amer. **53** (1963) 298.
[7.38m] Tawde, N. R. and B. S. Patil, Indian J. Phys. **33** (1959) 505.
[7.38n] Tawde, N. R., B. S. Patil, N. S. Murthy and M. R. Katti, Physica **23** (1957) 154.
[7.38o] Tawde, N. R. and M. J. Savadatti, Proc. Phys. Soc. Lond. **A 71** (1958) 514.
[7.38p] Tckenzie, R. D., Diss. Abstracts **25** (1964) 1970.
[7.39] Treanor, C. E. and W. H. Wurster, J. chem. Phys. **32** (1960) 758.
[7.39a] Tyte, D. C. and G. R. Hébert, Proc. Phys. Soc. **84** (1964) 830.
[7.39b] Vaidya, W. M., D. D. Desai and R. G. Bidaye, J. quant. Spectrosc. Radiat. Transfer **4** (1964) 353.
[7.39c] Vaidya, W. M. and K. C. Nagpal, Proc. Phys. Soc. Lond. **81** (1963) 682.
[7.39d] Wacks, M. E., J. chem. Phys. **41** (1964) 930.
[7.39e] Wallace, L. V. and R. W. Nicholls, J. atm. Terr. Phys. **7** (1955) 101.
[7.39f] Watson, R., J. quant. Spectrosc. Radiat. Transfer **2** (1962) 301.
[7.40] Watson, R., J. quant. Spectrosc. Radiat. Transfer **3** (1963) 255.
[7.41] Watson, R., J. quant. Spectrosc. Radiat. Transfer **4** (1964) 1.
[7.41a] Watson, R., Appl. Optics **5** (1965) 215.
[7.41b] Watson, R. and W. P. Ferguson, J. quant. Spectrosc. Radiat. Transfer **5** (1965) 595.
[7.41c] Watson, M. D., H. I. S. Ferguson and R. W. Nicholls, Canad. J. Phys. **41** (1963) 1405.
[7.42] Weber, D. and S. S. Penner, J. chem. Phys. **26** (1957) 860.
[7.42a] Wentink, T., L. Isaacson and J. Morreal, J. chem. Phys. **41** (1964) 278.
[7.42b] Wray, J., J. quant. Spectrosc. Radiat. Transfer **9** (1969) 255.
[7.42c] Wray, K. L. and T. J. Connolly, J. quant. Spectrosc. Radiat. Transfer **5** (1965) 111.

[7.42d] Wray, K. L. and T. J. Connolly, J. quant. Spectrosc. Radiat. Transfer 5 (1965) 633.
[7.43] Wurster, W. H., J. chem. Phys. 36 (1962) 2111.
[7.44] Wurster, W. H., J. quant. Spectrosc. Radiat. Transfer 3 (1963) 355.
[7.44a] Wyller, A., Mém. Soc. Sci. Liège 13 (1953) 97.
[7.44b] Wyller, A., Astrophys. J. 127 (1958) 763.
[7.44c] Yakubov, I. T., Optika i Spektroskopiya 9 (1960) 409.
[7.44d] Yerkovich, S. P., Optika i Spektroskopiya 6 (1959) 297.
[7.44e] Zare, R. N., Rep. Univ. Calif., Lawrence Radiation Lab., No. 10706 (1963) 84.
[7.44f] Zare, R. N., J. chem. Phys. 40 (1964) 1934.
[7.44g] Zare, R. N., E. O. Lasson and R. A. Berg, J. Mol. Spectrosc. 15 (1965) 117.
[7.44h] Zhirnov, N. I. and L. A. Kronrod, Optika i Spektroskopiya 19 (1965) 871.
[7.44i] Zinman, W. G. and S. I. Bogdan, J. chem. Phys. 40 (1964) 588.
[8.1] Ahmed, N., B. C. Gale and M. H. Key, J. Phys. **B** 2 (1969) 403.
[8.1a] Alcock, A. J., E. Panarella and S. A. Ramsden, Proc. 7. internat. Conf. Phenom. Ionized Gases, Belgrad 1965, Vol. 3, p. 224.
[8.1b] Allen, C. W., Astrophysical Quantities. Athlone Press, London 1955.
[8.2] Alpher, R. A. and D. R. White, Phys. Fluids 1 (1958) 452.
[8.2a] Alpher, R. A. and D. R. White, Phys. Fluids 2 (1959) 153.
[8.2b] Andelfinger, C., U. Igenbergs and W. von Jaskowsky, Proc. 6. internat. Congr. High-Speed Photography, Haarlem 1963, p. 238; Discussion p. 245, 642, 687.
[8.2c] Ascoli-Bartoli, U., in: Plasma Physics, IAEA, Vienna 1965, p. 287.
[8.3] Ascoli-Bartoli, U., A. de Angelis and S. Martellucci, Nuovo Cim. 18 (1960) 1116.
[8.4] Ascoli-Bartoli, U., A. de Angelis and S. Martellucci, Proc. 5. internat. Conf. Ionization Phenomena in Gases, Vol. 1, North Holland Publ. Co., Amsterdam 1962, p. 376.
[8.4a] Ascoli-Bartoli, U., J. Katzenstein and L. Lovisetto, Nature 204 (1964) 672.
[8.5] Ascoli-Bartoli, U. and S. Martellucci, Nuovo Cim. 27 (1963) 475.
[8.6] Ascoli-Bartoli, U. and F. Rasetti, Nuovo Cim. 13 (1959) 1296.
[8.7] Ascoli-Bartoli, U. and F. Rasetti, Proc. 4. internat. Conf. Ionization Phenomena in Gases, Vol. 2, North Holland Publ. Co., Amsterdam 1960, p. 839.
[8.7a] Ashby, D. E. T. F., D. F. Jephcott, A. Malein and F. A. Raynor, J. appl. Phys. 36 (1965) 29.
[8.7b] Baconnet, J. P., G. Cesari, A. Condeville and J. P. Watteau, Proc. 9. internat. Conf. Phenom. Ionized Gases, Bucharest 1969, p. 643.
[8.8] Barber, P. B., D. A. Swift and B. A. Tozer, Brit. J. appl. Phys. 14 (1963) 207.
[8.8a] Barnes, N. F., J. SMPTE 61 (1953) 487.
[8.8b] Bartoli, C. and A. DeAngelis, Proc. 7. internat. Conf. Phenom. Ionized Gases, Belgrad 1965, Vol. 3, p. 328.
[8.8c] Beach, A. D., J. sci. Instrum. 44 (1967) 690.
[8.8d] Besse, A. L. and J. G. Kelley, Rev. sci. Instrum. 37 (1966) 1497.
[8.8e] Besshanoshnikov, A. A., E. A. Voloshin, I. Ch. Kuchuberiya and N. V. Simonova, Zh. prikl. Spektrosk. 4 (1966) 458.
[8.8f] Bobin, J. L., J. C. Buges, P. Rouzaud and A. Terneaud, Proc. 9. internat. Conf. Phenom. Ionized Gases, Bucharest 1969, p. 638.
[8.8g] Brinkschulte, H. and H. Muntenbruch, Z. Naturforsch. 20a (1965) 196.
[8.9] Brown, S. C., G. Bekefi and R. E. Whitney, J. Opt. Soc. Amer. 53 (1963) 448.
[8.9a] Cary, B., Phys. Fluids 8 (1965) 26.
[8.9b] Chashchina, G. I., V. I. Gladushchak and E. Y. Shreider, Optika i Spektroskopiya 24 (1968) 1008.

[8.9c] CONSOLI, T., C. GORMEZANO and L. SLAMA, Compt. Rend. Acad. Sci., Paris 261 (1965) 86.

[8.9d] CONSOLI, T., C. GORMEZANO and L. SLAMA, Bull. inform. sci. et techn. Commissar. energie atom. No. 102 (1966) 81.

[8.9e] DANIEL, E. V., Optika i Spektroskopiya 25 (1968) 179.

[8.9f] DAVID, C., P. V. AVIZONIS, H. WEICHEL, C. BRUCE and K. D. PYATT, J. Quantum Electronics QE-2 (1966) 493.

[8.9g] DAVID, C. D. and H. WEICHEL, J. appl. Phys. **40** (1969) 674.

[8.9h] DAVIES, W. E. R. and S. A. RAMSDEN, Phys. Letters 8 (1964) 179.

[8.9i] DECROISETTE, M. and F. CABANNES, J. quant. Spectrosc. Radiat. Transfer 8 (1968) 619.

[8.9j] DEUCHARS, W. M., D. E. KIDD, J. IRVING and A. C. C. WARNOCK, Proc. 7. internat. Conf. Phenom. Ionized Gases, Belgrade 1965, Vol. 3, p. 218.

[8.9k] DING, A., G. HERZIGER and H. WEBER, Z. angew. Phys. **25** (1968) 168.

[8.9l] DMITRIYEV, A. K. and S. A. NIKIFOROVA, Zh. prikl. Spektrosk. **11** (1969) 545.

[8.10] DOUGAL, A. A., 6. internat. Conf. Ionization Phenomena in Gases, Paris 1963, Paper VIII, 22.

[8.10a] DOUGAL, A. A., J. P. CRAIG and R. F. GRIBBLE, Phys. Rev. Lett. **13** (1964) 156.

[8.10b] DUSHIN, L. A. and O. S. PAVLICHENKO, Issledovanie plazmy z pomoshchyu Lazerov, Atomizdat 1968.

[8.11] DYSON, J., Proc. 3. internat. Congr. on High-Speed Photography, Butterworths sci. Publ., London 1957, p. 285.

[8.11a] DYSON, J., J. Opt. Soc. Amer. **53** (1963) 690.

[8.12] DYSON, J., R. V. WILLIAMS and K. M. YOUNG, Nature, Lond. **195** (1962) 1291.

[8.12a] DYSON, J., R. V. WILLIAMS and K. M. YOUNG, Plasma Physics **6** (1964) 105.

[8.12b] ENGELHARDT, W. and W. KÖPPENDÖRFER, Rep. Inst. Plasmaphys., Garching near Munich, IPP 1/59, March 1967.

[8.12c] FALCONER, I. S., R. BENESCH and S. A. RAMSDEN, Phys. Lett. **14** (1965) 38.

[8.13] FOX, M. D. and F. J. WEINBERG, Brit. J. appl. Phhs. **11** (1960) 269.

[8.13a] FREUND, J., Rep. Inst. Plasmaphys., Garching near Munich, IPP 1/88, ca. 1968.

[8.14] FÜNFER, E., B. KRONAST and H. J. KUNZE, Phys. Letters **5** (1963) 125.

[8.14a] GERARDO, S., J. T. VERDEYEN and M. A. GUSINOW, J. appl. Phys. **36** (1965) 2146.

[8.14b] GERARDO, J. B., J. T. VERDEYEN and M. A. GUSINOW, J. appl. Phys. **36** (1965) 3526.

[8.14c] GERMAN, J. D. and P. W. SCHREIBER, Proc. 9. internat. Conf. Phenom. Ionized Gases, Bucharest 1969, p. 640.

[8.14d] GERRY, E. T. and R. M. PATRICK, Phys. Fluids 8 (1965) 208.

[8.14e] GRASSMANN, P. and H. WULFF, Proc. 6. internat. Conf. Ion. Phenom. Gases, Paris 1963, Vol. 4, p. 113.

[8.14f] GRIBBLE, R. F., E. M. LITTLE, R. L. MORRE and W. E. QUINN, Phys. Fluids **11** (1968) 1221.

[8.14g] HEARNE, K. R., H. EDELS and G. HOWARTH, Proc. 6. internat. Conf. Ion. Phenom. Gases, Paris 1963, Vol. 2, p. 253.

[8.14h] HEARNE, K. R. and N. KONJEVIĆ, Z. Physik **204** (1967) 443.

[8.14i] HEARNE, K. R. and N. KONJEVIĆ, Z. Physik **208** (1968) 65.

[8.14j] HEFFLINGER, L. O., R. F. WUERKER and R. E. BROOKS, J. appl. Phys. **37** (1966) 642.

[8.14k] HEISS, A., Interferometrische Messungen an Isar I, Rep. Inst. Plasmaphys., Garching near Munich, IPP 1/64, June 1967.

[8.14l] HOOPER, E. B. and G. BEKEFI, Proc. 7. internat. Conf. Phenom. Ionized Gases, Belgrade 1965, Vol. 3, p. 209.

[8.14m] HOOPER, E. B. and G. BEKEFI, J. appl. Phys. **37** (1966) 4083; Erratum: J. appl. Phys. **38** (1967) 1998.

[8.14n] HORA, H., Absorption coefficient of ruby laser radiation in fully ionized light elements. Rep. Inst. Plasmaphys. Garching near Munich, IPP-6/27, 1964.

[8.14o] HUG, W. F., D. EVANS, R. S. TANKIN and A. B. CAMBEL, Phys. Rev. **162** (1967) 117.

[8.14p] HUG, W. F., R. S. TANKIN and A. B. CAMBEL, Aerospace Res. Lab. Rep. ARL 67-0218, 1967 (USA).

[8.15] HUGHES, T. P., Nature, Lond. **194** (1962) 268.

[8.15a] IZAWA, Y., Y. NAKANISHI, M. YOKOYAMA and C. YAMANAKA, J. Phys. Soc. Japan **23** (1967) 1185.

[8.15b] JAHODA, F. C., R. A. JEFFRIES and G. SAWYER, Appl. Optics **6** (1967) 1407.

[8.15c] JAHODA, F. C., E. M. LITTLE, W. E. QUINN, F. L. RIBE and G. A. SAWYER, J. appl. Phys. **35** (1964) 2351.

[8.15d] JANNITTI, E. and G. TONDELLO, Proc. 8. internat. Conf. Phenom. Ionized Gases, Vienna 1967, p. 510.

[8.15e] JOHNSON, W. B., A. B. LARSEN and T. P. SOSNOWSKI, Proc. 7. internat. Conf. Phenom. Ionized Gases, Belgrade 1965, Vol. 3, p. 220.

[8.15f] JOHNSTONE, R. K. M. and W. SMITH, J. sci. Instrum. **42** (1965) 231.

[8.15g] KANTROWITZ, A. and R. L. TRIMPI, J. Aeronaut. Sci. **17** (1950) 311.

[8.15h] KAVEESHWAR, V. G., K. T. CHUNG and R. P. HURST, Phys. Rev. **172** (1968) 35.

[8.15i] KEGEL, W. H., Rep. Inst. Plasmaphys. Garching near Munich, IPP 6/9, Oktober 1963.

[8.15j] KIEFER, J. H. and R. W. LUTZ, Phys. Fluids **8** (1965) 1393.

[8.15k] KINDER, W., Optik **1** (1946) 413.

[8.15l] KINGSTON, A. E., J. Opt. Soc. Amer. **54** (1964) 1145.

[8.15m] KLEIN, A. F., Phys. Fluids **6** (1963) 310.

[8.15n] KOMISAROVA, I. I., G. V. OSTROVSKAYA and L. L. SHAPIRO, Zh. tekhn. Fiz. **38** (1968) 1369.

[8.15o] KOMISAROVA, I. I., G. V. OSTROVSKAYA, L. L. SHAPIRO and A. N. ZAIDEL, Proc. 9. internat. Conf. Phenom. Ionized Gases, Bucharest 1969, p. 641.

[8.15p] KOROBKIN, V. V. and A. A. MALUTIN, Zh. tekhn. Fiz. **38** (1968) 1095.

[8.15q] KUNZE, H.-J., Z. Naturforsch. **20a** (1965) 801.

[8.15r] KUNZE, H.-J., The laser as a tool for plasma diagnostics, in: Plasma Diagnostics, North Holland Publ. Co., Amsterdam 1968, p. 550.

[8.15s] KUNZE, H.-J., A. EBERHAGEN and E. FÜNFER, Phys. Letters **13** (1964) 38.

[8.15t] KUNZE, H.-J., E. FÜNFER, B. KRONAST and W. H. KEGEL, Phys. Letters **11** (1964) 42.

[8.15u] KUNZE, H.-J., E. FÜNFER and H. RÖHR, Phys. Letters **19** (1965) 11.

[8.15v] KUNZE, H.-J., A. H. GABRIEL and H. R. GRIEM, Phys. Rev. **165** (1968) 267.

[8.15w] KUNZE, H.-J., and W. D. JOHNSTON, Proc. 9. internat. Conf. Phenom. Ionized Gases, Bucharest 1969, p. 28.

[8.16] KÜPPER, F. P., Z. Naturforsch. **18a** (1963) 895.

[8.17] LADENBURG, R., DANIEL and D. BERSHADER, Physical Measurements in Gasdynamics and Combustion, Part I, Section 1, 3, Oxford Univ. Press, Oxford 1955.

[8.17a] LANDAU, L. D. and E. M. LIFSCHITZ, Lehrbuch der theoretischen Physik, Vol. 2, Klassische Feldtheorie, Akad. Verlag Berlin 1963, p. 238 (Translated from Russian).

[8.17b] LLEWELLYN-JONES, D. T., S. C. BROWN and G. BEKEFI, Proc. 6. internat. Conf. Ion. Phenom. Gases, Paris 1963, Vol. 4, p. 157.

[8.17c] LOVBERG, R. H., Phys. Fluids **8** (1965) 177.

[8.17d] MCLEAN, E. A. and S. A. RAMSDEN, Phys. Rev. **140** (1965) A 1122.

[8.17e] MAIGNAN, J., E. FABRE and P. VASSEUR, Phys. Letters **26 A** (1968) 135.

[8.18] MANDELSHTAM, S. L., Proc. Coll. spectrosc. internat. VI., Pergamon Press London 1957, p. 245.
[8.19] MANDELSHTAM, S. L., Spectrochim. Acta **15** (1959) 255.
[8.20] MEDFORD, R. D., A. L. T. POWELL, A. G. HUNT and J. K. WRIGHT, Proc. 5. internat. Conf. Ionization Phenomena in Gases, Vol. 2, North Holland Publ. Co., Amsterdam 1962, p. 2000.
[8.20a] METZ, S., W. H. BLOSS and J. A. SATKOWSKI, Proc. 9. internat. Conf. Phenom. Ionized Gases, Bucharest 1969, p. 642.
[8.20b] MILATZ, J. W. M., H. A. VREEDENBERG and J. W. BRASK, Physica **10** (1943) 433.
[8.20c] MIYOSHI, S., S. SHIOBARA and T. SHIMIZU, Jap. J. appl. Phys. 8 (1969) 383.
[8.20d] OERTEL, H., Shock tubes, Springer Verlag Wien-New York 1966, p. 390.
[8.20e] OETTINGER, P. E., AIAA-J. **6** (1968) 150.
[8.20f] PANARELLA, E., Proc. 8. internat. Conf. Phenom. Ionized Gases, IAEA Vienna 1967, p. 466.
[8.20g] PYATNITSKI, L. N., G. P. KHAUSMOVICH and V. V. KOROBKIN, Teplofizika **6** (1968) 751.
[8.20h] PATRICK, R. M., Phys. Fluids 8 (1965) 1985.
[8.20i] PAUL, J. W. M., G. C. GOLDENBAUM, A. LIYOSHI, L. S. HOLMES and R. A. HARDCASTLE, Nature **216** (1967) 363.
[8.20j] QUINN, J. M. P., Plasma Phys. 7 (1965) 113.
[8.20k] RAMSDEN, S. A. and W. E. R. DAVIES, Phys. Rev. Lett. **16** (1966) 303.
[8.20l] RAMSDEN, S. A. and E. A. MCLEAN, Nature, Lond. **194** (1962) 761.
[8.20m] RÖHR, H., Z. Physik **209** (1968) 295.
[8.20n] RÖHR, H. and G. DECKER, Z. Physik **214** (1968) 157.
[8.20o] RÖHR, H. and G. DECKER, Rep. Inst. Plasmaphys. Garching near Munich, IPP 1/78, March 1968.
[8.20p] ROZANOV, A. G., N. V. KHEBURKIN and N. M. SHVINDT, Optika i Spektroskopiya **21** (1966) 761.
[8.20q] ROZKWITALSKI, Z., Postepy Fiziyki **19** (1968) 383.
[8.20r] RUDDER, R. R. and D. R. BACH, J. Opt. Soc. Amer. 58 (1968) 1260.
[8.20s] RYE, B. J., J. W. WALLER, A. S. V. MCKENZIE and J. IRVING, Brit. J. appl. Phys. **16** (1965) 1404.
[8.20t] SALPETER, E. E., Phys. Rev. **120** (1960) 1528.
[8.20u] SALZMANN, H., On the use of Schlieren Optics for rotationally symmetric high density gradient Plasmas. Rep. Lab. Gas Ionizzati, Frascati, Roma, LGI 68/25, July 1968.
[8.21] SCHMITZ, G., Z. Physik **126** (1949) 1.
[8.21a] SCHWARZ, S. E., J. appl. Phys. **36** (1965) 1836.
[8.22] SHUCHTIN, A. M., Optika i Spektroskopiya **10** (1961) 436; German Translation: Phys. Abh. SU **5** (1961) 215.
[8.22a] DESILVA, A. W., D. E. EVANS and M. J. FORREST, Nature, Lond. **203** (1964) 1321.
[8.22b] DESILVA, A. W. and H.-J. KUNZE, J. appl. Phys. **39** (1968) 2458.
[8.23] SPERLING, J., Z. Physik **128** (1950) 269.
[8.23a] TATARINOV, V. V., Zh. prikl. Spektrosk. **9** (1968) 369.
[8.24] TEMPLE, E. B., J. Opt. Soc. Amer. **47** (1957) 91.
[8.24a] THOMPSON, E. and G. FIOCCO, Proc. 6. internat. Conf. Ion. Phenom. Gases, Paris 1963, Vol. 4, p. 111.
[8.24b] TUMAKAYEV, G. K. and V. R. LAZOVSKAYA, Aerofiz. Issled. Sverchsvuk. tekhenii, Nauka, Moscow-Leningrad 1967, p. 74.
[8.24c] ULMER, W., Proc. 7. internat. Conf. Phenom. Ionized Gases, Belgrade 1965, Vol. 3, p. 263.

[8.24d] ULMER, W. and W. THIELO, Z. angew. Phys. 19 (1965) 236.
[8.24e] VERDEYEN, J. T. and B. E. CHERRINGTON, Appl. Phys. Lett. 9 (1966) 360.
[8.24f] VASILYEV, L. A., A. G. GALANIN, I. V. YERSHOV and G. N. SUNTSOV, Pribory Tekhn. Eksp. No. 3 (1964) 195.
[8.25] WEINBERG, F. J., Optics of Flames, Butterworths & Co. Ltd., London 1963.
[8.26] WHITE, D. R., Phys. Fluids 4 (1961) 40.
[8.26a] WITKOWSKI, S., Rep. Inst. Plasmaphys. Garching near Munich, IPP 3/80, Oct. 1968.
[8.26b] WOLTER, H., Schlieren, Phasenkontrast und Lichtschnittverfahren, in: Handb. Phys. 24, Springer Verlag Berlin 1956.
[8.26c] WRIGHT, J. K., R. D. MEDFORD, A. G. HUNT and J. D. HERBERT, Proc. Phys. Soc. Lond. 78 (1961) 1439.
[8.26d] YASUDA, A., A. FUNAHSHI and S. TAKEDA, Japan. J. appl. Phys. 8 (1969) 497.
[8.26e] ZAIDEL, A. N., G. M. MALYSHEV and G. V. OSTROWSKI, Diagnostika Plazmy, Gosatomizdat, Moscow 1963, S. 31.
[8.26f] ZAIDEL, A. N., G. V. OSTROVSKAYA and YU. I. OSTROVSKII, Zh. tekhn. Fiz. 38 (1968) 1405.
[8.26g] ZICKERT, D. and H. SCHNEIDER, Proc. 8. internat. Conf. Phenom. Ionized Gases, IAEA Vienna 1967, p. 3.
[8.26h] ZIMMERMANN, W., Measurement of shock fronts in argon-filled membrane shocktubes made with a differential interferometer. Rep. Inst. Plasmaphys. Garching near Munich, IPP 3/57, Sept. 1967.
[9.1] ABRAMOV, V. A. and YU. A. TARASOV, Teplofizika 2 (1964) 160.
[9.1a] AGNEW, L. and C. SUMMERS, Proc. 7. internat. Conf. Phenom. Ionized Gases, Belgrade 1965, Vol. 2, p. 574.
[9.1b] AKCASU, A. Z. and L. H. WALD, Phys. Fluids 10 (1967) 1327.
[9.1c] ALCOCK, A. J., P. P. PASHININ and S. A. RAMSDEN, Phys. Rev. Lett. 17 (1966) 528.
[9.1d] ALEKSANDROV, V. YA., D. E. GUREVICH and I. V. PODMOSHENSKI, Teplofizika 5 (1967) 751.
[9.1e] ALEKSANDROV, V. YA., D. V. GUREVICH and I. V. PODMOSHENSKI, Optika i Spektroskopiya 25 (1968) 674.
[9.1f] ALEKSANDROV, V. YA., D. V. GUREVICH and I. V. PODMOSHENSKI, Optika i Spektroskopiya 26 (1969) 36.
[9.1g] ALEKSIN, V. F., V. A. SUPRUNENKO, YE. A. SUCHOMLIN and N. I. REVA, Zh. tekhn. Fiz. 36 (1966) 620.
[9.1h] ALESKOVSKI, Y. M. and V. L. GRANOVSKI, Zh. eksp. teor. Fiz. 41 (1961) 363; German Translation: Phys. Abh. SU 6 (1962) 182.
[9.1i] ALLEN, R. A., R. L. TAYLOR and J. C. CAMM, Proc. 6. internat. Conf. Ion. Phenom. Gases, Paris 1963, Vol. 3, p. 381.
[9.1j] ALLEN, R. A. and A. TEXTORIS, J. chem. Phys. 40 (1964) 3445.
[9.1k] ALLEN, R. A., A. TEXTORIS and and J. WILSON, J. quant. Spectrosc. Radiat. Transfer 5 (1965) 95.
[9.1l] ALPHER, R. A. and D. R. WHITE, Phys. Fluids 7 (1964) 1239.
[9.1m] ANDERSON, A. D. and H. R. GRIEM, US Naval Res. Lab., Washington, D. C., Technical Report N. 313, July 1963.
[9.1n] ANDERSON, A. D. and H. R. GRIEM, Proc. 6. internat. Conf. Ion. Phenom. Gases, Paris 1963, Vol. 3, p. 293.
[9.1o] ARMSTRONG, B. H., J. quant. Spectrosc. Radiat. Transfer 4 (1964) 207.
[9.1p] ARMSTRONG, B. H., J. SOKOLOFF, R. W. NICHOLLS, D. H. HOLLAND and R. E. MEYEROTT, J. quant. Spectrosc. Radiat. Transfer 1 (1961) 143.
[9.1q] ASAM, P. M., Inst. Plasmaphys. GmbH, Garching near Munich, Lab. Rep. IPP 1/22, May 1964.

[9.1r] Ashkin, M., Phys. Rev. **141** (1966) 41.
[9.1s] Asinovski, E. I. and V. M. Batenin, Teplofizika **3** (1965) 530.
[9.1t] Asinovski, E. I., A. V. Kirillin and G. A. Kobtsev, Teplofizika **6** (1968) 746.
[9.1u] Ashkin, M., Phys. Rev. **141** (1966) 41.
[9.1v] Bakeyev, A. A., R. E. Rovinsky and I.-P. Shirokova, Optika i Spektroskopiya **27** (1969) 215.
[9.2] Balazs, N., in: Optical spectrometric measurements of high Temperatures, Univ. of Chicago Press, Chicago 1961, p. 39.
[9.2a] Batenin, V. M. and V. F. Chinnov, Teplofizika **7** (1969) 832.
[9.2b] Bauder, U., Z. Physik **205** (1967) 303.
[9.3] Bearden, A. J., F. L. Ribe, G. A. Sawyer and T. F. Stratton, Phys. Rev. Letters **6** (1961) 257.
[9.3a] Bekefi, G., Radiation Processes in Plasmas, J. Wiley & Sons Inc., New York, London, Sydney 1966.
[9.3b] Bell, K. L. and A. E. Kingston, Proc. Phys. Soc. Lond. **90** (1967) 895.
[9.4] Berg, H. F., A. W. Ali, R. Lincke and H. R. Griem, Phys. Rev. **125** (1962) 199.
[9.4a] Berg, H. F. and G. Tondello, Inst. Plasmaphys., Kernforsch.anlage Jülich, Report JÜL-182-PP, 1964.
[9.4b] Berge, O. E., A. Böhm and L. Rehder, Z. Naturforsch. **20a** (1965) 120.
[9.4c] Berge, O. E. and W. Bötticher, Z. Naturforsch. **19a** (1964) 1460.
[9.5] Berger, J. M., Astrophys. J. **124** (1956) 550.
[9.5a] Berry, R. S. and C. W. Reimann, J. chem. Phys. **38** (1963) 1540.
[9.5b] Biberman, L. M. and G. E. Norman, Optika i Spektroskopiya **8** (1960) 433.
[9.5c] Biberman, L. M. and G. E. Norman, J. quant. Spectrosc. Radiat. Transfer **3** (1963) 221.
[9.5d] Biberman, L. M. and G. E. Norman, Usp. Fiz. Nauk **91** (1967) 193.
[9.6] Biberman, L. M., G. E. Norman and K. N. Ulyanov, Optika i Spektroskopiya **10** (1961) 297.
[9.6a] Biberman, L. M., G. E. Norman and K. N. Ulyanov, Astronomicheski Zhurnal **39** (1962) 107.
[9.7] Biberman, L. M. and V. Y. Romanov, Optika i Spektroskopiya **3** (1957) 646.
[9.8] Bogen, P., X-ray diagnostics of plasmas, in: Plasma Diagnostics, North Holland Publ. Co., Amsterdam 1968, p. 424.
[9.8a] Bogen, P. and D. Rusbüldt, Phys. Fluids **11** (1968) 2022.
[9.8b] Bogen, P. and J. Schlüter, Proc. 6. internat. Conf. Ion. Phenom. Gases, Paris 1963, Vol. 2, p. 561.
[9.8c] Böhm, A. and L. Rehder, Z. Naturforsch. **20a** (1965) 114.
[9.8d] Böhnlein, A., Z. Naturforsch. **21a** (1966) 1660.
[9.8e] Boland, B. C., F. E. Irons and R. W. P. McWhirter, United Kingdom Atomic Energy Authorities, Culham Lab., Abingdon, Berksh., Rep. CLM-P176, May 1, 1968.
[9.8f] Boland, B. C., F. E. Irons and R. W. P. McWhirter, J. Phys. **B 1** (1968) 1180.
[9.9] Boldt, G., Z. Physik **154** (1959) 319.
[9.10] Boldt, G., Z. Physik **154** (1959) 330.
[9.10a] Boldt, W., Beitr. Plasmaphys. **5** (1965) 143.
[9.10b] Bott, J. F., Phys. Fluids **9** (1966) 1540.
[9.10c] Bötticher, W., Z. Physik **150** (1958) 336.
[9.11] Breene, R. G., Planetary Space Sci. **2** (1959) 10.
[9.11a] Breene, R. G., J. quant. Spectrosc. Radiat. Transfer **5** (1965) 449.
[9.11b] Breene, R. G. and M. C. Nardone, J. Opt. Soc. Amer. **50** (1960) 1111.
[9.12] Breene, R. G. and M. C. Nardone, J. Opt. Soc. Amer. **51** (1961) 692.
[9.12a] Breene, R. G. and M. C. Nardone, J. quant. Spectrosc. Radiat. Transfer **2** (1962) 273.

[9.13] Brode, R., Proc. Roy. Soc. Lond. **A 125** (1929) 134.
[9.13a] Burakov, V. S., P. A. Naumenkov and V. P. Ivanov, Zh. prikl. Spektrosk. 8 (1968) 738.
[9.13b] Burgess, A. and M. J. Seaton, Rev. mod. Phys. **30** (1958) 992.
[9.13c] Burgess, A. and M. J. Seaton, Monthly Notices Roy. Astron. Soc. **120** (1960) 121.
[9.14] Busz, G. and W. Finkelnburg, Z. Physik **139** (1954) 212.
[9.15] Chandrasekhar, S., Astrophys. J. **102** (1945) 223.
[9.16] Chandrasekhar, S., Astrophys. J. **102** (1945) 395.
[9.17] Chandrasekhar, S., Astrophys. J. **128** (1958) 114.
[9.18] Chandrasekhar, S. and F. H. Breen, Astrophys. J **104** (1946) 430.
[9.18a] Chandrasekhar, S. and D. D. Elbert, Astrophys. J. **128** (1958) 633.
[9.18b] Chapelle, J., Proc. 8. internat. Conf. Phenom. Ionized Gases, Contributed Pap., IAEA Vienna 1967, p. 446.
[9.18c] Chapelle, J. and F. Cabannes, Chim. haut. temperat. Paris, CNRS 1963, p. 79.
[9.18d] Chapelle, J. and F. Cabannes, Proc. 7. Internat. Conf. Phenom. Ionized Gases, Belgrade 1965, Vol. 2, p. 589.
[9.18e] Chapelle, J. and F. Cabannes, in: Nizkotemperaturnaya Plazma, Izd. Mir, Moscow 1967, p. 102.
[9.18f] Chapelle, J. and F. Cabannes, J. quant. Spectrosc. Radiat. Transfer 9 (1969) 889.
[9.18g] Chapelle, J., M. Decroisette and F. Cabannes, J. quant. Spectrosc. Radiat. Transfer 7 (1967) 571.
[9.18h] Charnaya, F. A., Optika i Spektroskopiya **25** (1968) 24.
[9.18i] Churchill, D. R., B. H. Armstrong, R. R. Johnston and K. G. Müller, J. quant. Spectrosc. Radiat. Transfer **6** (1966) 371.
[9.18j] Conrads, H., Ber. Kernforsch.anlage Jülich No. 360 (1966).
[9.18k] Conrads, H., Z. Physik **200** (1967) 444.
[9.18l] Cooper, W. S. and W. B. Kunkel, Phys. Fluids 8 (1965) 482.
[9.18m] Cooper, J. W. and J. B. Martin, Phys. Rev. **126** (1962) 1482.
[9.18n] Costa, S. and G. Tondello, Ric. sci. 35, Ser. 2, Rend. Fis. Vol. 8 (1965) 949.
[9.18o] Dawson, J. and C. Oberman, Phys. Fluids **5** (1962) 517.
[8.18p] Döbele, H. F., R. Wienecke and S. Witkowski, Z. Naturforsch. **21a** (1966) 390.
[9.18q] Dobrynin, V. M. and V. A. Shingarkina, in: Aerofiz. issled. sverchsvuk. tekhenii, Nauka, Moscow-Leningrad 1967, p. 105.
[9.18r] Donohue, R. J. and R. F. Majkowski, J. appl. Phys. **33** (1962) 3.
[9.18s] Doughty, N. A. and P. A. Fraser, in: Atomic Collision Processes, North Holland Publ. Co., Amsterdam 1964, p. 527.
[9.18t] Doughty, N. A. and P. A. Fraser, Monthly Notices Roy. Astron. Soc. **132** (1966) 267.
[9.18u] Doughty, N. A., P. A. Fraser and R. P. McEachran, Monthly Notices Roy. Astron. Soc. **132** (1966) 255.
[9.19] Drawin, H. W., Z. Physik **146** (1956) 295.
[9.19a] Dreusicke, M. and W. Neumann, Beitr. Plasmaphys. **6** (1966) 175.
[9.20] Dronov, A. P., A. G. Sviridov and N. N. Sobolev, Optika i Spektroskopiya **12** (1962) 677.
[9.20a] Dyuzhev, G. A., V. B. Kaplan, B. Y. Moizhes and G. E. Pikus, Zh. tekhn. Fiz. **38** (1968) 963.
[9.20b] Eberhagen, A., Z. angew. Phys. **20** (1966) 244.
[9.20c] Eberhagen, A. and M. Keilhacker, Proc. 6. internat. Conf. Ion. Phenom. Gases, Paris 1963, Vol. 2, p. 577.

[9.20d] Eberhagen, A. and W. Lünow, Inst. Plasmaphys. GmbH Garching near Munich, Report IPP 1/23, June 1964.
[9.20e] Elton, R. C., L. J. Palumbo and H. R. Griem, Phys. Rev. Lett. **20** (1968) 783.
[9.20f] Fidone, L., Nuovo Cim. **24** (1962) 823.
[9.20g] Filippov, O. K. and V. M. Pivovarov, Zh. prikl. Spektrosk. **4** (1966) 64.
[9.21] Finkelnburg, W. and T. Peters, Hand. Phys. 28, Springer Verlag Berlin-Göttingen-Heidelberg 1957, p. 79.
[9.21a] Firsov, A. B. and M. I. Chibisov, Zh. eksp. teor. Fiz. **39** (1960) 1770.
[9.21b] Frank, H., G. Mück, M. Neiger and H.-P. Popp, Proc. 9. internat. Conf. Phenom. Ionized Gases, Bucharest 1969, p. 4.
[9.21c] Fucks, W., W. L. Bohn, G. Heinrich and P. Platz, Z, Physik **170** (1962) 409.
[9.22] Gaunt, J. A., Proc. Roy. Soc. Lond. **A 126** (1930) 654.
[9.23] Geltman, S., Phys. Rev. **104** (1956) 346.
[9.23a] Geltman, S., Astrophys. J. **141** (1965) 376.
[9.23b] Glaser, H., Inst. Plasmaphys. Garching near Munich, Report IPP 1/38, June 1965.
[9.24] Göing, W., H. Meier and H. Meinen, Z. Physik **140** (1955) 376.
[9.24a] Goldfarb, W. M., Optika i Spektroskopiya **19** (1965) 284.
[9.24b] Goldfarb, W. M. and S. W. Dresvin, Teplofizika **3** (1965) 333.
[9.24c] Goldfarb, V. M., E. V. Ilina, I. E. Kostygova, G. A. Lukyanov and V. A. Silantyev, Zh. prikl. mekhan. tekhn. Fiz. No. 1 (1967) 111.
[9.24d] Goldman, R. and L. Oster, Phys. Rev. **136** (1964) A 602.
[9.24e] Golubev, V. S., G. A. Kasabov and V. F. Konach, Teplofizika **2** (1964) 493.
[9.24f] Golubovski, Yu. B., Yu. M. Kagan and R. I. Lyagushchenko, Optika i Spektroskopiya **24** (1968) 286.
[9.24g] Grant, I. P., Monthly Notices Roy. Astron. Soc. **118** (1958) 241.
[9.24h] Green, T. S., D. L. Fisher, A. H. Gabriel, F. J. Morgan and A. A. Newton, Phys. Fluids **10** (1967) 1663.
[9.25] Greene, J., Astrophys. J. **130** (1959) 693.
[9.25a] Grenishin, S. G., L. S. Nikolaevski and A. F. Simonenko, Zh. prikl. Spektrosk. **8** (1968) 803.
[9.25b] Grenishin, S. G., A. F. Simonenko and G. P. Startsev, Optika i Spektroskopiya **25** (1968) 825.
[9.26] Griem, H. R., Proc. 5. internat. Conf. Ionization Phenomena in Gases, Vol. 2, North Holland Publ. Co., Amsterdam 1962, p. 1857.
[9.26a] Griem, H. R., in: Temperature, its Measurement and Control in Science and Industry, Vol. 3, Pt. 1, Reinhold Publ. Corp., New York 1962, p. 615.
[9.26b] Griem, H. R., Plasma Spectroscopy, McGraw Hill Book Co., New York 1964.
[9.27] Griem, H. R., A. C. Kolb, W. H. Lupton and D. T. Phillips, Nuclear Fusion **2**, Suppl. Part 2 (1962) 543.
[9.27a] Gubler, W. and M. J. O. Strutt, Z. Naturforsch. **20a** (1965) 1156; Appl. Optics **5** (1966) 659.
[9.27b] Gündel, H. and W. Neumann, Beitr. Plasmaphys. **7** (1967) 221.
[9.27c] Gündel, H. and W. Neumann, Monatsber. Dtsch. Akad. Wiss. Berlin **9** (1967) 604.
[9.27d] Hamberger, S. M. and A. W. Johnson, J. quant. Spectrosc. Radiat. Transfer **5** (1965) 683.
[9.28] Harding, G. N. and V. Roberts, Proc. 5. internat. Conf. Ionization Phenom. in Gases, Vol. 2, North Holland Publ. Co. Amsterdam 1962, p. 1977.
[9.28a] Hemmerich, H., Inst. Plasmaphys. GmbH Garching near Munich, Lab. Report IPP 1/16, January 1964.
[9.29] Henning, H., Z. Physik **169** (1962) 467.
[9.29a] Henning, H., Z. Astrophys. **62** (1965) 109.

[9.30] HETTNER, G., Z. Physik **150** (1958) 182.
[9.30a] HETTNER, G., Z. Physik **211** (1968) 483.
[9.31] ISLER, R. C. and D. E. KERR, Phys. Fluids **8** (1965) 1176.
[9.32] JAHODA, F. C., E. M. LITTLE, W. E. QUINN, O. A. SAWYER and T. F. STRATTON, Phys. Rev. **119** (1960) 843.
[9.33] JAHODA, F. C., F. L. RIBE, G. A. SAWYER and T. F. STRATTON, Proc. 5. internat. Conf. Ionization Phenomena in Gases, Vol. 2, North Holland Publ. Co., Amsterdam 1962, p. 1987.
[9.34] JANES, G. S. and H. KORITZ, J. appl. Phys. **31** (1960) 525.
[9.35] JANKOW, W. H., Optika i Spektroskopiya **14** (1963) 29.
[9.35a] JOHN, T. L., Monthly Notices Roy. Astron. Soc. **131** (1966) 315.
[9.36] JÜRGENS, G., Z. Physik **134** (1952) 21.
[9.36a] KAGAN, YU. M. and N. N. KHRISTOV, Optika i Spektroskopiya **27** (1969) 710.
[9.36b] KANDEL, R., Compt. Rend. Acad. Sci. AB **262** (1966) B373.
[9.36c] KARZAS, W. J. and R. LATTER, Astrophys. J. Suppl. **6** (1961) 167.
[9.36d] KASYANOV, V. and A. STAROSTIN, Zh. eksp. teor. Fiz. **48** (1965) 295.
[9.36e] KIMMITT, M. F. and B. F. NIBLETT, Proc. Phys. Soc. Lond. **82** (1963) 938.
[9.36f] KIVEL, B., J. quant. Spectrosc. Radiat. Transfer **7** (1967) 27.
[9.36g] KIVEL, B., J. quant. Spectrosc. Radiat. Transfer **7** (1967) 51.
[9.36h] KLEIN, L., Appl. Optics **7** (1968) 677.
[9.37] KOGAN, V. I., Fizika plazmy i problema upravlyayemych reaktsii, Vol. 3, Izd. Akad. Nauk SSSR, Moscow 1958, p. 99; German Translation: Phys. Abh. SU **1** (1959) 462.
[9.38] KOLB, A. C., H. R. GRIEM, W. H. LUPTON, D. T. PHILLIPS, S. A. RAMSDEN, F. A. MCLEAN, W. R. FAUST and M. A. SWARTZ, Nuclear Fusion **2**, Suppl. Part 2 (1962) 553.
[9.38a] KRIENER, H. and D. SCHLÜTER, Z. Astrophys. **67** (1967) 165.
[9.38b] KULSRUD, R. M., Astrophys. J. **119** (1954) 386.
[9.38c] KUTSENKO, A. M. and A. V. KORTNEV, Ukrain. Fiz. Zh., Kiev **10** (1965) 1033.
[9.39] KUMMERER, K., Z. Physik **147** (1957) 373.
[9.40] LARENZ, R. W., Z. Physik **129** (1951) 343.
[9.41] LIN, S. C., AVCO EVERETT Res. Lab., Res. Rep. **33** (1958).
[9.42] LIN, S. C., Rep. 18. Annual Conf. Phys. Electronics, Mass. Inst. Technol., Dep. Phys., March 1958.
[9.42a] LITVAK, M. M. and D. F. EDWARDS, J. appl. Phys. **37** (1966) 4462.
[9.43] LOCHTE-HOLTGREVEN, W. and W. NISSEN, Z. Physik **133** (1952) 124.
[9.43a] LOOKE, M., W. NEUMANN and K. RADEMACHER, Beitr. Plasmaphys. **4** (1964) 239.
[9.43b] LÜNOW, W., Inst. Plasmaphys. Garching near Munich, Lab. Rep. IPP 6/25, August 1964.
[9.44] MCLEAN, E. A., C. E. FANEUFF, A. C. KOLB and H. R. GRIEM, Phys. Fluids **3** (1960) 843.
[9.44a] MAECKER, H., Z. Physik **116** (1940) 257.
[9.44b] MAECKER, H., Ann. Physik **18** (1956) 441.
[9.45] MAECKER, H. and T. PETERS, Z. Physik **139** (1954) 448.
[9.46] MAGNUS-OBERHETTINGER, Formeln und Sätze für die speziellen Funktionen der mathematischen Physik. Springer-Verlag, Berlin 1948.
[9.47] MAK, A. A., Zh. tekhn. Fiz. **31** (1961) 94.
[9.47a] MANDELSHTAM, S. L., P. P. PASHININ, A. M. PROKHOROV, YU. P. RAIZER and N. K. SUKHODREV, Zh. eksp. teor. Fiz. **49** (1965) 127.
[9.47b] MARR, G. V., Photoionization processes in gases, Ac. Press New York-London 1967.
[9.48] MASTRUP, F., J. Opt. Soc. Amer. **50** (1960) 32.

[9.49] Maue, A. W., Ann. Physik **13** (1932) 161.
[9.49a] Meiners, D., Proc. 9. internat. Conf. Phenom. Ionized Gases, Bucharest 1969, p. 582.
[9.50] Menzel, D. H. and C. L. Pekeris, Monthly Notices Roy. Astron. Soc. **96** (1936) 77.
[9.50a] Mies, F. H., J. chem. Phys. **37** (1962) 1101.
[9.50b] Mjolsness, R. C. and H. M. Ruppel, J. quant. Spectrosc. Radiat. Transfer **7** (1967) 423.
[9.50c] Mjolsness, R. C. and H. M. Ruppel, Phys. Rev. **154** (1967) 98.
[9.50d] Mnatsakanyan, A. Kh., Teplofizika **6** (1968) 236.
[9.50e] Kriener, H. and D. Schlüter, Z. Astrophys. **67** (1967) 165.
[9.50f] McLean, E. A. and S. A. Ramsden, Phys. Rev. **140** (1965) A 1122.
[9.50g] Morgulis, N. D., I. L. Babich, Yu. A. Dyadyun and V. P. **18** (1965) 931.
[9.50h] Morris, J. C. and G. R. Bach, Trans. Nucl. Sci. NS-11 (1964) 119.
[9.50i] Morris, J. C., G. R. Bach, R. U. Krey, R. W. Liebermann and J. M. Yos, AIAA-J. **4** (1966) 165.
[9.50j] Morris, J. C., G. R. Bach, R. U. Krey, R. W. Liebermann and J. M. Yos, AIAA-J. **4** (1966) 1223.
[9.50k] Morris, J. C. and R. L. Garrison, J. quant. Spectrosc. Radiat. Transfer **6** (1966) 899.
[9.50l] Morris, J. C. and R. U. Krey, J. quant. Spectrosc. Radiat. Transfer **9** (1969) 1633.
[9.50m] Morris, J. C. and R. U. Krey, AVCO CORPORATION, Wilmington, Mass.: Argon continuum radiation. Manuscript, forseen for publication in: J. quant. Spectrosc. Radiat. Transfer.
[9.50n] Morris, J. C., R. U. Krey and G. R. Bach, J. quant. Spectrosc. Radiat. Transfer **6** (1966) 727.
[9.50o] Morris, J. C., R. U. Krey and G. R. Bach, Phys. Rev. **159** (1967) 113.
[9.50p] Morris, J. C., R. U. Krey and R. L. Garrison, Office of Aerospace Res., US Air Force, Aerospace Res. Labs., Report ARL 68-0103, May 1968.
[9.50q] Morris, J. C., R. U. Krey and R. L. Garrison, Phys. Rev. **180** (1969) 167.
[9.50r] Moskvin, Yu. V. and N. N. Chesnokova, Teplofizika **3** (1965) 370.
[9.50s] Mück, G. and H.-P. Popp, Z. Naturforsch. **23a** (1968) 1213.
[9.50t] Myerscough, V. P., Astrophys. J. **152** (1968) 1115.
[9.51] Neumann, W., Ann. Physik **17** (1956) 146.
[9.52] Neumann, W., Beitr. Plasmaphys. **1** (1961) 119.
[9.53] Neumann, W., Beitr. Plasmaphys. **2** (1962) 80.
[9.54] Neumann, W., Exp. Techn. Phys. **10** (1962) 124.
[9.54a] Neumann, W., Exp. Techn. Phys. **13** (1965) 157.
[9.55] Neumann, W. and H. Reimann, Ann. Phys. **3** (1959) 211.
[9.56] Neynaber, R. H., L. L. Marino, E. W. Rothe and S. M. Trujillo. Phys. Rev. **129** (1963) 2070.
[9.56a] Nikolayevskii, L. S., A. F. Simonenko and S. G. Grenishin, Zh. prikl. Spektrosk. **4** (1966) 485.
[9.57] Nissen, W., Z. Phys. **139** (1954) 638.
[9.57a] Norcross, D. W. and P. M. Stone, J. quant. Spectrosc. Radiat. Transfer **6** (1966) 277.
[9.57b] Ogurtsova, N. N., I. V. Podmoshenski and V. M. Shelemina, Optika i Spektroskopiya **16** (1964) 949.
[9.58] Ohmura, T. and H. Ohmura, Astrophys. J. **131** (1960) 8.
[9.58a] Ohmura, T. and H. Ohmura, Phys. Rev. **118** (1960) 154.
[9.59] Ohmura, T. and H. Ohmura, Phys. Rev. **121** (1961) 513.
[9.59a] Olfe, D. B., J. quant. Spectrosc. Radiat. Transfer **1** (1961) 104.

[9.60] OLSEN, H. N., Phys. Rev. **124** (1961) 1703.

[9.60a] OLSEN, H. N., in: Temperature, its Measurement and Control in Science and Industry, Vol. III Pt. 1, Reinhold Publ. Corp., New York 1962, p. 593.

[9.60b] ORRALL, F. Q., Temperature, its Measurement and Control in Science and Industry, Vol. 3, Pt. 1, Reinhold Publ. Corp. New York 1962, p. 723.

[9.60c] OSTER, L., Z. Astrophys. **47** (1959) 169.

[9.60d] PEACH, G., Monthly Notices Roy. Astron. Soc. **124** (1962) 371.

[9.60e] PEACH, G., Mem. Roy. Astron. Soc. **71** (1967) 1 (Pt. 1).

[9.60f] PEACH, G., Mem. Roy. Astron. Soc. **71** (1967) 13.

[9.60g] PEACH, G., Mem. Roy. Astron. Soc. **71** (1967) 29.

[9.60h] PENNER, S. S. and M. THOMAS, AIAA-J. **2** (1964) 1572.

[9.60i] PETERS, T., Z. Flugwiss. **14** (1966) 89.

[9.61] PETSCHEK, H. E., P. H. ROSE, H. S. GLICK, A. KANE and A. KANTROWITZ, J. appl. Phys. **26** (1955) 83.

[9.61a] PFAU, S. and A. RUTSCHER, Beitr. Plasmaphys. **8** (1968) 73.

[9.61b] PFAU, S. and A. RUTSCHER, Beitr. Plasmaphys. **8** (1968) 85.

[9.61c] PFAU, S. and A. RUTSCHER, Proc. 9. internat. Conf. Phenomena Ionized Gases, Bucharest 1969, p. 586.

[9.61d] PLASTININ, YU. A., in: Fiz. Gazodim. ion. khim. reag. gazov, Nauka, Moscow 1968, p. 29.

[9.61e] POPP, H.-P., Z. Naturforsch. **20a** (1965) 642.

[9.61f] POPP, H.-P., Z. Naturforsch. **22a** (1967) 254.

[9.61g] POPP, H.-P., Proc. 8. internat. Conf. Phenom. Ionized Gases, Contributed Pap., IAEA Vienna 1967, p. 448.

[9.61h] REHDER, L., W. BÖTTICHER, H. CARLS and V. GRAAP, Phys. Fluids **12** (1969) Suppl. I S. 50 (No. 5 Pt. II).

[9.61i] RICHTER, J., Z. Astrophys. **61** (1965) 57.

[9.61j] ROBERTS, J. R., P. A. VOIGT and J. J. NAGY, Proc. 9. internat. Conf. Phenom. Ionized Gases, Bucharest 1969, p. 595.

[9.61k] ROBINSON, E. J. and S. GELTMAN, Phys. Rev. **153** (1967) 4.

[9.61l] ROTHE, D. E., J. quant. Spectrosc. Radiat. Transfer **9** (1969) 49.

[9.61m] ROTHE, D. E., Phys. Rev. **177** (1969) 93.

[9.61n] ROVINSKI, R. E., V. A. GRUSDEV, T. M. GUTENMACHER and A. P. SOBOLEV, Teplofizika **5** (1967) 557.

[9.61o] RUTSCHER, A. and S. PFAU, Beitr. Plasmaphys. **8** (1968) 315.

[9.61p] SAWYER, G. A., F. C. JAHODA, F. L. RIBE and T. F. STRATTON, J. quant. Spectrosc. Radiat. Transfer **2** (1962) 467.

[9.61q] SCHIRMER, H., Z. angew. Phys. **11** (1959) 357.

[9.62] SCHIRMER, H., Zur Theorie der Plasmen unter besonderer Berücksichtigung der Xenon-Hochdruckentladungen. Habilitationsschrift TH, Berlin 1961.

[9.63] SCHIRMER, H., Z. angew. Phys. **13** (1961) 56.

[9.63a] SCHLEY, U., H. MAGDEBURG and M. PIWONSKI, PTB-Mitt. **76** (1966) 405.

[9.64] SCHLÜTER, D., Z. Astrophys. **56** (1962) 43.

[9.64a] SCHLÜTER, D., Z. Astrophys. **61** (1965) 67.

[9.64b] SCHLÜTER, D., Z. Physik **210** (1968) 80.

[9.64c] SCHNAPAUFF, Z., Astrophys. **68** (1968) 431.

[9.64d] SEKA, W., C. BRETON, J.-L. SCHWOB and C. MINIER, Plasma Phys. **12** (1970) 73.

[9.64e] SIMONENKO, A. F., Zh. prikl. Spektrosk. **9** (1968) 332.

[9.64f] SORRELL, F. Y., Phys. Fluids **12** (1969) 1218.

[9.64g] STICKFORTH, J., Z. Physik **164** (1961) 1.

[9.65] STRATTON, T. F., in: Temperature, its Measurement and Control in Science and Industry, Vol. III, 1, Reinhold Publ. Corp., New York 1962, p. 663.

[9.65a] Streissle, H. P., Absolute spektrosk. Strahlungsmessung im Vakuum-Ultraviolett bei Argon- und Stickstoffhochdruckplasmen, Diss. T. H. Stuttgart 1965.
[9.66] Taylor, R. L., AVCO-EVERETT Res. Lab., Res. Rep. 154 (1963).
[9.67] Taylor, R. L., J. chem. Phys. **39** (1963) 2354.
[9.67a] Taylor, R. L., J. chem. Phys. **39** (1963) 2354.
[9.67b] Taylor, R. L. and G. Caledonia, J. quant. Spectrosc. Radiat. Transfer **9** (1969) 657.
[9.67c] Taylor, R. L. and G. Caledonia, J. quant. Spectrosc. Radiat. Transfer **9** (1969) 681.
[9.68] Taylor, R. L. and B. Kivel, J. quant. Spectrosc. Radiat. Transfer **4** (1964) 239.
[9.68a] Theimer, O., Ann. Physics **22** (1963) 102.
[9.68b] Tietz, T., Acta phys. hung. **14** (1962) 1.
[9.69] Unsöld, A., Ann. Physik **33** (1938) 607.
[9.70] Unsöld, A., Physik der Sternatmosphären, Springer-Verlag, Berlin-Göttingen-Heidelberg 1955.
[9.71] Vitense, E., Himmelswelt **56** (1949) 33.
[9.72] Vitense, E., Z. Astrophys. **28** (1951) 81.
[9.72a] Vitkovitzsky, I. M., Phys. Fluids **7** (1964) 612.
[9.72b] DeVore, R. V., Phys. Rev. **136** (1964) A 666.
[9.73] Wahl, B. W., R. J. Gould and J. W. McKee, in: Magnetohydrodynamics, Evanston, Northwestern Univ. Press 1962, p. 109.
[9.73a] Weber, O., Erzeugung und Nachweis des H^--Kontinuums im Stoßwellenrohr, Diss. Univ. Kiel 1957.
[9.74] Weber, O., Z. Physik **152** (1958) 281.
[9.74a] Weinberg, M. and R. S. Berry, Proc. 4. internat. Conf. Phys. Electronic Atomic Collisions, Sci. Bookcrafters Inc., N. Y. 1965, DB3, p. 108.
[9.74b] Weinberg, M. and R. S. Berry, Phys. Rev. **144** (1966) 75; Erratum: Phys. Rev. **165** (1968) 334.
[9.74c] Wende, B., Spektrale Strahlungsemission und Temperatur eines Argon-Bogenplasmas, Diss. T. H. Karlsruhe 1966, Fak. Elektrotechn.
[9.74d] Wende, B., Z. Physik **198** (1967) 1.
[9.75] Wiese, W., H. F. Berg and H. R. Griem, Phys. Rev. **120** (1960) 1079.
[9.75a] Wiese, W. L., J. F. McClelland, D. E. Kelleher and D. R. Paquette, Proc. 9. internat. Conf. Phenom. Ionized Gases, Bucharest 1969, p. 596.
[9.75b] Wilkins, R. L., J. quant. Spectrosc. Radiat. Transfer **4** (1964) 775.
[9.75c] Wilson, K. H. and W. E. Nicolet, J. quant. Spectrosc. Radiat. Transfer **7** (1967) 891.
[9.76] Wulff, H., Nuclear Instrum. and Methods **4** (1959) 352.
[9.76a] Ya'akobi, B., Proc. Phys. Soc. Lond. **92** (1967) 100.
[9.76b] Ya'akobi, J. quant. Spectrosc. Radiat. Transfer **10** (1970) 61.
[9.77] Yamamoto, M., J. phys. Soc. Japan **14** (1959) 1739.
[9.77a] Yos, J. M., Experimental determination of electric and thermal conductivity in a laminar electric arc column. AVCO Technical Rep. RAD-TR-7, May 1965.
[9.77b] Zwicker, H., Z. Physik **178** (1964) 189.
[9.77c] Zwicker, H. and U. Schumacher, Proc. 6. internat. Conf. Ion. Phenom. Gases, Paris 1963, Vol. 2, p. 593.
[10.1] Agnew, L. and C. Summers, Proc. 7. internat. Conf. Ion. Phenom. Ionized Gases, Belgrade 1965, Vol. 2, p. 574.
[10.1a] Ahlborn, B. and A. J. Barnard, AIAA-J. **4** (1966) 1136.
[10.1b] Aleksandrov, E. B. and V. V. Khromov, Optika i Spektroskopiya **18** (1965) 545.

[10.1c] ALEKSANDROV, V. YA., D. B. GUREVICH and I. V. PODMOSHENSKI, Optika i Spektroskopiya 23 (1967) 521.
[10.1d] ALEKSEYEV, V. A., M. A. MAZING, P. D. SERAPINAS, I. I. SOBELMAN and L. A. VAINSHTEIN, Proc. 5. internat. Conf. Phys. Electronic and Atomic Collisions, Nauka, Leningrad 1967, p. 528.
[10.1e] ALEKSEYEV, V. A. and I. I. SOBELMAN, Zh. eksp. teor. Fiz. 49 (1965) 1274.
[10.1f] ALEKSEYEV, V. A. and E. A. YUKOV, Optika i Spektroskopiya 25 (1968) 645.
[10.1g] ALEKSEYEV, V. A. and E. A. YUKOV, Phys. Inst. Acad. Sci. USSR, Moscow, Reprint No. 87, 1968.
[10.1h] ALI, A. W. and H. R. GRIEM, Phys. Rev. 140 (1965) A1044; Erratum: Phys. Rev. 144 (1966) 366.
[10.1i] ALYAMOVSKI, V. N., A.-P. DRONOV, V. F. KITAYEVA, A. G. SVIRIDOV and N. N. SOBOLEV, in: Vopr. magnitn. gidrodinamiki i dinamiki plazmy, Vol. 2, Acad. Sci. Lett. SSR, Riga 1962, p. 379.
[10.2] ALYAMOVSKI, V. N. and V. F. KITAYEVA, Optika i Spektroskopiya 8 (1960) 152.
[10.2a] ANDO, K., J. Phys. Soc. Japan 23 (1967) 1183; 24 (1968) 1366.
[10.2b] ARMSTRONG, B. H., J. quant. Spectrosc. Radiat. Transfer 4 (1964) 207.
[10.3] ASAGOE, K., Sci. Pap. Inst. phys. chem. Res. Japan 11 (1929) 243.
[10.3a] ASINOVSKI, E. I. and V. M. BATENIN, Teplofizika 3 (1965) 530.
[10.3b] ATTAL, G. and F. SCHULLER, Compt. Rend. Acad. Sci. B 262 (1966) 807.
[10.4] AULEHLA, E. and L. HERMAN, Compt. Rend. Cad. Sci., Paris 246 (1958) 1676.
[10.5] BABCOCK, H. W., Astrophys. J. 118 (1953) 387.
[10.5a] BACON, M. E. and D. E. EDWARDS, Phys. Rev. 170 (1968) 125.
[10.5b] BACON, M. E. and D. F. EDWARDS, J. quant. Spectrosc. Radiat. Transfer 9 (1969) 951.
[10.5c] BAKANOVICH, G. I. and L. I. GRECHIKHIN, Teplofizika 3 (1965) 520.
[10.5d] BAKANOVICH, G. I. and L. I. GRECHIKHIN, Teplofizika 7 (1969) 338.
[10.6] BARANGER, M., in: Atomic and molecular Processes, Academic Press, New York and London 1962, S. 493.
[10.6a] BARDÓCZ, Á. and T. VÖRÖS, J. quant. Spectrosc. Radiat. Transfer 6 (1966) 351.
[10.7] BARDÓCZ, Á., T. VÖRÖS and U. M. VANYEK, Spectrochim. Acta 17 (1961) 642.
[10.8] BARDÓCZ, Á., T. VÖRÖS and U. M. VANYEK, Z. angew. Phys. 14 (1962) 581.
[10.8a] BARGER, R. L., Phys. Rev. 154 (1967) 94.
[10.8b] BARNARD, A. J., J. COOPER and L. J. SHAMEY, Astron. Astrophys. 1 (1969) 28.
[10.8c] BARNARD, A. J., J. COOPER and L. J. SHAMEY, Mém. Soc. Sci. Liège 17 (1969) 89.
[10.8d] BARNARD, A. J., H. G. JAMES and C. R. NEUFELD, Canad. J. Phys. 46 (1968) 1083.
[10.8e] BATES, D. R. and A. DAMGAARD, Phil. Trans. A 242 (1949) 101.
[10.8f] BAYER-HELMS, F., Z. angew. Phys. 16 (1963) 44.
[10.9] BEHMENBURG, W., J. quant. Spectrosc. Radiat. Transfer 4 (1964) 177.
[10.9a] BEHMENBURG, W., Z. Astrophys. 69 (1968) 368.
[10.10] BEHMENBURG, W. and H. KOHN, J. quant. Spectrosc. Radiat. Transfer 4 (1964) 163.
[10.10a] BENGTSON, R. D., M. H. MILLER, W. D. DAVIS and J. R. GREIG, Astrophys. J. 157 (1969) 957.
[10.11] BERESIN, A. B., A. N. SAIDEL and G. M. MALYSHEV, Zh. tekhn. Fiz. 33 (1963) 291.
[10.11a] BERG, H. F., Z. Physik 191 (1966) 503.
[10.12] BERG, H. F., A. W. ALI, R. LINCKE and H. R. GRIEM, Phys. Rev. 125 (1962) 199.
[10.12a] BERG, H. F., W. ERVENS and B. FURCH, Proc. 8. internat. Conf. Phenom. Ionized Gases, Contributed Pap., IAEA Vienna 1967, p. 439.
[10.12b] BERG, H. F., W. ERVENS and B. FURCH, Z. Physik 206 (1967) 309.
[10.12c] BERGEON, R. and B. VODAR, Compt. Rend. Acad. Sci., Paris 240 (1955) 172.

[10.13] BERGSTEDT, K., E. FERGUSON, H. SCHLÜTER and H. WULFF, Proc. 5. internat. Conf. Ionization Phenom. in Gases, Vol. 1, North Holland Publ. Co., Amsterdam 1962, p. 437.

[10.13a] BETH, M.-U., M. G. KLING, G. NEDDER and W. L. BOHN, Proc. 8. internat. Conf. Phenom. Ionized Gases, Contributed Pap., IAEA Vienna 1967, p. 545.

[10.14] BETHE, H. A. and E. E. SALPETER, Quantum Mechanics of One and Two Electron Atoms, Monograph, or in Handb. Phys. 35, Springer-Verlag, Berlin-Göttingen-Heidelberg 1957.

[10.14a] BIENEWSKI, T. M., in: Atomic Collision Processes, North Holland Publ. Co., Amsterdam 1964, p. 1055.

[10.14b] BIRD, P. F. and G. L. SCHOTT, J. quant. Spectrosc. Radiat. Transfer 5 (1965) 783.

[10.14c] BIRKELAND, J. W., J. P. OSS and W. G. BRAUN, Proc. 8. internat. Conf. Phenom. Ionized Gases, Contributed Pap., IAEA Vienna 1967, p. 438.

[10.15] BLAMONT, J. E., Ann. Physique 2 (1957) 551.

[10.15a] BLAMONT, J. E., Thesis Paris 1957.

[10.15b] BLAMONT, J. E. and J. BROSSEL, Compt. Rend. Acad. Sci., Paris **243** (1956) 2038.

[10.16] BLOCHINZEW, D. I., Grundlagen der Quantenmechanik. Deutscher Verl. d. Wiss. Berlin 1953 (Translation from Russian).

[10.16a] BOESCHOTEN, F. and L. J. DEMETER, Plasma Physics **10** (1968) 391.

[10.16b] BÖTTICHER, W., O. RODER and K. H. WOBIG, Z. Physik **175** (1963) 480.

[10.17] BOGEN, P., Z. Physik **149** (1957) 62.

[10.17a] BOGEN, P., H. CONRADS, G. GATTI and W. KOHLHAAS, J. Opt. Soc. Amer. **58** (1968) 203.

[10.17b] BOGEN, P. and J. SCHLÜTER, Nuclear Fusion **5** (1965) 251.

[10.17c] BOHN, W. L., M.-U. BETH and G. NEDDER, J. quant. Spectrosc. Radiat. Transfer 7 (1967) 661.

[10.17d] BOLDT, G. and W. S. COOPER, Z. Naturforsch. **19a** (1964) 968.

[10.17e] BONCH-BRUYEVICH, A. M. and V. A. CHODOWOI, Usp. Fiz. Nauk **93** (1967) 71.

[10.18] BOVEY, L. and A. H. GABRIEL, Spectrochim. Acta 18 (1962) 631.

[10.18a] BRÉCHOT, S., Phys. Letters **24 A** (1967) 476.

[10.18b] BRÉCHOT, S. and N. FEAUTRIER, Compt. Rend. Acad. Sci. Paris **265** (1967) B 39.

[10.18c] BRÉCHOT, S. and H. VAN REGENMORTER, Ann. Astrophys. **27** (1964) 739.

[10.18d] BRÉCHOT-SAHAL, S., Z. Astrophys. **69** (1968) 74.

[10.19] BREENE, R. G., Rev. mod. Phys. **29** (1957) 94.

[10.20] BREENE, R. G., The Shift and Shape of Spectral Lines. Pergamon Press, Oxford-London-New York-Paris 1961.

[10.20a] BREENE, R. G., in: Handb. Phys. **27** (1964) p. 1.

[10.20b] BREENE, R. G., Appl. Optics **6** (1967) 141.

[10.20c] BREENE, R. G., J. quant. Spectrosc. Radiat. Transfer 8 (1968) 1013.

[10.20d] BRIDGES, J. M., Diss. Univ. Maryland 1966.

[10.20e] BRIDGES, J. M. and W. L. WIESE, Proc. 7. internat. Conf. Phenom. Ionized Gases, Belgrade 1965, Vol. 3, p. 165.

[10.20f] BRIDGES, J. M. and W. L. WIESE, Phys. Rev. **159** (1967) 31.

[10.21] BROSE, E., Ann. Physik 58 (1919) 731.

[10.21a] BURHORN, F., Z. Physik **140** (1955) 440.

[10.21b] BUTAUX, J., and R. LENNUIER, Compt. Rend. Acad. Sci. Paris **265** (1967) B43.

[10.21c] BUTAUX, J. and R. LENNUIER, Compt. Rend. Acad. Sci., Paris **267** (1968) B36.

[10.21d] BUTTREY, D. E., J. quant. Spectrosc. Radiat. Transfer 9 (1969) 1227.

[10.21e] BUTTREY, D. E., H. R. MCCHESNEY and L. A. HOCKER, J. quant. Spectrosc. Radiat. Transfer 8 (1968) 717.

[10.21f] Cayless, M. A., Proc. 7. internat. Conf. Phenom. Ionized Gases, Belgrade 1965, Vol. 1, p. 651.

[10.21g] Chapelle, J., A. Sy, F. Cabannes and J. Blandin, Compt. Rend. Acad. Sci. **264** (1967) B853.

[10.21h] Chapelle, J., A. Sy, F. Cabannes and J. Blandin, Compt. Rend. Acad. Sci. Paris **266** (1968) B1513.

[10.21i] Ch'en, S. Y., Phys. Rev. **58** (1940) 844.

[10.21j] Ch'en, S. Y., Phys. Rev. **58** (1940) 1051.

[10.21k] Ch'en, S. Y., Phys. Rev. **73** (1948) 1470.

[10.21l] Ch'en, S. Y. and R. O. Garrett, Phys. Rev. **144** (1966) 59.

[10.21m] Ch'en, S. Y. and D. A. Kohler, Phys. Rev. **90** (1953) 1019.

[10.21n] Ch'en, S. Y., E. C. Looi and R. O. Garrett, Phys. Rev. **155** (1967) 38.

[10.21o] Ch'en, S. Y. and C. S. Pao, Phys. Rev. **58** (1940) 1058.

[10.22] Ch'en, S. Y. and W. J. Parker, J. Opt. Soc. Amer. **45** (1955) 22.

[10.23] Ch'en, S. Y. and M. Takeo, Rev. mod. Phys. **29** (1957) 20.

[10.23a] Choudhury, M. H., Phys. Rev. **186** (1969) 66.

[10.23b] Clayton, E. D. and S. Y. Ch'en, Phys. Rev. **85** (1952) 68.

[10.23c] Connor, T. R. and M. A. Biondi, Phys. Rev. **140** (1965) A778.

[10.23d] Cooper, W. S. and W. B. Kunkel, Phys. Fluids 8 (1965) 482.

[10.23e] Cooper, J. and G. K. Oertel, Phys. Rev. Lett. **18** (1967) 985.

[10.23f] Cooper, J. and G. K. Oertel, Phys. Rev. **180** (1969) 286.

[10.23g] Costa, S. and G. Tondello, Ric. sci. **35**, Ser. 2, Rend. Fis. Vol. 8 (1965) 949.

[10.23h] Dahmen, M. and H. J. Kusch, Z. Astrophys. **68** (1968) 445.

[10.23i] Damaschini, R. and J. Brochard, Compt. Rend. Acad. Sci. **264 B** (1967) 1350.

[10.23j] Davis, J. and D. E. Roberts, Proc. Phys. Soc. Lond. **92** (1967) 889.

[10.23k] Davis, J. and D. E. Roberts, J. Physics **B 1** (1968) 245.

[10.23l] Davis, J. and D. E. Roberts, J. Physics **B 1** (1968) 317.

[10.23m] Day, R. A., Phys. Fluids **12** (1969) I-47, No. 5.

[10.23n] Day, R. A. and H. R. Griem, Phys. Rev. **140** (1965) A1129.

[10.23o] Derevshchikov, V. A. and M. A. Deryabina, Teplofizika **5** (1967) 14.

[10.23p] Deutsch, C., H. W. Drawin, L. Herman and Nguyen-Hoe, J. quant. Spectrosc. Radiat. Transfer 8 (1968) 1027.

[10.23q] Deutsch, C., H. W. Drawin, L. Herman and H. Tittel, Proc. 9. internat. Conf. Phenom. Ionized Gases, Bucharest 1969, p. 587.

[10.24] Dickerman, P. J., J. appl. Phys. **29** (1958) 598.

[10.25] Dickerman, P. J., Proc. Conf. on extremely high temperatures. John Wiley & Sons Inc., New-York-London 1958, p. 77.

[10.25a] Dollfus, A., Compt. Rend. Acad. Sci. **236** (1953) 996.

[10.25b] Donohue, R. J. and R. F. Majkowski, J. appl. Phys. **33** (1962) 3.

[10.25c] Drawin, H.-W., Z. Physik **172** (1963) 181.

[10.25d] Drawin, H.-W. and M. Fumelli, Proc. Phys. Soc. Lond. **85** (1965) 997.

[10.25e] Drawin, H. W., H. Henning, L. Herman and Nguyen-Hoe, J. quant. Spectrosc. Radiat. Transfer **9** (1969) 317.

[10.26] Drobyshevski, E. M., Pribory Tekhn. Eksp. No. 6 (1963) 145.

[10.27] Durand, J., Z. Naturforsch. **18a** (1963) 281.

[10.27a] Eberhagen, A. und M. J. Bernstein, Z. Naturforsch. **20a** (1965) 1375.

[10.28] Eckarth, H., Z. Physik **107** (1937) 182.

[10.29] Ecker, G., Z. Physik **148** (1957) 593.

[10.29a] Ecker, G. and W. Kröll, Forsch.-Ber. Nordrhein-Westf. No. 1221 (1962).

[10.30] Ecker, G. and W. Kröll, Phys. Fluids **6** (1963) 62.

[10.30a] Ecker, G. and W. Kröll, Z. Naturforsch. **21a** (1966) 2023.

[10.31] Edels, H. and J. D. Craggs, Proc. Phys. Soc. Lond. **A 64** (1951) 574.

[10.31a] ELIESEV, V. V., Optika i Spektroskopiya **26** (1969) 481.
[10.31b] ELYASHEVICH, M. A. and V. I. TKACHYOVA, Optika i Spektroskopiya **25** (1968) 12.
[10.31c] ELYASHEVICH, M. A. and V. I. TKACHYOVA, Optika i Spektroskopiya **26** (1969) 3.
[10.31d] ENGLEMAN, R., J. quant. Spectrosc. Radiat. Transfer **9** (1969) 391.
[10.31e] ERVENS, W., Diss. Rhein-Westfl. Techn. Hochsch. Aachen 1969.
[10.31f] ESPENHAIN, M., H. J. KUSCH and W. LOCHTE-HOLTGREVEN, Z. Astrophys. **61** (1965) 77.
[10.31g] EVANS, D. L. and R. S. TANKIN, Phys. Fluids **10** (1967) 1137.
[10.31h] EVDOKIMOV, YU. V., Optika i Spektroskopiya **24** (1968) 832.
[10.31i] EVDOKIMOV, YU. V., M. P. CHAIKA and V. A. CHERENKOVSKI, Optika i Spektroskopiya **27** (1969) 184.
[10.31j] EVTUSHENKO, T. P., A. N. ZAIDEL, G. V. OSTROVSKAYA and T. YA. CHELIDZE, Zh. tekhn. Fiz. **36** (1966) 1506.
[10.31k] FAGG, L. W., J. quant. Spectrosc. Radiat. Transfer **5** (1965) 321.
[10.31l] FAROUX, J.-P., Compt. Rend. Acad. Sci. **264** (1967) B1573.
[10.31m] FEAUTRIER, N., S. BRECHOT and H. VAN REGEMORTER, Proc. 5. internat. Conf. Phys. Electronic Atomic Collisions, Nauka, Leningrad 1967, p. 525.
[10.31n] FELDHAUSEN, H. and H. J. KUSCH, Z. Astrophys. **67** (1967) 123.
[10.31o] FERGUSON, E. and H. SCHLÜTER, Ann. Physics **22** (1963) 351.
[10.31p] FINKELNBURG, W. and O. TH. HAHN, Phys. Z. **39** (1938) 98.
[10.32] FOSTER, J. S., Nature, Lond. **116** (1925) 135.
[10.33] FOSTER, J. S., Proc. Roy. Soc. Lond. **114** (1927) 47.
[10.34] FOSTER, J. S., Proc. Roy. Soc. Lond. **A 117** (1928) 137.
[10.35] FRISH, S. E. and YU. M. KAGAN, Zh. eksp. teor. Fiz. **17** (1947) 577.
[10.36] FRISH, S. E. and YU. M. KAGAN, Zh. eksp. teor. Fiz. **18** (1948) 519.
[10.36a] FRISH, S. E. and A. P. MAKAROV, Optika i Spektroskopiya **24** (1968) 831.
[10.36b] FUCKS, W., W. L. BOHN, G. HEINRICH and P. PLATZ, Z. Physik **170** (1962) 409.
[10.36c] FÜCHTBAUER, C. and F. GÖSSLER, Naturwiss. **21** (1933) 315.
[10.36d] FÜCHTBAUER, C. and F. GÖSSLER, Z. Physik **87** (1934) 89.
[10.36e] FÜCHTBAUER, C. and W. HEESEN, Z. Physik **113** (1939) 323.
[10.37] FÜCHTBAUER, C. and G. HEIMANN, Z. Phys. **110** (1938) 1.
[10.37a] FÜCHTBAUER, C., G. JOOS and O. DINKELACKER, Ann. Physik **71** (1923) 204.
[10.37b] FÜCHTBAUER, C. and C. S. PAO, Phys. Rev. **58** (1940) 1058.
[10.37c] FÜCHTBAUER, C. and P. SCHULZ, Z. Physik **97** (1935) 699.
[10.37d] FUJIOKA, Y., Sci. Pap. Inst. Phys.-Chem. Res., Tokyo **5** (1926) 45.
[10.38] FUJIOKA, J. and S. NAKAMURA, Astrophys. J. **65** (1927) 201.
[10.39] FURSSOW, W. and A. WLASSOW, J. Physics, Moskau **1** (1939) 335.
[10.39a] GABLER, F., Phys. Z. **44** (1943) 108.
[10.39b] GALATRY, L., Phys. Rev. **122** (1961) 1218.
[10.39c] GARDNER, A. L., in: Engineering aspects of Magnetohydrodynamics, Columbia Univ. Press, New York, London 1962, S. 438; also: UCRL-Rep. 6232-T-1962.
[10.40] GARDNER, A. L., W. L. BARR, R. L. KELLY and N. L. OLESON, Phys. Fluids **5** (1962) 794.
[10.40a] GARRETT, R. O. and S. Y. CH'EN, Phys. Rev. **144** (1966) 66.
[10.40b] GARRETT, R. O., S. Y. CH'EN and E. C. LOOI, Phys. Rev. **156** (1967) 48.
[10.40c] GEBAUER, R., Physikertagung Hamburg 1963, Physik-Verlag Mosbach/Baden 1964, p. 71.
[10.41] GEBAUER, R. and H. STEUL, Z. Physik **126** (1949) 465.
[10.41a] GERARDO, J. B. and R. A. HILL, Phys. Rev. Lett. **17** (1966) 623.
[10.42] GERICKE, W. E., Z. Astrophys. **53** (1961) 68.
[10.42a] GERMAN, K. R. and R. N. ZARE, Phys. Rev. **186** (1969) 9.

[10.43] GERTHSEN, P., Z. Physik **138** (1954) 515.
[10.44] GIBBONS, R. A. and R. J. MACKIN, Proc. 5. internat. Conf. Ionization Phenomena in Gases, Vol. II, North Holland Publ. Co., Amsterdam 1962, p. 1769.
[10.44a] GIESKE, H. A. and H. R. GRIEM, Astrophys. J. **157** (1969) 963.
[10.44b] GOLDFARB, V. M. and S. V. DRESVIN, Teplofizika **3** (1965) 333.
[10.44c] GOLDFARB, V. M. and YE. V. ILINA, Optika i Spektroskopiya **17** (1964) 302.
[10.44d] GOLDFARB, V. M., E. V. ILINA, I. E. KOSTYGOVA, G. A. DUKYANOV and V. A. SILANTYEV, Zh. prikl. mekhan. tekhn. Fiz. No. 1 (1967) 111.
[10.44e] GOLOBORODKO, V. T., V. L. KASHTANYER and I. V. ORFANOV, Zh. prikl. Spektrosk. **10** (1969) 13.
[10.44f] GOLUBOVSKI, YU. B. V. M. ZAKHAROVA and YU. M. KAGAN, Optika i Spektroskopiya **17** (1964) 333.
[10.44g] GRANIER, R. and J. GRANIER, Compt. Rend. Acad. Sci. **257** (1963) 2627.
[10.44h] GRANIER, R. and J. GRANIER, Compt. Rend. Acad. Sci. **260** (1965) 92.
[10.44i] GRANIER, R. and J. GRANIER, Compt. Rend. Acad. Sci., Paris, **AB 262** (1966) B 605.
[10.44j] GRANIER, R. and J. GRANIER, Compt. Rend. Acad. Sci., Paris, **B 262** (1966) 761.
[10.44k] GRANIER, R. and J. GRANIER, Compt. Rend. Acad. Sci., Paris, **B 262** (1966) 1502.
[10.44l] GRANIER, R., J. GRANIER and B. VODAR, J. quant. Spectrosc. Radiat. Transfer **6** (1966) 741.
[10.44m] GRECHIKHIN, L. I., Doklady Akad. Nauk Beloruss. SSR **10** (1966) 232.
[10.44n] GRECHIKHIN, L. I., Optika i Spektroskopiya **22** (1967) 548.
[10.44o] GRECHIKHIN, L. I., Optika i Spektroskopiya **25** (1968) 655.
[10.44p] GRECHIKHIN, L. I., Teplofizika **6** (1968) 507.
[10.44q] GRECHIKHIN, L. I., Zh. prikl. Spektrosk. **11** (1969) 221.
[10.44r] GRECHIKHIN, L. I. and M. A. ELYASHEVICH, Zh. prikl. Spektrosk. **3** (1965) 201.
[10.44s] GRECHIKHIN, L. I., M. A. ELYASHEVICH and L. I. KISELEVSKI, in: Nizkotemperaturnaya Plazma, Mir, Moscow 1967, p. 287.
[10.44t] GRECHIKHIN, L. I., L. YA. MINKO and N. I. NAGORNAYA, Zh. tekhn. Fiz. **36** (1966) 1207.
[10.44u] GRECHIKHIN, L. I. and D. K. SKUTOV, Zh. prikl. Spektrosk. **4** (1966) 381.
[10.44v] GRECHIKHIN, L. I. and E. S. TYUNINA, Teplofizika **1** (1963) 399.
[10.44w] GRECHIKHIN, L. I. and E. S. TYUNINA, Teplofizika **2** (1964) 689.
[10.44x] GRECHIKHIN, L. I. and E. S. TYUNINA, Zh. prikl. Spektrosk. **5** (1966) 426.
[10.44y] GRECHIKHIN, L. I. and E. S. TYUNINA, in: Fizika i Khimiya Obrabotki materialov No. **3** (1967) 3.
[10.44z] GRECHIKHIN, L. I. and E. S. TYUNINA, Zh. prikl. Spektrosk. **7** (1967) 7.
[10.45] GRECHIKHIN, L. I. and E. S. TYUNINA, Teplofizika **5** (1967) 764.
[10.45a] GREGORY, C., Phys. Rev. **61** (1942) 465.
[10.45b] GREIG, J. R., C. P. LIM, G. A. MOO-YOURIG, G. PALUMBO and H. R. GRIEM, Phys. Rev. **172** (1968) 148.
[10.45c] GRIDNEVA, S. M. and G. A. KASABOV, Proc. 7. internat. Conf. Phenom. Ionized Gases, Belgrade 1965, Vol. 2, p. 581.
[10.45d] GRIEM, H. R., Z. Physik **137** (1954) 280.
[10.46] GRIEM, H. R., Astrophys. J. **132** (1960) 883.
[10.47] GRIEM, H. R., Astrophys. J. **136** (1962) 422.
[10.47a] GRIEM, H. R., Plasma Spectroscopy, McGraw Hill Book Co., New York 1964.
[10.47b] GRIEM, H. R., Phys. Rev. **140** (1965) A1140; Erratum in: Phys. Rev. **144** (1966) 366.
[10.47c] GRIEM, H. R., Phys. Rev. Lett. **17** (1966) 509.
[10.47d] GRIEM, H. R., Proc. 8. internat. Conf. Phenom. Ionized Gases, Contributed Pap., IAEA Vienna 1967, p. 433.

[10.47e] GRIEM, H. R., Astrophys. J. **154** (1968) 1111.
[10.47f] GRIEM, H. R., Phys. Rev. **165** (1968) 258.
[10.48] GRIEM, H. R., M. BARANGER, A. C. KOLB and G. OERTEL, Phys. Rev. **125** (1962) 177.
[10.49] GRIEM, H. R., A. C. KOLB and K. Y. SHEN, Phys. Rev. **116** (1959) 4.
[10.50] GRIEM, H. R., A. C. KOLB and K. Y. SHEN, U. S. Naval Res. Lab., Rep. No. 5455, Washington 1960.
[10.51] GRIEM, H. R., A. C. KOLB and K. Y. SHEN, Astrophys. J. **135** (1962) 272.
[10.52] GRIEM, H. R., A. C. KOLB and K. Y. SHEN, U. S. Naval Res. Lab., Rep. No. 5805, Washington 1962.
[10.53] GRIEM, H. R. and K. Y. SHEN, Phys. Rev. **122** (1961) 1490.
[10.54] GRIFFIN, P. M., J. R. MCNALLY and G. K. WERNER, Temperature, its Measurement and Control in Science and Industry, Vol. III, 1, Reinhold Publ. Co., New York 1962, p. 651.
[10.54a] GRIGORYEV, V. N., Zh. prikl. mekhan. tekhn. Fiz. No. 2 (1965) 35.
[10.54b] GROSSWENDT, Z. Naturforsch. **24a** (1969) 1424.
[10.54c] GROTRIAN, W. and G. RAMSAUER, Phys. Z. **28** (1927) 846.
[10.54d] GRYCUK, T., M. KUBIAK and J. PROCHOROW, Bull. acad. polon. sci. ser. sci. math. astron. phys. **12** (1964) 517.
[10.54e] GÜNDEL, H., Diss. Ernst-Moritz-Arndt Univ. Greifswald 1970. Also: Beitr. Plasmaphys. **10** (1970) 455 and **11** (1971) 1.
[10.54f] HAGEN, S. and R. RITSCHL, Spectrochim. Acta **1** (1959) 477.
[10.54g] HAMMOND, G. L., Astrophys. J. **136** (1962) 431.
[10.55] HATTENBURG, A. T. and H. J. KOSTKOWSKI, in: Temperature, its Measurement and Control in Science and Industry, Vol. 3, 1, Reinhold Publ. Co., New York 1962, p. 587.
[10.55a] HAUN, R. D. and J. R. ZACHARIAS, Phys. Rev. **107** (1957) 107.
[10.55b] HEAD, R. M., Proc. 6. internat. Conf. Ion. Phenom. Gases, Paris 1963, Vol. 3, p. 395.
[10.56] HEARN, A. G., B. B. JONES and S. A. RAMSDEN, Plasma Physics **4** (1962) 22.
[10.56a] HEMMERLICH, H., Inst. Plasmaphys., Garching near Munich, Rep. IPP 1/16, 1964.
[10.57] HENKEL, W. D., Z. Phys. **137** (1954) 295.
[10.57a] HENNING, H., Z. Astrophys. **62** (1965) 109.
[10.58] HEPNER, G., Proc. Coll. spectrosc. internat. VI., Pergamon Press, London 1957, p. 356.
[10.59] HERMAN, L., Proc. 4. internat. Conf. Ionization Phenomena in Gases, Vol. 1, North Holland Publ. Co., Amsterdam 1960, p. 403.
[10.59a] HERMAN, R. M. and R. G. BREENE, J. Mol. Spectrosc. **23** (1967) 343.
[10.59b] HERMEL, A., Beitr. Plasmaphys. **9** (1969) 261.
[10.59c] HERMEL, A. and K. SELIGER, Monatsber. Dtsch. Akad. Wiss. Berlin **8** (1966) 286.
[10.59d] HESSBERG, H. and W. BÖTTICHER, Z. Naturforsch. **22a** (1967) 316.
[10.60] HEY, P., Z. Astrophys. **52** (1961) 254.
[10.60a] VON DER HEYDE, R. and H. J. KUSCH, Z. Astrophys. **68** (1968) 1.
[10.60b] HILL, R. A., J. quant. Spectrosc. Radiat. Transfer **4** (1964) 857.
[10.60c] HILL, R. A., J. quant. Spectrosc. Radiat. Transfer **7** (1967) 401.
[10.60d] HILL, R. A. and J. B. GERARDO, Phys. Rev. **162** (1967) 45.
[10.61] HINDMARSH, W. R., Proc. Phys. Soc. Lond. **72** (1958) 279.
[10.62] HINDMARSH, W. R., Mon. Notices Roy. astron. Soc. **119** (1959) 11.
[10.63] HINDMARSH, W. R., Mon. Notices Roy. astron. Soc. **121** (1960) 48.
[10.63a] HINDMARSH, W. R., Sympos. Molec. Struct. and Spectrosc., Columbus 1963.
[10.63b] HINDMARSH, W. R., A. D. PETFORD and G. SMITH, Proc. Roy. Soc. Lond. **A 297** (1967) 296.

[10.63c] HINDMARSH, W. R. and K. A. THOMAS, Monthly Notices Roy. Astron. Soc. **119** (1959) 21.
[10.63d] HINDMARSH, W. R. and K. A. THOMAS, Proc. Phys. Soc. Lond. 77 (1961) 1193.
[10.63e] HIRAMOTO, T., S. MATSUNAGA, H. SHIRAKATA and S. YANA, Electricity from MHD, IAEA Vienna 1966, Vol. I, p. 85.
[10.63f] HIRSCHBERG, J. G., Phys. Fluids 7 (1964) 543. Erratum in: Phys. Fluids 8 (1965) 772.
[10.63g] HIRSCHBERG, J. G. and E. HINNOV, J. chem. Phys. **45** (1966) 2233.
[10.63h] HOLLANDER, T. and H. P. BROIDA, J. quant. Spectrosc. Radiat. Transfer 7 (1967) 965.
[10.63i] HOLMES, R. and I. L. FREESTON, Proc. 9. internat. Conf. Phenom. Ionized Gases, Bucharest 1969, p. 631.
[10.63j] HUDIS, M., K. CHUNG and D. J. ROSE, J. appl. Phys. **39** (1968) 3297.
[10.64] HÜBNER, K., Proc. 6. internat. Conf. Ionization Phenomena in Gases, Paris 1963, Vol. 4, p. 31.
[10.64a] HÜBNER, K., Z. Naturforsch. **19a** (1964) 1111.
[10.65] HULDT, L. and E. KNALL, Z. Naturforsch. **9a** (1954) 663.
[10.65a] HULL, G. F., Phys. Rev. **50** (1936) 1148.
[10.66] INGLIS, D. E. and E. TELLER, Astrophys. J. **90** (1939) 439.
[10.66a] IOVA, J., Optika i Spektroskopiya **20** (1966) 27.
[10.66b] IOVA, J., Proc. 8. internat. Conf. Phenom. Ionized Gases, Contributed Pap., IAEA Vienna 1967, p. 440.
[10.66c] IOVA, J., Proc. 9. internat. Conf. Phenom. Ionized Gases, Bucharest 1969, p. 627.
[10.66d] ISHIDA, Y. and M. FUKUSHIMA, Sci. Pap. Inst. Phys.-Chem. Res., Tokyo 9 (1928) 141.
[10.67] ISHIDA, J. and M. FUKUSHIMA, Sci. Pap. Inst. Phys.-Chem. Res. Japan **14** (1930) 123.
[10.68] ISHIDA, Y. and S. HIYAMA, Sci. Pap. Inst. Phys.-Chem. Res. Tokyo **36** (1939) 390.
[10.68a] ISHIDA, Y., T. TAMURA and M. FUKUSHIMA, Sci. Pap. Inst. Phys.-Chem. Res. Tokyo **36** (1939) 417.
[10.69] JAHN, R. E., Proc. 5. internat. Conf. Ionization Phenomena in Gases, Vol. 1, North Holland Publ. Co., Amsterdam 1962, p. 955.
[10.70] JAHODA, F. C., F. L. RIBE and G. A. SAWYER, Phys. Rev. **131** (1963) 24.
[10.70a] JALUFKA, N. W., G. K. OERTEL and G. S. OFELT, Phys. Rev. Lett. **16** (1966) 1073.
[10.70b] JEFIMENKO, O., J. chem. Phys. **30** (1959) 1112.
[10.70c] JEFIMENKO, O., J. chem. Phys. **39** (1963) 2457.
[10.70d] JEFIMENKO, O., J. Opt. Soc. Amer. **54** (1964) 129.
[10.70e] JEFIMENKO, O. and S.-Y. CH'EN, J. chem. Phys. **26** (1957) 913.
[10.70f] JEFIMENKO, O. and W. CURTIS, J. chem. Phys. **27** (1957) 953.
[10.70g] JEFIMENKO, O. and G. M. WILLIAMS, J. chem. Phys. **42** (1965) 207.
[10.71] JOHN, R. M. ST. and J. G. WINANS, Phys. Rev. **98** (1955) 562 and 1664.
[10.72] JOOS, G. and A. SAUR, in: Landolt-Börnstein, Zahlenwerte u. Funktionen Vol. I, 1, Springer-Verlag Berlin-Göttingen-Heidelberg 1950.
[10.73] JÜRGENS, G., Z. Physik **134** (1952) 21.
[10.74] JUNG, M., Z. Astrophys. **58** (1963) 93.
[10.74a] KAGAN, YU. M. and N. N. KHRISTOV, Optika i Spektroskopiya **26** (1969) 886.
[10.74b] KAGAN, YU. M. and N. N. KHRISTOV, Optika i Spektroskopiya **27** (1969) 573.
[10.75] KAGAN, YU. M., L. A. LUISOVA, R. I. LYAGUSTSHENKO and A. D. CHACHAYEV, Optika i Spektroskopiya **15** (1963) 446.
[10.75a] KARIMOV, M. G. and V. G. KURT, Doklady Akad. Nauk SSSR **117** (1957) 207.

[10.75b] KEILHACKER, M., H. HEROLD, J. COOPER and D. E. ROBERTS, in: Plasma Physics and Controlled Nuclear Fusion Res., IAEA Vienna 1966, Vol. 1, p. 315.
[10.75c] KEPPLE, P. and H. R. GRIEM, Phys. Rev. **173** (1968) 317.
[10.75d] KHADJAVI, A., W. HAPPER and A. LURIO, Phys. Rev. Lett. **17** (1966) 463.
[10.75e] KHADJAVI, A., A. LURIO and W. HAPPER, Phys. Rev. **167** (1968) 128.
[10.76] KIMURA, M. and G. NAKAMURA, Jap. J. Phys. **2** (1923) 61.
[10.76a] KIELKOPF, J. F. and J. A. GWINN, J. chem. Phys. **48** (1968) 5570.
[10.76b] KISELEVSKI, L. I. and V. D. SHIMANOVICH, Teplofizika **2** (1964) 48.
[10.77] KITAYEVA, V. F., Izvest. Akad. Nauk SSSR Ser. Fiz. **19** (1955) 27.
[10.77a] KITAYEVA, V. F., Trudy Fiz. Inst. P. N. Lebedeva Vol. **11** (1959) 3.
[10.78] KITAYEVA, V. F., V. V. OBUCHOV-DENISOV and N. N. SOBOLEV, Optika i Spektroskopiya **12** (1962) 178.
[10.79] KITAYEVA, V. F., V. V. OBUCHOV-DENISOV and N. N. SOBOLEV, in: Fizicheskie problemy spektroskopii Vol. 1, Izd. Akad. Nauk SSSR, Moscow 1962, p. 116.
[10.79a] KITAYEVA, V. F., YU. I. OSIPOV and N. N. SOBOLEV, J. Quantum Electronics **2** (1966) 635.
[10.80] KITAYEVA, V. F. and N. N. SOBOLEV, Optika i Spektroskopiya **1** (1956) 302.
[10.81] KITAYEVA, V. F. and N. N. SOBOLEV, Doklady Akad. Nauk SSSR **137** (1961) 1091.
[10.82] KITAYEVA, V. F. and N. N. SOBOLEV, Proc. 5. internat. Conf. Ionization Phenomena in Gases, Vol. 2, North Holland Publ. Co., Amsterdam 1962. p. 1897.
[10.83] KIUTI, M., K. OCHAIAI, and Y. NISHIMURA, Jap. J. Phys. **5** (1929) 139.
[10.84] KIVEL, B., Phys. Rev. **98** (1955) 1055.
[10.84a] KLEIN, L. and H. MARGENAU, J. chem. Phys. **30** (1959) 1556.
[10.84b] KLEMAN, B. and E. LINDHOLM, Ark. Mat. Astron. Fys. **32** (1946) 1.
[10.84c] KLING, M. and M.-U. BETH, Proc. 9. internat. Conf. Phenom. Ionized Gases, Bucharest 1969, p. 341.
[10.84d] KNAUER, W. and M. A. LUTZ, Appl. Phys. Lett. **2** (1963) 109.
[10.85] KOCH, O. and H. DUNSTÄDTER, Z. Physik **129** (1951) 140.
[10.86] KOGAN, W. I., J. quant. Spectrosc. Radiat. Transfer **4** (1964) 243.
[10.87] KOLB, A. C., Diss. Univ. Michigan 1957.
[10.88] KOLB, A. C. and H. GRIEM, Phys. Rev. **111** (1958) 514.
[10.88a] KOLOSHNIKOV, V. G., Zh. tekhn. Fiz. **34** (1964) 34.
[10.88b] KOMISAROV, N. M., Optika i Spektroskopiya **25** (1968) 829.
[10.88c] KONJEVIC, N., J. LABAT, L. CIRKOVIC and J. PURIO, Proc. 9. internat. Conf. Phenom. Ionized Gases, Bucharest 1969, p. 593.
[10.89] KONOVALOV, Y. N. and V. V. MAGIDSON, Zh. tekhn. Fiz., **32** (1962) 450.
[10.89a] KOPFERMANN, H. and W. PAUL, Z. Physik **120** (1943) 545.
[10.90] KOPFERMANN, Atomspektren, in: Fiat Review, Physics of the Electron Shell, Wiesbaden 1948.
[10.91] KOSYREVA, Y. F., Optika i Spektroskopiya **2** (1957) 43.
[10.91a] KREFFT, H., Phys. Z. **32** (1931) 948.
[10.91b] KREFFT, H. and R. ROMPE, Z. Physik **73** (1932) 682.
[10.91c] KREFFT, H. and R. ROMPE, Tech.-wiss. Abh. Osram **3** (1934) 52.
[10.91d] KRYLOVA, S. I., L. A. LUIZOVA and A. D. KHAKHAEV, Optika i Spektroskopiya **25** (1968) 165.
[10.91e] KRYLOVA, S. I., L. A. LUIZOVA and V. A. SOLYANIKOVA, Optika i Spektroskopiya **27** (1969) 391.
[10.92] KUDRIN, L. P. and G. V. SHOLIN, Optika i Spektroskopiya **14** (1963) 607.
[10.92a] KUHN, H. G. and E. L. LEWIS, Proc. Roy. Soc. Lond. **A 299** (1967) 423.
[10.92b] KUHN, H. G. and J. M. VAUGHAN, Proc. Roy. Soc. Lond. **A 277** (1964) 297.

[10.92c] KUNZE, H.-J., H. R. GRIEM, A. W. DESILVA, G. C. GOLDENBAUM and I. J. SPALDING, Phys. Fluids **12** (1969) 2669.
[10.93] KUSCH, H. J., Z. Astrophys. **45** (1958) 1.
[10.93a] KUSCH, H. J., Z. Astrophys. **67** (1967) 64.
[10.93b] KUSCH, H. J. and G. MEINHOLD, Z. Astrophys. **66** (1967) 364.
[10.93c] KUSCH, H. J. and E. OBERSCHELP, Z. Astrophys. **67** (1967) 77.
[10.93d] KUSCH, H. J. and E. OBERSCHELP, Z. Astrophys. **67** (1967) 85.
[10.93e] KWIATKOWSKI, S., Bull. acad. polon. sci. ser. sci. math. astron. phys. **12** (1964) 125.
[10.94] LALOS, G. T. and G. L. HAMMOND, Astrophys. J. **135** (1962) 616.
[10.94a] LANG, K., Acta phys. austr. **11** (1957) 76.
[10.95] LAPORTE, O. and E. B. TURNER, in: Fundamental Data obtained from Shock-Tube Experiments. Pergamon Press, Oxford 1961, p. 386.
[10.96] LAPWORTH, K. C., Nature, Lond. **192** (1961) 252.
[10.96a] LAURENT, J. and S. WENIGER, Proc. 9. internat. Conf. Phenom. Ionized Gases, Bucharest 1969, p. 588.
[10.97] LENZ, W., Z. Physik **80** (1933) 423.
[10.97a] LEWIS, E. L., Proc. Phys. Soc. Lond. **92** (1967) 817.
[10.97b] LEWIS, E. L., M. M. REBBECK and J. M. VAUGHAN, Phys. Letters **30 A** (1969) 50.
[10.97c] LEYCURAS, Y., Ann. Physique **8** (1963) 577.
[10.97d] LEYCURAS, Y., Compt. Rend. Acad. Sci. Paris **267 B** (1968) 515.
[10.98] LIDÉN, K., Ark. Mat. Astron. Fys. **35 A** (1948) No. 23.
[10.98a] LIFSHITS, E. V., A. M. EGOROV and O. G. ZAGORODNOV, Ukrain. Fiz. Zh., Kiev **6** (1961) 793.
[10.98b] LIFSHITS, E. V., S. S. KRIVULYA and V. S. US, Ukrain. Fiz. Zh., Kiev **8** (1963) 1328.
[10.98c] LINCKE, R. and H. R. GRIEM, Phys. Rev. **143** (1966) 66.
[10.99] LINDHOLM, E., Ark. Mat. Astron. Fys. **28 B** (1941) No. 3.
[10.100] LINDHOLM, E., Diss. Upsala 1942.
[10.101] LINDHOLM, E., Ark. Mat. Astron. Fys. **32 A** (1946) Nr. 17.
[10.101a] LIN-TANG, C. and C. TUNG-CHING, Acta phys. sinica **21** (1965) 1591.
[10.101b] LITVAK, M. M. and D. F. EDWARDS, J. appl. Phys. **37** (1966) 4462.
[10.101c] LLOYD, P. E. and D. S. HUGHES, Phys. Rev. **52** (1937) 1251.
[10.101d] LOCHTE-HOLTGREVEN, W., in: A survey of Phenomena in Ionized Gases, IAEA, Vienna 1968, p. 489.
[10.102] LORENTZ, H. A., Versl. Amsterdam Acad. **14** (1905) 518, 577.
[10.103] LO SURDO, A., Phys. Z. **15** (1914) 122, 193.
[10.103a] LUDWIG, D., Inst. Plasmaphys., Garching near Munich, Report IPP 3/88, March 1969.
[10.104] LUISOVA, L. A., Optika i Spektroskopiya **15** (1963) 835.
[10.105] LUKYANOV, S. Y. and V. I. SINITSIN, Zh. eksp. teor. Fiz. **34** (1958) 849.
[10.106] MCCHESNEY, M. and J. D. CRAGGS, J. Electronics **4** (1958) 481.
[10.106a] MCLEAN, E. A. and S. A. RAMSDEN, Phys. Rev. **140** (1965) A1122.
[10.107] MCNALLY, J. R. and M. R. SKIDMORE, Appl. Optics **2** (1963) 699.
[10.107a] MCNALLY, J. R., R. SKIDMORE, J. E. FRANCIS, P. M. JENKINS and D. A. GRIFFIN, J. Opt. Soc. Amer. **53** (1963) 495.
[10.108] MAISSEL, L., J. Opt. Soc. Amer. **48** (1958) 853.
[10.108a] MAJKOWSKI, R. F. and R. J. DONOHUE, Phys. Rev. **173** (1968) 177.
[10.109] MAK, A. A., Zh. eksp. teor. Fiz. **37** (1959) 1488.
[10.109a] MAKAROV, A. P., Optika i Spektroskopiya **26** (1969) 341.
[10.110] MANDELSHTAM, S. L., in: Optik und Spektroskopie aller Wellenlängen, Akademie-Verlag, Berlin 1962, p. 372.

[10.111] MANDELSHTAM, S. L. and M. A. MAZING, Izvest. Akad. Nauk SSSR Ser. Fiz. **23** (1959) 1017; German Translation: Phys. Abh. SU **2** (1960) 228.
[10.112] MANDELSHTAM, S. L. and M. A. MAZING, Proc. 8. Coll. spectrosc. internat., Verlag H. R. Sauerländer & Co., Aarau 1960, p. 184.
[10.112a] MANDELSHTAM, S. L., M. A. MAZING, I. I. SOBELMAN and L. A. VAINSHTEIN, Proc. 6. internat. Conf. Ion. Phenom. Gases, Paris 1963, Vol. 3, p. 331.
[10.112b] MANDELSHTAM, S. L., P. P. PASHININ, A. V. PROKHINDEYEV, A. M. PROKHOROV and N. K. SUKHODREV, Zh. eksp. teor. Fiz. **47** (1964) 2003.
[10.113] MARGENAU, H., Proc. 4. internat. Conf. Ionization Phenomena in Gases, Vol. 2, North Holland Publ. Co., Amsterdam 1960, p. 799.
[10.114] MARGENAU, H. and M. LEWIS, Rev. mod. Phys. **31** (1959) 569.
[10.114a] MARGENAU, H. and W. WATSON, Phys. Rev. **44** (1933) 92.
[10.114b] MARRUS, R., D. MCCOLM and J. YELLIN, Phys. Rev. **147** (1966) 55.
[10.114c] MAZING, M. A., Trudy Fiz. Inst. P. N. Lebedeva **15** (1961) 55.
[10.115] MAZING, M. A. and S. L. MANDELSHTAM, Probleme des Plasmas in Physik und Astronomie, Akademie-Verlag, Berlin 1958, p. 150.
[10.116] MAZING, M. A. and S. L. MANDELSHTAM, Materialy 10. soveshchaniya po spektroskopii, Vol. 2, Lvov 1958, p. 305.
[10.117] MAZING, M. A. and S. L. MANDELSHTAM, Zh. eksp. teor. Fiz. **36** (1959) 1329.
[10.117a] MAZING, M. A. and P. D. SERAPINIS, Phys. Inst. Akad. Nauk USSR, Moscow 1969, Reprint No. 155.
[10.118] MAZING, M. A. and N. A. VRUBLEVSKAYA, Optika i Spektroskopiya **13** (1962) 308.
[10.119] MAZING, M. A. and N. A. VRUBLEVSKAYA, Optika i Spektroskopiya **16** (1964) 11.
[10.119a] MEJAŠKI-TONEJC, A. and V. VUJNOVIĆ, Proc. 9. internat. Conf. Phenom. Ionized Gases, Bucharest 1969, p. 629.
[10.119b] MEWE, R. and R. F. DE VRIES, Plasma Physics **6** (1964) 591.
[10.119c] MEYER. J., Z. Astrophys. **60** (1964) 94.
[10.120] MICHELS, A. and H. DE KLUIVER, Physica **22** (1956) 919.
[10.121] MICHELS, A., H. DE KLUIVER and B. CASTLE, Physica **23** (1957) 1131.
[10.121a] MICHELS, A., H. DE KLUIVER and B. CASTLE, Coll. internat. Centre nat. rech. sci. **77** (1959) 221.
[10.122] MICHELS, A., H. DE KLUIVER and D. MIDDELKOOP, Physica **24** (1958) 543.
[10.123] MICHELS, A., H. TJIN A DJIE, H. DE KLUIVER and C. A. TEN SELDAM, Physica **23** (1957) 1115.
[10.123a] MICHNENKO, G. A. and E. D. PROTSENKO, Optika i Spektroskopiya **26** (1969) 668.
[10.123b] MILIYANTSHUK, V. S., Doklady Akad. Nauk SSSR **67** (1949) 1001.
[10.123c] MILLER, M. H. and R. D. BENGTSON, Phys. Rev. **A 1** (1970) 983.
[10.123d] MING-YUAN, H., Y. MOW-FOH and S. HSIANG, Acta phys. sinica **21** (1965) 1606.
[10.124] MINNHAGEN, L., Diss. Lund 1944.
[10.125] MINNHAGEN, L., Ark. Mat. Astron. Fys. **35 A** (1948) Nr. 16.
[10.126] MINNHAGEN, L., Ark. Fys. **1** (1950) 425.
[10.127] MINNHAGEN, L. and U. B. NORDBERG, Ark. Fys. **8** (1954) 197.
[10.128] MISJUNAS, A. W., Fizicheskie problemy spektroskopii, Vol. 1, Izd. Akad. Nauk SSSR, Moscow 1962, p. 94.
[10.128a] MITCHELL, A. C. G. and M. W. ZEMANSKY, Resonance radiation and excited atoms. Cambridge Univ. Press, New York 1961.
[10.128b] MITTLEMAN, M. H., Phys. Rev. **162** (1967) 81; Erratum in: Phys. Rev. **166** (1968) 206.
[10.128c] MIYAMOTO, K., M. OTSUKA, J. FUJITA, S. KAWASAKI, N. INOUE, Y. SUZUKI, T. UCHIDA, K. MORI, K. ANDO and T. ISHIMURA, Japan. J. appl. Phys. **5** (1966) 970.

[10.129] Mitsuk, V. T., Zh. tekhn. Fiz. Moskau **28** (1958) 1316.
[10.130] Mohler, F. L., Astrophys. J. **90** (1939) 429.
[10.130a] Moore, C., Atomic Energy Levels, U. S. Nat. Bur. Stand. Circ. 467 (1949).
[10.130b] Morris, R. N. and B. Ahlborn, Univ. Brit. Columbia, Dep. Phys., UVA Rep. No. 3 (1968).
[10.130c] Morris, J. C. and R. L. Garrison, Phys. Rev. **188** (1969) 112.
[10.130d] Morris, J. C. and R. U. Krey, Phys. Rev. Lett. **21** (1968) 1043.
[10.130e] Morris, J. C., R. U. Krey and R. L. Garrison, US Air Force, Office of Aerospace Res., Aerospace Res. Labs., Rep. ARL 68-0103, May 1968.
[10.130f] Mozer, B. and M. Baranger, Phys. Rev. **118** (1960) 626.
[10.130g] Muntenbruch, H., Spectrochim. Acta **16** (1960) 1031.
[10.131] Murakawa, K. and S. Mizuno-Hashimoto, J. phys. Soc. Japan **14** (1959) 1235.
[10.131a] Murakawa, K. and M. Yamamoto, J. phys. Soc. Japan **20** (1965) 1057.
[10.131b] Murakawa, K. and M. Yamamoto, J. Phys. Soc. Japan **23** (1967) 416.
[10.131c] Murakawa, K., M. Yamamoto and S. Hashimoto, Proc. 7. internat. Conf. Phenom. Ionized Gases, Belgrade 1965, Vol. 2, p. 594.
[10.131d] Mustafin, K. S., Vestnik Leningrad Univ. Ser. Fiz. i Khim. No. 22 (1960) 130.
[10.131e] Nadler, M. and W. E. Kaskan, J. quant. Spectrosc. Radiat. Transfer **10** (1970) 25.
[10.131f] Nguyen-Hoe, E. Banerjea, H.-W. Drawin and L. Herman, J. quant. Spectrosc. Radiat. Transfer **5** (1965) 835.
[10.131g] Nguyen-Hoe, H.-W. Drawin and L. Herman, J. quant. Spectrosc. Radiat. Transfer **4** (1964) 847.
[10.131h] Nguyen-Hoe, H.-W. Drawin and L. Herman, Association EURATOM, Fontenay-aux-Roses, Rapp. CEA No. 2912, 1965.
[10.131i] Nguyen-Hoe, H.-W. Drawin and L. Herman, Ass. EURATOM, Fontenay-aux-Roses, Rapp. CEA-R-3057, 1966.
[10.131j] Nguyen-Hoe, H.-W. Drawin and L. Herman, Z. Naturforsch. **21a** (1966) 1515.
[10.131k] Nguyen-Hoe, H.-W. Drawin and L. Herman, J. quant. Spectrosc. Radiat. Transfer **7** (1967) 429.
[10.131l] Nguyen-Hoe, H.-W. Drawin and L. Herman, Associat. EURATOM, Fontenay-aux-Roses, Rapp. EUR-CEA-FC-367, ca 1967.
[10.131m] Nguyen-Hoe, H.-W. Drawin and L. Herman, Associat. EURATOM, Fontenay-aux-Roses, Rapp. CEA-R 3161, 1967.
[10.131n] Nikolayevski, L. S., A. F. Simonenko and S. G. Grenishin, Zh. prikl. Spektrosk. **4** (1966) 485.
[10.132] Nissen, W., Z. Physik **139** (1954) 638.
[10.132a] Nubbemeyer, H. and B. Wende, Z. angew. Physik **27** (1969) 214.
[10.132b] Ny, T. Z. and S. Y. Ch'en, Phys. Rev. **51** (1937) 567.
[10.132c] Ny, T. Z. and S. Y. Ch'en, Phys. Rev. **52** (1937) 1158.
[10.132d] Odintsova, G. A., Zh. prikl. Spektrosk. **7** (1967) 414.
[10.132e] Odintsova, G. A. and E. P. Vaulin, Zh. prikl. Spektrosk. **3** (1965) 105.
[10.132f] Oertel, G. K., Proc. 8. internat. Conf. Phenom. Ionized Gases, Contributed Pap., IAEA Vienna 1967, p. 279.
[10.132g] Oettinger, P. E., J. quant. Spectrosc. Radiat. Transfer **10** (1970) 127.
[10.132h] Oettinger, P. E. and J. Cooper, J. quant. Spectrosc. Radiat. Transfer **9** (1969) 591.
[10.133] Ohkawa, T., H. K. Forsen, A. A. Schupp and D. W. Kerst, Phys. Fluids **6** (1963) 846.
[10.134] Olsen, H. N. and W. S. Huxford, Phys. Rev. **87** (1952) 922.
[10.134a] Patel, C. K. N., Phys. Rev. **131** (1963) 1582.
[10.134b] Pavlov, M. and A. N. Prasad, Z. Physik **209** (1968) 244.

[10.134c] PEACOCK, N. J., J. COOPER and J. R. GREIG, Proc. Phys. Soc. Lond. **83** (1964) 803.
[10.134d] PENKIN, N. P. and L. N. SHABANOVA, Optika i Spektroskopiya **23** (1967) 22.
[10.134e] PENKIN, N. P. and L. N. SHABANOVA, Optika i Spektroskopiya **25** (1968) 795.
[10.134f] PENKIN, N. P. and L. N. SHABANOVA, Optika i Spektroskopiya **26** (1969) 346.
[10.134g] PESCHMANN, K. R., Z. Physik **201** (1967) 79.
[10.134h] PESCHMANN, K. R., Z. Physik **215** (1968) 228.
[10.135] PESIN, M. S., Zap. Voroneshsky selskokhozyajistrenny institut **28** (1959) 413.
[10.135a] PETROVA, G. P., Isledovaniya po fizicheskoi gazodinamike, Izd. Nauka, Moscow 1966, p. 96.
[10.136] PETSCHEK, H. E., P. H. ROSE, H. S. GLICK, A. KANE and A. KANTROWITZ, J. appl. Phys. **26** (1955) 83.
[10.136a] PFENNIG, H., J. quant. Spectrosc. Radiat. Transfer **6** (1966) 549.
[10.136b] PFENNIG, H., Z. Naturforsch. **21a** (1966) 1648.
[10.136c] PFENNIG, H. and E. TREFFTZ, Z. Naturforsch. **21a** (1966) 697.
[10.136d] PFENNIG, H. and E. TREFFTZ, Z. Physik **190** (1966) 253.
[10.136e] PITTACK, U., Z. Astrophys. **60** (1964) 190.
[10.136f] PLATZ, P. and J. G. HIRSCHBERG, Compt. Rend. Acad. Sci. **261** (1965) 1207.
[10.137] POIROT, A., Ann. Physique **4** (1935) 533.
[10.138] POIROT, A., Ann. Physique **4** (1935) 627.
[10.138a] POPENOE, C. H. and J. B. SHUMAKER, J. Res. Nat. Bur. Stand. **A 69** (1965) 495.
[10.138b] POWELL, W. R., Appl. Optics 8 (1969) 1041.
[10.138c] PRANOTO, S. and P. SCHULZ, Z. angew. Phys. **17** (1964) 539.
[10.138d] PRANOTO, S. and P. SCHULZ, Beitr. Plasmaphys. **5** (1965) 455.
[10.138e] PRASAD, A. N. and K. ABD EL-KARIM, J. Physics **B 1** (1968) 77.
[10.138f] PRESNYANKOV, L. P., Phys. Lebedev Inst., Akad. Wiss. UdSSR, Moscow, Preprint No. 156, 1969.
[10.138g] RANK, D. H. and T. A. WIGGINS, J. quant. Spectrosc. Radiat. Transfer **3** (1963) 377.
[10.138h] RASMADSE, N. A., G. V. SHOLIN and A. K. PAVLOVSKI, Zh. prikl. Spektrosk. **9** (1968) 43.
[10.138i] RAUTIAN, S. G. and I. I. SOBELMAN, Usp. Fiz. Nauk **90** (1966) 209.
[10.138j] REBANE, V. N., Optika i Spektroskopiya **26** (1969) 673.
[10.138k] RECK, G. P., H. TAKEBE and C. A. MEAD, Phys. Rev. **137** (1965) A 683.
[10.138l] RICHTER, J., Z. Astrophys. **61** (1965) 57.
[10.138m] RITTER, M., Ann. Physik **59** (1919) 170.
[10.138n] ROBERTS, D. E., Phys. Letters **22** (1966) 417.
[10.138o] ROBERTS, D. E., J. Physics **B 1** (1968) 53.
[10.138p] ROBERTS, D. E., J. quant. Spectrosc. Radiat. Transfer 8 (1968) 1241.
[10.138q] ROBERTS, D. E., Phys. Letters **27 A** (1968) 716.
[10.138r] ROBERTS, D. E. and F. L. CURZON, J. Physics **B 1** (1968) 973.
[10.138s] ROBERTS, D. E. and J. DAVIS, J. Physics **B 1** (1968) 48.
[10.138t] ROBERTS, J. R. and K. L. ECKERLE, Phys. Rev. **159** (1967) 104.
[10.138u] ROBIN, S., J. Chim. Phys. **49** (1951) 1.
[10.138v] ROBIN, J., Compt. Rend. Acad. Sci. Paris **238** (1954) 1491.
[10.138w] ROBIN, J. and S. ROBIN, Compt. Rend. Acad. Sci., Paris **233** (1951) 928.
[10.138x] ROBIN, J. and S. ROBIN, Compt. Rend. Acad. Sci. **233** (1951) 1019.
[10.138y] ROBIN, S. and S. ROBIN, J. Phys. Radium **17** (1956) 143.
[10.138z] ROBIN, J., S. ROBIN, R. BERGEON and B. VODAR, Coll. internat. Centre nat. rech. sci. 77 (1959) 237.
[10.139] ROBIN, S., J. ROBIN and B. VODAR, Compt. Rend. Acad. Sci. **232** (1951) 1754.
[10.139a] ROBINSON, E. J., J. Opt. Soc. Amer. **59** (1969) 782.
[10.139b] RODER, O. and A. STAMPA, Z. Physik **178** (1964) 348.

[10.139c] ROGERS, W. A. and M. A. BIONDI, Phys. Rev. **134** (1964) A1215.
[10.139d] ROMANOV, N. P. and A. S. PASYUK, Optika i Spektroskopiya **24** (1968) 337.
[10.139e] ROMPE, R., Z. Physik **74** (1932) 175.
[10.139f] ROMPE, R., Acta phys. polon. **5** (1936) 151.
[10.140] ROMPE, R. and P. SCHULZ, Z. Physik **108** (1938) 654.
[10.141] ROMPE, R. and P. SCHULZ, Z. Physik **110** (1938) 223.
[10.142] ROMPE, R. and P. SCHULZ, Z. Physik **118** (1941) 269.
[10.143] ROTHER, H., Ann. Physik **2** (1958) 326.
[10.144] RÜHMKOPF, H., Ann. Physik **33** (1938) 21.
[10.145] SADJIAN, H., in: Temperature, its Measurement and Control in Science and Industry, Vol. III, 1, Reinhold Publ. Co., New York 1962, p. 629.
[10.146] SAWYER, G. A., A. J. BEARDEN, I. HENINS, F. C. JAHODA and F. L. RIBE, Phys. Rev. **131** (1963) 1891.
[10.147] SCHLÜTER, H., Z. Naturforsch. **16a** (1961) 972.
[10.148] SCHLÜTER, H. and C. AVILA, Astrophys. J. **144** (1966) 785.
[10.149] SCHLÜTER, H., C. AVILA and J. DURHAM, Proc. 7. internat. Conf. Phenom. Ionized Gases, Belgrade 1965, Vol. **3**, p. 160.
[10.150] SCHMALJOHANN, P., Druckverbreiterung von Wasserstofflinien nach der Holtsmarkschen Theorie. Staatsexamensarbeit Kiel 1936.
[10.151] SCHMIDT, K., Proc. 6. internat. Conf. Ion. Phenom. Gases, Paris 1963, Vol. 3, p. 323.
[10.152] SCHOTT, L., Z. Physik **183** (1965) 521.
[10.152a] SCHOTT, L., Z. Physik **189** (1966) 449.
[10.152b] SCHOTT, L., Z. Physik **198** (1967) 165.
[10.153] SCHPOLSKI, E. W., Atomphysik, Part I, VEB Deutscher Verlag der Wiss., Berlin 1954 (translated from Russian).
[10.154] SCHRÖDINGER, E., Ann. Physik **80** (1926) 437.
[10.154a] SCHULZ, P., Phys. Z. **39** (1938) 412.
[10.154b] SCHULZ, G. and W. STOPP, Z. Physik **210** (1968) 223.
[10.154c] SCHULZ, P. and B. WENDE, Z. Physik **208** (1968) 116.
[10.155] SEAY, G. E., L. B. SEELY and R. G. FOWLER, J. appl. Phys. **32** (1961) 2439.
[10.155a] SEMENOVA, O. P., T. M. GORBUKOVA, N. A. BOKOVA and G. B. SUKHANOVA, Zh. prikl. Spektrosk. **9** (1968) 937.
[10.155b] SHOLIN, G. V., Proc. 8. internat. Conf. Phenom. Ionized Gases, Contributed Pap., IAEA Vienna 1967, p. 435.
[10.155c] SHOLIN, G. V., Optika i Spektroskopiya **26** (1969) 489.
[10.156] SHUMAKER, J. B. and W. L. WIESE, in: Temperature, its Measurement and Control in Science and Industry, Vol. III, 1, Reinhold Publ. Co., New York 1962, p. 575.
[10.156a] DE SILVA, A. W. and H.-J. KUNZE, J. appl. Phys. **39** (1968) 2458.
[10.156b] SKVORTSOV, YU. V., V. S. KOMELKOV and V. N. TERESHCHENKO, Zh. tekhn. Fiz. **34** (1964) 1790.
[10.156c] SMITH, E. W., Phys. Rev. Lett. **18** (1967) 990.
[10.156d] SMITH, E. W., J. COOPER and C. R. VIDAL, Phys. Rev. **185** (1969) 140.
[10.156e] SMITH, E. W. and C. F. HOOPER, J. quant. Spectrosc. Radiat. Transfer **8** (1968) 1617.
[10.156f] SNIDER, J. L., Phys. Letters **21** (1966) 172.
[10.157] SOBELMAN, I. I., Fortschr. Phys. **5** (1957) 175.
[10.157a] SOBELMAN, I. I., Vvedenie v teoriyu atomnych spektrov, Fizmatgis, Moscow 1963, Chapter 10.
[10.158] SOBOLEV, N. N., Proc. Coll. spectrosc. internat. VI., Pergamon Press, London 1957, p. 310.
[10.158a] SPIEGELHALTER, F., Beitr. Plasmaphys. **8** (1968) 1.

[10.158b] STACEY, D. N. and J. M. VAUGHAN, Phys. Letters **11** (1964) 105.
[10.159] STAMPA, A., Z. Astrophys. 58 (1963) 82.
[10.159a] STARK, J., R. RITSCHL and H. VERLEGER, Phys. Z. **39** (1938) 502.
[10.160] STEUBING, W. and R. GÜNTHER, Ann. Physik **15** (1955) 345.
[10.161] STEUBING, W. and J. HENGEVOSS, Ann. Physik **2** (1958) 191.
[10.161a] STONE, P. M., Phys. Rev. **127** (1962) 1151.
[10.162] STONE, P. M. and L. AGNEW, Phys. Rev. **127** (1962) 1157.
[10.163] STRIJLAND, J. C. and A. J. NANASSY, Physica **24** (1958) 935.
[10.163a] SUEOKA, S. and M. SATO, J. Phys. Soc. Japan **6** (1951) 444.
[10.163b] SULZMANN, K. G. P., F. BIEN and S. S. PENNER, J. quant. Spectrosc. Radiat. Transfer 7 (1967) 969.
[10.164] SVENSSON, B., Z. Physik **107** (1937) 485.
[10.165] SVENSSON, B., Diss. Lund 1940.
[10.165a] SYKOV, V. G., I. A. STEPANENKO, L. A. DUSHIN, I. K. NIKOLSKI, O. S. PAVLITSHENKO and V. T. TOLOK, Fiz. plazmy i problemy upravlyaemogo termoyadernogo sinteza, Vyp. 4, Naukova Dumka, Kiev 1965, p. 532.
[10.165b] SYKOV, V. G., I. A. STEPANENKO, L. A. DUSHIN, I. K. NIKOLSKI, O. S. PAVLITSHENKO and V. T. TOLOK, Zh. tekhn. Fiz. **35** (1965) 56.
[10.165c] SYKOVA, T. F. and G. E. SOLOTUKHIN, Zh. prikl. Spektrosk. **5** (1966) 417.
[10.165d] SZUDY, J., Acta phys. polon. **29** (1966) 605.
[10.165e] SZUDY, J., Bull. acad. polon. sci. ser. sci. math. astron. phys. **15** (1967) 297.
[10.165f] THORSON, W. R. and R. M. BADGER, J. chem. Phys. **27** (1957) 609.
[10.165g] TOMISER, J., Acta phys. austr. 8 (1954) 198; 276.
[10.165h] TOMISER, J., Acta phys. austr. 8 (1954) 323.
[10.165i] TOMISER, J., Acta phys. austr. 9 (1954) 18.
[10.166] TRAVING, G., Über die Theorie der Druckverbreiterung von Spektrallinien, Verlag G. Braun, Karlsruhe 1960.
[10.166a] TRAVING, G., Line broadening and line shift. in: Plasmadiagnostics, North Holland Publ. Co., Amsterdam 1968, p. 66.
[10.166b] TRIMMIER, J. R. and A. MILLER, Phys. Fluids **10** (1967) 1997.
[10.166c] TRYNKOWSKA, Bull. acad. polon. sci. ser. sci math. astron. phys. **13** (1965) 437.
[10.167] UNDERHILL, A. B. and J. H. WADDELL, Stark broadening functions for the hydrogen lines, National Bureau of Standards Circular No. 603, Washington 1959.
[10.168] UNSÖLD, A., Z. Astrophys. **23** (1944) 75.
[10.169] UNSÖLD, A., Z. Astrophys. **24** (1948) 355.
[10.170] UNSÖLD, A., Physik der Sternatmosphären, Springer-Verlag, Berlin-Göttingen-Heidelberg 1955.
[10.170a] VAINSHTEIN, L. A. and I. I. SOBELMAN, Optika i Spektroskopiya **6** (1959) 440; German Translation in: Phys. Abh. SU **1** (1959) 363.
[10.170b] VANHAUWERMEIREN, R., Plasma Physics 7 (1965) 433.
[10.170c] VAUGHAN, J. M., Phys. Letters **21** (1966) 153.
[10.170d] VAUGHAN, J. M., Proc. Roy. Soc. Lond. **A 295** (1966) 164.
[10.170e] VAUGHAN, J. M. and G. SMITH, Phys. Rev. **166** (1968) 17.
[10.171] VERWEY, S., Diss. Amsterdam 1936.
[10.171a] VIDAL, C.-R., Z. Naturforsch. **19a** (1964) 947.
[10.171b] VIDAL, C. R., Proc. 7. internat. Conf. Phenom. Ionized Gases, Belgrade 1965, Vol. 3, p. 168.
[10.171c] VIDAL, C. R., Max Planck Inst. Phys. Astrophys., Munich, Rep. MPI-PAE/Pl 13/65, August 1965.
[10.171d] VIDAL, C.-R., J. quant. Spectrosc. Radiat. Transfer **6** (1966) 461.
[10.171e] VIDAL, C.-R., J. quant. Spectrosc. Radiat. Transfer **6** (1966) 575.
[10.171f] VOITOVICH, A. P., Optika i Spektroskopiya **27** (1969) 556.

[10.171g] VOSLAMBER, D., Z. Naturforsch. **24a** (1969) 1458.
[10.171h] DE VRIES, R. F., Proc. 7. internat. Conf. Phenom. Ionized Gases, Belgrade 1965, Vol. 3, S. 173.
[10.171i] VUJNOVIĆ, V., Period. math.-phys. astr., Zagreb **19** (1964) 97.
[10.172] VUJNOVIĆ, V., J. A. HARRISON and J. D. CRAGGS, Proc. Phys. Soc. Lond. **80** (1962) 516.
[10.173] WARNER, B., Z. Astrophys. **69** (1968) 161.
[10.173a] WATANABE, K., Phys. Rev. **59** (1941) 151.
[10.173b] WATSON, W. and H. MARGENAU, Phys. Rev. **44** (1933) 748.
[10.174] WEISSKOPF, V., Z. Physik **75** (1932) 287.
[10.174a] WENDE, B., Diss. Techn. Hochsch. Karlsruhe 1966.
[10.174b] WENDE, B., Z. angew. Physik **22** (1967) 181.
[10.175] WENDT, G. and R. A. WETZEL, Ann. Physik **50** (1916) 429.
[10.175a] WENIGER, S., J. Physique **25** (1964) 946.
[10.175b] WHITE, A. D., Appl. Phys. Lett. **10** (1967) 24.
[10.176] WIESE, W. L. and P. W. MURPHY, Phys. Rev. **131** (1963) 2108.
[10.177] WIESE, W. L., D. R. PAQUETTE, and J. E. SOLARSKI, Proc. 5. internat. Conf. Ionization Phenomena in Gases, Vol. 1, North Holland Publ. Co., Amsterdam 1962, p. 907.
[10.178] WIESE, W. L., D. R. PAQUETTE and J. E. SOLARSKI, Phys. Rev. **129** (1963) 1225.
[10.178a] WIESE, W. L. and J. B. SHUMAKER, J. Opt. Soc. Amer. **51** (1961) 937.
[10.178b] WILSON, K. H. and W. E. NICOLET, J. quant. Spectrosc. Radiat. Transfer **7** (1967) 891.
[10.178c] WOJTCZAK, L., Acta phys. polon. **24** (1963) 723.
[10.178d] WULFF, H., Diss. Univ. Kiel 1957.
[10.179] WULFF, H., Z. Physik **150** (1958) 614.
[10.180] WULFF, H., Nuclear Instrum. **4** (1959) 352.
[10.180a] YA'AKOBI, B., Phys. Rev. **176** (1968) 227.
[10.180b] YA'AKOBI, B., J. quant. Spectrosc. Radiat. Transfer **9** (1969) 137.
[10.180c] YA'AKOBI, B., J. quant. Spectrosc. Radiat. Transfer **9** (1969) 143.
[10.180d] YA'AKOBI, B., J. quant. Spectrosc. Radiat. Transfer **9** (1969) 309.
[10.180e] YA'AKOBI, B., J. quant. Spectrosc. Radiat. Transfer **9** (1969) 1097.
[10.180f] YAO, Y. T., Z. Physik 77 (1932) 307.
[10.180g] YEN, M. P., H. SUN and M. Y. HONG, Acta phys. sinica **21** (1965) 1606.
[10.180h] YI, C. S. and T. MAKOTO, Propriétés optiques et acoustiques fluides comprimes et actions intermolecul., Paris, CNRS (1959) 401.
[10.180i] ZAGORYANSKAYA, E. V. and P. S. KIREYEV, Izvest. VUZ. Fiz. Tomsk, No. 4 (1961) 163.
[10.181] ZAIDEL', A. N., V. A. BELYAYEVA, A. B. BEREZIN, V. I. GLADUSHAK et al., in: Fizicheskie problemy spektroskopii Vol. 1, Izd. Akad. Nauk SSSR, Moscow 1962, p. 96.
[10.182] ZAIDEL', A. N., G. M. MALYSHEV, A. B. BEREZIN and G. T. RAZDOBARIN, Zh. tekhn. Fiz. **30** (1960) 1437.
[10.183] ZAIDEL', A. N., G. M. MALYSHEV, Y. I. MOSKALEV et al., Zh. tekhn. Fiz. **30** (1960) 1433.
[10.184] ZAIDEL', A. N., G. M. MALYSHEV and Y. Y. SHREIDER, Zh. tekhn. Fiz. **30** (1960) 1422.
[10.185] ZAKHAROVA, V. M., Optika i Spektroskopiya **1** (1956) 636.
[10.186] ZAKHAROVA, V. M., YU. M. KAGAN and V. I. PEREL, Optika i Spektroskopiya **11** (1961) 777; German Translation in: Phys. Abh. SU **6** (1962) 416.
[11.1] ALEKSANDROV, V. YA, D. V. GUREVICH and I. V. PODMOSHENSKI, Optika i Spektroskopiya **23** (1967) 521.

[11.1a] Aleksandrov, V. Ya. D. B. Gurevich and I. V. Podmoshenski, Teplofizika 5 (1967) 751.
[11.1b] Averin, V. G., M. A. Mazing and A. I. Pisanko, Zh. tekhn. Fiz. 34 (1964) 767.
[11.1c] Bates, D. R. and A. Dalgarno, in: Atomic and Molecular Processes, Acad. Press New York 1962, p. 245.
[11.1d] Bekefi, G., J. L. Hirshfield and S. C. Brown, Phys. Fluids 4 (1961) 173.
[11.1e] Berg, H. F., Z. Physik 207 (1967) 404.
[11.2] Biberman, L. M. and K. N. Ulyanov, Optika i Spektroskopiya 16 (1964) 394.
[11.2a] Bibermann, L. M., V. S. Vorobyev and I. T. Yakubov, Teplofizika 5 (1967) 201.
[11.2b] Biberman, L. M., V. S. Vorobyev and I. T. Yakubov, in: Magnitogidrodinamicheskii metod polucheniya elektroenergii, Izd. Energiya 1968, p. 209.
[11.3] Blaha, M., Probleme des Plasmas in Physik und Astronomie, Akademie-Verlag, Berlin 1958, 48.
[11.3a] Bott, J. F., J. quant. Spectrosc. Radiat. Transfer 6 (1966) 807.
[11.4] Burhorn, F., Z. Physik 140 (1955) 440.
[11.4a] Deputat, G. G. and V. G. Konovalov, Optika i Spektroskopiya 24 (1968) 806.
[11.5] Dewan, E. M., Generalization of the Saha equation, Air Force Cambridge Res. Labs., AFCRL-Report 42, 1961.
[11.6] Dewan, E. M., Phys. Fluids 4 (1961) 759.
[11.7] Drawin, H. W., Atomic Cross Sections for inelastic electronic Collisions, Association-EURATOM-C. E. A., Fontenay-aux-Roses, France, Report EUR-CEA-FC 236, 1963.
[11.8] Drawin, H. W., Z. Physik 172 (1963) 429.
[11.8a] Drawin, H. W., Report EUR-CEA-383, March 1966, revised version January 1967, Assoc. EURATOM, Fontenay-aux-Roses, France.
[11.8b] Drawin, H. W., Validity conditions for local thermodynamic equilibrium. German translation in: Ergebnisse der Plasmaphysik und der Gaselektronik, Band 3, Akademie-Verlag, Berlin 1972, p. 363; Original: Akademie-Verlag, Berlin, in this volume, p. 591. See also: Z. Physik 228 (1969) 99.
[11.8c] Drawin, H. W., J. quant. Spectrosc. Radiat. Transfer 10 (1970) 33.
[11.8d] Drawin, H. W. and H. Henning, Z. Naturforsch. 22a (1967) 587.
[11.9] Durand, J., Z. Naturforsch. 18a (1963) 281.
[11.10] Edlen, B., Z. Astrophys. 22 (1942) 30.
[11.11] Elwert, G., Z. Naturforsch. 7a (1952) 432.
[11.12] Elwert, G., Z. Naturforsch. 7a (1952) 703.
[11.13] Finkelnburg, W. and H. Maeckeg, Handb. Phys. 22, Springer-Verlag, Berlin-Göttingen-Heidelberg 1956, p. 254.
[11.14] Fite, W. L., in: Atomic and Molecular Processes, Academic Press, New York and London 1962, p. 421.
[11.15] Frish, S. E., Uspekhi fiz. Nauk 61 (1957) 461.
[11.16] Frish, S. E. and V. F. Revald, Optika i Spektroskopiya 15 (1963) 726.
[11.16a] Goldfarb, V. M., E. V. Ilina and A. I. Kasakov, in: Proc. 14. Coll. Spectrosc. Internat., Debrecen, Hungary 1967, p. 603.
[11.16b] Goldman, L. M. and R. W. Kilb, Plasma Physics 6 (1964) 217.
[11.16c] Grankova, D. A. and L. I. Kiselevski, Zh. prikl. Spektrosk. 4 (1966) 342.
[11.16d] Grechikhin, L. I., Teplofizika 6 (1968) 507.
[11.17] Greene, E. F. and J. P. Toennis, Chemische Reaktionen in Stoßwellen, Verlag D. Steinkopff, Darmstadt 1959, Chap. 4 and 7.
[11.17a] Greig, J. R., Brit. J. appl. Phys. 16 (1965) 957.
[11.17b] Gridneva, S. M. and G. A. Kasabov, in: Electricity from MHD, LAEA Vienna 1966, Vol. 1, p. 73.
[11.18] Griem, H. R., Phys. Rev. 131 (1963) 1170.

[11.19] GUREVICH, D. B. and I. V. PODMOSHENSKI, Optika i Spektroskopiya 15 (1963) 587.

[11.20] HATTENBURG, A. T. and H. J. KOSTKOWSKI, in: Temperature, its Measurement and Control in Science and Industry, Vol. III, 1, Reinhold Publ. Co., New York 1962, p. 587.

[11.20a] HEROUX, L., Proc. Phys. Soc. Lond. **83** (1964) 121.

[11.21] HINNOV, E., Report MATT 194, May 1963, Princeton Univ., Plasmaphysics Laboratory, Princeton, N. Jersey.

[11.22] HOLSTEIN, T., Phys. Rev. **83** (1951) 1159.

[11.23] KAGAN, YU. M., R. I. LYAGUSTSHENKO and A. D. CHACHAYEV, Optika i Spektroskopiya **14** (1963) 598.

[11.24] KAGAN, YU. M., R. I. LYAGUSTSHENKO and A. D. CHACHAYEV, Optika i Spektroskopiya **15** (1963) 13.

[11.25] KAPZOW, N. A., Elektrische Vorgänge in Gasen und im Vakuum. VEB Deutscher Verlag der Wissenschaften, Berlin 1955, p. 275; German Translation from: N. A. Kaptsov, Elektricheskie yavleniya v gazach i vakuume, Moscow-Leningrad 1950.

[11.26] KAUFMAN, S. and R. V. WILLIAMS, Proc. 4. internat. Conf. Ionization Phenomena in Gases, Vol. II, North Holland Publ. Co., Amsterdam 1960, p. 824.

[11.27] KISELEVSKI, L. I. and V. N. SNOPKO, Optika i Spektroskopiya **17** (1964) 637.

[11.27a] KISELEVSKI, L. I. and V. N. SNOPKO, Zh. prikl. Spektrosk. **2** (1965) 207.

[11.27b] KOLESNIKOV, V. N., in: Fizicheskaya Optika, Moscow 1964, p. 66.

[11.28] KOLESNIKOV, V. N. and N. N. SOBOLEV, in: Fizicheskii problemy spektroskopii, Vol. 1, Izd. Akad. Nauk SSSR, Moscow 1962, p. 119.

[11.28a] KOZAK, O. V., N. A. MIKHATSKAYA and L. L. PASECHNIK, Ukrain. Fiz. Zh. **11** (1966) 253.

[11.28b] KRYSMANSKI, K. H., Beitr. Plasmaphys. 7 (1967) 349.

[11.28c] KRYSMANSKI, K. H. and U. BERNDT, Beitr. Plasmaphys. 8 (1968) 101.

[11.29] LOSEV, S. A. and A. I. OSIPOV, Usp. Fiz. Nauk **74** (1961) 393.

[11.29a] LOTZ, W., Astrophys. J., Suppl. **14** (1967) No. 128, 32 pp.

[11.29b] LOTZ, W., Report IPP 1/62, May 1967, Inst. Plasmaphys. Garching near Munich.

[11.30] LYOT, B., Mon. Not. Roy. Astron. Soc. **99** (1939) 580.

[11.31] MCNALLY, J. R., in: Optical spectrometric Measurements of high Temperatures. Univ. of Chicago Press, Chicago 1961, p. 170.

[11.31a] MAHN, C., Report IPP 3/52, Febr. 1967, Inst. Plasmaphys. Garching near Munich.

[11.31b] MANDELSHTAM, S. L. and N. K. SUKHODREV, Izv. Akad. Nauk SSSR Ser. Fiz. **19** (1955) 11.

[11.32] MASSEY, H. S. W., Handb. Phys. 36 (1956) p. 307 (Springer-Verlag, Berlin-Göttingen-Heidelberg).

[11.32a] MOTZ, R. O., J. quant. Spectrosc. Radiat. Transfer 8 (1970) 961.

[11.33] NEUMANN, W., Beitr. Plasmaphys. 2 (1962) 252.

[11.34] NEUMANN, W., Beitr. Plasmaphys. **3** (1963) 12.

[11.34a] NEUMANN, W., Beitr. Plasmaphys. **11** (1971) 249.

[11.34b] PODGORNY, I. M., Lektsii po diagnostike plazmy. Atomizdat Moskau 1968.

[11.35] PROKOFYEV, V. K., D. B. GUREVICH, I. M. BELOUSOVA and Y. A. SNIGIREV, Optika i Spektroskopiya 7 (1959) 14; German Translation: Phys. Abh. SU **1** (1959) 518.

[11.36] RASUMOVSKAYA, L. P., Optika i Spektroskopiya **14** (1963) 189.

[11.37] ROBBEN, F., W. B. KUNKEL and L. TALBOT, Phys. Rev. **132** (1963) 2363.

[11.38] ROTHER, H., Z. Physik **173** (1963) 441.

[11.39] SCHLÜTER, H., Z. Naturforsch. **16a** (1961) 972.

[11.40] SEATON, M. J., Monthly Notices Roy. Astron. Soc. **119** (1959) 81.

[11.41] SEATON, M. J., in: Atomic and Molecular Processes, Academic Press, New York and London 1962, p. 374.
[11.42] SEMENOVA, O. P. and V. V. KOKHANENKO, Izv. Akad. Nauk SSSR, Ser. Fiz. **14** (1950) 727.
[11.42a] SUCKEWER, S., Phys. Letters **25 A** (1967) 284.
[11.42b] SUCKEWER, S., Phys. Letters **25 A** (1967) 405.
[11.42c] VILENSKAYA, T. V., Izv. VUZ Fizika, No. 6 (1962) 111.
[11.43] VOLKOV, YA. F., V. G. DYATKOV and I. N. MITINA, Teplofizika **3** (1965) 659.
[11.44] WALDMEIER, M., Die Sonnenkorona, Birkhäuser Verlag Basel 1951.
[11.45] WILLIAMS, R. V. and S. KAUFMAN, Proc. Phys. Soc. Lond. **75** (1960) 329.
[11.45a] WIMEL-PECKER, C., JILA-Report No. 90, April 1967, Joint Inst. Laboratory Astrophysics, Boulder, Colorado.
[11.46] WITTE, H., Z. Physik **88** (1934) 415.
[11.46a] ZAIDEL, A. N., G. M. MALYSHEV and YE. A. PTITSYNA, Zh. tekhn. Fiz. **33** (1963) 200.

K. Baumgärtel and G. Wallis

Magnetohydrodynamic Waves

(Waves in Ideal Media)

This article is translated from the German original without additions.

CHAPTER 1

Introduction

In the first section we shall study the propagation phenomena of small disturbances in an ideal magnetohydrodynamic medium. "Ideal" will refer to a medium which is a classical, compressible, and dissipationless continuum of infinite electrical conductivity.

In the presence of a magnetic field the reaction of a continuum of this type to small disturbances is more complicated than in hydrodynamics. An ordinary compressible medium transmits only one type of disturbance, namely sound waves which are compression waves. These propagate (locally) isentropically. Disturbances, which are not related to pressure changes, are at rest relative to the fluid. In a magneto-hydrodynamic medium however, noncompressive disturbances can also propagate (Alfvén-waves). These are brought about by the magnetic stresses and produce transverse oscillations of the magnetic field lines. Compressions of the medium are generally linked with disturbances of the magnetic pressure. Both can produce similar effects (fast magnetoacoustic waves or f-waves), or partially compensate one another (slow magnetoacoustic waves or s-waves).

The above three wave modes are characterized by different phase velocities and polarizations. The magnetic field produces a strong anisotropy: the phase velocities depend on the direction of propagation relative to the magnetic field direction. Only the f-waves can propagate normal to the direction of the magnetic field.

The equations describing these disturbances are linear, or more precisely, they comprise a system of hyperbolic linear partial differential equations of the first order. This system is obtained by linearization from the basic magneto-hydrodynamic (MHD) equations (section 2.2).

When linearizing it is assumed that a solution to the basic equations is known, i.e., for the "undisturbed" state. The propagation properties of the disturbances are essentially dependent on the structure of this state. The simplest relationships are obtained when initially considering a uniform, steady, infinitely long flow. In this case we can assume plane periodic waves, and the above three wave modes will propagate independently of one another. Phase velocities polarization, and group velocities can be studied in a straightforward manner (sections 2.3—2.5).

We shall then discuss the reflection and refraction of these waves at a plane discontinuity front (section 3.1). The different wave modes transform into one another: an incident wave in general produces all other modes as reflected and refracted waves. We shall derive the analog of Snell's Law and present a geometrical method for the construction of the angles of reflection and refraction (section 3.2).

Upon calculating the amplitudes of the reflected and refracted waves we can reach certain conclusions concerning the stability of MHD-discontinuity fronts (section 3.3). If the disturbed medium is nonuniform, the different wave modes will be linked to one another. (A discontinuity can in this sense be considered as a special type of inhomogeneity).

The precise description of the disturbed state is thus given by the linearized equations. Solving these equations is inconvenient and in most cases unnecessary. It is more convenient to limit ourselves to an approximate description of the propagation process, as for example, is done in geometrical optics or in "geometrical acoustics". Instead of the system of linearized MHD-equations we solve only the relevant characteristic equation, which is determined by the basic solution and no longer depends on the disturbances. We are dealing with a first-order partial differential equation of the Hamilton-Jacobi type, which describes spatial surfaces changing with time (initial conditions are previously given) (section 4.1).

These characteristic surfaces can be physically interpreted (besides being boundaries between disturbed and undisturbed regions) in the following way:

1. They are identical to the so-called weak shock waves or weak discontinuities (surfaces at which the state parameters can undergo small jumps or along which short-time weak pulse disturbances can propagate).
2. They represent the wave surfaces of periodic, local plane waves in the limiting case $\omega \to \infty$ ($\lambda \to 0$) (the linearized MHD-equations are related to the corresponding characteristic equation in the same way as Maxwell's equations are related to the eikonal equation of geometrical optics).

We can construct the wavefronts, which emanate from a surface or point excitation (periodic or pulsed). In general, there are seven such surfaces fronts which propagate independently of one another with different velocities.

The characteristic equation itself possesses characteristic properties (the so-called bicharacteristics of the initial system). These determine the spatial curves ("rays") along which the characteristic surface elements move and by means of which we can construct the characteristic surfaces (planes). Their direction at every point corresponds to the direction of energy propagation at that point.

As examples the propagation of Alfvén disturbances in a dipole field and the propagation of f-waves in the magnetosphere of the earth are discussed.

Furthermore, one can obtain additional information on the variation of the excitation intensity (i.e., the amplitudes) of disturbances along the rays.

Among the references listed (p. 589 ff.) the following are of particular importance: J. Bazer and J. Hurley [8], J. Bazer and O. Fleischmann [9], M. Kantorovich [13], as well as the book by A. Jeffrey and T. Taniuti, Nonlinear Wave Propagation with Application to Physics and Magnetohydrodynamics [1]. In several sections we shall make frequent reference to these studies. Further details and comprehensive literature can also be found in them.

CHAPTER 2

Propagation of Plane Waves in Homogeneous Media

2.1. Basic MHD-Equations for Ideal Media

The basic magnetohydrodynamic equations for ideal media are the following (in rationalized mks-units)

$$\left.\begin{aligned} \operatorname{rot} \boldsymbol{E} &= -\frac{\partial \boldsymbol{B}}{\partial t}, && (2.1)\\ \operatorname{div} \boldsymbol{E} &= \frac{\varrho_0}{\varepsilon_0}, && (2.2)\\ \operatorname{rot} \boldsymbol{B} &= \mu \boldsymbol{j}, && (2.3)\\ \operatorname{div} \boldsymbol{B} &= 0, && (2.4) \end{aligned}\right\} \text{Maxwell's equations*}$$

$$\boldsymbol{E} = -(\boldsymbol{v} \times \boldsymbol{B}), \quad \text{Ohm's law} \tag{2.5}$$

$$\left.\begin{aligned} &\frac{\partial \varrho}{\partial t} + \operatorname{div}(\varrho\, \boldsymbol{v}) = 0, && (2.6)\\ &\varrho\left\{\frac{\partial \boldsymbol{v}}{\partial t} + (\boldsymbol{v} \operatorname{grad})\, \boldsymbol{v}\right\} = -\operatorname{grad} p + \varrho_e \boldsymbol{E} + (\boldsymbol{j} \times \boldsymbol{B}), && (2.7)\\ &\frac{\partial s}{\partial t} + (\boldsymbol{v} \operatorname{grad} s) = 0, && (2.8) \end{aligned}\right\} \text{hydrodynamic equations}$$

$$p = p(\varrho, s) \quad \text{equation of state} \tag{2.9}$$

where

$\boldsymbol{E}, \boldsymbol{B}, \boldsymbol{j}, \varrho_e$ respectively denote the electronic field strength, magnetic induction, current density, and the electric charge density;

$p, \boldsymbol{v}, \varrho, s$ are respectively the pressure, current velocity, mass density, and the entropy per unit mass;

μ, ε_0 are the permeability and the dielectric constant in a vacuum, respectively.

The above system links Maxwell's equations with those of mechanics for the corresponding material continuum. The connection is essentially due to the term $(\boldsymbol{j} \times \boldsymbol{B})$ appearing in the equation of motion (2.7) and Ohm's law (2.5). The latter states that an electric field cannot exist in a coordinate system, which moves with the medium (due to the infinite conductivity).

Equation (2.4) on account of (2.1) is only useful as an initial condition. Apart from (2.6), we are dealing with 16 equations in 16 scalar variables.

* The displacement current has been left out.

The following energy equation is equivalent to (2.8)

$$\frac{\partial}{\partial t}\left(\frac{\varrho \boldsymbol{v}^2}{2} + \varrho e + \frac{\boldsymbol{B}^2}{2\mu}\right) = -\operatorname{div}\left\{\boldsymbol{v}\left(\frac{\varrho \boldsymbol{v}^2}{2} + \varrho e\right) + p\boldsymbol{v} + \frac{1}{\mu}(\boldsymbol{E}\times\boldsymbol{B})\right\}. \quad (2.10)$$

where $e = e(\varrho, s)$ is the integral energy per unit mass. $\boldsymbol{E}$ and $\boldsymbol{j}$ can be eliminated from all the equations so that we are left only with relationships between $\boldsymbol{B}$ and the hydrodynamic and thermodynamic quantities $\boldsymbol{v}$, ϱ, p, s, and e

$$\operatorname{div}\boldsymbol{B} = 0\,, \quad (2.11)$$

$$\frac{\partial \boldsymbol{B}}{\partial t} - \operatorname{rot}(\boldsymbol{v}\times\boldsymbol{B}) = 0\,, \quad (2.12)$$

$$\frac{\partial \varrho}{\partial t} + \operatorname{div}(\varrho\boldsymbol{v}) = 0\,, \quad (2.13)$$

$$\varrho\left\{\frac{\partial \boldsymbol{v}}{\partial t} + (\boldsymbol{v}\operatorname{grad})\,\boldsymbol{v}\right\} = -\operatorname{grad} p + \frac{1}{\mu}(\operatorname{rot}\boldsymbol{B}\times\boldsymbol{B})\,, \quad (2.14)$$

$$\frac{\partial s}{\partial t} + (\boldsymbol{v}\operatorname{grad} s) = 0\,, \quad (2.15)$$

$$p = p(\varrho, s)\,, \quad (2.16)$$

$$\varrho_e = -\varepsilon_0 \operatorname{div}(\boldsymbol{v}\times\boldsymbol{B})\,. \quad (2.17)$$

In the equation of motion (2.14) we neglected the term $\varepsilon_0(\boldsymbol{v}\times\boldsymbol{B})\operatorname{div}(\boldsymbol{v}\times\boldsymbol{B})$, which results from the term $\varrho_e\boldsymbol{E}$. This is permissible when $v^2 \ll c^2$, in which case we have

$$|\varepsilon_0(\boldsymbol{v}\times\boldsymbol{B})\operatorname{div}(\boldsymbol{v}\times\boldsymbol{B})| + \left|\frac{1}{\mu}(\operatorname{rot}\boldsymbol{B}\times\boldsymbol{B})\right| \approx \frac{B^2}{\mu}\left(\frac{v^2}{c^2} + 1\right).$$

In a system which moves locally with the medium $\operatorname{div}\boldsymbol{E} = 0$, since $\boldsymbol{E} = 0$. However, the (formal) space charge

$$\varrho_e = -\varepsilon_0 \operatorname{div}(\boldsymbol{v}\times\boldsymbol{B})$$

in general does not vanish.

The energy equation, equivalent to (2.15), is given by

$$\frac{\partial}{\partial t}\left(\frac{\varrho \boldsymbol{v}^2}{2} + \varrho e + \frac{\boldsymbol{B}^2}{2\mu}\right) = -\operatorname{div}\left\{\boldsymbol{v}\left(\frac{\varrho \boldsymbol{v}^2}{2} + \varrho e\right) + \boldsymbol{v}p + \frac{1}{\mu}(\boldsymbol{B}\times(\boldsymbol{v}\times\boldsymbol{B}))\right\}. \quad (2.18)$$

Equations (2.11)—(2.16) represent a quasilinear system of differential equations.

We shall write this system in another way and subsequently transform it (except for (2.11)), into the so-called conservation form, which can be applied after determination of the boundary conditions at the discontinuity planes (see section 3.1).

$$\frac{\partial a_i}{\partial t} + \operatorname{div}\boldsymbol{F}_i = 0\,, \quad i = 1, \ldots, 8\,, \quad (2.19)$$

where $\boldsymbol{F}_i = F_{i1}\,\boldsymbol{e}_1 + F_{i2}\,\boldsymbol{e}_2 + F_{i3}\,\boldsymbol{e}_3$; $F_{ij} = F_{ij}(a_1, \ldots, a_8)$. (Equations (2.13) and (2.18) are already in this form.) From

a_i	$\boldsymbol{F}_i$	
ϱ	$\varrho\,\boldsymbol{v}$,	(2.20)
$\varrho\,v_i$	$\sum_{j=1}^{3}\left(p\,\delta_{ij} + \varrho\,v_i\,v_j + \frac{1}{\mu}\left(\frac{\boldsymbol{B}^2}{2}\delta_{ij} - B_i\,B_j\right)\right)\boldsymbol{e}_j$,	(2.21)
B_i	$\sum\,(B_i\,v_j - B_j\,v_i)\,\boldsymbol{e}_j$,	(2.22)
$\mathcal{E} = \varrho\frac{\boldsymbol{v}^2}{2} + \varrho\,e + \frac{\boldsymbol{B}^2}{2\,\mu}$	$\boldsymbol{v}\left(\frac{\varrho\,\boldsymbol{v}^2}{2} + \varrho\,e + p\right) + \frac{1}{\mu}\left(\boldsymbol{B}^2\,\boldsymbol{v} - (\boldsymbol{v}\,\boldsymbol{B})\,\boldsymbol{B}\right)$,	(2.23)

follow the continuity equation, the three components of the equation of motion, the three components of the "induction equation" (2.12), and the energy equation. $\boldsymbol{e}_1$, $\boldsymbol{e}_2$, and $\boldsymbol{e}_3$ are unit vectors directed along the axes of a cartesian coordinate system.

With the vector

$$\boldsymbol{a} = (a_1, \ldots, a_8) = (\varrho, \varrho\,v_1, \varrho\,v_2, \varrho\,v_3, B_1, B_2, B_3, \mathcal{E}) \tag{2.24}$$

we can arrive at the following condensed form

$$\frac{\partial\boldsymbol{a}}{\partial t} + \sum_{j=1}^{3} A_j\,\frac{\partial\boldsymbol{a}}{\partial x_j} = 0\,. \tag{2.25}$$

where

$$A_j = \begin{pmatrix} \frac{\partial F_{1j}}{\partial a_1}, & \ldots, & \frac{\partial F_{1j}}{\partial a_8} \\ \vdots & & \vdots \\ \frac{\partial F_{8j}}{\partial a_1}, & \ldots, & \frac{\partial F_{8j}}{\partial a_8} \end{pmatrix}, \qquad j = 1, 2, 3\,. \tag{2.26}$$

The "standard form" is similarly obtained. For

$$\boldsymbol{u} = (\varrho, v_1, v_2, v_3, B_1, B_2, B_3, s) \tag{2.27}$$

we get

$$\frac{\partial\boldsymbol{u}}{\partial t} + \sum_{j=1}^{3} U_j\,\frac{\partial\boldsymbol{u}}{\partial x_j} = 0\,. \tag{2.28}$$

The matrices A_j and U_j are given in the appendix, pp. 582—584. Systems (2.25) and (2.28) are complemented by the expression

$$\operatorname{div}\boldsymbol{B} = 0\,.$$

The quasilinearity lies herein that the coefficients of matrices A_j and U_j are still functions of the a_i and u_i, respectively. The relationship between the two vectorial representations is given by

$$a_i = a_i(u_1, \ldots, u_8)\,, \qquad i = 1, \ldots, 8\,. \tag{2.29}$$

Consequently, we have

$$\frac{\partial\boldsymbol{a}}{\partial t} = M\,\frac{\partial\boldsymbol{u}}{\partial t}\,, \qquad \frac{\partial\boldsymbol{a}}{\partial x_j} = M\,\frac{\partial\boldsymbol{u}}{\partial x_j} \tag{2.30}$$

where

$$M = \frac{\partial(a_1, \ldots, a_8)}{\partial(u_1, \ldots, u_8)} = \begin{pmatrix} \frac{\partial a_1}{\partial u_1}, \ldots, \frac{\partial a_1}{\partial u_8} \\ \vdots \qquad \vdots \\ \frac{\partial a_8}{\partial u_1}, \ldots, \frac{\partial a_8}{\partial u_8} \end{pmatrix} \tag{2.31}$$

and

$$U_j = M^{-1} A_j M \,. \; *$$

The matrix M is also given in the appendix.

2.2. Linearization and the Equation for Plane Waves

Let us assume that we know a solution

$$\boldsymbol{u}(\boldsymbol{x}, t) = \big(\varrho(\boldsymbol{x}, t), \ldots, s(\boldsymbol{x}, t)\big)$$

of the basic equations and that it describes the "undisturbed" state. We shall now investigate the influence of superimposing small disturbances

$$\delta\boldsymbol{u}(\boldsymbol{x}, t) = \big(\delta\varrho(\boldsymbol{x}, t), \ldots, \delta s(\boldsymbol{x}, t)\big)$$

on this equation. The linearized system for the disturbances has the following form

$$L_1(\delta\boldsymbol{B}) = \operatorname{div} \delta\boldsymbol{B} = 0 \,, \tag{2.32}$$

$$L_2(\delta\varrho, \delta\boldsymbol{v}) = \frac{\partial\delta\varrho}{\partial t} + \operatorname{div}(\delta\varrho\, \boldsymbol{v}) + \operatorname{div}(\varrho\, \delta\boldsymbol{v}) = 0 \,, \tag{2.33}$$

$$\begin{aligned} L_3(\delta\varrho, \delta\boldsymbol{v}, \delta\boldsymbol{B}) = {} & \varrho \frac{\partial\delta\boldsymbol{v}}{\partial t} + \varrho\,(\boldsymbol{v} \operatorname{grad})\, \delta\boldsymbol{v} + \varrho\,(\delta\boldsymbol{v} \operatorname{grad})\, \boldsymbol{v} + \delta\varrho\,(\boldsymbol{v} \operatorname{grad} \boldsymbol{v}) \\ & + \delta\varrho \frac{\partial\boldsymbol{v}}{\partial t} + \operatorname{grad} \frac{\partial p}{\partial\varrho}\, \delta\varrho + \operatorname{grad} \frac{\partial p}{\partial s}\, \delta s - \frac{1}{\mu} (\operatorname{rot} \delta\boldsymbol{B} \times \boldsymbol{B}) \\ & - \frac{1}{\mu} (\operatorname{rot} \boldsymbol{B} \times \delta\boldsymbol{B}) = 0 \,, \end{aligned} \tag{2.34}$$

$$L_4(\delta\boldsymbol{v}, \delta\boldsymbol{B}) = \frac{\partial\delta\boldsymbol{B}}{\partial t} - \operatorname{rot}(\delta\boldsymbol{v} \times \boldsymbol{B}) - \operatorname{rot}(\boldsymbol{v} \times \delta\boldsymbol{B}) = 0 \,, \tag{2.35}$$

$$L_5(\delta\boldsymbol{v}, \delta s) = \frac{\partial\delta s}{\partial t} + (\boldsymbol{v} \operatorname{grad} \delta s) + (\delta\boldsymbol{v} \operatorname{grad} s) = 0 \,. \tag{2.36}$$

Except for (2.32) these equations can be written in matrix form:

$$\frac{\partial\delta\boldsymbol{u}}{\partial t} + \sum_{j=1}^{3} U_j \frac{\partial\delta\boldsymbol{u}}{\partial x_j} + \sum_{j=1}^{3} \left(\frac{\partial U_j}{\partial u_1} \delta u_1 + \cdots + \frac{\partial U_j}{\partial u_8} \delta u_8 \right) \frac{\partial\boldsymbol{u}}{\partial x_j} = 0 \,. \tag{2.37}$$

The matrix elements of U_j are given by the above basic solution. Equation (2.37) assumes a particularly simple form when we set

$$\boldsymbol{u} = \text{const} \,,$$

i.e., by taking a steady, infinitely long flow of matter in a homogeneous magnetic field as the basis of an undisturbed state (ϱ = const, $\boldsymbol{v}$ = const, $\boldsymbol{B}$ = const, s = const).

* Remembering that div $\boldsymbol{B} = 0$.

In this case $\frac{\partial \boldsymbol{u}}{\partial x_j} = \frac{\partial \boldsymbol{u}}{\partial t} = 0$, and the linearized system becomes

$$\operatorname{div} \delta \boldsymbol{B} = 0 \,, \tag{2.38}$$

$$\frac{\partial \delta \boldsymbol{u}}{\partial t} + \sum_{j=1}^{3} U_j \frac{\partial \delta \boldsymbol{u}}{\partial x_j} = 0 \,, \tag{2.39}$$

or written out in full

$$\operatorname{div} \delta \boldsymbol{B} = 0 \,, \tag{2.40}$$

$$\frac{\partial \delta \varrho}{\partial t} + (\boldsymbol{v} \operatorname{grad} \delta \varrho) + \varrho \operatorname{div} \delta \boldsymbol{v} = 0 \,, \tag{2.41}$$

$$\varrho \frac{\partial \delta \boldsymbol{v}}{\partial t} + \varrho\, (\boldsymbol{v} \operatorname{grad})\, \delta \boldsymbol{v} + \frac{1}{\mu} (\boldsymbol{B} \times \operatorname{rot} \delta \boldsymbol{B}) + a^2 \operatorname{grad} \delta \varrho + p_s \operatorname{grad} \delta s = 0 \,, \tag{2.42}$$

$$\frac{\partial \delta \boldsymbol{B}}{\partial t} + (\boldsymbol{v} \operatorname{grad})\, \delta \boldsymbol{B} - (\boldsymbol{B} \operatorname{grad})\, \delta \boldsymbol{v} + \boldsymbol{B} \operatorname{div} \delta \boldsymbol{v} = 0 \,, \tag{2.43}$$

$$\frac{\partial \delta s}{\partial t} + (\boldsymbol{v} \operatorname{grad} \delta s) = 0 \,, \tag{2.44}$$

where $p_s = \frac{\partial p(\varrho, s)}{\partial s}$, $a^2 = \frac{\partial p}{\partial \varrho}$, and a is the sound velocity. Since the undisturbed parameters are dependent on time and location, we can use the variables $\frac{a}{\varrho} \delta \varrho$, $\frac{\partial B_j}{\sqrt{\mu \varrho}} = \delta b_j$, and $\frac{1}{a \varrho}\, p_s \delta s$ instead of $\delta \varrho$, δB_j, and δs, respectively. Thus all the disturbed parameters have the dimensions of velocity.

From

$$\delta \boldsymbol{u}^* = \left(\frac{a}{\varrho} \delta \varrho, \delta v_1, \delta v_2, \delta v_3, \delta b_1, \delta b_2, \delta b_3, \frac{1}{a \varrho} p_s\, \delta s \right) = I^* \delta \boldsymbol{u} \,, \tag{2.45}$$

$$I^* = \begin{pmatrix} \frac{a}{\varrho} & & & & & & & \\ & 1 & & & & & & \\ & & 1 & & & & & \\ & & & 1 & & & & \\ & & & & \frac{1}{\sqrt{\mu \varrho}} & & & \\ & & & & & \frac{1}{\sqrt{\mu \varrho}} & & \\ & & & & & & \frac{1}{\sqrt{\varrho \mu}} & \\ & & & & & & & \frac{p_s}{a \varrho} \end{pmatrix}$$

it follows that

$$\frac{\partial \delta \boldsymbol{u}^*}{\partial t} + \sum_{j=1}^{3} U_j^* \frac{\partial \boldsymbol{u}^*}{\partial x_j} = 0 , \qquad U_j^* = I^* U_j I^{*-1} . \tag{2.46}$$

The matrices U_j^* (see appendix) are thus to within one element symmetric.

We shall assume a plane-wave form for the disturbations. With the equation

$$\delta \boldsymbol{u}(\boldsymbol{x}, t) = \delta \boldsymbol{u}\, \mathrm{e}^{i(\omega t - (\boldsymbol{k}\boldsymbol{x}))} \tag{2.47}$$

($\delta \boldsymbol{u}$ is a constant amplitude vector) equations (2.38) and (2.39) are transformed into the linear system

$$(\boldsymbol{k}\, \delta \boldsymbol{B}) = 0 , \tag{2.48}$$

$$\left\{\omega I - \sum_{j=1}^{3} k_j U_j\right\} \delta \boldsymbol{u} = 0 \tag{2.49}$$

(I is the unity matrix). Setting

$$\boldsymbol{n} = (n_1, n_2, n_3) = \frac{\boldsymbol{k}}{k} , \qquad C = \frac{\omega}{k} , \qquad L_u = \sum_{j=1}^{3} n_j U_j$$

we can rewrite the above system in the following way

$$(\boldsymbol{n}\, \delta \boldsymbol{B}) = 0 , \tag{2.50}$$

$$L_u\, \delta \boldsymbol{u} = C\, \delta \boldsymbol{u} . \tag{2.51}$$

The secular equation of L_u

$$||L_u - C I|| = 0 \tag{2.52}$$

gives the dispersion relation; the eigenvalues C define the phase velocities of the allowed plane waves. The eigenvectors represent the relevant amplitudes of the disturbed parameters. Calculations of the secular equation and the eigenvectors will be made on the basis of the coordinate system specified by $\boldsymbol{n}$ and $\boldsymbol{B}$ (Figure 2.1):

$$\boldsymbol{e}_1 = \boldsymbol{n} , \quad \boldsymbol{e}_2 = \boldsymbol{r} = \frac{\boldsymbol{n} \times \boldsymbol{B}}{|\boldsymbol{n} \times \boldsymbol{B}|} , \quad \boldsymbol{e}_3 = \boldsymbol{t} = (\boldsymbol{r} \times \boldsymbol{n}) = \frac{\boldsymbol{B} - (\boldsymbol{n}\, \boldsymbol{B})\, \boldsymbol{n}}{|\boldsymbol{B} - (\boldsymbol{n}\, \boldsymbol{B})\, \boldsymbol{n}|} . \tag{2.53}$$

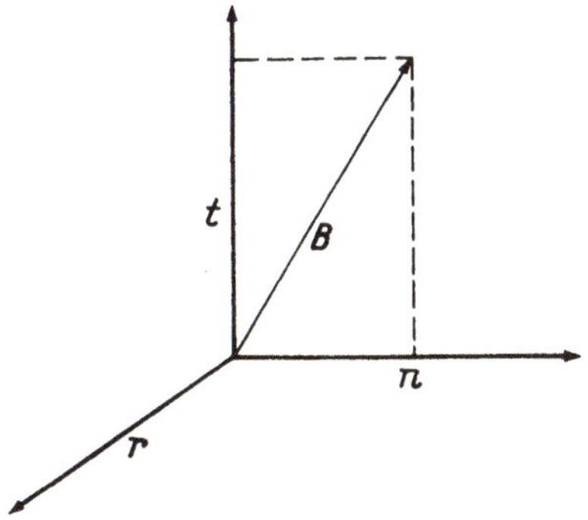

Figure 2.1. Orientation of the eigenvectors of the *nrt*-coordinate system

In addition, we shall consider the variables in (2.45) in the following order

$$\delta \boldsymbol{w} = \left(\frac{a}{\varrho}\,\delta\varrho,\ \delta v_n,\ \delta v_t,\ \delta b_t,\ \frac{1}{a\,\varrho}\,p_s\,\delta s,\ \delta v_r,\ \delta b_r,\ \delta b_n\right). \tag{2.54}$$

Thus, $\delta \boldsymbol{w} = N\,\delta \boldsymbol{u}$, where

$$N = \begin{array}{|c|c|c|c|c|c|c|c|}
\hline
\frac{a}{\varrho} & & & & & & & \\ \hline
 & 1 & & & & & & \\ \hline
 & & & 1 & & & & \\ \hline
 & & & & & & \frac{1}{\sqrt{\mu\,\varrho}} & \\ \hline
 & & & & & & & \frac{p_s}{a\,\varrho} \\ \hline
 & & 1 & & & & & \\ \hline
 & & & & & \frac{1}{\sqrt{\mu\,\varrho}} & & \\ \hline
 & & & & \frac{1}{\sqrt{\mu\,\varrho}} & & & \\ \hline
\end{array}$$

Consequently, we obtain from (2.51) the system of equations

$$L_w\,\delta \boldsymbol{w} = C\,\delta \boldsymbol{w} \tag{2.55}$$

where

$$L_w = N\,L_u\,N^{-1} = \begin{array}{|c|c|c|c|c|c|c|c|}
\hline
v_n & a & & & & & & \\ \hline
a & v_n & & b_t & a & & & \\ \hline
 & & v_n & -b_n & & & & \\ \hline
 & b_t & -b_n & v_n & & & & \\ \hline
 & & & & v_n & & & \\ \hline
 & & & & & v_n & -b_n & \\ \hline
 & & & & & -b_n & v_n & \\ \hline
 & & & & & & & v_n \\ \hline
\end{array} \tag{2.56}$$

Here we have

$$v_n = (\boldsymbol{v}\,\boldsymbol{n})\,, \quad b_n = (\boldsymbol{b}\,\boldsymbol{n})\,, \quad b_t = (\boldsymbol{b}\,\boldsymbol{t})\,, \quad \text{where} \quad \boldsymbol{b} = \frac{\boldsymbol{B}}{\sqrt{\mu\,\varrho}}\,.$$

L_w reduces to a quasidiagonal matrix and is symmetric except for the element a in the fifth column. The secular equation is

$$||L_w - C\,I|| = c^2\,(c^2 - b_n^2)\,\{c^4 - (a^2 + b^2)\,c^2 + a^2\,b_n^2\} = 0\,, \tag{2.57}$$

where $c = C - v_n$ is the phase velocity relative to the medium, $b = |\boldsymbol{b}|$ has the dimension of velocity and is called the Alfvén velocity.

Equation (2.57) is equivalent to (2.52), since $L_w = N\,L_u\,N^{-1}$. Let us once again write the system (2.55) explicitly:

$$c \cdot \frac{a}{\varrho}\,\delta\varrho - a\,\delta v_n = 0\,, \tag{2.58}$$

$$c\,\delta v_n - b_t\,\delta b_t - \frac{a^2}{\varrho}\,\delta\varrho - \frac{1}{\varrho}\,p_s\,\delta s = 0\,, \tag{2.59}$$

$$c\,\delta v_r + b_n\,\delta b_r = 0\,, \tag{2.60}$$

$$c\,\delta v_t + b_n\,\delta b_t = 0\,, \tag{2.61}$$

$$c\,\delta b_n = 0\,, \tag{2.62}$$

$$c\,\delta b_r + b_n\,\delta v_r = 0\,, \tag{2.63}$$

$$c\,\delta b_t + b_n\,\delta v_t - b_t\,\delta v_n = 0\,, \tag{2.64}$$

$$c\,\delta s = 0\,. \tag{2.65}$$

Allowing for (2.50) $\big((\boldsymbol{n}\,\delta\boldsymbol{B}) = \delta b_n = 0\big)$, equation (2.62) is identically satisfied.

As a result a factor c cancels out in (2.57).

We can also linearize the system starting with the conservation equation. The equations analogous to (2.39) and (2.51) are respectively

$$\frac{\partial\delta\boldsymbol{a}}{\partial t} + \sum_{j=1}^{3} A_j\,\frac{\partial\delta\boldsymbol{a}}{\partial x_j} = 0 \tag{2.66}$$

and

$$L_A\,\delta\boldsymbol{a} = C\,d\boldsymbol{a} \quad \text{where} \quad L_A = \sum_{j=1}^{3} n_j\,A_j\,. \tag{2.67}$$

Since

$$L_A = M\,L_u\,M^{-1}$$

the secular equation of L_a agrees with that of L_u, and thus also with (2.57). The eigenvectors $\delta\boldsymbol{a}$, $\delta\boldsymbol{u}$, and $\delta\boldsymbol{w}$ are linked by the following relation

$$\delta\boldsymbol{a} = M\,\delta\boldsymbol{u} = M \cdot N^{-1}\,\delta\boldsymbol{w}\,. \tag{2.68}$$

2.3. The Dispersion Equation

The roots of the secular equation are

$$c = C - (\boldsymbol{v}\,\boldsymbol{n}) = \pm\, c_i\,, \qquad i = 0, 1, 2, 3\,,$$

where

$$c_0 = 0 \qquad \text{“entropy wave”}\,; \tag{2.69}$$

$$c_s = c_1 = \sqrt{\frac{1}{2}\,\{a^2 + b^2 - \sqrt{(a^2 + b^2) - 4\,a^2\,b_n^2}\}} = b\,\sqrt{\frac{1}{2}\,(1 + r - q)} \tag{2.70}$$

“slow magnetoacoustic wave” or s-wave (slow wave);

$$c_A = c_2 = |b_n| = b\,|\cos\vartheta| \qquad \text{“Alfvén wave”}\,; \tag{2.71}$$

$$c_f = c_3 = \sqrt{\frac{1}{2}\,\{a^2 + b^2 + \sqrt{(a^2 + b^2)^2 - 4\,a^2\,b_n^2}\}} = b\,\sqrt{\frac{1}{2}\,(1 + r + q)} \tag{2.72}$$

“fast magnetoacoustic wave” of f-wave (fast wave).
where ϑ $(0 \le \vartheta < \pi)$ is the angle between $\boldsymbol{n}$ and $\boldsymbol{B}$. $r = \frac{a^2}{b^2}$ is (to within a constant factor) the ratio of the internal energy density to the magnetic energy density and

$$q^2 = r^2 - 2\,r\cos 2\,\vartheta + 1\,. \tag{2.73}$$

Magnetoacoustic waves will be subsequently denoted as MA-waves.

The c_i are all real and distinct, provided $\boldsymbol{n}$ is not parallel or perpendicular to $\boldsymbol{B}$. The anisotropy produced by the magnetic field is expressed by the ϑ-dependence of the phase velocity. The dependence of c_s, c_A, and c_f on ϑ is given in Figures 2.2a and 2.2b for three different values of r.

From (2.70)–(2.72) we obtain the following inequalities

$$0 \le c_s \le \min\,(a\cos\vartheta,\, b\cos\vartheta) \le \max\,(a, b) \le c_f \le \sqrt{a^2 + b^2}\,; \tag{2.74}$$

and in particular

$$c_s \le c_A \le c_f\,. \tag{2.75}$$

In the limiting cases where the propagation is respectively parallel and perpendicular to the magnetic field direction $\left(\vartheta = 0,\ \frac{\pi}{2}\right)$, we have

$$\vartheta = 0: \quad c_s = \min\,(a, b)\,, \qquad c_A = b\,, \qquad c_f = \max\,(a, b)\,; \tag{2.76}$$

$$\vartheta = \frac{\pi}{2}: \quad c_s = 0\,, \qquad c_A = 0\,, \qquad c_f = \sqrt{a^2 + b^2}\,. \tag{2.77}$$

The phase velocity of the s-wave and the Alfvén-wave vanish upon propagation normal to the field direction.

The dispersion is purely spatial in nature. The phase velocities depend only on the direction of propagation. Thus

$$c = c(\boldsymbol{n})\,, \quad \text{and consequently} \quad C = C(\boldsymbol{n})$$

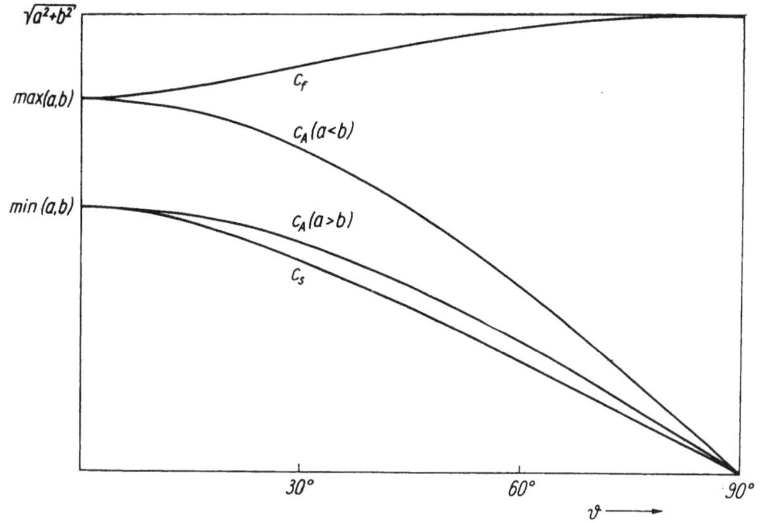

a)

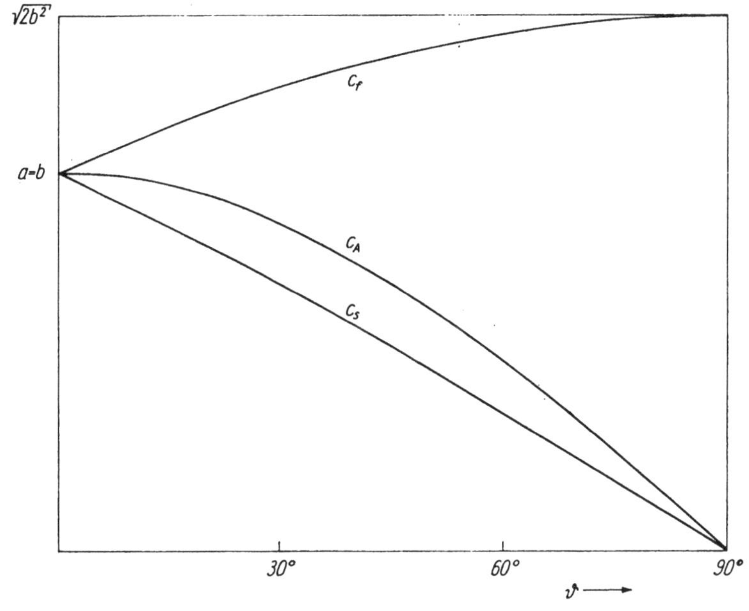

b)

Figure 2.2. c_s, c_A, and c_f as functions of the angle ϑ between the direction of propagation and the magnetic field for the values $r = 1/2$, $r = 2$ (a), and $r = 1$ (b)

or

$$\omega = \omega(\boldsymbol{k}) = k \cdot C\left(\frac{\boldsymbol{k}}{k}\right). \tag{2.78}$$

More specifically

$$c = c(\boldsymbol{n}\,\boldsymbol{m}) = c\,(\cos\vartheta)\,, \qquad \boldsymbol{m} = \frac{\boldsymbol{B}}{|\boldsymbol{B}|}\,.$$

This expresses the rational symmetry about the $\boldsymbol{B}$-axis.

Of interest are the two limiting cases, in which one of the velocities a, b predominates

When

$$r \gg 1\,, \quad \text{i.e.,} \quad a^2 \gg b^2$$

it follows that

$$c_f \approx a\left(1 + \frac{1}{2\,r}\sin^2\vartheta\right) \approx a\,, \tag{2.79}$$

$$c_s \approx b\cos\vartheta\,.$$

Similarly, when

$$r \ll 1\,, \quad \text{i.e.,} \quad a^2 \ll b^2$$

we have

$$c_f \approx b\left(1 + \frac{r}{2}\sin^2\vartheta\right) \approx b\,, \tag{2.80}$$

$$c_s \approx a\cos\vartheta\,.$$

The f-wave propagates nearly isotropically in both cases. The phase velocity of the slow wave is in each case small compared to that of the fast wave.

2.4. Polarization Conditions

When $\boldsymbol{n}$ is neither parallel nor perpendicular to $\boldsymbol{B}$, the roots of the dispersion equation are distinct. The associated eigenvectors are then fixed (to within a constant factor). Let us write these in the form

$$\delta\boldsymbol{w} = \varepsilon \cdot \frac{1}{W} \cdot \boldsymbol{w}\,,$$

where ε is a parameter, which ensures the "smallness" of the disturbance and represents a measure of its intensity. $\boldsymbol{w}$ is a dimensionless vector and $\frac{1}{W}$ a normalization factor, which is chosen in such a way that

$$\frac{\varrho}{2} \cdot \delta\boldsymbol{w}^2 = \varepsilon^2$$

(ε^2 and W^2 have the dimensions of energy and density, respectively). In the nrt-system we have

$$\delta\boldsymbol{w} = \left(\frac{a}{\varrho}\,\delta\varrho,\, \delta v_n,\, \delta v_t,\, b_t,\, \frac{1}{a\,\varrho}\,p_s\,\delta s,\, \delta v_r,\, \delta b_r\right). \;* \tag{2.81}$$

* We omit the component $\varrho\, b_n$, since $\delta b_n = 0$ according to (2.62).

It clearly follows from the form of the L_w-matrix (2.56) that two orthogonal characteristic spaces exist: a 5-dimensional space Ω_1 (MA- and entropy waves) and a two-dimensional space Ω_2, (Alfvén-waves).

Entropy waves:

In this case

$$c_0 = 0$$

and it follows that

$$\boldsymbol{w}_0 = (-1, 0, 0, 0, 1, 0, 0)\,, \qquad W = \sqrt{\varrho}\,, \tag{2.82}$$

i.e.,

$$\delta \boldsymbol{v} = \delta \boldsymbol{B} = 0\,, \qquad p = a^2\,\delta\varrho + p_s\,\delta s = 0\,.$$

These are not waves in the actual sense. Here we are dealing with changes of the density and entropy at a constant pressure, which are moving with the medium, and which also occur in ordinary hydrodynamics. When $(\boldsymbol{n}\,\boldsymbol{B}) = 0$, changes in $\boldsymbol{v}$ and $\boldsymbol{B}$ at a constant total pressure $p^* = p + \dfrac{\boldsymbol{B}^2}{2\mu}$ can also occur. For all other waves $c \neq 0$ and thus $\delta s = 0$, because of (2.65).

To determine the eigenvectors $\delta\boldsymbol{w}_i\ (i = 1, 2, 3)$ we shall only use the symmetric part of L_w. The six $\delta\boldsymbol{w}_i$ are therefore mutually orthogonal and the characteristic space spanned by the eigenvalues has the dimension of the multiplicity of the eigenvalue. In addition, the c_i are real.

Alfvén-waves:

Here we have

$$c = \pm\, c_2 = \pm\, |b_n|$$

so that

$$\boldsymbol{w}_2^{+,-} = (0, 0, 0, 0, 0, 1, \pm\ \mathrm{sgn}\cos\vartheta)\,, \qquad W = \sqrt{\varrho}\,. \tag{2.83}$$

This means

$$\delta\varrho = \delta p = 0\,, \qquad \delta v_n = \delta v_t = \delta B_t = 0 \tag{2.84}$$

and

$$\varrho\,\delta v_r = \pm\ \mathrm{sgn}\cos\vartheta\,\frac{\partial B_r}{\sqrt{\mu}}\,. \tag{2.85}$$

The thermodynamic quantities ϱ, p, and s remain constant, while $\delta\boldsymbol{v}$ and $\delta\boldsymbol{B}$ are oriented perpendicular to the $\boldsymbol{n}\,\boldsymbol{B}$-plane. Thus, Alfvén-waves are not associated with compressions or changes in the magnetic pressure.

$$(\boldsymbol{n}\,\delta\boldsymbol{v}) = \frac{1}{\mu}\,(\boldsymbol{B}\,\delta\boldsymbol{B}) = 0\,.$$

The change in $\boldsymbol{E}$ can be calculated in accordance with (1.5). In the system moving with the medium, we have

$$\delta\boldsymbol{E} = -\,(\delta\boldsymbol{v}\times\boldsymbol{B}) = B_t\,\delta v_r\,\boldsymbol{n} - B_n\,\delta v_r\,\boldsymbol{t}\,. \tag{2.86}$$

$(\boldsymbol{n}\,\delta\boldsymbol{E}) = 0$ only if $B_t = 0$, i.e., when propagation takes place parallel to the magnetic field. With respect to $\boldsymbol{E}$ the Alfvén-waves are thus in general not transverse.

Magnetoacoustic waves:

In this case

$$c = \pm c_i \qquad (i = 1, 3)$$

and we obtain

$$\left.\begin{aligned} \boldsymbol{w}_i^{+,-} &= \left(\frac{a}{\pm c_i}\left(\frac{c_i^2}{b^2} - \cos^2\vartheta\right), \frac{c_i^2}{b^2} - \cos^2\vartheta\,, \; -\sin\vartheta\cos\vartheta\,, \frac{\pm c_i}{b}\sin\vartheta, 0, 0, 0\right), \\ W &= \sqrt{\varrho\left\{\frac{c^4}{b^4} + \cos^2\left(1 - \frac{2\,c^2}{\varrho^2}\right)\right\}}. \end{aligned}\right\} \tag{2.87}$$

From this equation it follows that $\delta B_r = \delta v_r = 0$. In contrast to the Alfvén-waves, $\delta \boldsymbol{v}$ and $\delta \boldsymbol{B}$ in this case lie in the $\boldsymbol{n}\,\boldsymbol{B}$-plane. These waves are thus associated with compressions and changes in the magnetic pressure. δp and $\delta p_m = \frac{1}{\mu}(\delta\,\boldsymbol{B}\,\boldsymbol{B})$ are of like sign for fast waves and of opposite sign for slow waves.

According to (2.74) we see that

$$\frac{\partial p_m}{\partial p} = \frac{\frac{1}{\mu}(\delta \boldsymbol{B}\,\boldsymbol{B})}{\delta p} = \frac{c_i^2 - a^2\cos^2\vartheta}{c_i^2\cdot r} = \begin{cases} >0 & \text{for} \quad i = 3\,, \\ <0 & \text{for} \quad i = 1\,. \end{cases} \tag{2.88}$$

The changes in the magnetic pressure and the gas pressure have a comparable effect in the case of f-waves, whereas they partly compensate each other for s-waves.

Both MA-waves become degenerate upon propagation parallel to the magnetic field direction. We obtain

$$\boldsymbol{w}_i^{+,-} = (0, 0, 1, \pm 1, 0, 0, 0) \quad \text{when} \quad \vartheta \to 0 \quad \text{and} \quad c_i \to b\,;$$

furthermore

$$\delta\, v_n = \delta\varrho = 0\,.$$

We are dealing with an Alfvén-wave, which is polarized perpendicular to the wave in (2.83).*

When

$$\vartheta \to 0 \quad \text{and} \quad c \to a$$

we have

$$\boldsymbol{w}_i^{+,-} = (\pm 1, 1, 0, 0, 0, 0, 0)\,; \quad \text{i.e.,} \quad \delta B_t = \delta v_t = 0\,.$$

This represents a pure sound wave.

For fast waves propagating perpendicular to the magnetic field, we obtain

$$\boldsymbol{w}_i^{+,-} = \left(\pm\frac{a}{\sqrt{a^2 + b^2}}, 1, 0, \pm\frac{b}{\sqrt{a^2 + b^2}}, 0, 0, 0\right).$$

* Thus circular polarization is possible upon propagation parallel to the magnetic field.

Here also we are dealing with a longitudinal wave with respect to $\delta \boldsymbol{v}$. We may consider this wave as a sound wave modified by the field. The propagation speed $c_f = \sqrt{a^2 + b^2}$ is higher than the pure gasdynamical propagation speed. The magnetic field reduces the compressibility of the medium in a direction perpendicular to itself.

When propagation takes place in a direction oblique to the magnetic field, $\delta \boldsymbol{v}$ is composed of longitudinal as well as transverse components. However, with respect to $\delta \boldsymbol{E}$, the MA-waves are transverse, i.e.,

$$(\boldsymbol{n}\,\delta\boldsymbol{E}) = -\,\boldsymbol{n}\,(\delta\boldsymbol{v}\times\boldsymbol{B}) = 0\,.$$

since $\boldsymbol{n}$, $\delta\boldsymbol{v}$, and $\boldsymbol{B}$ lie in the same plane.

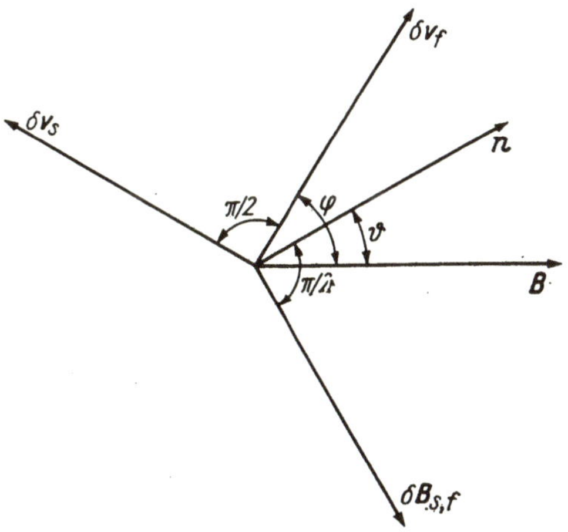

Figure 2.3. Orientation of the amplitude vectors $\delta\boldsymbol{v}$ and $\delta\boldsymbol{B}$ for MA-waves

Figure 2.3 illustrates the orientation of the vectors $\delta\boldsymbol{v}$, $\delta\boldsymbol{B}$, $\boldsymbol{n}$, and $\boldsymbol{B}$ for the case $(\boldsymbol{n}\,\boldsymbol{B}) > 0$. In accordance with (2.87), we can show that

$$(\delta\boldsymbol{v}_f\,\delta\boldsymbol{v}_s) = 0 \quad \text{and} \quad (\delta\boldsymbol{B}\,\delta\boldsymbol{v}) \begin{cases} < 0 \\ > 0 \end{cases} \quad \text{when} \quad (\boldsymbol{n}\,\boldsymbol{B}) \begin{matrix} > 0\,, \\ < 0\,. \end{matrix}$$

It is possible to derive a relationship between the angles φ (between $\delta\boldsymbol{v}$ and $\boldsymbol{B}$) and ϑ. From equations (2.58) to (2.64) for $c \neq b_n$, after eliminating $\delta\varrho$, δb_r, and δb_t, and by virtue of the coplanar property of the vectors $\delta\boldsymbol{v}$, $\delta\boldsymbol{B}$, $\boldsymbol{n}$, and $\boldsymbol{B}$, we obtain

$$(\boldsymbol{m}\,\delta\boldsymbol{v})\,(\boldsymbol{m}\times\delta\boldsymbol{v}) = r(\boldsymbol{n}\,\delta\boldsymbol{v})\,(\boldsymbol{n}\times\delta\boldsymbol{v})\,, \qquad \boldsymbol{m} = \frac{\boldsymbol{B}}{|\boldsymbol{B}|}\,.$$

Rewriting in terms of φ and ϑ

$$\sin 2\,\varphi = r \sin 2\,(\varphi - \vartheta)\;;$$

and solving for $\sin 2\varphi$

$$\sin 2\varphi = \frac{r}{q} \sin 2\vartheta .$$

To each given ϑ correspond two solutions for φ consistent with the remaining equations, and which differ by a factor of $\frac{\pi}{2}$. In Figure 2.4 φ is shown as a function of ϑ (continuous curves represent the f-wave, the dotted curves the s-wave; r is the curve parameter).

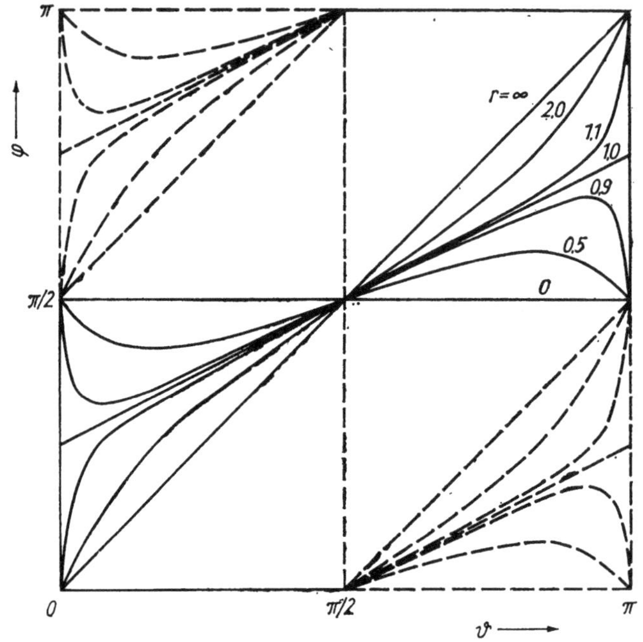

Figure 2.4. The angle φ between $\delta \boldsymbol{v}$ and $\boldsymbol{B}$ as a function of ϑ for different values of r. The continuous and dotted curves represent the f-wave and s-wave, respectively (Kunkel [15])

We can determine the polarization conditions from the limiting cases (2.79) and (2.80).

When $r \to \infty$, the f-wave becomes longitudinal and changes to a sound wave.

When $r \to 0$, $\varphi \to \frac{\pi}{2}$ is independent of ϑ. The change in the velocity is perpendicular to $\boldsymbol{B}$, independent of the direction of propagation.

2.5. Group Velocity and Energy Transport

From crystal optics it is known that for plane waves in an anisotropic medium the direction in which energy is propagated (ray direction) in general does not conform with the direction of the wave normals. Similarly, the phase velocity is in general different from the velocity at which the energy is transported.

In optics the energy flux is represented by the Poynting vector $\boldsymbol{S}$. If U is the local energy density of the electromagnetic field, then

$$\frac{\partial U}{\partial t} = -\operatorname{div} \boldsymbol{S} . \tag{2.89}$$

The velocity at which the energy is propagated is then given by the vector

$$\boldsymbol{s} = \frac{1}{U} \cdot \boldsymbol{S} .$$

This corresponds to the group velocity in optics

$$\boldsymbol{v}_g = \operatorname{grad}_k \omega(k) .$$

Similar relations hold for MHD-waves.

We can derive an energy equation for the changes analogous to (2.89) from the linearized equations. Next we shall disregard the entropy waves and let $\delta s = 0$. Performing a scalar multiplication of (2.46) with $\varrho \cdot \delta \boldsymbol{u}^*$, we obtain after some rearranging

$$\frac{\partial \delta \mathscr{E}}{\partial t} = -\operatorname{div} \delta \boldsymbol{F} \tag{2.90}$$

where

$$\delta \mathscr{E} = \frac{\varrho}{2} (\delta \boldsymbol{u}^*)^2 = \frac{\varrho}{2} (\delta \boldsymbol{v})^2 + \frac{a^2}{2\varrho} (\delta \varrho)^2 + \frac{1}{2\mu} (\delta \boldsymbol{B})^2 \tag{2.91}$$

and

$$\delta \boldsymbol{F} = \boldsymbol{v} \cdot \delta \mathscr{E} + a^2 \delta \varrho \, \delta \boldsymbol{v} + \frac{1}{\mu} \big(\delta \boldsymbol{B} \times (\delta \boldsymbol{v} \times \boldsymbol{B})\big) . \tag{2.92}$$

From (2.90) it clearly follows that $\delta \mathscr{E}$ and $\delta \boldsymbol{F}$ respectively represent the local energy density and the energy flux, which are connected with a disturbance. When $\boldsymbol{B} = 0$, equation (2.90) transforms in to energy equation for the propagation of a sound wave; the term $a^2 \, \delta p \, \delta \boldsymbol{v}$ is the acoustic energy flux. The energy transport by convection is expressed by $\boldsymbol{v}\, \delta \mathscr{E}$ and $\frac{1}{\mu} \big(\delta \boldsymbol{B} \times (\delta \boldsymbol{v} \times \boldsymbol{B})\big) = \frac{1}{\mu} (\delta \boldsymbol{E} \times \delta \boldsymbol{B})$ can be considered as the Poynting vector. The vector

$$\boldsymbol{s} = \frac{\partial \boldsymbol{F}}{\partial \mathscr{E}} \tag{2.93}$$

can be regarded as the velocity at which energy is propagated.

From expressions (2.83) and (2.87) for the amplitudes we can calculate the energy density $\delta \mathscr{E}_i^{+,-}$ and the energy flux $\delta \boldsymbol{F}_i^{+,-}$ for the three wave modes.

Because of the normalization $\frac{\varrho}{2} (\delta \boldsymbol{w})^2 = \varepsilon^2$ we have

$$\delta \mathscr{E}_i^{+,-} = (\varepsilon_i^{+,-})^2 . \tag{2.94}$$

The intensity of the corresponding wave is denoted by $\varepsilon_i^{+,-}$. We still have to verify the following relationship:

$$\frac{1}{2} \varrho (\delta \, \boldsymbol{v})^2 = \frac{1}{2} \frac{a^2}{\varrho} (\delta \varrho)^2 + \frac{(\partial \boldsymbol{B})^2}{2\mu} . \tag{2.95}$$

The total energy density associated with the wave is equally made up of the kinetic energy density and the density of the internal and magnetic energy $\left(\text{for Alfvén-waves } \frac{1}{2} \frac{a_2}{\varrho} (\delta\varrho)^2 = 0\right)$. Letting $\delta s \neq 0$, an additional term p_s $(\delta \boldsymbol{v} \cdot \operatorname{grad} \delta s)$ enters into (2.90), which cannot be written as a divergence. However, from the last equation of system (2.39), which is independent of the other equations, we can obtain an expression quadratic in

$$\frac{\partial}{\partial t} (\delta s)^2 = - \operatorname{div} \{ \boldsymbol{v} (\delta s)^2 \} . \tag{2.96}$$

Let us now calculate the energy transport velocity for the different wave modes.

Entropy waves:

In this case $\delta \boldsymbol{v} = 0$ and $\delta p \sim \delta s \neq 0$. Equations (2.90) and (2.96) are equivalent, i.e.,

$$\mathbf{s} = \boldsymbol{v} . \tag{2.97}$$

Thus, as we expected, energy transport can take place by convection only.

Alfvén-waves:

Here we have $\delta \varrho = 0$; $\delta \boldsymbol{B}, \delta \boldsymbol{v} \perp \boldsymbol{B}$, so that

$$\frac{1}{\mu} \left(\delta \boldsymbol{B} \times (\delta \boldsymbol{v} \times \boldsymbol{B}) \right) = - \frac{1}{\mu} \boldsymbol{B} \delta v_r \, \delta B_r$$

and allowing for (2.83) we obtain

$$\mathbf{s}^{+,-} = \boldsymbol{v} \pm \boldsymbol{b} \operatorname{sgn} \cos \vartheta . \tag{2.98}$$

Magnetoacoustic waves:

Let us calculate by first determining its components. We obtain

$$s_n = v_n + \frac{1}{\mathscr{E}} \left\{ \delta v_n \left(a^2 \, \delta \varrho + \frac{1}{\mu} (\delta B_t \, B_t) \right) - \frac{B_n}{\mu} \delta v_t \, \delta B_t \right\} . \tag{2.99}$$

Substituting the expressions (2.87), we can transform the parenthetical expression with the help of the dispersion equation for the MA-waves $\left(c^4 - (a^2 + b^2) \times c^2 + a^2 b^2 \cos^2 \vartheta = 0\right)$, giving

$$s_n = v_n + c . \tag{2.100}$$

where here and in the following c is set $\pm c_1$, $\pm c_3$. Furthermore,

$$s_r = v_r \quad \text{since} \quad \delta v_r = 0 \quad \text{and} \quad \delta B_r = 0 \tag{2.101}$$

and

$$s_t = v_t + \frac{1}{\mathscr{E}} \left\{ \delta v_t \left(a^2 \, \delta \varrho + \frac{1}{\mu} (\delta B_t \, B_t) \right) - \frac{B_t}{\mu} \delta v_t \, \delta B_t \right\} .$$

After substituting (2.87), the parenthetical expression can once again be transformed when allowing for the dispersion relation:

$$s_t = v_t - c \cdot \frac{\sin \vartheta \cos \vartheta \left(\frac{c^2}{b^2} - 1\right)}{\frac{c^4}{b^4} + \cos^2 \vartheta \left(1 - \frac{2\,b^2}{c^2}\right)} . \tag{2.102}$$

Let us now calculate the group velocity of the individual waves. Due to (2.87) the phase and group velocity are connected by the expression

$$\boldsymbol{v}_g = \mathrm{grad}_k\, \omega(k) = C\, \boldsymbol{n} + \mathrm{grad}_n\, C(\boldsymbol{n}) - (\boldsymbol{n}\, \mathrm{grad}_n\, C)\, \boldsymbol{n} . \tag{2.103}$$

In general $\boldsymbol{v}_g$ and $\boldsymbol{C} = C\,\boldsymbol{n}$ are no longer in agreement when $\frac{\partial C}{\partial n_j} \neq 0$. For an arbitrary direction

$$(\boldsymbol{v}_g\, \boldsymbol{n}) = C \tag{2.104}$$

holds, i.e., the phase velocity in a direction $\boldsymbol{n}$ is equal to the projection of the group velocity along $\boldsymbol{n}$.

In a stationary coordinate system the following holds for MHD-waves:

$$C = (\boldsymbol{v} \cdot \boldsymbol{n}) + c\,, \qquad c = c\,(\boldsymbol{n} \cdot \boldsymbol{m}) = c(\vartheta)\,.$$

Calculating $\boldsymbol{v}_g$ in accordance with (2.103) and expressing the components in the nrt-system (2.53), we obtain

$$v_{g\,n} = v_n + c\,, \tag{2.105}$$

$$v_{g\,r} = v_r\,, \tag{2.106}$$

$$v_{g\,t} = v_t - \frac{\partial c}{\partial \vartheta} . \tag{2.107}$$

Let us compare these expressions with those of $\boldsymbol{s}$ for the different wave modes.

Entropy waves:

$$\boldsymbol{v}_g = \boldsymbol{v} \quad \text{since} \quad c = 0\,.$$

Alfvén-waves:

From

$$c_2 = \pm\, b\, |\cos \vartheta|\,, \quad \frac{\partial c_2}{\partial \vartheta} = \pm\, b \sin \vartheta \;\mathrm{sgn} \cos \vartheta$$

it follows that

$$\boldsymbol{v}_g = \boldsymbol{v} \pm \boldsymbol{b}\; \mathrm{sgn} \cos \vartheta$$

in agreement with (2.98).

Thus, energy is transported relative to the medium independent of $\boldsymbol{n}$ only in the direction $\pm\boldsymbol{B}$ with velocity b*. This is the characteristic property of Alfvén-waves.

* See footnote on page 564.

Magnetoacoustic waves:

The agreement between $v_{g\,n}$, $v_{g\,r}$ and s_n, s_r, respectively, follows immediately. We can calculate $\frac{\partial c}{\partial \vartheta}$ with the help of the dispersion relation. We obtain

$$\frac{\partial c}{\partial \vartheta} = \frac{c \sin \vartheta \cos \vartheta \left(\frac{c^2}{b^2} - 1\right)}{\frac{c^4}{b^4} + \cos^2 \vartheta \left(1 - \frac{2\,b^2}{c^2}\right)}. \tag{2.108}$$

Thus, $v_{g\,t}$ and s_t are also identical.

$\boldsymbol{s}$ lies in the $\boldsymbol{n}\,\boldsymbol{B}$-plane of the coordinate system, in which the medium is at rest. The component of $\boldsymbol{s}$ in the direction of propagation is c (in agreement with (2.104)), and the component perpendicular to this direction (in the t-direction) is $-\frac{\partial c}{\partial \vartheta}$.

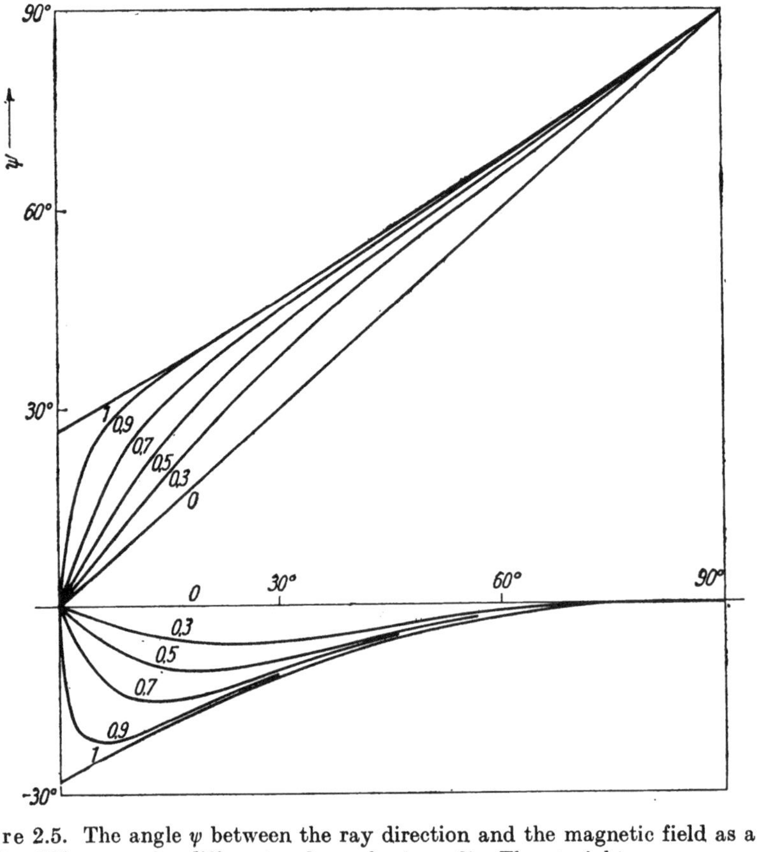

Figure 2.5. The angle ψ between the ray direction and the magnetic field as a function of ϑ for MA-waves for different values of r ($\boldsymbol{v} = 0$). The straight curves correspond to the parameter values r and $1/r$. Top curves: fast waves, bottom curves: slow waves

In Figure 2.5 the **s**-direction (angle ψ between **s** and $\boldsymbol{B}$) is shown as a function of ϑ for the two MA-waves when $\boldsymbol{v} = 0$ (the curve parameter is r; the straight curves correspond to r and $\frac{1}{r}$). We see that the f-waves become isotropic when $r \to \infty$ and $r \to 0$ ($\psi = \vartheta$).

In the case of the s-wave there is for every r a ψ_m, such that

$$\psi(\vartheta) \leq \psi_m .$$

For arbitrary ϑ, the rays lie inside a cone with vertex angle ψ_m. When $r = 1$, ψ_m attains its maximum value

$$\psi_m(1) = \tan^{-1} \frac{1}{2} \approx 26°30' ,$$

when $r \to \infty$ and $r \to 0$, $\psi_m \to 0$, and the rays of the s-waves are then directed along $\boldsymbol{B}$.

For Alfvén-waves, $\psi \equiv 0$, independent of ϑ.

Geometric interpretation

The relationship between the phase and group velocity can be readily shown geometrically. If in some coordinate system we plot the phase velocity C in every direction (from the origin) we obtain a surface N, which is called the surface of normal velocity. This is also known as the reciprocal normal surface. The radius vector from the origin to the points on the surface is

$$\boldsymbol{x} = C(\boldsymbol{n})\, \boldsymbol{n} .$$

We can express this radius in the form $\boldsymbol{x} = \boldsymbol{x}(u, v)$, where u and v are independent variables.

To every direction $\boldsymbol{n}$ corresponds the ray direction **s**. If we plot

$$\mathbf{s} = C\, \boldsymbol{n} + \text{grad}_n\, C - (\boldsymbol{n} \cdot \text{grad}_n\, C)\, \boldsymbol{n} \tag{2.109}$$

letting $\boldsymbol{n}$ assume all directions, the so-called ray surface S is obtained. Figures 2.6 and 2.7 show the surfaces of normal velocity and the ray surfaces of the Alfvén-wave and the two MA-waves for the case $\boldsymbol{v} = 0$.

N is the pedal surface of S relative to the origin.

The general relationship between a surface $\mathbf{s} = \mathbf{s}(u, v)$ and the corresponding pedal surface $\boldsymbol{x} = \boldsymbol{x}(u, v)$ is given by the following expression

$$\mathbf{s}(u, v) = \frac{\boldsymbol{x}^2(\boldsymbol{x}_u \times \boldsymbol{x}_v) - 2\,(\boldsymbol{x}\, \boldsymbol{x}_u)\,(\boldsymbol{x} \times \boldsymbol{x}_v) - 2\,(\boldsymbol{x}\, \boldsymbol{x}_v)\,(\boldsymbol{x}_u \times \boldsymbol{x})}{\boldsymbol{x} \cdot (\boldsymbol{x}_u \times \boldsymbol{x}_v)} , \tag{2.110}$$

where

$$\boldsymbol{x}_u = \frac{\partial}{\partial u}\, \boldsymbol{x}(u, v) , \qquad \boldsymbol{x}_v = \frac{\partial}{\partial v}\, \boldsymbol{x}(u, v) .$$

By means of these formulas we can check the relationship between N and S. The required parametric representation of the surface of normal velocity is

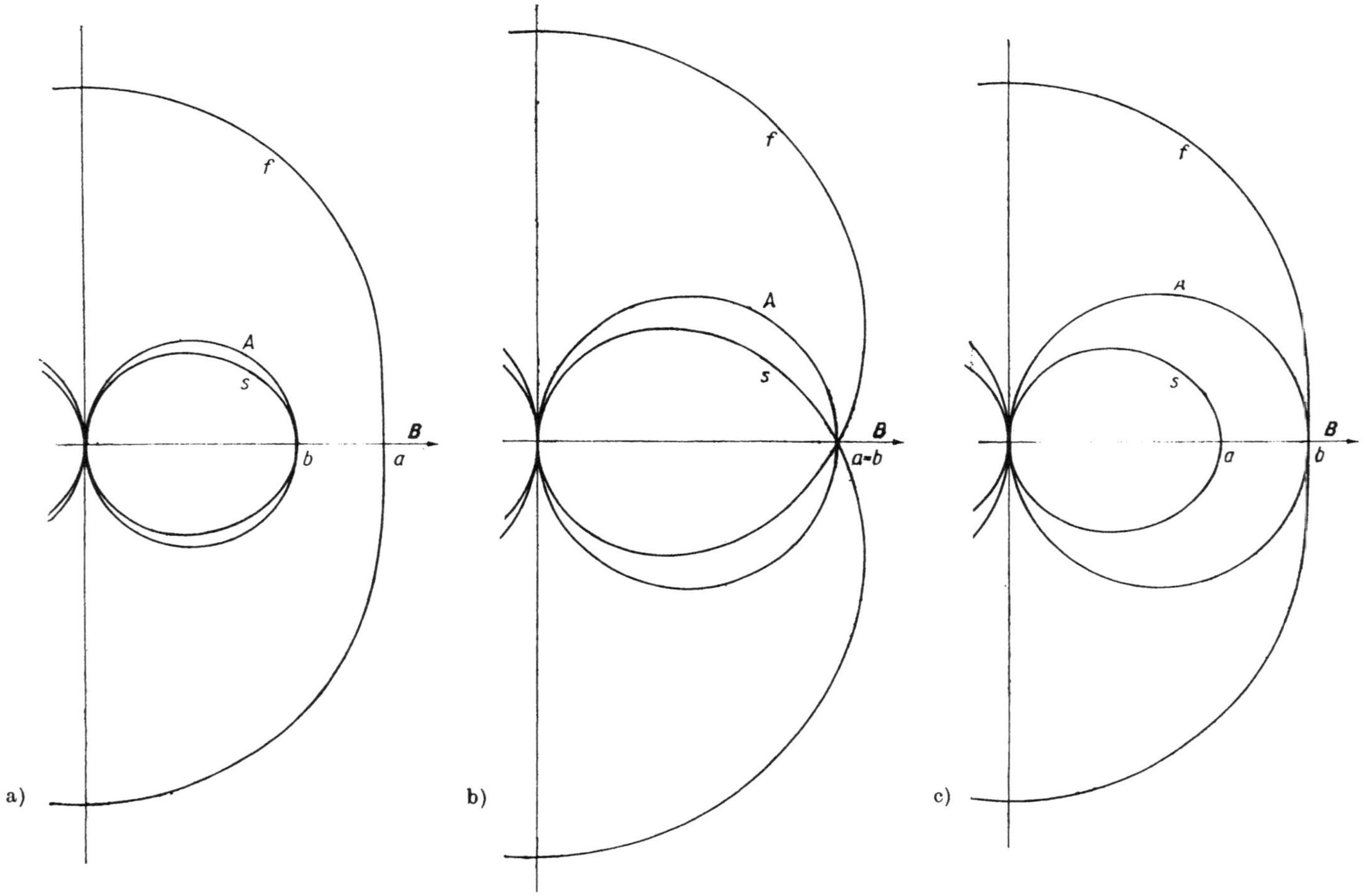

Figure 2.6. Surfaces of normal velocity of an Alfvén-wave and the MA-waves for the values $r = 2\ (a)$, $r = 1\ (b)$, and $r = 1/2\ (c)$

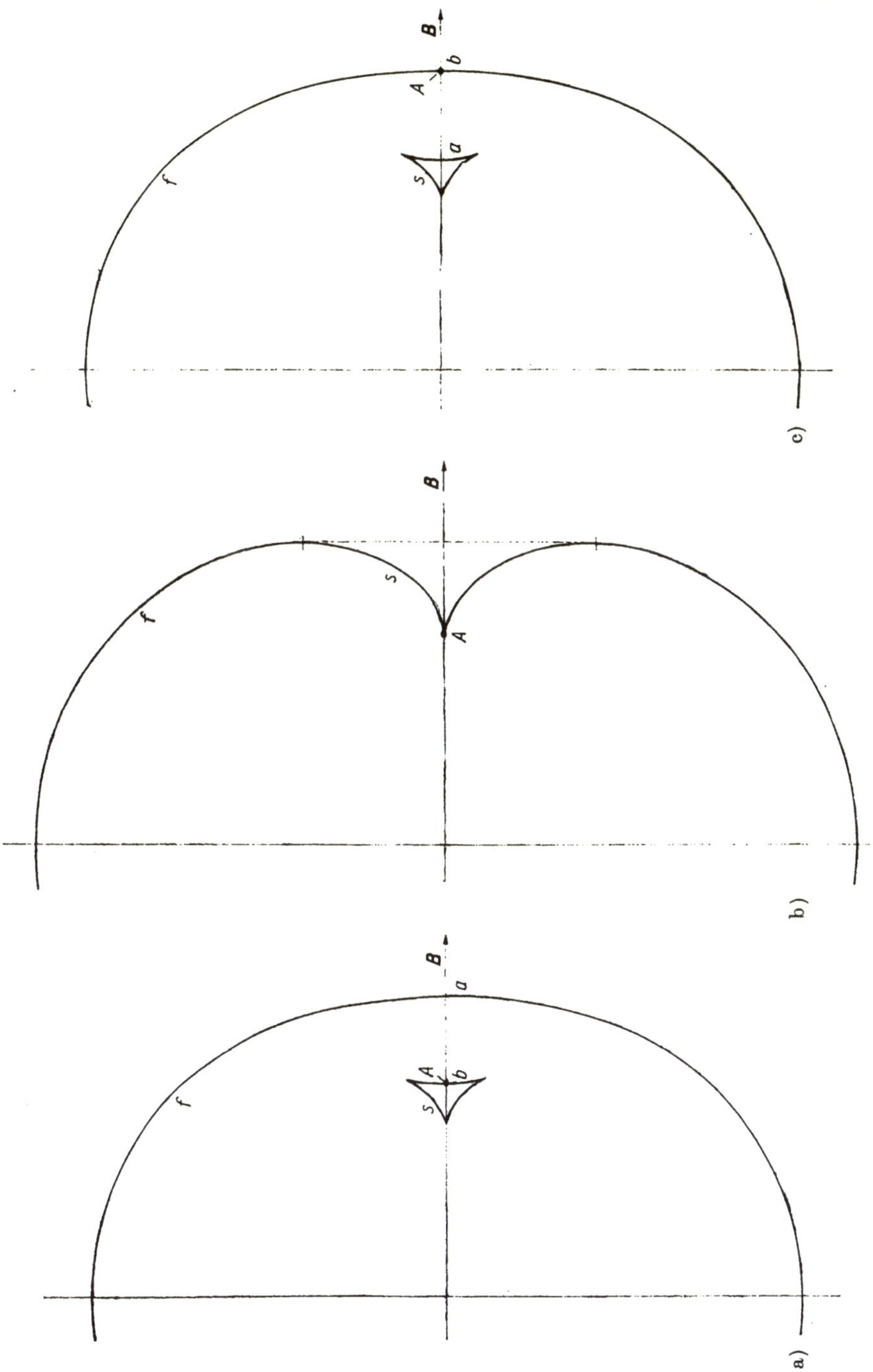

Figures 2.7. Ray surfaces of the Alfvén-wave and the MA-waves (Friedrichs diagrams) for the values $r = 1/2$ (*a*), $r = 1$ (*b*), and $r = 2$ (*c*), when $\boldsymbol{v} = 0$

obtained upon introducing spherical polar coordinates. Thus,

$$\boldsymbol{x}(\vartheta', \varphi') = \begin{pmatrix} C \sin \vartheta' \cos \varphi' \\ C \sin \vartheta' \sin \varphi' \\ C \cos \vartheta' \end{pmatrix}, \qquad C = C(\vartheta', \varphi') .$$

Considering (2.109) and (2.110) and allowing for the above, we obtain in both cases the same expressions

$$s_x = \cos \varphi' \left(C \sin \vartheta' + \frac{\partial C}{\partial \vartheta'} \cos \vartheta' \right) - \frac{\partial C}{\partial \varphi'} \cdot \frac{\sin \varphi'}{\sin \vartheta'},$$

$$s_y = \sin \varphi' \left(C \sin \vartheta' + \frac{\partial C}{\partial \vartheta'} \cos \vartheta' \right) + \frac{\partial C}{\partial \varphi'} \cdot \frac{\sin \varphi'}{\sin \vartheta'},$$

$$s_z = C \cos \vartheta' - \frac{\partial C}{\partial \vartheta'} \sin \vartheta' .$$

This gives the parametric representation of ray surfaces of the MA-waves (for $\boldsymbol{v} = 0$). Taking the z-axis along the direction of $\boldsymbol{B}$, we have $\vartheta = \vartheta'$ and $c = c(\vartheta)$. It follows from (2.70) and (2.72) that

$$\frac{\partial c_s}{\partial \vartheta} = \frac{b^2 r \sin \vartheta \cos \vartheta}{c_s\, q}; \quad \frac{\partial c_f}{\partial \vartheta} = - \frac{b^2\, r \sin \vartheta \cos \vartheta}{c_f\, q}, \tag{2.111}$$

and, for example, for the s-wave we obtain

$$s_x = \sin \vartheta \cos \varphi' \left(c_s - \frac{b^2\, r \cos^2 \vartheta}{c_s\, q} \right), \tag{2.112}$$

$$s_y = \sin \vartheta \sin \varphi' \left(c_s - \frac{b^2\, r \cos^2 \vartheta}{c_s\, q} \right), \tag{2.113}$$

$$s_z = \cos \vartheta \left(c_s + \frac{b^2\, r \sin^2 \vartheta}{c_s\, q} \right). \tag{2.114}$$

The formulas for the f-wave follow from the above when the sign of the second expression in the parenthesis is reversed and c_s is replaced by c_f. The parametric representations of the curves in Figure 2.7 are obtained from (2.112) to (2.114) when setting $\varphi' = 0$.

There exists a closer relationship between the ray surface S and the surface of phase normals P. This is defined by

$$\boldsymbol{p} = \frac{\boldsymbol{n}}{c(\boldsymbol{n})}. \tag{2.115}$$

This surface is obtained by reflection of N in the unit circle. Figure 2.8 illustrates this surface for the Alfvén-wave and both MA-waves (again for the case $\boldsymbol{v} = 0$). The surface of phase normals and the ray surface are dual in the following sense:

$\boldsymbol{s}(\boldsymbol{n})$ is normal to P at $p(\boldsymbol{n})$,

$\boldsymbol{p}(\boldsymbol{n})$ is normal to S at $s(\boldsymbol{n})$.

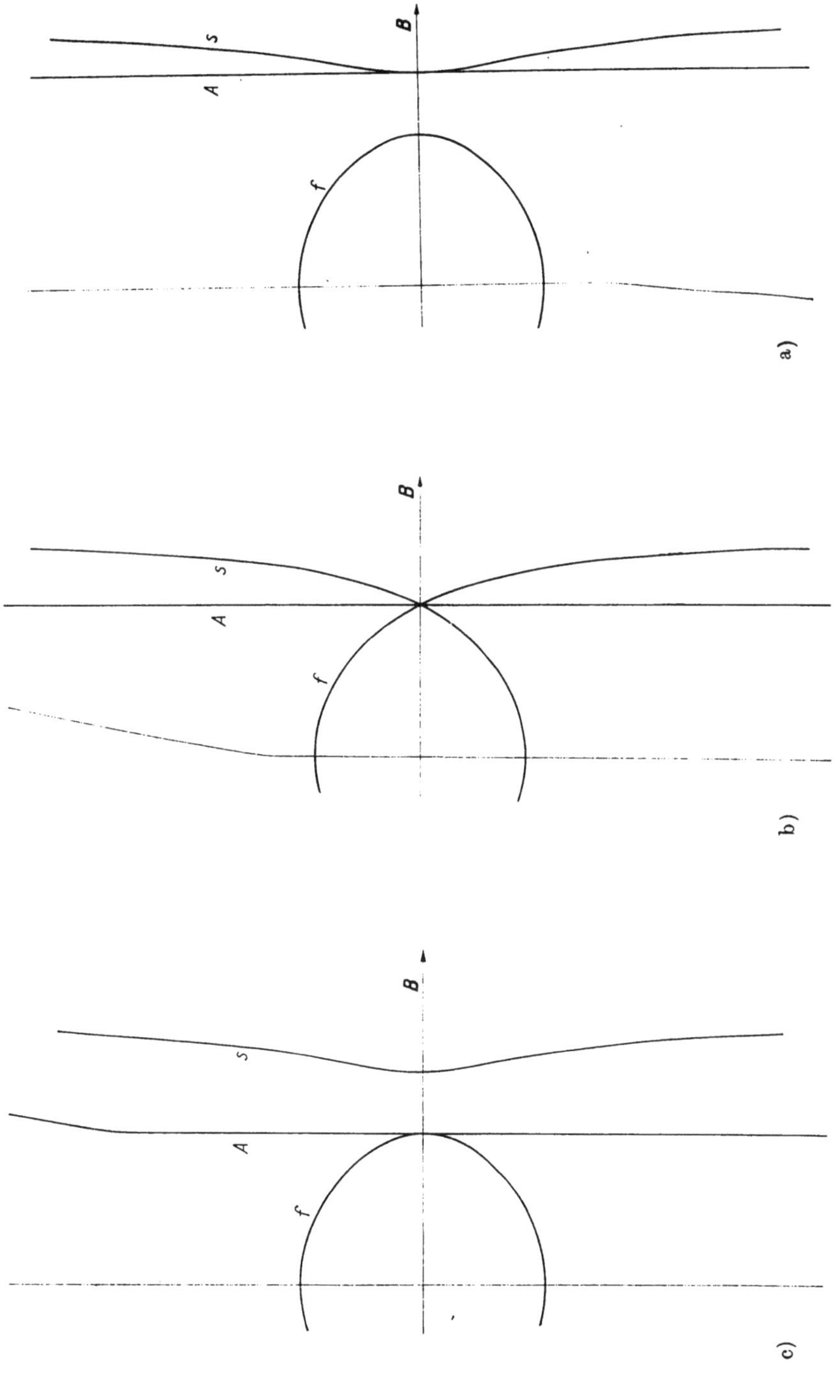

Figure 2.8. Surfaces of phase normals of the Alfvén-wave and the MA-waves for the values $r = 1/2$ (a), $r = 1$ (b), and $r = 2$ (c) (for $\boldsymbol{v} = 0$)

Thus

$$(\boldsymbol{s}\boldsymbol{p}) = 1 \qquad \text{or} \qquad (\boldsymbol{s}\,\boldsymbol{n}) = C\,. \tag{2.116}$$

These surfaces can be constructed from one another in the following manner. Given the surface of phase normals $\boldsymbol{p} = \boldsymbol{p}(u, v)$, the ray surface is obtained by performing the operation

$$\boldsymbol{s} = \frac{\boldsymbol{N}_p}{(\boldsymbol{N}_p\,\boldsymbol{p})}\,,$$

where $\boldsymbol{N}_p$ is some vector normal to P at (u, v). Conversely,

$$\boldsymbol{p} = \frac{\boldsymbol{N}_s}{(\boldsymbol{N}_s\,\boldsymbol{s})}\,,$$

where $\boldsymbol{N}_s$ is a vector normal to S at (u, v). S is then called the reciprocal polar surface of P relative to the unit circle and vice versa.

CHAPTER 3

Interaction of MHD-Waves with Surfaces of Discontinuity. Propagation in Inhomogeneous Media

In many cases the conclusions we arrived at in the first chapter (based on a homogeneous medium) are not applicable, since the undisturbed quantities (magnetic field, density, velocity) can be functions of space and time. The equation of the plane waves for the disturbances does not hold and the different wave modes can no longer propagate independently of one another. The disturbed state is in this case expressed by equation (2.37). An approximate solution of this equation is obtained in an analogous manner to the treatment of wave propagation in geometrical optics based on the already mentioned (chapter 1) geometric theory of MHD wave propagation.

However, before we examine this technique more closely (sections 4.1—4.3), we shall consider reflection and refraction of MHD-waves at surfaces of discontinuity. These surfaces of discontinuity can be regarded as special inhomogeneities: the discontinuous change of the state parameters leads to an interaction of the different wave modes.

Surfaces of discontinuity are known from hydrodynamics and gasdynamics (for example, interfaces between two different media or shock fronts). They can displace themselves relative to the medium and the discontinuities of the state parameters satisfy conditions, which follow from the basic equations (section 3.1).

We are interested in the following problem: A discontinuity front separates two steadily flowing homogeneous media. A plane MHD-wave of the type i and intensity ε_i is incident at an angle Θ_i upon one of the sides of the discontinuity. We want to know

1. the laws of reflection and refraction, i.e., the phase fronts of all the reflected and refracted waves;
2. the intensities of the reflected and refracted waves as a function of the intensity of the incident wave.

Due to the relatively complicated dispersion law for MHD-waves (double root in the expression for c_1 and c_3) the angles of reflection and refraction (and therefore also the intensities) can be specified analytically only in special cases. However, it is possible to arrive at a simple geometric method, with which we can obtain a general idea concerning the conditions upon reflection and refraction (section 3.2.2). When calculating the intensities we obtain statements on the stability of the discontinuity under consideration (section 3.3)

3.1. Boundary Conditions at Surfaces of Discontinuity

To derive the boundary conditions we make use of the conservation form of the basic equations (2.19)

$$\frac{\partial a_i}{\partial t} + \operatorname{div} \boldsymbol{F}_i = 0 . \tag{3.1}$$

We shall now consider a plane discontinuity U having a normal $\boldsymbol{N}$, and employ a coordinate system whose z-axis is directed along $\boldsymbol{N}$. Let the velocity of the discontinuity in the direction of the normal be V, and the position of U at a time t on the z-axis be $z(t)$ $\big(\dot{z}(t) = V\big)$.

The integrals

$$\frac{\mathrm{d}}{\mathrm{d}t} \int\limits_{z(t)-\varepsilon}^{z(t)+\varepsilon} a_i(x, y, z)\, \mathrm{d}z \tag{3.2}$$

will then vanish when $\varepsilon \to 0$, provided that the a_i have no singularities (for example, sources of mass, momentum, energy, etc) at the discontinuity.

The integral can be rewritten as

$$\lim_{\varepsilon\to 0} \frac{\mathrm{d}}{\mathrm{d}t} \int\limits_{z(t)-\varepsilon}^{z(t)+\varepsilon} a_i\, \mathrm{d}z = V\, (a_i^{(2)} - a_i^{(1)}) + \lim_{\varepsilon\to 0} \int\limits_{z(t)-\varepsilon}^{z(t)+\varepsilon} \frac{\partial}{\partial t}\, a_i\, \mathrm{d}z = 0 . \tag{3.3}$$

where

$$a_i^{(1,\,2)} = \lim_{\varepsilon\to 0} a_i\big(x, y, z(t) \pm \varepsilon\big)$$

are the values of a_i on both sides of the surface of discontinuity. The remaining integral does not vanish due to the singularity of $\frac{\partial a_i}{\partial t}$ at the discontinuity. It can be transformed with the help of (3.1), giving

$$\lim_{\varepsilon\to 0} \int\limits_{z(t)-\varepsilon}^{z(t)+\varepsilon} \frac{\partial a}{\partial t}\, \mathrm{d}z = -\, (F_{iz}^{(2)} - F_{iz}^{(1)}) - \lim_{\varepsilon\to 0} \int\limits_{z(t)-\varepsilon}^{z(t)+\varepsilon} \left(\frac{\partial F_{ix}}{\partial x} + \frac{\partial F_{iy}}{\partial y}\right) \mathrm{d}z . \tag{3.4}$$

Assuming that the derivatives of F_{ix} and F_{iy} "along" U have no singularities, the integral on the right-hand side of (3.4) vanishes. Together with (3.3) we obtain the expressions

$$V[a_i] - [F_{iz}] = 0 , \qquad [x] = (x^{(2)} - x^{(1)}) , \tag{3.5}$$

or, in general

$$V[a_i] - [(\boldsymbol{F}_i\, \boldsymbol{N})] = 0 . \tag{3.6}$$

Comparing this to (3.1) shows that the boundary conditions at a discontinuity plane follow from the basic equations in the conservation form, when the operators $\frac{\partial}{\partial t}$ and div are respectively replaced by $V \cdot [\;]$ and $-\,[(\boldsymbol{N} \cdot\,)]$.

If we now successively substitute the values of a_i and F_i given in (2.20) to (2.23), we obtain the following transition relations:

$$[m] = [\varrho\,(v_z - V]) = 0\,, \tag{3.7}$$

$$\left[\varrho\,(v_z - V)\,v_x - \frac{B_z\,B_x}{\mu}\right] = 0\,, \tag{3.8}$$

$$\left[\varrho\,(v_z - V)\,v_y - \frac{B_z\,B_y}{\mu}\right] = 0\,, \tag{3.9}$$

$$\left[\varrho\,(v_z - V)\,v_z + p + \frac{\boldsymbol{B}^2}{2\,\mu}\right] = 0\,, \tag{3.10}$$

$$[(v_z - V)\,B_x - B_z\,v_x] = 0\,, \tag{3.11}$$

$$[(v_z - V)\,B_y - B_z\,v_y] = 0\,, \tag{3.12}$$

$$\left[(v_z - V)\left\{\frac{\varrho\,\boldsymbol{v}^2}{2} + \varrho\,e + \frac{\boldsymbol{B}^2}{2\,\mu}\right\} + v_z\left(p + \frac{\boldsymbol{B}^2}{2\,\mu}\right) - \frac{B_z}{\mu}\,(\boldsymbol{B}\,\boldsymbol{v})\right] = 0\,. \tag{3.13}$$

Equations (3.11) and (3.12) express the constancy of the tangential component of $\boldsymbol{E}$ at the front in the coordinate system moving with the discontinuity.

In the system fixed in space we have

$$\boldsymbol{E} = -\,(\boldsymbol{v}\times\boldsymbol{B})\,;$$

in the system moving with the front ($v_z \to v_z - V$, all other components of $\boldsymbol{v}$ and $\boldsymbol{B}$ are unchanged)

$$\boldsymbol{E} = -\,\big((\boldsymbol{v} - V\,\boldsymbol{N})\times\boldsymbol{B}\big)\,.$$

These two equations can be written as

$$[E_x] = 0 \quad \text{and} \quad [E_y] = 0\,.$$

Conditions (3.7)—(3.13) are supplemented by the expression

$$[B_z] = 0\,, \tag{3.14}$$

which follows from div $\boldsymbol{B} = 0$. (Upon substituting the z-component of the induction equation (2.22), we only obtain $V[B_z] = 0$.) Without referring to the coordinate system we can write the front conditions in a slightly more condensed form:

$$[B_N] = 0\,, \tag{3.15}$$

$$[m] = 0\,, \tag{3.16}$$

$$\left[m\,\boldsymbol{v} + \left(p + \frac{\boldsymbol{B}^2}{2\,\mu}\right)\boldsymbol{N} - \frac{1}{\mu}\,B_N\,\boldsymbol{B}\right] = 0\,, \tag{3.17}$$

$$[m\,\tau\,\boldsymbol{B} - B_N\,\boldsymbol{v}] = 0\,, \tag{3.18}$$

$$m\left[\frac{\boldsymbol{v}^2}{2} + e + \frac{\boldsymbol{B}^2}{2\,\mu}\,\tau\right] + \left[v_N\left(p + \frac{\boldsymbol{B}^2}{2\,\mu}\right) - \frac{B_N}{\mu}\,(\boldsymbol{B}\,\boldsymbol{v})\right] = 0\,. \tag{3.19}$$

where $B_N = (\boldsymbol{B}\,\boldsymbol{N})$, $v_N = (\boldsymbol{v}\,\boldsymbol{N})$, and m is the mass flux per cm² per s through the front and $\tau = 1/\varrho$ is the specific volume. A conservation law for the entropy cannot be formulated.

We must require that

$$m[s] \geq 0 . \tag{3.20}$$

The front conditions are in general represented by eight conditions for the variables

$$\Gamma = (\varrho, p, \boldsymbol{v}, \boldsymbol{B}, m)$$

on both sides of the front (9 scalar quantities for each side).

This means that for a given $\Gamma^{(1)}$ there exists a one-parameter family of states $\Gamma^{(2)}$ behind the front:

$$\Gamma^{(2)} = \Gamma^{(2)}(\Gamma^{(1)}, \lambda) , \qquad \lambda\text{: parameter} . \tag{3.21}$$

(λ, for example, can be one of the state parameters behind the front.)

In certain cases the number of front conditions can be considerably reduced, and the states on both sides of the front are then largely independent of one another (e.g., when $m = 0$ and $B_N = 0$).

When $\boldsymbol{B} = 0$, equations (3.7)–(3.13) become the *hydrodynamic jump conditions*:

$$[m] = 0 , \tag{3.22}$$

$$m[v_x] = m[v_y] = 0 , \tag{3.23}$$

$$[p + m\, v_N] = 0 , \tag{3.24}$$

$$\left[m\left(\frac{\boldsymbol{v}^2}{2} + e\right) + v_N\, p\right] = 0 . \tag{3.25}$$

We can distinguish between two classes of discontinuities.

1) If $m = 0$, i.e., if mass is transferred through the front, then $[v_x] = [v_y] = 0$; the tangential component of the velocity is constant. To this class belong the *shock waves*. According to the way in which the density jumps ($[\varrho] > 0$ or $[\varrho] < 0$) we can divide these waves into compression and expansion shock waves.

2) If $[v_y] \neq 0$ or $[v_x] \neq 0$, m must vanish. This case we are dealing with the so-called *contact discontinuities*, at which the pressure must be constant.

For shock waves ($m = 0$) the dependence expressed in (3.21) holds on both sides of the front. For example, we have the so-called Hugoniot relation

$$[e(p, \tau)] + \frac{p^{(1)} + p^{(2)}}{2}\,[\tau] = 0 . \tag{3.26}$$

The corresponding representation $p^{(2)} = p^{(2)}(\varrho^{(2)})$ is the shock adiabatic. We can express this relationship in another way

$$M^{(2)} = M^{(2)}(M^{(1)}) , \qquad M = \frac{v}{a} . \tag{3.27}$$

Equation (3.27) is represented in Figure 3.1 for a polytropic gas.

MHD-discontinuities consist of a considerably greater number of types. A comprehensive discussion on this subject is given by Bazer and Ericson [16]. We shall be content with a less detailed characterization.

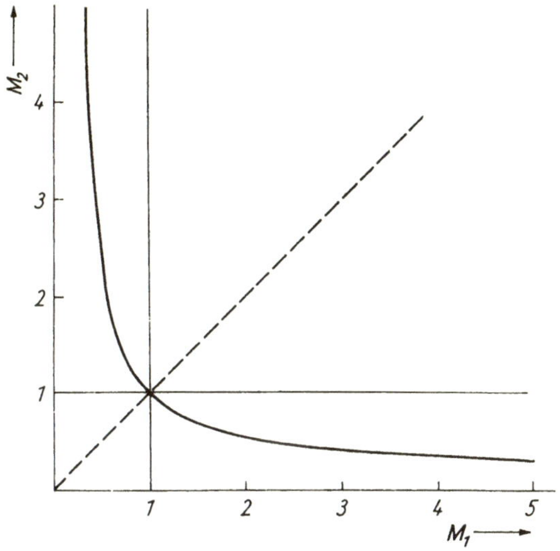

Figure 3.1. Dependence of the Mach numbers $M = v/a$ in front of and behind a gasdynamic shock front (polytropic gas)

First we again make the distinction between the two cases

1) $m = 0$ (contact discontinuities),

2) $m = 0$ (shock waves in a somewhat more general sense).

In the case of contact discontinuities the front conditions can be simplified. We have

$$[\boldsymbol{v}] = [\boldsymbol{B}] = [p] = 0\,, \qquad [\varrho], [s] \text{ are arbitrary if } B_N \neq 0$$

and

$$[v_N] = 0\,, \qquad \left[p + \frac{\boldsymbol{B}^2}{2\,\mu}\right] = 0\,,$$

$$[\boldsymbol{v}_t]\,,\ [\boldsymbol{B}_t]\,,\ [\varrho]\,,\ [s] \qquad \text{are arbitrary if } B_N = 0\,.$$

$\boldsymbol{v}_t$ and $\boldsymbol{B}_t$ are respectively the tangential components of $\boldsymbol{v}$ and $\boldsymbol{B}$. In the first case we are dealing with discontinuities which are at rest relative to the medium, and where the density and entropy can undergo arbitrary jumps. The second type corresponds to the contact discontinuities of hydrodynamics, except that here the total pressure $p + \frac{\boldsymbol{B}^2}{2\,\mu}$ must be constant.

We shall classify MHD shock waves into compressive ($[\varrho] > 0$), noncompressive or Alfvén ($[\varrho] = 0$), and rarefaction ($[\varrho] < 0$) shock waves.

The equation analogous to (3.26) is

$$[e(p, \tau)] + \frac{p^{(1)} + p^{(2)}}{2} [\tau] = - [\tau] \frac{[\boldsymbol{B}]^2}{2 \mu} \tag{3.28}$$

(generalized Hugoniot relation).

The representation

$$M^{(2)} = M^{(2)}(M^{(1)}) , \qquad M = \frac{v_N}{b_N} \tag{3.29}$$

is given in Figure 3.2.

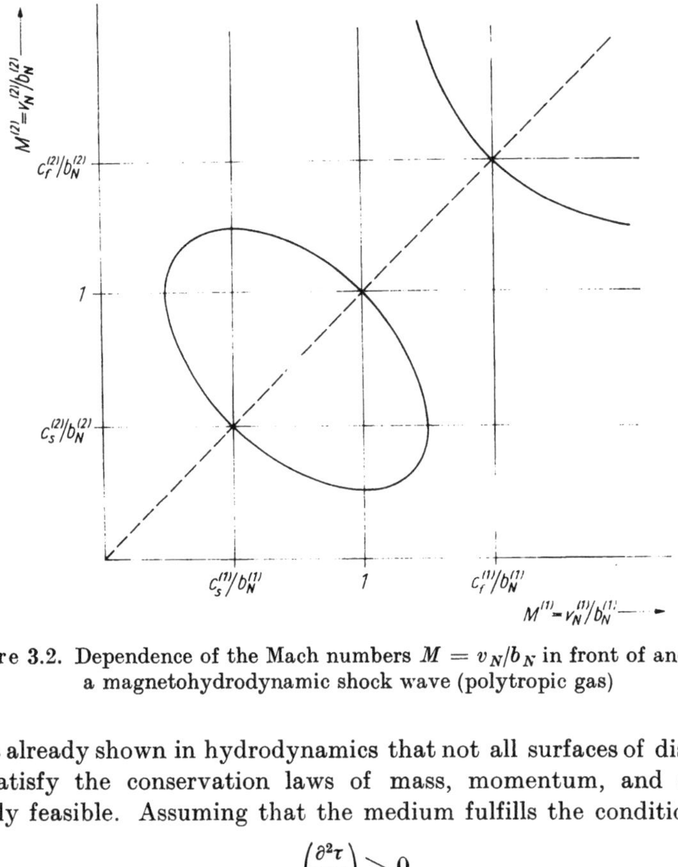

Figure 3.2. Dependence of the Mach numbers $M = v_N/b_N$ in front of and behind a magnetohydrodynamic shock wave (polytropic gas)

It was already shown in hydrodynamics that not all surfaces of discontinuity which satisfy the conservation laws of mass, momentum, and energy are physically feasible. Assuming that the medium fulfills the condition

$$\left(\frac{\partial^2 \tau}{\partial p^2}\right) > 0 \tag{3.30}$$

which is usually the case, we can exclude the rarefaction waves with the requirement $m[s] \geq 0$ (entropy increases upon passage of gas throught the front) on the basis of purely thermodynamic arguments. This means, that only the part of the curve located in the region of $M_2 < M_1$ of Figure 3.1 applies. (By similar reasoning we can exclude the MHD rarefaction shock waves for the case when (3.30) is valid). When condition (3.30) no longer applies, it is no longer possible by requiring that $m[s] > 0$, to obtain information on the variation of ϱ and p upon passage through the front.

We can nevertheless retain the unequality $m[\varrho] > 0$ in an entirely different manner (independent of thermodynamic considerations), by requiring that in some reflection and refraction process at the discontinuity under consideration the system of equations for the amplitudes of the wave has a unique solution. A front, which satisfies this condition, will be called evolutionary. The requirement of evolutionarity in the case of gas-dynamic shock waves leads to the condition, that the medium must move relative to the front with a speed above that of sound on one side and a speed lower than the sound speed on the other side of the discontinuity. This is self evident when $m[\varrho] > 0$.

For MHD-discontinuities the entropy condition carries less weight than the evolutionarity condition: not all MHD-discontinuities satisfying the condition $m[\varrho] \geq 0$ are evolutionary.

3.2. Reflection and Refraction

3.2.1. The Laws of Reflection and Refraction

When a MHD-wave is incident upon a plane discontinuity, the state parameters at the front undergo small disturbances. These disturbances cause deformations of the discontinuity plane and a slight increase in the velocity. (In the case of perpendicular incidence the front is displaced as a whole, whereas upon oblique incidence slight deviations in the front occur). Consequently, the state parameters on the other side of the front will also experience small changes, i.e., waves will be reflected from the front. In order to describe the process of reflection and refraction, we must derive certain relationships between the disturbances on both sides of the front. We can obtain these if we linearize the transition conditions. When linearizing we should allow for the above mentioned change in the speed and position of the front. (The linearized conditions are, strictly speaking, valid at the deformed front). However, due to the smallness of the disturbances we can (as in the case of reflection of refraction of sound waves) disregard the latter effect and formulate the linearized front conditions at the undisturbed (plane) front.

Consequently, upon incidence of a plane wave we have:

1. The frequencies of the reflected and refracted waves are equal to the frequency of the incident wave.
2. The projection of $\boldsymbol{k}$ on the discontinuity plane is the same for all waves (in particular, the wave normals of the reflected and refracted waves lie in the plane of incidence).

These two conditions are sufficient to derive the laws of reflection and refraction.

Consider a discontinuity front U, which separates two flowing homogeneous media 1 and 2. Let the velocities on both sides of the front be $\boldsymbol{v}^{(1)}$ and $\boldsymbol{v}^{(2)}$

with normal components $v_z^{(1)}$ and $v_z^{(2)}$, respectively. Since $[m] = 0$, $v_z^{(1)}$ and $v_z^{(2)}$ have the same sign; let us assume that $m > 0$ and that the normal $\boldsymbol{N}$ points to medium 2. This means that the medium flows from 1 to 2. Consider a wavefront of type i with wave normals $\boldsymbol{n}_i = \frac{\boldsymbol{k}_i}{k_i}$ $(i = 0, 1, 2, 3)$ moving in 1. The velocity of the wavefront in the coordinate system at rest in 1 is

$$\boldsymbol{c}_i = c_i \, \boldsymbol{n}_i , \qquad c_i \geq 0 ,$$

For the system, in which the discontinuity is at rest, we have on the other hand

$$\boldsymbol{C}_i = \big((\boldsymbol{v}^{(1)} \, \boldsymbol{n}_i) + c_i\big) \, \boldsymbol{n}_i = C_i \, \boldsymbol{n}_i \, . \tag{3.31}$$

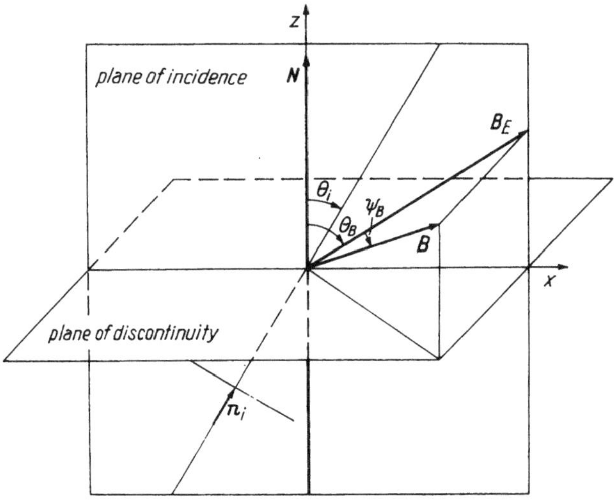

Figure 3.3. Geometric relationships at a plane of discontinuity

C_i can also be negative, i.e., the wavefront in the system associated with U can also move in the opposite direction of $\boldsymbol{n}_i$. This occurs, when

$$(\boldsymbol{v}^{(1)} \, \boldsymbol{n}_i) < - \, c_i < 0 \, .$$

Figures 3.3 and 3.4 illustrate the geometrical relationships at the discontinuity. The x-axis is taken along the direction of $\boldsymbol{q}$, $\boldsymbol{k}_i$ is projected on the discontinuity; $\boldsymbol{e}$ is a unit vector in this direction. Θ_i is the angle between $\boldsymbol{N}$ and $\boldsymbol{n}_i$ (positive when measured clockwise from $\boldsymbol{N}$).

The wave front i moves toward or away from the discontinuity, according to whether

$$(\boldsymbol{C}_i \, \boldsymbol{N}) = C_i \cos \Theta_i > 0 \quad \text{or} \quad C_i \cos \Theta_i < 0 \, . \tag{3.32}$$

In Medium 2 the opposite holds.

We have

$$\boldsymbol{q} = k_x \boldsymbol{e} = \frac{(\boldsymbol{n}_i \boldsymbol{e}) \boldsymbol{e}}{C_i} = \frac{\sin \Theta_i}{C_i} \boldsymbol{e} .$$

Since $\boldsymbol{q}$ is the same for all waves participating in a reflection or refraction process with the wave i, we have

$$\frac{\sin \Theta_{ij}^{(1)}}{C_{ij}^{(1)}} = \frac{\sin \Theta_{ij}^{(2)}}{C_{ij}^{(2)}} = \frac{\sin \Theta_i}{C_i} . \tag{3.33}$$

Here $\Theta_{ij}^{(1)}$, $C_{ij}^{(1)}$, $\Theta_{ij}^{(2)}$, and $C_{ij}^{(2)}$ are respectively the angles of reflection and refraction and the phase velocities of the participating waves in 1 and 2.

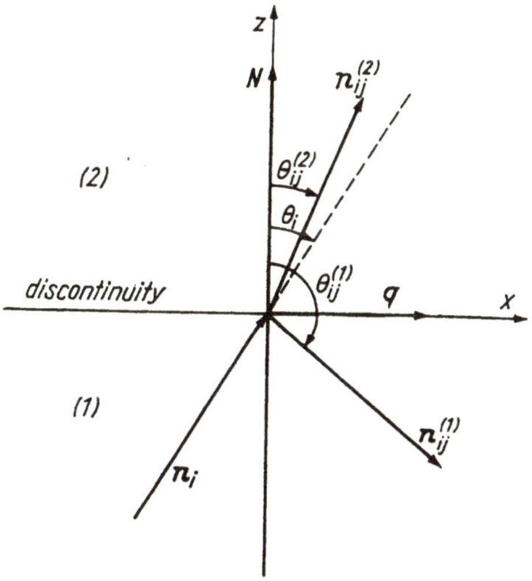

Figure 3.4. Orientation of the phase normal of the incident wave (n_i) and those of a possible reflected and refracted wave $\left(n_{ij}^{(1)}\right.$ and $\left.n_{ij}^{(2)}\right)$ in the plane of incidence

Expression (3.33) is the *analog of Snell's law of refraction.* Substituting the expression for $C_{ij}^{(1)}$, $C_{ij}^{(2)}$, and C_i, we obtain for the wavefronts in 1

$$\frac{v_z^{(1)} \cos \Theta_{ij}^{(1)} + c_{ij}^{(1)}}{\sin \Theta_{ij}^{(1)}} = \frac{v_z^{(1)} \cos \Theta_i^{(1)} + c_i^{(1)}}{\sin \Theta_i^{(1)}} = \text{const} \quad (\text{``law of reflection''}) \tag{3.34}$$

and for those in 2

$$\frac{v_z^{(2)} \cos \Theta_{ij}^{(2)} + c_{ij}^{(2)}}{\sin \Theta_{ij}^{(2)}} = (v_x^{(1)} - v_x^{(2)}) + \frac{v_z^{(1)} \cos \Theta_i^{(1)} + c_i^{(1)}}{\sin \Theta_i^{(1)}} = \text{const} \quad (\text{``law of refraction''}) \tag{3.35}$$

("Reflection" and "refraction" are now to be understood in a more general sense; formulas (3.34) and (3.35) describe the incident and reflected waves on both sides; in particular, the wave i can be a reflected wave.)

Let us first consider the law of reflection. Considering (3.34), we also have

$$\frac{v_z^{(1)} \cos \Theta_{ij}^{(1)} + c_{ij}^{(1)}}{v_z^{(1)} \sin \Theta_{ij}^{(1)}} = \frac{v_z^{(1)} \cos \Theta_i^{(1)} + c_i^{(1)}}{v_z^{(1)} \sin \Theta_i^{(1)}} . \tag{3.36}$$

The quotient on the right-hand side can be interpreted as the cotangent of the angle ψ_i between the vectors

$$v_z^{(1)} \boldsymbol{N} + c_i \boldsymbol{n}_i \quad \text{and} \quad c_i \boldsymbol{n}_i$$

If we take ψ_i to be positive when measured in a clockwise direction from $v_z^{(1)} \boldsymbol{N} + c_i \boldsymbol{n}_i$, this relation is valid for arbitrary orientations of both vectors.

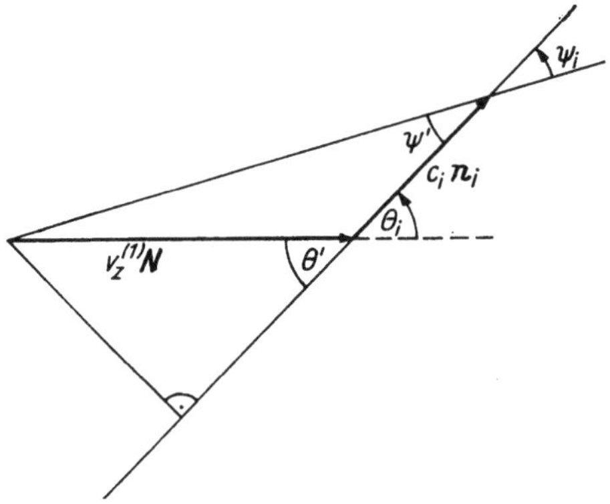

Figure 3.5. Meaning of the angle ψ

In Figure 3.5 the connection for a special case is evident. When $\psi', \Theta' > 0$ we have

$$\cot \psi' = \frac{v_z^{(1)} \cos \Theta' + c_i}{v_z^{(1)} \sin \Theta'} .$$

Since $\psi_i' = -\psi_i$ and $\Theta_i' = -\Theta_i$, equation (3.36) follows. The law of reflection can therefore be written in the form

$$\cot \psi_{ij} = \cot \psi_i . \tag{3.37}$$

This states that

$$\psi_{ij} = \psi_i \quad \text{or} \quad \psi_{ij} \begin{cases} \psi_i - \pi & \text{when} \quad \psi_i > 0 , \\ \psi_i + \pi & \text{when} \quad \psi_i < 0 . \end{cases}$$

The geometric locus of all points for which

$$\cot \varphi = \cot \psi_i$$

is valid for given $v_z^{(1)}$, Θ_i, and c_i, is a circle through the endpoints of the vectors $v_z^{(1)} \boldsymbol{N}$ and $v_z^{(1)} \boldsymbol{N} + c_i \boldsymbol{n}_i$ (Figure 3.6).

For the points lying on the upper half of the circle we have

$$\psi = \psi_i ,$$

and for those on the lower half

$$\psi = \psi_i + \pi \quad (\psi_i < 0) .$$

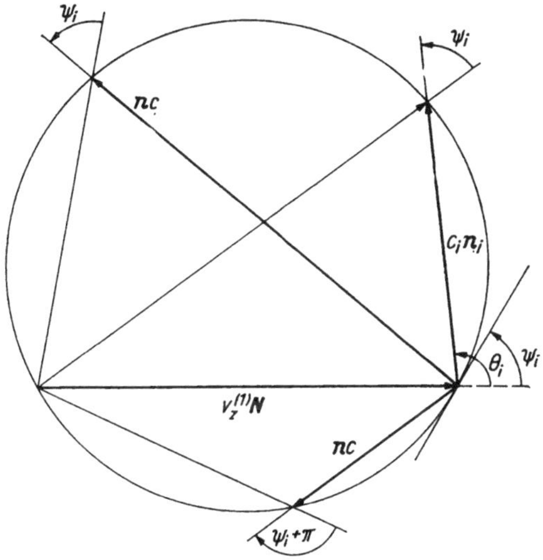

Figure 3.6. The circle ψ = const

3.2.2. Geometric Construction of the Angle of Reflection and the Angle of Refraction

Figure 3.6 clearly suggests the following method to find all angles Θ_{ij} and simultaneously the phase velocities c_{ij}, which are connected by (3.36) with the wave i: we take the origin of a system of polar coordinates (**N**-direction is along the axis $\Theta = 0$) at the endpoint of the vector $v_z^{(1)}$ **N** and introduce into this system the curves

$$c = c_i(\Theta) .$$

These are the curves of the reciprocal normal surfaces intersecting the plane of incidence. We obtain the representation of $c_i(\Theta)$ from expressions (2.70) to (2.72), when setting

$$\cos \vartheta = \cos \psi_B \cdot \cos (\Theta_B - \Theta) , \tag{3.38}$$

where Θ_B is the angle between **N** and $\boldsymbol{B}_E$ (the projection of **B** on the plane of incidence), ψ_B the angle between **B** and $\boldsymbol{B}_E$. (When $\psi_B = 0$, the magnetic field is directed along the plane of incidence). One obtains expression (3.38) from the spherical right-angular triangle, which fixes the directions of **n**, **B**, and $\boldsymbol{B}_E$ (Figure 3.3). The $c_i(\Theta)$ contain as parameters the sound speed and the Alfvén speed of medium 1.

The points where the curves $c_i(\Theta)$ intersect the circle $\psi = \text{const}$ give the desired angles Θ_{ij} and phase velocities c_{ij}. These are (by construction) independent of the tangential component of $\boldsymbol{v}$. Whether the "reflected" wave fronts move in the direction of $\boldsymbol{n}_{ij}$ or $-\boldsymbol{n}_{ij}$, is determined by the sign of C_{ij}. According to (3.31), the tangential component of $\boldsymbol{v}$ enters into the calculation of C_{ij}.

If $v_z^{(1)} = 0$, we clearly have

$$\operatorname{sgn} C_{ij} = \operatorname{sgn} \{(\boldsymbol{v}^{(1)} \boldsymbol{n}_{ij}) + C_{ij}\} = \operatorname{sgn} (v_z^{(1)} \boldsymbol{N} + c_{ij} \boldsymbol{n}_{ij}) \boldsymbol{n}_{ij} = \operatorname{sgn} \cos \psi_{ij} ,$$

i.e.,

$$C_{ij} \begin{Bmatrix} > 0 \\ < 0 \end{Bmatrix} \text{ for points on the } \begin{Bmatrix} \text{upper} \\ \text{lower} \end{Bmatrix} \text{ half of the circle } \psi = \text{const} .$$

The "refracted" waves in the Medium 2 can be constructed in an analogous manner.

From (3.35) we also have

$$\frac{v_z^{(2)} \cos \Theta_{ij}^{(2)} + c_{ij}}{v_z^{(2)} \sin \Theta_{ij}^{(2)}} = \frac{1}{v_z^{(2)}} \left\{ v_x^{(1)} - v_x^{(2)} + \frac{v_z^{(1)} \cos \Theta_i + c_i}{\sin \Theta_i} \right\} .$$

The left-hand side can be expressed, as before, as the cotangent of the angle ψ_{ij} between $v_z^{(1)} \boldsymbol{N} + c_{ij} \boldsymbol{n}_{ij}$ and $c_{ij} \boldsymbol{n}_{ij}$, and the right-hand side is known for given $\boldsymbol{n}_i$ and c_i. We obtain an equation of the form

$$\cot \psi_{ij} = \text{const} = \cot \psi_i' .$$

The values $a^{(2)}$ and $b^{(2)}$ of medium 2 are now substituted into the curves $c_i(\Theta)$.

When $c_0 = 0$, the entropy waves are not included in the formulation. It clearly follows from (3.34) and (3.35) that for these waves we have respectively

$$\cot \Theta_{0i}^{(1)} = \cot \psi_i$$

and

$$\cot \Theta_{0i}^{(2)} = \cot \psi_i'$$

Both solutions of each of the equations give the same C_{0i}, because $c_0 = 0$. Thus, as expected, there can only be an entropy wave on each side of the discontinuity.

The above construction becomes more clear upon reflection of the loci in Figure 3.7 in the unit circle. The curves $c_i(\Theta)$ are then transformed into the curves $p_i(\Theta)$ of the normal surfaces (section 2.5) intersecting with the plane of incidence, and the circle $\psi = \text{const}$ degenerates into a straight line l. Which intersects the axis $\Theta = \pi$ at an angle of ψ $\left(\text{when } p = \dfrac{1}{v_z^{(1)}}\right)$. Figure 3.8 shows the relationships when $\psi_B = 0$ (the magnetic field is directed along the plane of incidence).

A change in Θ causes a rotation of l about the point of intersection with the $\boldsymbol{N}$-axis, and a change in $v_z^{(1)}$ produces a shift in the position of the intersection point. A variation in the direction of $\boldsymbol{B}$ signifies a rotation of the curves $p_i(\Theta)$ about the origin. It is found that the angle of reflection is invariant under a change of $\boldsymbol{B}$ bis $-\boldsymbol{B}$. Furthermore, negative angles of incidence are not equivalent to positive one (except for specific orientations of the magnetic field, e.g., $\boldsymbol{B}$ parallel to $\boldsymbol{N}$ or $\boldsymbol{B}$ perpendicular to $\boldsymbol{N}$).

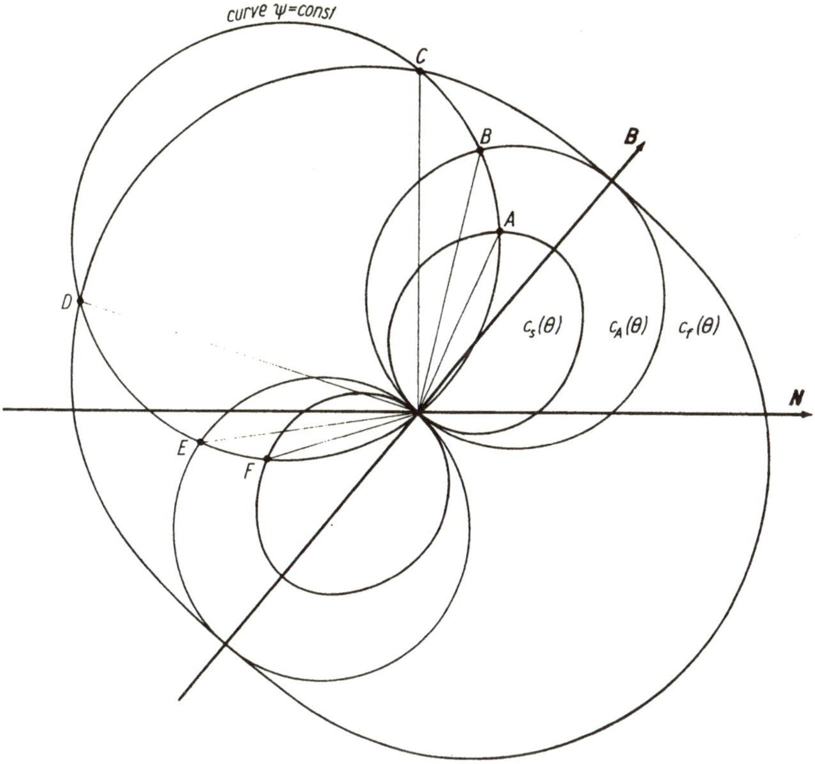

Figure 3.7. Points of intersection of the curves $c_i(\Theta)$ with the curve $\psi = \text{const}$ for a specific case ($\psi_B = 0$, $\Theta_B = 50°$, $r = 1/2$)

For normal incidence $\psi \to 0$, and the line l will be identical to the $\boldsymbol{N}$-axis. In this case we have $\Theta_{ij} = \Theta_i = 0$.

For contact fronts ($m = 0$, i.e., $v_z^{(1)} = v_z^{(2)} = 0$) the line l intersects the $\boldsymbol{N}$-axis at infinity. Here the law of reflection is given by

$$\frac{c_{ij}}{\sin \Theta_{ij}} = \frac{c_i}{\sin \Theta_i}. \tag{3.39}$$

The line l becomes parallel to the $\boldsymbol{N}$-axis at a distance $c_i \sin \Theta_i$ from the origin. We can construct the angle of reflection in a similar manner.

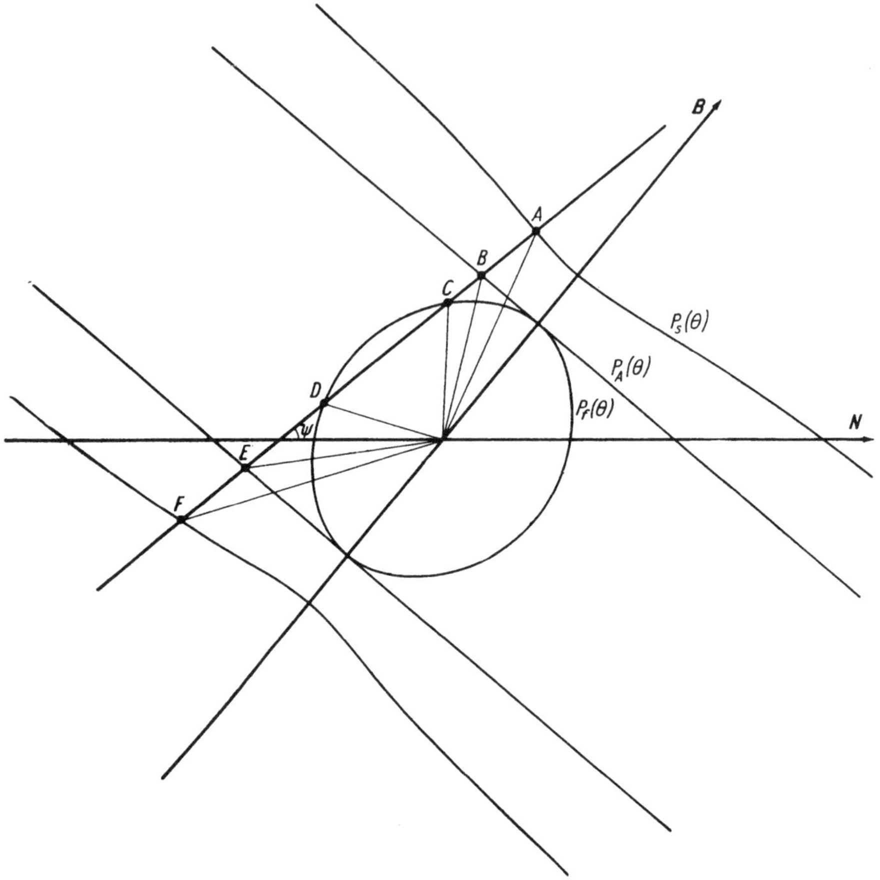

Figure 3.8. Points of intersection of the straight line l with the curves $p_i(\Theta)$ for a specific case ($\psi_B = 0$, $\Theta_B = 50°$, $r = 1/2$)

For Alfvén-waves equation (3.39) can be evaluated analytically. When

$$c_2 = b \cos \psi_B \, |\cos (\Theta_2 - \Theta_B)|$$

we obtain from (3.39)

$$\cot \bar{\Theta}_{22} = \cot \Theta_2 + 2 \tan \Theta_B \,, \tag{3.40}$$

where

$$\bar{\Theta}_{22} = \begin{cases} \pi - \Theta_{22} & \text{when} \quad \Theta_{22} > 0 \,, \\ -(\pi + \Theta_{22}) & \text{when} \quad \Theta_{22} < 0 \,. \end{cases}$$

In Figure 3.9 $\bar{\Theta}_{22}$ is plotted against Θ_2 $\left(\Theta_B = 0, \frac{\pi}{4}, -\frac{\pi}{4}\right)$. $\bar{\Theta}_{22}$ does not depend on ψ_B.

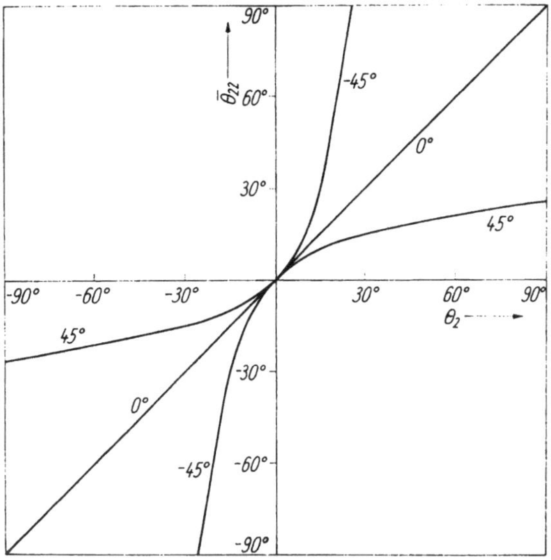

Figure 3.9. The angle of reflection $\bar{\Theta}_{22}$ as a function of the angle of incidence Θ_2 for Alfvén-waves ($\Theta_B = 0°, \pm 45°$) (Bazer and Hurley [8])

3.2.3. Significance of the Group Velocity in Reflection

Hitherto we have distinguished between "incoming" and "outgoing" waves according to the sign of the projection $\boldsymbol{C}$ on $\boldsymbol{N}$ (equation (3.32)). It would be physically more informative if we were to make this distinction in accordance with the group velocity. In the case of perpendicular incidence both methods lead to the same result, since the component of the group velocity in the direction of the normal is just the phase velocity (equation (2.104)). For oblique incidence this is in general not true, since there are angular ranges for which the projections of the phase and group velocity on $\boldsymbol{N}$ possess a different sign ("anomalous reflection").

We shall consider a wave in Medium 1 as "incoming" or "outgoing", respectively, when

$$(\boldsymbol{v}_g^{(1)}\,\boldsymbol{N}) > 0 \qquad \text{bzw.} \qquad (\boldsymbol{v}_g^{(2)}\,\boldsymbol{N}) < 0\,. \tag{3.41}$$

In medium 2 the opposite applies.

We can calculate the z-component of the group velocity of a wave with normal n according to (2.103). In the selected coordinate system, wen have

$$\boldsymbol{n} = n_z\,\boldsymbol{N} + n_z\,\boldsymbol{z}$$

where

$$n_z = \cos\Theta\,, \qquad n_x = \sin\Theta; \qquad C = (\boldsymbol{v}^{(1)}\,\boldsymbol{n}) + c(\boldsymbol{n})\,,$$

and

$$(\boldsymbol{v}_g^{(1)}\,\boldsymbol{N}) = v_{g\,z}^{(1)} = C\,n_z + \frac{\partial C}{\partial n_z} - n_z\left(n_z\frac{\partial C}{\partial n_z} + n_x\frac{\partial C}{\partial n_x}\right) = v_z^{(1)} + c\cos\Theta - \sin\Theta\,\frac{\partial c}{\partial \Theta}\,.$$

On the other hand, it follows from

$$\cot\psi = \frac{v_z^{(1)}\cos\Theta + c(\Theta)}{v_z^{(1)}\sin\Theta}$$

that

$$\frac{\partial\psi}{\partial\Theta} = \frac{\sin^2\psi}{v_z^{(1)}\sin^2\Theta}\left[v_v^{(1)} + c\cdot\cos\Theta - \sin\Theta\,\frac{\partial c}{\partial\Theta}\right].$$

From this we have

$$\operatorname{sgn} v_{g\,z}^{(1)} = \operatorname{sgn}\frac{\partial\psi}{\partial\Theta} \quad (\text{since } v_z^{(1)} > 0)\,.$$

The sign of $\frac{\partial\psi}{\partial\Theta}$ for the different points of intersection can be readily obtained from Figure 3.8. Allowing for the direction in which Θ and ψ are measured (positive in the clockwise direction), we have

$$\frac{\partial\psi}{\partial\Theta} > 0 \quad \text{for} \quad A, B, C, E, F$$

and

$$\frac{\partial\psi}{\partial\Theta} < 0 \quad \text{für} \quad D\,.$$

In particular, the f-wave, which corresponds to the point C, is thus incident ("incoming"), despite the fact that the phase front is moving away from the discontinuity. In this case the components of $\boldsymbol{v}_g$ and $\boldsymbol{C}$ along the normal $\boldsymbol{N}$ have different signs.

A further decrease in ψ causes the intersection points D and C to approach one another, and at ψ_0 these points coincide. At this point we have

$$\left.\frac{\partial\psi}{\partial\Theta}\right|_{\psi=\psi_0} = 0\,,$$

i.e., the group velocity has only a tangential component. If ψ is further reduced, there will be no points at which the curves $p_f(\Theta)$ intersect the line l. We obtain two complex angles of refraction and two corresponding complex velocities. The exponent in expression (2.47) is made up of a real and an imaginary part. We can exclude one solution (that which describes the exponential growth of the amplitude). We are left with a surface wave, whose intensity descrases exponentially with increasing distance from the front.

These considerations can also be applied to medium 2. We have for the entropy waves

$$v_{g\,z}^{(1),\,(2)} = v_z^{(1),\,(2)}\,.$$

Since $v_z^{(1)}, v_z^{(2)} > 0$, they are always incoming in medium 1 and outgoing in medium 2.

We can further deduce from Figure 3.8, that the number of outgoing and incoming waves, respectively, is independent of ψ (except for the above chosen limiting case). In particular, this number is equal to that corresponding to normal incidence ($\psi = 0$). If we count the surface wave as an outgoing wave, the above strictly applies to the outgoing waves.

The number of outgoing and incoming waves in medium 1 for normal incidence is readily found. We form the expressions

$$v_z^{(1)} \pm c_i\,, \qquad i = 0, 1, 2, 3\,; \tag{3.42}$$

the number of positive sums is equal to that of the incoming waves, and the number of negative sums corresponds to the number of outgoing waves. In all there are always seven waves.

The above method allows us to determine the number of positive angles of reflection and refraction for an arbitrary incident wave. Whether or not the corresponding waves are produced, depends on the relationships between the amplitudes, i.e., on the linearized front conditions.

3.3. Relations between the Amplitudes. Stability of Discontinuity Fronts (Evolution Conditions)

Consider a MHD-discontinuity front U, at which the transition conditions (3.7)—(3.14) are satisfied. It is assumed that the state parameters on both sides of the front are subject to small changes. If we again make use of the coordinate system whose z-axis is directed along the normal, and if we replace a_i in (3.6) by $a_i + \delta a_i$ and V by $V + \delta V$, where

$$V[a_i] = [F_{iz}(a_i, \ldots, a_8)]\,, \qquad i = 1, \ldots, 8\,, \tag{3.43}$$

we have

$$(V + \delta V)\,[a_i + \delta a_i] = [F_{iz}\,(a_i + \delta a_1, \ldots, a_8 + \delta a_8)]$$
$$= \left[F_{iz}(a_1, \ldots, a_8) + \sum_{\nu=1}^{8} \frac{\partial F_{iz}}{\partial a_\nu}\,\delta a_\nu\right], \qquad i = 1, \ldots, 8\,,$$

or, due to (3.43)

$$\delta V[a_i] = \left[\sum_{\nu=1}^{8} \frac{\partial F_{iz}}{\partial a_\nu}\,\delta a_\nu - V\,\delta a_i\right], \qquad i = 1, \ldots, 8\,. \tag{3.44}$$

Vectorially, this can be written

$$\delta V[\boldsymbol{a}] = [(A_z - V\mathrm{I})\,\delta \boldsymbol{a}]\,. \tag{3.45}$$

The matrix A_z can be identified with the matrix A_3 (2.26). The seventh equation of the system (3.45) ($V[\delta B_z] = 0$) also holds for $V = 0$, since $[B_z] = 0$.

If we refer to the coordinate system, in wich the front is at rest ($V = 0$, front is located at $z = 0$), equation (3.45) becomes

$$\delta V[\boldsymbol{a}] = [A_z\,\delta \boldsymbol{a}]\,. \tag{3.46}$$

The disturbances $\delta \boldsymbol{a}$ on both sides of the front must again satisfy the linearized equations (2.66)

$$\frac{\partial \delta \boldsymbol{a}}{\partial t} + \sum_{j=1}^{3} A_j \frac{\partial \delta \boldsymbol{a}}{\partial \delta_j} = 0 .$$

Let us assume that we are dealing with periodic plane waves which are participating in a reflection and refraction process at U.

On each side of U we then obtain a representation of the form

$$\begin{aligned}\delta \boldsymbol{a} &= \delta \boldsymbol{a}_0 + \sum_{i=1}^{3} (\delta \boldsymbol{a}_i^+ + \delta \boldsymbol{a}_i^-) \\ &= \varepsilon_0 \, \boldsymbol{a}_0 \, e^{i(\omega_0 t - (\boldsymbol{k}_0 \boldsymbol{x}))} + \sum_{i=1}^{3} \{ \varepsilon_i^+ \, \boldsymbol{a}_i^+ \, e^{i(\omega_i^+ t + (\boldsymbol{k}_i^+ \boldsymbol{x}))} + \varepsilon_i^- \, \boldsymbol{a}_i^- \, e^{i(\omega_i^- t - (\boldsymbol{k}_i^- \boldsymbol{x}))} \} . \end{aligned} \quad (3.47)$$

The frequency ω and the projection $\boldsymbol{q}$ of $\boldsymbol{k}$ on the discontinuity must be the same for all waves. If we bear this in mind upon substitution in (3.46), we obtain the following expression at the location of the front:

$$\delta V[\boldsymbol{a}] = \left[A_z \left\{ \varepsilon_0 \, \boldsymbol{a}_0 + \sum_{i=1}^{3} (\varepsilon_i^+ \, \boldsymbol{a}_i^+ + \varepsilon_i^- \, \boldsymbol{a}_i^-) \right\} \right] e^{i(\omega t - (\boldsymbol{q}\boldsymbol{x}))}.$$

The disturbance δV must therefore have the same dependence on time and space:

$$\delta V = \varepsilon_V \cdot e^{i(\omega t - (\boldsymbol{q}\boldsymbol{x}))} .$$

We are thus left with a linear system of equations for the intensities ε_V, $\varepsilon_i^{\pm(1)}$, and $\varepsilon_i^{\pm(2)}$.

For periodic disturbances the linearized front conditions are not independent of one another. Equations 5 and 6 of (3.46) can be written in the following form

$$[\delta\{(v_z - V) B_x - B_z v_x\}|_{V=0}] = [\delta\{(\boldsymbol{v} - V \boldsymbol{N}) \times \boldsymbol{B}\}_y|_{V=0}] = 0 ,$$

$$[\delta\{(v_z - V) B_y - B_z v_y\}|_{V=0}] = -[\delta\{(\boldsymbol{v} - V \boldsymbol{N}) \times \boldsymbol{B}\}_x|_{V=0}] = 0 .$$

They express the constancy of the tangential components of

$$\delta \boldsymbol{E} = -\delta \{(\boldsymbol{v} - V \boldsymbol{N}) \times \boldsymbol{B}\}|_{V=0}$$

at the discontinuity front. Due to the periodicity, we now have

$$\omega \, \delta B_s = - k_x \cdot \delta E_y + k_y \cdot \delta E_x .$$

ω, k_x, and k_y are constant, and from $[\delta E_y] = [\delta E_x] = 0$ it follows that $[\delta B_z] = 0$.

The system (3.46) reduces to 7 independent conditions. Altogether 14 wave modes can occur on the two sides of U. Of these μ_e and μ_a are respectively incoming and outgoing from the standpoint of group velocity ($\mu_e + \mu_a = 14$).

Let us bring all the outgoing waves in (3.47) to the left-hand side. In addition we shall renumber the ε_0, $\varepsilon_i^{+,-}$ by

$$\varepsilon_j^e\,, \quad j = 1, \ldots, \mu_e$$
$$\varepsilon_j^a\,, \quad j = 1, \ldots, \mu_a$$

where the ε_j^e are the intensities of the incoming and ε_j^a the intensities of the outgoing waves, respectively. (Outgoing waves an appear in (1) and (2)).

From (3.47) we obtain the following system:

$$\varepsilon_V[\boldsymbol{a}] + \sum_{j=1}^{\mu_a} \varepsilon_j^a\, A_z\, \boldsymbol{a}_j^a = \sum_{j=1}^{\mu_e} \varepsilon_j^e\, A_z\, \boldsymbol{a}_j^e\,. \tag{3.48}$$

(A_z is different in 1 and 2, since the elements depend on the undisturbed parameters).

The intensities of the incoming waves can be assumed to be known, so that ε_V and the ε_j^a, $j = 1, \ldots, \mu_a$ remain to be determined. To solve (3.48), we require the angles of reflection and refraction. They define the wave normals n and, consequently, the *nrt*-systems associated with the different waves, in which the amplitudes are expressed. The amplitudes must then be transformed into the *xyz*-system, in which the discontinuity is expressed and the matrix A_z is defined.

The system (3.48) does not have a unique solution in all cases. Thus, a necessary condition for such a solution to exist is that the number of equations is equal to the number of unknowns, i.e.,

$$\mu_a + 1 = 7 \tag{3.49}$$

(condition I). There must be 6 outgoing waves.*

Furthermore, the determinant of the system must be nonzero, i.e., the vectors $[\boldsymbol{a}]$ and $A_z\,\boldsymbol{a}_j^a, j = 1, \ldots, \mu_a$ must be linearly independent (condition II).

The discontinuity is called evolutionary (section 3.1) when both conditions are fulfilled.

When a discontinuity is not evolutionary, i.e., if (3.48) has no unique solution, this means that the change a small disturbance undergoes at the discontinuity does not remain "small", and can no longer be described by the linearized equations (e.g., the discontinuity splits up into several discontinuity fronts). In other words, the discontinuity under consideration is not stable with respect to the effect caused by small disturbances.

In principle, one can examine the evolutionary of arbitrary MHD-discontinuities with the help of conditions I and II.

The number of outgoing waves is readily found in accordance with (3.42). It is fixed for a given discontinuity and is independent of the specific reflection and refraction process.

* More precisely: $\mu_a + 1$ must be equal to the number of independent equations of the system (3.46).

In Figure 3.10 is given a $v_z^{(1)}$ $v\,s$ $v_z^{(2)}$ plot. The number of possible outgoing waves is in each case indicated in the squares (the first number corresponds to Alfvén-waves, the second to MA and entropy waves). The condition $\mu_a = 6$ is satisfied only in the three shaded areas.

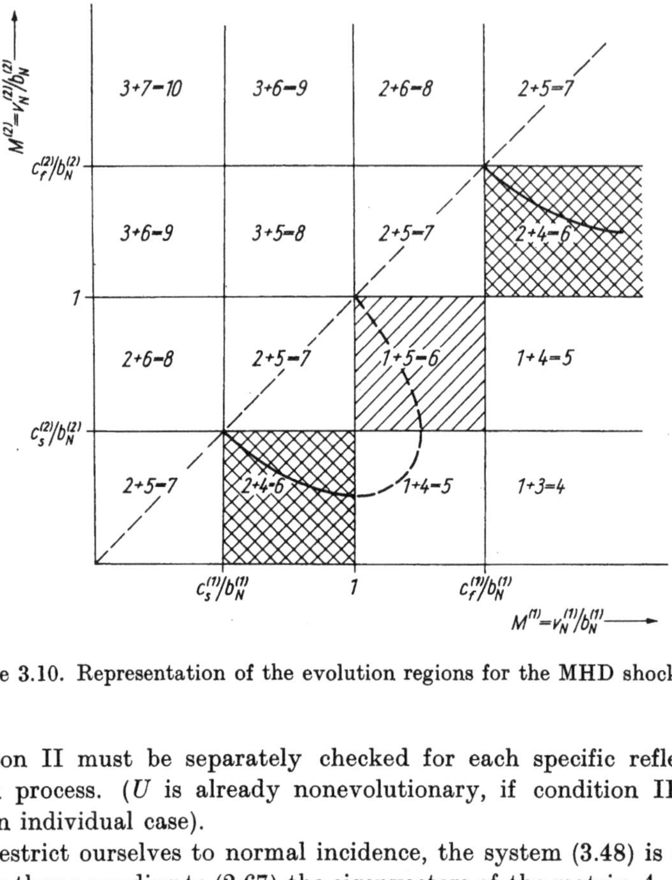

Figure 3.10. Representation of the evolution regions for the MHD shock waves

Condition II must be separately checked for each specific reflection and refraction process. (U is already nonevolutionary, if condition II does not hold in an individual case).

If we restrict ourselves to normal incidence, the system (3.48) is simplified. The $\boldsymbol{a}_j^a$ are then according to (2.67) the eigenvectors of the matrix A_z. Equation (3.48) becomes

$$\varepsilon_V[\boldsymbol{a}] + \sum_{j=1}^{\mu_a} \varepsilon_j^a\, C_j^a\, \boldsymbol{a}_j^a = \sum_{j=1}^{\mu_e} \varepsilon_j^e\, C_j^e\, \boldsymbol{a}_j^e\,. \tag{3.50}$$

Condition II is reduced upon requiring that the vectors $[\boldsymbol{a}]$, $\boldsymbol{a}_j^a$, $j = 1, \ldots, 6$ are linearly independent.

Let us consider the specific case, in which U represents a compressive MHD shock wave ($m \neq 0$, $[\varrho] \neq 0$).

Compressive MHD shock waves haves the following property: the normals and the magnetic field lie in the same plane on both sides ($\boldsymbol{B}_t^{(1)} \sim \boldsymbol{B}_t^{(2)}$).

Taking the x-axis along the direction of $\boldsymbol{B}_t$, then $B_y^{(1)} = B_y^{(2)} = 0$, from which again follows $[v_y] = 0$. We can choose a coordinate system, in which $v_y^{(1)}$ and $v_y^{(1)}$ vanish.

The *nrt*-systems of the individual waves coincide with each other to within an interchange of the axial direction.

The a_j^a result from the corresponding $\boldsymbol{w}_j^a$ through the transformation

$$\boldsymbol{a}_j^a = M\, N^{-1}\, \boldsymbol{w}_j^a \, .$$

Since $B_y = 0$ and $v_y = 0$, M has the form

$$M = \begin{array}{|c|c|c|c|c|c|c|c|}
\hline
1 & & & & & & & \\
\hline
v_x & \varrho & & & & & & \\
\hline
 & & \varrho & & & & & \\
\hline
v_z & & & \varrho & & & & \\
\hline
 & & & & 1 & & & \\
\hline
 & & & & & 1 & & \\
\hline
 & & & & & & 1 & \\
\hline
\dfrac{\boldsymbol{v}^2}{\varrho} + \dfrac{\partial(e\varrho)}{\partial\varrho} & \varrho\, v_x & & \varrho\, v_z & \dfrac{B_x}{\mu} & & \dfrac{B_z}{\mu} & \dfrac{\partial(e\,\varrho)}{\delta s} \\
\hline
\end{array}$$

This means that the characteristic spaces Ω_1 and Ω_2 are preserved under the transformation and coincide on both sides of U.

The vector $[\boldsymbol{a}]$ has the following form:

$$[\boldsymbol{a}] = ([\varrho], [\varrho\, v_x], 0, [\varrho\, v_z] \cdot [B_x], 0, 0, [\mathscr{E}]) \, .$$

Thus, this vector belongs to Ω_1.

We thus find that of the three shaded areas in Figure 3.10 the middle one falls out of the evolution region. To this region correspond 5 outgoing waves, which belong to Ω_1, and together with $[\boldsymbol{a}]$ they are linearly dependent. The two remaining regions are found to be the evolution regions for the MHD shock waves. (In each case there are two outgoing Alfvén-waves). The MHD shock waves corresponding to the two evolution regions are designated as "fast" (+) and "slow" (−) shock waves. They are distinguished by the jump in the tangential component of the magnetic field. For fast and slow shock waves we have, respectively, $[B_x] > 0$ and $[B_x] < 0$.

The fact that upon normal incidence of disturbances on MHD shock waves the characteristic spaces Ω_1 and Ω_2 on both sides coincide, has a further consequence. Incident Alfvén-waves cannot produce reflected or refracted magneto-acoustic and entropy waves. Conversely, if an MA or entropy wave is incident, then no Alfvén-waves are generated. This also still holds when the wave normals of the incident wave lie in the $\boldsymbol{n}\,\boldsymbol{B}$-plane.

In principal the above considerations do not change if one of the phase velocities $v_2 \pm v$ vanishes on one side (or on both sides) of a discontinuity. There are some MHD discontinuities, for which this is the case e.g., the Alfvén shock waves of the contact discontinuities). Thus, one (or two) of the 14 possible waves are neither incoming nor outgoing.

The evolutionarity can be investigated in a smiliar way. The Alfvén shock waves as well as the contact discontinuities are evolutionary.

CHAPTER 4

Propagation in Inhomogeneous Media

4.1. Separation of the Linearized Equation in the Homogeneous Case, coupling in the Inhomogeneous Case

We have already seen in section 2.2, that the linearized MHD-equations separate into two groups when a homogeneous undisturbed medium is initially considered and the disturbances are assumed to be plane waves. (This fact was expressed through the reduction of L_{iv} to a quasidiagonal matrix).

This separation is a consequence of the homogeneity of the basic solution, and not due to an effect which is a result of the assumption if plane waves. This becomes clear upon introducing the new scalar variables

$$\delta\varrho,\ \mathrm{rot}_z\,\delta\boldsymbol{v},\ \mathrm{div}_\perp\,\delta\boldsymbol{v},\ \delta v_z,\ \mathrm{rot}_z\,\delta\boldsymbol{B},\ \mathrm{div}_\perp\,\delta\boldsymbol{B},\ \mathrm{d}B_z,\ \delta s$$

instead of $\delta\boldsymbol{u} = (\delta\varrho, \ldots, \delta s)$. (The z-axis is in the $\boldsymbol{B}$-direction, $\mathrm{div}_\perp\,\delta\boldsymbol{v} = \frac{\partial\delta v_x}{\partial x} + \frac{\partial\delta v}{\partial y}$, etc.) Two separate systems are readily obtained, and in fact we have the two groups of variables

$$\underbrace{\mathrm{rot}_z\,\delta\boldsymbol{v},\ \mathrm{rot}_z\,\delta\boldsymbol{B}}_{A} \quad \text{and} \quad \underbrace{\mathrm{div}_\perp\,\delta\boldsymbol{v},\ \delta v_z,\ \mathrm{div}_\perp\,\delta\boldsymbol{B},\ \delta\varrho_s,\ \delta B,\ \delta s}_{B}\ .$$

The corresponding equations are obtained in the following fashion: operating with rot_z on (2.42) and (2.43) gives

$$\frac{\mathrm{d}}{\mathrm{d}t}\,\mathrm{rot}_z\,\delta\boldsymbol{v} - \frac{B}{\mu\,\varrho}\,\frac{\partial}{\partial z}\,(\mathrm{rot}_z\,\delta\boldsymbol{B}) = 0 \tag{4.1}$$

and

$$\frac{\mathrm{d}}{\mathrm{d}t}\,\mathrm{rot}_z\,\delta\boldsymbol{B} - B\,\frac{\partial}{\partial z}\,(\mathrm{rot}_z\,\delta\boldsymbol{v}) = 0\ , \tag{4.2}$$

where we have set $\frac{\mathrm{d}}{\mathrm{d}t} = \frac{\partial}{\partial t} + (\boldsymbol{v}\ \mathrm{grad})$.

Equations (2.40) and (2.41) as well as div operating on (2.42) and (2.43) the z-components of these two equations, and finally (2.44) form the second system for the group of variables B:

$$\mathrm{div}_\perp\,\delta\boldsymbol{B} + \frac{\partial\delta B_z}{\partial z} = 0\ , \tag{4.3}$$

$$\frac{\mathrm{d}}{\mathrm{d}t}\,\delta\varrho + \varrho\left(\mathrm{div}_\perp\,\delta\boldsymbol{v} + \frac{\partial\delta v_z}{\partial z}\right) = 0\ , \tag{4.4}$$

$$\frac{\mathrm{d}}{\mathrm{d}t}\left(\operatorname{div}_\perp \delta\boldsymbol{v} + \frac{\partial \delta v_z}{\partial z}\right) + \frac{1}{\varrho}\,\Delta\,(a^2\,\delta\varrho + p_s\,\delta s) + \frac{B}{\mu\,\varrho}\left\{\frac{\partial}{\partial z}\operatorname{div}_\perp \delta\boldsymbol{B} - \left(\Delta - \frac{\partial^2}{\partial z^2}\right)\delta B_z\right\} = 0\,, \tag{4.5}$$

$$\frac{\mathrm{d}}{\mathrm{d}t}\,\delta v_z + \frac{1}{\varrho}\,\frac{\partial}{\partial z}\,(a^2\,\delta\varrho + p_s\,\delta s) = 0\,, \tag{4.5}$$

$$\frac{\mathrm{d}}{\mathrm{d}t}\,\operatorname{div}_\perp \delta\boldsymbol{B} - B\,\frac{\partial}{\partial z}\,\operatorname{div}_\perp \delta\boldsymbol{v} = 0\,, \tag{4.7}$$

$$\frac{\mathrm{d}}{\mathrm{d}t}\,\delta B_z + B\,\operatorname{div}_\perp \delta\boldsymbol{v} = 0\,, \tag{4.8}$$

$$\frac{\mathrm{d}}{\mathrm{d}t}\,\delta s = 0\,. \tag{4.9}$$

Both systems can be reduced by repeated application of the appropriate differentiation operations to a single differential equation (corresponding higher order).

For group A we get

$$\left\{\frac{\mathrm{d}^2}{\mathrm{d}t^2} - b^2\,\frac{\partial^2}{\partial z^2}\right\}\operatorname{rot}_z \delta\boldsymbol{v} = 0 \tag{4.10}$$

($\operatorname{rot}_z \delta\boldsymbol{B}$ satisfies the same equation). When $\boldsymbol{v} = 0$, this equation represents a one-dimensional wave equation, which describes the propagation of disturbances along the direction of the magnetic field (or in the opposite direction) with velocity b* (Alfvén-type disturbances). In this case the variables in group B can vanish identically.

From (4.3)–(4.9) we obtain for group B (e.g., for $\delta\varrho$)

$$\frac{\mathrm{d}^2}{\mathrm{d}t^2}\left\{\frac{\mathrm{d}^4}{\mathrm{d}t^4} - (a^2 + b^2)\,\Delta\,\frac{\mathrm{d}^2}{\mathrm{d}t^2} + a^2\,b^2\,\Delta\,\frac{\partial^2}{\partial z^2}\right\}\delta\varrho = 0\,. \tag{4.11}$$

The following solutions are allowable:

1) $\delta\varrho, \delta s \neq 0$ when $\dfrac{\mathrm{d}\delta\varrho}{\mathrm{d}t} = \dfrac{\mathrm{d}\delta s}{\mathrm{d}t} = 0$; all other disturbance variables are identically zero (along with the density and entropy disturbances associated with the medium),

2) $\delta s = 0$, all other variables $\neq 0$ with

$$\left\{\frac{\mathrm{d}^4}{\mathrm{d}t^4} - (a^2 + b^2)\,\Delta\,\frac{\mathrm{d}^2}{\mathrm{d}t^2} + a^2\,b^2\,\Delta\,\frac{\partial^2}{\partial z^2}\right\}\delta\varrho = 0$$

(disturbances of a magnetoacoustic nature).

Assuming plane waves for the disturbance variables, we again obtain the previously stated dispersion relations.

* $b = \dfrac{B}{\sqrt{\mu\,\varrho}}$ can become arbitrarily large in a sufficiently strong or sufficiently weak magnetic field, and can even exceed c, the velocity of light. However, when allowing for the displacement current in the initial equations, we obtain from the propagation speed for Alfvén-waves $b^* = \dfrac{b}{\sqrt{1 + b^2/c^2}}$. Now $b^* \to c$ when $b \to \infty$. When $b \approx c$, we must allow for the displacement current.

Equations (4.10) and (4.11) completely describe the propagation of disturbances in an homogeneous medium. In particular, they are appropriate for the "radiation" produced by (e.g., localized) sources of disturbances. These equations must then be solved with a nonvanishing right-hand side.

The system of linearized MHD-equations no longer follows when the initial basic solution is inhomogeneous. The terms, in which the gradients of the undisturbed quantities appear $\big((\delta \boldsymbol{v}\ \mathrm{grad})\, \boldsymbol{v}$, $(\delta \boldsymbol{B}\ \mathrm{grad})\, \boldsymbol{v}$, etc$\big)$ have the effect that, for example, all components of the perturbation vectors $\delta \boldsymbol{v}$ and $\delta \boldsymbol{B}$ are coupled throughout.

We must then refer to equation (2.37); plane waves no longer satisfy this equation.

Due to the large number of basic solutions it is not possible to obtain general statements concerning exact solutions of (2.37). (The problem already involves difficulties for simple inhomogeneities, for example, for a one-dimensional stratified medium).

We therefore would like to consider an approximate description of the wave propagation by analogy with geometrical optics or geometrical acoustics. The realm of problems associated herewith is collectively denoted as "geometric magnetohydrodynamics" (after Bazer and Hurley).

It is assumed that the basic solution is time independent. The expression for plane monochromatic waves is generalized in the usual manner by assuming an unknown phase function $W(\boldsymbol{x})$. The task of the geometrical theory is then to determine the time and spatial behavior of the phase surfaces.

An alternate way to describe the propagation of disturbances (nonperiodic) in inhomogeneous media is to consider the surfaces of a weak discontinuity ("weak shock waves"). These are surfaces, at which the state parameters can undergo small jumps, or surfaces of the excitation originating from pulse-like disturbances. It can be shown, that the phase surfaces of the periodic harmonic waves correspond to the surface of a weak discontinuity. Both can be again considered as different interpretations of the characteristic surfaces of (2.37); then the conditions, which are satisfied at both surfaces, are just the characteristic equations (2.37). We shall derive them below.

4.2. Characteristics of the linearized MHD-Equations

A characteristic surface of the basic equations (2.28) is defined as a surface $\boldsymbol{F}$ $\big(\Phi(x_1, x_2, x_3, t) = 0\big)$, at which any first or higher order derivative of u_i can become discontinuous, whereas the u_i themselves are constant. With the symbol [ ], such as used in section 3.1, we obtain the following conditions at every point on a characteristic surface:

$$[\boldsymbol{u}] = 0 \,, \tag{4.12}$$

$$\frac{\partial[\boldsymbol{u}]}{\partial t} + \sum_{j=1}^{3} U_j \frac{\partial[\boldsymbol{u}]}{\partial x_j} = 0 \,. \tag{4.13}$$

The last equation simply follows from requirement that the basic equations (2.28) must be satisfied on both sides of $\boldsymbol{F}$. For $[\boldsymbol{u}] = 0$

$$\mathrm{d}[\boldsymbol{u}] = \frac{\partial[\boldsymbol{u}]}{\partial t}\,\mathrm{d}t + \frac{\partial[\boldsymbol{u}]}{\partial x_1}\,\mathrm{d}x_1 + \cdots + \frac{\partial[\boldsymbol{u}]}{\partial x_3}\,\mathrm{d}x_3 = 0 \tag{4.14}$$

holds also "along" $\boldsymbol{F}$, i.e., provided the displacement $\mathrm{d}x_i, \ldots, \mathrm{d}t$ is chosen in such a way that

$$\frac{\partial\Phi}{\partial x_1}\,\mathrm{d}x_1 + \cdots + \frac{\partial\Phi}{\partial x_3}\,\mathrm{d}x_3 + \frac{\partial\Phi}{\partial t}\,\mathrm{d}t = 0 \tag{4.15}$$

is satisfied.

Comparing (4.14) and (4.15) gives

$$\begin{pmatrix} \dfrac{\partial[\boldsymbol{u}_i]}{\partial x_1} \\ \vdots \\ \dfrac{\partial[\boldsymbol{u}_i]}{\partial t} \end{pmatrix} = \alpha_i \begin{pmatrix} \dfrac{\partial\Phi}{\partial x_1} \\ \vdots \\ \dfrac{\partial\Phi}{\partial t} \end{pmatrix}. \tag{4.16}$$

The α_i are proportionality factors, which can vary from point to point on $\boldsymbol{F}$. For $\vec{\alpha} = (\alpha_1, \ldots, \alpha_8)$ we can also write

$$\frac{\partial[\boldsymbol{u}]}{\partial x_1} = \vec{\alpha}\,\frac{\partial\Phi}{\partial x_1}, \ldots, \frac{\partial[\boldsymbol{u}]}{\partial t} = \vec{\alpha}\,\frac{\partial\Phi}{\partial t}. \tag{4.17}$$

Inserting (4.17) into (4.13) yields

$$\left\{\frac{\partial\Phi}{\partial t}\,I + \sum_{j=1}^{3} U_j\,\frac{\partial\Phi}{\partial x_j}\right\}\vec{\alpha} = 0 \tag{4.18}$$

(I = unit matrix), or for

$$\boldsymbol{n} = \frac{\operatorname{grad}\Phi}{|\operatorname{grad}\Phi|} \quad \text{and} \quad C = -\frac{\dfrac{\partial\Phi}{\partial t}}{|\operatorname{grad}\Phi|}$$

(local normal to $\boldsymbol{F}$) (velocity of $\boldsymbol{F}$ in the direction of the normal)

$$\left\{\sum_{j=1}^{3} n_j\,U_j - C\,I\right\}\vec{\alpha} = \{L_u - C\,I\}\,\vec{\alpha} = 0\,. \tag{4.19}$$

At every point on $\boldsymbol{F}$ this is a system of linear homogeneous equations for the $\alpha_1, \ldots, \alpha_8$. Nontrivial solutions exist only when the corresponding determinant vanishes:

$$\|L_u - C\,I\| = 0\,. \tag{4.20}$$

Owing to the expressions for n and C this condition is a partial differential equations of the first order in $\Phi(x, y, z, t)$ and represents the equations for the characteristic surfaces. Via the L_u the u_j enter into (4.20), and consequently the characteristic surfaces are dependent on the solutions.

The characteristic equation of the linearized system is identical to (4.20), only now the u_i are the given functions of position and time. The characteristic surfaces are in this case not dependent on the solutions (the δu_i), but are given by the unperturbed configuration.

Expression (4.20) has exactly the same form as the dispersion equation (2.52). This can be interpreted in the following way: over small ranges of space and time, in which the time and space dependence of the basic equation can be disregarded, a characteristic surface element moves as a part of a plane wave front.

a) *The "geometrical magnetohydrodynamic" approximation*

This consists of generalizing the expression (2.47) in the following manner (under the assumption of a time-independent basic solution)

$$\delta \boldsymbol{u}(\boldsymbol{x}, t) = e^{i\omega\{W(\boldsymbol{x})-(t-t_0)\}} \sum_{\nu=0}^{\infty} \left(\frac{1}{i\,\omega}\right)^{\nu} \delta \boldsymbol{u}^{(\nu)}\,, \tag{4.21}$$

and considering the limiting case $\omega \to \infty$ (The $\delta \boldsymbol{u}^{(\nu)}$ are only dependent on location). Comparing (4.21) and (2.37), we obtain the following recursive system

$$(\boldsymbol{u}\,\delta \boldsymbol{B}^{(0)}) = 0\,, \tag{4.22}$$

$$(L_W - C\,I)\,\delta \boldsymbol{w}^{(0)} = 0\,, \tag{4.23}$$

$$(\boldsymbol{n}\,\delta \boldsymbol{B}^{(k)}) = -\,C\,L_1\,(\delta \boldsymbol{B}^{(k-1)})\,, \tag{4.24}$$

$$(L_W - C\,I)\,\delta \boldsymbol{w}^{(k)} = P(\delta\,\boldsymbol{u}^{(k-1)})\,, \tag{4.25}$$

where

$$\boldsymbol{n} = \frac{\operatorname{grad} W}{|\operatorname{grad} W|}\,, \qquad C = \frac{1}{|\operatorname{grad} W|} \tag{4.26}$$

and

$$P(\delta \boldsymbol{u}^{(k-1)}) = -\,C\left\{\frac{a}{\varrho}\,\overline{L}_2^{(n)}, \frac{1}{\varrho}\,\overline{L}_{3,n}^{(n)}, \frac{1}{\varrho}\,\overline{L}_{3,t}^{(n)}, \frac{1}{\sqrt{\mu\,\varrho}}\,\overline{L}_{4,t}^{(n)}, \overline{L}_5^{(n)}, \frac{1}{\varrho}\,\overline{L}_{3,r}^{(n)}, \frac{1}{\sqrt{\mu\,\varrho}}\,\overline{L}_{4,r}^{(n)}\right\}. \tag{4.27}$$

The $\overline{L}_i$ are identical to the operators in (2.32)—(2.38) without the time derivative*.

(The second subscript designates the component associated with the vector operator of $\overline{L}_i$.) Restricting ourselves to the limiting case $\omega \to \infty$, the two equations (4.22) and (4.23) are retained. In particular we must have

$$||L_W - C\,I|| = 0\,, \qquad \text{i.e.,} \tag{4.28}$$

the phase surfaces $W(\boldsymbol{x}) = 0$ are characteristic surfaces.

* For example, $\overline{L}_2^{(n)} = L_2(\delta\varrho^{(n)}, \varrho \boldsymbol{v}^{(n)})$, etc.

b) *Weak discontinuities*

In Section 3.1 we derived the conditions, which were satisfied at "strong" discontinuities. From the derivation it is clear that the flatness of the surface is only locally required. The conditions (3.6) must therefore be valid at every point on a curved discontinuity surface.

If $\varphi(x_1, \ldots, t)$ represents the equation of such a surface, then

$$N_j = \frac{\dfrac{\partial \varphi}{\partial x_j}}{|\text{grad}\ \varphi|}, \qquad V = \frac{-\dfrac{\partial \varphi}{\partial t}}{|\text{grad}\ \varphi|}.$$

When $[a_j] \ll a_i^{(1),(2)}$, i.e., in the limiting case of small jumps, we can set

$$a_i^{(2)} = a_i^{(1)} + \delta a_i .$$

Consequently, the boundary conditions become

$$\begin{aligned} V\,\delta a_i &= \sum_{j=1}^{3} N_j\,[F_{ij}(a_1^{(1)} + \delta a_1, \ldots, a_8^{(1)} + \delta a_8) - F(a_1^{(1)}, \ldots, a_8^{(1)})] \\ &= \sum_{\nu=1}^{3} N_j \sum_{\nu=1}^{8} \cdot \frac{\partial F_{ij}}{\partial a_\nu}\,\delta a_\nu , \end{aligned}$$

or

$$\left\{\sum_{j=1}^{3} N_j\,A_j - V\,I\right\} \delta \boldsymbol{a} = 0 . \tag{4.29}$$

Therefore, we must have

$$\left\| \sum_{j=1}^{3} N_j\,A_j - V\,I \right\| = 0 \quad \text{or} \quad \|L_A - V\,I\| = 0 . \tag{4.30}$$

This system is identical to (4.28), since $L_A = M\,L_u\,M^{-1}$. While there was no limitation with regard to V for strong discontinuities, the velocity of a weak discontinuity in the direction of its normal can not assume arbitrary values. It moves with one of the velocities $\mp c_i$ relative to the medium.

The surfaces of a weak discontinuity satisfy the same conditions as the characteristic surfaces.

4.3. Construction of the Characteristic Surfaces by Means of Rays

The equation

$$\|L_u - C\,I\| = 0 \tag{4.31}$$

is a partial differential equation for $\Phi(x_1, \ldots, x_0)$ $(x_0 = t)$. If we set

$$\frac{\partial \Phi}{\partial x_j} = p_j , \quad \text{i.e.,} \quad \text{grad}\ \Phi = \boldsymbol{p} , \quad \frac{\partial \Phi}{\partial t} = p_0 ,$$

we can write (4.31), utilizing the roots c_i, in the following way:

$$||L_u - C\,I|| = H(x_j, p_j) = \prod_{i=0}^{3} H_i^+ \, H_i^- = 0\,, \tag{4.32}$$

where

$$H_i^{+,-} = p_0 + (\boldsymbol{v}\,\boldsymbol{p}) \pm p \cdot c_i; \qquad \left(c_i = c_i\left(\frac{\boldsymbol{p}}{p}\right) = c_i(\boldsymbol{n})\right), \tag{4.33}$$

H is a homogeneous polynomial in the p_i. For $H = 0$ we therefore also have

$$\sum_{j=0}^{3} p_j \frac{\partial H}{\partial p_j} = (\boldsymbol{p} \operatorname{grad}_p H) + p_0 \frac{\partial H}{\partial p_0} = 0\,. \tag{4.34}$$

To construct the solutions of $H = 0$ we can again make use of the characteristic surfaces. The characteristic surfaces corresponding to (4.31) (the bicharacteristics of the system (2.37)) are curves $x_j(s)$ and $p_j(s)$ in $x_j\,p_j$-space. They satisfy the equations

$$\frac{\mathrm{d}x_j}{\mathrm{d}s} = \frac{\partial H}{\partial p_j}\,; \quad \frac{\partial p_j}{\partial s} = -\frac{\partial H}{\partial x_j} - p_j \frac{\partial H}{\partial \Phi}\,; \quad \frac{\partial \Phi}{\partial s} = \sum_{j=0}^{3} p_j \frac{\partial H}{\partial p_j} \tag{4.35}$$

(s = parameter). We now have

$$\frac{\partial H}{\partial \Phi} = 0 \quad \text{and} \quad \sum_{j=0}^{3} p_j \frac{\partial H}{\partial p_j} = 0; \tag{4.36}$$

leaving us the system

$$\frac{\partial \boldsymbol{x}}{\partial s} = \operatorname{grad}_{\boldsymbol{p}} H\,, \qquad \frac{\partial t}{\partial s} = \frac{\partial H}{\partial p_0}\,,$$

$$\frac{\partial \boldsymbol{p}}{\partial s} = -\operatorname{grad}_{\boldsymbol{x}} H\,, \qquad \frac{\partial p_0}{\partial s} = \frac{\partial H}{\partial t}\,.$$

Because of (4.32), it is sufficient to consider the equations

$$H_i^{+,-} = 0$$

separately. In all cases we have $\frac{\mathrm{d}t}{\mathrm{d}s} = 1$; thus, we can set $s = t$.

If the undisturbed state is time independent (this assumption was already employed in the geometric MHD approximation), we have

$$\frac{\mathrm{d}p_0}{\mathrm{d}s} = \frac{\mathrm{d}p_0}{\mathrm{d}t} \equiv 0\,, \qquad \text{i.e.,} \qquad p_0 = \text{const}\,(= -1)\,.$$

(It is found that this assumption enables us to write the expression $\Phi = W - (t - t_0)$.)

The following equations are now obtained

$$\frac{\mathrm{d}\boldsymbol{x}_i^{+,-}}{\mathrm{d}t} = \operatorname{grad}_{\boldsymbol{p}} H_i^{+,-}(\boldsymbol{x}, \boldsymbol{p})\,, \tag{4.37}$$

$$\frac{\mathrm{d}\boldsymbol{p}_i^{+,-}}{\mathrm{d}t} = -\operatorname{grad}_{\boldsymbol{x}} H_i^{+,-}(\boldsymbol{x}, \boldsymbol{p})\,. \tag{4.38}$$

This system has the same form as the canonical equations of motion for the Hamiltonian $H_i^{+,-}(\boldsymbol{x},\boldsymbol{p})$.

$H_i^{+,-}(\boldsymbol{x},\boldsymbol{p}) = 0$ gives at every point in space a plane in $\boldsymbol{p}$-space, which is the normal locus mentioned in section 2.5.

The curves in space described by (4.37) are the rays. The right-hand side of (4.37) is in agreement with the expression for the (local) velocity of the energy transport, since

$$\operatorname{grad}_{\boldsymbol{p}} H_i^{+,-}(\boldsymbol{x},\boldsymbol{p}) = \boldsymbol{v} \pm \operatorname{grad}_{\boldsymbol{p}} (p\, c_i) .$$

We therefore have

$$\frac{\mathrm{d}\boldsymbol{x}_i^{+,-}}{\mathrm{d}t} = \mathbf{s}_i^{+,-} ; \tag{4.39}$$

the rays everywhere have the direction of the local energy flux. The directions of $\boldsymbol{p}$ and $\frac{\mathrm{d}\boldsymbol{x}}{\mathrm{d}t}$ in general do not coincide; i.e., in general, the rays do not intersect (as in crystal optics) the constant-phase surfaces orthogonally.

In accordance with (4.34) and (4.37) we have

$$\left(\boldsymbol{p}\,\frac{\mathrm{d}\boldsymbol{x}}{\mathrm{d}t}\right) = -\,p_0 = 1 . \tag{4.40}$$

If the basic solution is homogeneous, we have

$$\frac{\mathrm{d}\boldsymbol{x}}{\mathrm{d}t} = \text{const} , \quad \frac{\mathrm{d}\boldsymbol{p}}{\mathrm{d}t} = 0 .$$

The rays are straight lines; the characteristic surface elements move along the rays without altering their shape.

With the rays we can follow the time and spatial development of a given characteristic surface. Let $\boldsymbol{x}^0 = \boldsymbol{x}^0(\xi_1, \xi_2)$ be a surface in space in parametric representation and $\boldsymbol{p}^0 = \boldsymbol{p}^0(\xi_1, \xi_2)$ the corrresponding normal at every point on the surface. The values $\boldsymbol{x}^0$ and $\boldsymbol{p}^0$ will be considered as initial conditions for the system (4.37/4.38). This yields a solution of the form

$$\boldsymbol{x}_i^{+,-} = \boldsymbol{x}_i^{+,-}(\xi_1, \xi_2; t) , \tag{4.41}$$

$$\boldsymbol{p}_i^{+,-} = \boldsymbol{p}_i^{+,-}(\xi_1, \xi_2; t) , \tag{4.42}$$

when $t \geq t_0$, this is the parametric representation of the characteristic surfaces at time t. (The same solutions are obtained when $\boldsymbol{p}^0$ is taken as the initial value for $\boldsymbol{p}$.)

We shall specify the ray equations for the different wave modes.

Entropy rays

$$\frac{\mathrm{d}\boldsymbol{x}}{\mathrm{d}t} = \boldsymbol{v} , \quad \frac{\mathrm{d}\boldsymbol{p}}{\mathrm{d}t} = -\operatorname{grad}_{\boldsymbol{x}} (\boldsymbol{v}\,\boldsymbol{p}) = -(\boldsymbol{p}\operatorname{grad})\,\boldsymbol{v} - (\boldsymbol{p}\times\operatorname{rot}\boldsymbol{v}) . \tag{4.43}$$

Alfvén rays

$$\frac{\mathrm{d}\boldsymbol{x}^{+,-}}{\mathrm{d}t} = \boldsymbol{v} \pm \boldsymbol{b}\ \mathrm{sgn}\,(\boldsymbol{n}\boldsymbol{b})\,, \tag{4.44}$$

$$\frac{\mathrm{d}\boldsymbol{p}^{+,-}}{\mathrm{d}t} = -(\boldsymbol{p}\ \mathrm{grad})\,\boldsymbol{v} - (\boldsymbol{p}\times\mathrm{rot}\,\boldsymbol{v}) \mp p\ \mathrm{grad}\, c_A\,. \tag{4.45}$$

s-rays

$$\frac{\mathrm{d}\boldsymbol{x}^{+,-}}{\mathrm{d}t} = \boldsymbol{v} \pm \left\{c_s\,\boldsymbol{n} + \frac{b^2\,r\sin\vartheta\cos\vartheta}{c_s\,q}\,\boldsymbol{t}\right\}, \tag{4.46}$$

$$\frac{\mathrm{d}\boldsymbol{p}^{+,-}}{\mathrm{d}t} = -(\boldsymbol{p}\ \mathrm{grad})\,\boldsymbol{v} - (\boldsymbol{p}\times\mathrm{rot}\,\boldsymbol{v}) \mp p\cdot\mathrm{grad}\, c_s\,. \tag{4.47}$$

f-rays

$$\frac{\mathrm{d}\boldsymbol{x}^{+,-}}{\mathrm{d}t} = \boldsymbol{v} \pm \left\{c_f\,\boldsymbol{n} - \frac{b^2\,r\sin\vartheta\cos\vartheta}{c\,q}\,\boldsymbol{t}\right\}, \tag{4.48}$$

$$\frac{\mathrm{d}\boldsymbol{p}^{+,-}}{\mathrm{d}t} = -(\boldsymbol{p}\ \mathrm{grad})\,\boldsymbol{v} - (\boldsymbol{p}\times\mathrm{rot}\,\boldsymbol{v}) \mp p\cdot\mathrm{grad}\, c_f\,, \tag{4.49}$$

when $\boldsymbol{n} = \frac{\boldsymbol{p}}{p}$, $\vartheta = \sphericalangle\,(\boldsymbol{B},\boldsymbol{p})$ and $\boldsymbol{t}$ is the unit vector defined in (2.53). The $c_{s,f}$ and q are functions of $\boldsymbol{x}$ and $\boldsymbol{n}$.

The ray equations can be directly integrated when the basic solution is homogeneous. Assuming further that $\boldsymbol{v} = 0$, we obtain (the entropy waves do not enter here):

$$\boldsymbol{x}(t) - \boldsymbol{x}^{(0)} = \pm\,\boldsymbol{b}\ \mathrm{sgn}\cos\vartheta\cdot(t - t_0)\,, \tag{4.50}$$

$$\boldsymbol{x}(t) - \boldsymbol{x}^{(0)} = \pm\left(c_s\,\boldsymbol{n}^0 + \frac{b^2\,r\sin\vartheta\cos\vartheta}{c_s\,q}\,\boldsymbol{t}^0\right)(t - t_0)\,, \tag{4.51}$$

$$\boldsymbol{x}(t) - \boldsymbol{x}^{(0)} = \pm\left(c_f\,\boldsymbol{n}^0 - \frac{b^2\,r\sin\vartheta\cos\vartheta}{c_f\,q}\,\boldsymbol{t}^0\right)(t - t_0) \tag{4.52}$$

and $p_i = \pm\,p_i^0\ (i = 1, 2, 3)$.

$\boldsymbol{n}$ is constant along every ray and the rays themselves are straight lines. From an initial surface element $\delta\gamma(t_0)$ emanate 6 different characteristic surface elements, which displace themselves along the rays.

Explicit formulas for the solutions can be specified when the initial surface is fixed.

Simple expressions are obtained, for example, when we choose a spherical surface of radius R as the initial surface. We then have

$$\boldsymbol{x}^0 = R\,\boldsymbol{n}^0(\vartheta,\varphi) \qquad \text{where} \qquad \boldsymbol{n}^0(\vartheta,\varphi) = \begin{pmatrix}\sin\vartheta\cos\varphi\\ \sin\vartheta\sin\varphi\\ \cos\vartheta\end{pmatrix}$$

and

$$\boldsymbol{t}^0 = \frac{1}{\sin\vartheta}\,(\boldsymbol{m} - \boldsymbol{n}^0\cos\vartheta)\,.$$

The parametric representations of the characteristic surfaces propagating from the spherical surface are

1. *Alfvén surfaces:*

$$x(t) = x^0\,,\ y(t) = y^0\,,\ z(t) = z^0 \pm b \operatorname{sgn} \cos\vartheta\,(t - t_0)\,. \tag{4.53}$$

Two congruent spherical surfaces propagating from the initial surface in the direction of $\boldsymbol{B}$ and $-\boldsymbol{B}$ with velocity b (Figure 4.1)

2. *s-surfaces:*

$$x(t) = \sin\vartheta\cos\varphi\left\{R \pm \left(c_s - \frac{b^2\, r\cos^2\vartheta}{c_s\, q}\right)\cdot(t - t_0)\right\}, \tag{4.54}$$

$$y(t) = \sin\vartheta\sin\varphi\left\{R \pm \left(c_s - \frac{b^2\, r\cos^2\vartheta}{c_s\, q}\right)\cdot(t - t_0)\right\}, \tag{4.55}$$

$$z(t) = \cos\vartheta\left\{R \pm \left(c_s + \frac{b^2\, r\sin^2\vartheta}{c_s\, q}\right)\cdot(t - t_0)\right\}. \tag{4.56}$$

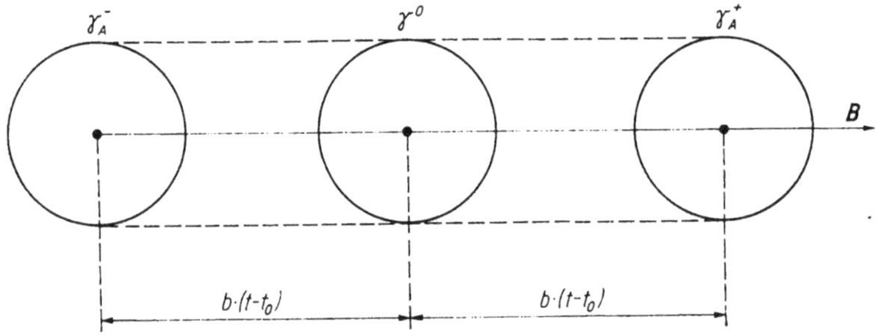

Figure 4.1. Alfvén-wave fronts propagating from a spherical surface

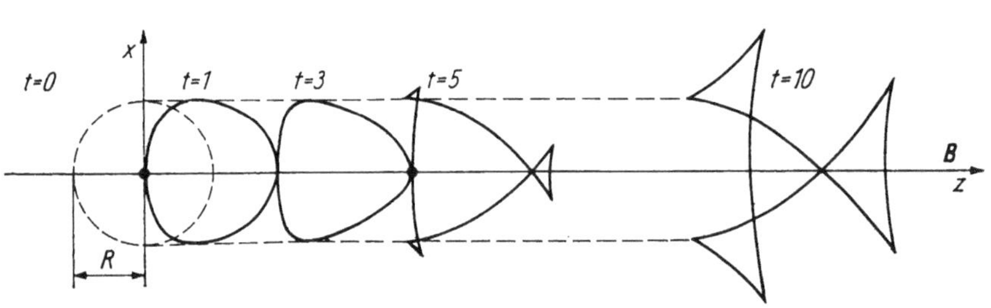

Figure 4.2. s-waves propagating from a spherical surface (after Bazer and Fleischmann [9])

The surfaces progressing in one of the directions are shown in Figure 4.2 (at different points of time). They possess rotational symmetry about the $\boldsymbol{B}$-axis. The surfaces progressing in the opposite direction are obtained by reflection with respect to the x-axis.

3. The *f-surfaces* can be determined in an analogous fashion (in (4.54)–(4.56) c_s is replaced by c_f and the sign of the second expression in the small parantheses is changed).

If we let $R \to 0$ in the above formulas, we obtain the parametric representations of the wavefronts, which emanate from a point-source excitation. Setting $t - t^0 = 1$ in these expressions gives the parametric representations of the ray surfaces (2.112)—(2.114). They represent the wave surfaces produced by a point source after a time $t = 1$.

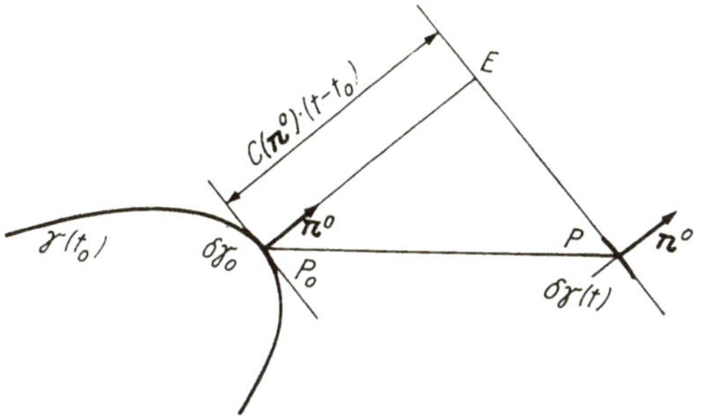

Figure 4.3. Geometric construction of the surfaces $\gamma(t)$ from the initial surface γ^0 for a homogeneous undisturbed medium

Figure 4.3 illustrates once more the geometric correlations (for the homogeneous case). $\gamma(t_0)$ is the initial surface at time $t = t_0$, $\delta\gamma^{(0)}$ is a surface element at $\boldsymbol{x}^0$ with normal $\boldsymbol{n}^0$. After a time t, $\delta\gamma^{(0)}$ has progressed along the ray until P. Since

$$\boldsymbol{n} \cdot \frac{\mathrm{d}\boldsymbol{x}}{\mathrm{d}t} = c(\boldsymbol{n}^0) ,$$

it has therefore moved a distance $c(n^0)\,(t - t^0)$ in the direction of n^0. The EP-plane is tangential to $\delta\gamma(t)$. We obtain $\gamma(t)$ by plotting at all points the vectors $c(\boldsymbol{n}^0)\,(t - t^0)\,\boldsymbol{n}^0$, constructing planes normal to their end points, and finally, by forming their envelope.

4.4. Examples

4.4.1. Propagation of Alfvén Waves in a Dipole Field

Let us consider a dipole field in a medium at rest and of constant density as the undisturbed configuration.

In polar coordinates the ray equation for the Alfvén rays for $\boldsymbol{v} = 0$ is

$$\frac{\mathrm{d}r}{\mathrm{d}t} = b_r \operatorname{sgn}(\boldsymbol{p}\,\boldsymbol{b}) , \tag{4.57}$$

$$\frac{\mathrm{d}\vartheta}{\mathrm{d}t} = \frac{b_\vartheta}{r} \operatorname{sgn}(\boldsymbol{p}\,\boldsymbol{b}) , \tag{4.58}$$

$$\frac{\mathrm{d}\varphi}{\mathrm{d}t} = \frac{b_\varphi}{r \sin\vartheta} \operatorname{sgn}(\boldsymbol{p}\,\boldsymbol{b}) . \tag{4.59}$$

We have hereby restricted ourselves to the positive sign in (4.44) ("plus" rays).

Since $\left(\frac{\mathrm{d}\boldsymbol{x}}{\mathrm{d}t}\boldsymbol{p}\right) = 1 = (\boldsymbol{p}\,\boldsymbol{b})\,\mathrm{sgn}(\boldsymbol{p}\,\boldsymbol{b})$, $(\boldsymbol{p}\,\boldsymbol{b}) \neq 0$ and thus $\mathrm{sgn}(\boldsymbol{p}\,\boldsymbol{b})$ is constant along a ray. Equations (4.57)–(4.59) are no longer interconnected with those for $\boldsymbol{p}$. Let the initial surface be given in the form

$$r = r_0(\vartheta, \varphi)\,.$$

If we take the z-axis of the polar coordinate system along the direction of the dipole, the magnetic field is given by

$$\boldsymbol{B} = -M \operatorname{grad}\left(\frac{\cos\vartheta}{r^2}\right)$$

or

$$B_r = \frac{2M\cos\vartheta}{r^3}\,, \qquad B_\vartheta = \frac{M\sin\vartheta}{r^3}\,, \qquad B_\varphi = 0 \tag{4.60}$$

$$(M = \text{dipole moment})\,.$$

From (4.59) it thus follows that $\frac{\mathrm{d}\vartheta}{\mathrm{d}t} = 0$ and we can restrict ourselves to the meridional plane $\varphi = 0$.

With (4.60) the ray equations are

$$\frac{\mathrm{d}r}{\mathrm{d}t} = \frac{2M\cos\vartheta}{\sqrt{\mu\varrho}\cdot r^3}\,\mathrm{sgn}(\boldsymbol{p}\,\boldsymbol{b})\,, \tag{4.61}$$

$$\frac{\mathrm{d}\vartheta}{\mathrm{d}t} = \frac{M\sin\vartheta}{\sqrt{\mu\varrho}\,r^4}\,\mathrm{sgn}(\boldsymbol{p}\,\boldsymbol{b})\,. \tag{4.62}$$

Taking the quotient of both equations, we obtain the following expression

$$\frac{1}{r}\frac{\mathrm{d}r}{\mathrm{d}\vartheta} = 2\cot\vartheta\,; \tag{4.63}$$

or after integration

$$r(\vartheta) = \frac{r_0}{\sin^2\vartheta_0}\sin^2\vartheta = k_0\sin^2\vartheta\,. \tag{4.64}$$

It further follows from (4.61) that

$$\frac{\mathrm{d}t}{\mathrm{d}r} = \sqrt{\mu\varrho}\,\mathrm{sgn}(\boldsymbol{p}\,\boldsymbol{b})\,\frac{r^3}{2M\,(\pm)\sqrt{1-\frac{r}{k_0}}}\,. \tag{4.65}$$

$\left(\cos\vartheta = \pm\sqrt{1-\frac{r}{k_0}}\right.$ in accordance with (4.64)$\left.\right)$. The positive sign pertains to $0 \leq \vartheta < \frac{\pi}{2}$, the negative sign for $\frac{\pi}{2} \leq \vartheta < \pi$). Integration yields

$$t - t_0 = \frac{\sqrt{\mu\varrho}\,\mathrm{sgn}(\boldsymbol{p}\,\boldsymbol{b})}{2M}\int\limits_{r_0}^{r}\frac{r'^3}{\sqrt{1-\frac{r'}{k_0}}}\,\mathrm{d}r' = \frac{\sqrt{\mu\varrho}\,\mathrm{sgn}(\boldsymbol{p}\,\boldsymbol{b})\,k_0^4}{2M}\int\limits_{r_0/k_0}^{r/k_0}\frac{x^3}{\sqrt{1-x}}\,\mathrm{d}x\,. \tag{4.66}$$

Equations (4.64) and (4.66) represent the solution. The function

$$I(x) = \frac{1}{B} \int_0^x \frac{x^3}{\sqrt{1-x}}\, dx\,, \qquad \text{where} \qquad B = \int_0^1 \frac{x^3}{\sqrt{1-x}}\, dx$$

can be found in the tables.

Figure 4.4 shows the phase surfaces (in the meridional plane $\varphi = 0$) emanating from the plane

$$r_0 = \frac{d}{\sin\vartheta_0 \cos\varphi}\,,$$

which is the initial surface γ^0. (d is the distance of the plane from the dipole; time points were chosen at $t = 0.1;\ 0.2;\ 0.3;\ 0.4$.)

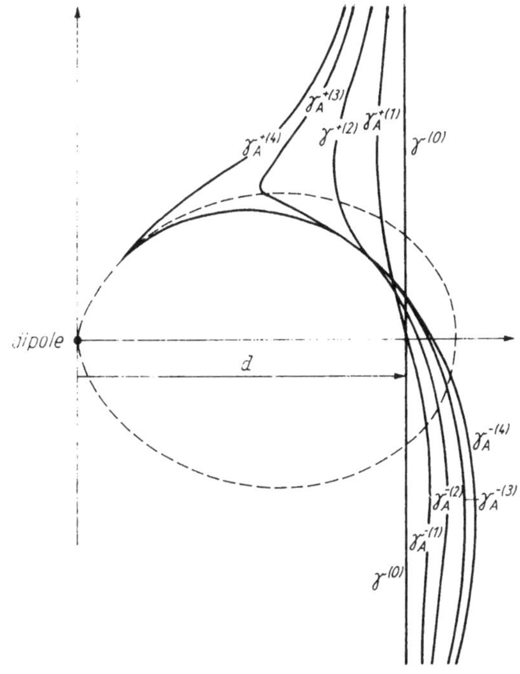

Figure 4.4. Phase surfaces of Alfvén disturbances propagating from a plane in a dipole field (after Bazer and Hurley [8])

The normal to γ^0 was taken as directed toward the left. The surfaces in the upper part of Figure 4.4 are therefore produced by plus rays, those in the lower part by minus rays. The complete picture of all surfaces emanating from γ^* is obtained by reflecting the curves in Figure 4.4 with respect to the equatorial plane. As the surfaces $\gamma(t)$ approach the dipole they form a point. The dotted curve designates the position at which the point developes.

The surfaces $\gamma(t)$ can be interpreted according to section 4.2 either as weak shock waves, diverging from an instantaneous disturbance of γ; or as phase surfaces of waves, which originate in a localized harmonic disturbance of γ^0. With the solution of (4.64), equation (4.66) approximately describes the propagation of Alfvén disturbances in the earth's magnetic field.

4.4.2. Propagation of *f*-Waves in the Magnetosphere of the Earth

Compared to the Alfvén speed ($r \ll 1$), the sound speed is very small in the magnetosphere. Disregarding terms of the order of r, we obtain in accordance with (2.80)

$$c_f = b\,, \qquad c_s = 0\,.$$

Further assuming that $\boldsymbol{v} = 0$ the ray equations for the f-rays reduce to

$$\frac{\mathrm{d}\boldsymbol{x}}{\mathrm{d}t} = \pm\, b^2 \boldsymbol{p}\,, \tag{4.67}$$

$$\frac{\mathrm{d}\boldsymbol{p}}{\mathrm{d}t} = \mp\, p \,\mathrm{grad}\, b\,. \tag{4.68}$$

We again drop the double sign and consider only the plus rays. We can eliminate p:

$$\frac{\mathrm{d}}{\mathrm{d}t}\left\{\frac{1}{b^2}\frac{\mathrm{d}\boldsymbol{x}}{\mathrm{d}t}\right\} = -\frac{1}{b}\,\mathrm{grad}\, b \qquad \left(p = \frac{1}{b}\right). \tag{4.69}$$

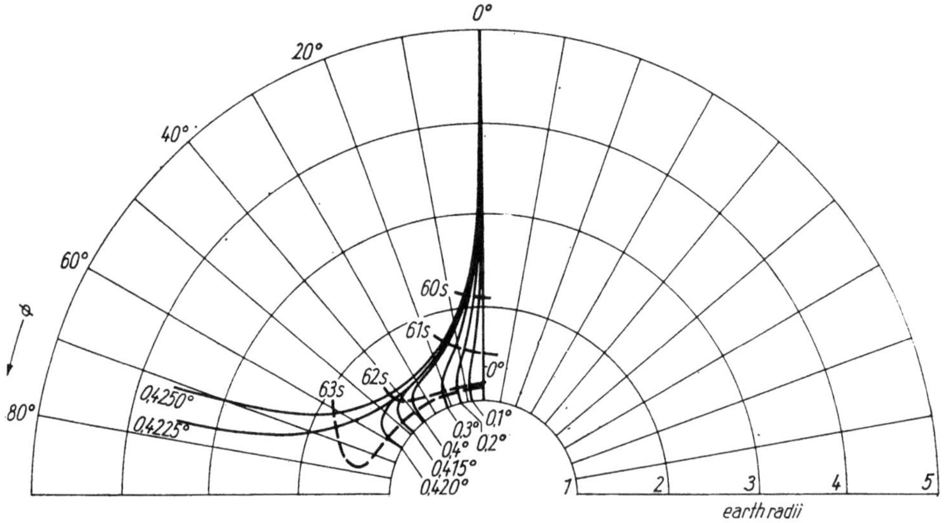

Figure 4.5a. f-rays in the earth's magnetosphere (Stegelmann and Kenschitzki [19])

The different rays in a and b correspond to the different initial directions. The deviations from the line connecting the source to the center of the earth are designated as well as the travel times of the rays from the source to the dotted lines.

(a) course of f-rays in the geomagnetic equatorial plane,

(b) course of f-rays in a geomagnetic meridional plane

In terms of polar coordinates:

$$\ddot{r} - \frac{2\,r}{b}\left(\frac{\partial b}{\partial r}\,\dot{r} + \frac{\partial b}{\partial \vartheta}\,\dot{\vartheta}\right) + \frac{1}{b}\,\frac{\partial b}{\partial r} - \frac{r}{b^2}\,\dot{\vartheta}^2 - \frac{r\sin^2\vartheta}{b^2}\,\dot{\varphi}^2 = 0\,, \tag{4.70}$$

$$\ddot{\vartheta} - \frac{2\,\dot{\vartheta}}{b}\left(\frac{\partial b}{\partial r}\,\dot{r} + \frac{\partial b}{\partial \vartheta}\,\dot{\vartheta}\right) + \frac{2\,\dot{r}\,\dot{\vartheta}}{r} + \frac{b}{r^2}\,\frac{\partial b}{\partial \vartheta} - \sin\vartheta\cos\vartheta\,\dot{\varphi}^2 = 0\,, \tag{4.71}$$

$$\ddot{\varphi} - \frac{2}{b}\,\dot{\varphi}\left(\frac{\partial b}{\partial r}\,\dot{r} + \frac{\partial b}{\partial \vartheta}\,\dot{\vartheta}\right) + \frac{2\,\dot{r}\,\dot{\varphi}}{r} + \frac{2\cos\vartheta}{\sin\vartheta}\,\dot{\varphi}\,\dot{\vartheta} = 0\,, \tag{4.72}$$

where we remembered that $\frac{\partial b}{\partial \varphi} = 0$ (dipole field).

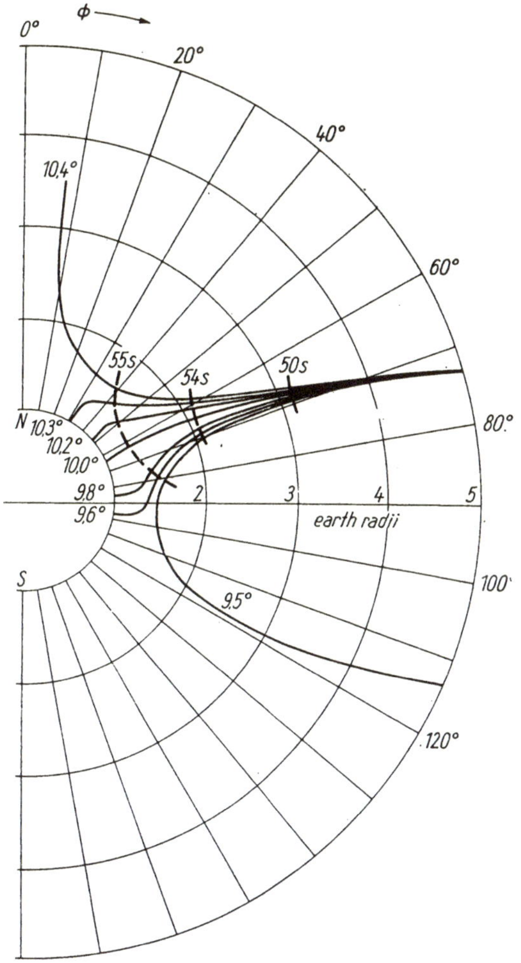

Figure 4b

To integrate the equations we make use of the density profile $\varrho = \varrho(r)$ in the magnetosphere, since it enters into the Alfvén speed $\left(b = B/\sqrt{\mu\,\varrho}\right)$.

In [17], [18], [19] ray paths in the magnetosphere diverging from point sources have been calculated with equations (4.70)—(4.72) using an empirical distribution $\varrho = \varrho(r)$. These were used to interpret the "sudden commencements" in geomagnetic storms, and in particular to explain the travel times of the disturbances from the position of the excitation (4—10 earthradii) to the earth's surface.

The relationships are the following: the constant plasma flow (solar wind) from the sun interacts with the earth's magnetic field, which in turn penetrates the plasma of the magnetosphere. The dipole field of the earth is therefore heavily distorted on the side facing the sun. Thus a relatively sharp boundary surface between the earth's magnetic field and the solar wind is formed, at which MHD-waves (f-waves) can be produced by the disturbances due to the solar flux of corpuscles.

From these surface sources the disturbances travel along the corresponding rays to the earth's surface. The travel time to the earth depends on the particular point from which the ray emanates. In Figures 4.5a and b are shown different ray paths in the equatorial plane and a meridional plane, respectively, which emanate from point sources with different initial starting directions. The corresponding travel times and the deviations of the starting directions from the line connecting the source to the earth's—center are marked on the rays.

4.5 Variation of intensities along the Rays*

Hitherto our considerations were based on determining the phase surfaces $W(x)$ diverging from an initial excitation surface. This was achieved with the help of rays. We shall now discuss the variation in the excitation intensities at the points of $W(x)$ during propagation.

Let $\gamma(t)$ be a phase surface, which emanates from $\gamma(t_0) = \gamma^0$ $\left(\gamma(t) = \gamma_i^{+,-}(t),\ i = 1, 2, 3\right)$ and let

$$\delta \boldsymbol{w}^0(\boldsymbol{x}) = \varepsilon(\boldsymbol{x}) \cdot \boldsymbol{w}(\boldsymbol{x}, \boldsymbol{n})$$

be the excitation at a point $\boldsymbol{x}$ of $\gamma(t)$ $\left(\boldsymbol{n}\text{: normal to } \gamma(t) \text{ at the point } \boldsymbol{x}\right)$. It is now of interest to examine how $\delta\boldsymbol{w}(\boldsymbol{x})$ varies as $\gamma(t)$ progresses, when, for example, the initial excitation

$$\delta \boldsymbol{w}^0(\boldsymbol{x}) = \varepsilon^0(\boldsymbol{x})\, \boldsymbol{w}_0(\boldsymbol{x}, \boldsymbol{n})$$

is given on the surface γ^0.

It is sufficient to know the variation of $\delta\boldsymbol{w}$ along the rays diverging from γ^0: $\boldsymbol{n}(t) = \dfrac{\boldsymbol{p}(t)}{p(t)}$ is known from (4.38) along the rays, so that $\boldsymbol{w}(\boldsymbol{r}, \boldsymbol{n})$ is also known. The problem reduces to determining how $\varepsilon(\boldsymbol{x})$ varies along the rays. It is possible to derive a differential equations for $\varepsilon\left(\boldsymbol{x}(t)\right)$.

* We are essentially following the work of Bazer and Hurley [8].

For this purpose we once again consider the recursive system in section 4.2 and in particular equation (4.23) and the first approximation of (4.25):

$$(L_W - C\,I)\,\delta\boldsymbol{w}^{(0)} = 0\,,$$
$$(L_W - C\,I)\,\delta\boldsymbol{w}^{(1)} = P(\delta\boldsymbol{u}^{(0)})\,.$$

We shall exclude entropy waves, so that we can make use of the symmetric part of $(L_W - C\,I)$. For this reason the equation for the first approximation has only one solution, when $\delta\boldsymbol{w}^{(0)}$ and $P(\delta\boldsymbol{u}^{(0)})$ are mutually orthogonal:

$$\delta\boldsymbol{w}^{(0)}\,P(\delta\boldsymbol{u}^{(0)}) = 0\,. \tag{4.73}$$

We can satisfy this condition with (4.27), (2.54), and (2.32)–(2.36). After rearranging (in a similar way as was necessary to derive (2.90)) we obtain

$$\operatorname{div}\delta\boldsymbol{F} + \varrho\,\delta\boldsymbol{v}\,(\delta\boldsymbol{v}\operatorname{grad})\,\boldsymbol{v} - \frac{1}{2}\frac{(\partial\varrho)^2}{\varrho}\,(\boldsymbol{v}\operatorname{grad}a^2)$$
$$+\frac{1}{2\,\mu}\,(\delta\boldsymbol{B})^2\operatorname{div}\boldsymbol{v} - \frac{1}{\mu}\,\delta\boldsymbol{B}\,(\delta\boldsymbol{B}\operatorname{grad})\,\boldsymbol{v} = 0\,. \tag{4.74}$$

In this derivation it was remembered that for Alfvén-waves as well as for the MA-waves the vectors $\delta\boldsymbol{v}$ and $\delta\boldsymbol{B} - \frac{\partial\varrho}{\varrho}\,\boldsymbol{B}$ are coplanar. (This can be readily verified with (2.83) and (2.87).) $\delta\boldsymbol{F}$ is the energy flux vector. Since the amplitudes $\delta\boldsymbol{v}$, $\delta\varrho$, and $\delta\boldsymbol{B}$ are linear in ε, we can condense the last four terms of (4.74) to $\varepsilon^2\,I$, where I is independent of ε and is obtained by the successive substitution of (2.83) and (2.87) into (4.24).

From (2.93) and (2.94) it follows that

$$\delta\boldsymbol{F} = \varepsilon^2\,\mathbf{s}\,.$$

With (4.74) we have

$$(\mathbf{s}\operatorname{grad}\varepsilon^2) + \varepsilon^2\,(\operatorname{div}\mathbf{s} + I) = 0\,. \tag{4.75}$$

The first term denotes the time variation of ε^2 along the ray $x(t)$. We have $\frac{\mathrm{d}\varepsilon^2(\boldsymbol{x}(t))}{\mathrm{d}t} = \left(\operatorname{grad}\varepsilon^2\,\frac{\mathrm{d}\boldsymbol{x}}{\mathrm{d}t}\right) = (\operatorname{grad}\varepsilon^2\,\mathbf{s})$. Furthermore,

$$\operatorname{div}\mathbf{s} = \frac{\mathrm{d}}{\mathrm{d}t}\log A\cdot\frac{C}{C_0}\bigg|_{\boldsymbol{x}=\boldsymbol{x}(t)}\,. \tag{4.76}$$

$C = 1/p$ is known and $A\big(\boldsymbol{x}(t)\big)$ is the so-called expansion relation along the ray, which is defined as follows:

Let $\mathrm{d}\sigma_0$ be a surface element of γ_0 and $\mathrm{d}\sigma(t)$ the surface element on $\gamma(t)$, which is bounded by the tubular ray bundle emanating from $\mathrm{d}\sigma^0$. We then obtain

$$A\big(\boldsymbol{x}(t)\big) = \frac{\mathrm{d}\sigma(t)}{\mathrm{d}\sigma^0}\,. \tag{4.77}$$

Expression (4.76) is obtained as follows:

We integrate div $\boldsymbol{s}$ over the volume ΔV, formed by the two surfaces $d\sigma(t)$ and $d\sigma(t + \Delta t)$ and the corresponding ray tube. Let $\boldsymbol{n}(t)$ and $\boldsymbol{n}(t + \Delta t)$ be the normals to $d\sigma(t)$ and $d\sigma(t + \Delta t)$, respectively. We have

$$\int_{\Delta V} \operatorname{div} \boldsymbol{s} \, d\tau = \int_{d\sigma(t)} -(\boldsymbol{n}\,\boldsymbol{s}) \, df + \int_{d\sigma(t+\Delta t)} (\boldsymbol{n}\,\boldsymbol{s}) \, df . \tag{4.78}$$

(The tubular surface does not contribute, since $\boldsymbol{s}$ is here parallel to the surface.) After dividing by Δt and letting $\Delta t \to 0$ we obtain from (4.78), remembering that $(\boldsymbol{n}\,\boldsymbol{s}) = C$ and $\Delta V = \Delta t\, C \cdot d\sigma$,

$$C \cdot d\sigma \operatorname{div} \boldsymbol{s} = \frac{d}{dt}\, C \cdot d\sigma ,$$

from which (4.76) immediately follows.

Consequently, we obtain the sought differential equation for the variation of ε^2 along aray $\boldsymbol{x}(t)$

$$\frac{d\varepsilon^2}{dt} + \varepsilon^2 \cdot I \Big|_{\boldsymbol{x}=\boldsymbol{x}(t)} + \varepsilon^2 \frac{d}{dt} \log \left\{ A \frac{C}{C_0} \right\} \Big|_{\boldsymbol{x}=\boldsymbol{x}(t)} = 0 . \tag{4.79}$$

Integration gives

$$\varepsilon^2(\boldsymbol{x}(t))\, A(\boldsymbol{x}(t))\, C(\boldsymbol{x}(t)) = \varepsilon_0^2\, C_0 \exp \int_{t_0}^{t} - I\, dt , \tag{4.80}$$

or since $A(\boldsymbol{x}(t)) = \dfrac{d\sigma(\boldsymbol{x}(t))}{d\sigma_0}$

$$\varepsilon^2\, d\sigma\, C\big|_{\boldsymbol{x}=\boldsymbol{x}(t)} = \varepsilon_0^2\, d\sigma_0\, C_0 \cdot \exp - \int_{t_0}^{t} I\, dt . \tag{4.81}$$

Since $\delta \boldsymbol{F} = \varepsilon^2\, \boldsymbol{s}$ and $(\delta \boldsymbol{F}\, \boldsymbol{n}) = \varepsilon^2 (\boldsymbol{s}\, \boldsymbol{n}) = \varepsilon^2\, C$, we may also write

$$(\delta \boldsymbol{F}\, \boldsymbol{n})\, d\sigma\big|_{\boldsymbol{x}=\boldsymbol{x}(t)} = (\delta \boldsymbol{F}\, \boldsymbol{n})\, d\sigma_0\big|_{\boldsymbol{x}=\boldsymbol{x}_0} \cdot \exp - \int_{t_0}^{t} I\, dt . \tag{4.82}$$

The expressions I for the individual waves modes are

$$I_A = -\frac{1}{2} (\boldsymbol{v} \operatorname{grad} \log \varrho) , \tag{4.83}$$

$$I_i = \frac{\lambda_i}{\sqrt{(a^2 + b^2)^2 - 4\, a^2 b^2 \cos^2 \vartheta}} \left\{ \frac{1}{\mu \varrho} \boldsymbol{B} (\boldsymbol{B} \operatorname{grad}) \boldsymbol{v} - c_i^2\, \boldsymbol{n} (\boldsymbol{n} \operatorname{grad}) \boldsymbol{v} \right.$$
$$\left. - \frac{c_i^2 - a^2}{2} (\boldsymbol{v} \operatorname{grad} \log \varrho) + \frac{c_i^2 - b^2 \cos^2 \vartheta}{2\, c_i^2} (\boldsymbol{v} \operatorname{grad} a^2) \right\} , \tag{4.84}$$

$$\lambda_i = \begin{cases} 1 & i = s , \\ -1 & i = f . \end{cases}$$

We can interpret (4.82) in the following manner:
$(\delta \boldsymbol{F}\, \boldsymbol{n})\, d\sigma|_{\boldsymbol{x}=\boldsymbol{x}(t)}$ and $(\delta \boldsymbol{F}\, \boldsymbol{n})\, d\sigma_0|_{\boldsymbol{x}=\boldsymbol{x}_0}$ are respectively the normal components of the energy flux through the surface elements $d\sigma(t)$ and $d\sigma_0$, which are connected

to one another by the ray tube. Both expressions are proportional to one another. The proportionality factor $\exp - \int_{t_0}^{t} I \, dt$ is unity, when the basic solution is homogeneous, or more precisely when

$$\boldsymbol{v} \equiv 0 \quad \text{or} \quad \boldsymbol{v} = \text{const} \quad \text{and} \quad \varrho = \text{const}. \tag{4.85}$$

Then the I_i vanish identically and we have

$$(\delta \boldsymbol{F} \, \boldsymbol{n}) \, \mathrm{d}\sigma\big|_{\boldsymbol{x}=\boldsymbol{x}(t)} = (\delta \boldsymbol{F} \, \boldsymbol{n}) \, \mathrm{d}\sigma_0\big|_{\boldsymbol{x}=\boldsymbol{x}_0} \tag{4.86}$$

irrespective of whether the magnetic field is constant or not. If $\boldsymbol{B}$ is constant, the rays are straight lines, $C(\boldsymbol{x}(t)) = 1/p$, is constant along the rays and (4.86) becomes

$$A(\boldsymbol{x}(t)) \cdot \varepsilon^2(\boldsymbol{x}(t)) = \varepsilon_0^2 \, . \tag{4.87}$$

In this case

$$A(\boldsymbol{x}(t)) = \frac{K(\boldsymbol{x}(t_0))}{K((t\boldsymbol{x}))} \, , \tag{4.88}$$

where $K(\boldsymbol{x}(t))$ is the Gaussian curvature of the surface $\gamma(t)$ at the point $\boldsymbol{x}(t)$. ($K(\boldsymbol{x}(t)) \neq 0$ for all t is a prerequisite for the validity of (4.88).)

If (4.85) is not satisfied, the factor

$$\exp - \int_{t_0}^{t} I \, \mathrm{d}t$$

describes the energy coupling of the disturbances to the basic flux.

CHAPTER 5

Appendix: Matrices

1. The Matrices U_j $(j = 1, 2, 3)$

$$
U_1 = \begin{array}{|c|c|c|c|c|c|c|c|}
\hline
v_1 & \varrho & & & & & & \\
\hline
\frac{a^2}{\varrho} & v_1 & & & & \frac{B_2}{\mu\,\varrho} & \frac{B_3}{\mu\,\varrho} & \frac{1}{\varrho}\,\frac{\partial p}{\partial s} \\
\hline
 & & v_1 & & & -\frac{B_1}{\mu\,\varrho} & & \\
\hline
 & & & v_1 & & & -\frac{B_1}{\mu\,\varrho} & \\
\hline
 & & & & v_1 & & & \\
\hline
 & B_2 & -B_1 & & & v_1 & & \\
\hline
 & B_3 & & -B_1 & & & v_1 & \\
\hline
 & & & & & & & v_1 \\
\hline
\end{array}
$$

$$U_2 = \begin{pmatrix} v_2 & & \varrho & & & & & \\ & v_2 & & & -\frac{B_2}{\mu\,\varrho} & & & \\ \frac{a^2}{\varrho} & & v_2 & & \frac{B_1}{\mu\,\varrho} & & \frac{B_3}{\mu\,\varrho} & \frac{1}{\varrho}\,\frac{\partial p\,\varrho}{\partial s} \\ & & & v_2 & & & -\frac{B_2}{\mu\,\varrho} & \\ & -B_2 & B_1 & & v_2 & & & \\ & & & & & v_2 & & \\ & & B_3 & -B_2 & & & v_2 & \\ & & & & & & & v_2 \end{pmatrix}$$

$$U_3 = \begin{pmatrix} v_3 & & & \varrho & & & & \\ & v_3 & & & -\frac{B_3}{\mu\,\varrho} & & & \\ & & v_3 & & & -\frac{B_3}{\mu\,\varrho} & & \\ \frac{a^2}{\varrho} & & & v_3 & \frac{B_1}{\mu\,\varrho} & \frac{B_2}{\mu\,\varrho} & & \frac{1}{\varrho}\,\frac{\partial p}{\partial s} \\ & -B_3 & & B_1 & v_3 & & & \\ & & -B_3 & B_2 & & v_3 & & \\ & & & & & & v_3 & \\ & & & & & & & v_3 \end{pmatrix}$$

2. The Matrices A_j $(j = 1, 2, 3)$ as Functions of the u_i

It means $\varkappa = \dfrac{\partial s(\varrho, \varrho e)}{\partial \varrho}$, $\quad \lambda = \dfrac{\partial s(\varrho, \varrho e)}{\partial (\varrho e)}$.

$A_1 =$

	1						
$a^2 + \frac{\partial p}{\partial s}\left(\varkappa + \lambda \frac{\boldsymbol{v}^2}{2}\right) - v_1^2$	$v_1\left(2 - \frac{\partial p}{\partial s}\lambda\right)$	$- v_2 \frac{\partial p}{\partial s}\lambda$	$- v_3 \frac{\partial p}{\partial s}\lambda$	$-\frac{B_1}{\mu} \times \left(1 + \frac{\partial p}{\partial s}\lambda\right)$	$\frac{B_2}{\mu}\left(1 - \frac{\partial p}{\partial s}\lambda\right)$	$\frac{B_3}{\mu}\left(1 - \frac{\partial p}{\partial s}\lambda\right)$	$\frac{\partial p}{\partial s}\lambda$
$- v_1 v_2$	v_2	v_1		$-\frac{B_2}{\mu}$	$-\frac{B_1}{\mu}$		
$- v_1 v_3$	v_3		v_1	$-\frac{B_3}{\mu}$		$-\frac{B_1}{\mu}$	
$-\frac{1}{\varrho}(v_1 B_2 - B_1 v_2)$	$\frac{B_2}{\varrho}$	$-\frac{B_1}{\varrho}$		$- v_2$	v_1		
$-\frac{1}{\varrho}(v_1 B_3 - B_1 v_3)$	$\frac{B_3}{\varrho}$		$-\frac{B_1}{\varrho}$	$- v_3$		v_1	
$-\frac{v_1}{\varrho}\varepsilon - \frac{v_1}{\varrho}\left(p + \frac{\boldsymbol{B}^2}{\mu}\right) + v_1\left(a^2 + \frac{\partial p}{\partial s}\left(\varkappa + \frac{\lambda \boldsymbol{v}^2}{2}\right)\right) + \frac{B_1}{\mu\varrho}(\boldsymbol{v}\cdot\boldsymbol{B})$	$\frac{\varepsilon}{\varrho} + \frac{1}{\varrho}\left(p + \frac{\boldsymbol{B}^2}{2\mu}\right) - \frac{B_1^2}{\mu\varrho} - v_1^2 \frac{\partial p}{\partial s}\lambda$	$- v_1 v_2 \frac{\partial p}{\partial s}\lambda - \frac{B_1 B_2}{\mu\varrho}$	$- v_1 v_3 \frac{\partial p}{\partial s}\lambda - \frac{B_1 B_3}{\mu\varrho}$	$- v_1 B_1 \frac{\partial p}{\partial s}\lambda - \frac{1}{\mu}(\boldsymbol{B}\cdot\boldsymbol{v})$	$\frac{1}{\mu}\left(v_1 B_2\left(1 - \frac{\partial p}{\partial s}\lambda\right) - B_1 v_2\right)$	$\frac{1}{\mu}\left(v_1 B_3\left(1 - \frac{\partial p}{\partial s}\lambda\right) - B_1 v_3\right)$	$v_1\left(1 + \frac{\partial p}{\partial s}\lambda\right)$

$$
A_2 = \left(\begin{array}{c|c|c|c|c|c|c|c}
 & & 1 & & & & & \\
\hline
- v_1 v_2 & v_2 & v_1 & & -\frac{B_2}{\mu} & -\frac{B_1}{\mu} & & \\
\hline
\begin{array}{c} a^2 + \frac{\partial p}{\partial s}\left(\varkappa + \lambda \frac{\boldsymbol{v}^2}{2}\right) \\ - v_2^2 \end{array} & - v_1 \frac{\partial p}{\partial s} \lambda & v_2 \left(2 - \frac{\partial p}{\partial s} \lambda\right) & - v_3 \frac{\partial p}{\partial s} \lambda & \frac{B_1}{\mu}\left(1 - \frac{\partial p}{\partial s} \lambda\right) & \begin{array}{c} -\frac{B_2}{\mu} \times \\ \times \left(1 + \frac{\partial p}{\partial s} \lambda\right) \end{array} & \frac{B_3}{\mu}\left(1 - \frac{\partial p}{\partial s} \lambda\right) & \frac{\partial p}{\partial s} \lambda \\
\hline
- v_2 v_3 & & v_3 & v_2 & & -\frac{B_3}{\mu} & -\frac{B_2}{\mu} & \\
\hline
-\frac{1}{\varrho}(v_2 B_1 - B_2 v_1) & -\frac{B_2}{\varrho} & \frac{B_1}{\varrho} & & v_2 & - v_1 & & \\
\hline
 & & & & & & & \\
\hline
-\frac{1}{\varrho}(v_2 B_3 - B_2 v_3) & & \frac{B_3}{\delta} & -\frac{B_2}{\varrho} & & - v_3 & v_2 & \\
\hline
\begin{array}{c} -\frac{v_2}{\varrho}\varepsilon - \frac{v_2}{\varrho}\left(p + \frac{\boldsymbol{B}^2}{2\mu}\right) \\ + v_2 \left(a^2 + \frac{\partial p}{\partial s} \right. \\ \left. \times \left(\varkappa + \lambda \frac{\boldsymbol{v}^2}{2}\right)\right) \\ + \frac{B_2}{\mu \varrho}(\boldsymbol{v} \cdot \boldsymbol{B}) \end{array} & \begin{array}{c} - v_1 v_2 \frac{\partial p}{\partial s} \lambda \\ - \frac{B_2 B_1}{\mu \varrho} \end{array} & \begin{array}{c} \frac{\varepsilon}{\varrho} + \frac{1}{\varrho}\left(p + \frac{\boldsymbol{B}^2}{2\mu}\right) \\ - v_2^2 \frac{\partial p}{\partial s} \lambda - \frac{B_2^2}{\mu \varrho} \end{array} & \begin{array}{c} - v_2 v_3 \frac{\partial p}{\partial s} \lambda \\ - \frac{B_2 B_3}{\mu \varrho} \end{array} & \begin{array}{c} \frac{1}{\mu}\left(v_2 B_1 \left(1 - \frac{\partial p}{\partial s} \lambda\right) \right. \\ \left. - B_3 v_1 \right) \end{array} & \begin{array}{c} - \frac{v_2 B_2}{\mu} \frac{\partial p}{\partial s} \lambda \\ - \frac{1}{\mu}(\boldsymbol{v} \cdot \boldsymbol{B}) \end{array} & \begin{array}{c} \frac{1}{\mu}\left(v_2 B_3 \left(1 - \frac{\partial p}{\partial s} \lambda\right) \right. \\ \left. - B_2 v_3 \right) \end{array} & v_2 \left(1 + \frac{\partial p}{\partial s} \lambda\right)
\end{array}\right)
$$

$$
A_3 = \begin{pmatrix}
 & & & 1 & & & & \\
-v_3 v_1 & v_3 & & v_1 & -\frac{B_3}{\mu} & & -\frac{B_1}{\mu} & \\
-v_3 v_2 & & v_3 & v_2 & & -\frac{B_3}{\mu} & -\frac{B_2}{\mu} & \\
\begin{matrix} a^2 + \frac{\partial p}{\partial s}\left(\varkappa + \lambda \frac{\boldsymbol{v}^2}{2}\right) \\ - v_3^2 \end{matrix} & -v_1 \frac{\partial p}{\partial s}\lambda & -v_2 \frac{\partial p}{\partial s}\lambda & v_3\left(2 - \frac{\partial p}{\partial s}\lambda\right) & \frac{B_1}{\mu}\left(1 - \frac{\partial p}{\partial s}\lambda\right) & \frac{B_2}{\mu}\left(1 - \frac{\partial p}{\partial s}\lambda\right) & \begin{matrix} -\frac{B_3}{\mu} \times \\ \times\left(1 + \frac{\partial p}{\partial s}\lambda\right) \end{matrix} & \frac{\partial p}{\partial s}\lambda \\
-\frac{1}{\varrho}(v_3 B_1 - B_3 v_1) & -\frac{B_3}{\varrho} & & \frac{B_1}{\varrho} & v_3 & & -v_1 & \\
-\frac{1}{\varrho}(v_3 B_2 - B_3 v_2) & & -\frac{B_3}{\varrho} & \frac{B_2}{\varrho} & & v_3 & -v_2 & \\
 & & & & & & & \\
\begin{matrix} -\frac{v_3}{\varrho}\varepsilon - \frac{v_3}{\varrho}\left(p + \frac{\boldsymbol{B}^2}{2\mu}\right) \\ + v_3\left(a^2 + \frac{\partial p}{\partial s}\left(\varkappa + \lambda\frac{\boldsymbol{v}^2}{2}\right)\right) \\ + \frac{B_3}{\mu\varrho}(\boldsymbol{v}\cdot\boldsymbol{B}) \end{matrix} & \begin{matrix} -v_3 v_1 \frac{\partial p}{\partial s}\lambda \\ -\frac{B_3 B_1}{\mu\varrho} \end{matrix} & \begin{matrix} -v_3 v_2 \frac{\partial p}{\partial s}\lambda \\ -\frac{B_3 B_2}{\mu\varrho} \end{matrix} & \begin{matrix} \frac{\varepsilon}{\varrho} + \frac{1}{\varrho}\left(p + \frac{\boldsymbol{B}^2}{2\mu}\right) \\ -v_3^2 \frac{\partial p}{\partial s}\lambda - \frac{B_3^2}{\mu\varrho} \end{matrix} & \frac{1}{\mu}\left(\begin{matrix} v_3 B_1\left(1 - \frac{\partial p}{\partial s}\lambda\right) \\ - B_3 v_1 \end{matrix}\right) & \frac{1}{\mu}\left(\begin{matrix} v_3 B_2\left(1 - \frac{\partial p}{\partial s}\lambda\right) \\ - B_3 v_2 \end{matrix}\right) & \begin{matrix} -v_3 B_3 \frac{\partial p}{\partial s}\lambda \\ -\frac{1}{\mu}(\boldsymbol{v}\cdot\boldsymbol{B}) \end{matrix} & v_3\left(1 + \frac{\partial p}{\partial s}\lambda\right)
\end{pmatrix}
$$

3. The Matrix $M = \frac{\partial(u_1, \ldots u_8)}{\partial(a_1, \ldots a_8)}$

$M =$

1							
v_1	ϱ						
v_2		ϱ					
v_3			ϱ				
				1			
					1		
						1	0
$\frac{\boldsymbol{v}^2}{2} + \frac{\partial(\varrho\, e)}{\partial \varrho}$	$\varrho\, v_1$	$\varrho\, v_2$	$\varrho\, v_3$	$\frac{B_1}{\mu}$	$\frac{B_2}{\mu}$	$\frac{B_3}{\mu}$	$\frac{\partial(\varrho\, e)}{\partial s}$

4. The Matrices $U_j^* = I^*\, U_j\, I^{*-1}$

$U_1^* =$

v_1	a						
a	v_1				$\frac{B_2}{\sqrt{\mu\,\varrho}}$	$\frac{B_3}{\sqrt{\mu\,\varrho}}$	a
		v_1			$-\frac{B_1}{\sqrt{\mu\,\varrho}}$		
			v_1			$-\frac{B_1}{\sqrt{\mu\,\varrho}}$	
				v_1			
	$\frac{B_2}{\sqrt{\mu\,\varrho}}$	$-\frac{B_1}{\sqrt{\mu\,\varrho}}$			v_1		
	$\frac{B_3}{\sqrt{\mu\,\varrho}}$		$-\frac{B_1}{\sqrt{\mu\,\varrho}}$			v_1	
							v_1

$$U_2^* = \left|\begin{array}{cccccccc}
v_2 & & a & & & & & \\
 & v_2 & & & -\frac{B_2}{\sqrt{\mu\varrho}} & & & \\
a & & v_2 & & \frac{B_1}{\sqrt{\mu\varrho}} & & \frac{B_3}{\sqrt{\mu\varrho}} & a \\
 & & & v_2 & & & -\frac{B_2}{\sqrt{\mu\varrho}} & \\
 & -\frac{B_2}{\sqrt{\mu\varrho}} & \frac{B_1}{\sqrt{\mu\varrho}} & & v_2 & & & \\
 & & & & & v_2 & & \\
 & & \frac{B_3}{\sqrt{\mu\varrho}} & -\frac{B_2}{\sqrt{\mu\varrho}} & & & v_2 & \\
 & & & & & & & v_2
\end{array}\right|$$

$$U_2^* = \left|\begin{array}{cccccccc}
v_3 & & & a & & & & \\
 & v_3 & & & -\frac{B_3}{\sqrt{\mu\varrho}} & & & \\
 & & v_3 & & & -\frac{B_3}{\sqrt{\mu\varrho}} & & \\
a & & & v_3 & \frac{B_1}{\sqrt{\mu\varrho}} & \frac{B_2}{\sqrt{\mu\varrho}} & & a \\
 & -\frac{B_3}{\sqrt{\mu\varrho}} & & \frac{B_1}{\sqrt{\mu\varrho}} & v_3 & & & \\
 & & -\frac{B_3}{\sqrt{\mu\varrho}} & \frac{B_2}{\sqrt{\mu\varrho}} & & v_3 & & \\
 & & & & & & v_3 & \\
 & & & & & & & v_3
\end{array}\right|$$

Bibliography

[1] A. JEFFREY and T. TANIUTI, Nonlinear Wave Propagation with Applications to Physics and Magnetohydrodynamics, Academic Press, New York, London, 1964.
[2] S. I. SYROVATSKII, "Magnetohydrodynamics", Uspekhi fiz. Nauk **62** (1958) 247.
[3] J. BRANDSTATTER, An Introduction to Waves, Rays and Radiation in Plasma Media, McGraw-Hill Book Company, 1963.
[4] J. E. ANDERSON, Magnetohydrodynamic Shock Waves, M.I.T. Press, Cambridge, Massachusetts, 1963.
[5] R. COURANT and D. HILBERT, Methods of Mathematical Physics, Interscience Publishers, Inc., New York, London, 1962.
[6] R. COURANT and K. O. FRIEDRICHS, Supersonic Flow and Shock Waves, Interscience Publishers, Inc., New York, London 1948.
[7] A. G. KULIKOVSKII and B. A. LYUBIMOV, "Magnetohydrodynamics", Moscow 1962.
[8] J. BAZER and J. HURLEY, Geometrical Hydromagnetics, J. geophys. Res. **68** (1963) 147.
[9] J. BAZER and J. FLEISCHMANN, Propagation of Weak Hydromagnetic Discontinuities, Phys. Fluids **2** (1959) 366.
[10] H. GRAD, Propagation of Magnetohydrodynamic Waves without Radial Attenuation, in: The Magnetodynamics of Conducting Fluids, D. BERSHADER, Stanford University Press, Stanford (California) 1959.
[11] H. GRAD, Reducible Problems in Magneto-Fluid Dynamic Steady Flows, Rev. mod. Physics **32** (1960) 830.
[12] G. S. S. LUDFORD, The Propagation of Small Disturbances in Hydromagnetics, J. Fluid Mechan. **5** (1959) 387.
[13] V. M. KONTOROVICH, "On the interaction of small disturbances with discontinuity surfaces in magnetohydrodynamics and the stability of shock waves", JETP **35** (1958) 1216.
[14] M. J. LIGHTHILL, Studies on Magneto-Hydrodynamic Waves and Other Anisotropic Wave Motions, Philos. Trans. Roy. Soc. London, Ser. A, **252** (1960) 397.
[15] W. B. KUNKEL, Remarks on Magnetoacoustic Waves, Phys. Fluids **5** (1962) 867.
[16] J. BAZER and W. B. ERICSON, Hydromagnetic Shocks, Astrophysic. J. **129** (1959) 758.
[17] W. E. FRANCIS, M. I. GREEN and A. J. DESSLER, Hydromagnetic Propagation of Sudden Commencements of Magnetic Storms, J. Geophys. Res. **64** (1959) 1643.
[18] A. J. DESSLER, W. E. FRANCIS and E. N. PARKER, Geomagnetic Storm Sudden Commencement Rise Times, J. Geophys. Res. **65** (1960) 2715.
[19] E. J. STEGELMANN and C. H. VON KENSCHITZKI, On the Interpretation of Sudden Commencement of Geomagnetic Storms, J. Geophys. Res. **69** (1964) 139.
[20] R. LUST, "Über die Ausbreitung von Wellen in einem Plasma", Fortschritte der Physik **7** (1959) 503.
[21] H. KAUTZLEBEN and O. LUCKE, "Über Stoßwellen", Beitr. Plasmaphysik **3** (1962) 143.
[22] R. V. POLOVIN, "Shock waves in magneto-hydrodynamics", Uspekhi fiz. Nauk **9** (1961) 526.
[23] R. V. POLOVIN, Evolution Conditions in Hydrodynamics and Magneto-Hydrodynamics in: "Plasma physics", Vol. 3, Kiev 1963.
[24] A. I. AKHIESER, G. I. LYUBARSKII and R. V. POLOVIN, "Stability of shock waves in magnetohydrodynamics", JETP **35** (1958) 731.

H. W. Drawin

Validity Conditions for Local Thermodynamic Equilibrium

CHAPTER 1

Introduction

The concept of (complete) local thermodynamic equilibrium (complete L.T.E.) plays an important role in the field of plasma spectroscopy. Indeed, many plasma spectroscopical measurements and evaluations are based on the assumption that complete L.T.E. exists throughout the plasma. When complete L.T.E. prevails the plasma state at any spatial point r at time t can be calculated from very simple and very general equations. More generally speaking, all laws and thermodynamic relations for *complete thermodynamic* equilibrium also hold in the case of *complete local thermodynamic equilibrium,* the only exception from this rule makes Planck's radiation law. The laws valid for plasmas in complete thermodynamic equilibrium (C.T.E.) and those valid for complete local thermodynamic equilibrium have been listed in Table 1.1.

Plasmas in C.T.E. only exist under very special conditions, e.g. the plasma state in the interior of stars is in C.T.E. The core of plasmas formed during the short time interval of strong explosions can also assumed to be in C.T.E. In the latter case the temperatures are relatively low and the densities are high. Low-temperature plasmas in C.T.E. can be obtained by enclosing them in a black-body box of wall temperature T_{wall}. The kinetic energy distribution of all plasma particles is then given by a Maxwell-Boltzmann distribution function of kinetic temperature T_{kin}, the radiation field has a distribution according to Planck's law with a radiation temperature T_{rad}, the excited particles are populated according to a Boltzmann distribution of excitation temperature T_{exc}, and the molecules, atoms, ions and electrons are distributed according to the mass action law, where the mass action constant depends on the reaction temperature $T_{\text{react.}}$. In C.T.E., all these temperatures are equal to each other, hence

$$T = T_{\text{wall}} = T_{\text{kin}} = T_{\text{rad}} = T_{\text{excit}} = T_{\text{react.}} \,. \tag{1.1}$$

However, under laboratory conditions the temperature, and hence the electron density of C.T.E. plasmas can not be very high, since all material walls already evaporate at moderately high temperatures. Plasmas in C.T.E. as defined in Table 1.1 have no practical importance, and laboratory plasmas in C.T.E. are practically non existent, due to the following three reasons:

1. all actual plasmas show a temperature distribution, the outer parts are always at lower temperatures than the inner regions. A unique thermodynamic temperature T does not exist. It follows from this that

Table 1.1. Laws for Plasmas in C.T.E. and complete L.T.E.

Distributions	Complete thermodynamic equilibrium (C.T.E.)	Complete local thermo-dynamic equilibrium (L.T.E.)
Temperature	$\nabla T = 0$	$\nabla T \neq 0$
Photons	Planck law $0 < \nu < \infty$ $T = T_{\text{radiation}}$	no Planck law for $0 < \nu < \infty$ $T \neq T_{\text{radiation}}$
Energy, velocity	Maxwell-Boltzmann $T = T_{\text{kinetic}}$	Maxwell-Boltzmann $T = T_{\text{kinetic}}$
Dissociation products	Guldberg-Waage $T = T_{\text{react.}} = T_{\text{kinetic}}$	Guldberg-Waage $T = T_{\text{react.}} = T_{\text{kinetic}}$
Ionisation products	Saha-Eggert $T = T_{\text{react.}} = T_{\text{kinetic}}$	Saha-Eggert $T = T_{\text{react.}} = T_{\text{kinetic}}$
Population of discrete energy levels	Boltzmann $T = T_{\text{excit}} = T_{\text{react.}} = T_{\text{kinetic}}$	Boltzmann $T = T_{\text{excit}} = T_{\text{react.}} = T_{\text{kinetic}}$
Thermodynamic Functions	All formulas valid for C.T.E. also hold for L.T.E. plasmas with $T = T_{\text{excit}} = T_{\text{react.}} = T_{\text{kinetic}}$	

2. there does not exist a radiation field of unique radiation temperature, there does not exist a unique kinetic temperature.
3. very often the walls enclosing plasmas are transparent over a wide spectral range. Then radiation escapes, and consequently the radiation field will be diluted compared to the field fulfilling the Planck radiation law. Due to the escape of radiation it is then not possible to describe the radiation field by a mean radiation temperature valid for all frequencies. Only in small frequency regions the Planck law can be fulfilled. These frequencies correspond to the resonance transitions between bound energy states.

Although most plasmas are not in the state of C.T.E. there exists a large class of plasmas which fulfill for each volume element all thermodynamic distribution laws derived for complete thermodynamic equilibrium, except Planck's radiation law. Such plasmas are said to be in *local* thermodynamic equilibrium (L.T.E.). The laws valid for L. T.E. plasmas have been listed in the third column of Table 1.1. One sees that the concept of local thermodynamic equilibrium is a very powerful one. All formulae valid for C.T.E. plasmas also apply to L.T.E. plasmas, except Planck's radiation law. The radiation, however, plays an important role in the establishment of local thermodynamic equilibrium at low electron densities (see later). The interaction of the radiation field (produced by the plasma) with the plasma particles is treated via the radiative transfer equations. When the radiation field is diluted deviations from L.T.E. have to be expected for electron densities $n_e < 10^{17}$ cm^{-3} in the case of hydrogen.

L.T.E. plasmas have been treated in many books and articles, an excellent review has been given by Neumann earlier in this book where the reader may find corresponding references.

All plasmas not fulfilling the L.T.E. laws — as given in the third column of Table 1 — are said to be non-equilibrium plasmas, often referred to as non-L.T.E. plasmas. Their properties can only be described by so-called rate equations containing the Einstein coefficients A_{nm}, B_{nm} for spontaneous emission and absorption and the collisional rate coefficients $n_k \langle \sigma v \rangle_k^l$ for collisions between particles of kind k and l

$$\langle \sigma v \rangle_k^l = \frac{1}{n_k} \int_\omega \sigma_k^l(\boldsymbol{w}_k^l)\, f_k(\boldsymbol{w}_k^l)\, \boldsymbol{w}_k^l \, \mathrm{d}\boldsymbol{w} \tag{1.2}$$

as a parameter. In this last equation, $f_k(\boldsymbol{w}_k^l)$ represents the velocity distribution function of the relative velocity $\boldsymbol{w}_k^l$ between the plasma particles of kind k and l, and $\sigma_k^l(\boldsymbol{w}_k^l)$ denotes the cross section as a function of $\boldsymbol{w}_k^l$. For collision processes and rate coefficients the reader is referred to the special literature. A review may for instance be found in the articles of Drawin [1], [2] and Shahin [3], and in the book of McDaniel [4]*. The rate coefficients can only then be calcualted when σ *and* f are known. The distribution function f is obtained by solving the Boltzmann collision equation with the proper initial and boundary conditions. The Boltzmann collision equation contains the collision frequency per phase-space-element $\mathrm{d}\boldsymbol{r}\,\mathrm{d}\boldsymbol{w}$. This means that in a consistent treatment of non-equilibrium problems the rate equations for the excited levels must be coupled with the Boltzmann equation in the μ-space, and the rate equations and the Boltzmann equation must simultaneously be solved for each spatial volume element $\mathrm{d}V$ $(= \mathrm{d}x_1\, \mathrm{d}x_2\, \mathrm{d}x_3)$ centered at $\boldsymbol{r}$. The number of particles in this $\mathrm{d}V - \mathrm{d}w$ element is equal to $f(\boldsymbol{r}, \boldsymbol{w}, t)\, \mathrm{d}V\, \mathrm{d}w$. Further, one has to account for quasi-neutrality of each volume element, and finally one has to account for radiative transfer. A rigorous treatment of such a coupled system of equations poses considerable mathematical difficulties. In practice, one introduces many simplifying assumptions, otherwise the mathematical and numerical evaluations would be impossible or take so much time so that the efforts would be in no relation to the results obtained. For more details the reader is refered to a recent review article by Drawin [5].

For practical applications it is very useful to have validity criteria for *complete* and for *partial L.T.E.* There is a vast amount of literature dealing with this subject; only a few authors may be quoted: Giovanelli [6] showed that for a given electron density and temperature all excited levels lying between the ionization limit and a lowest bound state with principal quantum number i'

* The reader is also referred to the new edition of the former book of Massey and Burhop. The new edition appears in four volumes under the title: H. S. W. Massey, E. H. S. Burhop, "Electronic and Ionic Impact Phenomena", Clarendon Press, Oxford. Vol. 1 and Vol. 2 appeared in 1969, Vol. 3 in 1971, Vol. 4 is in preparation.

always show a Saha-population with respect to the ground state of the next higher ionization stage. For hydrogen, for instance, he found $i' = 11$ for $n_e = 1 \cdot 10^{10}\,\text{cm}^{-3}$ and $T_e = 5 \cdot 10^3\,°\text{K}$. For $i < i'$ the population densities $n_{i<i'}$ lie below those given by the Saha-Eggert equation in the form

$$\frac{n_e\, n_{z,1}^{(\text{Saha})}}{n_{z-1,i}^{(\text{Saha})}} = \frac{2\, g_{z,1}}{g_{z-1,i}} \frac{(2\pi m_e k T_e)^{3/2}}{h^3} \exp\left(-\frac{E_{z-1,z}^{(i)} - \Delta E_{z-1,z}}{kT_e}\right), \tag{1.3}$$

where $n_{z-1,i}^{(\text{Saha})}$ is the number density of quantum state i, and $n_{z,1}^{(\text{Saha})}$ that of the ground state of the next higher ionization stage. $\Delta E_{z-1,z}$ is the lowering of the ionization energy due to the electric microfield, see for instance the section by Neumann.

Deviations from L.T.E. in star atmospheres, especially that of the sun has been discussed by Unsöld [7] and more recently by Pagel [8]*. Griem [9], [10] using generalized classical cross sections for electronic collisions derived simple formulas giving an idea wether the applicability of the L.T.E. concept is justified or not. Biberman and Ulyanov [11], Vorobev [12], and Lagarkov [13] discuss in length the departure from L.T.E. caused by the radiative transport in plasmas showing spatial density and temperature gradients. The same problem is treated by Thomas [14], Jefferies and Thomas [15], [16], Hummer [17], [18], and many others in connection with departures from L.T.E. in star atmospheres. Sampson [19] using quantum mechanical cross sections derived a relation between the electron density n_e, the electron temperature T_e and the ionization energy $E_{z-1,z}$, which must be fulfilled in order to ensure that an ionized gas is collision-dominated in all bound-bound transitions. The corresponding relation equals the condition for the establishment of complete L.T.E. Further references are Wilson [20], McWhirter [21], Drawin [22], Deutsch [23], and many others.

The problem of possible departures from L.T.E. has already briefly been discussed by Neumann in this book. The subject is treated here in more detail and on a general scale taking recent publications into account. Possible departures from a Maxwell distribution will not be discussed.** When a plasma is close to the state of L.T.E. the number density is always sufficiently high to assure a Maxwellian velocity distribution with $T_e = T_{\text{atoms}} = T_{\text{ions}}$.

Molecules will not be discussed in this chapter.

* The problem is also treated in the book of J. T. Jefferies: "Spectral Line Formation", Blaisdell Publ. Comp., Waltham/Massachusetts (U.S.A.) 1968.

** The problem has briefly been treated by Neumann (see his article in this volume). The reader is also referred to the following recent papers: J. Oxenius, Z. Naturforschg. **25a** (1970) 101, **25a** 1302; J. F. Shaw, M. Mitchner and L. H. Krüger, Phys. Fluids **13** (1970) 325, **13** (1970) 339. Departures from a Maxwellian distribution are possible when the degree of ionization is smaller than approximately 0.5%. The degree of this departure is not only a function of the degree of ionization, it also depends on other quantities such as the electron and neutral particle densities and the electron and gas temperatures. — The problem is treated elswhere (H. W. Drawin, High temperatures—High pressures **2** (1970) 359; S. Suckewer, Physics Letters **33 A** (1970) 110; H. W. Drawin, Physics Letters **34 A** (1971) 151).

CHAPTER 2

Partial L. T. E. in Time-Independent and Homogeneous Optically Thin Plasmas

We consider the most unfavorable case, namely an optically thin plasma. In low-density plasmas the radiative decay rates of the lower-lying levels are much larger than those for superelastic collisions. This is the reason why the lowerlying states are underpopulated with respect to the ground state, compared to the equilibrium population. For the high-lying levels the collision rates become larger and the radiative decay rates smaller so that one finally arrives at a level for which the collisional rates are more frequent than the radiative ones. However, one can not conclude that a level, say i (see figure 2.1),

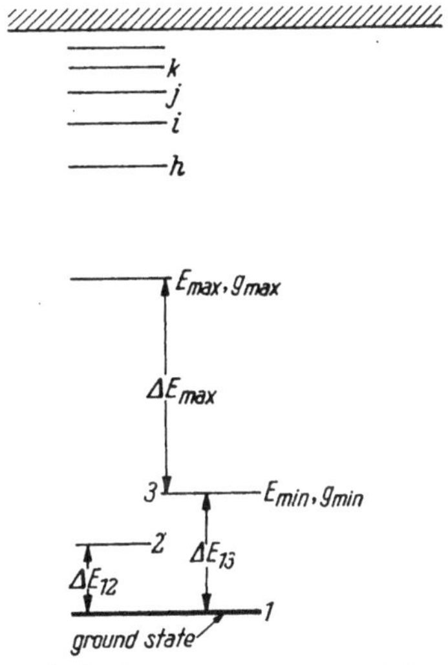

Figure 2.1. Illustration of the largest energy gap $\Delta E_{\max}$ within a level system

is already in partial L.T.E. when its excitation rate is much larger than the radiative decay rate down to lower lying levels. The fact that the collision rate $n_e\, n_i\, \langle \sigma^{(e)}\, v_e \rangle_{i\to j}$ (with $i < j$) is larger than the radiative decay rate $n_i \sum_{h<i} A_{ih}$, at a first glance, is of signification as regards partial L.T.E. In a complicated level system one may eventually find higher excited levels $k > j$ having radiative decay rates which are larger than the collisional de-excitation rates, see e.g. the level system of O II. Due to the mutual coupling of all states, the level i

may eventually not be in partial L.T.E., although a first estimate of the collisional and radiative rates seemed to indicate partial L.T.E.

Denoting by $n_i^{(\text{actual})} = n_i$ the population densities calculated on the basis of the collisional-radiative model, and by $n_i^{(\text{Saha})}$ the corresponding Saha-density as given by eq. (1.3) one may define a coefficient b_i by the relation (Menzel [24])

$$n_i = b_i \, n_i^{(\text{Saha})} , \tag{2.1}$$

where n_i denotes the actual number density. A level is said to be in partial L.T.E. when b_i is equal to unity within 10% or better. When this condition is fulfilled one may be sure that spectroscopically measured population densities n_i permit a determination of the electron density and of the electron temperature by means of the Saha-Eggert eq. (1.3).

In order to define a validity condition which ensures that $b_i = 1$ (within 10% or better) one may climb down the level system beginning at the ionization limit. Due to the large two-body ionization and three-body collisional recombination rates the highly excited states will be in a collision-dominated equilibrium with the free electrons. Thus, the condition $b_i = 1$ is fulfilled.

In climbing further downward the direct collision coupling of the excited states with the continuum of the free electrons will become looser, however, the collision excitation and de-excitation rates between the closely spaced levels are still sufficiently large, so that a Saha-population is accomplished via these many closely spaced levels. The collision-dominated coupling between the excited states ceases when one arrives at a level $j = i + 1$ for which the spontaneous decay rate $n_j \sum\limits_{h<j} A_{jh}$ is of the same order as or larger than the collisional decay rate $n_j \, n_e \, \langle \sigma^{(e)} \, v_e \rangle_{j \to i = j-1}$. It depends now on the level $i = j - 1$ whether the enhanced radiative decay of level j can yet be compensated by a sufficient number of excitation processes $i \to j$, or not. It is evident that the population densities n_j and n_i can be stabilized by collisions when the following relations are fulfilled

$$n_e \, n_j \, \langle \sigma^{(e)} \, v_e \rangle_{j \to i} \gg n_j \sum_{h<j} A_{jh} , \tag{2.2a}$$

$$n_e \, n_i \, \langle \sigma^{(e)} \, v_e \rangle_{i \to j} \gg n_i \sum_{h<i} A_{ih} . \tag{2.2b}$$

By putting the collision rates ten times larger than the radiative ones, one may be sure that partial L.T.E. will be fulfilled within 10% or better for all levels equal to or larger than i, hence the conditions:

$$n_e \geq 10 \, \frac{\sum\limits_{h<j} A_{jh}}{\langle \sigma^{(e)} \, v_e \rangle_{j \to i}} \tag{2.3a}$$

and

$$n_e \geq 10 \, \frac{\sum\limits_{h<i} A_{ih}}{\langle \sigma^{(e)} \, v_e \rangle_{i \to j}} \tag{2.3b}$$

In simple level systems (such as hydrogen) with "wellbehaving" atomic constants the condition (2.3a) is already contained in the condition (2.3b) generally used for estimating the electron density necessary for the establishment of partial L.T.E. in completely stationary homogeneous optically thin plasmas.— However, in quasi-stationary plasmas which show recombination or ionization relaxation this may not be true (see Section VII).

The right-hand sides of the relations (2.3a) and (2.3b) are functions of the quantum numbers of the levels involved and of the electron temperature. Under actual conditions the electron density may be slightly smaller than required by the above conditions due to the simultaneous mutual coupling of the excited states, and also due to radiative absorption (see Section V).

Using for neutral atoms and ions two different types of cross sections (the first type being zero or nearly zero at threshold, the other one giving finite values there, see Van Regemorter [25]), one will obtain two different expressions for the validity of partial L.T.E.

The general relation for the excitation cross section (allowed transitions) can be put into the following form (Seaton [26], Drawin [1]):

$$\sigma_{ij}^{(e)}(U_{ij}) = 4\pi a_0^2 \left(\frac{E_1^H}{E_j - E_i}\right)^2 f_{ij}\, g(U_{ij}), \tag{2.4}$$

where U_{ij} equals the translational energy E of the electrons in unit $E_j - E_i$, hence

$$U_{ij} = \frac{E}{E_j - E_i} \equiv \frac{E}{E_{ij}}, \qquad E = \frac{1}{2} m_e v_e^2 .$$

f_{ij} is the absorption oscillator strength, and E_1^H is the ionization energy of the hydrogen ground state atom ($E_1^H = 13.58$ eV). The function $g(U_{ij})$ can be written in the following forms (Drawin [1]).

For neutral atoms:

$$g(U_{ij}) \cong \frac{U_{ij} - 1}{U_{ij}^2} \ln(1.25\, \beta_{ij}\, U_{ij}) . \tag{2.5a}$$

For ions:

$$g(U_{ij}) \cong \begin{cases} 0.302 & \text{for} \quad 1 \le U_{ij} \le 3.85\,, \\ \dfrac{U_{ij} - 1}{U_{ij}^2} \ln(1.25\, \beta_{ij}\, U_{ij}) & \text{for} \quad U_{ij} > 3.85\,. \end{cases} \tag{2.5b}$$

The coefficients β_{ij} take values between 1 and 5, for the numerical evaluation one may put $\beta_{ij} = 2$ for neutrals and $\beta_{ij} = 1$ for ions. The formulas (2.5a) and (2.5b) agree well with the quantum mechanical calculations of other authors at the low and the high energy side in the energy dependence and in the absolute values.

Integrating eq. (2.4) over a Maxwellian velocity distribution of the electrons yields for the rate coefficient for excitation $i \to j$

$$\langle \sigma^{(e)} v_e \rangle_{i \to j} = 8\pi a_0^2 \left(\frac{E_1^H}{E_j - E_i}\right)^2 \left(\frac{2kT_e}{\pi m_e}\right)^{1/2} f_{ij}\, u_{ij}\, \Psi(\beta_{ij}, u_{ij}) \tag{2.6a}$$

and numerically in units [cm³ s⁻¹]

$$\langle\sigma^{(e)} v_e\rangle_{i\to j} = 8.685\cdot 10^{-8} \frac{u_1^{3/2}}{z^3} \frac{f_{ij}}{u_{ij}} \cdot \begin{cases} \Psi_1(\beta_{ij}, u_{ij}) & \text{for atoms ,} \\ \Psi_2(\beta_{ij}, u_{ij}) & \text{for ions .} \end{cases} \qquad (2.6\text{b})$$

with

$$u_1 = z^2 E_1^H/kT_e \; ; \qquad u_{ij} = (E_j - E_i)/kT_e \equiv E_{ij}/kT_e \; ;$$

$E_1^H = 13.58$ eV = ionization energy of hydrogen ground state atoms;

$$\Psi_1(\beta_{ij}, u_{ij}) = \int_{u=u_{ij}}^{\infty} \left[\left(1 - \frac{u_{ij}}{u}\right) e^{-u} \ln\left(1.25\, \beta_{ij} \frac{u}{u_{ij}}\right)\right] du \; ;$$

$$\Psi_2(\beta_{ij}, u_{ij}) = \frac{0.302}{u_{ij}} \int_{u=u_{ij}}^{3.85\, u_{ij}} u\, e^{-u}\, du + \int_{3.85\, u_{ij}}^{\infty} \left[\left(1 - \frac{u_{ij}}{u}\right) e^{-u} \ln\left(1.25\, \beta_{ij} \frac{u}{u_{ij}}\right)\right] du \, .$$

Table 2.1. Numerical values of the functions $\Psi_1(\beta, x)$ and $\Psi_2(\beta, x)$

x	$\Psi_1(1, x)$	$\Psi_1(2, x)$	$\Psi_2(1, x)$	x	$\Psi_1(1, x)$	$\Psi_1(2, x)$	$\Psi_2(1, x)$
1.10⁻⁵	1.11, +1	1.18, +1	1.12, +1	8	1.43, −5	3.87, −5	1.14, −4
1.10⁻⁴	8.85, +0	9.54, +0	8.85, +0	10	1.48, −6	4.22, −6	1.51, −5
1.10⁻³	6.53, +0	7.22, +0	6.53, +0	12	1.61, −7	4.77, −7	2.01, −6
1.10⁻²	4.16, +0	4.82, +0	4.16, +0	14	1.81, −8	5.54, −8	2.69, −7
1.10⁻¹	1.76, +0	2.26, +0	1.79, +0	16	2.10, −9	6.57, −9	3.61, −8
0.4	5.90, −1	8.59, −1	7.29, −1	18	2.47, −10	7.91, −10	4.86, −9
1.0	1.55, −1	2.58, −1	2.21, −1	20	2.96, −11	9.64, −11	6.53, −10
1.5	6.37, −2	1.14, −1	1.10, −1	22	3.58, −12	1.18, −11	8.81, −11
2.0	2.87, −2	5.47, −2	6.13, −2	24	4.39, −13	1.47, −12	1.19, −11
3.0	6.78, −3	1.42, −2	2.07, −2	26	5.42, −14	1.84, −13	1.60, −12
4.0	1.79, −3	4.01, −3	6.91, −3	28	6.73, −15	2.31, −14	2.16, −13
6.0	1.48, −4	3.71, −4	8.73, −4	30	8.43, −16	2.92, −15	2.92, −14

Read: 1.11, +1 = 1.11 · 10¹, etc. ...

For the numerical values of the functions $\Psi_1(\beta, x)$ and $\Psi_2(\beta, x)$, see Table 2.1. To see the influence of the parameter β on Ψ_1 we chose two different values for β. The function Ψ_2 is nearly independent of β, if $x > 1$. In the following, $\Psi_1(2, x) \equiv \Psi_1(x)$ and $\Psi_2(1, x) \equiv \Psi_2(x)$. For practical purposes we additionally introduce the functions

$$\Phi_1(\beta, x) = [\Psi_1(\beta, x) \exp(+x)]^{-1} \; ; \qquad \Phi_2(\beta, x) = [\Psi_2(\beta, x) \exp(+x)]^{-1}$$

of which the numerical values have been listed in Table 2.2. Using the method of detailed balancing one obtains for the rate coefficient for de-excitation $j \to i$ (with $i < j$):

$$\langle\sigma^{(e)} v_e\rangle_{j\to i} = 8.685\cdot 10^{-8} \frac{g_i}{g_j} \frac{u_1^{3/2}}{z^3} \frac{f_{ij}}{u_{ij}} \cdot \begin{cases} \Phi_1^{-1}(u_{ij}) & \text{for atoms} \\ \Phi_1^{-1}(u_{ij}) & \text{for ions} \end{cases} \quad \text{cm}^3\,\text{s}^{-1} \qquad (2.7)$$

The g_i and g_j are the statistical weights of the lower and upper states, respectively.

Table 2.2. Numerical values of the functions $\Phi_1(\beta = 2, x)$ and $\Phi_2(\beta = 1, x)$

x	$\Phi_1(2, x) \equiv \Phi_1(x)$	$\Phi_2(1, x) \equiv \Phi_2(x)$	x	$\Phi_1(x)$	$\Phi_2(x)$
1.10^{-5}	8.44, −2	8.96, −2	12.0	1.31, 1	3.06, 0
1.10^{-4}	1.05, −1	1.13, −1	14.0	1.53, 1	3.09, 0
1.10^{-3}	1.38, −1	1.53, −1	16.0	1.75, 1	3.12, 0
1.10^{-2}	2.05, −1	2.38, −1	18.0	1.96, 1	3.14, 0
1.10^{-1}	4.00, −1	5.06, −1	20.0	2.18, 1	3.15, 0
0.4	7.80, −1	1.02, 0	22.0	2.40, 1	3.17, 0
1.0	1.43, 0	1.66, 0	24.0	2.61, 1	3.18, 0
2.0	2.47, 0	2.22, 0	26.0	2.83, 1	3.19, 0
3.0	3.52, 0	2.48, 0	28.0	3.05, 1	3.20, 0
4.0	4.57, 0	2.65, 0	30.0	3.27, 1	3.21, 0
5.0	5.62, 0	2.76, 0	35.0	3.81, 1	3.22, 0
6.0	6.69, 0	2.84, 0	40.0	4.36, 1	3.23, 0
7.0	7.75, 0	2.90, 0	45.0	4.90, 1	3.24, 0
8.0	8.82, 0	2.94, 0	50.0	5.45, 1	3.25, 0
9.0	9.90, 0	2.98, 0	60.0	6.54, 1	3.26, 0
10.0	1.10, 1	3.01, 0	100.0	4.14, 2	3.28, 0

Read: 8.44, − 2 = $8.44 \cdot 10^{-2}$, etc. ...

Putting the relations (2.6) and (2.7) into the inequalities (2.3a, b) leads to the following conditions for n_e

$$n_e \geq 1.15 \cdot 10^8 \, z^3 \frac{g_j}{g_i} \frac{u_{ij}}{u_1^{3/2}} \frac{\sum\limits_{h<j} A_{jh}}{f_{ij}} \cdot \begin{cases} \Phi_1(u_{ij}) \text{ cm}^{-3} & \text{for atoms } (z = 1), \\ \Phi_2(u_{ij}) \text{ cm}^{-3} & \text{for ions } \quad (z < 1) \end{cases} \tag{2.8a}$$

and

$$n_e \geq 1.15 \cdot 10^8 \, z^3 \frac{u_{ij}}{u_1^{3/2}} \frac{\sum\limits_{h<j} A_{jh}}{f_{ij}} \cdot \begin{cases} [\Psi_1(u_{ij})]^{-1} \text{ cm}^{-3} & \text{for atoms } (z = 1), \\ [\Psi_2(u_{ij})]^{-1} \text{ cm}^{-3} & \text{for ions } \quad (z < 1), \end{cases} \tag{2.8b}$$

which must be satisfied when partial L.T.E. is to exist within 10% or better down to the level i. A_{jh} is in s^{-1}, and u_{ij} equals the energy difference (in units kT_e) between two adjacent levels i and $j = i + 1$, hence $u_{ij} = (E_j - E_i)/kT_e$. In this form, the validity condition for partial L.T.E. is very general and may be applied to any level i in any atomic or ionic level system. As has already been mentioned, the eqs. (2.3a) or (2.8a) are sufficiently good conditions for the homogeneous stationary state provided the temperature is not too low. Thus, we only consider eq. (2.8a) for the further discussion.

Eq. (2.8a) can be put into a simpler form by making the following approximations for *sufficiently* excited hydrogen and hydrogen-like levels of principal quantum numbers i and $j = i + 1$:

$$\sum_{h<j} A_{jh} \cong 1.6 \cdot 10^{10} \frac{z^4}{i^{9/2}} \text{ s}^{-1}; \quad f_{ij} \cong 0.5\, i; \quad u_{ij} \cong \frac{2\, u_1}{i^3}; \quad \frac{g_j}{g_i} \cong 1.$$

Thus, one obtains the condition

$$n_e \geq 7.4 \cdot 10^{18} \frac{z^7}{i^{17/2}} \left(\frac{kT_e}{Z^2 E_1^H}\right)^{1/2} \cdot \begin{cases} \Phi_1(x_i) \text{ cm}^{-3} & \text{for atoms ,} \\ \Phi_2(x_i) \text{ cm}^{-3} & \text{for ions .} \end{cases} \tag{2.8c}$$

For sufficiently large values of i the argument x_i of the function $\Phi_1(x_i)$ and $\Phi_2(x_i)$ can also be put equal to $2\,u_1/i^3$. For low principal quantum numbers, however, one should use the true energy difference, hence

$$x_i = z^2 E_1^H \left(\frac{1}{i^2} - \frac{1}{(i+1)^2}\right) \Big/ kT_e \,.$$

Eq. (2.8c) differs from Griem's formula by a factor $\Phi_1(x_i)$ for atoms and by a factor $\Phi_2(x_i)$ for ions. The practical consequence is that formula (2.8c) gives slightly larger values for n_e than Griem's formula when x_i is large. For small values of the quantity x_i eq. (2.8c) leads to smaller values than Griem's expression.

We now compare eq. (2.8c) with the values as obtained when the population densities are calculated from a coupled system of rate equations. In deriving eq. (2.8c) no collisional-radiative coupling between the more distant levels has been taken into account. It is interesting to see whether these coupling processes favor the establishment of partial L.T.E. or not.

It has been shown (Mc Whirter and Hearn [29], Drawin and Emard [41]) that the coefficient $b_{z-1,i}$ can be put into the following form when electronic collision processes and radiative transitions only are taken into account

$$b_{z-1,i} = r^{(0)}_{z-1,i} + r^{(1)}_{z-1,i} \frac{n_{z-1,1}}{n^{(\mathrm{Saha})}_{z-1,1}} \,. \tag{2.9}$$

The population coefficients $r^{(0)}_{z-1,i}$ and $r^{(1)}_{z-1,i}$ are complicated functions of n_e, T_e, i, and the radiative absorption processes. $n_{z-1,1}$ denotes the actual number density of the ground state particles, and $n^{(\mathrm{Saha})}_{z-1,1}$ is the corresponding equilibrium number density for the same values of n_e and T_e Numerical values of $r^{(0)}_{z-1,i}$ and $r^{(1)}_{z-1,i}$ may be found in the following publications:

for hydrogen atoms	Ref. [41], [72], [73],
hydrogen-like ions	Refs. [29], [41],
helium atoms	Ref. [41].

The coefficient $r^{(0)}_{z-1,i}$ represents the Saha decrement which is obtained when collision excitation from the ground level is assumed to be absent. (This case is, for instance, realized in a cold recombining plasma). The population coefficient $r^{(1)}_{z-1,i}$ accounts for collision excitation from the ground level, with the additional condition that recombination processes are absent. At low principal quantum numbers the second term of eq. (2.9) can only be omitted when the temperatures are low. In the case of a stationary homogeneous plasma it must always be taken into account. For large values of i, however, $b_{z-1,i}$ is practically equal to $r^{(0)}_{z-1,i}$, since $r^{(1)}_{z-1,i}$ rapidly decreases with increasing values of i.

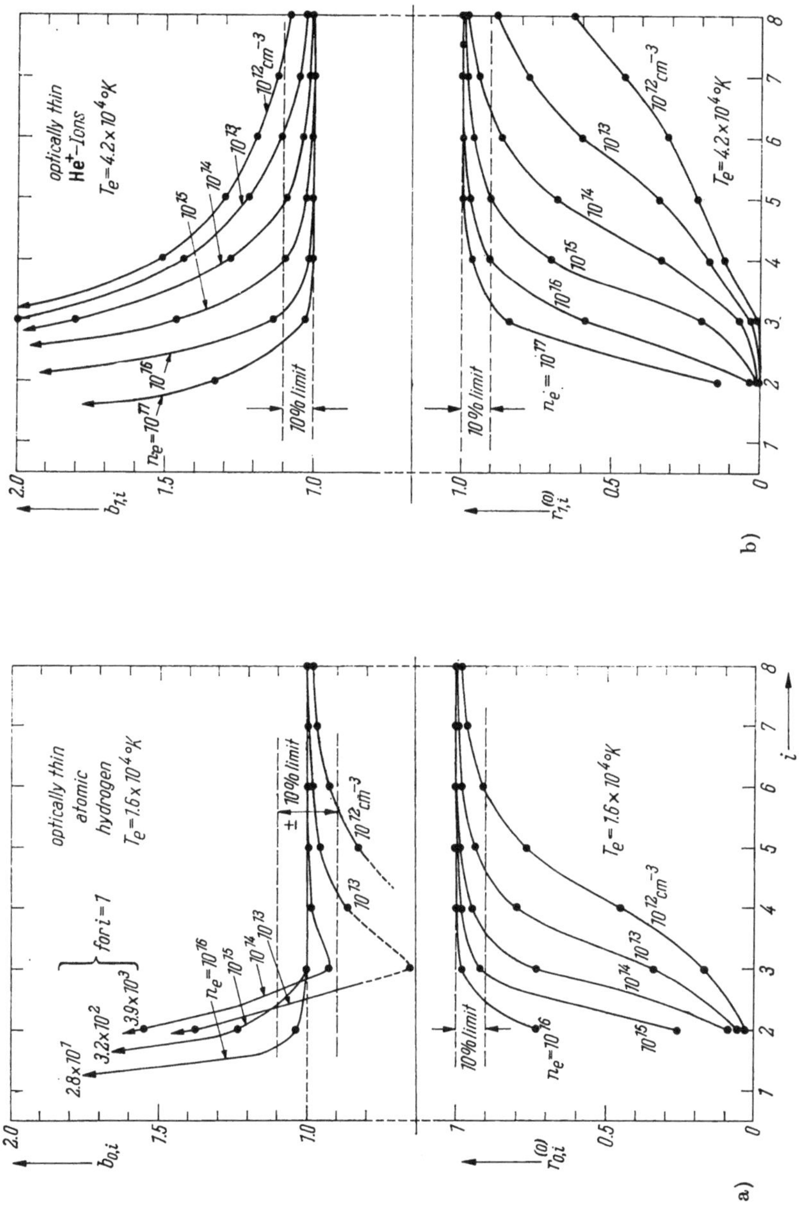

Figures 2.2a, b. Saha-decrement $b_{z-1,\,i}$ and population coefficient $r^{(0)}_{z-1,\,i}$

a) for atomic hydrogen ($z = 1$) b) for He$^+$-ions ($z = 2$)

All transitions are assumed to be optically thin. The coefficient $b_{z-1,\,i}$ (upper curves) refers to a completely homogeneous stationary plasma. The lower curves represent the relative population when collision excitation from the ground level is negligibly small (so-called recombination-dominated régime)

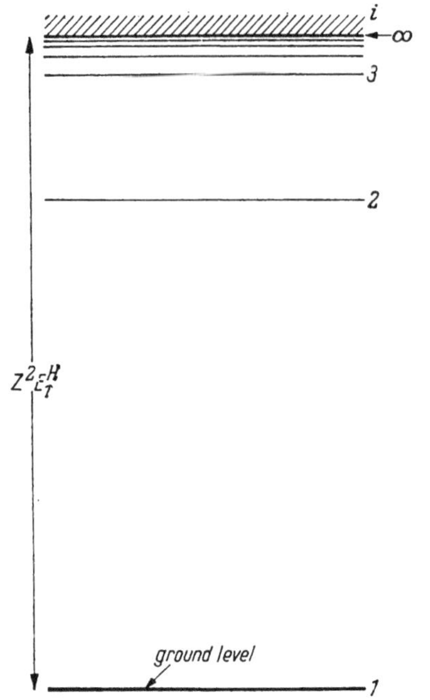

Figure 2.3. Model used for calculating the population coefficients $r_{z-1,i}^{(0)}$ and $r_{z-1,i}^{(1)}$ of H and He^+

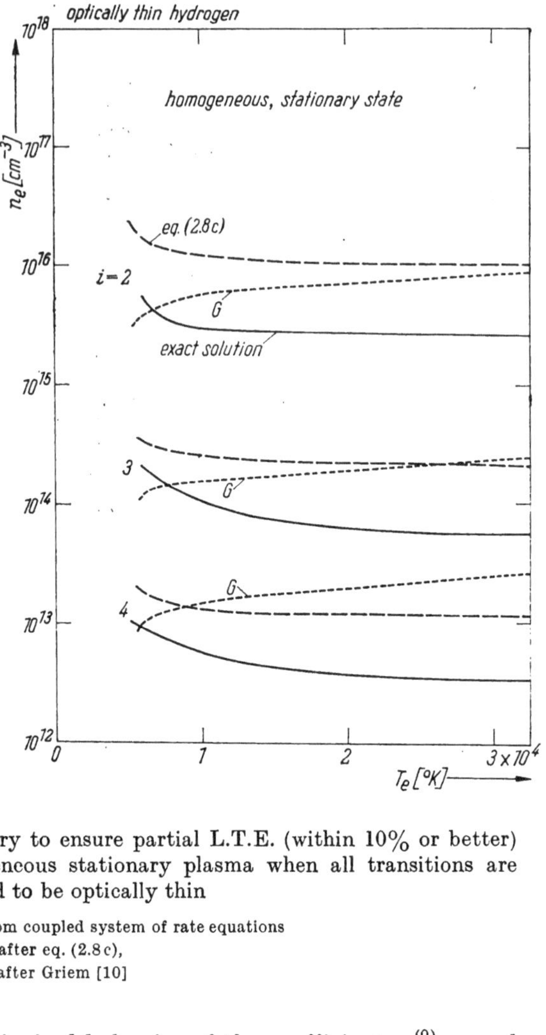

Figure 2.4. Electron density necessary to ensure partial L.T.E. (within 10% or better) for the hydrogen levels in a homogeneous stationary plasma when all transitions are assumed to be optically thin

- solid curves: from coupled system of rate equations
- broken curves: after eq. (2.8c),
- dotted curves: after Griem [10]

The figures 2.2a, b show the principal behavior of the coefficients $r_{z-1,i}^{(0)}$ and $b_{z-1,i}$ for the levels of hydrogen atoms and He^+-ions for an electron temperature of $1.6 \cdot 10^4$ °K and $4.2 \cdot 10^4$ °K, respectively. The upper part of the figures shows the values for $b_{z-1,i}$ when the plasma is assumed to be homogeneous and stationary. The difference between the lower and upper curves is caused by the second term of eq. (2.9). For the calculations of these coefficients we used a model as shown in figure 2.3.

Since we are dealing with a homogeneous stationary plasma we have to consider the 10%-limit of the upper part of figure 2.2. All population densities of which the coefficients $b_{z-1,i}$ fall into the 10%-limit are said to be in partial L.T.E. The result of the numerical evaluation is shown in figures 2.4 and 2.5

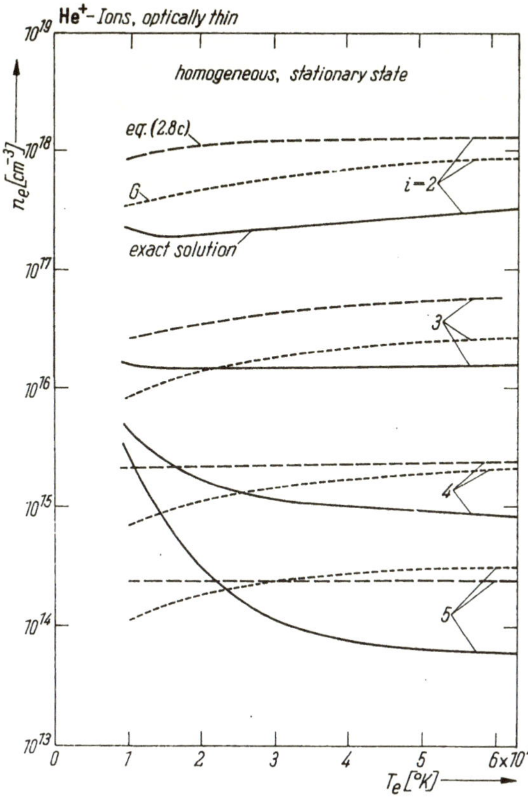

Figure 2.5. Electron density necessary to ensure partial L.T.E. (within 10% or better for the levels of He^+-ions ($z = 2$) in a homogeneous stationary plasma when all transition are assumed to be optically thin

– solid curves: from coupled system of rate equations,
– broken curves: after eq. (2.8c),
– dotted curves: after Griem [10]

by the solid curves. These curves give the electron density as a function of the electron temperature to ensure partial L.T.E. within 10% or better. For comparison we have indicated the validity condition as given by eq. (2.8c) (broken curves) and by Griem's expression (dotted curves). One sees, that collisional-radiative coupling between the excited levels and with the ground level favors the establishment of partial L.T.E. compared to the validity criteria as obtained by considering only two adjacent levels. For high temperatures, such validity conditions largely overestimate the electron density necessary for establishing partial L.T.E. At very low temperatures the validity criteria will lead to an underestimation of the electron density. It should be mentioned, however, that homogeneous optically thin steady-state plasmas occur very seldom in nature.

CHAPTER 3

Complete L. T. E. in Time-Independent and Homogeneous Optically Thin Plasmas

The method applied for obtaining the validity criteria for *partial* L.T.E. in neutral and ionic level systems may also be used for obtaining a condition for the validity of complete L.T.E. For *complete* L.T.E. to be established one has to require that the condition (2.8a) is fulfilled for any two adjacent levels with the ground state included. The most severe condition, however, does not originate from the energy gap between the ground and the first excited state in all cases, it is mostly determined by the largest possible energy gap which can ever be realized between two adjacent levels. In hydrogen and helium, for instance, this energy gap just corresponds to that between the ground and the first excited state. However, there are many level systems which do not follow this rule, e.g. O II showing two excited levels relatively close to the ground state (excitation energies $E_{1,2} = 2.2$ eV, and $E_{1,3} = 3.3$ eV, see e.g. fig. 2.1). Then follows a large energy gap of approximately $\Delta E_{\max} \simeq 7$ eV before one finds another excited level. It is obvious that the condition for complete L.T.E. should be applied to the energy gap with $\Delta E_{\max} \simeq 7$ eV rather than to the gaps $E_{1,2} \simeq 2.2$ eV, and $E_{1,3} \simeq 3.3$ eV, etc.

Modifying the condition (2.8a) in this sense we obtain the following general condition for the validity of complete L.T.E.

$$n_e \geq 1.15 \cdot 10^8 \, z^3 \, \frac{g_{\max}}{g_{\min}} \, \frac{E_{\max} - E_{\min}}{kT_e \, u_1^{3/2}} \, \frac{\Sigma A_{\max,i}}{f_{\min,\max}} \cdot \begin{cases} \Phi_1(\Delta u_{\max}) & \text{for atoms,} \\ \Phi_2(\Delta u_{\max}) & \text{for ions} \end{cases} \text{cm}^{-3} \tag{3.1a}$$

where we used the relation $\Delta u_{\max} = (E_{\max} - E_{\min})/kT_e$.

The index "min" indicates the level at the low-energy side of the largest possible energy gap (fig. 2.1) and the index "max" means the corresponding level on the high-energy side. The ratio $\sum\limits_{i<\max} A_{\max i}/f_{\max,\min}$ may be scaled with respect to the energy gap $E_{\max} - E_{\min}$ using the relation

$$\frac{\sum\limits_{i<\max} A_{\max,i}}{f_{\min,\max}} \cong \frac{A^H_{2,1}}{f^H_{1,2}} \left(\frac{E_{\max} - E_{\min}}{E^H_1} \right)^2 = 1.13 \cdot 10^9 \left(\frac{E_{\max} - E_{\min}}{E^H_1} \right)^2 \text{s}^{-1} ,$$

with $A^H_{2,1} = 4.7 \cdot 10^8 \text{ s}^{-1}$, $f^H_{1,2} = 0.42$. Thus, one finally obtains the following general condition for the validity of *complete* L.T.E. in an optically thin plasma

$$n_e \geq 6.5 \cdot 10^{16} \frac{g_{\max}}{g_{\min}} \left(\frac{E_{\max} - E_{\min}}{E_1^H} \right)^3 \left(\frac{kT_e}{E_1^H} \right)^{1/2} \cdot \begin{cases} \Phi_1(\Delta u_{\max}) & \text{for atoms } (z=1), \\ \Phi_2(\Delta u_{\max}) & \text{for ions } (z>1). \end{cases} \text{cm}^{-3}. \tag{3.1b}$$

For hydrogen and hydrogen-like ions, $g_{\min}/g_{\max}$ is equal to $1/2^2$. The functions $\Phi_1(x)$ and $\Phi_2(x)$ have the same meaning as in eq. (2.8a) with the argument $2\,u_1/j^3$ replaced by $\Delta u_{\max}$. The numerical values for n_e calculated from relation (3.1b) differ from those of other authors by not more than a factor three to four. According to relation (3.1b) n_e decreases with decreasing temperature, i.e. lower electron temperatures fevor the establishment of complete L. T. E.. The temperature dependence is different for atoms and ions, however. Consider relation (3.1b) for atoms with the values Φ_1 as listed in Table 2.2. One sees that the function $T_e^{1/2}\,\Phi_1$ decreases very slightly when T_e decreases. For ions, the temperature dependence of relation (3.1b) is larger than for atoms, since the function Φ_2 is nearly temperature-independent. Thus, $T_e^{1/2}\,\Phi_2$ varies approximately proportional to $T_e^{1/2}$. This general temperature behaviour agrees very well with the results obtained from the solutions of the corresponding systems of collisional-radiative rate equations (see Drawin and Emard [41]).

CHAPTER 4

Influence of Resonance Absorption on the Establishment of Complete L. T. E. in Homogeneous Stationary Plasmas

One generally argues that radiative absorption in the first members of the resonance lines favours the establishment of complete L.T.E., and that due to the large optical thickness of the resonance lines complete L.T.E. prevails down to rather small electron densities. This may partly be true for the inner regions of stellar plasmas, however, for laboratory plasmas this argument does not hold. In general, only the first two or three members of the resonance

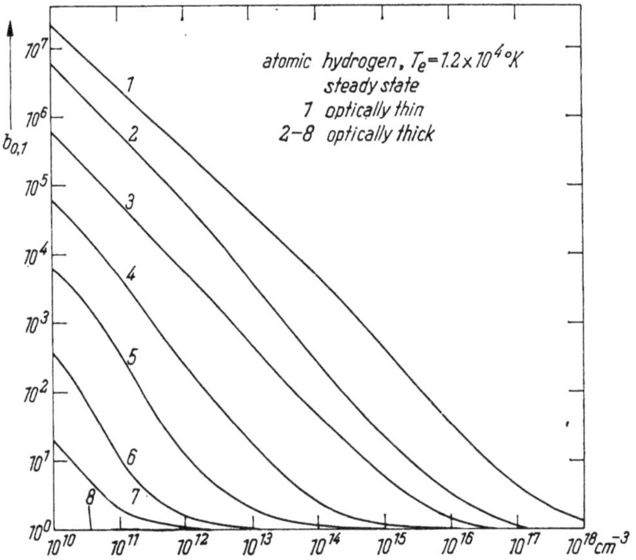

Figure 4.1. $b_{0,1}$-values for the ground state of hydrogen without and with radiative absorption

curve *1*: all transitions optically thin ($\Lambda_i = \Lambda_{ji} = 1$),
curve *2*: Lyman-α optically opaque ($\Lambda_{21} = 0$),
curve *3*: Lyman series optically opaque ($\Lambda_{j_1} = 0$),
curve *4*: Lyman series, Lyman continuum, and $H\alpha$-line optically opaque ($\Lambda_1 = \Lambda_{j_1} = \Lambda_{32} = 0$),
curve *5*: Lyman series, Lyman continuum, Balmer series and Balmer continuum opaque ($\Lambda_1 = \Lambda_{j_1} = \Lambda_2 = \Lambda_{j_2} = 0$),
curve *6*: Lyman series, Lyman continuum, Balmer lines, Balmer continuum, Paschen lines, and Paschen continuum optically opaque ($\Lambda_1 = \Lambda_{j_1} = \Lambda_2 = \Lambda_{j_2} = \Lambda_3 = \Lambda_{j_3} = 0$),
curve *7*: Lyman series, Lyman continuum, Balmer lines, Balmer continuum, Paschen lines, Paschen continuum, Pfund lines, and Pfund continuum optically opaque ($\Lambda_1 = \Lambda_{j_1} = \Lambda_2 = \Lambda_{j_2} = \Lambda_3 = \Lambda_{j_3} = \Lambda_4 = \Lambda_{j_4} = 0$)
curve *8*: All free-free and free-bound transitions optically opaque

series are completely optically opaque. All other transitions are, in general, optically thin or nearly optically thin.

In order to show quantitatively the influence of resonance absorption on the establishment of complete L.T.E. we have calculated the population densities of the hydrogen and of the helium levels from a coupled system of rate equations for the stationary state assuming different degrees of reabsorption. The method consisted in reducing successively the transition probabilities for the radiative transitions. Although not consistent, this method shows nevertheless how serious may be the departure from complete L.T.E. in laboratory plasmas of small geometrical dimensions. (In a rigorous treatment one has to solve simultaneously the coupled system of rate equations and the corresponding radiative transfer equations for the geometrical problem given).

Figure 4.1 shows a typical result for the ground state density of neutral hydrogen. The coefficient $b_{0,1}$ denotes the calculated number density $n_{0,1}$ divided by the number density $n_{0,1}^{(\mathrm{Saha})}$ calculated from eq. (1.3), hence

$$b_{0,1} = n_{0,1}/n_{0,1}^{(\mathrm{Saha})} .$$

Curve 1 is for an optically thin plasma (all photons escape from the plasma). By reducing successively the radiative transition probabilities one obtains the curves 2 to 8. There is only a small difference between the b_1-values for an optically thin plasma and a plasma with the resonance lines completely absorbed. Only when the transitions of the higher spectral series are optically thick complete L.T.E. may exist at correspondingly lower electron densities. However, in laboratory plasmas one never obtains substantial reabsorption in the Balmer-, Paschen-, ... etc. series.* (A similar result has been obtained for He I, He II, and for Li). It follows from this, that even in the case of extremely strong absorption in the resonance lines the right-hand side of the relations (3.1 a) and (3.1 b) may be reduced by not more than a factor 10. The effect of reabsorption can quantitatively be taken into account by multiplying the right-hand sides of the relations (3.1 a) and (3.1 b) by a coefficient Λ which may take values lying between $1 \geq \Lambda \geq 0.1$. The numerical value of Λ may be estimated by considering the number of photons absorbed when traversing a homogeneous layer of characteristic length l, l may for instance represent the mean radius R of a discharge, $l = R$. For a transition $i \rightleftarrows j$ one has for the reduction factor the general expression

$$\Lambda_{ji} = \int P_{ji}(\Delta\lambda) \exp\left(-\bar{n}_i l \sigma_{ji}(\Delta\lambda)\right) \mathrm{d}\lambda \tag{4.1}$$

where $\bar{n}_i$ is the average number density of particles in the (lower) level i, and $\sigma_{ji}(\lambda\Delta)$ the optical absorption cross section for photons belonging to the transition $i \rightleftarrows j$. Assuming only Doppler broadening to be effective, the absorption

* An exception make spectral line series containing frequencies higher than the plasma frequency, i.e. series originating from highly excited levels can strongly be absorbed. This case has also been treated in the frame of the collisional-radiative model (H. W. Drawin and F. Emard, to be publ.). It was found that this kind of reabsorption has a negligibly small influence on the establishment of C.L.T.E. It can in almost all cases be neglected.

cross section as a function of the wave-length distance can be written

$$\sigma_{ji}(\Delta\lambda) = \frac{\pi^{1/2} e_0^2 \lambda_{ij}^2}{m_e c^2 \Delta\lambda_D} f_{ij} \exp - \left(\frac{\Delta\lambda}{\Delta\lambda_D}\right)^2 = \frac{\pi e_0^2 \lambda_{ij}^2}{m_e c^2} P_{ji}(\Delta\lambda) . \qquad (4.2)$$

The *average* optical absorption cross section* is, thus, estimated by

$$\langle\sigma_{ij}\rangle \simeq \frac{1}{2} \frac{\pi^{1/2} e_0}{m_e c} \left(\frac{M}{2 R T_M}\right)^{1/2} \lambda_{ij} f_{ij} = 5.13 \cdot 10^{-15} f_{ij} \lambda_{ij} \left(\frac{M}{T_M}\right)^{1/2} \text{cm}^2$$

with the wave-length λ_{ij} in Angstrom units, the temperature T_M of the emitters in °K, and the mass number M in relative mass units (e.g., $M = 1$ for hyrdogen)' When eq. (4.1) applied to a resonance transition $1 \rightleftarrows 2$ yields a value $\Lambda_{21} < 0.1$ the reduction factor for the electron density becomes approximately equal to 0.1 for the species considered. — Table 4.1 shows some numerical values for

Table 4.1. Numerical values of the coefficient Λ_{j1} for the hydrogen Lyman lines as a function of $\bar{n}_1$ and T_M; $\bar{n}_1$ denotes the number density of hydrogen ground state atoms

T_M [°K]	$\bar{n}_1 l$ [cm²]	Λ_{j1} Ly − α $i = 1 \to j = 2$	Ly − β $i = 1 \to j = 2$	Ly − γ $i = 1 \to j = 3$	Ly − δ $i = 1 \to j = 4$
$1.0 \cdot 10^4$	1.10^{14}	7.2, −2	5.4, −1	8.1, −1	9.0, −1
	1.10^{15}	5.1, −3	4.1, −2	1.5, −1	4.4, −1
	1.10^{16}	4.1, −4	3.0, −3	9.7, −3	2.3, −2
	1.10^{17}	3.6, −5	2.5, −4	7.7, −4	1.8, −3
$1.6 \cdot 10^4$	1.10^{14}	1.7, −1	7.3, −1	9.0, −1	9.5, −1
	1.10^{15}	1.1, −2	9.5, −2	6.7, −1	9.2, −1
	1.10^{16}	8.7, −4	6.5, −3	2.1, −2	5.2, −2
	1.10^{17}	7.5, −5	5.2, −4	1.6, −3	3.8, −2

Note: Read 7.2, − 2 = $7.2 \cdot 10^{-2}$, etc. . .

the first Lyman lines of hydrogen calculated from eq. (4.1). For the estimate of the electron density necessary for the establishment of complete L.T.E. one only needs the Λ_{21}-value for the Ly α-line. Table 4.1 shows that under most actual conditions a reduction factor $\Lambda = 0.1$ is permissible by which the right-hand side of the relations (3.1 a) and (3.1 b) may be multiplied. Thus, one obtains the following condition for complete L.T.E.:

$$n_e \geq 6.5 \cdot 10^{15} \frac{g_{\max}}{g_{\min}} \left(\frac{E_{\max} - E_{\min}}{E_1^H}\right)^3 \left(\frac{kT_e}{E_1^H}\right)^{1/2} \cdot \begin{cases} \Phi_1(\Delta u_{\max}) & \text{for atoms} \\ \Phi_2(\Delta u_{\max}) & \text{for ions} \end{cases} \text{cm}^{-3} . \qquad (4.3)$$

It follows from these considerations that wall-stabilized arcs operated at atmospheric pressure and sufficiently high currents can fulfill the condition for complete L.T.E. in the central part of the arc column, however, in the outer regions serious departures from complete L.T.E. may exist. All types of discharges operated below atmospheric pressure may often deviate from complete L.T.E. accross the entire cross section of the plasma column.

* averaged over frequency ν.

CHAPTER 5

Influence of Resonance Absorption on the Establishment of Partial L. T. E. in Homogeneous Stationary Plasmas

Absorption of resonance radiation $h\nu_{1j}$ may influence the establishment of partial L.T.E. This will be the case when the levels j and $j-1$ considered partly decay down to the ground state. The corresponding transition probabilities appearing in the sum ΣA_{ji} of eqs. (2.8) should thus be multiplied by a reduction factor Λ_{ji}. As a rule, all A_{ji} should be multiplied by the corresponding coefficient Λ_{ji}, however, only the Λ_{ji} coefficients associated with the ground state will have an influence, the number densities $n_{i>1}$ of the excited states are generally too small to give values $\Lambda_{ji} < 1$. The Λ_{ji} values for the hydrogen resonance transitions may be taken from Table 4.1. Fortunately, there is only a small difference between the two cases $\Lambda_{ji} = 1$ and $\Lambda_{ji} < 1$. As an example, let us consider the hydrogen levels $j = 4$ and $j = 3$. We assume strong reabsorption in the transitions $1 \rightleftarrows 4$ and $1 \rightleftarrows 3$, so that Λ_{41} and Λ_{31} become very small. We first consider the level $j = 4$: we assume $\Lambda_{41} = 10^{-2}$. Since the other Λ_{ji} values associated with the excited levels $i = 2$ and $i = 3$ are equal to 1 one obtains for the sum $\sum\limits_{i<4} \Lambda_{ji} A_{ji} = 1.75 \cdot 10^7\ \text{s}^{-1}$, compared to $\sum\limits_{i<4} A_{ji} = 3.02 \cdot 10^7\ \text{s}^{-1}$ in the optically thin case. We consider now the level $j = 3$ and assume $\Lambda_{31} = 10^{-5}$. One obtains $\sum\limits_{i<3} \Lambda_{ji} A_{ji} = 4.41 \cdot 10^7\ \text{s}^{-1}$, compared to $\sum\limits_{i<3} A_{ji} = 9.98 \cdot 10^7\ \text{s}^{-1}$ in the optically thin case. For levels $j > 4$ resonance absorption has still a smaller influence.

One concludes from these considerations that strong resonance absorption does not change appreciably the conditions (2.8a) and (2.8c) for partial L.T.E. When strong resonance absorption exists, the right-hand sides of the relations (2.8a) and (2.8c) should be multiplied by a coefficient not smaller than 0.5. In the special case of hydrogen and hydrogen-like levels instead of eq. (2.8c) one thus obtains the condition

$$n_e \geq 3.7 \cdot 10^{18} \frac{z^7}{i^{17/2}} \left(\frac{kT_e}{z^2 E_1^H}\right)^{1/2} \cdot \begin{cases} \Phi_1(x_i)\ \text{cm}^{-3} & \text{for atoms}\,, \\ \Phi_2(x_i)\ \text{cm}^{-3} & \text{for ions}\,. \end{cases} \tag{5.1}$$

The values derived from this formula are shown in figure 5.1 (broken curves). The solid curves are the results which have been obtained by solving the corresponding system of coupled rate equations. For comparison, we also indicated the electron densities as given by Griem's formula (dotted curves). One sees that the enalytical expressions of the validity criteria are very close to the results of the complicated numerical calculations. This will only be true, however, as long as reabsorption in the higher spectral series is negligibly small.

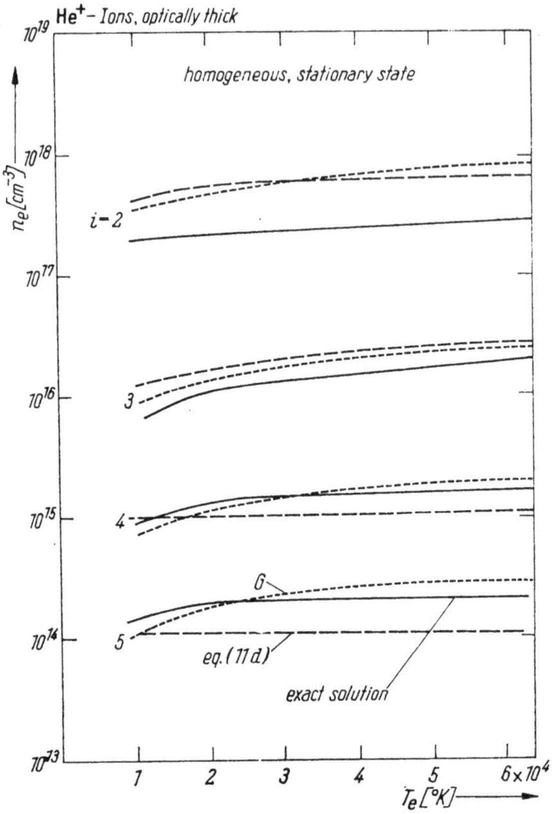

Figures 5.1. Electron density necessary to ensure partial L.T.E. (within 10% or better) for the levels of He^+-ions in an optically thick homogeneous stationary plasma

– solid curves: from coupled system of rate equations. In these calculations the following assumptions have been made: 99.99% reabsorption in the Ly α-line (λ = 304 Å); 99.9% in the Ly β-line (λ = 256 Å), 99% in the Ly γ-line (λ = 244 Å), 90% in the Ly δ-line (λ = 238 Å), and 33% reabsorption in the Ly ε-line; all other transitions are assumed to be optically thin,
– broken curves: after eq. (5.1),
– dotted curves: after Griem [10]

There is another important problem related to the population densities $n_{i>1}$ of the excited states, namely the order of magnitude of the ratio n_i/n_1, i.e. the population densities of the excited states relative to that of the number density n_1 of the ground level.

In the case of complete L.T.E. this ratio equals the Boltzmann ratio $(g_i/g_1) \exp\left(-(E_i - E_1)\right)/kT$. In non-L.T.E. plasmas, however, it is much smaller. The departure of the actual ratio from the Boltzmann ratio may be characterized by a new quantity a_i defined by $a_i = b_i/b_1$, hence for particles of type $(z-1)$:

$$\frac{n_{z-1,i}}{n_{z-1,1}} = a_{z-1,i}\,\frac{n_{z-1,i}^{(\text{Saha})}}{n_{z-1,1}^{(\text{Saha})}} \equiv \frac{b_{z-1,i}}{b_{z-1,1}}\,\frac{n_{z-1,i}^{(\text{Saha})}}{n_{z-1,1}^{(\text{Saha})}}, \tag{5.2}$$

where the b_i's are given by eq. (2.1). For complete L.T.E. $a_{z-1,i} = 1$ holds; for the non-equilibrium state $a_{z-1,i} < 1$.

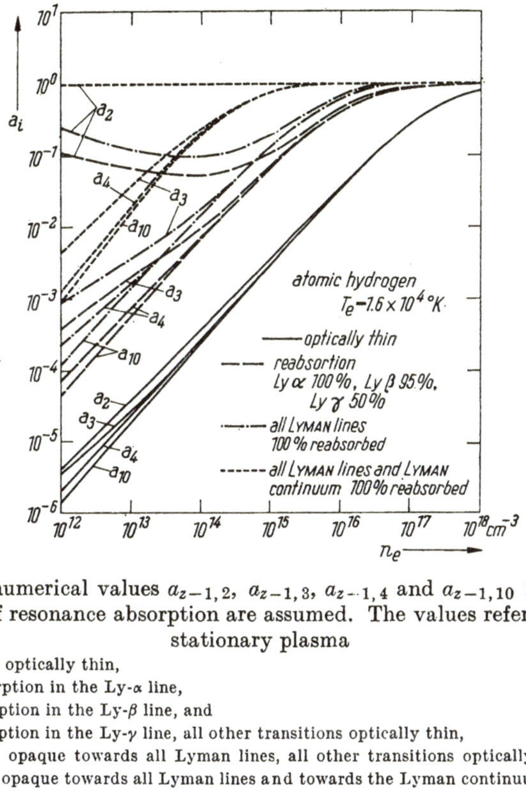

Figure 5.2. The numerical values $a_{z-1,2}$, $a_{z-1,3}$, $a_{z-1,4}$ and $a_{z-1,10}$ for hydrogen when different degrees of resonance absorption are assumed. The values refer to a homogeneous stationary plasma

curve *1*: all transitions optically thin,

curve *2*: 100% reabsorption in the Ly-α line,
95% reabsorption in the Ly-β line, and
50% reabsorption in the Ly-γ line, all other transitions optically thin,

curve *3*: Plasma 100% opaque towards all Lyman lines, all other transitions optically thin,

curve *4*: Plasma 100% opaque towards all Lyman lines and towards the Lyman continuum, all other transitions optically thin

In the figure, the subscript $z-1$ has been dropped

Figure 5.2 shows numerical values $a_{z-1,i}$ for the hydrogen levels with principal quantum numbers $i = 2$, 3, 4, and 10, as a function of the electron density and for different degrees of resonance absorption. The $a_{z-1,i}$-values have been calculated from a coupled system of rate equations. (For details, the reader is referred to the original publication Ref. [27]). One sees from Fig. 5.2 that even in the case of complete optically thick resonance lines the excited levels will never be in a Boltzmann equilibrium with the ground level for electron densities $n_e < 1 \cdot 10^{16}\ \text{cm}^{-3}$.

For the calculation of the $b_{z-1,i}$- and $a_{z-1,i}$-values we neglected the heavy particle collisions, only electronic collisions have been taken into account.

CHAPTER 6

Influence of Heavy Particle Collisions on the Equilibrium Population of Highly Excited Levels

We come now to the important problem of the influence of the heavy particle collisions on the equilibrium distribution of highly excited states. When heavy particle collisions are important the excited states will show population densities which are determined by the gas temperature T_a rather than by the electron temperature T_e, provided that $T_e \neq T_a$.

The heavy particles may either be ions or neutral particles. Since the number density of the neutrals in the ground state is very often one or several orders of magnitudes larger than that of the ions, neutral particle collisions may be much more efficient than ions, although the ionic cross sections are larger (by a factor 2 to 5) than those for neutral particle encounters in the region of interest. In completely ionized plasmas, however, the influence of the ion collisions may be important.

The relative influence of the electronic and neutral particle collisions may be treated by considering the de-excitation rates from a level j to an adjacent level $i = j - 1$ due to electronic and neutral particle encounters. The de-excitation rate due to neutral particle encounters can be calculated form the cross section $\sigma_{ij}^{(a)}$ for excitation $i \rightarrow j$. We use for $\sigma_{ij}^{(a)}$ a relation (in the center-of-mass system of identical particles, see Drawin [28]) which is similar to that for ionization. For excitation, this formula is

$$\sigma_{ij}^{(a)} = 4\,\pi\,a_0^2 \left(\frac{E_1^H}{E_j - E_i}\right)^2 f_{ij}\,\frac{2\,m_e}{m_H}\,\frac{W_{ij} - 1}{\left[1 + \frac{2\,m_e}{m_a}(W_{ij} - 1)\right]^2}, \tag{6.1}$$

where f_{ij} is the oscillator strength for the transition $i \rightarrow j$, $E_j - E_i$ is the energy difference of the two levels i and j, and m_H is the mass of the hydrogen atom, m_a the mass of the colliding particles in question, m_e the mass of the electrons, and W_{ij} has the meaning of a reduced translational energy of the atoms:

$$W_{ij} = \frac{E^{(a)}}{E_{ij}}\,; \quad E_{ij} \equiv |E_j - E_i| = |E_i - E_j|$$

$\left(E^{(a)} = (1/4)\,m_a\,v_a^2;\ E_{ij} = \text{threshold energy of levels } (i, j) \text{ in the laboratory system}\right)$. The threshold for the atom-atom collisions lies in the center-of-mass system at $W_{ij} = 1$.

Integrating eq. (6.1) over a Maxwellian velocity distribution for the heavy particles (in the center-of-mass system) yields the following formula for the excitation coefficient [28]

$$\langle\sigma^{(a)} v_a\rangle_{j\to i} = 32\,\pi\, a_0^2 \left(\frac{E_1^H}{E_{ij}}\right)^2 \left(\frac{kT_a}{\pi\, m_a}\right)^{1/2} f_{ij}\, \frac{m_e}{m_H}\, \Psi_{m_a}(w_{ij})\,, \tag{6.2}$$

with $w_{ij} = (E_j - E_i)/kT_a$. The function $\Psi_{m_a}(w_{ij})$ may be approximated by

$$\Psi_{m_a}(w_{ij}) \cong \left(1 + \frac{2}{w_{ij}}\right) \left(\frac{1}{1 + \left[\dfrac{2\, m_e}{m_a\, \omega_{ij}}\right]^2}\right) \exp\,(-\, w_{ij})\,.$$

Thus, the method of detailed balancing yields for the rate coefficient for de-excitation the formula

$$\langle\sigma^{(a)} v_a\rangle_{j\to i} \cong 32\,\pi\, a_0^2 \left(\frac{E_1^H}{E_{ij}}\right)^2 \left(\frac{kT_a}{\pi\, m_a}\right)^{1/2} \frac{g_i}{g_j}\, f_{ij}\, \frac{m_e}{m_H}\, \frac{1 + \dfrac{2}{\omega_{ij}}}{1 + \left(\dfrac{2\, m_e}{m_a\, w_{ij}}\right)^2}\,. \tag{6.3}$$

Heavy particle collisions will be more efficient than those by electrons if

$$\frac{n_j\, n_a\, \langle\sigma^{(a)} v_a\rangle_{j\to(j-1)}}{n_j\, n_e\, \langle\sigma^{(e)} v_e\rangle_{j\to(j-1)}} \gg 1\,. \tag{6.4}$$

Inserting the equations for $\langle\sigma^{(e)} v_e\rangle_{j\to(j-1)}$ and $\langle\sigma^{(a)} v_a\rangle_{j\to(j-1)}$ into relation (6.4) yields

$$\frac{4}{\sqrt{2}}\, \frac{n_a}{n_e}\, \frac{m_e}{m_H} \left(\frac{m_e}{m_a}\right)^{1/2} \left(\frac{T_a}{T_e}\right)^{1/2} \frac{\Phi_1(u_{j,\,j-1})}{u_{j,\,j-1}}\, \frac{1 + \dfrac{2}{\omega_{j,\,j-1}}}{1 + \left(\dfrac{2\, m_e}{m_a\, \omega_{j,\,j-1}}\right)^2} \gg 1\,. \tag{6.4a}$$

Using for $\Psi_1(\beta, x)$ a simple analytical approximation, (see Ref. [28]) the function $\Phi_1(\beta, x)$ may be written as follows

$$\Phi_1(\beta, x) \cong (1 + x) \Big/ \left(\frac{1}{20 + x} + \ln\left(1.25\,\beta\left[1 + \frac{1}{x}\right]\right)\right) \tag{6.5}$$

and further for sufficiently excited levels

$$u_{j,\,j-1} \cong \frac{2\, E_1^H}{kT_e\, j^3} = \frac{2\, u_1}{j^3}\,; \quad w_{j,\,j-1} \cong \frac{2\, E_1^H}{kT_a\, j^3}\,,$$

where j is identified with the principal quantum number in the hydrogenic approximation. With these quantities the relation (6.4a) is fulfilled for all levels $k > j$. This means that the population densities n_k belonging to levels of principal quantum numbers $k > j$ are determined by the gas temperature rather than by the electron temperature. As long as $T_a = T_e$, this is not important. However, when the gas temperature is very different from T_e spectroscopic measurements will give the gas temperature rather than the electron temperature when spectral lines originating from highly excited levels are evaluated.

The physical reason of this effect originates from the quite different energy behaviour of the electronic and atomic cross sections. It would surely be very interesting to check the relation (6.4 a) for well-defined experimental conditions.*

One can also show that for highly excited levels three-body collisional recombination with an atom as the third body may eventually be much more important than the usual three-body collisional recombination process with two electrons involved, provided that $n_a \gg n_e$, [28].

* Corresponding measurements have recently been performed (see Drawin, Klan, Ringler: Z. Naturforschg. **26a** (1970) 186). Unfortunately, the exact determination of the population densities of the highly excited levels (up to principal quantum number 25) was not possible due to too large error bars. It was thus impossible to show wether or not the Boltzmann plot was determined by the gas or the electron temperature. It could be shown, however, that the determination of the electron density n_e from the population densities of levels with medium principal quantum numbers yields wrong values when the calculation of n_e is based on the Saha equation in which the temperature T equals the electron temperature T_e. This showed that at low degrees of ionization and different electron and gas temperatures the population of the excited levels are determined by both temperatures! Quantitatively can this only be treated in the frame of a collisional-radiative model in which all processes are correctly taken into account. For further details see: Collins, C. B., Phys. Rev. **158** (1967) 94, Phys. Rev. **186** (1969) 113; Drawin, H. W., Z. Naturforschg. **25A** (1970) 145; Drawin, H. W., and F. Emard, Z. Physik **254** (1972) 202.

CHAPTER 7

Validity of Partial L.T.E. in Quasi-Stationary Quasi-Homogeneous Plasmas (optically thin and thick)

A stationary and homogeneous plasma is characterized by the fact that temporal as well as spatial variations of the particle densities are non existent, hence

$$\nabla_r \cdot (n_{z-1,i}\langle \boldsymbol{v}_{z-1,i}\rangle) = 0 \; ; \quad \frac{\partial}{\partial t} n_{z-1,i} = 0 \, .$$

It often happens that these equations are fulfilled for all excited states but not for the ground state particles. One then says that the particles are in a quasi-homogeneous quasi-stationary state. Such a state is, for instance, obtained when the actual number density of the ground state particles, $n_{z-1,1}$, is smaller than the number density for the completely homogeneous stationary state. (The relaxation time for the establishment of a quasi-stationary state is given by eq. (8.5b), the relaxation length for establishing a quasi-homogeneous state by eq. (9.17).)

We consider the completely homogeneous stationary state in more detail: As long as complete L.T.E. is not fulfilled the ground state number densities as obtained from the coupled systems of rate equations are always larger than the corresponding equilibrium number densities, hence

$$b_{z-1,1} = \frac{n_{z-1,1}}{n_{z-1,1}^{(\mathrm{Saha})}} > 1 \, .$$

A typical example is shown in figure 4.1. The number densities themselves can be calculated from such systems of equations as soon as the electron density, the ion density, and the electron temperature are known (provided that only electronic collision processes are assumed to be important and that radiative absorption can be estimated). The numerical solutions for different cases of radiative absorption are shown in figures 7.1a to 7.3c for the special case

$$\varrho_z = \frac{n_e}{n_z} = 1 \, .$$

If the number density n_z is different from n_e (which is the case for all ions with $Z = 2$ and for heterogenes plasmas) one obtains the ground state number densities by dividing the values of these curves by ϱ_z, hence

$$n_{z-1,1} = n'_{z-1,1}/\varrho_z \, .$$

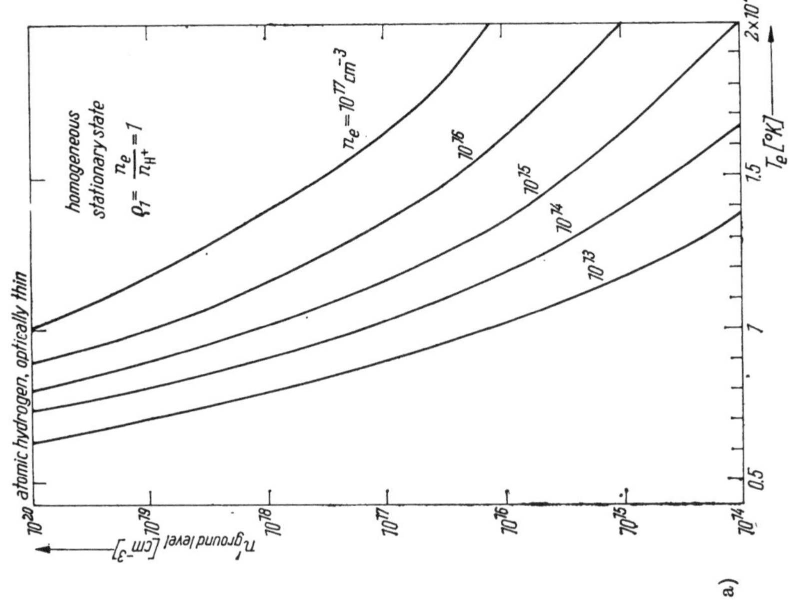
atomic hydrogen, optically thin
homogeneous stationary state
$\varrho_1 = \frac{n_e}{n_{H^+}} = 1$
$n_e = 10^{17}\,cm^{-3}$
10^{16}
10^{15}
10^{14}
10^{13}
$n_{ground\,level}\,[cm^{-3}]$
10^{20}
10^{19}
10^{18}
10^{17}
10^{16}
10^{15}
10^{14}
0.5
1
1.5
2×10^4
$T_e\,[°K]$

a)

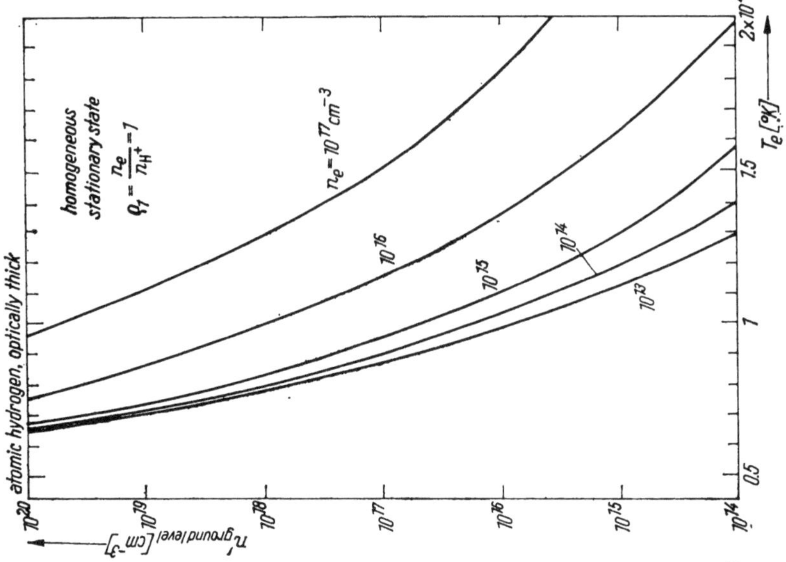
atomic hydrogen, optically thick
homogeneous stationary state
$\varrho_1 = \frac{n_e}{n_{H^+}} = 1$
$n_e = 10^{17}\,cm^{-3}$
10^{16}
10^{15}
10^{14}
10^{13}
$n_{ground\,level}\,[cm^{-3}]$
10^{20}
10^{19}
10^{18}
10^{17}
10^{16}
10^{15}
10^{14}
0.5
1
1.5
2×10^4
$T_e\,[°K]$

b)

c)

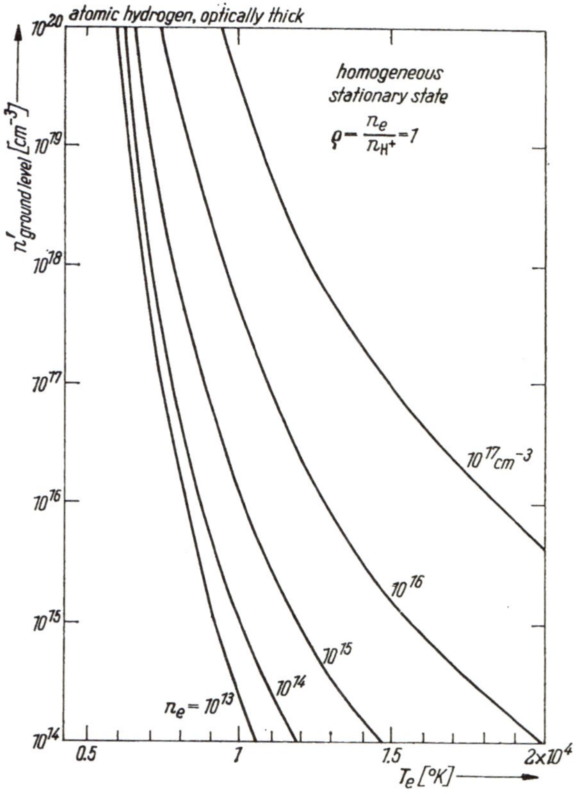

Figures 7.1. The number density $n'_{\text{ground level}}$ of hydrogen atoms in the ground state $1\,s\,^2S_{1/2}$ in a homogeneous stationary plasma when $\varrho_1 = 1$. The curves represent the solutions of a coupled system of rate equations (after Ref. [41])

Figure 7.1a: all transitions optically thin

Figure 7.1b 99% reabsorption in the Ly α-line, 90% in the Ly β-line, 33% in the Ly γ-line; all other transitions optically thin

Figure 7.1c. 99.99% reabsorption in the Ly α-line, 99.9% in the Ly β-line, 99% in the Ly γ-line, 90% in the Ly δ-line, 33% in the Ly ε-line; all other transitions optically thin

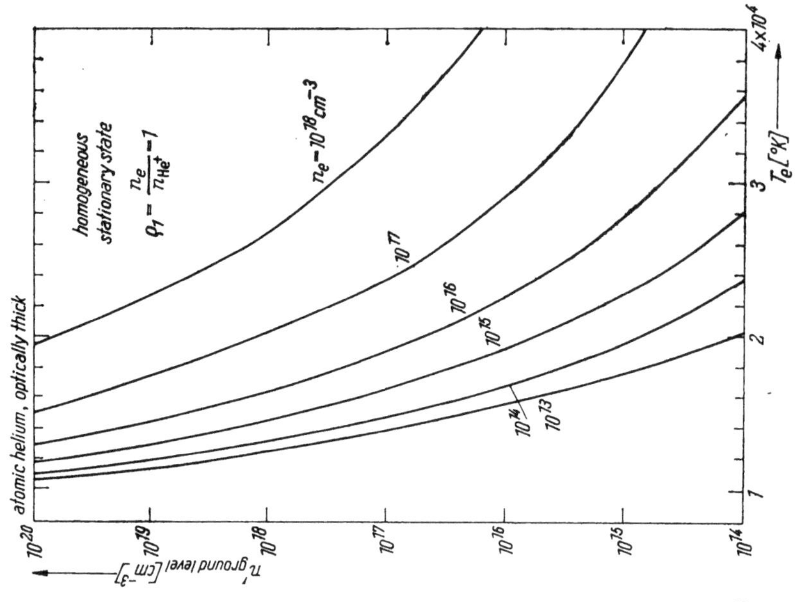
atomic helium, optically thick
homogeneous stationary state
$\varrho_1 = \frac{n_e}{n_{He^+}} = 1$
$n_e = 10^{18}\,cm^{-3}$
10^{17}
10^{16}
10^{15}
10^{14}
10^{13}
$n_{ground\,level}\,[cm^{-3}]$
10^{20}
10^{19}
10^{18}
10^{17}
10^{16}
10^{15}
10^{14}
1
2
3
4×10^4
$T_e\,[°K]$

b)

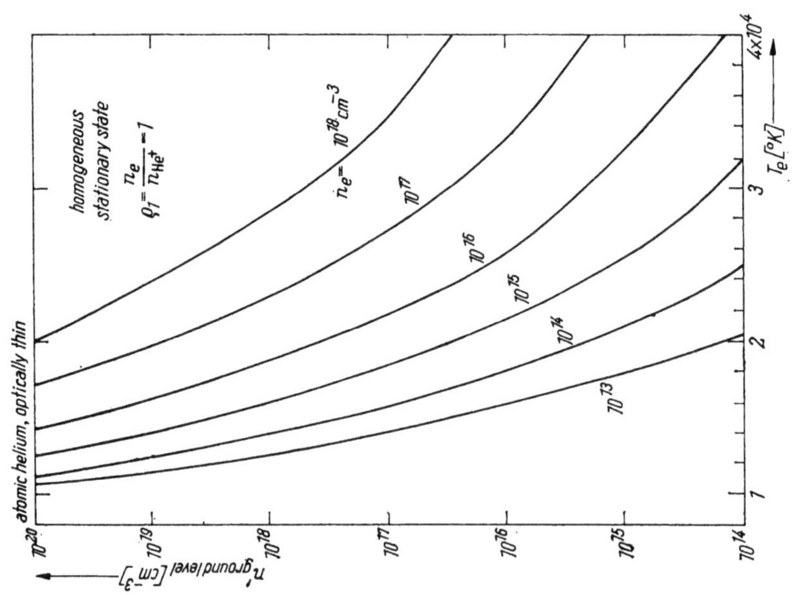
atomic helium, optically thin
homogeneous stationary state
$\varrho_1 = \frac{n_e}{n_{He^+}} = 1$
$n_e = 10^{18}\,cm^{-3}$
10^{17}
10^{16}
10^{15}
10^{14}
10^{13}
$n_{ground\,level}\,[cm^{-3}]$
10^{20}
10^{19}
10^{18}
10^{17}
10^{16}
10^{15}
10^{14}
1
2
3
4×10^4
$T_e\,[°K]$

a)

c)

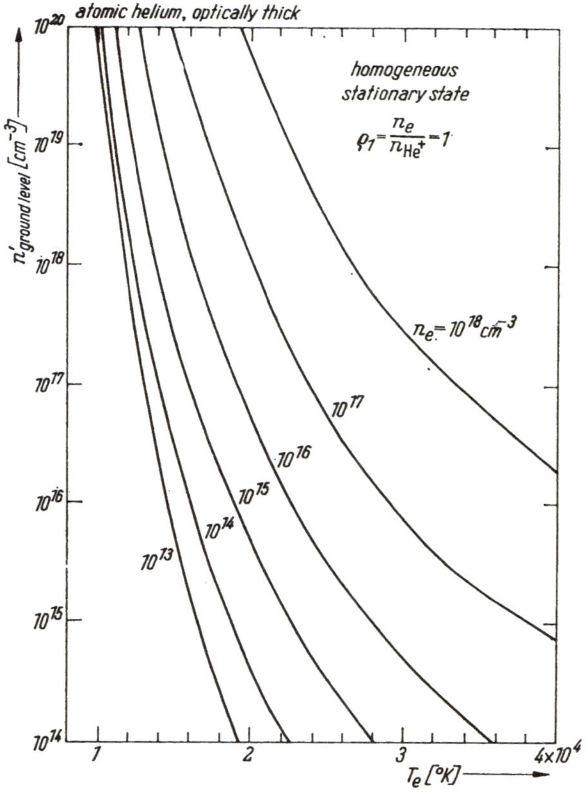

Figures 7.2a, b, c. The number density $n'_{\text{ground level}}$ of helium atoms in the ground state $(1s)^2\,{}^1S_0$ in a homogeneous stationary plasma when $\varrho_1 = 1$ (after Ref. [41])

Figure 7.2a. all transitions optically thin

Figure 7.2b. 99.99% reabsorption in the line $\lambda = 584$ Å

Figure 7.2c. 99.9% reabsorption in the line $\lambda = 537$ Å, 99% in the line $\lambda = 522$ Å, 90% in the line $\lambda = 516$ Å, 33% in the line $\lambda = 512$ Å

all other transitions within the level system of He are assumed to be optically thin

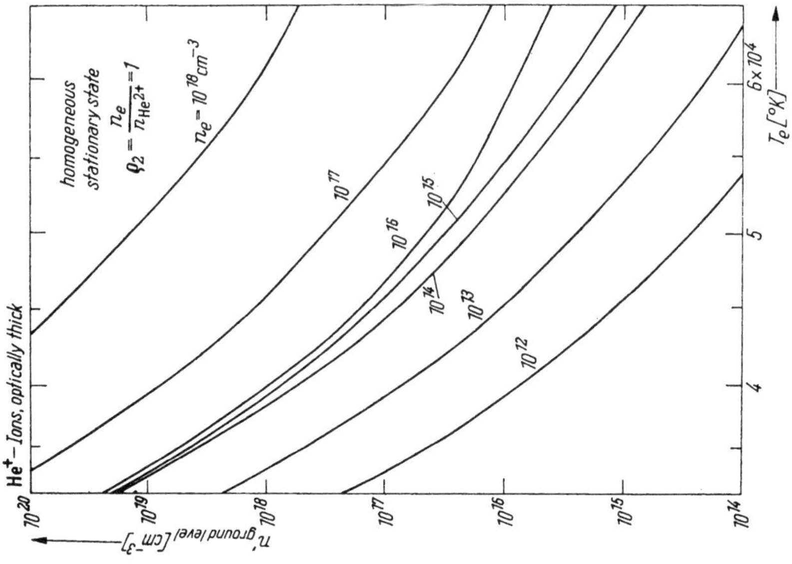
He+ – Ions, optically thick
homogeneous stationary state
$\varrho_2 = \frac{n_e}{n_{He^{2+}}} = 1$
$n_e = 10^{18}\,cm^{-3}$
10^{17}
10^{16}
10^{15}
10^{14}
10^{13}
10^{12}
$n_{ground\ level}\,[cm^{-3}]$
10^{20} 10^{19} 10^{18} 10^{17} 10^{16} 10^{15} 10^{14}
$T_e\,[°K]$
4 5 6×10^4

b)

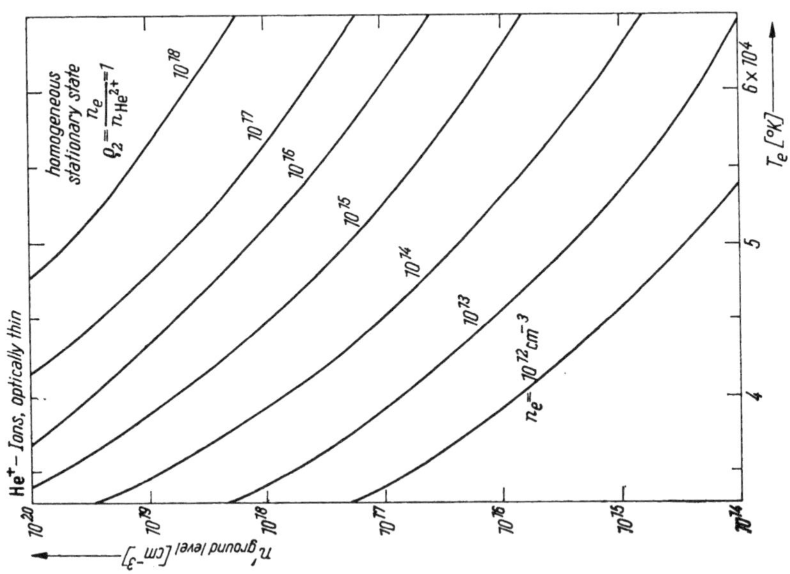
He+ – Ions, optically thin
homogeneous stationary state
$\varrho_2 = \frac{n_e}{n_{He^{2+}}} = 1$
10^{18}
10^{17}
10^{16}
10^{15}
10^{14}
10^{13}
$n_e = 10^{12}\,cm^{-3}$
$n_{ground\ level}\,[cm^{-3}]$
10^{20} 10^{19} 10^{18} 10^{17} 10^{16} 10^{15} 10^{14}
$T_e\,[°K]$
4 5 6×10^4

a)

c)

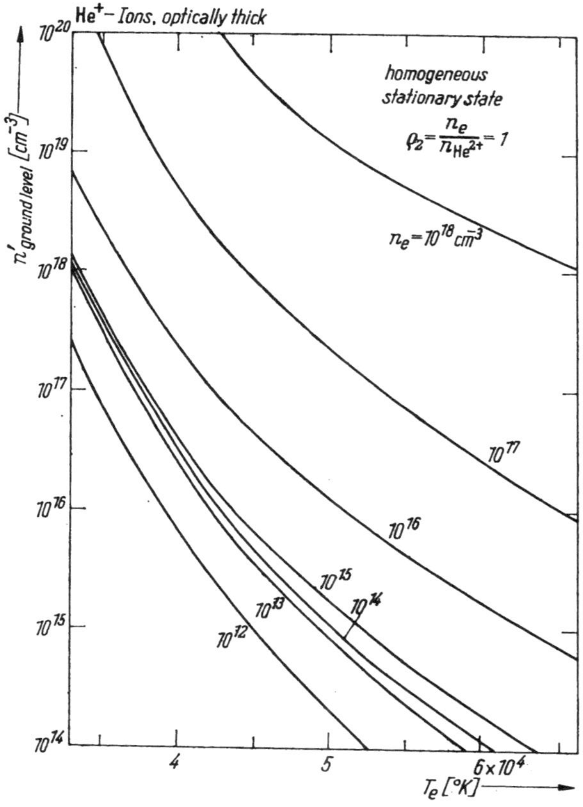

Figures 7.3a, b, c. The number density $n'_{\text{ground level}}$ of He^+-ions in the ground state $1s\,^2S_{1/2}$ in a homogeneous stationary plasma when $\varrho_2 = 1$ (after Ref. [41])

Figure 7.3a: all transitions optically thin

Figure 7.3.b: 99% reabsorption in the Ly α-line, 90% in the Ly β-line, and 33% in the Ly γ-line; all other transitions optically thin

Figure 7.3.c: 99.99% reabsorption in the Ly α-line, 99.9% in the Ly β-line, 99% in the Ly γ-line, 90% in the Ly δ-line, and 33% reabsorption in the Ly ε-line; all other transitions are assumed to be optically thin

For a pure hydrogen and helium plasma $\varrho_z = 1$ holds, thus, figures 7.1a to 7.2c represent the solutions for the stationary homogeneous state. One sees that for given values of T_e, n_e, the experimentally given ground state number densities may very often differ from the values as given by these curves. Especially in recombining plasmas the actual ground state number densities will be much smaller than those belonging to the homogeneous stationary state at the same temperature and the same electron density. Thus, the second term of eq. (2.9) has no further influence on the coefficient $b_{z-1,1}$. When this happens the conditions (2.3a), (2.8a), (2.8c) and (5.1) do not further represent good validity conditions for the establishment of partial L.T.E.

7.1. Optically Thin Case

A minimum condition for establishing partial L.T.E. must be the fulfillment of the conditions (2.3b) (2.8b). Making the same approximations as in Section II for eq. (2.8c) one obtains for the electron density the condition

$$n_e \geq 7.4 \cdot 10^{18} \frac{z^7}{i^{17/2}} \left(\frac{kT_e}{z^2 E_1^H}\right)^{1/2} \cdot \begin{cases} [\Psi_1(x_i)]^{-1} \text{ cm}^{-3} & \text{for atoms ,} \\ [\Psi_2(x_i)]^{-1} \text{ cm}^{-3} & \text{for ions ,} \end{cases} \tag{7.1a}$$

which must be satisfied when partial L.T.E. is to exist within 10% or better down to the level i.

In figure 7.4 and 7.5, we compare the values of the condition (7.1a) with the solutions of a coupled system of rate equations. The solid curves represent the electron density which necessary to give $r_{z-1,1}^{(0)}$ values lying between 0.9 and 1. The broken curves are the values of eq. (7.1a). One sees that the condition (7.1a) is too optimistic. At low temperatures in particular the discrepancies are rather large. A better agreement with the solid curves of Fig. 7.4 is obtained when the r. h. s. of (7.1a) is multiplied by $i^{1/2}$.

7.2. Optically Thick Case

Resonance absorption has a very complicated influence on the population coefficients $r_{z-1,i}^{(0)}$ and $r_{z-1,i}^{(1)}$ of the lower lying levels, see e.g. Ref. [41]: At low electron densities these coefficients are inversely proportional to the reduction coefficients Λ_{j1} already discussed in Sections IV and V. With increasing electron density the influence of Λ_{ji} upon $r_{z-1,i}^{(0)}$ and $r_{z-1,i}^{(1)}$ becomes smaller. It is higher at high temperatures than at low ones. At very low temperatures (i.e. $u_1 < 100$) resonance absorption has very little influence upon the population coefficients for all electron densities.

On this account it is practically impossible to give a simple and well-founded validity condition for partial L.T.E. of the lower-lying quantum levels.

For levels with high principal quantum number the situation is different, since resonance absorption is less important for these levels. The population densities are mainly determined by the electronic collisions and the radiative

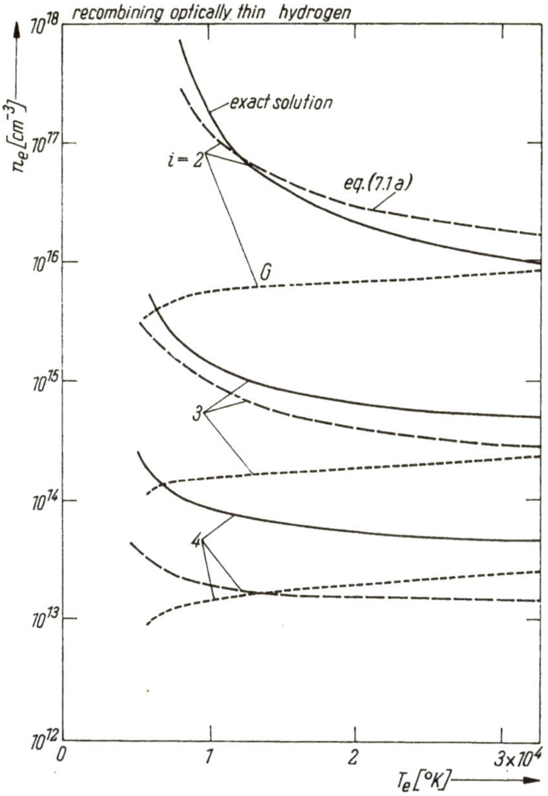

Figure 7.4. The electron density which is necessary to ensure partial L.T.E. (within 10% or better) for the hydrogen levels in a recombining optically thin plasma

- solid curves: from the coupled system of rate equations,
- broken curves: after eq. (7.1a),
- dotted curves: are for the corresponding homogeneous stationary state using Griem's [10] criterium.

de-excitation processes. To obtain an appropriate validity condition one may argue as in Chapter 5. Thus, one gets instead of relation (7.1a) the following condition for the electron density

$$n_e \geq 3.7 \cdot 10^{18} \frac{z^7}{i^{17/2}} \left(\frac{kT_e}{z^2 E_1^H}\right)^{1/2} \cdot \begin{cases} [\Psi_1(x_i)]^{-1} \text{ cm}^{-3} & \text{for atoms} \\ [\Psi_2(x_i)]^{-1} \text{ cm}^{-3} & \text{for ions,} \end{cases} \tag{7.1b}$$

It should only be applied when n_e is smaller than or equal to 10^{16} cm^{-3}.

Figure 7.6 shows as an example the electron densities necessary to attain $r^{(0)}_{z-1,\,i}$-values within the 10%-limit of unity for the lower lying levels of He$^+$-ions in the case of strong resonance absorption. These curves have been obtained by solving the corresponding system of coupled rate equations. One sees that

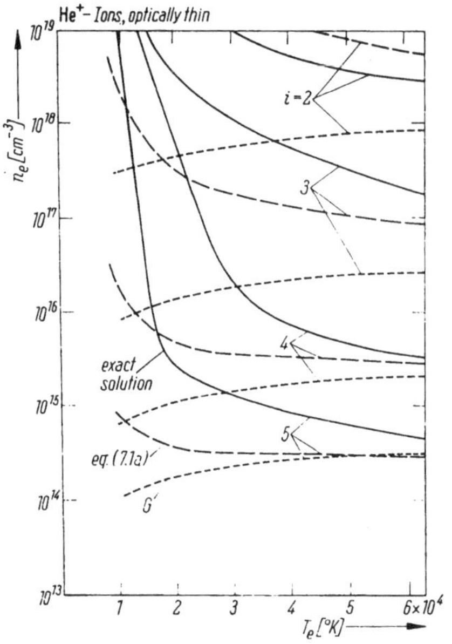

Figure 7.5. The electron density which is necessary to ensure partial L.T.E. (within 10% or better) for the levels of He^+-ions in a recombining optically thin plasma
The meaning of the curves is as in Figure 7.4

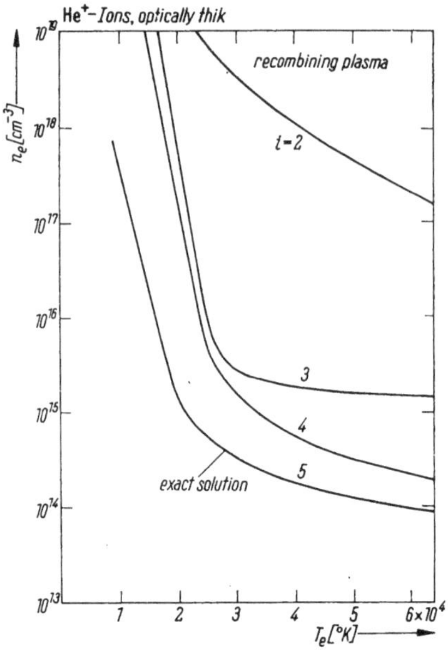

Figure 7.6. The electron density which is necessary to ensure partial L.T.E. (within 10% or better) for the levels of He^+-ions in a recombining optically thick plasma. Results obtained from the solutions of the system of coupled rate equations

relatively high n_e-values are needed to ensure partial L.T.E. at comparatively low electron temperatures. Further, there is not much difference between the optically thin and thick case at low T_e-values. This result is in so far of interest as high-density low-temperature plasmas immediately recombine due to the short recombination time which is of the order (e-fold decay time)

$$\tau^{(\text{recombination})} \cong 1/n_e \alpha_z = 1/z\, n_e \alpha^{CR} ,$$

where α_z is the collisional-radiative recombination coefficient (see tables 7.1 and 7.2). For instance, He^{2+}-ions at a density of $n_e = 1.28 \cdot 10^{16}$ cm^{-3} and a temperature of $T_e = z^2\, \Theta_z = 1.6 \cdot 10^4$ °K have disappeared in a time of

$$\tau^{(\text{recombination})} \cong (2 \cdot 1.28 \cdot 10^{16} \cdot 5 \cdot 10^{-11})^{-1} \approx 10^{-6}\ \text{s} .$$

Thus, one sees that a validity condition for partial L.T.E. of the lower-lying levels in a highly ionized low-temperature recombining plasma is only of little interest.

Table 7.1. Reduced collisional-radiative recombination coefficient α^{CR} in [cm^3 s^{-1}] for optically thin hydrogen (upper value) and hydrogenic ion plasma (lower-value) as a function of (reduced) electron density η_z and (reduced) electron temperature Θ_z (after Ref. [32])

η_z [cm^{-3}]	Θ_z [°K] 250	500	1000	2000	4000	8000	16000	32000	64000
10^4	4.3, −12	2.4, −12	1.7, −12	1.1, −12	7.0, −13	4.3, −13	2.6, −13	1.5, −13	8.8, −14
	3.9, −12	2.6, −12	1.7, −12	1.1, −12	7.0, −13	4.3, −13	2.6, −13	1.5, −13	8.8, −14
10^8	1.5, −11	6.8, −12	2.9, −12	1.4, −12	8.0, −13	4.6, −13	2.7, −13	1.6, −13	8.8, −14
	1.5, −11	6.4, −12	2.9, −12	1.4, −12	8.0, −13	4.6, −13	2.7, −13	1.6, −13	8.8, −14
10^9	1.0, −10	2.6, −11	6.4, −12	2.2, −12	9.9, −13	5.1, −13	2.8, −13	1.6, −13	8.9, −14
	1.0, −10	2.6, −11	6.5, −12	2.2, −12	9.9, −13	5.1, −13	2.8, −13	1.6, −13	8.9, −14
10^{10}	9.0, −10	1.3, −10	1.8, −11	4.1, −12	1.4, −12	6.2, −13	3.1, −13	1.7, −13	9.2, −14
	9.5, −10	1.4, −10	1.9, −11	4.2, −12	1.4, −12	6.2, −13	3.1, −13	1.7, −13	9.2, −14
10^{11}	8.7, − 9	8.8, −10	6.6, −11	9.3, −12	2.3, −12	8.2, −13	3.6, −13	1.8, −13	9.5, −14
	9.4, − 9	1.1, − 9	7.7, −11	9.8, −12	2.3, −12	8.2, −13	3.6, −13	1.8, −13	9.5, −14
10^{12}	8.6, − 8	7.1, − 9	3.1, −10	2.5, −11	4.2, −12	1.2, −12	4.4, −13	2.0, −13	1.0, −13
	9.4, − 8	1.0, − 8	4.4, −10	3.0, −11	4.5, −12	1.2, −12	4.5, −13	2.0, −13	1.0, −13
10^{13}	8.6, − 7	6.7, − 8	2.1, − 9	1.0, −10	1.0, −11	1.9, −12	6.0, −13	2.4, −13	1.1, −13
	9.4, − 7	1.0, − 7	3.6, − 9	1.5, −10	1.2, −11	2.1, −12	6.1, −13	2.4, −13	1.1, −13
10^{14}	8.6, − 6	6.6, − 7	1.8, − 8	5.4, −10	3.3, −11	4.2, −12	9.3, −13	3.1, −13	1.3, −13
	9.4, − 6	1.0, − 6	3.5, − 8	1.1, − 9	5.0, −11	4.9, −12	9.8, −13	3.1, −13	1.3, −13
10^{15}	—	—	1.7, − 7	4,1, − 9	1.3, −10	1.0, −11	1.8, −12	4.5, −13	1.5, −13
	—	—	3.5, − 7	1.0, − 8	3.0, −10	1.7, −11	2.1, −12	4.7, −13	1.5, −13
10^{16}	—	—	1.7, − 6	3.4, − 8	7.8, −10	4.6, −11	4.0, −12	6.4, −13	1.8, −13
	—	—	3.5, − 6	9.9, − 8	2.6, − 9	8.8, −11	5.1, −12	7.2, −13	1.9, −13
10^{17}	—	—	1.4, − 5	3.3, − 7	6.5, − 9	1.4, −10	6.7, −12	1.0, −12	3.1, −13
	—	—	2.6, − 6	9.7, − 7	2.5, − 8	4.8, −10	1.7, −11	1.8, −12	4.1, −13
[illegible]	[illegible]	[illegible]	[illegible]	2.0, − 6	4.8, − 8	4.7, −10	2.4, −11	4.6, −12	1.6, −12
	[illegible]	[illegible]	[illegible]	[illegible]	[illegible]	[illegible]	[illegible] −10	1.2, −11	2.6, −12

Note: $\alpha_z(n_e, T_e) = z\,\alpha^{CR}$; $S_{z-1}(n_e, T_e) = z^{-3}\,S^{CR}$; $\Theta_z = T_e/z^2$; $\eta_z = n_e/z^7$

... combination coefficient α^{CR} in [cm^3 s^{-1}] for optically thick hydrogen (upper value) and hydro- ... transition optically thin (after Ref. [32])

η_z [cm^{-3}]	Θ_z [°K] 250	500	1000	2000	4000	8000	16 000	32 000	64 000
10^4	4.3, −12 3.9, −12	2.9, −12 2.6, −12	1.7, −12 1.7, −12	1.1, −12 1.1, −12	6.9, −13 7.0, −13	3.4, −13 4.2, −13	1.4, −13 1.9, −13	8.2, −14 9.0, −14	5.1, −14 5.2, −14
10^8	1.5, −11 1.5, −11	6.6, −12 6.3, −12	2.8. −12 2.8, −12	1.4, −12 1.4, −12	7.8, −13 7.8, −13	3.5, −13 4.4, −13	1.4, −13 2.0, −13	8.2, −14 9.0, −14	5.1, −14 5.2, −14
10^9	1.0, −10 1.0, −10	2.4, −11 2.5, −11	5.8, −12 5.8, −12	2.0, −12 2.0, −12	9.2, −13 9.2, −13	3.7, −13 4.8, −13	1.4, −13 2.0, −13	8.2, −14 9.0, −14	5.1, −14 5.2, −14
10^{10}	8.9, −10 9.5, −10	1.3, −10 1.4, −10	1.6, −11 1.7, −11	3.6, −12 3.7, −12	1.2, −12 1.3, −12	4.1, −13 5.5, −13	1.4, −13 2.0, −13	8.2, −14 9.0, −14	5.1, −14 5.2, −14
10^{11}	8.7, − 9 9.4, − 9	8.5, −10 1.1, − 9	6.0, −11 7.0, −11	7.9, −12 8.4, −12	1.9, −12 2.0, −12	4.8, −13 6.7, −13	1.4, −13 2.0, −13	8.2, −14 8.9, −14	5.0, −14 5.1, −14
10^{12}	8.6, − 8 9.4, − 8	7.0, − 9 1.0, − 8	2.9, −10 4.4, −10	2.1, −11 2.6, −11	3.3, −12 3.6, −12	5.3, −13 8.3, −13	1.4, −13 1.9, −13	8.1, −14 8.7, −14	5.0, −14 5.1, −14
10^{13}	8.6, − 7 9.4, − 7	6.7, − 8 1.0, − 7	1.1, − 9 3.6, − 9	8.6, −11 1.3, −10	7.2, −12 9.3, −12	4.3, −13 9.3, −13	1.3, −13 1.6, −13	8.0, −14 8.3, −14	5.0, −14 5.0, −14
10^{14}	8.6, − 6 9.4, − 6	6.6, − 7 1.0, − 6	1.8, − 8 3.5, − 8	4.9, −10 1.1, − 9	1.9, −11 3.9, −11	3.7, −13 1.2, −12	1.3, −13 1.6, −13	8.0, −14 8.3, −14	5.0, −14 5.0, −14
10^{15}	— —	— —	1.7, − 7 3.6, − 7	3.8, − 9 1.0, − 8	6.0, −11 2.6, −10	6.7, −13 4.9, −12	1.5, −13 2.7, −13	8.4, −14 9.4, −14	5.1, −14 5.2, −14
10^{16}	— —	— —	1.7, − 6 3.5, − 6	3.3, − 8 9.9, − 8	4.1, −10 2.5, − 9	3.7, −12 4.1, −11	3.2, −13 1.4, −12	1.2, −13 2.0, −13	6.4, −14 7.4, −14
10^{17}	— —	— —	— —	3.3, − 7 9.7, − 7	3.9, − 9 2.4, − 8	3.4, −11 4.0, −10	2.0, −12 1.3, −11	4.7, −13 1.3, −12	1.9, −13 2.9, −13
10^{18}	— —	— —	— —	3.0, − 6 7.6, − 6	3.9, − 8 2.4, − 7	3.4, −10 4.0, − 9	1.9, −11 1.3, −10	4.0, −12 1.2, −11	1.5, −12 2.4, −12

Table 7.3. Reduced collisional-radiative ionization coefficient S^{CR} in [cm³ s⁻¹] for optically thin hydrogen (upper value) and hydrogenic ion plasma (lower value) as a function of (reduced) electron density η_z and (reduced) electron temperature Θ_z (after Ref. [32])

η_z [cm⁻³]	Θ_z [°K] 4000	7000	8000	9000	10000	11000	12000	16000	32000	64000	128000	256000
10^4	1.9, −26 2.6, −26	6,2, −19 6.4, −19	1.1, −17 1.1, −17	1.1, −16 1.1, −16	6.9, −16 6.9, −16	3.1, −15 3.1, −15	1.1, −14 1.1, −14	3.6, −13 3.6, −13	8.1, −11 8.1, −11	1.5, −9 1.5, −9	7.5, −9 7.5, −9	1.8, −8 1.8, −8
10^8	1.9, −26 8.7, −27	6.2, −19 6.7, −19	1.1, −17 1.2, −17	1.1, −16 1.2, −16	6.9, −16 7.2, −16	3.1, −15 3.2, −15	1.1, −14 1.1, −14	3.6, −13 3.7, −13	8.2, −11 8.2, −11	1.5, −9 1.5, −9	7.5, −9 7.5, −9	1.8, −8 1.8, −8
10^9	1.9, −26 3.0, −26	6.3, −19 7.5, −19	1.1, −17 1.3, −17	1.1, −16 1.3, −16	6.9, −16 7.8, −16	3.1, −15 3.5, −15	1.1, −14 1.2, −14	3.6, −13 3.9, −13	8.2, −11 8.4, −11	1.5, −9 1.5, −9	7.5, −9 7.5, −9	1.8, −8 1.8, −8
10^{10}	2.0, −26 3.9, −26	6.3, −19 9.0, −19	1.2, −17 1.6, −17	1.1, −16 1.5, −16	7.0, −16 8.8, −16	3.1, −15 3.9, −15	1.1, −14 1.3, −14	3.6, −13 4.2, −13	8.3, −11 8.7, −11	1.5, −9 1.6, −9	7.6, −9 7.6, −9	1.8, −8 1.8, −8
10^{11}	2.1, −26 6.9, −26	6.4, −19 1.3, −18	1.2, −17 2.1, −17	1.2, −16 1.9, −16	7.2, −16 1.1, −15	3.2, −15 4.9, −15	1.1, −14 1.6, −14	3.7, −13 4.8, −13	8.4, −11 9.4, −11	1.6, −9 1.6, −9	7.7, −9 7.8, −9	1.8, −8 1.9, −8
10^{12}	2.1, −26 1.6, −25	6.8, −19 2.3, −18	1.3, −17 3.6, −17	1.2, −16 3.1, −16	7.5, −16 1.7, −15	3.4, −15 7.2, −15	1.2, −14 2.4, −14	3.9, −13 6.4, −13	8.7, −11 1.1, −10	1.6, −9 1.7, −9	7.9, −9 8.2, −9	1.9, −8 1.9, −8
10^{13}	2.9, −26 6.6, −25	8.1, −19 5.9, −18	1.5, −17 8.5, −17	1.4, −16 6.8, −16	8.6, −16 3.6, −15	3.8, −15 1.4, −14	1.3, −14 4.4, −14	4.3, −13 1.1, −12	9.5, −11 1.4, −10	1.7, −9 2.1, −9	8.6, −9 9.1, −9	2.0, −8 2.1, −8
10^{14}	7.7, −26 5.1, −24	1.4, −18 2.7, −17	2.4, −17 3.5, −16	2.2, −16 2.5, −15	1.3, −15 1.2, −14	5.6, −15 4.5, −14	1.9, −14 1.3, −13	5.9, −13 2.6, −12	1.2, −10, 2.6, −10	2.2, −9 3.0, −9	1.0, −8 1.2, −8	2.5, −8 2.6, −8
10^{15}	4.9, −25 5.3, −23	5.1, −18 1.9, −16	7.8, −17 2.2, −15	6.6, −16 1.5, −14	3.7, −15 6.8, −14	1.5, −14 2.3, −13	5.0, −14 6.5, −13	1.4, −12 1.1, −11	2.3, −10 6.7, −10	3.5, −9 5.7, −9	1.5, −8 1.9, −8	3.5, −8 3.7, −8
10^{16}	4.4, −24 3.7, −22	3.3, −17 1.0, −15	4.5, −16 1.0, −14	3.4, −15 6.2, −14	1.7, −14 2.6, −13	6.2, −14 8.0, −13	1.8, −13 2.0, −12	3.6, −12 2.6, −11	3.6, −10 1.1, − 9	4.5, −9 7.6, −9	1.8, −8 2.3, −8	4.0, −8 4.3, −8
10^{17}	3.5, −23 8.8, −22	1.2, −16 1.8, −15	1.2, −15 1.7, −14	7.6, −15 9.4, −14	3.3, −14 3.6, −13	1.1, −13 1.1, −12	2.9, −13 2.7, −12	4.6, −12 3.1, −11	3.9, −10 1.2, − 9	4.6, −9 7.9, −9	1.9, −8 2.3, −8	4.1, −8 4.4, −8
10^{18}	1.2, −22 1.0, −21	1.6, −16 2.0, −15	1.5, −15 1.8, −14	8.7, −15 1.0, −13	3.6, −14 3.8, −13	1.2, −13 1.1, −12	3.1, −13 2.7, −12	4.8, −12 3.1, −11	3.9, −10 1.2, − 9	4.6, −9 7.9, −9	1.9, −8 2.3, −8	4.1, −8 4.4, −8

Table 7.4. Reduced collisional-radiative ionization coefficient S^{CR} in [cm^3 s^{-1}] for optically thick hydrogen (upper value) and hydrogenic ion plasma (lower value) as a function of (reduced) electron density η_z and (reduced) electron temperature Θ_z. All lines of Lyman series completely optically thick, all other lines optically thin (after Ref. [32])

η_z [cm^{-3}]	Θ_z [°K] 4000	7000	8000	9000	10000	11000	12000	16000	32000	64000	128000	256000
10^4	1.6, −24 1.3, −24	4.1, −17 5.2, −17	6.2, −16 9.5, −16	4.8, −15 9.0, −15	2.3, −14 5.3, −14	8.4, −14 2.2, −13	2.4, −13 7.3, −13	4.3, −12 1.6, −11	3.8, −10 1.1, − 9	4.6, −9 7.6, −9	1.9, −8 2.3, −8	4.1, −8 4.4, −8
10^8	1.6, −24 4.0, −25	4.1, −17 5.5, −17	6.2, −16 9.8, −16	4.8, −15 9.2, −15	2.3, −14 5.5, −14	8.5, −14 2.3, −13	2.4, −13 7.5, −13	4.3, −12 1.7, −11	3.8, −10 1.1, − 9	4.6, −9 7.6, −9	1.9, −8 2.3, −8	4.1, −8 4.4, −8
10^9	1.6, −24 2.4, −24	4.2, −17 6.0, −17	6.3, −16 1.0, −15	4.8, −15 9.9, −15	2.4, −14 5.8, −14	8.5, −14 2.4, −13	2.4, −13 7.8, −13	4.3, −12 1.7, −11	3.8, −10 1.1, − 9	4.6, −9 7.7, −9	1.9, −8 2.3, −8	4.1, −8 4.4, −8
10^{10}	1.8, −24 3.6, −24	4.5, −17 8.3, −17	6.6, −16 1.4, −15	5.0, −15 1.3, −14	2.4, −14 7.1, −14	8.7, −14 2.9, −13	2.5, −13 9.0, −13	4.3, −12 1.8, −11	3.8, −10 1.1, − 9	4.6, −9 7.7, −9	1.9, −8 2.3, −8	4.1, −8 4.4, −8
10^{11}	2.2, −24 7.5, −24	5.0, −17 1.3, −16	7.2, −16 2.1, −15	5.3, −15 1.7, −14	2.6, −14 9.4, −14	9.0, −14 3.6, −13	2.5, −13 1.1, −12	4.4, −12 2.0, −11	3.8, −10 1.1, − 9	4.6, −9 7.7, −9	1.9, −8 2.3, −8	4.1, −8 4.4, −8
10^{12}	3.7, −24 2.3, −23	6.6, −17 2.7, −16	8.8, −16 3.9, −15	6.2, −15 3.0, −14	2.8, −14 1.5, −13	9.7, −14 5.4, −13	2.7, −13 1.5, −12	4.5, −12 2.4, −11	3.8, −10 1.1, − 9	4.6, −9 7.8, −9	1.9, −8 2.3, −8	4.1, −8 4.4, −8
10^{13}	2.0, −23 1.8, −22	1.2, −16 1.0, −15	1.3, −15 1.1, −14	8.0, −15 7.1, −14	3.4, −14 3.0, −13	1.1, −13 9.3, −13	3.0, −13 2.4, −12	4.7, −12 3.0, −11	3.9, −10 1.2, − 9	4.6, −9 7.9, −9	1.9, −8 2.3, −8	4.1, −8 4.4, −8
10^{14}	9.6, −23 7.1, −22	1.6, −16 1.8, −15	1.5, −15 1.7, −14	8.8, −15 9.6, −14	3.6, −14 3.7, −13	1.2, −13 1.1, −12	3.1, −13 2.7, −12	4.8, −12 3.1, −11	3.9, −10 1.2, − 9	4.6, −9 7.9, −9	1.8, −8 2.3, −8	4.1, −8 4.4, −8
10^{15}	1.6, −22 1.0, −21	1.6, −16 2.0, −15	1.5, −15 1.8, −14	8.9, −15 1.0, −13	3.6, −14 3.8, −13	1.2, −13 1.1, −12	3.1, −13 2.8, −12	4.8, −12 3.2, −11	3.9, −10 1.2, − 9	4.6, −9 7.9, −9	1.9, −8 2.3, −8	4.1, −8 4.4, −8
10^{16}	1.7, −22 1.0, −21	1.6, −16 2.0, −15	1.5, −15 1.8, −14	8.9, −15 1.0, −13	3.6, −14 3.8, −13	1.2, −13 1.1, −12	3.1, −13 2.8, −12	4.8, −12 3.2, −11	3.9, −10 1.2, − 9	4.6, −9 7.9, −9	1.9, −8 2.3, −8	4.1, −8 4.4, −8
10^{17}	1.7, −22 1.0, −21	1.6, −16 2.0, −15	1.5, −15 1.8, −14	8.9, −15 1.0, −13	3.6, −14 3.8, −13	1.2, −13 1.1, −12	3.1, −13 2.8, −12	4.8, −12 3.2, −11	3.9, −10 1.2, − 9	4.6, −9 7.9, −9	1.9, −8 2.3, −8	4.1, −8 4.4, −8
10^{18}	1.7, −22 1.0, −21	1.6, −16 2.0, −15	1.5, −15 1.8, −14	8.9, −15 1.0, −13	3.6, −14 3.8, −13	1.2, −13 1.1, −12	3.1, −13 2.8, −12	4.8, −12 3.2, −11	3.9, −10 1.2, − 9	4.6, −9 7.9, −9	1.9, −8 2.3, −8	4.1, −8 4.4, −8

CHAPTER 8

Relaxation Times for the Establishment of Partial and Complete L.T.E. in Homogeneous Transient Plasmas

Relaxation processes characterized by so-called relaxation times τ_i play an important role when plasmas suffer fast temporal changes. The different bound states of the plasma particles follow these changes in different manners, the excited states much faster than the ground state. The higher the temperature and the electron density the shorter is the relaxation time, since the collision rates increase like $n_e \exp(-E_{ij}/kT_e)$. One generally assumes that relaxation times are considerably reduced in optically thick plasmas. This is true for small and medium electron densities, however, the higher the electron density the smaller is the radiative influence on the establishment of L.T.E.

It is extremely difficult to give an estimate of the relaxation times valid for optically thin and optically thick plasmas for all electron densities, as may be seen from the following equations.

When the relaxation times are to be calculated one has to consider all collisional and radiative processes within a level system. The spatial and temporal change of the number density $n_{z-1,i}$ of level i of a $(z-1)$-fold charged particle is governed by the following fundamental equation

$$\frac{\partial n_{z-1,i}}{\partial t} + \nabla \cdot (n_{z-1,i} \langle \boldsymbol{v}_{z-1,i} \rangle) = \left(\frac{\partial n_{z-1,i}}{\partial t}\right)_{\substack{\text{collision,}\\ \text{radiation}}}, \qquad i = 1, 2, \ldots \tag{8.1}$$

For homogeneous plasmas, the diffusion term is equal to zero. Expressing the right-hand side in terms of the collisional-radiative rate coefficients leads to the following relation

$$\begin{aligned} \frac{\partial n_{z-1,i}}{\partial t} = \left(\frac{\partial n_{z-1,i}}{\partial t}\right)_{\substack{\text{collision,}\\ \text{radiation}}} &= \sum_{h<i}^{i-1} n_h \left[n_e C_{hi}^{(e)} + \frac{n_i}{n_h} (1 - \Lambda_{ih}) A_{ih}\right] \\ &+ \sum_{k>i}^{p} n_k [n_e F_{ki}^{(e)} + A_{ki}] + n_z n_e [R_i^{(e)} + n_e Q_i^{(e)}] \\ &- n_i \sum_{h<i}^{i-1} [n_e F_{ih}^{(e)} + A_{ih}] \\ &- n_i \sum_{k>i}^{p} \left[n_e C_{ik}^{(e)} + \frac{n_k}{n_i} (1 - \Lambda_{ki}) A_{ki}\right] \\ &- n_i \left[n_e S_i^{(e)} + \frac{n_z n_e}{n_i} (1 - \Lambda_i) R_i^{(e)}\right], \end{aligned} \tag{8.2}$$

where the $n_{z-1,i}$ and n_z denote the number densities of particles of two successively following ionization stages (i.e. $z = 1$ for hydrogen). $C_{ij}^{(e)}$ and $F_{ji}^{(e)}$ are the temperature-dependent rate coefficients for excitation ($i \to j$) and de-excitation ($j \to i$), respectively, the A_{ji} the spontaneous transition probabilities, $R_i^{(e)}$ and $n_e\, Q_i^{(e)}$ the rate coefficients for radiative and three-body collisional recombination, respectively, $S_i^{(e)}$ is the collisional ionization coefficient, and Λ_j and Λ_{ji} are reduction coefficients due to radiative absorption of free-bound and bound-bound radiation, respectively. ($\Lambda_j, \Lambda_{ji} = 1$ for optically thin radiation; $\Lambda_j, \Lambda_{ji} < 1$ when re-absorption takes place). On the right-hand side of eq. (8.2) the index $z - 1$ has been dropped for the number densities $n_{z-1,h}$, $n_{z-1,i}$, etc.

8.1. Excited Levels $i > 1$

One sees that the return to a quasi-stationary state with respect to the actual number densities of ground state particles, $n_{z-1,1}$, and of electrons, n_e, after a sudden change of n_e and/or T_e takes place in a time of order $\tau_{z-1,i}$, where

$$\frac{1}{\tau_{z-1,i}} \cong n_e\, S_i^{(e)} + \frac{n_z\, n_e}{n_{z-1,i}} (1 - \Lambda_i)\, R_i^{(e)} + \sum_{h<i}^{i-1} [n_e\, F_{ih}^{(e)} + A_{ih}] + \sum_{k>i}^{p} \left[n_e\, C_{ik}^{(e)} + \frac{n_{z-1,k}}{n_{z-1,i}} (1 - \Lambda_{ki})\, A_{ki} \right]. \tag{8.3}$$

This equation, together with the condition for the homogeneous stationary state

$$\left(\frac{\partial n_{z-1,i}}{\partial t} \right)_{\substack{\text{collision} \\ \text{radiation}}} \simeq 0\,, \quad \text{with} \quad i = 1, 2, \ldots, p = i^* \tag{8.4}$$

yields the relaxation time $\tau_{z-1,i}$ for the level i.

Numerical values for hydrogen and hydrogen-like ions calculated from eqs. (8.3), (8.4) may be found in the papers of McWhirter and Hearn [29] and Drawin [30]. The reader is also referred to the paper [72].

A simple analytical formula is obtained by retaining in eq. (8.3) only the most important terms. This leads to the following relation

$$\tau_{z-1,i} \approx \frac{1}{n\, \langle \sigma^{(e)}\, v_e \rangle_{i \to (i-1)} + \sum_{h<i}^{i-1} A_{ih}}\,, \tag{8.5a}$$

where $\langle \sigma^{(e)}\, v_e \rangle_{i \to (i-1)} \equiv F_{i,i-1}^{(e)}$ is the rate coefficient for collisional de-excitation. At low electron densities the relaxation is essentially determined by the spontaneous transition probabilities at least for the lower-lying levels. However, when the electron density additionally fulfills the conditions (2.8b) or (2.8c), respectively, the radiative terms of eq. (8.5a) may be neglected and we obtain for the relaxation time necessary to establish partial L.T.E. for the level of principal quantum number i (in the hydrogenic approximation) the formula:

$$\tau_{z-1,i} \cong \frac{4.6\cdot 10^7}{n_e\, i^4}\, z^3 \left(\frac{kT_e}{E_1 H}\right)^{1/2} \cdot \begin{cases} \Phi_1(x_i)\ \mathrm{s} & \text{for atoms ,} \\ \Phi_2(x_i)\ \mathrm{s} & \text{for ions ,} \end{cases} \tag{8.5b}$$

with $x_i \cong 2\, z^2\, E_1^H/k\, T_e\, i^3$, and $z = 1$ for neutral atoms, and $z > 1$ for ions, n_e in cm^{-3}. Eqs. (8.5a, b) do, practically, not depend on radiative absorption, see Ref. [30]. When the ground state density $n_{z-1,1}$ is much larger than the value $n_{z-1,1}$ as calculated from eq. (8.4) the relaxation time for the establishment of partial L.T.E. is determined by the relaxation time for the ground level.

8.2. Ground Level $i = 1$

For the ground level $i = 1$ the situation is different compared to that of the excited states. If one would apply eq. (8.3) to the ground level one would neglect collisional de-excitation from the excited states (term $\sum_{k>1}^{p} n_k\, n_e\, F_{k_1}^{(e)}$ in eq. (8.2)) and the three-body collisional recombination (term $n_z\, n_e\, Q_1^{(e)}$ in eq. (8.2)). Both processes may give non-negligible contributions to the actual relaxation time when the plasma is already close to the steady state. If one would apply eq. (8.3) to the ground level one would obtain the relaxation time for ionization only. To see this more clearly one may write eq. (8.2) in terms of the effective collisional-radiative ionization and recombination coefficients*, S_{z-1} and α_z. (Numerical values of S_{z-1} and α_z may for instance be found in Refs. [31] and [32], some values have been listed in Tables 7.1 to 7.4). One obtains instead of eq. (8.1) for the ground level

$$\frac{\partial n_{z-1,1}}{\partial t} = \left(\frac{\partial n_{z-1,1}}{\partial t}\right)_{\substack{\text{collision}\\ \text{radiation}}} = n_z\, n_e\, \alpha_z - n_{z-1,1}\, n_e\, S_{z-1}\,. \tag{8.6}$$

The return to a steady state (complete ionization) takes place in a time of order

$$\tau_{z-1,1}^{(\text{ionization})} \approx \frac{1}{n_e\, S_{z-1}}\,. \tag{8.7}$$

This formula should, thus, be applied when the ground state number density at the beginning of the ionization is much larger than the density belonging to the steady state at given values n_e and T_e.

An appreciable change of the ground state number density always leads to an equivalent change of the number density n_z of the next higher ionized species. Thus, one can also write

$$-\frac{\partial n_{z-1,1}}{\partial t} \cong \frac{\partial n_z}{\partial t} = -\, n_z\, n_e\, \alpha_z + n_{z-1,1}\, n_e\, S_{z-1}\,. \tag{8.8}$$

* S_{z-1} is the rate coefficient for transitions $(z-1) \to z$ leading to ionization of particles of type $(z-1)$. The coefficient α_z characterizes transitions in the opposite direction, i.e. $z \to (z-1)$.

We now assume that the second term on the right-hand side of eq. (8.8) becomes so small (i.e. by a sudden lowering of the temperature) that it can be neglected. Thus, the plasma recombines with a time constant given by

$$\tau_{z,1}^{(\text{recombination})} \approx \frac{1}{n_e \alpha_z}, \tag{8.9}$$

which has the meaning of a "relaxation time for complete recombination". Eq. (8.9) should therefore be applied during the recombination phase, when ionizing collisions are negligibly small.

When the plasma is already very close to the steady state, ionization as well as recombination processes take place simultaneously with about the same frequency, thus, both processes will contribute to the actual relaxation time. Therefore, the appropriate relation will be

$$\frac{1}{\tau_{z-1,1}} \approx \frac{1}{\tau_{z-1,1}^{(\text{ionization})}} + \frac{1}{\tau_{z,1}^{(\text{recombination})}}. \tag{8.10}$$

The actual relaxation time for the ionization phase may be put into another useful form: Let $\Delta n_{z-1,1}/n_{z-1,1}\,(t=0)$ be the relative deviation from the equilibrium density, hence

$$\Delta n_{z-1,1} = n_{z-1,1}\,(t=0) - n_{z-1,1}^{(\text{equilib})}\,(t=\tau)\,.$$

The particles of initial number density $n_{z-1,1}$ $(t=0)$ will reach the equilibrium density $n_{z-1,1}^{(\text{equilib})}$ $(t=\tau)$ after a time approximately given by

$$\boxed{\tau_{z-1,1} \cong \frac{\Delta n_{z-1,1}}{n_{z-1,1}\,(t=0)}\,\tau_{z-1,1}^{(\text{ionization})} \cong \left(\frac{n_z}{n_{z-1,1}+n_z}\right)_{t=\tau} \tau_{z-1,1}^{(\text{ionization})}}, \tag{8.11}$$

where $\tau_{z-1,1}^{(\text{ionization})}$ is the time for complete ionization. If $n_{z-1,1}$ $(t=0)$ $\gg n_{z-1,1}^{(\text{equilib})}$ $(t=\tau)$ the eq. (8.11) leads back to eq. (8.7). The equilibrium density is given by the Saha-Eggert equation, for sufficiently high electron densities it may alternatively be calculated from the relation

$$n_z\, n_e \alpha_z - n_{z-1,1}\, n_e\, S_{z-1} = 0\,. \tag{8.12}$$

When the temperature is high the final state of the gas is reached after the particles have passed through several ionization stages. To obtain the relaxation time for ions of charge number $z'-1$, with respect to $t=0$, one has to sum over all individual relaxation times, hence

$$\tau_{z'-1,1} \approx \sum_{z=1}^{z'} \tau_{z-1,1}\,.$$

For all further discussion we use eq. (8.11).

To obtain a simple analytical expression for $\tau_{z-1,1}$ one has to find an appropriate relation for $\tau_{z-1,1}^{(\text{ionization})}$. It can be seen from eq. (8.2) that the following relation holds

$$\frac{1}{\tau_{z-1,1}^{(\text{ionization})}} \cong n_e\, S_1^{(e)} + \frac{n_z\, n_e}{n_{z-1,1}}\,(1-\Lambda_1)\, R_1^{(e)} + \sum_{k>1}^{p}\left[n_e\, C_{1k}^{(e)} + \frac{n_{z-1,k}}{n_{z-1,1}}\,(1-\Lambda_{k1})\, A_{k1}\right]. \tag{8.13}$$

In the optically thin case $\Lambda_1 = \Lambda_{k1} = 1$, thus, all terms containing the coefficients $1 - \Lambda_1$ and $1 - \Lambda_{k1}$ vanish. We now approximate the remaining terms by appropriate expressions:

In most cases holds

$$\frac{1}{10} \langle \sigma^{(e)} v_e \rangle_{1\to 2} < S_1^{(e)} < \frac{1}{2} \langle \sigma^{(e)} v_e \rangle_{1\to 2} .$$

We take

$$n_e S_1^{(e)} = 0.3\, n_e \langle \sigma^{(e)} v_e \rangle_{1\to 2} = 0.3\, n_e C_{12}^{(e)} . \tag{8.14a}$$

One further has

$$\sum_{k>1}^{p} n_e C_{1k}^{(e)} \equiv n_e \sum_{k>1}^{p} \langle \sigma^{(e)} v_e \rangle_{1\to 2} \approx n_e \langle \sigma^{(e)} v_e \rangle_{1\to 2} \equiv n_e C_{12}^{(e)} . \tag{8.14b}$$

With these approximations follows

$$n_e S_1^{(e)} + \sum_{k>1}^{p} n_e C_{1k}^{(e)} \approx 1.3\, n_e \langle \sigma^{(e)} v_e \rangle_{1\to 2} \cong \frac{1}{\tau_{z-1,1}^{(\text{ionization})} (\Lambda_{21} = 1)} , \tag{8.14c}$$

where $\langle \sigma^{(e)} v_e \rangle_{1\to 2}$ is the rate coefficient for collisional excitation of the first excited level of the resonance series in the hydrogen-like approximation. According to eq. (2.6b) one obtains two different analytical expressions, the one being valid for neutral atoms, the other one for ions. The relaxation time for complete ionization can finally be put into the following analytical form valid for the optically thin case ($\Lambda_{21} = 1$):

$$\tau_{z-1,1}^{(\text{ionization})} (\Lambda_{21} = 1) \cong \frac{0.89 \cdot 10^7}{n_e f_{12}} \frac{E_2 - E_1}{E_1^H} \left(\frac{kT_e}{E_1^H} \right)^{1/2} \cdot \left\{ \begin{array}{ll} \dfrac{1}{\Psi_1\left(\dfrac{E_2 - E_1}{kT_e}\right)} & \text{for atoms}, \\ \dfrac{1}{\Psi_2\left(\dfrac{E_2 - E_1}{kT_e}\right)} & \text{for ions}, \end{array} \right\} \text{s.} \tag{8.15}$$

The numerical values of the functions $1/\Psi_1(x)$ and $1/\Psi_2(x)$ may be taken from Table 2.1. f_{12} is the oscillator strength for the resonance line, and $E_2 - E_1$ is the energy difference between the ground and the first resonance level. In non-hydrogenlike level systems $E_2 - E_1$ has to be replaced by $E_{\max} - E_{\min}$ and f_{12} by $f_{\min,\max}$ (see e. g. eq. (3.1a)).

In the optically thick case ($\Lambda_1 = 1$, $\Lambda_{21} < 1$) for any degree of reabsorption) one may write

$$\tau_{z-1,1}^{(\text{ionization})} (\Lambda_{21} \neq 1) \cong \frac{1}{1.3\, n_e \langle \sigma^{(e)} v_e \rangle_{1\to 2} + \dfrac{n_{z-1,2}}{n_{z-1,1}} (1 - \Lambda_{21}) A_{21}} , \tag{8.16}$$

where the ratio $n_{z-1,2}/n_{z-1,1}$ has to be calculated from the coupled system of rate equations (8.4). However, when the system is very close to or in the state of complete L.T.E., this ratio can be replaced by the Boltzmann ratio times a coefficient a_2 accounting for eventual deviations from this ratio ($a_2 \leq 1$, see eq. (5.1). One obtains

$$\frac{1}{\tau_{z-1,1}^{(\text{ionization})}\,(\Lambda_{21} \neq 1)} \cong 1.3\, n_e \langle \sigma^{(e)} v_e \rangle_{1\to 2} + (1-\Lambda_{21})\, A_{21}\, a_2 \frac{g_2}{g_1} \exp\left(-\frac{E_2-E_1}{kT_e}\right). \tag{8.17 a}$$

With eq. (8.14c) follows

$$\frac{1}{\tau_{z-1,1}^{(\text{ionization})}\,(\Lambda_{21} \neq 1)} \cong \frac{1}{\tau_{z-1,1}^{(\text{ionization})}\,(\Lambda_{21} = 1)} + (1-\Lambda_{21})\, A_{21}\, a_2 \frac{g_2}{g_1} \exp\left(-\frac{E_2-E_1}{kT_e}\right). \tag{8.17 b}$$

When the resonance line is completely optically opaque one has $\Lambda_{21} = 0$! Under certain experimental conditions the second term in eq. (8.17 a, b) may be more important than the first one, the relaxation process is then determined by radiative rather than by collisional processes. For $\Lambda_{21} = 1$ (optically thin case) eqs. (8.17) reduce to eq. (8.15).

Numerical values for $\tau_{z-1,1}^{(\text{ionization})}$ calculated from the coupled system of rate equations may be found in the papers of McWhirter and Hearn [29] and Drawin [30], [33].

For an estimate of the relaxation time necessary for establishing complete L.T.E. it suffices to use eq. (8.11) in connection with eqs. (8.15) or (8.17) with the additional condition that the electron density has simultaneously to fulfill the relations (3.1) and/or (4.2). In other words: The ground state particles will be in a stationary state with respect to the instantaneous value of T_e when — or a given value of n_e — the relative temporal change of T_e obeys the criterion

$$\frac{T_e(t+\tau_{z-1,1}) - T_e(t)}{T_e(t)} \ll 1\,,$$

and complete L.T.E. is fulfilled when n_e additionally fulfills the conditions (3.1) and/or (4.2).

It may happen that n_e is not sufficiently high to fulfill the eqs. (3.1) or (4.2). In this case $\tau_{z-1,1}$ represents the relaxation time for the establishment of a stationary state in a homogeneous non-L.T.E. plasma. Especially $\tau_{z-1}^{(\text{ionization})}$ is the ionization time of a non-L.T.E. plasma.

In plasmas of high neutral gas density and low degree of ionization the relaxation times are mainly determined by atom-atom rather than by electron-atom collisions as treated here. Mon-atomic and poly-atomic gases behave quite differently. Atom-atom collisions play a role during the initial phase of a shock-heated gas. Some indicatians may be found in [59]—[65].

CHAPTER 9

Relaxation Lengths for the Establishment of Partial and Complete L.T.E. in Stationary Non-Uniform Plasmas

In non-uniform plasmas the density and temperature gradients will cause diffusion fluxes of various kinds. When the diffusion velocities are high and the spatial extensions of the plasma are small relaxation phenomena may lead to a spatial relaxation displacement of the particle densities and, thus, of the degree of ionization compared to the equilibrium state (Griem [10], Drawin et al. [34], Drawin [35], Artmann [36], Bergstedt [37], Tichmann [38]). Most sensitive are the ground state particles, since their cross sections are much smaller than those for the highly excited states. In order to obtain a validity condition for L.T.E. we start with the "zero-moment" of the Boltzmann collision equation for particles of chemical species $z-1$ in the i-th state

$$\frac{\partial n_{z-1,1}}{\partial t} + \nabla_r \cdot (n_{z-1,1}\langle \boldsymbol{v}_{z-1,i}\rangle) = \left(\frac{\partial n_{z-1,i}}{\partial t}\right)_{\substack{\text{collision}\\ \text{radiation}}}, \qquad i = 1, 2, \ldots, \tag{9.1}$$

where $\langle \boldsymbol{v}_{z-1,i}\rangle$ denotes the average diffusion velocity of particles $(z-1, i)$ in a co-ordinate frame fixed in space. The collision term on the right-hand side provides the coupling with the other particles and with the radiation field. A similar equation holds for the electrons. For time-independent plasmas, the first term on the left-hand side is equal to zero.

The remaining term is composed of a particle flux due to a density gradient and a particle flux due to a temperature gradient. However, $\langle \boldsymbol{v}_{z-1,i}\rangle$ does not equal the mean thermal velocity, it has in principle to be calculated from a closed system of equations in which the higher moments of the Boltzmann collision equation interfer. The collisional-radiative term on the right-hand side contains all types of collisions, superelastic and inelastic ones. Coupling with the photon field must also be taken into account.

To obtain a simple relation between the spatial dimension of the plasma and the diffusion coefficients we consider only the diffusion equation for the ground state particles ($i = 1$). The diffusion terms in eq. (9.1) may be approximated by

$$\nabla_r \cdot (n_{z-1,1}\langle \boldsymbol{v}_{z-1,1}\rangle) \approx \nabla \cdot (D_{z-1,1} \nabla_r n_{z-1,1}) \approx D_{z-1,1} \frac{\Delta n_{z-1,1}}{(\Delta l)^2}, \tag{9.2}$$

where $D_{z-1,1}$ is the effective diffusion coefficient of particles of type $(z-1, 1)$. For neutrals ($z = 1$), $D_{z-1,1}$ is the ordinary diffusion coefficient; for ions ($z > 1$), $D_{z-1,1}$ equals the ambipolar diffusion coefficient for diffusion of ion-electron pairs. The quantity $\Delta n_{z-1,1}$ is the change of the number density over the distance Δl.

We now express the right-hand side of eq. (9.1) in terms of the effective collisional-radiative recombination and ionization coefficients, α_z and S_{z-1}, see e.g. eq. (8.6). Putting the corresponding relations into eq. (9.2) yields for the stationary state (i.e. $\partial/\partial t \to 0$) the following expressions

for particles of kind $(z-1, 1)$:

$$\nabla_r \cdot (n_{z-1,1}\langle \boldsymbol{v}_{z-1,1}\rangle) \cong D_{z-1,1}\,\frac{\Delta n_{z-1,1}}{(\Delta l)^2} \cong +\, n_{z,1}\, n_e \alpha_z - n_{z-1,1}\, n_e\, S_{z-1}\,, \tag{9.3a}$$

(for particles of kind $(z, 1)$)

$$\nabla_r \cdot (n_{z,1}\langle \boldsymbol{v}_{z,1}\rangle) \cong D_{z,1}\,\frac{\Delta n_{z,1}}{(\Delta l)^2} \cong -\, n_{z,1}\, n_e \alpha_z + n_{z-1,1}\, n_e\, S_{z-1}\,. \tag{9.3b}$$

These equations show: *First*, in a sufficiently homogeneous plasma (i.e. $\nabla_r \cdot (\div) \to 0$) the ratio of the particle densities of two successively following ionization stages is inversely proportional to the corresponding collisional radiative rate coefficients

$$\left.\frac{n_{z-1,1}}{n_z}\right|_{\substack{\text{stationary}\\ \text{homogeneous}}} = \frac{\alpha_z}{S_{z-1}}\,.$$

This is the solution for the so-called "stationary-homogeneous case".

Second, in regions of strong diffusion fluxes (i.e. $\nabla_r \cdot (\div) \neq 0$) the right-hand sides of eqs. (9.3a), (9.3b) can be so different from zero that the actual population densities — especially those of the ground levels — considerably deviate from the solutions for the "stationary-homogeneous case", hence

$$\frac{n_{z-1,1}}{n_z} \neq \left.\frac{n_{z-1,1}}{n_z}\right|_{\substack{\text{stationary}\\ \text{homogeneous}}}$$

In the diffusion zone, the particle densities may be governed either by spatial recombination relaxation or by spatial ionization relaxation, or by both effects together.

Considering the local values of n_e and T_e as given quantities one can say that a plasma region shows (spatial) recombination relaxation if

$$n_z\, n_e \alpha_z > n_{z-1,1}\, n_e\, S_{z-1} \tag{9.4a}$$

and (spatial) ionization relaxation if

$$n_z\, n_e \alpha_z < n_{z-1,1}\, n_e\, S_{z-1}\,. \tag{9.4b}$$

It follows from this that in regions of recombination relaxation

$$\frac{n_{z-1,1}}{n_z} < \left.\frac{n_{z-1,1}}{n_{z,1}}\right|_{\substack{\text{stationary}\\ \text{homogeneous}}}\,.$$

holds, i.e. the number density of particles of type $(z-1, 1)$ is generally closer to a Saha-population than in stationary, homogeneous non-L.T.E.

plasmas. This is due to the fact that in stationary homogeneous non-L.T.E. plasmas

$$\left.\frac{n_{z-1,1}}{n_{z,1}}\right|_{\substack{\text{stationary}\\ \text{homogeneous}}} > \left.\frac{n_{z-1,1}}{n_{z,1}}\right|_{\text{Saha}}$$

holds. — In regions of (spatial) ionization relaxation the actual ground state density $n_{z-1,1}$ obeys the in-equality

$$\frac{n_{z-1,1}}{n_z} > \left.\frac{n_{z-1,1}}{n_{z,1}}\right|_{\substack{\text{stationary}\\ \text{homogeneous}}} .$$

This can easily be proved by comparing, for instance, the population densities for the stationary homogeneous state (for given values of n_e, T_e) with the number density given by the gas equation

$$n_{z-1,1} \cong \frac{p_{z-1}}{kT_{z-1}} . \tag{9.5}$$

In a region where ionization relaxation prevails one will find

$$\frac{p_{z-1,1}}{kT_{z-1}} > n_z \left(\frac{n_{z-1,1}}{n_{z,1}}\right)_{\substack{\text{stationary}\\ \text{homogeneous}}} .$$

In wall-stabilized as well as in magnetically confined plasmas the diffusion fluxes can be so strong that neither the particles in the ground level nor the excited particle densities will reach the equilibrium (Saha) population densities with respect to the local values of n_e and T_e. This is a consequence of eq. (2.9) which may be put into the following form

$$n_{z-1,i} = g^{(0)}_{z-1,i} + g^{(1)}_{z-1,i}\, n_{z-1,1} , \quad i > 1 .$$

One sees that the actual population densities $n_{z-1,i}$ can exceed the population densities $(n_{z-1,i})_{\text{hom. stat.}}$ for the homogeneous stationary state when the population density $n_{z-1,1}$ of ground state particles is much larger than the number density $(n_{z-1,1})_{\text{hom. stat.}}$ of the homogeneous stationary state for the same values of n_e and T_e.

In an inhomogeneous plasma, the particle diffusion influences the usual condition for complete L.T.E. if the terms due to electronic collisions and radiative processes are much smaller than the diffusion rates. On the other hand, diffusion does not influence the population densities if the collisional-radiative rates are larger than the diffusion rates, hence the condition

$$\left(\frac{\partial n}{\partial t}\right)_{\substack{\text{collision}\\ \text{radiation}}} \geq D\,\frac{\Delta n}{(\Delta l)^2} . \tag{9.6}$$

We now express the recombination and ionization rates in terms of the recombination and ionization times*: $n_z\, n_e\, \alpha_z = n_z/\tau_z^{(\text{recombination})}$, $n_{z-1,\,1}\, n_e\, S_{z-1} = n_{z-1}/\tau_{z-1}^{(\text{ionization})}$. Putting these relations into the expression (9.6) yields with the conditions (9.4a) and (9.4b) the following general conditions for the spatial extension of the diffusion zone which must be fulfilled so that diffusion processes, and hence spatial relaxation in the region Δl around l are negligible:

for recombination relaxation:

$$\Delta l \geq \left(\frac{\Delta n_z}{n_z}\right)^{1/2} (\tau_z^{(\text{recombination})}\, D_{z,\,1})^{1/2} \tag{9.7a}$$

for ionization relaxation:

$$\Delta l \geq \left(\frac{\Delta n_{z-1,1}}{n_{z-1,1}}\right)^{1/2} (\tau_{z-1,}^{(\text{ionization})}\, D_{z-1,\,1})^{1/2}\,. \tag{9.7b}$$

It is clear that the electron density appearing in eqs. (9.7a), (9.7b) has, in addition, to fulfill the conditions (3.1) or (4.2), respectively, when complete local thermodynamic equilibrium shall be ensured.

The right-hand sides of eqs. (9.7a), (9.7b) have the meaning of "mean diffusion lengths" $\lambda_{Z,\,1}^{(\text{diffusion})}$ and $\lambda_{Z-1,\,1}^{(\text{diffusion})}$, respectively, which the particles traverse before they come into chemical equilibrium. The quantity $(\tau_{Z-1,\,1}^{(\text{ionization})}\, D_{Z-1,\,1})^{1/2}$, for instance, is the diffusion length for complete ionization, $(\tau_{Z-1}^{(\text{ionization})}\, D_{Z-1,\,1})^{1/2} = \lambda_{z-1,\,1}^{(\text{ionization})}$. $\Delta n_{z-1,\,1}/n_{z-1,\,1}$ is the mean relative change of $n_{z-1,\,1}$ across the plasma column.

In the following we only consider eq. (9.7b). To obtain an analytical expression for the diffusion length one has to know $D_{z-1,\,1}$ and $\tau_{z-1}^{(\text{ionization})}$, and one has to make an estimate of the ratio $\Delta n_{z-1,\,1}/n_{z-1,\,1}$.

The relaxation time $\tau_{z-1}^{(\text{ionization})}$ for ionization has already been calculated in Chapter 8. The ratio $\Delta\, n_{z-1,\,1}/n_{z-1,\,1}$ can be considered as a known quantity (e.g. from measurements). Thus, we only need an analytical expression for $D_{z-1,\,1}$.

The diffusion coefficient $D_{z-1,\,1}$ is a function of the cross sections for heavy particle encounters and of the relative velocity between the collision partners. When no neutrals are present, $D_{z-1,\,1}$ may be approximated by the usual ambipolar diffusion coefficient for charged particles (elastic Coulomb encounters). In the presence of neutral atoms, however, collisions between neutral particles have also to be considered.

* Note that n_e, $n_{z,1}$, and $n_{z-1,\,1}$ are the actual mean number densities within the interval Δl around l. The coefficients α_z and S_{z-1} are functions of n_e and T_e.

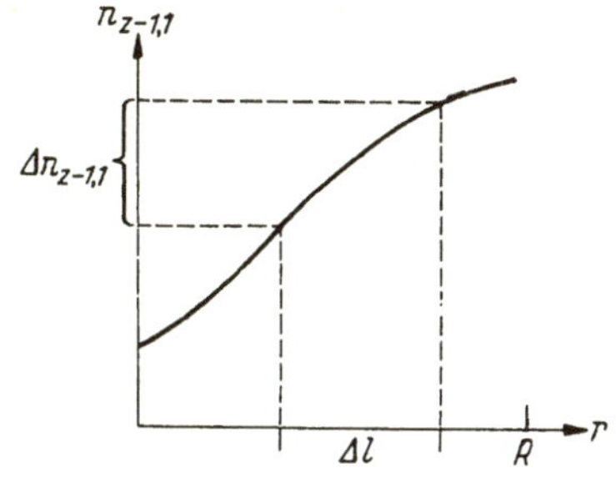

Figure 9.1. Illustration of the definition of $\Delta n_{z-1,\,1}$

The first approximation to the coefficient of diffusion of a binary mixture consisting of the two chemical species (*a*) and (*b*) is (see Ref. [39])

$$D^{(a,\,b)} = \frac{3\,(m_a + m_b)}{16\,n\,m_a\,m_b}\,\frac{kT}{\Omega^{(1,\,1)}_{a,\,b}} \approx \frac{3}{16}\left(\frac{\pi\,kT}{m_{a\,b}}\right)^{1/2}\frac{1}{n\,\langle\sigma^b_a(T)\rangle}\,, \tag{9.8}$$

where $\Omega^{(1,\,1)}_{a,\,b}$ denotes the collision integral for binary mixtures, in reduced units, $m_{a\,b}$ is the reduced mass, and $\langle\sigma^b_a(T)\rangle$ the average cross section at temperature T.

9.1. Collisions between Neutral Particles

When the degree of ionization is low diffusion is mainly determined by collisions between neutral particles. For mixtures consisting of two chemical species (*a*) and (*b*) the diffusion coefficient for the rigid sphere model is (see e.g. Hirschfelder et al. [39])

$$D^{(a,\,b)} = 2.63\cdot 10^{-3}\,\frac{T^{3/2}}{p\,d^2_{a\,b}}\left(\frac{M_a + M_b}{2\,M_a\,M_b}\right)^{1/2}\frac{\text{cm}^2}{\text{s}}\,, \tag{9.9}$$

with the neutral particle pressure p in Atm., the temperature T in °K, M in relative molecular weight (e.g. $M = 1$ for hydrogen), the effective molecular diameter $d_{a\,b}$ in Ångstrom units $\big(d_{a\,b} = (1/2)\,(d_a + d_b)\big)$. When only one chemical species is present one has $M_a = M_b$, $d_a = d_b$, and eq. (9.9) reduces to the so-called self-diffusion coefficient, $D^{(a,\,b)}$. Thus, at low degrees of ionization x ($x < 1\%$), one may identify $D_{z-1,\,1}$ with $D^{(a,\,b)}$ or $D^{(a,\,a)}$. In connection with the validity condition for complete L.T.E., however, this case is of minor importance, since complete L.T.E. demands high electron densities. Thus, collisions with ions may be much more important.

9.2. Collisions between Neutral Atoms and Singly Charged Ions

At moderately high degrees of ionization resonance charge transfer between atoms and ions (of the same chemical species) may play an important role.

Identifying $\langle\sigma(T)\rangle$ in eq. (9.8) with the cross section for resonance charge transfer one obtains with

$$\langle\sigma_{z-1,1}^{(\text{ch-transf})}\rangle \cong \pi\, a_0^2 \left(\frac{m_z}{2\,m_e}\right)^{1/2} \left(\frac{E_1^H}{kT_z}\right)^{1/2} \tag{9.10}$$

the following estimate for the diffusion coefficient of the neutral atoms $(z-1, 1)$

$$D_{z-1,1}^{(\text{ch-transf.})} \cong \frac{3\sqrt{2}}{16\,\pi\, e_0^2\, n_{z,1}} \left(\frac{m_e}{m_z}\right)^{1/2} \left(\frac{\pi\, kT_z}{m_z}\right)^{1/2} \left(\frac{kT_z}{E_1^H}\right)^{1/2}, \tag{9.11}$$

where $n_{z,1}$ is the number density of the charged particles (in cm^{-3}), m_z the mass (in g) of the collisions partners.

For ions, the diffusion coefficient is

$$D_{z,1}^{(\text{ch-transf.}} \cong \frac{3\sqrt{2}}{16\,\pi\, a_0^2\, n_{z-1,1}} \left(\frac{m_e}{m_{z-1}}\right)^{1/2} \left(\frac{\pi\, kT_{z-1}}{m_{z-1}}\right)^{1/2} \left(\frac{kT_{z-1}}{E_1^H}\right)^{1/2}. \tag{9.12}$$

$n_{z-1,1}$ is now the number density of the ground state atoms, further $m_{z-1} = m_z = m_a$. For L.T.E. plasmas holds $T_{z-1} = T_z = T_a$.

9.3. Collisions between Ions and Ions

In fully ionized plasmas the collisions are due to the long-range Coulomb interactions. Using the "90 degree" Coulomb scattering cross section one obtains from eq. (9.8) for singly ionized particles the diffusion coefficient

$$D_{1,1} \cong \frac{3}{16}\,\frac{3^2}{8\,\pi} \left(\frac{\pi}{m_Z}\right)^{1/2} \frac{(kT)^{5/2}}{e_0^4\, n_z \ln \Lambda_s}, \qquad \Lambda_s = \varrho_{\max}/\varrho_{\min}, \tag{9.13}$$

where n_z is the number density of the ions and m_z the ion mass. $\varrho_{\max}$ and $\varrho_{\min}$ denote the maximum and minimum cut-off distances in the theoretical treatment of Coulomb scattering.

9.4. Effective Diffusion Lengths $\lambda^{(\text{diffusion})}$

The effective diffusion coefficient may now be calculated from

$$\frac{1}{D_{\text{eff}}} \cong \sum_k \frac{1}{D_k}, \tag{9.14}$$

where D_k denotes the individual diffusion coefficient due to one type of interaction. Using further the analytical expressions for $\tau_{z-1,1}^{(\text{ionization})}$ (see eqs. (8.15), (8.17)) one obtains the following formulas for the mean diffusion lengths:

— diffusion of atoms ($z - 1 = 0$) in the ground level ($i = 1$):

$$\lambda^{(\text{diffusion})}_{\text{atoms},\,1} \cong \left(\frac{\Delta n_{0,1}}{n_{0,1}}\right)^{1/2} \frac{1.7 \cdot 10^{11}\,(m_e/m_a)^{1/2}\, n_e^{-1}}{\left[1 + \dfrac{n_a \pi r^2_{a,a}}{n_e \pi a_0^2}\left(\dfrac{m_e\, kT_a}{m_a\, E_1^H}\right)^{1/2}\right]^{1/2}} \left(\frac{E_2 - E_1}{f_{12}\, E_1^H}\right)^{1/2} \times \left(\frac{kT_e}{E_1^H}\right)^{1/4} \left(\frac{k\,T_a}{E_1^H}\right)^{1/4} \left(\frac{8\,kT_a}{\pi\, m_a}\right)^{1/4} \left(\frac{1}{\Psi_1\left(\dfrac{E_2 - E_1}{kT_e}\right)}\right)^{1/2} \text{cm} \tag{9.15}$$

— diffusion of ions ($z > 1$) in the ground level ($i = 1$):

$$\lambda^{(\text{diffusion})}_{z-1,\,1} \cong \left(\frac{\Delta n_{z-1,1}}{n_{z-1,1}}\right)^{1/2} \frac{6.2 \cdot 10^7\, T_z^{5/4}\, M_a^{-1/4}\, n_e^{-1}\,(\ln \Lambda_s)^{-1/2}}{\left[1 + \dfrac{2.1 \cdot 10^{-7}\, T_z^{5/2}}{M_a \ln \Lambda_s}\, \dfrac{n_a}{n_e} \left(\dfrac{m_a}{m_e}\right)^{1/2} \left(\dfrac{\pi\, m_a}{8\, kT_a}\right)^{1/2} \left(\dfrac{E_1^H}{kT_a}\right)^{1/2}\right]^{1/2}} \times \left(\frac{E_2 - E_1}{f_1\, E_1^H}\right)^{1/2} \left(\frac{kT_e}{E_1^H}\right)^{1/4} \left(\frac{1}{\Psi_2\left(\dfrac{E_2 - E_1}{kT_e}\right)}\right)^{1/2} \text{cm} \tag{9.16}$$

with n_a and n_e in cm^{-3}, T in °K, M_a in relative atomic weight units. The quantity $\pi\, r^2_{a,a} \equiv \langle \sigma^a_a(T) \rangle$ denotes the (classical) atomic gas kinetic cross section for the special type of atomic collisions considered, n_a is the number density of neutral atoms, T_ε the electron temperature, T_z the ion temperature, T_a the temperature of the neutral atoms. The quantity $(8\, kT_a/\pi\, m_a)^{1/4}$ appearing in eqs. (9.15) and (9.16) equals the square root of the mean thermal velocity of the neutral atoms (in units (cm/s)$^{1/2}$). The numerical value of $\ln \Lambda_s$ lies generally between 10 and 20.

One sees from eqs. (9.15) and (9.16) that the L.T.E. condition

$$\Delta l \geq \lambda^{(\text{diffusion})} \tag{9.17}$$

may be severe for neutral particles as well as for ions. The diffusion length of ions will especially be large in regions close to the boundary where the gradient of the electron density can reach large values.

Table 9.1. Values for $(n_{0,1}/\Delta n_{0,1})^{1/2}\, \lambda^{(\text{diffusion})}_{\text{atom},\,1}$ in cm for ground state atoms when the electron density is $n_e = 1 \cdot 10$ cm^{16} cm^{-3}, after eq. (9.15)

gas	$T_e = T_a = 8000$ °K	12000°K	16000 °K	24000 °K	48000 °K
H	$1.2 \cdot 10^1$	$7.9 \cdot 10^0$	$1.9 \cdot 10^{-1}$	$3.4 \cdot 10^{-2}$	—
He	$1.8 \cdot 10^5$	$1.9 \cdot 10^3$	$1.4 \cdot 10^{+1}$	$9.8 \cdot 10^0$	$7.8 \cdot 10^{-1}$
Ar	$6.2 \cdot 10^0$	$4.5 \cdot 10^0$	$9.2 \cdot 10^{-2}$	$2.1 \cdot 10^{-2}$	—

In table 9.1 we give some values of the quantity $(n_{0,1}/\Delta n_{0,1})^{1/2}$ times $\lambda^{(\text{diffusion})}_{\text{atoms},\,1}$ for the ground state particles of H, He, and Ar atoms at an electron density of $n_e = 10^{16}$ cm^{-3} as calculated from eq. (9.15) The bracket in the denominator of eq. (9.15) has been put equal to unity. At low degrees of ionization, however, the second term in the bracket may contribute to the value of $\lambda^{(\text{diffusion})}_{\text{atoms},1}$. The condition for complete L.T.E. is (see eq. (9.17)) $\Delta l \geq \lambda^{(\text{diffusion})}$. It should be emphasized, however, that the diffusion length of the higher excited particles is much smaller than that of the ground state atoms. The diffusion length of excited atoms in a state of principal quantum number i may be estimated from eq. (9.7) when $\tau_{z-1,1}$ and $D_{z-1,1}$ are replaced by $\tau_{z-1,i}$ and $D_{z-1,i}$, respectively. One obtains for neutral particles in a level of principal quantum number i the formula

$$\lambda^{(\text{diffusion})}_{\text{atoms},\,i} \cong \frac{8.3 \cdot 10^{11}\,(m_e/m_a)^{1/4}\,n_e^{-1}\,i^{-2}}{\left[1 + \frac{n_a\,\pi\,r^2_{a,a}}{n_e\,\pi\,a_0^2}\,i^4\left(\frac{m_e\,kT_a}{m_a\,E_1^H}\right)^{1/2}\right]^{1/2}} \left(\frac{kT_e}{E_1^H}\right)^{1/4}\left(\frac{kT_a}{E_1^H}\right)^{1/4} \times \left(\frac{8\,kT_a}{\pi\,m_a}\right)^{1/4}\left(\frac{1}{\Psi_1\left(\frac{2\,E_1^H}{kT_e\,i^3}\right)}\right)^{1/2}\text{cm} \tag{9.18}$$

The function $(\Psi_1(x))^{-1/2}$ may be calculated from the values listed in Table 2.1. It follows from this relation that partial L.T.E. can easily be established, and even in extremely inhomogeneous plasmas diffusion processes will have a small influence on the relative equilibrium population of the higher excited states, provided that the electron density is sufficiently high. It follows from this that the spectroscopic temperature determination based on a measurement of the relative intensities originating from the same parent (ion) atoms is practically not influenced by diffusion. However, the measurement of relative intensities of spectral lines originating from two successively following ionization stages may eventually lead to temperatures which are not consistent with the assumption of complete L.T.E. required for this kind of measurement.

The eqs. (9.15) to (9.18) are based on the cross sections for symmetrical or resonant charge transfer of the type $A + \vec{A}^+ = A^+ + \vec{A}$. The corresponding cross sections are large at low impact energies, see eq. (9.10). In gas mixtures one has also to consider the asymmetric charge transfer collisions $A + \vec{B}^+ = A^+ + \vec{B}$. These cross sections are extremely small at low impact energies (see, e.g. Hasted [40]). When small traces are added to a main gas, e.g. a small percentage of hydrogen in argon, the diffusion of the ground state particles of the tracer gas can not be calculated from eq. (9.15). Due to the small values of the cross sections for asymmetric charge transfer the diffusion of the tracer gas is governed by the elastic atom-atom collisions rather than by the charge transfer collisions. With eq. (9.8), this leads to the following diffusion length for the ground state atoms of the tracer gas or the impurities

$$\lambda^{(\text{diffusion})}_{\text{impurity}} \cong \frac{1.6 \cdot 10^{11}}{n_e^{1/2}\,n_a^{1/2}\,(\langle\sigma_a^b(T)\rangle)^{1/2}}\left(\frac{T_{ab}}{M_{ab}}\right)^{1/4}\left(\frac{E_2 - E_1}{f_{12}\,E_1^H}\right)^{1/2}\left(\frac{kT_e}{E_1^H}\right)^{1/4}\left(\frac{1}{\Psi_1\left(\frac{E_2 - E_1}{kT_e}\right)}\right)^{-1/2}\text{cm}, \tag{9.19}$$

where n_a in cm^{-3} is the number density of the main gas, $\langle\sigma_a^b(T)\rangle$ in cm^2 the elastic cross section for atom-atom collisions (tracer gas $\rightarrow$ main gas), and $E_2 - E_1$ the excitation energy of the resonance level of the tracer gas. This equation can lead to large diffusion lengths.

The results may now be summarized as follows: Any level i of particles of type $(z - 1)$ shows the equilibrium population (i.e. partial L.T.E.) with respect to the local value of the electron temperature T_e if the spatial variation of T_e obeys the criterion,

$$\frac{T_e(l) - T_e\left(l + \lambda_{z-1,i}^{(\text{diffusion})}\right)}{T_e(l)} \ll 1$$

with the additional requirement that n_e fulfills the conditions (2.8b), (2.8c), or (5.1), respectively. For the fulfillment of complete L.T.E. with respect to the local value of T_e the electron temperature has to obey the criterion

$$\frac{T_e(l) - T_e\left(l + \lambda_{z-1,1}^{(\text{diffusion})}\right)}{T_e(l)} \ll 1\,,$$

with the additional requirement that n_e fulfills the conditions (3.1a), (3.1b) or (4.2), respectively.

It should be pointed out that the diffusion length $\lambda_{z-1,1}^{(\text{diffusion})}$ for a given type of particle can directly be estimated when for this type of particle the collisional-radiative recombination (α_z) and ionization coefficients (S_{z-1}) are known from the solutions of the system of compled rate equations. Especially the relaxation length for ionization can be given the following analytical form

$$\lambda_{z-1,1}^{(\text{ionization})} \simeq \left(\frac{D_{z-1,1}}{n_e S_{z-1,1}}\right)^{1/2}. \tag{9.20}$$

CHAPTER 10

Perturbation of the Electron Temperature by Externally Applied Electric Fields

Since the fundamental papers of Druyvesteyn [43], [44] and Smit [45] many articles have been published in which the influence of an externally applied electric field upon the velocity distribution of the electrons is treated. Concerning more recent papers the following authors may be mentioned: Boyd and Twiddy [46], Thompson [47], Twiddy [48], Carleton and Megilll [49], Sampson and Enoch [50], Wojaczek [51], Swift [52], Dugan, Lyman and Albers [53], Morrone [54], Kagan and Millenin [55], Kagan et al. [56], Peyraud [57], Postma [58], and many others. A detailed description of these papers is out of the scope of this article. We only discuss very briefly the influence of an externally applied electric d.c. field on the mean thermal energy of the electrons and heavy particles.

In order to get equal mean thermal kinetic energies, i.e. equal temperatures for electrons and ions, one has to require that the energy gain of the electrons due to the presence of the electric field remains small over the mean free path (Maecker and Finkelnburg [42]). In other words: To obtain at any instant $T_e = T_{\text{ion}} = T_{\text{gas}}$ one has to ensure that the rate for energy transfer by electronic collisions $(\partial\langle E\rangle/\partial t)_{\substack{\text{electronic}\\ \text{collisions}}}$ is much larger than the rate with which the electrons gain kinetic energy from the electric field, i.e. the condition

$$\left(\frac{\partial\langle E\rangle}{\partial t}\right)_{\substack{\text{electronic}\\ \text{collisions}}} \gg \left(\frac{\partial\langle E\rangle}{\partial t}\right)_{\substack{\text{electric}\\ \text{field}}} . \tag{10.1}$$

The term on the left-hand side accounts for collisions of electrons with neutrals and with ions. With the assumption that $T_{\text{ion}} = T_{\text{gas}}$ follows

$$\left(\frac{\partial\langle E\rangle}{\partial t}\right)_{\substack{\text{electronic}\\ \text{collisions}}} \cong \frac{3}{2}\, k\,(T_e - T_{\text{ion}})\, \frac{2\, m_e}{m_z\, \tau_c} . \tag{10.2}$$

The coefficient $2\, m_e/m_z$ accounts for the energy transfer per collision. τ_c is the time between two effective collisions, $\nu_c = 1/\tau_c$ is the collision frequency. The right-hand side of (10.1) can be written as follows

$$\left(\frac{\partial\langle E\rangle}{\partial t}\right)_{\substack{\text{electric}\\ \text{field}}} = e_0|\varepsilon|\, \frac{d_e}{\tau_c} = e_0^2|\varepsilon|^2\, \frac{\tau_c}{2\, m_e} , \tag{10.3}$$

where $|\varepsilon|$ is the absolute value of the electric field strength, d_e is the distance that travels an electron between two collisions. With $\tau_c = \lambda_e/\langle v_e\rangle$, $\langle v_e\rangle =$

$= (8\,kT_e/\pi\, m_e)^{1/2}$ follows after division by T_e the condition

$$\frac{\Delta T}{T} = \frac{T_e - T_{\text{ion}}}{T_e} = \frac{\pi\, m_z}{3\, m_e} \left(\frac{\lambda_e\, e_0\, |\varepsilon|}{4\, kT_e} \right)^2 \ll 1\,, \tag{10.4}$$

which has already been given by Finkelnburg and Maecker [42] in a slightly different form. Relation (10.4) can only by fulfilled when λ_e and ε are sufficiently small, otherwise the electron temperature lies above the gas temperature. When λ_e and $|\varepsilon|$ are known one can calculate the percentage increase of the electron temperature $\Delta T/T$.

A relation similar to (10.4) has been given by Griem [10]:

$$|\varepsilon|^2 \ll \left(5.5 \cdot 10^{-12}\, n_e\, \frac{E_1^H}{kT_e} \right)^2 \frac{m_e}{m_z} \qquad (\text{V/m})^2\,, \tag{10.5}$$

with n_e in cm^{-3}. In this inequality only electron-ion collisions have been taken into account, electron-atom encounters have been disregarded. Thus, relation (10.4) is more general than relation (10.5).

Concerning the equality of ion and gas temperatures one can say that under almost all experimental conditions the differences between T_{ion} and T_{gas} are generally so small that they need not be taken into account.

CHAPTFR 11

Calculation of Optical Reduction Factors Λ_{ji} and Λ_i

The Fig. 4.1 shows that reabsorption has a significant influence on the establishment of complete local thermodynamic equilibrium. Reabsorption has been accounted for by optical reduction factors Λ_{ji} for bound-bound radiation and by optical reduction factors Λ_i for free-bound transitions. For resonance lines, such reduction factors have already been calculated by Holstein [66] who thermed them "photon escape factors g". Holstein's g-factors and our Λ_{ji}-factors are physically equivalent.

Denoting by $\hat{\sigma}_{ji}$ the absorption cross section for photons in the center of a spectral line $i \rightleftarrows j$ $(j > i)$ and by

$$\hat{\sigma}_{ji}\,\bar{n}_i\,l = \hat{\varkappa}_{ji}\,l = \hat{\tau}_{ji} \tag{11.1}$$

the optical depth in the center of this line, Holstein's method yields for particles situated on the axis of an infinitely long cylinder of radius $l = R$

— in the case of a Doppler-(Gauss)-profile:

$$\Lambda_{ji}^{(G)} \cong \frac{1.6}{\hat{\sigma}_{ji}\,\bar{n}_i\,R(\pi \ln(\hat{\sigma}_{ji}\,\bar{n}_i\,R))^{1/2}} = \frac{1.6}{\hat{\tau}_{ji}(\pi \ln \hat{\tau}_{ji})^{1/2}}, \tag{11.2a}$$

— in the case of a dispersion profile:

$$\Lambda_{ji}^{(D)} \cong \frac{1.115}{(\pi\,\hat{\sigma}_{ji}\,\bar{n}_i\,R)^{1/2}} = \frac{1.115}{(\pi\,\hat{\tau}_{ij})^{1/2}}. \tag{11.2b}$$

These formulas are only valid for large optical depths. In practice it is sufficient to put $\hat{\tau}_{ji} > 3$. Holstein did not calculate reduction factors for free-bound transitions.

For practical applications it is important to estimate the order of magnitude of Λ_{ji} and Λ_i. Therefore, we give a brief mathematical derivation of these factors. It is based on an application of the equation of radiative transfer and the rate requations determining the population densities.

1. The Equation of Radiative Transfer

The equation of transfer for the specific radiation intensity I_{ji} in the direction of $\boldsymbol{r}$ in the frequency range $\nu, \ldots, \nu + \mathrm{d}\nu$ within a spectral line $i \rightleftarrows j$ is given by

$$\frac{\mathrm{d}I_{ji}(\boldsymbol{r},\nu)}{\mathrm{d}\boldsymbol{r}} = -\,I_{ji}(\boldsymbol{r},\nu)\,[B_{ij}\,n_i\,\varphi_{ij}(\nu) - B_{ji}\,n_j\,\psi_{ji}(\nu)]\,\frac{h\,\nu}{4\,\pi} + A_{ji}\,n_j\,\Phi_{ji}(\nu)\,\frac{h\,\nu}{4\,\pi}. \tag{11.3}$$

with the normalized line profiles

$$\int \varphi_{ij}(\nu)\,d\nu = \int \psi_{ji}(\nu)\,d\nu = \int \Phi_{ji}(\nu)\,d\nu = 1\,. \tag{11.4}$$

The A_{ji}, B_{ij} and B_{ji} are the Milne-Einstein-coefficients:

$$B_{ji} = \frac{c^2}{2\,h\,\nu_{ji}^3}\,A_{ji}\,, \tag{11.5a}$$

$$B_{ij} = \frac{c^2}{2\,h\,\nu_{ji}^3}\,\frac{g_j}{g_i}\,A_{ji}\,, \tag{11.5b}$$

$$B_{ji}\,g_j = B_{ij}\,g_i\,. \tag{11.5c}$$

They are defined in terms of the mean radiation intensity $I_{ji}(\boldsymbol{r}', \nu_{ji})$ in the spatial point $\boldsymbol{r}'$:

$$4\,\pi\,\bar{J}_{ji}(\boldsymbol{r}', \nu_{ji}) = \int\limits_{\Omega} I_{ji}(\boldsymbol{r}', \Omega, \nu_{ji})\,d\Omega\,. \tag{11.6}$$

Integration is over all directions $\{\Omega\}$ surrounding $\boldsymbol{r}'$.

By introducing the emission and absorption coefficients, ε_{ji} and $\varkappa_{ji}$, (11.3) can be transformed in a simple transport equation. One defines (see e.g. [67])

$$\varepsilon_{ji}(\boldsymbol{r}, \nu) = \frac{h\,\nu}{4\,\pi}\,A_{ji}\,n_j(\boldsymbol{r})\,\Phi_{ji}(\nu)\,, \tag{11.7a}$$

$$\varkappa_{ji}(\boldsymbol{r}, \nu) = \frac{h\,\nu}{4\,\pi}\,[B_{ij}\,n_i(\boldsymbol{r})\,\varphi_{ij}(\nu) - B_{ji}\,n_j(\boldsymbol{r})\,\psi_{ji}(\nu)]\,. \tag{11.7b}$$

We make the assumption that the line profiles for spontaneous emission, true absorption and stimulated emission are equal:

$$\Phi_{ji}(\nu) = \varphi_{ij}(\nu) = \psi_{ji}(\nu) = P_{ji}(\nu)\,. \tag{11.8}$$

Introducing now (11.7a, b) into (11.3) yields the simple relation

$$\frac{dI_{ji}(\boldsymbol{r}, \nu)}{d\boldsymbol{r}} = -\,\varkappa_{ji}(\boldsymbol{r}, \nu)\,[I_{ji}(\boldsymbol{r}, \nu) - S_{ji}(\boldsymbol{r}, \nu)]\,, \tag{11.9}$$

where S_{ji} is termed "source function" which — due to (11.5) — can be put into the following analytical form:

$$S_{ji}(\boldsymbol{r}, \nu) = \frac{\varepsilon_{ji}(\boldsymbol{r}, \nu)}{\varkappa_{ji}(\boldsymbol{r}, \nu)} = \frac{2\,h\,\nu_{ji}^3}{c^2}\cdot\frac{1}{g_j\,n_i(\boldsymbol{r})/g_i\,n_j(\boldsymbol{r}) - 1}\,. \tag{11.10}$$

For the absorption coefficient follows

$$\varkappa_{ji}(\boldsymbol{r}, \nu) = \frac{c^2}{8\,\pi\,\nu_{ji}^2}\,A_{ji}\left[1 - \frac{n_j(\boldsymbol{r})\,g_i}{n_i(\boldsymbol{r})\,g_j}\right]\frac{g_j}{g_i}\,n_i(\boldsymbol{r})\,P_{ji}(\nu)\,. \tag{11.11}$$

We consider first a plasma in complete thermodynamic equilibrium. In this case $n_j(\boldsymbol{r})/n_i(\boldsymbol{r})$ is equal to the Boltzmann ration. One obtains

$$S_{ji}(\boldsymbol{r}, \nu) = S_{ji}(\boldsymbol{r}, \nu_{ji}) = \frac{2\,h\,\nu_{ji}^3}{c^2}\,[e^{h\nu_{ji}/kT} - 1]^{-1}\,, \tag{11.12}$$

i.e. the source function does not depend on the line profile. It equals the Planck-function at temperature T. It is also independent of $\boldsymbol{r}$, since T was assumed to be independent of $\boldsymbol{r}$. We now define a differential optical depth $d\tau_{ji}(\nu) = \varkappa_{ji}(\boldsymbol{r}, \nu)\, d\boldsymbol{r}$. Thus, (11.9) leads to

$$\frac{dI_{ji}(\tau_{ji}, \nu)}{d\tau_{ji}} = S_{ji}(\tau_{ji}, \nu) - I_{ji}(\tau_{ji}, \nu)\,. \qquad (11.13)$$

Inserting (11.12) into (11.13) and integrating over τ_{ji} yields with the boundary condition $I_{ji} = 0$ at $\tau_{ji} = 0$ the following expression for the specific intensity

$$I_{ji}(\tau_{ji}, \nu) = \frac{2\,h\,\nu_{ji}^3}{c^2}\,\frac{1}{\exp(h\,\nu_{ji}/kT) - 1}\,[1 - e^{-\tau_{ji}}]\,. \qquad (11.14)$$

For $\tau_{ji} \to \infty$, the specific intensity approaches the Planck intensity!

We consider now a non-equilibrium plasma, i.e. a plasma in which $n_j(\boldsymbol{r})/n_i(\boldsymbol{r})$ is not further given by a Boltzmann ratio. We make the assumption that the source function remains constant across the whole plasma, i.e.

$$S_{ji}(\tau_{ji}, \nu) = S_{ji}(\nu_{ji}) = \text{constant}\,. \qquad (11.15)$$

Under actual conditions (11.15) will not be fulfilled. We use this relation only as a formal hypothesis. Putting (11.15) into (11.13) and integrating over τ_{ji} yields with the boundary condition $I_{j0} = 0$ at $\tau_{ji} = 0$ the relation

$$I_{ji}(\tau_{ji}, \nu_{ji}) = S_{ji}(\nu_{ji})\,[1 - e^{-\tau_{ji}})]\,. \qquad (11.16)$$

For $\tau_{ji} \to \infty$, the specific intensity approaches now the non-equilibrium source function! This result is in obvious contradiction to a fundamental law of thermodynamics which says that a gas enclosed in an infinitely large box ($\tau_{ji} \propto \boldsymbol{r} \to \infty$) must be in equilibrium at least at large optical depths, i.e. the source function must reach there the value given by the Planck function.

We have chosen this example in order to demonstrate that the calculation of Λ_{ji} and Λ_i which follows below should only be considered as a rough estimate! It does not represent an exact solution. Our estimate will be based on the assumption that the source function does not vary in space. When one makes this assumption the specific intensity calculated from a two-level line source function must approach the line source function at large optical depths. This is the analogue to (11.14). However, by making the assumption of a constant source function we can become free from the special geometry of plasma and only in this case a calculation of the optical reduction factors can easily be performed. It should be emphasized that the influence which radiative absorption has on the local thermodynamic state of a plasma can only exactly be treated in the frame of a model in which the Boltzmann collision equations are simultaneously solved with the equations of radiative transfer. The solutions depend on the special geometry treated.

The estimates for the Λ_{ji} and Λ_i which follow below may nevertheless be considered as relatively realistic as concerns their application to laboratory plasmas. This has the following physical reason: Laboratory plasmas always have small spatial dimensions and therefore spectral lines and continua are never completely optically thick. Even in cases in which the optical depth in the center of a resonance line becomes very large (i.e. 10^4 or 10^6) many photons can still escape through the far line wings and, thus, leading to a ratio of the population densities (involved in this transition) which may deviate from the equilibrium ratio. This effect is still enhanced due to the fact that lines between excited states are generally optically thin. Since all states are coupled by many collision and radiation processes, the ratio $n_{\text{excited}}/n_{\text{ground}}$ is further pushed towards non-equilibrium values. It follows from this that even at large optical depths in the line center of a resonance line the source function can strongly deviate from the equilibrium value, i.e. from the Planck function. Even completely optically thick resonance lines and a completely optically thick resonance continuum are generally not sufficient to ensure at low electron densities the establishment of complete local thermodynamic equilibrium in the corresponding level system. For details see [27], [70].

2. The Reduction Factors Λ_{ji}

The number of effective spontaneous de-excitation processes $j \to i$ per unit volume and unit time at spatial position $\boldsymbol{r}'$ is defined by

$$N_{ji}(\boldsymbol{r}') = \Lambda_{ji}(\boldsymbol{r}')\, A_{ji}\, n_j(\boldsymbol{r}') \,. \tag{11.17}$$

This quantity is equal to the actual number of spontaneous plus stimulated de-excitation processes, minus the actual number fo true absorption processes, also per unit volume and unit time at position $\boldsymbol{r}'$:

$$N_{ji}(\boldsymbol{r}') = [A_{ji} + B_{ji}\, \bar{J}_{ji}(\boldsymbol{r}')]\, n_j(\boldsymbol{r}') - B_{ij}\, \bar{J}_{ji}(\boldsymbol{r}')\, n_i(\boldsymbol{r}') \,. \tag{11.18}$$

Comparison of (11.17) and (11.18) yields with (11.5a, b) the relation

$$\Lambda_{ji}(\boldsymbol{r}') = 1 - \left[\frac{g_j}{g_i}\, \frac{n_i(\boldsymbol{r}')}{n_j(\boldsymbol{r}')} - 1\right] \frac{c^2}{2\, h\, \nu_{ji}^3}\, \bar{J}_{ji}(\boldsymbol{r}') \,. \tag{11.19}$$

The coefficient in front of $\bar{J}_{ji}(\boldsymbol{r}')$ is the inverse of the source function (see (11.10)). We thus obtain for the reduction factor

$$\Lambda_{ji}(\boldsymbol{r}') = 1 - \frac{\bar{J}_{ji}(\boldsymbol{r}')}{S_{ji}(\boldsymbol{r})} \,. \tag{11.20}$$

We make now the assumption that the source function is constant in space. We obtain for the mean radiation intensity (compare (11.6) with (11.16)):

$$\bar{J}_{ji}(\boldsymbol{r}', \nu) = \frac{1}{4\pi} \int\limits_{\Omega} I_{ji}\, (r', \Omega, \nu)\, \mathrm{d}\Omega = \frac{1}{4\pi}\, S_{ji}(\nu_{ji}) \int\limits_{\Omega} [1 - \mathrm{e}^{-\tau_{ji}(r', \Omega, \nu)}]\, \mathrm{d}\Omega \,. \tag{11.21}$$

Integration over the whole sphere Ω surrounding $\boldsymbol{r}'$ can only be performed when the geometry is known. To become free of the special geometry we define with the help of (11.21) a mean optical depth $\bar{\tau}_{ji}(\boldsymbol{r}', \nu)$ at spatial position $\boldsymbol{r}'$:

$$\bar{J}_{ji}(\boldsymbol{r}', \nu) = S_{ji}(\nu_{ji})\, [1 - \mathrm{e}^{-\bar{\tau}_{ji}(\boldsymbol{r}', \nu)}] \,. \tag{11.22}$$

It is convenient to write

$$\bar{\tau}_{ji}(\boldsymbol{r}', \nu) = \sigma_{ji}(\boldsymbol{r}', \nu)\, \bar{n}_i\, l\,, \tag{11.23}$$

where $\sigma_{ji}(\boldsymbol{r}', \nu)$ is the absorption cross section for a line $i \rightleftarrows j$ in a region around $\boldsymbol{r}'$, l is an effective or a mean length (e.g. the distance between position $\boldsymbol{r}'$ within the plasma and the plasma boundary), and $\bar{n}_i$ denotes a mean population density of particles in state i, averaged over the length l. We introduce now a normalized line profile $P_{ji}(\Delta\nu)$ by the equation

$$\sigma_{ji}(\Delta\nu)\, \mathrm{d}\nu = \sigma_{ji}(\nu_{ji})\, P_{ji}(\Delta\nu)\, \mathrm{d}\nu\,; \qquad \int\limits_{-\infty}^{+\infty} P_{ji}(\Delta\nu)\, \mathrm{d}\nu = 1\,. \tag{11.24}$$

We thus obtain from (11.22) at a frequency distance $\Delta\nu = \nu - \nu_{ji}$ from the line center

$$\bar{J}_{ji}(\boldsymbol{r}', \Delta\nu) = S_{ji}(\nu_{ji})\, [1 - \exp(-\sigma_{ji}\, \bar{n}_i\, l\, P_{ji}(\Delta\nu)]\,. \tag{11.25}$$

In order to evaluate (11.20) we need the mean intensity averaged over the whole frequency spectrum of the line. It is obtained from

$$\bar{J}_{ji}(\boldsymbol{r}') = \frac{\int P_{ji}(\Delta\nu)\, \bar{J}_{ji}(r', \nu)\, \mathrm{d}\nu}{\int P_{ji}(\Delta\nu)\, \mathrm{d}\nu} = \int\limits_{-\infty}^{+\infty} P_{ji}(\Delta\nu)\, \bar{J}_{ji}(\boldsymbol{r}', \Delta\nu)\, \mathrm{d}\nu\,. \tag{11.26}$$

Inserting (11.25) into (11.26) and using the normalization condition for the line profile yields

$$\bar{J}_{ji}(\boldsymbol{r}') = S_{ji}\left[1 - \int\limits_{-\infty}^{+\infty} P_{ji}(\Delta\nu)\, \mathrm{e}^{-\sigma_{ji}(\nu_{ji})\, \bar{n}_i\, l\, P_{ji}(\Delta\nu)}\, \mathrm{d}\nu\right]. \tag{11.27}$$

When this expression is inserted into (11.20) we obtain for Λ_{ji} the general relation

$$\boxed{\Lambda_{ji}(\boldsymbol{r}') = \int\limits_{-\infty}^{+\infty} P_{ji}(\Delta\nu) \exp\left(-\sigma_{ji}(\nu_{ji})\, \bar{n}_i\, l\, P_{ji}(\Delta\nu)\right) \mathrm{d}\nu}\,. \tag{11.28}$$

We use now the Ladenburg-relation for the absorption cross section:

$$\sigma_{ji}(\Delta\nu)\, \mathrm{d}\nu = \frac{\pi\, e_0^2}{m_e\, c} f_{ij}\, P_{ji}(\Delta\nu)\, \mathrm{d}\nu\,, \tag{11.29}$$

where f_{ij} denotes the absorption oscillator strength. $P_{ji}(\Delta\nu)$ is given by the following relations:

— For a Gauss-(Doppler)-profile

$$P_{ji}^{(\mathrm{G})}(\Delta\nu) = \frac{1}{N_{ji}^{\mathrm{G}}} \exp - \left(\frac{\Delta\nu}{\Delta\nu_{\mathrm{D}}}\right)^2; \qquad N_{ji}^{\mathrm{G}} = \Delta\nu_{\mathrm{D}}\, \pi^{1/2}\,. \tag{11.30}$$

— For a dispersion profile

$$P_{ji}^{(\mathrm{D})}(\Delta\nu) = \frac{1}{N_{ji}^{\mathrm{D}}} \cdot \frac{a_{ji}}{a_{ji}^2 + (\Delta\nu)^2}\,; \qquad N_{ji}^{\mathrm{D}} = \pi\,. \tag{11.31}$$

The N_{ji} denote normalization coefficients.

Λ_{ji}-values calculated after (11.28) for some Doppler-broadened Lyman lines of atomic hydrogen have been listed in Tab. 4.1.

The quantities

$$\hat{\sigma}_{ji}^{(G)} = \frac{1}{\Delta\nu_D \pi^{1/2}} \frac{\pi e_0^2}{m_e c} f_{ij}, \tag{11.32a}$$

$$\hat{\sigma}_{ji}^{(D)} = \frac{1}{\pi a_{ji}} \frac{\pi e_0^2}{m_e c} f_{ij}, \tag{11.32b}$$

represent the absorption cross sections in the center ($\Delta\nu = 0$) of the spectral lines. When one introduces the effective optical depth $\hat{\tau}_{ji}$ in the line center (11.28) can also be written as follows:

a — For a Gauss-profile

$$\Lambda_{ji}^{(G)} = \frac{1}{\Delta\nu_D \pi^{1/2}} \int_{-\infty}^{+\infty} e^{-(\Delta\nu/\Delta\nu_D)^2} \exp\left[-\hat{\tau}_{ji} \exp - (\Delta\nu/\Delta\nu_D)^2)\right] d\nu, \tag{11.33}$$

where we have to put $\hat{\tau}_{ji} = \hat{\sigma}_{ji}^{(G)} \bar{n}_i l$. For $\hat{\tau}_{ji} \geq 3$ one obtains as a good approximation to (11.33)

$$\Lambda_{ji}^{(G)} \cong \frac{1}{\hat{\tau}_{ji}(\pi \ln \tau_{ji})^{1/2}}. \tag{11.34}$$

This corresponds to Holstein's result (11.2a) when the geometry is omitted.

b — For a dispersion-profile

$$\Lambda_{ji}^{(D)} = \frac{1}{\pi} \int_{-\infty}^{+\infty} \frac{a_{ji}}{a_{ji}^2 + (\Delta\nu)^2} \exp\left[-\hat{\tau}_{ji} \frac{a_{ji}^2}{a_{ji}^2 + (\Delta\nu)^2}\right] d\nu, \tag{11.35}$$

where we have now to put $\hat{\tau}_{ji} = \hat{\sigma}_{ji}^{D} \bar{n}_i l$. For $\hat{\tau}_{ji} \geq 5$ we obtain as a good approximation to (11.35)

$$\Lambda_{ji}^{(D)} \cong \frac{1}{(\pi \hat{\tau}_{ji})^{1/2}}. \tag{11.36}$$

This is Holstein's result (11.2b) when the geometry is neglected. Holstein obtained his results in quite another way.

When a line profile follows a Voigt-function, the Λ_{ji}-values lie between those given by (11.33) and (11.35). Corresponding values are given in [71].

We see that the reduction factors Λ_{ji} — by which the Einstein coefficients A_{ji} for spontaneous emission have to be multiplied — depend in a sensitive manner on the form of the line profile. For large values $\hat{\tau}_{ji}$, (11.33), (11.34) on the one hand and (11.35), (11.36) on the other hand give values which deviate by orders of magnitude from each other.

Within a spectral series the Λ_{ji} approach the value unity with increasing principal quantum number j. This has the following reason: First, the oscillator strengths f_{ij} decrease approximately like $1/j^3$. Second, broadening effects become more and more important for the higher series members. Both effects together lead to a rapid approach to unity of the Λ_{ji}-values. This effect can already be seen from the numerical values listed in Tab. 4.1. Under actual conditions the Λ_{ji}-values of the higher members tend still more rapid to unity than those of Tab. 4.1, since the line contours are Stark rather than Doppler profiles. Numerical exemples for helium lines, for instance, may be found in [71].

3. The Reduction Factors Λ_i

The number of effective photorecombination processes into level i, per unit volume and unit time at a spatial point $\boldsymbol{r}'$ is defined by

$$N_i(\boldsymbol{r}') = \Lambda_i(\boldsymbol{r}')\, n_+(\boldsymbol{r}')\, n_e(\boldsymbol{r}')\, R_i(\boldsymbol{r}') \tag{11.37}$$

where R_i denotes the rate coefficient

$$R_i = \langle \sigma_i\, v_e \rangle\,. \tag{11.38}$$

n_+ and n_e are the ion and electron densities respectively. $N_i(\boldsymbol{r}')$ is equal to the actual number of spontaneous plus stimulated photorecombination processes into level i, minus the actual number of true photoionization processes starting from level i, per unit volume and unit time at spatial point $\boldsymbol{r}'$:

$$N_i(\boldsymbol{r}') = n_+(\boldsymbol{r}')\, n_e(\boldsymbol{r}')\, R_i(\boldsymbol{r}') + \begin{pmatrix}\text{stimulated}\\ \text{processes}\end{pmatrix} - n_i(\boldsymbol{r}')\, P_i(\boldsymbol{r}')\,. \tag{11.39}$$

The coefficient for photoionization from level i is obtained from the general relation

$$P_i(\boldsymbol{r}') = \int\limits_{\nu=\nu_i}^{\infty} \sigma_i(\nu)\, \frac{c}{h\,\nu}\, u_i(\boldsymbol{r}', \nu)\, \mathrm{d}\nu\,, \tag{11.40}$$

where $\sigma_i(\nu)$ is the corresponding cross section for photoionization and $u_i(\boldsymbol{r}', \nu)$ the radiation density within the i-th recombination continuum. Neglecting the stimulated processes and comparing (11.37) with (11.39) yields for $\Lambda_i(\boldsymbol{r}')$ at position $\boldsymbol{r}'$ in the plasma

$$\Lambda_i(\boldsymbol{r}') = 1 - \frac{n_i(\boldsymbol{r}')}{n_+(\boldsymbol{r}')\, n_e(\boldsymbol{r}')}\, \frac{P_i(\boldsymbol{r}')}{R_i(\boldsymbol{r}')}\,. \tag{11.41}$$

When one uses the relation for detailed balancing in the form

$$n_+\, n_e\, R_i = n_i^{(\mathrm{Saha})}\, P_i^{(\mathrm{Saha})}\,, \tag{11.42}$$

we obtain

$$\Lambda_i(\boldsymbol{r}') = 1 - b_i(\boldsymbol{r}')\, \frac{P_i(\boldsymbol{r}')}{P_i^{(\mathrm{Saha})}(\boldsymbol{r}')}\,, \tag{11.43}$$

where $b_i(\boldsymbol{r}')$ denotes the Saha decrement of level i (see (2.1).) Relation (11.43) is the analogue to (11.20).

$P_i^{(\mathrm{Saha})}(\boldsymbol{r}')$ is obtained from (11.40) when one introduces for $u_i(\boldsymbol{r}, \nu)$ the Planck function

$$u(\nu) = \frac{8\,\pi\, h\, \nu^3}{c^3}\, [\exp(h\,\nu/kT_e) - 1]^{-1} \cong \frac{8\,\pi\, h\, \nu^3}{c^3}\, \exp(-\,h\,\nu/kT_e)\,, \tag{11.44}$$

whereas $P_i(\boldsymbol{r}')$ can only be calculated from (11.40) when one inserts the true radiation density which depends on the actual thermodynamic state of the plasma. We are thus faced with the problem of transfer of radiation in the recombination continuum.

We define now a mean radiation intensity in the frequency range $\nu \cdots \nu + d\nu$ in the i-th recombination continuum at spatial position $\boldsymbol{r}'$:

$$\bar{J}_i(\boldsymbol{r}', \nu) = S_i(\nu)\,[1 - e^{-\bar{\tau}_i(\boldsymbol{r}', \nu)}] . \tag{11.45}$$

This is the analogue relation to (11.22). The quantity $\bar{\tau}_i(\boldsymbol{r}', \nu)$ can be interpreted as the effective or mean optical depth in the frequency range $\nu \ldots \nu + d\nu$ of the i-th recombination continuum at spatial point $\boldsymbol{r}'$. We put in analogy to (11.23):

$$\bar{\tau}_i(\boldsymbol{r}', \nu) = \sigma_i(\nu)\,\bar{n}_i\,l . \tag{11.46}$$

We try now to make an estimate of $P_i(\boldsymbol{r}')$ after (11.40). The radiation density is given by

$$u_i(\boldsymbol{r}', \nu) = \frac{4\pi}{c}\,\bar{J}_i(\boldsymbol{r}', \nu) , \tag{11.47}$$

where we have to insert $\bar{J}_i(\boldsymbol{r}', \nu)$ from (11.45). We put the source function $S_i(\nu)$ equal to the source function in equilibrium (i.e. equal to the Planck intensity in Wien's approximation) divided by the Saha decrement for level i:

$$S_i(\nu) = \frac{1}{b_i(\boldsymbol{r}')}\,\frac{2\,h\,\nu^3}{c^2}\,\exp\left(-\frac{h\,\nu}{kT_e}\right). \tag{11.48}$$

It follows

$$P_i(\boldsymbol{r}') = \frac{1}{b_i(\boldsymbol{r}')}\,\frac{8\,\pi}{(h\,c)^2}\int_{\nu=\nu_i}^{\infty}(h\,\nu)^2\,\sigma_i(\nu)\,e^{-h\nu/kT_e}\,[1 - e^{-\bar{\tau}_i(\boldsymbol{r}', \nu)}]\,d\nu . \tag{11.49}$$

$P_i^{(\mathrm{Saha})}(\boldsymbol{r}')$ is obtained from (11.40) when one inserts the Planck radiation density (also in Wien's approximation). It follows

$$P_i^{(\mathrm{Saha})}(\boldsymbol{r}') = \frac{8\,\pi}{(h\,c)^2}\int_{\nu=\nu_i}^{\infty}(h\,\nu)^2\,\sigma_i(\nu)\,\exp\left(-\frac{h\,\nu}{kT_e}\right)d\nu . \tag{11.50}$$

Putting now (11.49) and (11.50) into (11.43) we obtain the general relation

$$\boxed{\Lambda_i(\boldsymbol{r}') = \frac{\int_{\nu=\nu_i}^{\infty}(h\nu)^2\,\sigma_i(\nu)\,e^{-(h\nu/kT_e+\bar{\tau}_i(\boldsymbol{r}', \nu))}\,d\nu}{\int_{\nu=\nu_i}^{\infty}(h\nu)^2\,\sigma_i(\nu)\,e^{-h\nu/kT_e}\,d\nu}} . \tag{11.51}$$

Until now we have not yet specified the frequency dependence of the cross section for photoionization. Therefore (11.51) can be applied to any atomic or ionic system.

Consider now the cross section $\sigma_i(\nu)$ in the hydrogenic approximation which can be put into the following analytical form

$$\sigma_i(\nu) = \frac{2^3}{3\sqrt{3}} \frac{h^2 Z^{-2}}{\pi^2 m_e^2 e_0^2 c} \frac{1}{i^5} \left(\frac{Z^2 E_1^{\mathrm{H}}}{h\nu}\right)^3 . \tag{11.52}$$

We put $u = h\nu/kT_e$, $u_i = h\nu_i/kT_e = Z^2 E_1^{\mathrm{H}}/i^2 kT_e$ and introduce the effective optical depth $\hat{\tau}_i$ at the series limit $\nu = \nu_i$:

$$\hat{\tau}_i(\boldsymbol{r}') = \hat{\sigma}_i \bar{n}_i l . \tag{11.53}$$

$\hat{\sigma}_i$ denotes the cross section for photoionization at the series limit:

$$\hat{\sigma}_i \equiv \sigma_i(\nu = \nu_i) = \frac{2}{3\sqrt{3}} \frac{h^3 Z^{-2} i}{\pi^2 m_e^2 e_0^2 c} . \tag{11.54}$$

We obtain from (11.51)

$$\boxed{\Lambda_i(\boldsymbol{r}') = \frac{1}{[-Ei(-u_i)]} \int\limits_{u=u_i}^{\infty} \frac{e^{-\left(u + \left(\frac{u}{u_i}\right)^3 \tau_i(\boldsymbol{r}')\right)}}{u} \, du} . \tag{11.55}$$

In this form the reduction factor has recently been given by Gehren [68].

Λ_i is a sensitive function of T_e and of the frequency dependence of the cross section for photoionization. Small changes of T_e and/or $\sigma_i(\nu)$ can change Λ_i by orders of magnitudes in the region of low temperatures. With increasing T_e/ν_i the factor Λ_i approaches unity.

Figs. 11.1a, b show the numerical values of Λ_i calculated after (11.55). Under actual conditions Λ_i lies mostly within the limits $10^{-2} < \Lambda_i \leq 1$.

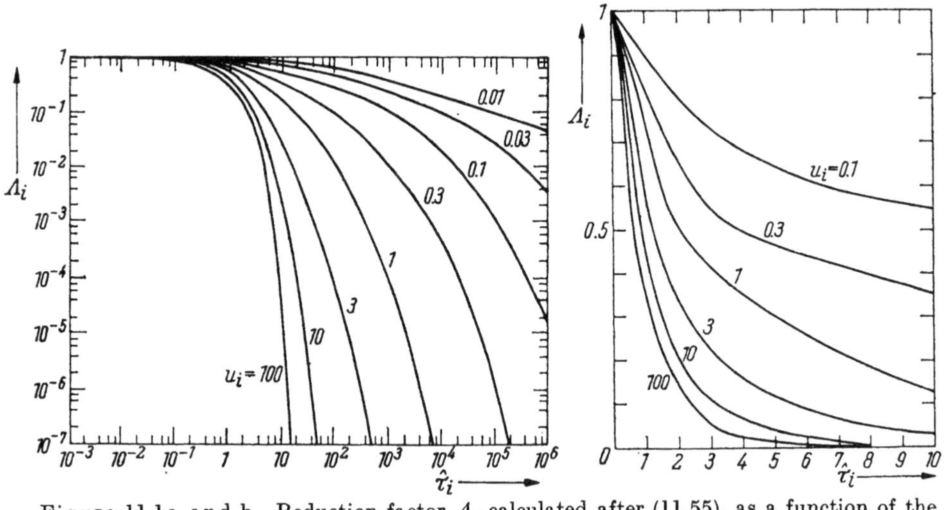

Figure 11.1a and b. Reduction factor Λ_i calculated after (11.55), as a function of the effective optical depth $\hat{\tau}_i$ at the series limit. The parameter u_i is given by $u_i = h\nu_i/kT_e = Z^2 E_1^{\mathrm{H}}/i^2 kT_e$

CHAPTER 12

Concluding Remarks

In this section a number of criteria have been given which permit a rather precise estimate whether local thermodynamic equilibrium can be established under special conditions, or not. All validity criteria contain a temperature dependence, in all relations the temperature is specified, i.e. they refer to a special type of particle responsible for the collision rates. When the condition for equal temperatures for all types of particles is fulfilled (i.e. in the case of thermal equilibrium for the translational degrees of freedom) the specification of the temperature disappears. Then, all criteria depend on one single temperature, namely the true or thermodynamic temperature T.

All criteria have been checked by solving the corresponding coupled systems of rate equations*. The numerical computer calculations have shown that coupling processes between the many levels of a level system favour the establishment of L.T.E. compared to the simple estimates on which the criteria are based. This means that the validity criteria generally lie on the "unfavorable side". However, the differences between the validity criteria as given in this article and the numerical evaluations of the coupled systems of rate equations are relatively small, as far as hydrogen and helium are concerned and provided the experimentel conditions are not too extreme. It should be emphasized, however, that all L.T.E. criteria are based on very simplified level systems, simplified collisional-radiative rate equations (especially for transfer of radiation), and extremely simplified relations for the collision cross sections. Details of the plasma compositrons have not been taken into account nor have plasma-dynamical processes been considered. Due to this situation the L.T.E. criteria should be supplemented by numerical studies of collisional-radiative models applied to the specific situations considered.

* The numerical calculations have been performed on the IBM 360/50 360/75 computers at Fontenay-aux-Roses and Saclay.

Bibliography

[1] DRAWIN, H. W., Collision and Transport Cross Sections, Report EUR-CEA-FC-383, Fontenay-aux-Roses 1966/67.
[2] DRAWIN, H. W., "Collision and Transport Cross Sections", in: Plasma Diagnostics, Ed. W. Lochte-Holtgreven, North-Holland Publ. Comp., Amsterdam (1968).
[3] SHAHIN, M. M., "Fundamental Processes in Plasmas", in: Chemical Reactions Under Plasma Conditions, Ed. M. Venugopalan, John Wiley & Sons Interscience Publ. New York.
[4] McDANIEL, W. E., Collision Phenomena in Ionized Gases, John Wiley & Sons, New York, 1964.
[5] DRAWIN, H. W., "Thermodynamic properties of the equilibrium and non-equilibrium states of plasmas", in: Chemical Reactions Under Plasma Conditions, Ed. M. Venugopalan, John Wiley & Sons Interscience Publ. New York.
[6] GIOVANELLI, R. G., Austral. J. Sci. Research **1** (1948) 275.
[7] UNSÖLD, A., Z. Physik **171** (1963) 44.
[8] PAGEL, B. E. J., Proc. Roy. Soc. (London) **306 A** (1968) 91.
[9] GRIEM, H. R., Phys. Rev. **131** (1963) 1170.
[10] GRIEM, H. R., Plasma Spectroscopy, McGraw Hill, New York (1964).
[11] BIBERMAN, L. M. and K. N. ULYANOV, Optics and Spectroscopy **16** (1964) 216.
[12] VOROBEV, V. S., High Temperature **4** (1966) 468.
[13] LAGARKOV, A. N., High Temperature **4** (1966) 301.
[14] THOMAS, R. N., Astrophys. J. **123** (1957) 260; **131** (1960) 429.
[15] JEFFERIES, J. T. and R. N. THOMAS, Astrophys. J. **129** (1959) 401.
[16] JEFFERIES, J. T., Astrophys. J. **132** (1960) 775.
[17] HUMMER, D. G., Monthly Notic. Roy. Astron. Soc. **125** (1963) 461.
[18] HUMMER, D. G., J. Quant. Spectr. Rad. Transf. **8** (1968) 193.
[19] SAMPSON, D. H., Astroph. J. **144** (1966) 96.
[20] WILSON, R., J. Quant. Specty. Radiative Trans. **2** (1962) 477.
[21] McWHIRTER, R. W. P., "Spectral Intensities", in: Plasma Diagnostic Techniques, Eds. R. H. Huddlestone and S. L. Leonard, Academic Press, New York, London (1965).
[22] DRAWIN, H. W., Z. Physik **228**, (1969) 99.
[23] DEUTSCH, C., Physics Letters **28 A** (1968) 525 and **28 A** (1969) 752.
[24] MENZEL, D. H., Astrophys. J. **85** (1937) 330; **86** (1937) 70.
[25] VAN REGEMORTER, H., Monthly Not. Roy. Astron. Soc. **121** (1960) 213.
[26] SEATON, M. J., Article in the book Atomic and Molecular Processes, Ed. D. R. Bates, Academic Press New York, London 1962.
[27] DRAWIN, H. W., Report EUR-CEA-FC-516, May (1969), Z. Naturfg. **24a** (1969) 1492.
[28] DRAWIN, H. W., Z. Physik **211** (1968) 404.
[29] McWHIRTER, R. W. P. and A. G. HEARN, Proc. Phys. Soc. **82** (1963) 641.
[30] DRAWIN, H. W., J. Quantitative Specty. Radiative Transf. **10** (1970) 33.
[31] BATES, D. R., A. E. KINGSTON and R. W. P. McWHIRTER, Proc. Roy. Soc. (London) **267 A** (1962) 297 and **270 A** (1963) 155.

[32] DRAWIN, H. W., Z. Physik **225** (1969) 470 and **225** (1969) 483.
[33] DRAWIN, H. W., Z. Naturforschg. **19a** (1964) 1451.
[34] DRAWIN, H. W., M. FUMELLI and G. WESTE, Z. Naturf. **20a** (1965) 184.
[35] DRAWIN, H. W., Z. Physik **186** (1965) 99.
[36] ARTMANN, J., Z. Naturf. **23a** (1968) 2132.
[37] BERGSTEDT, K., Report IPP 3/64, Institut für Plasmaphysik, Garching (1967), Z. Naturf. **24a** (1969) 299.
[38] TICHMANN, K., Report IPP 3/77, Institut für Plasmaphysik, Garching (1968).
[39] HIRSCHFELDER, J. O., C. F. CURTISS and R. B. BIRD, "Molecular Theory of Gases and Liquids", John Wiley & Sons, New York (1954), 4th-ed. 1967.
[40] HASTED, J. B., "Physics of Atomic Collisions", Butterworths, London 1964.
[41] DRAWIN, H. W. and F. EMARD, Report EUR-CEA-FC-534 (1970). Instantaneous population densities of the excited levels of hydrogen atoms, hydrogen-like ions, and helium atoms in optically thin and thick non-L.T.E. plasmas.
[42] FINKELNBURG, W. and H. MAECKER, "Elektrische Bögen und thermisches Plasma", article in: Handbuch der Physik, Vol. 22, Ed. S. Flügge, Springer Verlag, Berlin-Göttingen-Heidelberg (1956).
[43] DRUYVESTEYN, M. J., Physica (old) **10** (1930) 61, Physica (new) **3** (1936) 65.
[44] DRUYVESTEYN, M. J., Phys. Zs. **33** (1932) 856.
[45] SMIT, J. A., Physica **3** (1936) 543.
[46] BOYD, R. L. F. and N. D. TWIDDY, Proc. Roy. Soc. (Lond.) **A 250** (1959) 53 and **A 259** (1960) 145.
[47] THOMPSON, J. B., Proc. Roy. Soc. (Lond.) **A 262** (1961) 503.
[48] TWIDDY, N. D., Proc. Roy. Soc. (Lond.) **A 275** (1963) 338.
[49] CARLETON, N. P. and L. R. MEGILL, Phys. Rev. **126** (1962) 2089.
[50] SAMPSON, D. H. and J. ENOCH, Phys. Fluids **6** (1963) 28.
[51] WOJACZEK, K., Beiträge Plasmaphysik **5** (1965) 181.
[52] SWIFT, J. D., Brit. J. Applied Phys. **16** (1965) 837.
[53] DUGAN, J. V., F. A. LYMAN and L. U. ALBERS, in: Electricity from MHD, Vol. II, p. 85–99, Eds. Intern. Atomic Energy Agency, Vienna 1966.
[54] MORRONE, T., Phys. Rev. **141** (1966) 27.
[55] KAGAN, YU. M. and V. M. MILLENIN, Soviet Phys.-Techn. Phys. **10** (1966) 1470.
[56] KAGAN, YU. M., V. M. MILLENIN and N. K. MITRIFANOV, Soviet Phys.-Techn. Phys. **11** (1967) 1661.
[57] PEYRAUD, N., J. de Physique **29** (1968) 201, **29** (1968) 747, **29** (1968) 997.
[58] POSTMA, A. J., Physica **43** (1969) 229, **43** (1969) 581, **45** (1970) 609, **57** (1972) 624.
[59] HOBSON, R. M., in: A survey of phenomena in ionized gases, invited paper, Int. Conf. on Phenomena in Ionized Gases, Vienna 1967, published by the Intern. Atomic Energy Agency, Vienna 1968.
[60] KELLY, A. J., J. Chem. Physics **45** (1966) 1723.
[61] MCLAREN, T. I. and R. M. HOBSON, Phys. Fluids **11** (1968) 2162.
[62] MCELWAIN, D. L. S., L. WAGSCHALL and H. O. PRITCHARD, Phys. Fluids **13** (1970) 2200.
[63] BELOZEROV, A. N. and R. M. MEASURES, J. Fluid Mech. **36** (1969) 695.
[64] KALRA, S. P. and R. M. MEASURES, Phys. Fluids **14** (1971) 2544.
[65] DRAWIN, H. W. and F. EMARD, Z. Physik **254** (1972) 202 and Physics Letters **43 A** (1973) 333.
[66] HOLSTEIN, T., Phys. Review **72** (1947) 1212; **83** (1951) 1159.
[67] JEFFERIES, J. T., Spectral Line Formation, Blaisdell Publ. Comp. Walthasam (U.S.A.) 1968.
[68] GEHREN, T., Diplomarbeit, University Kiel (1970).

[69] NEUMANN, W., Spectroscopic Methods of Plasma Diagnostics, in: Progress in Plasmas and Gas Electronics, Eds. R. Rompe and M. Steenbeck, Akademie-Verlag, Berlin 1974. See Article in this volume.
[70] DRAWIN, H. W., Physics Letters **42A** (1972) 423.
[71] DRAWIN, H. W. and F. EMARD, Beiträge aus der Plasmaphysik **13** (1973) 143.
[72] JOHNSON, L. C. and L. C. HINNOV, J. Quant. Spectroscopy Radiative Transfer **13** (1973) 333.
[73] FUJIMOTO, T., SUGIYAMA, I. and FUKUDA, K., Memoirs Faculty of Engineering, Kyoto University, Japan, Vol. 34, part 2 (1972) 249.

Subject Index

Absorption coefficients, determination of 228
— —, molecular, references 235
—, continuous, data 316ff.
Absorptive power, of plasma, increasing 202
Alfvén velocity 523
— waves, in energy transport 532
— —, phase velocity 524
— —, polarization 527
Anode crater, of carbon arc 23
Applications of plasmas 3
Arcs, types of stabilized 151ff., 156
Atomic spectra, classification 87

Band intensities, plasma data from 240
— strengths 254
Bichromatic methods, for temperature measurement 204
Black body radiation, for temperature measurement 209
Bolometers 19
Boltzmann distribution 4
— —, validity 408ff.

Carbon arc 23
Chemical equilibria, from thermodynamic data 110
Collision broadening, by Stark effect 363
— theories, classical 348ff.
Collisions, and diffusion coefficient 641ff.
Composition, of atomic plasma, as function of temperature and pressure 101
—, — — —, calculation 95
—, — — —, simplifications 99
—, plasma, from emission coefficients 125
—, —, references on 115
Conductivity, electrical, from infrared radiation 321
Continuous radiation, kinds of 289
Correction of line profiles 68

Densities, particle, ionization temperature for low 432
Density, electron, from emission coefficient, data 320
—, —, — forbidden lines 391
—, —, infrared bremsstrahlung 321
—, —, rarefied plasmas 435
Description 3
Diatomic molecules, intensities in spectra 237
Discontinuity surfaces, boundary conditions at 542
Distortion, of line profiles 66
Doppler broadening 325
— shift 332
— width 167, 170

Eberhard effect 52
Echelle grating 11
Electric field strength, macroscopic 402
Electron-atom continuum, and electron-ion continuum 308
—-— —, characteristics 304
— density, from relative intensities of spectral lines 139
— —, — rotation of laser beams 283
Electron-ion continuum 308
—-— —, types 291
—, temperature, and external electric fields 646
Emission coefficient, and ionization energy 301
— coefficients, determination of, with optical probes 80
— —, plasma parameters from 119, 310
— —, temperature from ratio of 129
—, continuous, data 316ff.
Emissivity, of tungsten 28
Energies of dissociation, data 112
Energy transport, by MHD waves 530ff.

Fabry-Perot interferometer 167
Film curve, characteristic 54
Fortrat diagram 94

Framing devices, spectroscopic 44
Free-bound radiation 296
Free-free radiation 291

Gaseous discharges, as standard sources 29
Gaunt-factor 295

Heavy particle collisions 614
High-speed recording, of line profils 49
High-speed spectroscopy 34
Holtsmark distribution 354
Hönl-London factor 238
"Hook" method 168ff.

Ideal media, MHD-equations for 516
Inhomogeneous media, propagation in 563ff.
— plasmas, temperature, determination in 210
Integral line intensities, determination 62
Intensities, line 62
Intensity distribution, transformation of 72
— measurements, photographic 50
Interferometers 11
—, laser 271
—, multiples beam 272
—, two-beam 270
Interferometry 260ff.
—, for temperature and density, summary 274
—, high-speed 269
—, holographic 273
—, laser 276
International partition functions, of molecules 109
Ionization energy, of atomic plasma 96
— —, reduction 393, 399
— temperature for low particle densities 432

King furnace 164
Kirchhoff's law 199
— —, validity 419

Lamp, tungsten 24
Lamps, hydrogen discharge 31
—, pulse discharge arc, as sources 30
—, transmission factors 25
Laser radiation, non resonant scattering 284
— scattering experiments, data 288
Light intensity, control of 58
Line broadening, and particle interaction 370
— —, due to particle interaction 347
— —, — — Stark effect 365
— —, — — — —, data 381, 384
— —, electron and ion 367
— —, factors in 388
— —, natural 346
— —, statistical theory 353
—, natural, broadening 346
— profiles, correction of 68
— —, distortion of 66
— -reversal method, data 211
—-— — for temperature determination 202
— width photographic 61
Lines, fusion of, at series limit 393
Low levels, excitation of 425

Magnetic field strengths, from Zeeman effect 338
Magnetosphere, propagation of f-waves in 576
Maxwell distribution 3
— velocity distribution, of atoms, molecules, and ions 408
— — —, — electrons 405
MHD-equations, characteristics of linearized 565
MHD-waves, at discontinuity surfaces 542ff.
—-—, reflection and refraction 547ff.
Microfield ionization, weakening of lines by 395
— probability, for emitters 357
Molecular spectra, classification 89
Monochromatic pictures, with spectrograph 45
Monochromatically resolved radiation, temperature determination by 210
Multichannel systems, for spectrum registration 47
Multiplet strengths, data 179

Optical reduction factors, and thermodynamics 648
— refractivity, and particle number density 258
Oscillator strengths 254
— —, at high pressures 175
— —, determination 149
— —, from absorption measurements 164, 166
— —, — broadening of resonance level 171
— —, — dispersion 167

Parameters, plasma, from relative line intensities 127
Particle densities, determination of 227
— —, fluorescence methods for 235
— —, from absorption measurements 232
— —, — electron beam fluorescence 236
— —, — resonance line scattering 235
— density, and optical refractivity 258
— —, by interferometry 264
— —, determination 264
— —, from relative line intensities 127
— —, measurement, Schlieren technique 277
Partition function 395
— —, of atomic plasma 96
— —, references on, table 114, 115
Photoelectric detectors 62
Photographic blackening 51
— developing 52
— intensity 50
— materials 18
Photoionization, atomic, references 233
Plane waves, dispersion equation for 524
— —, equation for 519
— —, group velocity and energy transport in 530
— —, polarization conditions for 526
Plasma absorption 81
— —, effect of 74
— diagnosis by continuous radiation 310
— —, spectroscopic methods 407
—, homogeneous stationary, thermodynamics 608ff.
—, optically thick, thermodynamics 617ff.
—, — thin, thermodynamics 597ff., 617ff.
— parameters, from absorption 254
— —, — continuous radiation 322
— —, — emission coefficients 310
— —, — line profiles 324ff.
—, thermodynamic equilibrium in 593ff.
Plasmas, inhomogeneous, Boltzmann equilibrium in 422
Polychromators 47
Pressure, plasma, from intensity and composition 123

Quantum defect method, for emission coefficients 298
Quasistationarity 419ff.

Radiation, blackbody 21
—, continuous 289
— detectors 18
— —, chemical 21
— —, electrical 19
—, free-bound 296
—, free-free 291
—, standard sources of 21, 32
—, surface sources 24
— thermocouples 19
—, total continuous 296
Reflection and refraction 547ff.
Refractive index, by interferometry 260
— —, measurement, Schlieren technique 277
Relaxation lengths, and thermodynamics 637ff.
— times, and thermodynamics 631ff.

Schlieren technique 277, 279
Schwarzschild effect 55
Self-absorption 81, 83
Self-reversed spectral lines, profile correction 225
Slit width, effect of 57
— widths, adjustment 53
Spectra, atomic, classification 87
—, continuous time-resolved 41
—, instanteneous recording 34
—, molecular 89
Spectral lines, self-reversed 225
— tables 17
Spectrograph, quartz 87
Spectrographs, commercial 8
—, grating 8
—, prism 5
-, resolution factors 6
Spectrometer, crystal 15
Spectroscopes, calibration of 16
Spectroscopic instrumentation 5
— methods, conspectus 406
— —, plasma diagnosis 407
— —, prerequisites 423
Spectroscopy, high speed 34
—, nonthermal, techniques 423
—, thermal, validity 404ff.
Stark constants, determination 356
— —, references 389
— effect 363ff.
— —, and field strength 402
— —, and line profiles 370
— —, and oscillator strength 175
Surface sources 24

Temperature by norm method 140
— determination, Bartel's method 219ff., 226

– –, by line-reversal method 199
– –, from decay of intensity 314
– –, – molecular spectra 248
– –, – self-reversed lines 218
– –, in inhomogeneous plasmas 210
– –, – monochromatically resolved radiation 210
– –, integrating over a spectral line 214
–, electron, for partly overlapping bands 252
–, –, from emission coefficient, data 320
–, –, in rarefied plasmas 428
–, from ratio of emission coefficients 129
–, – relative intensities of spectral lines 135, 136
–, plasma, from Doppler width, data 329
–, –, – intensity and composition 123
–, –, – pressure and composition 120
– profile, radial 215
–, rotational 241, 245
–, translational, in plasma 414
–, vibrational 244
–, –, determination, when bands not resolved 249

Thermodynamic data, chemical equilibrium from 110
– equilibrium, in plasma 593 ff.
Total continuous radiation 296
Transition probabilities 254
– –, absolute, determination 149
– – by f-sum rule 161
– –, calculation 176
– –, determination 149 ff., 175
– –, from lifetime of excited levels 172
– –, – one spectral line 162
– –, – spectral lines in optically thin layer 154
– –, reference data 183
Tungsten, emissivity of 27
Two-path method, for temperature measurement 207

Velocity, from Doppler shift, data 339
Voigt-profiles, data 71

Zeeman effect 338, 345
Zeeman effect, and line profiles 370

Bei Fragen zur Produktsicherheit wenden Sie sich bitte an:
If you have any questions regarding product safety,
please contact:

Walter de Gruyter GmbH
Genthiner Straße 13
10785 Berlin
productsafety@degruyterbrill.com